The *Blueprint for Problem Solving* is a general outline that you can use to help you solve application problems.

BLUEPRINT FOR PROBLEM SOLVING

STEP 1: *Read* the problem, and then mentally *list* the items that are known and the items that are unknown.

STEP 2: *Assign a variable* to one of the unknown items. (In most cases this will amount to letting x = the item that is asked for in the problem.) Then *translate* the other *information* in the problem to expressions involving the variable.

STEP 3: *Reread* the problem, and then *write an equation,* using the items and variables listed in steps 1 and 2, that describes the situation.

STEP 4: *Solve the equation* found in step 3.

STEP 5: *Write* your *answer* using a complete sentence.

STEP 6: *Reread* the problem, and *check* your solution with the original words in the problem.

Here is a *Blueprint for Problem Solving* that you can use specifically for problems about systems of equations (beginning in Chapter 4).

BLUEPRINT FOR PROBLEM SOLVING USING A SYSTEM OF EQUATIONS

STEP 1: *Read* the problem, and then mentally *list* the items that are known and the items that are unknown.

STEP 2: *Assign variables* to each of the unknown items; that is, let x = one of the unknown items and y = the other unknown item. Then *translate* the other *information* in the problem to expressions involving the two variables.

STEP 3: *Reread* the problem, and then *write a system of equations,* using the items and variables listed in steps 1 and 2, that describes the situation.

STEP 4: *Solve the system* found in step 3.

STEP 5: *Write* your *answers* using complete sentences.

STEP 6: *Reread* the problem, and *check* your solution with the original words in the problem.

ENHANCED

Web**Assign**

The perfect instructional tool to help you bridge the gap!

Elementary Algebra is infused with Pat McKeague's passion for teaching mathematics. His attention to detail and exceptionally clear writing style move students through each new concept with ease. This Eighth Edition of *Elementary Algebra* is enriched with new features and pedagogy that will help your students bridge the concepts. And now this best-selling text integrates **Enhanced WebAssign**—a dynamic homework management system that enables you and your students to work as partners in the teaching and learning process. McKeague's *Elementary Algebra* is a bridge to student learning, and **Enhanced WebAssign** builds a bridge between the text's proven content and powerful learning resources!

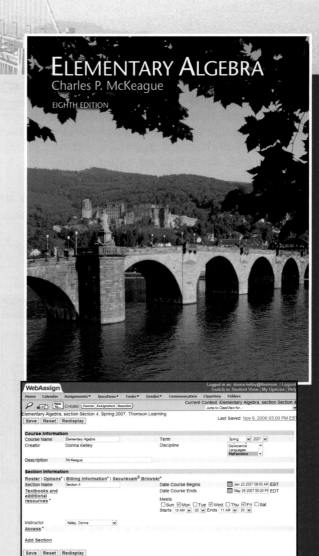

ELEMENTARY ALGEBRA
Charles P. McKeague
EIGHTH EDITION

Open here to learn more about **Enhanced WebAssign**!

Enhanced WebAssign
has the features you need!

* Gradebook with weighted categories and scores
* Student forums, bulletin boards, messaging
* Textbook and open source questions
* Textbook PDFs for quick, portable reference
* 1,500 randomized, algorithmic questions
* Proper display of mathematical expressions
* Timed, secure, password- and IP-restricted exercises
* Built-in calendar and communication tools
* Simulations, animations, and videos
* Windows® and Apple® Macintosh compatible
* Easy to use and reliable

Instructors: Take your free 45-Day Test Drive!

Visit **www.webassign.net/brookscole**

Enhanced WebAssign content is also available with these developmental mathematics texts from Pat McKeague:

**Intermediate Algebra,
Eighth Edition**
0-495-10840-5

**Elementary and Intermediate
Algebra, Third Edition**
0-495-10851-0

**Basic College Mathematics:
A Text/Workbook,
Second Edition**
0-495-01391-9

**Beginning Algebra:
A Text/Workbook,
Seventh Edition**
0-495-01252-1

**Intermediate Algebra:
A Text-Workbook,
Seventh Edition**
0-495-01250-5

THOMSON
™
BROOKS/COLE

*Contact your local Thomson Brooks/Cole
representative for packaging and ordering information.*

ELEMENTARY ALGEBRA

EIGHTH EDITION

Charles P. McKeague

CUESTA COLLEGE

THOMSON

BROOKS/COLE

Australia • Brazil • Canada • Mexico • Singapore • Spain
United Kingdom • United States

Executive Editor: *Charlie Van Wagner*
Development Editor: *Donald Gecewicz*
Assistant Editor: *Laura Localio*
Editorial Assistant: *Lisa Lee*
Technology Project Manager: *Rebecca Subity*
Marketing Manager: *Greta Kleinert*
Marketing Assistant: *Cassandra Cummings*
Marketing Communications Manager: *Darlene Amidon-Brent*
Project Manager, Editorial Production: *Cheryll Linthicum*
Creative Director: *Rob Hugel*
Art Director: *Vernon T. Boes*
Print Buyer: *Karen Hunt*

Permissions Editor: *Roberta Broyer*
Production Service: *Graphic World Publishing Services*
Text Designer: *Diane Beasley*
Photo Researcher: *Kathleen Olson*
Illustrator: *Graphic World Illustration Studio*
Cover Designer: *Diane Beasley*
Cover Image: Arch bridge across a river, Neckar River, Heidelberg Castle, Heidelberg, Germany, *Glowimages/Getty Images*
Cover Printer: *R.R. Donnelley/Willard*
Compositor: *Graphic World Inc.*
Printer: *R.R. Donnelley/Willard*

© 2008, 2004 Thomson Brooks/Cole, a part of The Thomson Corporation. Thomson, the Star logo, and Brooks/Cole are trademarks used herein under license.

ALL RIGHTS RESERVED. No part of this work covered by the copyright hereon may be reproduced or used in any form or by any means—graphic, electronic, or mechanical, including but not limited to photocopying, recording, taping, Web distribution, information networks, or information storage and retrieval systems—without the written permission of the publisher.

Printed in the United States of America
1   2   3   4   5   6   7   11   10   09   08   07

ExamView® and *ExamView Pro®* are registered trademarks of FSCreations, Inc. Windows is a registered trademark of the Microsoft Corporation used herein under license. Macintosh and Power Macintosh are registered trademarks of Apple Computer, Inc. Used herein under license.

Thomson Higher Education
10 Davis Drive
Belmont, CA 94002-3098
USA

Library of Congress Control Number: 2006937868

Student Edition:
ISBN-13: 978-0-495-10839-9
ISBN-10: 0-495-10839-1

For more information about our products, contact us at:
Thomson Learning Academic Resource Center
1-800-423-0563
For permission to use material from this text or product, contact us at: **http://www.thomsonrights.com.**
Any additonal questions about permissions can be submitted by e-mail to **thomsonrights@thomson.com.**

Brief Contents

Contents

Preface to the Instructor

I have a passion for teaching mathematics. That passion carries through to my textbooks. My goal is a textbook that is user-friendly for both students and instructors. For students, this book forms a bridge to intermediate algebra with clear, concise writing, continuous review, and foreshadowing of topics to come. For the instructor, I build features into the text that reinforce the habits and study skills we know will bring success to our students.

The eighth edition of *Elementary Algebra* builds upon these strengths. In this edition, renewal of the problem sets was my first priority. This renewal of the problem sets, along with the continued emphasis on foreshadowing of later topics, a program of continuous cumulative review, and the focus on applications, make this the best edition of *Elementary Algebra* yet.

Renewal and Reorganization of Exercise Sets

This edition of *Elementary Algebra* contains 1,400 new problems, roughly 30 percent of the problems in the reorganized problem sets. Most of these new problems have been used to shore up the midrange of exercises. Doing so gives our problem sets a better bridge from easy problems to more difficult problems. Many of these midrange of problems cover more than one concept or technique in a new and slightly more challenging way—in short, they start students thinking mathematically and working with algebra productively. This enhanced midrange of exercises also underscores our series' appeal to the middle level of rigor for the course. We think instructors will use many of these midrange problems as classroom examples, so we have labeled some of them as *chalkboard problems*.

As part of our revamping of the problem sets in this edition, we have also reordered the categories of problems to make a more logical bridge between sections.

A Better Progression of Categories of Problems

The categories in our problem sets now appear in the following order.

General (Undesignated) Exercises These exercises normally do not have labels. They involve a certain amount of the drill necessary to master basic techniques. These problems then progress in difficulty so that students can begin to put together more than one concept or idea. It is here that you will find the foreshadowing problems. Instead of drill for the sake of drill, we have students work the problems that they will need later in the course—hence the description *foreshadowing*. This is also where you will find the midrange of problems discussed earlier. As in previous editions, we have kept the odd-even similarity of the problems in this part of the problem set.

Applying the Concepts Students are always curious about how the algebra they are learning can be applied, so we have included applied problems in most of the problem sets in the book and have labeled them to show students the array of uses of mathematics. These applied problems are written in an inviting way, to help students overcome some of the apprehension associated with application problems. We have a number of new applications under the heading *Improving Your Quantitative Literacy* that are particularly inviting.

Maintaining Your Skills One of the major themes of our book is continuous review. We strive to continuously hone techniques learned earlier by keeping the important concepts in the forefront of the course. The *Maintaining Your Skills* problems review material from the previous chapter, or they review problems that form the foundation of the course—the problems that you expect students to be able to solve when they get to the next course.

Getting Ready for the Next Section Many students think of mathematics as a collection of discrete, unrelated topics. Their instructors know that this is not the case. The new *Getting Ready for the Next Section* problems reinforce the cumulative, connected nature of this course by showing how the concepts and techniques flow one from another throughout the course. These problems review all of the material that students will need in order to be successful, forming a bridge to the next section, gently preparing students to move forward.

The Elementary Course as a Bridge to Further Success

Elementary algebra is a bridge course. The course and its syllabus bring the student to the level of ability required to do quantitative work in their major, while getting them ready to make a successful start in intermediate algebra. The algebraic concepts are bridges linked cumulatively to other and more advanced algebra concepts.

Our Proven Commitment to Student Success

After seven successful editions, we have developed several interlocking, proven features that will improve students' chances of success in the course. We place practical, easily understood study skills in the first six chapters (look at the page after the chapter opener). Here are some of the other, important success features of the book.

Getting Ready for Class Just before each problem set is a list of four questions under the heading *Getting Ready for Class.* These problems require written responses from students and are to be done before students come to class. The answers can be found by reading the preceding section. These questions reinforce the importance of reading the section before coming to class.

Linking Objectives and Examples This feature at the end of each section helps students to understand how the section and its examples are built around objectives. We think that this feature helps to make the structure of exposition of concepts clearer

Blueprint for Problem Solving Found in the main text, this feature is a detailed outline of steps required to successfully attempt application problems. Intended as a guide to problem solving in general, the blueprint takes the student through the solution process to various kinds of applications.

Maintaining Your Skills We believe that students who consistently work review problems will be much better prepared for class than students who do not engage in continuous review. The *Maintaining Your Skills* problems cumulatively review the most important concepts from the previous chapter as well as concepts that form the foundation of the course.

Getting Ready for the Next Section At the ends of section problem sets, you will find *Getting Ready for the Next Section,* a category of problems that students can work to prepare themselves to navigate the next section successfully. These problems polish techniques and reinforce the idea that all topics in the course are built on previous topics.

End-of-Chapter Summary, Review, and Assessment

We have learned that students are more comfortable with a chapter that sums up what they have learned thoroughly and accessibly, and reinforces concepts and techniques well. To help students grasp concepts and get more practice, each chapter ends with the following features that together give a comprehensive reexamination of the chapter.

Chapter Summary The chapter summary recaps all main points from the chapter in a visually appealing grid. In the margin, next to each topic, is an example that illustrates the type of problem associated with the topic being reviewed. Our way of summarizing shows students that concepts in mathematics do relate—and that mastering one concept is a bridge to the next. When students prepare for a test, they can use the chapter summary as a guide to the main concepts of the chapter.

Chapter Review/Test Following the chapter summary in each chapter is the chapter review/test. It contains an extensive set of problems that review all the main topics in the chapter. This feature can be used flexibly, as assigned review, as a recommended self-test for students as they prepare for examinations, or as an in-class quiz or test.

Chapter Projects Each chapter closes with a pair of projects. One is a group project, suitable for students to work on in class. Group projects list details about number of participants, equipment, and time, so that instructors can determine how well the project fits into their classroom. The second project is a research project for students to do outside of class and tends to be open ended.

Additional Features of the Book

Facts from Geometry Many of the important facts from geometry are listed under this heading. In most cases, an example or two accompanies each of the facts to give students a chance to see how topics from geometry are related to the algebra they are learning.

Unit Analysis Chapter 7 contains problems requiring students to convert from one unit of measure to another. The method used to accomplish the conversions is the method they will use if they take a chemistry class. Since this method is similar to the method we use to multiply rational expressions, unit analysis is covered in Section 7.2 as an application of multiplying rational expressions.

Chapter Openings Each chapter opens with an introduction in which a real-world application is used to stimulate interest in the chapter. We expand on these opening applications later in the chapter.

Supplements

Test Bank (0495383082) Draw from hundreds of text-specific questions to easily create tests that target your course objectives. The Test Bank includes multiple tests per chapter as well as final exams. The tests are made up of a combination of multiple-choice, free-response, true/false, and fill-in-the-blank questions.

Text-Specific Videos (0495383090) These DVD sets created by Pat McKeague are available at no charge to qualified adopters of the text. They feature 10- to 20-minute problem-solving lessons that cover each section of every chapter.

Student Solutions Manual (0495383139) Provides worked-out solutions to the odd-numbered problems in the text.

Annotated Instructor's Edition (0495383295) The Instructor's Edition provides the complete student text with answers next to each respective exercise.

Complete Solutions Manual (0495383279) Provides worked-out solutions to all of the problems in the text.

Printed Access Card (ThomsonNOW™) (0495392480) This printed access card provides entrance to all the content that accompanies McKeague's *Elementary Algebra* within ThomsonNOW. This powerful and fully integrated teaching and learning system provides instructors and students with unsurpassed control, variety, and all-in-one utility. ThomsonNOW ties together the fundamental learning activities: diagnostics, tutorials, homework, personalized study, quizzing, and testing. The Personalized Study is a learning companion that helps students gauge their unique study needs and makes the most of their study time by building focused Personalized Study plans that reinforce key concepts. **Pre-Tests** give students an initial assessment of their knowledge. **Personalized Study** plans, based on the students' answers to the pre-test questions, outline key elements for review. **Post-Tests** assess student mastery of core chapter concepts; results can be e-mailed to the instructor!

JoinIn™ on TurningPoint® (0495383309) Book-specific content for classroom response systems allows you to transform your classroom and assess student progress with instant in-class quizzes and polls. Pose book-specific questions and display students' answers seamlessly within the Microsoft® PowerPoint® slides of your own lecture, in conjunction with the "clicker" hardware of your choice. Enhance how your students interact with you, your lecture, and each other. For college and university adopters only. Contact your local Thomson representative to learn more.

Enhanced WebAssign Instant feedback and ease of use are just two reasons why WebAssign is the most widely used homework system in higher education. WebAssign's homework delivery system lets you deliver, collect, grade, and record assignments via the web. And now, this proven system has been enhanced to include end-of-chapter problems from McKeague's *Elementary Algebra,* Eighth Edition—incorporating figures, videos, examples, PDF pages of the text, and quizzes to promote active learning and provide the immediate, relevant feedback students want.

ThomsonNOW with Personalized Study (0495392472) This powerful and fully integrated teaching and learning system provides instructors and students with unsurpassed control, variety, and all-in-one utility. ThomsonNOW ties together the fundamental learning activities: diagnostics, tutorials, homework, personalized study, quizzing, and testing. The Personalized Study is a learning companion that helps students gauge their unique study needs and makes the most of their study time by building focused Personalized Study plans that reinforce key concepts. **Pre-Tests** give students an initial assessment of their knowledge. **Personalized Study** plans, based on the students' answers to the pre-test questions, outline key elements for review. **Post-Tests** assess student mastery of core chapter concepts; results can be e-mailed to the instructor!

Blackboard ThomsonNOW Integration (0495383473) Combines easy-to-use course management tools with content from this text's rich companion website. Ready to use as soon as you log on—or, customize WebTutor ToolBox with web links, images, and other resources.

WebCT ThomsonNOW Integration (0495383481) Combines easy-to-use course management tools with content from this text's rich companion website. Ready to use as soon as you log on—or, customize WebTutor ToolBox with web links, images, and other resources.

Instant Access Code (ThomsonNOW) (0495392499) Gives students without a new copy of the text one access code to all available technology associated with this product. ThomsonNOW, a powerful and fully integrated teaching and learning system, provides instructors and students with unsurpassed control, variety, and all-in-one utility. ThomsonNOW ties together the fundamental learning activities: diagnostics, tutorials, homework, personalized study, quizzing, and testing. The Personalized Study is a learning companion that helps students gauge their unique study needs and makes the most of their study time by building focused Personalized Study plans that reinforce key concepts. **Pre-Tests** give students an initial assessment of their knowledge. **Personalized Study** plans, based on the students' answers to the pre-test questions, outline key elements for review. **Post-Tests** assess student mastery of core chapter concepts; results can be e-mailed to the instructor!

ExamView (0495383317) A computerized testbank on CD allows instructors to create exams quickly and easily.

Website www.thomsonedu.com/math/mckeague
Instant access to the Student Resource Center—a rich teaching and learning resources—including chapter-by-chapter online tutorial quizzes, a final exam, chapter outlines, chapter review, chapter-by-chapter web links, videos, flash cards, and more!

Acknowledgments

I would like to thank my editor at Brooks/Cole, Charlie Van Wagner, for his help and encouragement with this project. Many thanks also to Don Gecewicz, my developmental editor, for his suggestions on content, his proofreading, and his availability for consulting. This is a better book because of Don. Patrick McKeague, Tammy Fisher-Vasta, and Devin Christ assisted me with all parts of this revision, from manuscript preparation to proofreading page proofs and preparing the index. They are a fantastic team to work with, and this project could not have been completed without them. Susan Caire did an excellent job of proofreading the entire book in page proofs, and Jeff Brower of Chaffee Community College and Steve Marsden of Glendale Community College contributed their expertise to the accuracy check of page proofs. Mary Gentilucci, Shane Wilwand, and Annie Stephens assisted with error checking and proofreading. Thanks to Rebecca Subity and Laura Localio for handling the media and ancillary packages on this project. Cheryll Linthicum of Brooks/Cole and Carol O'Connell of Graphic World Publishing Services turned the manuscript into a book. Ross Rueger produced the excellent solutions manuals that accompany the book.

Thanks also to Diane McKeague and Amy Jacobs for their encouragement with all my writing endeavors.

Finally, I am grateful to the following instructors for their suggestions and comments: Jess L. Collins, McLennan Community College; Richard Drey, Northampton Community College; Peg Hovde, Grossmont College; Sarah Jackman, Richland College; Carol Juncker, Delgado Community College; Joanne Kendall, College of the Mainland; Harriet Kiser, Floyd College; Domíngo Javier Lítong, South Texas Community College; Cindy Lucas, College of the Mainland; Jan MacInnes, Florida Community College of Jacksonville; Rudolfo Maglio, Oakton Community College; Janice McFatter, Gulf Coast Community College; Nancy Olson, Johnson County Community College; John H. Pleasants, Orange County Community College; Barbara Jane Sparks, Camden County College; Jim Stewart, Jefferson Community College; David J. Walker, Hinds Community College; and Deborah Woods, University of Cincinnati.

Charles P. McKeague
January 2007

The Basics

Caryl Bryer Fallert

Much of what we do in mathematics is concerned with recognizing patterns. If you recognize the patterns in the following two sequences, then you can easily extend each sequence.

Sequence of odd numbers = 1, 3, 5, 7, 9, . . .

Sequence of squares = 1, 4, 9, 16, 25, . . .

Once we have classified groups of numbers as to the characteristics they share, we sometimes discover that a relationship exists between the groups. Although it may not be obvious at first, there is a relationship that exists *between* the two sequences shown. The introduction to *The Book of Squares,* written in 1225 by the mathematician known as Fibonacci, begins this way:

> I thought about the origin of all square numbers and discovered that they arise out of the increasing sequence of odd numbers.

The relationship that Fibonacci refers to is shown visually in Figure 1.

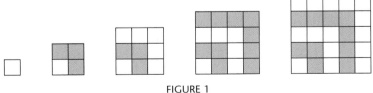

FIGURE 1

Many times we describe a relationship or pattern in a number of different ways. Figure 1 is a visual description of a relationship. In this chapter we will work on describing relationships numerically and verbally (in writing).

▶ Improve your grade and save time!
Go online to **www.thomsonedu.com/login**
where you can
- Watch videos of instructors working through the in-text examples
- Follow step-by-step online tutorials of in-text examples and review questions
- Work practice problems
- Check your readiness for an exam by taking a pre-test and exploring the modules recommended in your Personalized Study plan
- Receive help from a live tutor online through vMentor™

Try it out! Log in with an access code or purchase access at **www.ichapters.com**.

Some of the students enrolled in my elementary algebra classes develop difficulties early in the course. Their difficulties are not associated with their ability to learn mathematics; they all have the potential to pass the course. Students who get off to a poor start do so because they have not developed the study skills necessary to be successful in algebra. Here is a list of things you can do to begin to develop effective study skills.

1 Put Yourself on a Schedule

The general rule is that you spend 2 hours on homework for every hour you are in class. Make a schedule for yourself in which you set aside 2 hours each day to work on algebra. Once you make the schedule, stick to it. Don't just complete your assignments and stop. Use all the time you have set aside. If you complete an assignment and have time left over, read the next section in the book, and then work more problems.

2 Find Your Mistakes and Correct Them

There is more to studying algebra than just working problems. You must always check your answers with the answers in the back of the book. When you have made a mistake, find out what it is and correct it. Making mistakes is part of the process of learning mathematics. In the prologue to *The Book of Squares*, Leonardo Fibonacci (ca. 1170–ca. 1250) had this to say about the content of his book:

> I have come to request indulgence if in any place it contains something more or less than right or necessary; for to remember everything and be mistaken in nothing is divine rather than human . . .

Fibonacci knew, as you know, that human beings make mistakes. You cannot learn algebra without making mistakes.

3 Gather Information on Available Resources

You need to anticipate that you will need extra help sometime during the course. There is a form to fill out in Appendix A to help you gather information on resources available to you. One resource is your instructor; you need to know your instructor's office hours and where the office is located. Another resource is the math lab or study center, if they are available at your school. It also helps to have the phone numbers of other students in the class, in case you miss class. You want to anticipate that you will need these resources, so now is the time to gather them together.

Suppose you have a checking account that costs you $15 a month, plus $0.05 for each check you write. If you write 10 checks in a month, then the monthly charge for your checking account will be

$$15 + 10(0.05)$$

Do you add 15 and 10 first and then multiply by 0.05? Or do you multiply 10 and 0.05 first and then add 15? If you don't know the answer to this question, you will after you have read through this section.

Because much of what we do in algebra involves comparison of quantities, we will begin by listing some symbols used to compare mathematical quantities. The comparison symbols fall into two major groups: equality symbols and inequality symbols.

We will let the letters a and b stand for (represent) any two mathematical quantities. When we use letters to represent numbers, as we are doing here, we call the letters *variables.*

Variables: An Intuitive Look

When you filled out the application for the school you are attending, there was a space to fill in your first name. "First name" is a variable quantity because the value it takes depends on who is filling out the application. For example, if your first name is Manuel, then the value of "First Name" is Manuel. However, if your first name is Christa, then the value of "First Name" is Christa.

If we denote "First Name" as FN, "Last Name" as LN, and "Whole Name" as WN, then we take the concept of a variable further and write the relationship between the names this way:

$$FN + LN = WN$$

(We use the + symbol loosely here to represent writing the names together with a space between them.) This relationship we have written holds for all people who have only a first name and a last name. For those people who have a middle name, the relationship between the names is

$$FN + MN + LN = WN$$

A similar situation exists in algebra when we let a letter stand for a number or a group of numbers. For instance, if we say "let a and b represent numbers," then a and b are called *variables* because the values they take on vary. We use the variables a and b in the following lists so that the relationships shown there are true for all numbers that we will encounter in this book. By using variables, the following statements are general statements about all numbers, rather than specific statements about only a few numbers.

> **Comparison Symbols**
>
> | | | |
> |---|---|---|
> | *Equality:* | $a = b$ | a is equal to b (a and b represent the same number) |
> | | $a \neq b$ | a is not equal to b |
> | *Inequality:* | $a < b$ | a is less than b |
> | | $a \not< b$ | a is not less than b |
> | | $a > b$ | a is greater than b |
> | | $a \not> b$ | a is not greater than b |
> | | $a \geq b$ | a is greater than or equal to b |
> | | $a \leq b$ | a is less than or equal to b |

The symbols for inequality, $<$ and $>$, always point to the smaller of the two quantities being compared. For example, $3 < x$ means 3 is smaller than x. In this case we can say "3 is less than x" or "x is greater than 3"; both statements are correct. Similarly, the expression $5 > y$ can be read as "5 is greater than y" or as "y is less than 5" because the inequality symbol is pointing to y, meaning y is the smaller of the two quantities.

Next, we consider the symbols used to represent the four basic operations: addition, subtraction, multiplication, and division.

Note
In the past you may have used the notation 3×5 to denote multiplication. In algebra it is best to avoid this notation if possible, because the multiplication symbol $\times$ can be confused with the variable x when written by hand.

> **Operation Symbols**
>
> | | | |
> |---|---|---|
> | *Addition:* | $a + b$ | The *sum* of a and b |
> | *Subtraction:* | $a - b$ | The *difference* of a and b |
> | *Multiplication:* | $a \cdot b$, $(a)(b)$, $a(b)$, $(a)b$, ab | The *product* of a and b |
> | *Division:* | $a \div b$, a/b, $\dfrac{a}{b}$, $b\overline{)a}$ | The *quotient* of a and b |

When we encounter the word *sum,* the implied operation is addition. To find the sum of two numbers, we simply add them. *Difference* implies subtraction, *product* implies multiplication, and *quotient* implies division. Notice also that there is more than one way to write the product or quotient of two numbers.

> **Grouping Symbols** Parentheses ( ) and brackets [ ] are the symbols used for grouping numbers together. (Occasionally, braces { } are also used for grouping, although they are usually reserved for set notation, as we shall see.)

The following examples illustrate the relationship between the symbols for comparing, operating, and grouping and the English language.

 EXAMPLES

Mathematical Expression	English Equivalent
1. $4 + 1 = 5$	The sum of 4 and 1 is 5.
2. $8 - 1 < 10$	The difference of 8 and 1 is less than 10.
3. $2(3 + 4) = 14$	Twice the sum of 3 and 4 is 14.

4. $3x \geq 15$ The product of 3 and x is greater than or equal to 15.

5. $\dfrac{y}{2} = y - 2$ The quotient of y and 2 is equal to the difference of y and 2.

The last type of notation we need to discuss is the notation that allows us to write repeated multiplications in a more compact form—*exponents.* In the expression 2^3, the 2 is called the *base* and the 3 is called the *exponent.* The exponent 3 tells us the number of times the base appears in the product; that is,

$$2^3 = 2 \cdot 2 \cdot 2 = 8$$

The expression 2^3 is said to be in exponential form, whereas $2 \cdot 2 \cdot 2$ is said to be in expanded form. Here are some additional examples of expressions involving exponents.

EXAMPLES   Expand and multiply.

6. $5^2 = 5 \cdot 5 = 25$ Base 5, exponent 2

7. $2^5 = 2 \cdot 2 \cdot 2 \cdot 2 \cdot 2 = 32$ Base 2, exponent 5

8. $10^3 = 10 \cdot 10 \cdot 10 = 1{,}000$ Base 10, exponent 3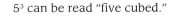

Notation and Vocabulary Here is how we read expressions containing exponents.

Mathematical Expression	Written Equivalent
5^2	five to the second power
5^3	five to the third power
5^4	five to the fourth power
5^5	five to the fifth power
5^6	five to the sixth power

We have a shorthand vocabulary for second and third powers because the area of a square with a side of 5 is 5^2, and the volume of a cube with a side of 5 is 5^3.

5^2 can be read "five squared." 5^3 can be read "five cubed."

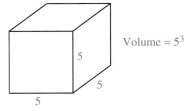

The symbols for comparing, operating, and grouping are to mathematics what punctuation symbols are to English. These symbols are the punctuation symbols for mathematics.

Consider the following sentence:

Paul said John is tall.

It can have two different meanings, depending on how it is punctuated.

1. "Paul," said John, "is tall."

2. Paul said, "John is tall."

Let's take a look at a similar situation in mathematics. Consider the following mathematical statement:

$$5 + 2 \cdot 7$$

If we add the 5 and 2 first and then multiply by 7, we get an answer of 49. However, if we multiply the 2 and the 7 first and then add 5, we are left with 19. We have a problem that seems to have two different answers, depending on whether we add first or multiply first. We would like to avoid this type of situation. Every problem like $5 + 2 \cdot 7$ should have only one answer. Therefore, we have developed the following rule for the order of operations.

> **Rule (Order of Operations)** When evaluating a mathematical expression, we will perform the operations in the following order, beginning with the expression in the innermost parentheses or brackets first and working our way out.
> 1. Simplify all numbers with exponents, working from left to right if more than one of these expressions is present.
> 2. Then do all multiplications and divisions left to right.
> 3. Perform all additions and subtractions left to right.

 EXAMPLES Simplify each expression using the rule for order of operations.

9. $5 + 8 \cdot 2 = 5 + 16$ Multiply $8 \cdot 2$ first
$\quad = 21$

10. $12 \div 4 \cdot 2 = 3 \cdot 2$ Work left to right
$\quad = 6$

11. $2[5 + 2(6 + 3 \cdot 4)] = 2[5 + 2(6 + 12)]$ Simplify within the innermost
$\quad = 2[5 + 2(18)]$ grouping symbols first
$\quad = 2[5 + 36]$ Next, simplify inside
$\quad = 2[41]$ the brackets
$\quad = 82$ Multiply

12. $10 + 12 \div 4 + 2 \cdot 3 = 10 + 3 + 6$ Multiply and divide left to right
$\quad = 19$ Add left to right

13. $2^4 + 3^3 \div 9 - 4^2 = 16 + 27 \div 9 - 16$ Simplify numbers with exponents
$\quad = 16 + 3 - 16$ Then, divide
$\quad = 19 - 16$ Finally, add and subtract
$\quad = 3$ left to right

Reading Tables and Bar Charts

The following table shows the average amount of caffeine in a number of beverages. The diagram in Figure 1 is a bar chart. It is a visual presentation of the information in Table 1. The table gives information in numerical form, whereas the chart gives the same information in a geometric way. In mathematics, it is important to be able to move back and forth between the two forms.

TABLE 1
Caffeine Content of Hot Drinks

Drink (6-ounce cup)	Caffeine (milligrams)
Brewed coffee	100
Instant coffee	70
Tea	50
Cocoa	5
Decaffeinated coffee	4

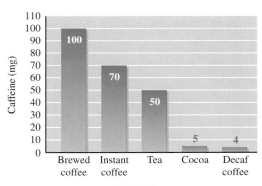

FIGURE 1

 EXAMPLE 14   Referring to Table 1 and Figure 1, suppose you have 3 cups of brewed coffee, 1 cup of tea, and 2 cups of decaf in one day. Write an expression that will give the total amount of caffeine in these six drinks, then simplify the expression.

SOLUTION   From the table or the bar chart, we find the number of milligrams of caffeine in each drink; then we write an expression for the total amount of caffeine:

$$3(100) + 50 + 2(4)$$

Using the rule for order of operations, we get 358 total milligrams of caffeine.

Number Sequences and Inductive Reasoning

Suppose someone asks you to give the next number in the sequence of numbers below. (The dots mean that the sequence continues in the same pattern forever.)

$$2, 5, 8, 11, \ldots$$

If you notice that each number is 3 more than the number before it, you would say the next number in the sequence is 14 because $11 + 3 = 14$. When we reason in this way, we are using what is called *inductive reasoning*. In mathematics we use inductive reasoning when we notice a pattern to a sequence of numbers and then use the pattern to extend the sequence.

 EXAMPLE 15   Find the next number in each sequence.
    a. 3, 8, 13, 18, . . .
    b. 2, 10, 50, 250, . . .
    c. 2, 4, 7, 11, . . .

SOLUTION   To find the next number in each sequence, we need to look for a pattern or relationship.

    a. For the first sequence, each number is 5 more than the number before it; therefore, the next number will be $18 + 5 = 23$.
    b. For the sequence in part (b), each number is 5 times the number before it; therefore, the next number in the sequence will be $5 \cdot 250 = 1,250$.

c. For the sequence in part (c), there is no number to add each time or multiply by each time. However, the pattern becomes apparent when we look at the differences between the numbers:

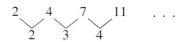

Proceeding in the same manner, we would add 5 to get the next term, giving us $11 + 5 = 16$.

In the introduction to this chapter we mentioned the mathematician known as Fibonacci. There is a special sequence in mathematics named for Fibonacci. Here it is.

Fibonacci sequence = 1, 1, 2, 3, 5, 8, . . .

Can you see the relationship among the numbers in this sequence? Start with two 1's, then add two consecutive members of the sequence to get the next number. Here is a diagram.

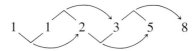

Sometimes we refer to the numbers in a sequence as *terms* of the sequence.

EXAMPLE 16 Write the first 10 terms of the Fibonacci sequence.

SOLUTION The first six terms are given above. We extend the sequence by adding 5 and 8 to obtain the seventh term, 13. Then we add 8 and 13 to obtain 21. Continuing in this manner, the first 10 terms in the Fibonacci sequence are

1, 1, 2, 3, 5, 8, 13, 21, 34, 55

GETTING READY FOR CLASS

LINKING OBJECTIVES AND EXAMPLES

Next to each **objective** we have listed the examples that are best described by that objective. Connecting the examples to the objectives gives us a more complete understanding of the overall structure of the section.

A	1–5
B	6–8
C	9–13
D	14
E	15, 16

Each section of the book will end with some problems and questions like the ones below. They are for you to answer after you have read through the section but before you go to class. All of them require that you give written responses in complete sentences. Writing about mathematics is a valuable exercise. If you write with the intention of explaining and communicating what you know to someone else, you will find that you understand the topic you are writing about even better than you did before you started writing. As with all problems in this course, you want to approach these writing exercises with a positive point of view. You will get better at giving written responses to questions as you progress through the course. Even if you never feel comfortable writing about mathematics, just the process of attempting to do so will increase your understanding and ability in mathematics.

After reading through the preceding section, respond in your own words and in complete sentences.

1. What is a variable?
2. Write the first step in the rule for order of operations.
3. What is inductive reasoning?
4. Explain the relationship between an exponent and its base.

Answers appear in the Instructor's Edition only.

For each sentence below, write an equivalent expression in symbols.

1. The sum of x and 5 is 14. $x + 5 = 14$

2. The difference of x and 4 is 8. $x - 4 = 8$

▶ **3.** The product of 5 and y is less than 30. $5y < 30$

4. The product of 8 and y is greater than 16. $8y > 16$

5. The product of 3 and y is less than or equal to the sum of y and 6. $3y \leq y + 6$

6. The product of 5 and y is greater than or equal to the difference of y and 16. $5y \geq y - 16$

7. The quotient of x and 3 is equal to the sum of x and 2. $\frac{x}{3} = x + 2$

8. The quotient of x and 2 is equal to the difference of x and 4. $\frac{x}{2} = x - 4$

Expand and multiply.

9. 3^2 9

10. 4^2 16

11. 7^2 49

12. 9^2 81

13. 2^3 8

14. 3^3 27

▶ **15.** 4^3 64

16. 5^3 125

17. 2^4 16

18. 3^4 81

19. 10^2 100

20. 10^4 10,000

21. 11^2 121

22. 111^2 12,321

Use the rule for order of operations to simplify each expression as much as possible.

23. a. $2 \cdot 3 + 5$ 11
   b. $2(3 + 5)$ 16

24. a. $8 \cdot 7 + 1$ 57
   b. $8(7 + 1)$ 64

▶ **25. a.** $5 + 2 \cdot 6$ 17
   b. $(5 + 2) \cdot 6$ 42

26. a. $8 + 9 \cdot 4$ 44
   b. $(8 + 9) \cdot 4$ 68

27. a. $5 \cdot 4 + 5 \cdot 2$ 30
   b. $5(4 + 2)$ 30

28. a. $6 \cdot 8 + 6 \cdot 3$ 66
   b. $6(8 + 3)$ 66

29. a. $8 + 2(5 + 3)$ 24
   b. $(8 + 2)(5 + 3)$ 80

30. a. $7 + 3(8 - 2)$ 25
   b. $(7 + 3)(8 - 2)$ 60

31. $20 + 2(8 - 5) + 1$ 27

32. $10 + 3(7 + 1) + 2$ 36

33. $5 + 2(3 \cdot 4 - 1) + 8$ 35

34. $11 - 2(5 \cdot 3 - 10) + 2$ 3

▶ **35.** $4 + 8 \div 4 - 2$ 4

36. $6 + 9 \div 3 + 2$ 11

37. $3 \cdot 8 + 10 \div 2 + 4 \cdot 2$ 37

38. $5 \cdot 9 + 10 \div 2 + 3 \cdot 3$ 59

39. a. $(5 + 3)(5 - 3)$ 16
   b. $5^2 - 3^2$ 16

40. a. $(7 + 2)(7 - 2)$ 45
   b. $7^2 - 2^2$ 45

41. a. $(4 + 5)^2$ 81
   b. $4^2 + 5^2$ 41

42. a. $(6 + 3)^2$ 81
   b. $6^2 + 3^2$ 45

43. $2 \cdot 10^3 + 3 \cdot 10^2 + 4 \cdot 10 + 5$ 2,345

44. $5 \cdot 10^3 + 6 \cdot 10^2 + 7 \cdot 10 + 8$ 5,678

▶ **45.** $10 - 2(4 \cdot 5 - 16)$ 2

46. $15 - 5(3 \cdot 2 - 4)$ 5

47. $4[7 + 3(2 \cdot 9 - 8)]$ 148

48. $5[10 + 2(3 \cdot 6 - 10)]$ 130

49. $3(4 \cdot 5 - 12) + 6(7 \cdot 6 - 40)$ 36

50. $6(8 \cdot 3 - 4) + 5(7 \cdot 3 - 1)$ 220

▶ **51.** $3^4 + 4^2 \div 2^3 - 5^2$ 58

52. $2^5 + 6^2 \div 2^2 - 3^2$ 32

53. $5^2 + 3^4 \div 9^2 + 6^2$ 62

54. $6^2 + 2^5 \div 4^2 + 7^2$ 87

Simplify each expression.

55. $20 \div 2 \cdot 10$ 100

56. $40 \div 4 \cdot 5$ 50

57. $24 \div 8 \cdot 3$ 9

58. $24 \div 4 \cdot 6$ 36

59. $36 \div 6 \cdot 3$ 18

60. $36 \div 9 \cdot 2$ 8

61. $16 - 8 + 4$ 12

62. $16 - 4 + 8$ 20

63. $24 - 14 + 8$ 18

64. $24 - 16 + 6$ 14

65. $36 - 6 + 12$ 42

66. $36 - 9 + 20$ 47

We are assuming that you know how to do arithmetic with decimals. Here are some problems to practice. Simplify each expression.

▶ **67.** $0.08 + 0.09$ 0.17

▶ **68.** $0.06 + 0.04$ 0.10

▶ **69.** $0.10 + 0.12$ 0.22

▶ **70.** $0.08 + 0.06$ 0.14

▶ **71.** $4.8 - 2.5$ 2.3

▶ **72.** $6.3 - 4.8$ 1.5

▶ **73.** $2.07 + 3.48$ 5.55

▶ **74.** $4.89 + 2.31$ 7.20

▶ **75.** $0.12(2,000)$ 240

▶ **76.** $0.09(3,000)$ 270

▶ **77.** $0.25(40)$ 10

▶ **78.** $0.75(40)$ 30

▶ **79.** $510 \div 0.17$ 3,000

▶ **80.** $400 \div 0.1$ 4,000

▶ **81.** $240 \div 0.12$ 2,000

▶ **82.** $360 \div 0.12$ 3,000

Use a calculator to find the following quotients. Round your answers to the nearest hundredth, if necessary.

83. $37.80 \div 1.07$ 35.33

84. $85.46 \div 4.88$ 17.51

85. $555 \div 740$ 0.75

86. $740 \div 108$ 6.85

87. $70 \div 210$ 0.33

88. $15 \div 80$ 0.19

89. $6,000 \div 22$ 272.73

90. $51,000 \div 17$ 3,000

= Videos available by instructor request

▶ = Online student support materials available at www.thomsonedu.com/login

Applying the Concepts

Food Labels In 1993 the government standardized the way in which nutrition information was presented on the labels of most packaged food products. Figure 2 shows a standardized food label from a package of cookies that I ate at lunch the day I was writing the problems for this problem set. Use the information in Figure 2 to answer the following questions.

Nutrition Facts	Amount/serving	%DV*	Amount/serving	%DV*
Serving Size 5 Cookies (about 43 g)	**Total Fat** 9 g	**15%**	**Total Carb.** 30 g	**10%**
Servings Per Container 2	Sat. Fat 2.5 g	**12%**	Fiber 1 g	**2%**
Calories 210	**Cholest.** less than 5 mg	**2%**	Sugars 14 g	
Fat Calories 90	**Sodium** 110 mg	**5%**	**Protein** 3 g	

* Percent Daily Values (DV) are based on a 2,000 calorie diet.

Vitamin A 0% • Vitamin C 0% • Calcium 2% • Iron 8%

Sandwich cremes

FIGURE 2

91. How many cookies are in the package?   10

92. If I paid $0.50 for the package of cookies, how much did each cookie cost? $0.05

93. If the "calories" category stands for calories per serving, how many calories did I consume by eating the whole package of cookies?   420

94. Suppose that, while swimming, I burn 11 calories each minute. If I swim for 20 minutes, will I burn enough calories to cancel out the calories I added by eating 5 cookies?   Yes, by 10 calories

95. Reading Tables and Charts The following table and bar chart give the amount of caffeine in five different soft drinks. How much caffeine is in each of the following?
 a. A 6-pack of Jolt   600 mg
 b. 2 Coca-Colas plus 3 Tabs   231 mg

Caffeine Content in Soft Drinks	
Drink	Caffeine (milligrams)
Jolt	100
Tab	47
Coca-Cola	45
Diet Pepsi	36
7UP	0

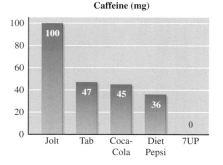

Caffeine (mg)

96. Reading Tables and Charts The following table and bar chart give the amount of caffeine in five different nonprescription drugs. How much caffeine is in each of the following?
 a. A box of 12 Excedrin   780 mg
 b. 1 Dexatrim plus 4 Excedrin   460 mg

Caffeine Content in Nonprescription Drugs	
Nonprescription Drug	Caffeine (milligrams)
Dexatrim	200
NoDoz	100
Excedrin	65
Triaminicin tablets	30
Dristan tablets	16

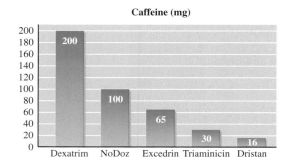

Caffeine (mg)

97. Reading Tables and Charts The following bar chart gives the number of calories burned by a 150-pound person during 1 hour of various exercises. The accompanying table should display the same information. Use the bar chart to complete the table.

Calories Burned by 150-Pound Person	
Activity	Calories Burned in 1 Hour
Bicycling	374
Bowling	265
Handball	680
Jogging	680
Skiing	544

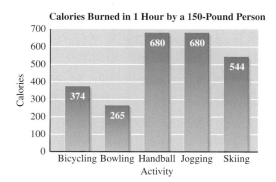

Calories Burned in 1 Hour by a 150-Pound Person

98. Reading Tables and Charts The following bar chart gives the number of calories consumed by eating some popular fast foods. The accompanying table should display the same information. Use the bar chart to complete the table.

Calories in Fast Food

Food	Calories
McDonald's Hamburger	270
Burger King Hamburger	260
Jack in the Box Hamburger	280
McDonald's Big Mac	510
Burger King Whopper	630

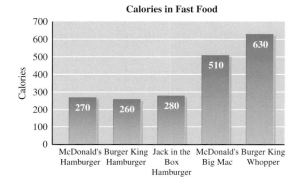

Calories in Fast Food

Find the next number in each sequence.

99. 1, 2, 3, 4, . . . (The sequence of counting numbers)
5

100. 0, 1, 2, 3, . . . (The sequence of whole numbers)
4

101. 2, 4, 6, 8, . . . (The sequence of even numbers)
10

102. 1, 3, 5, 7, . . . (The sequence of odd numbers)
9

103. 1, 4, 9, 16, . . . (The sequence of squares)
25

104. 1, 8, 27, 64, . . . (The sequence of cubes)
125

105. 2, 2, 4, 6, . . . (A Fibonacci-like sequence)
10

106. 5, 5, 10, 15, . . . (A Fibonacci-like sequence)
25

1.2 Real Numbers

OBJECTIVES

A Locate and label points on the number line.

B Change a fraction to an equivalent fraction with a new denominator.

C Simplify expressions containing absolute value.

D Identify the opposite of a number.

E Multiply fractions.

F Identify the reciprocal of a number.

G Find the value of an expression.

H Find the perimeter and area of squares, rectangles, and triangles.

Table 1 and Figure 1 give the record low temperature, in degrees Fahrenheit, for each month of the year in the city of Jackson, Wyoming. Notice that some of these temperatures are represented by negative numbers.

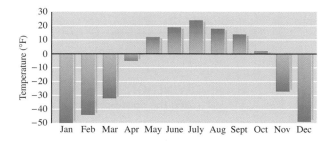

FIGURE 1

In this section we start our work with negative numbers. To represent negative numbers in algebra, we use what is called the *real number line*. Here is how we construct a real number line: We first draw a straight line and label a conve-

TABLE 1
Record Low Temperatures for Jackson, Wyoming

Month	Temperature (Degrees Fahrenheit)	Month	Temperature (Degrees Fahrenheit)
January	−50	July	24
February	−44	August	18
March	−32	September	14
April	−5	October	2
May	12	November	−27
June	19	December	−49

Note

If there is no sign (+ or −) in front of a number, the number is assumed to be positive (+).

nient point on the line with 0. Then we mark off equally spaced distances in both directions from 0. Label the points to the right of 0 with the numbers 1, 2, 3, . . . (the dots mean "and so on"). The points to the left of 0 we label in order, −1, −2, −3, ... Here is what it looks like.

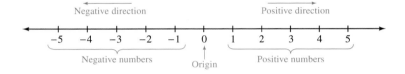

Note

There are other numbers on the number line that you may not be as familiar with. They are irrational numbers such as π, $\sqrt{2}$, $\sqrt{3}$. We will introduce these numbers later in the chapter.

The numbers increase in value going from left to right. If we "move" to the right, we are moving in the positive direction. If we move to the left, we are moving in the negative direction. When we compare two numbers on the number line, the number on the left is always smaller than the number on the right. For instance, −3 is smaller than −1 because it is to the left of −1 on the number line.

 EXAMPLE 1 Locate and label the points on the real number line associated with the numbers -3.5, $-1\frac{1}{4}$, $\frac{1}{2}$, $\frac{3}{4}$, 2.5.

SOLUTION We draw a real number line from −4 to 4 and label the points in question.

> **DEFINITION** The number associated with a point on the real number line is called the **coordinate** of that point.

In the preceding example, the numbers $\frac{1}{2}$, $\frac{3}{4}$, 2.5, -3.5, and $-1\frac{1}{4}$ are the coordinates of the points they represent.

> **DEFINITION** The numbers that can be represented with points on the real number line are called **real numbers**.

Real numbers include whole numbers, fractions, decimals, and other numbers that are not as familiar to us as these.

Fractions on the Number Line

As we proceed through Chapter 1, from time to time we will review some of the major concepts associated with fractions. To begin, here is the formal definition of a fraction.

> **DEFINITION** If a and b are real numbers, then the expression
> $$\frac{a}{b} \qquad b \neq 0$$
> is called a **fraction.** The top number a is called the **numerator,** and the bottom number b is called the **denominator.** The restriction $b \neq 0$ keeps us from writing an expression that is undefined. (As you will see, division by zero is not allowed.)

The number line can be used to visualize fractions. Recall that for the fraction $\frac{a}{b}$, a is called the numerator and b is called the denominator. The denominator indicates the number of equal parts in the interval from 0 to 1 on the number line. The numerator indicates how many of those parts we have. If we take that part of the number line from 0 to 1 and divide it into *three equal parts,* we say that we have divided it into *thirds* (Figure 2). Each of the three segments is $\frac{1}{3}$ (one third) of the whole segment from 0 to 1.

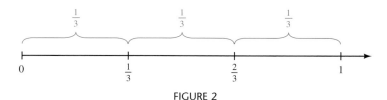

FIGURE 2

Two of these smaller segments together are $\frac{2}{3}$ (two thirds) of the whole segment. And three of them would be $\frac{3}{3}$ (three thirds), or the whole segment.

Let's do the same thing again with six equal divisions of the segment from 0 to 1 (Figure 3). In this case we say each of the smaller segments has a length of $\frac{1}{6}$ (one sixth).

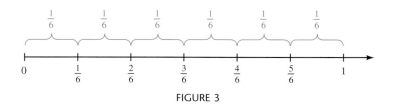

FIGURE 3

The same point we labeled with $\frac{1}{3}$ in Figure 2 is now labeled with $\frac{2}{6}$. Likewise, the point we labeled earlier with $\frac{2}{3}$ is now labeled $\frac{4}{6}$. It must be true then that

$$\frac{2}{6} = \frac{1}{3} \qquad \text{and} \qquad \frac{4}{6} = \frac{2}{3}$$

Actually, there are many fractions that name the same point as $\frac{1}{3}$. If we were to divide the segment between 0 and 1 into 12 equal parts, 4 of these 12 equal parts ($\frac{4}{12}$) would be the same as $\frac{2}{6}$ or $\frac{1}{3}$; that is,

$$\frac{4}{12} = \frac{2}{6} = \frac{1}{3}$$

Even though these three fractions look different, each names the same point on the number line, as shown in Figure 4. All three fractions have the same *value* because they all represent the same number.

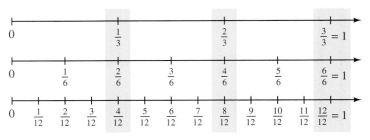

FIGURE 4

> **DEFINITION** Fractions that represent the same number are said to be **equivalent.** Equivalent fractions may look different, but they must have the same value.

It is apparent that every fraction has many different representations, each of which is equivalent to the original fraction. The next two properties give us a way of changing the terms of a fraction without changing its value.

> **Property 1** Multiplying the numerator and denominator of a fraction by the same nonzero number never changes the value of the fraction.

> **Property 2** Dividing the numerator and denominator of a fraction by the same nonzero number never changes the value of the fraction.

EXAMPLE 2   Write $\frac{3}{4}$ as an equivalent fraction with denominator 20.

SOLUTION   The denominator of the original fraction is 4. The fraction we are trying to find must have a denominator of 20. We know that if we multiply 4 by 5, we get 20. Property 1 indicates that we are free to multiply the denominator by 5 as long as we do the same to the numerator.

$$\frac{3}{4} = \frac{3 \cdot 5}{4 \cdot 5} = \frac{15}{20}$$

The fraction $\frac{15}{20}$ is equivalent to the fraction $\frac{3}{4}$.

Absolute Values and Opposites

Representing numbers on the number line lets us give each number two important properties: a direction from zero and a distance from zero. The direction from zero is represented by the sign in front of the number. (A number without a sign is understood to be positive.) The distance from zero is called the absolute value of the number, as the following definition indicates.

> **DEFINITION** The **absolute value** of a real number is its distance from zero on the number line. If x represents a real number, then the absolute value of x is written $|x|$.

 EXAMPLES   Write each expression without absolute value symbols.

3. $|5| = 5$ **The number 5 is 5 units from zero**
4. $|-5| = 5$ **The number -5 is 5 units from zero**
5. $\left|-\dfrac{1}{2}\right| = \dfrac{1}{2}$ **The number $-\frac{1}{2}$ is $\frac{1}{2}$ units from zero**

The absolute value of a number is *never* negative. It is the distance the number is from zero without regard to which direction it is from zero. When working with the absolute value of sums and differences, we must simplify the expression inside the absolute value symbols first and then find the absolute value of the simplified expression.

 EXAMPLES   Simplify each expression.

6. $|8 - 3| = |5| = 5$
7. $|3 \cdot 2^3 + 2 \cdot 3^2| = |3 \cdot 8 + 2 \cdot 9| = |24 + 18| = |42| = 42$
8. $|9 - 2| - |8 - 6| = |7| - |2| = 7 - 2 = 5$

Another important concept associated with numbers on the number line is that of opposites. Here is the definition.

> **DEFINITION** Numbers the same distance from zero but in opposite directions from zero are called **opposites.**

 EXAMPLES   Give the opposite of each number.

Number	Opposite	
9. 5	-5	**5 and -5 are opposites**
10. -3	3	**-3 and 3 are opposites**
11. $\dfrac{1}{4}$	$-\dfrac{1}{4}$	**$\frac{1}{4}$ and $-\frac{1}{4}$ are opposites**
12. -2.3	2.3	**-2.3 and 2.3 are opposites**

Each negative number is the opposite of some positive number, and each positive number is the opposite of some negative number. The opposite of a negative number is a positive number. In symbols, if a represents a positive number, then

$$-(-a) = a$$

Opposites always have the same absolute value. And, when you add any two opposites, the result is always zero:

$$a + (-a) = 0$$

Reciprocals and Multiplication with Fractions

The last concept we want to cover in this section is the concept of reciprocals. Understanding reciprocals requires some knowledge of multiplication with fractions. To multiply two fractions, we simply multiply numerators and multiply denominators.

 EXAMPLE 13   Multiply $\frac{3}{4} \cdot \frac{5}{7}$.

SOLUTION   The product of the numerators is 15, and the product of the denominators is 28:

$$\frac{3}{4} \cdot \frac{5}{7} = \frac{3 \cdot 5}{4 \cdot 7} = \frac{15}{28}$$

Note

In past math classes you may have written fractions like $\frac{7}{3}$ (improper fractions) as mixed numbers, such as $2\frac{1}{3}$. In algebra it is usually better to write them as improper fractions rather than mixed numbers.

 EXAMPLE 14   Multiply $7\left(\frac{1}{3}\right)$.

SOLUTION   The number 7 can be thought of as the fraction $\frac{7}{1}$:

$$7\left(\frac{1}{3}\right) = \frac{7}{1}\left(\frac{1}{3}\right) = \frac{7 \cdot 1}{1 \cdot 3} = \frac{7}{3}$$

EXAMPLE 15   Expand and multiply $\left(\frac{2}{3}\right)^3$.

SOLUTION   Using the definition of exponents from the previous section, we have

$$\left(\frac{2}{3}\right)^3 = \frac{2}{3} \cdot \frac{2}{3} \cdot \frac{2}{3} = \frac{8}{27}$$

We are now ready for the definition of reciprocals.

> **DEFINITION**   Two numbers whose product is 1 are called **reciprocals.**

 EXAMPLES   Give the reciprocal of each number.

	Number	Reciprocal	
16.	5	$\frac{1}{5}$	Because $5\left(\frac{1}{5}\right) = \frac{5}{1}\left(\frac{1}{5}\right) = \frac{5}{5} = 1$
17.	2	$\frac{1}{2}$	Because $2\left(\frac{1}{2}\right) = \frac{2}{1}\left(\frac{1}{2}\right) = \frac{2}{2} = 1$
18.	$\frac{1}{3}$	3	Because $\frac{1}{3}(3) = \frac{1}{3}\left(\frac{3}{1}\right) = \frac{3}{3} = 1$
19.	$\frac{3}{4}$	$\frac{4}{3}$	Because $\frac{3}{4}\left(\frac{4}{3}\right) = \frac{12}{12} = 1$

Although we will not develop multiplication with negative numbers until later in the chapter, you should know that the reciprocal of a negative number is also a negative number. For example, the reciprocal of -4 is $-\frac{1}{4}$.

Previously we mentioned that a variable is a letter used to represent a number or a group of numbers. An expression that contains any combination of numbers, variables, operation symbols, and grouping symbols is called an *algebraic expression* (sometimes referred to as just an *expression*). This definition includes the use of exponents and fractions. Each of the following is an algebraic expression.

$$3x + 5 \qquad 4t^2 - 9 \qquad x^2 - 6xy + y^2 \qquad -15x^2y^4z^5 \qquad \frac{a^2 - 9}{a - 3} \qquad \frac{(x - 3)(x + 2)}{4x}$$

In the last two expressions, the fraction bar separates the numerator from the denominator and is treated the same as a pair of grouping symbols; it groups the numerator and denominator separately.

The Value of an Algebraic Expression

An expression such as $3x + 5$ will take on different values depending on what x is. If we were to let x equal 2, the expression $3x + 5$ would become 11. On the other hand, if x is 10, the same expression has a value of 35:

	When	$x = 2$		When	$x = 10$
the expression		$3x + 5$	the expression		$3x + 5$
becomes		$3(2) + 5$	becomes		$3(10) + 5$
		$= 6 + 5$			$= 30 + 5$
		$= 11$			$= 35$

Table 2 lists some other algebraic expressions, along with specific values for the variables and the corresponding value of the expression after the variable has been replaced with the given number.

TABLE 2

Original Expression	Value of the Variable	Value of the Expression
$5x + 2$	$x = 4$	$5(4) + 2 = 20 + 2$
		$= 22$
$3x - 9$	$x = 2$	$3(2) - 9 = 6 - 9$
		$= -3$
$4t^2 - 9$	$t = 5$	$4(5^2) - 9 = 4(25) - 9$
		$= 100 - 9$
		$= 91$
$\dfrac{a^2 - 9}{a - 3}$	$a = 8$	$\dfrac{8^2 - 9}{8 - 3} = \dfrac{64 - 9}{8 - 3}$
		$= \dfrac{55}{5}$
		$= 11$

FACTS FROM GEOMETRY

Formulas for Area and Perimeter

A square, rectangle, and triangle are shown in the following figures. Note that we have labeled the dimensions of each with variables. The formulas for the perimeter and area of each object are given in terms of its dimensions.

Note
The vertical line labeled h in the triangle is its height, or altitude. It extends from the top of the triangle down to the base, meeting the base at an angle of 90°. The altitude of a triangle is always perpendicular to the base. The small square shown where the altitude meets the base is used to indicate that the angle formed is 90°.

A Square

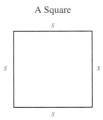

Perimeter = $4s$
Area = s^2

A Rectangle

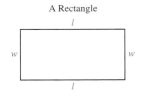

Perimeter = $2l + 2w$
Area = lw

A Triangle

Perimeter = $a + b + c$
Area = $\frac{1}{2}bh$

The formula for perimeter gives us the distance around the outside of the object along its sides, whereas the formula for area gives us a measure of the amount of surface the object has.

EXAMPLE 20   Find the perimeter and area of each figure.

a.

5 feet

b.

6 inches

8 inches

c.

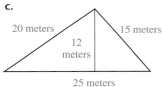

20 meters   15 meters

12 meters

25 meters

SOLUTION   We use the preceding formulas to find the perimeter and the area. In each case, the units for perimeter are linear units, whereas the units for area are square units.

a.   Perimeter = $4s = 4 \cdot 5$ feet = 20 feet
Area = $s^2 = (5 \text{ feet})^2 = 25$ square feet

b.   Perimeter = $2l + 2w = 2(8 \text{ inches}) + 2(6 \text{ inches}) = 28$ inches
Area = $lw = (8 \text{ inches})(6 \text{ inches}) = 48$ square inches

c.   Perimeter = $a + b + c = (20 \text{ meters}) + (25 \text{ meters}) + (15 \text{ meters})$
= 60 meters

Area = $\frac{1}{2}bh = \frac{1}{2}(25 \text{ meters})(12 \text{ meters}) = 150$ square meters

LINKING OBJECTIVES AND EXAMPLES

Next to each **objective** we have listed the examples that are best described by that objective.

A	1
B	2
C	3–8
D	9–12
E	13–15
F	16–19
G	Table 2
H	20

GETTING READY FOR CLASS

After reading through the preceding section, respond in your own words and in complete sentences.

1. What is a real number?
2. Explain multiplication with fractions.
3. How do you find the opposite of a number?
4. Explain how you find the perimeter and the area of a rectangle.

Draw a number line that extends from −5 to +5. Label the points with the following coordinates.

1. 5

2. −2

3. −4

4. −3

5. 1.5

6. −1.5

7. $\dfrac{9}{4}$

8. $\dfrac{8}{3}$

Write each of the following fractions as an equivalent fraction with denominator 24.

9. $\dfrac{3}{4}$ $\dfrac{18}{24}$

10. $\dfrac{5}{6}$ $\dfrac{20}{24}$

11. $\dfrac{1}{2}$ $\dfrac{12}{24}$

12. $\dfrac{1}{8}$ $\dfrac{3}{24}$

13. $\dfrac{5}{8}$ $\dfrac{15}{24}$

14. $\dfrac{7}{12}$ $\dfrac{14}{24}$

Write each fraction as an equivalent fraction with denominator 60.

15. $\dfrac{3}{5}$ $\dfrac{36}{60}$

16. $\dfrac{5}{12}$ $\dfrac{25}{60}$

17. $\dfrac{11}{30}$ $\dfrac{22}{60}$

18. $\dfrac{9}{10}$ $\dfrac{54}{60}$

Fill in the missing numerator so the fractions are equal.

19. $\dfrac{1}{2} = \dfrac{2}{4}$

20. $\dfrac{1}{5} = \dfrac{2}{10}$

21. $\dfrac{5}{9} = \dfrac{25}{45}$

22. $\dfrac{2}{5} = \dfrac{18}{45}$

23. $\dfrac{3}{4} = \dfrac{6}{8}$

24. $\dfrac{1}{2} = \dfrac{4}{8}$

For each of the following numbers, give the opposite, the reciprocal, and the absolute value. (Assume all variables are nonzero.)

25. 10 $-10, \frac{1}{10}, 10$

26. $\dfrac{3}{4}$ $-\frac{3}{4}, \frac{4}{3}, \frac{3}{4}$

27. −3 $3, -\frac{1}{3}, 3$

28. $-\dfrac{2}{5}$ $\frac{2}{5}, -\frac{5}{2}, \frac{2}{5}$

29. x $-x, \frac{1}{x}, |x|$

30. a $-a, \frac{1}{a}, |a|$

Place one of the symbols < or > between each of the following to make the resulting statement true.

31. −5 < −3

32. −8 < −1

33. −3 > −7

34. −6 < 5

35. $|-4| > -|-4|$

36. $3 > -|-3|$

37. $7 > -|-7|$

38. $-7 < |-7|$

39. $-\dfrac{3}{4} < -\dfrac{1}{4}$

40. $-\dfrac{2}{3} < -\dfrac{1}{3}$

41. $-\dfrac{3}{2} < -\dfrac{3}{4}$

42. $-\dfrac{8}{3} > -\dfrac{17}{3}$

Simplify each expression.

43. $|8 - 2|$ 6

44. $|6 - 1|$ 5

45. $|5 \cdot 2^3 - 2 \cdot 3^2|$ 22

46. $|2 \cdot 10^2 + 3 \cdot 10|$ 230

47. $|7 - 2| - |4 - 2|$ 3

48. $|10 - 3| - |4 - 1|$ 4

49. $10 - |7 - 2(5 - 3)|$ 7

50. $12 - |9 - 3(7 - 5)|$ 9

51. $15 - |8 - 2(3 \cdot 4 - 9)| - 10$ 3

52. $25 - |9 - 3(4 \cdot 5 - 18)| - 20$ 2

Multiply the following.

53. $\dfrac{2}{3} \cdot \dfrac{4}{5}$ $\frac{8}{15}$

54. $\dfrac{1}{4} \cdot \dfrac{3}{5}$ $\frac{3}{20}$

55. $\dfrac{1}{2}(3)$ $\frac{3}{2}$

56. $\dfrac{1}{5}(4)$ $\frac{4}{5}$

57. $\dfrac{4}{3} \cdot \dfrac{3}{4}$ 1

58. $\dfrac{5}{7} \cdot \dfrac{7}{5}$ 1

59. $3 \cdot \dfrac{1}{3}$ 1

60. $4 \cdot \dfrac{1}{4}$ 1

61. Multiply.

  a. $\dfrac{1}{2}(4)$ 2

  b. $\dfrac{1}{2}(8)$ 4

  c. $\dfrac{1}{2}(16)$ 8

  d. $\dfrac{1}{2}(0.06)$ 0.03

62. Multiply.

  a. $\dfrac{1}{4}(8)$ 2

  b. $\dfrac{1}{4}(24)$ 6

  c. $\dfrac{1}{4}(16)$ 4

  d. $\dfrac{1}{4}(0.20)$ 0.05

63. Multiply.

  a. $\dfrac{3}{2}(4)$ 6

  b. $\dfrac{3}{2}(8)$ 12

  c. $\dfrac{3}{2}(16)$ 24

  d. $\dfrac{3}{2}(0.06)$ 0.09

64. Multiply.

  a. $\dfrac{3}{4}(8)$ 6

  b. $\dfrac{3}{4}(24)$ 18

  c. $\dfrac{3}{4}(16)$ 12

  d. $\dfrac{3}{4}(0.20)$ 0.15

Expand and multiply.

65. $\left(\dfrac{3}{4}\right)^2$ $\frac{9}{16}$

66. $\left(\dfrac{5}{6}\right)^2$ $\frac{25}{36}$

67. $\left(\dfrac{2}{3}\right)^3$ $\frac{8}{27}$

68. $\left(\dfrac{1}{2}\right)^3$ $\frac{1}{8}$

69. Find the value of $2x - 6$ when
  a. $x = 5$ 4
  b. $x = 10$ 14
  c. $x = 15$ 24
  d. $x = 20$ 34

70. Find the value of $2(x - 3)$ when
  a. $x = 5$ 4
  b. $x = 10$ 14
  c. $x = 15$ 24
  d. $x = 20$ 34

71. Find the value of each expression when x is 10.
  a. $x + 2$ 12
  b. $2x$ 20
  c. x^2 100
  d. 2^x 1,024

72. Find the value of each expression when x is 3.
  a. $x + 3$ 6
  b. $3x$ 9
  c. x^2 9
  d. 3^x 27

73. Find the value of each expression when x is 4.
  a. $x^2 + 1$ 17
  b. $(x + 1)^2$ 25
  c. $x^2 + 2x + 1$ 25

74. Find the value of $b^2 - 4ac$ when
  a. $a = 2, b = 6, c = 3$ 12
  b. $a = 1, b = 5, c = 6$ 1
  c. $a = 1, b = 2, c = 1$ 0

Find the perimeter and area of each figure.

75.

1 inch

1 inch

4 inches; 1 square inch

76.

15 millimeters

15 millimeters

60 millimeters; 225 square millimeters

77.

0.75 inches

1.5 inches

4.5 inches; 1.125 square inches

78.

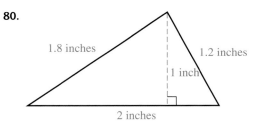

1.5 centimeters

4.5 centimeters

12 centimeters; 6.75 square centimeters

79.

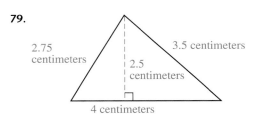

2.75 centimeters

3.5 centimeters

2.5 centimeters

4 centimeters

10.25 centimeters; 5 square centimeters

80.

1.8 inches

1.2 inches

1 inch

2 inches

5 inches; 1 square inch

Applying the Concepts

81. **Football Yardage** A football team gains 6 yards on one play and then loses 8 yards on the next play. To what number on the number line does a loss of 8 yards correspond? The total yards gained or lost on the two plays corresponds to what negative number? $-8, -2$

82. Checking Account Balance A woman has a balance of $20 in her checking account. If she writes a check for $30, what negative number can be used to represent the new balance in her checking account?   −$10

Temperature In the United States, temperature is measured on the Fahrenheit temperature scale. On this scale, water boils at 212 degrees and freezes at 32 degrees. To denote a temperature of 32 degrees on the Fahrenheit scale, we write

32°F, which is read "32 degrees Fahrenheit"

Use this information for Problems 83 and 84.

83. Temperature and Altitude Marilyn is flying from Seattle to San Francisco on a Boeing 737 jet. When the plane reaches an altitude of 35,000 feet, the temperature outside the plane is 64 degrees below zero Fahrenheit. Represent the temperature with a negative number. If the temperature outside the plane gets warmer by 10 degrees, what will the new temperature be?   −64°F; −54°F

84. Temperature Change At 10:00 in the morning in White Bear Lake, Minnesota, John notices the temperature outside is 10 degrees below zero Fahrenheit. Write the temperature as a negative number. An hour later it has warmed up by 6 degrees. What is the temperature at 11:00 that morning?   −10°F; −4°F

85. Scuba Diving Steve is scuba diving near his home in Maui. At one point he is 100 feet below the surface. Represent this number with a negative number. If he descends another 5 feet, what negative number will represent his new position?   −100 feet; −105 feet

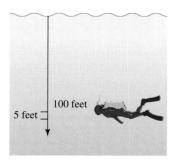

100 feet

5 feet

86. Reading a Chart The chart shows yields for certificates of deposit during one week in 2005. Write a mathematical statement using one of the symbols < or > to compare the following:

a. 6-month yield a year ago to 1-year yield last week   1.64% < 3.26%

b. 2½-year yield this week to 5-year yield a year ago   3.5% < 3.53%

c. 5-year yield last week to 6-month yield this week   3.91% > 2.81%

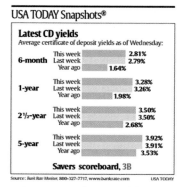

USA TODAY Snapshots®

Latest CD yields
Average certificate of deposit yields as of Wednesday:

6-month	This week	2.81%
	Last week	2.79%
	Year ago	1.64%
1-year	This week	3.28%
	Last week	3.26%
	Year ago	1.98%
2½-year	This week	3.50%
	Last week	3.50%
	Year ago	2.68%
5-year	This week	3.92%
	Last week	3.91%
	Year ago	3.53%

Savers scoreboard, 3B

Source: Bank Rate Monitor, 800-327-7717, www.bankrate.com   USA TODAY

From *USA Today*. Copyright 2005. Reprinted with permission.

87. Geometry Find the area and perimeter of an $8\frac{1}{2}$-by-11-inch piece of notebook paper.
Area = 93.5 square inches; perimeter = 39 inches

88. Geometry Find the area and perimeter of an $8\frac{1}{2}$-by-$5\frac{1}{2}$-inch piece of paper.
Area = 46.75 square inches; perimeter = 28 inches

Calories and Exercise Table 3 gives the amount of energy expended per hour for various activities for a person weighing 120, 150, or 180 pounds. Use Table 3 to answer questions 89–92.

TABLE 3
Energy Expended from Exercising

Activity	Calories per Hour		
	120 lb	**150 lb**	**180 lb**
Bicycling	299	374	449
Bowling	212	265	318
Handball	544	680	816
Horseback trotting	278	347	416
Jazzercise	272	340	408
Jogging	544	680	816
Skiing (downhill)	435	544	653

89. Suppose you weigh 120 pounds. How many calories will you burn if you play handball for 2 hours and then ride your bicycle for an hour?   1,387 calories

90. How many calories are burned by a person weighing 150 pounds who jogs for $\frac{1}{2}$ hour and then goes bicycling for 2 hours?   1,088 calories

91. Two people go skiing. One weighs 180 pounds and the other weighs 120 pounds. If they ski for 3 hours, how many more calories are burned by the person weighing 180 pounds?   654 more calories

92. Two people spend 3 hours bowling. If one weighs 120 pounds and the other weighs 150 pounds, how many more calories are burned during the evening by the person weighing 150 pounds?

159 more calories

93. Improving Your Quantitative Literacy Quantitative literacy is a subject discussed by many people involved in teaching mathematics. The person they are concerned with when they discuss it is you. We are going to work at improving your quantitative literacy; but before we do that, we should answer the question: What is quantitative literacy? Lynn Arthur Steen, a noted mathematics educator, has stated that quantitative literacy is "the capacity to deal effectively with the quantitative aspects of life."
  a. Give a definition for the word *quantitative*.
  b. Give a definition for the word *literacy*.
  c. Are there situations that occur in your life that you find distasteful, or that you try to avoid, because they involve numbers and mathematics? If so, list some of them here. (For example, some people find the process of buying a car particu-

larly difficult because they feel that the numbers and details of the financing are beyond them.)

94. Improving Your Quantitative Literacy Use the chart shown here to answer the following questions.
  a. How many millions of camera phones were sold in 2004?   93 million
  b. True or false? The chart shows projected sales in 2005 to be more than 155 million camera phones.   False
  c. True or false? The chart shows projected sales in 2007 to be less than 310 million camera phones.   True

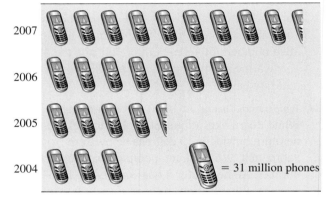

Camera Phone Growth:
Projected sales in millions of units

= 31 million phones

1.3   Addition of Real Numbers

OBJECTIVES

A Add any combination of positive and negative numbers.

B Simplify expressions using the rule for order of operations.

C Extend an arithmetic sequence.

Suppose that you are playing a friendly game of poker with some friends, and you lose $3 on the first hand and $4 on the second hand. If you represent winning with positive numbers and losing with negative numbers, how can you translate this situation into symbols? Because you lost $3 and $4 for a total of $7, one way to represent this situation is with addition of negative numbers:

$$(-\$3) + (-\$4) = -\$7$$

From this equation, we see that the sum of two negative numbers is a negative number. To generalize addition with positive and negative numbers, we use the number line.

Because real numbers have both a distance from zero (absolute value) and a direction from zero (sign), we can think of addition of two numbers in terms of distance and direction from zero.

Let's look at a problem for which we know the answer. Suppose we want to add the numbers 3 and 4. The problem is written 3 + 4. To put it on the number line, we read the problem as follows:

1. The 3 tells us to "start at the origin and move 3 units in the positive direction."
2. The + sign is read "and then move."
3. The 4 means "4 units in the positive direction."

To summarize, 3 + 4 means to start at the origin, move 3 units in the *positive* direction, and then move 4 units in the *positive* direction.

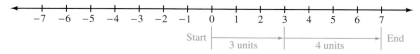

We end up at 7, which is the answer to our problem: 3 + 4 = 7.

Let's try other combinations of positive and negative 3 and 4 on the number line.

 EXAMPLE 1   Add 3 + (−4).

SOLUTION   Starting at the origin, move 3 units in the *positive* direction and then 4 units in the *negative* direction.

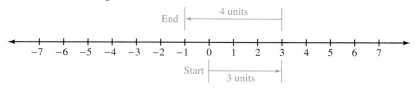

We end up at −1; therefore, 3 + (−4) = −1.

 EXAMPLE 2   Add −3 + 4.

SOLUTION   Starting at the origin, move 3 units in the *negative* direction and then 4 units in the *positive* direction.

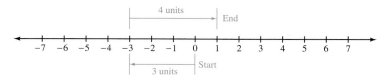

We end up at +1; therefore, −3 + 4 = 1.

EXAMPLE 3   Add −3 + (−4).

SOLUTION   Starting at the origin, move 3 units in the *negative* direction and then 4 units in the *negative* direction.

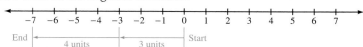

We end up at −7; therefore, −3 + (−4) = −7.

Here is a summary of what we have just completed:

$$3 + 4 = 7$$
$$3 + (-4) = -1$$
$$-3 + 4 = 1$$
$$-3 + (-4) = -7$$

Let's do four more problems on the number line and then summarize our results into a rule we can use to add any two real numbers.

EXAMPLE 4   Show that $5 + 7 = 12$.

SOLUTION

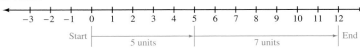

EXAMPLE 5   Show that $5 + (-7) = -2$.

SOLUTION

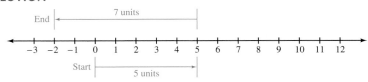

EXAMPLE 6   Show that $-5 + 7 = 2$.

SOLUTION

EXAMPLE 7   Show that $-5 + (-7) = -12$.

SOLUTION

If we look closely at the results of the preceding addition problems, we can see that they support (or justify) the following rule.

Note

This rule is what we have been working toward. The rule is very important. Be sure that you understand it and can use it. The problems we have done up to this point have been done simply to justify this rule. Now that we have the rule, we no longer need to do our addition problems on the number line.

Rule To add two real numbers with
1. The *same* sign: Simply add their absolute values and use the common sign. (Both numbers are positive, the answer is positive. Both numbers are negative, the answer is negative.)
2. *Different* signs: Subtract the smaller absolute value from the larger. The answer will have the sign of the number with the larger absolute value.

This rule covers all possible combinations of addition with real numbers. You must memorize it. After you have worked a number of problems, it will seem almost automatic.

 EXAMPLE 8   Add all combinations of positive and negative 10 and 13.

SOLUTION   Rather than work these problems on the number line, we use the rule for adding positive and negative numbers to obtain our answers:

$$10 + 13 = 23$$
$$10 + (-13) = -3$$
$$-10 + 13 = 3$$
$$-10 + (-13) = -23$$

 EXAMPLE 9   Add all possible combinations of positive and negative 12 and 17.

SOLUTION   Applying the rule for adding positive and negative numbers, we have

$$12 + 17 = 29$$
$$12 + (-17) = -5$$
$$-12 + 17 = 5$$
$$-12 + (-17) = -29$$

 EXAMPLE 10   Add $-3 + 2 + (-4)$.

SOLUTION   Applying the rule for order of operations, we add left to right:

$$-3 + 2 + (-4) = -1 + (-4)$$
$$= -5$$

 EXAMPLE 11   Add $-8 + [2 + (-5)] + (-1)$.

SOLUTION   Adding inside the brackets first and then left to right, we have

$$-8 + [2 + (-5)] + (-1) = -8 + (-3) + (-1)$$
$$= -11 + (-1)$$
$$= -12$$

EXAMPLE 12   Simplify $-10 + 2(-8 + 11) + (-4)$.

SOLUTION   First, we simplify inside the parentheses. Then, we multiply. Finally, we add left to right:

$$-10 + 2(-8 + 11) + (-4) = -10 + 2(3) + (-4)$$
$$= -10 + 6 + (-4)$$
$$= -4 + (-4)$$
$$= -8$$

Arithmetic Sequences

The pattern in a sequence of numbers is easy to identify when each number in the sequence comes from the preceding number by adding the same amount each time. This leads us to our next level of classification, in which we classify together groups of sequences with a common characteristic.

> **DEFINITION** An **arithmetic sequence** is a sequence of numbers in which each number (after the first number) comes from adding the same amount to the number before it.

Here is an example of an arithmetic sequence:

$$2, 5, 8, 11, \ldots$$

Each number is obtained by adding 3 to the number before it.

 EXAMPLE 13   Each sequence below is an arithmetic sequence. Find the next two numbers in each sequence.

a. $7, 10, 13, \ldots$
b. $9.5, 10, 10.5, \ldots$
c. $5, 0, -5, \ldots$

SOLUTION   Because we know that each sequence is arithmetic, we know to look for the number that is added to each term to produce the next consecutive term.

a. $7, 10, 13, \ldots$: Each term is found by adding 3 to the term before it. Therefore, the next two terms will be 16 and 19.
b. $9.5, 10, 10.5, \ldots$: Each term comes from adding 0.5 to the term before it. Therefore, the next two terms will be 11 and 11.5.
c. $5, 0, -5, \ldots$: Each term comes from adding -5 to the term before it. Therefore, the next two terms will be $-5 + (-5) = -10$ and $-10 + (-5) = -15$.

LINKING OBJECTIVES AND EXAMPLES

Next to each **objective** we have listed the examples that are best described by that objective.

A   1–9

B   10–12

C   13

GETTING READY FOR CLASS

After reading through the preceding section, respond in your own words and in complete sentences.

1. Explain how you would add 3 and -5 on the number line.
2. How do you add two negative numbers?
3. What is an arithmetic sequence?
4. Why is the sum of a number and its opposite always 0?

1. Add all combinations of positive and negative 3 and 5. (Look back to Examples 8 and 9.)
$3 + 5 = 8, 3 + (-5) = -2; -3 + 5 = 2, -3 + (-5) = -8$

2. Add all combinations of positive and negative 6 and 4.
$6 + 4 = 10, 6 + (-4) = 2, -6 + 4 = -2, -6 + (-4) = -10$

3. Add all combinations of positive and negative 15 and 20. $15 + 20 = 35, 15 + (-20) = -5,$
$-15 + 20 = 5, -15 + (-20) = -35$

4. Add all combinations of positive and negative 18 and 12. $18 + 12 = 30, 18 + (-12) = 6,$
$-18 + 12 = -6, -18 + (-12) = -30$

Work the following problems. You may want to begin by doing a few on the number line.

5. $6 + (-3)$ 3

6. $7 + (-8)$ -1

▶ **7.** $13 + (-20)$ -7

8. $15 + (-25)$ -10

9. $18 + (-32)$ -14

10. $6 + (-9)$ -3

▶ **11.** $-6 + 3$ -3

12. $-8 + 7$ -1

13. $-30 + 5$ -25

14. $-18 + 6$ -12

▶ **15.** $-6 + (-6)$ -12

16. $-5 + (-5)$ -10

17. $-9 + (-10)$ -19

18. $-8 + (-6)$ -14

19. $-10 + (-15)$ -25

20. $-18 + (-30)$ -48

Work the following problems using the rule for addition of real numbers. You may want to refer back to the rule for order of operations.

▶ **21.** $5 + (-6) + (-7)$ -8

22. $6 + (-8) + (-10)$ -12

23. $-7 + 8 + (-5)$ -4

24. $-6 + 9 + (-3)$ 0

25. $5 + [6 + (-2)] + (-3)$ 6

26. $10 + [8 + (-5)] + (-20)$ -7

27. $[6 + (-2)] + [3 + (-1)]$ 6

28. $[18 + (-5)] + [9 + (-10)]$ 12

29. $20 + (-6) + [3 + (-9)]$ 8

30. $18 + (-2) + [9 + (-13)]$ 12

31. $-3 + (-2) + [5 + (-4)]$ -4

32. $-6 + (-5) + [-4 + (-1)]$ -16

▶ **33.** $(-9 + 2) + [5 + (-8)] + (-4)$ -14

34. $(-7 + 3) + [9 + (-6)] + (-5)$ -6

35. $[-6 + (-4)] + [7 + (-5)] + (-9)$ -17

36. $[-8 + (-1)] + [8 + (-6)] + (-6)$ -13

37. $(-6 + 9) + (-5) + (-4 + 3) + 7$ 4

38. $(-10 + 4) + (-3) + (-3 + 8) + 6$ 2

The problems that follow involve some multiplication. Be sure that you work inside the parentheses first, then multiply, and, finally, add left to right.

39. $-5 + 2(-3 + 7)$ 3

40. $-3 + 4(-2 + 7)$ 17

41. $9 + 3(-8 + 10)$ 15

42. $4 + 5(-2 + 6)$ 24

43. $-10 + 2(-6 + 8) + (-2)$ -8

44. $-20 + 3(-7 + 10) + (-4)$ -15

45. $2(-4 + 7) + 3(-6 + 8)$ 12

46. $5(-2 + 5) + 7(-1 + 6)$ 50

Add.

▶ **47.** $3.9 + 7.1$ 11

▶ **48.** $4.7 + 4.3$ 9

▶ **49.** $8.1 + 2.7$ 10.8

▶ **50.** $2.4 + 7.3$ 9.7

▶ **51.** Find the value of $0.06x + 0.07y$ when $x = 7,000$ and $y = 8,000$.   980

▶ **52.** Find the value of $0.06x + 0.07y$ when $x = 7,000$ and $y = 3,000$.   630

▶ **53.** Find the value of $0.05x + 0.10y$ when $x = 10$ and $y = 12$.   1.70

▶ **54.** Find the value of $0.25x + 0.10y$ when $x = 3$ and $y = 11$.   1.85

Each sequence below is an arithmetic sequence. In each case, find the next two numbers in the sequence.

55. 3, 8, 13, 18, . . .   23, 28

56. 1, 5, 9, 13, . . .   17, 21

57. 10, 15, 20, 25, . . .   30, 35

58. 10, 16, 22, 28, . . .   34, 40

59. 20, 15, 10, 5, . . .   0, -5

60. 24, 20, 16, 12, . . .   8, 4

61. 6, 0, -6, . . . $-12, -18$

62. 1, 0, -1, . . . $-2, -3$

63. 8, 4, 0, . . . $-4, -8$

64. 5, 2, -1, . . . $-4, -7$

65. Is the sequence of odd numbers an arithmetic sequence?   Yes

66. Is the sequence of squares an arithmetic sequence?   No

Recall that the word *sum* indicates addition. Write the numerical expression that is equivalent to each of the following phrases and then simplify.

67. The sum of 5 and 9 $5 + 9 = 14$

68. The sum of 6 and -3

$6 + (-3) = 3$

69. Four added to the sum of -7 and -5

$[-7 + (-5)] + 4 = -8$

70. Six added to the sum of -9 and 1

$(-9 + 1) + 6 = -2$

71. The sum of -2 and -3 increased by 10

$[-2 + (-3)] + 10 = 5$

72. The sum of -4 and -12 increased by 2

$[-4 + (-12)] + 2 = -14$

Answer the following questions.

73. What number do you add to -8 to get -5?   3

74. What number do you add to 10 to get 4? -6

75. The sum of what number and -6 is -9? -3

76. The sum of what number and -12 is 8?   20

Applying the Concepts

77. **Temperature Change** The temperature at noon is 12 degrees below 0 Fahrenheit. By 1:00 it has risen 4 degrees. Write an expression using the numbers -12 and 4 to describe this situation.

$-12 + 4$

78. **Stock Value** On Monday a certain stock gains 2 points. On Tuesday it loses 3 points. Write an expression using positive and negative numbers with addition to describe this situation and then simplify. $2 + (-3) = -1$

79. **Gambling** On three consecutive hands of draw poker a gambler wins $10, loses $6, and then loses another $8. Write an expression using positive and negative numbers and addition to describe this situation and then simplify. $10 + (-6) + (-8) = -\$4$

80. **Number Problem** You know from your past experience with numbers that subtracting 5 from 8 results in 3 $(8 - 5 = 3)$. What addition problem that starts with the number 8 gives the same result? $8 + (-5) = 3$

81. **Checkbook Balance** Suppose that you balance your checkbook and find that you are overdrawn by $30; that is, your balance is $-\$30$. Then you go to the bank and deposit $40. Translate this situation into an addition problem, the answer to which gives the new balance in your checkbook. $-30 + 40 = 10$

82. **Improving Your Quantitative Literacy** One of our goals is for you to be able to read graphs and charts like the one you see here, and to recognize when the information in the chart is displayed correctly. Although the chart appears exactly as it did in *USA Today,* this does not mean it is correct. A problem surfaces when we ask ourselves "How many births does one baby carriage represent?" Explain what is wrong with the chart.

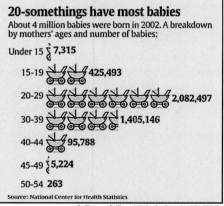

USA TODAY Snapshots

20-somethings have most babies

About 4 million babies were born in 2002. A breakdown by mothers' ages and number of babies:

Under 15   7,315

15-19   425,493

20-29   2,082,497

30-39   1,405,146

40-44   95,788

45-49   5,224

50-54   263

Source: National Center for Health Statistics

By Shannon Reilly and Alejandro Gonzalez, USA TODAY

1.4 Subtraction of Real Numbers

OBJECTIVES

A Subtract any combination of positive and negative numbers.

B Simplify expressions using the rule for order of operations.

C Translate sentences from English into symbols and then simplify.

D Find the complement and the supplement of an angle.

Suppose that the temperature at noon is 20° Fahrenheit and 12 hours later, at midnight, it has dropped to −15° Fahrenheit. What is the difference between the temperature at noon and the temperature at midnight? Intuitively, we know the difference in the two temperatures is 35°. We also know that the word difference indicates subtraction. The difference between 20 and −15 is written

$$20 - (-15)$$

It must be true that $20 - (-15) = 35$. In this section we will see how our definition for subtraction confirms that this last statement is in fact correct.

In the previous section we spent some time developing the rule for addition of real numbers. Because we want to make as few rules as possible, we can define subtraction in terms of addition. By doing so, we can then use the rule for addition to solve our subtraction problems.

> **Rule** To subtract one real number from another, simply add its opposite.
>
> Algebraically, the rule is written like this: If a and b represent two real numbers, then it is always true that
>
> $$\underbrace{a - b}_{\text{To subtract } b} = \underbrace{a + (-b)}_{\text{add the opposite of } b}$$

This is how subtraction is defined in algebra. This definition of subtraction will not conflict with what you already know about subtraction, but it will allow you to do subtraction using negative numbers.

 EXAMPLE 1 Subtract all possible combinations of positive and negative 7 and 2.

SOLUTION

$$\left.\begin{array}{l} 7 - 2 = 7 + (-2) = 5 \\ -7 - 2 = -7 + (-2) = -9 \end{array}\right\} \quad \text{Subtracting 2 is the same as adding } -2$$

$$\left.\begin{array}{l} 7 - (-2) = 7 + 2 = 9 \\ -7 - (-2) = -7 + 2 = -5 \end{array}\right\} \quad \text{Subtracting } -2 \text{ is the same as adding 2}$$

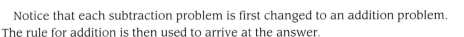

Notice that each subtraction problem is first changed to an addition problem. The rule for addition is then used to arrive at the answer.

We have defined subtraction in terms of addition, and we still obtain answers consistent with the answers we are used to getting with subtraction. Moreover, we now can do subtraction problems involving both positive and negative numbers.

As you proceed through the following examples and the problem set, you will begin to notice shortcuts you can use in working the problems. You will not always have to change subtraction to addition of the opposite to be able to get answers quickly. Use all the shortcuts you wish as long as you consistently get the correct answers.

EXAMPLE 2   Subtract all combinations of positive and negative 8 and 13.

SOLUTION

$$8 - 13 = 8 + (-13) = -5$$
$$-8 - 13 = -8 + (-13) = -21$$

Subtracting +13 is the same as adding −13

$$8 - (-13) = 8 + 13 = 21$$
$$-8 - (-13) = -8 + 13 = 5$$

Subtracting −13 is the same as adding +13

 EXAMPLES   Simplify each expression as much as possible.

3. $7 + (-3) - 5 = 7 + (-3) + (-5)$
$$= 4 + (-5)$$
$$= -1$$

Begin by changing all subtractions to additions
Then add left to right

4. $8 - (-2) - 6 = 8 + 2 + (-6)$
$$= 10 + (-6)$$
$$= 4$$

Begin by changing all subtractions to additions
Then add left to right

5. $-2 - (-3 + 1) - 5 = -2 - (-2) - 5$
$$= -2 + 2 + (-5)$$
$$= -5$$

Do what is in the parentheses first

The next two examples involve multiplication and exponents as well as subtraction. Remember, according to the rule for order of operations, we evaluate the numbers containing exponents and multiply before we subtract.

 EXAMPLE 6   Simplify $2 \cdot 5 - 3 \cdot 8 - 4 \cdot 9$.

SOLUTION   First, we multiply left to right, and then we subtract.

$$2 \cdot 5 - 3 \cdot 8 - 4 \cdot 9 = 10 - 24 - 36$$
$$= -14 - 36$$
$$= -50$$

 EXAMPLE 7   Simplify $3 \cdot 2^3 - 2 \cdot 4^2$.

SOLUTION   We begin by evaluating each number that contains an exponent. Then we multiply before we subtract:

$$3 \cdot 2^3 - 2 \cdot 4^2 = 3 \cdot 8 - 2 \cdot 16$$
$$= 24 - 32$$
$$= -8$$

 EXAMPLE 8   Subtract 7 from −3.

SOLUTION   First, we write the problem in terms of subtraction. We then change to addition of the opposite:

$$-3 - 7 = -3 + (-7)$$
$$= -10$$

EXAMPLE 9   Subtract −5 from 2.

SOLUTION   Subtracting −5 is the same as adding +5:

$$2 - (-5) = 2 + 5$$
$$= 7$$

 EXAMPLE 10   Find the difference of 9 and 2.

SOLUTION   Written in symbols, the problem looks like this:

$$9 - 2 = 7$$

The difference of 9 and 2 is 7.

 EXAMPLE 11   Find the difference of 3 and −5.

SOLUTION   Subtracting −5 from 3 we have

$$3 - (-5) = 3 + 5$$
$$= 8$$

In the sport of drag racing, two cars at the starting line race to the finish line $\frac{1}{4}$ mile away. The car that crosses the finish line first wins the race.

Jim Rizzoli owns and races an alcohol dragster. On board the dragster is a computer that records data during each of Jim's races. Table 1 gives some of the data from a race Jim was in. Figure 1 gives the same information visually.

TABLE 1
Speed of a Race Car

Time in Seconds	Speed in Miles/Hour
0	0
1	72.7
2	129.9
3	162.8
4	192.2
5	212.4
6	228.1

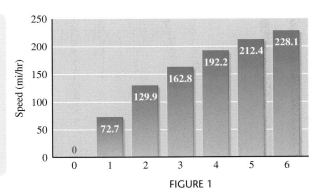

FIGURE 1

 EXAMPLE 12   Use Table 1 to find the difference in speed after 5 seconds and after 2 seconds have elapsed during the race.

SOLUTION   We know the word *difference* implies subtraction. The speed at 2 seconds is 129.9 miles per hour, whereas the speed at 5 seconds is 212.4 miles per hour. Therefore, the expression that represents the solution to our problem looks like this:

$$212.4 - 129.9 = 82.5 \text{ miles per hour}$$

FACTS FROM GEOMETRY

Complementary and Supplementary Angles
If you have studied geometry at all, you know that there are 360° in a full rotation—the number of degrees swept out by the radius of a circle as it rotates once around the circle.

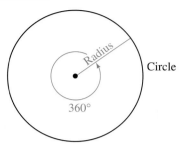

We can apply our knowledge of algebra to help solve some simple geometry problems. Before we do, however, we need to review some of the vocabulary associated with angles.

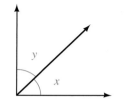

Complementary angles: $x + y = 90°$

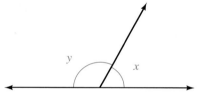

Supplementary angles: $x + y = 180°$

> **DEFINITION** In geometry, two angles that add to 90° are called **complementary angles.** In a similar manner, two angles that add to 180° are called **supplementary angles.** The diagrams at the left illustrate the relationships between angles that are complementary and between angles that are supplementary.

EXAMPLE 13   Find x in each of the following diagrams.

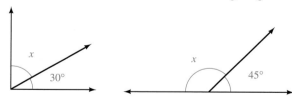

SOLUTION   We use subtraction to find each angle.

a. Because the two angles are complementary, we can find x by subtracting 30° from 90°:

$$x = 90° - 30° = 60°$$

We say 30° and 60° are complementary angles. The complement of 30° is 60°.

b. The two angles in the diagram are supplementary. To find x, we subtract 45° from 180°:

$$x = 180° - 45° = 135°$$

We say 45° and 135° are supplementary angles. The supplement of 45° is 135°.

Subtraction and Taking Away

For some people taking algebra for the first time, subtraction of positive and negative numbers can be a problem. These people may believe that $-5 - 9$ should be -4 or 4, not -14. If this is happening to you, you probably are thinking of subtraction in terms of taking one number away from another. Thinking of subtraction in this way works well with positive numbers if you always subtract the smaller number from the larger. In algebra, however, we encounter many situations other than this. The definition of subtraction, that $a - b = a + (-b)$, clearly indicates the correct way to use subtraction; that is, when working subtraction problems, you should think "addition of the opposite," not "take one number away from another." To be successful in algebra, you need to apply properties and definitions exactly as they are presented here.

LINKING OBJECTIVES AND EXAMPLES

Next to each **objective** we have listed the examples that are best described by that objective.

A	1, 2
B	3–7
C	8–12
D	13

GETTING READY FOR CLASS

After reading through the preceding section, respond in your own words and in complete sentences.

1. Why do we define subtraction in terms of addition?
2. Write the definition for $a - b$.
3. Explain in words how you would subtract 3 from -7.
4. What are complementary angles?

Problem Set 1.4

Online support materials can be found at www.thomsonedu.com/login

The following problems are intended to give you practice with subtraction of positive and negative numbers. Remember, in algebra subtraction is not taking one number away from another. Instead, subtracting a number is equivalent to adding its opposite.

Subtract.

1. $5 - 8$ -3

2. $6 - 7$ -1

3. $3 - 9$ -6

4. $2 - 7$ -5

5. $5 - 5$ 0

6. $8 - 8$ 0

▶ **7.** $-8 - 2$ -10

8. $-6 - 3$ -9

9. $-4 - 12$ -16

10. $-3 - 15$ -18

11. $-6 - 6$ -12

12. $-3 - 3$ -6

13. $-8 - (-1)$ -7

14. $-6 - (-2)$ -4

▶ **15.** $15 - (-20)$ 35

16. $20 - (-5)$ 25

▶ **17.** $-4 - (-4)$ 0

18. $-5 - (-5)$ 0

Simplify each expression by applying the rule for order of operations.

19. $3 - 2 - 5$ -4

20. $4 - 8 - 6$ -10

▶ **21.** $9 - 2 - 3$ 4

22. $8 - 7 - 12$ -11

23. $-6 - 8 - 10$ -24

24. $-5 - 7 - 9$ -21

25. $-22 + 4 - 10$ -28

26. $-13 + 6 - 5$ -12

27. $10 - (-20) - 5$ 25

28. $15 - (-3) - 20$ -2

29. $8 - (2 - 3) - 5$ 4

30. $10 - (4 - 6) - 8$ 4

▶ **31.** $7 - (3 - 9) - 6$ 7

32. $4 - (3 - 7) - 8$ 0

33. $5 - (-8 - 6) - 2$ 17

34. $4 - (-3 - 2) - 1$ 8

35. $-(5 - 7) - (2 - 8)$ 8

36. $-(4 - 8) - (2 - 5)$ 7

37. $-(3 - 10) - (6 - 3)$ 4

38. $-(3 - 7) - (1 - 2)$ 5

39. $16 - [(4 - 5) - 1]$ 18

40. $15 - [(4 - 2) - 3]$ 16

41. $5 - [(2 - 3) - 4]$ 10

42. $6 - [(4 - 1) - 9]$ 12

43. $21 - [-(3 - 4) - 2] - 5$ 17

44. $30 - [-(10 - 5) - 15] - 25$ 25

   = Videos available by instructor request

▶ = Online student support materials available at www.thomsonedu.com/login

The following problems involve multiplication and exponents. Use the rule for order of operations to simplify each expression as much as possible.

45. $2 \cdot 8 - 3 \cdot 5$ 1

46. $3 \cdot 4 - 6 \cdot 7$ -30

47. $3 \cdot 5 - 2 \cdot 7$ 1

48. $6 \cdot 10 - 5 \cdot 20$ -40

49. $5 \cdot 9 - 2 \cdot 3 - 6 \cdot 2$ 27

50. $4 \cdot 3 - 7 \cdot 1 - 9 \cdot 4$ -31

51. $3 \cdot 8 - 2 \cdot 4 - 6 \cdot 7$ -26

52. $5 \cdot 9 - 3 \cdot 8 - 4 \cdot 5$ 1

53. $2 \cdot 3^2 - 5 \cdot 2^2$ -2

54. $3 \cdot 7^2 - 2 \cdot 8^2$ 19

55. $4 \cdot 3^3 - 5 \cdot 2^3$ 68

56. $3 \cdot 6^2 - 2 \cdot 3^2 - 8 \cdot 6^2$ -198

Subtract.

▶ **57.** $-3.4 - 7.9$ -11.3

▶ **58.** $-3.5 - 2.3$ -5.8

▶ **59.** $3.3 - 6.9$ -3.6

▶ **60.** $2.2 - 7.5$ -5.3

61. Find the value of $x + y - 4$ when
 a. $x = -3$ and $y = -2$ -9
 b. $x = -9$ and $y = 3$ -10
 c. $x = -\dfrac{3}{5}$ and $y = \dfrac{8}{5}$ -3

62. Find the value of $x - y - 3$ when
 a. $x = -4$ and $y = 1$ -8
 b. $x = 2$ and $y = -1$ 0
 c. $x = -5$ and $y = -2$ -6

Rewrite each of the following phrases as an equivalent expression in symbols, and then simplify.

63. Subtract 4 from -7. $-7 - 4 = -11$

64. Subtract 5 from -19. $-19 - 5 = -24$

65. Subtract -8 from 12. $12 - (-8) = 20$

66. Subtract -2 from 10. $10 - (-2) = 12$

67. Subtract -7 from -5. $-5 - (-7) = 2$

68. Subtract -9 from -3. $-3 - (-9) = 6$

69. Subtract 17 from the sum of 4 and -5.
 $[4 + (-5)] - 17 = -18$

70. Subtract -6 from the sum of 6 and -3.
 $[6 + (-3)] - (-6) = 9$

Recall that the word *difference* indicates subtraction. The difference of a and b is $a - b$, in that order. Write a numerical expression that is equivalent to each of the following phrases, and then simplify.

71. The difference of 8 and 5 $8 - 5 = 3$

72. The difference of 5 and 8 $5 - 8 = -3$

73. The difference of -8 and 5 $-8 - 5 = -13$

74. The difference of -5 and 8 $-5 - 8 = -13$

75. The difference of 8 and -5 $8 - (-5) = 13$

76. The difference of 5 and -8 $5 - (-8) = 13$

Answer the following questions.

77. What number do you subtract from 8 to get -2?   10

78. What number do you subtract from 1 to get -5?   6

79. What number do you subtract from 8 to get 10? -2

80. What number do you subtract from 1 to get 5? -4

Applying the Concepts

81. Savings Account Balance A man with $1,500 in a savings account makes a withdrawal of $730. Write an expression using subtraction that describes this situation. $1,500 - 730$

First Bank Account No. 12345			
Date	Withdrawals	Deposits	Balance
1/1/99			1,500
2/2/99	730		

82. Temperature Change The temperature inside a Space Shuttle is 73°F before reentry. During reentry the temperature inside the craft increases 10°. On landing it drops 8°F. Write an expression using the numbers 73, 10, and 8 to describe this situation. What is the temperature inside the shuttle on landing? $73 + 10 - 8$, 75°F

83. Gambling A man who has lost $35 playing roulette in Las Vegas wins $15 playing blackjack. He then loses $20 playing the wheel of fortune. Write an expression using the numbers -35, 15, and 20 to describe this situation and then simplify it.
 $-35 + 15 - 20 = -\$40$

84. Altitude Change An airplane flying at 10,000 feet lowers its altitude by 1,500 feet to avoid other air traffic. Then it increases its altitude by 3,000 feet to clear a mountain range. Write an expression that describes this situation and then simplify it.
 $10,000 - 1,500 + 3,000 = 11,500$ feet

85. Checkbook Balance Bob has $98 in his checking account when he writes a check for $65 and then another check for $53. Write a subtraction problem that gives the new balance in Bob's checkbook. What is his new balance? $98 - 65 - 53 = -\$20$

86. Temperature Change The temperature at noon is 23°F. Six hours later it has dropped 19°F, and by midnight it has dropped another 10°F. Write a subtraction problem that gives the temperature at midnight. What is the temperature at midnight? $23 - 19 - 10 = -6°F$

87. Depreciation Stacey buys a used car for $4,500. With each year that passes, the car drops $550 in value. Write a sequence of numbers that gives the value of the car at the beginning of each of the first 5 years she owns it. Can this sequence be considered an arithmetic sequence? $4,500, $3,950, $3,400, $2,850, $2,300; yes

88. Depreciation Wade buys a computer system for $6,575. Each year after that he finds that the system is worth $1,250 less than it was the year before. Write a sequence of numbers that gives the value of the computer system at the beginning of each of the first four years he owns it. Can this sequence be considered an arithmetic sequence?
$6,575, $5,325, $4,075, $2,825; yes

Drag Racing The table shown here extends the information given in Table 1 of this section. In addition to showing the time and speed of Jim Rizzoli's dragster during a race, it also shows the distance past the starting line that his dragster has traveled. Use the information in the table shown here to answer the following questions.

Speed and Distance for a Race Car

Time in Seconds	Speed in Miles/Hour	Distance Traveled in Feet
0	0	0
1	72.7	69
2	129.9	231
3	162.8	439
4	192.2	728
5	212.4	1,000
6	228.1	1,373

89. Find the difference in the distance traveled by the dragster after 5 seconds and after 2 seconds.   769 feet

90. How much faster is he traveling after 4 seconds than he is after 2 seconds?   62.3 miles per hour

91. How far from the starting line is he after 3 seconds?   439 feet

92. How far from the starting line is he when his speed is 192.2 miles per hour?   728 feet

93. How many seconds have gone by between the time his speed is 162.8 miles per hour and the time at which he has traveled 1,000 feet?   2 seconds

94. How many seconds have gone by between the time at which he has traveled 231 feet and the time at which his speed is 228.1 miles per hour?   4 seconds

Find x in each of the following diagrams.

95.
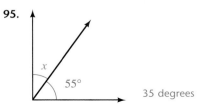
35 degrees

96.

45 degrees

97.
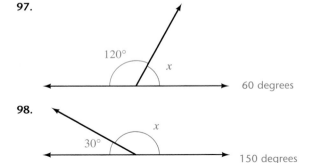
60 degrees

98.
150 degrees

99. Grass Growth The bar chart below shows the growth of a certain species of grass over a period of 10 days.
a. Use the chart to fill in the missing entries in the table.
b. How much higher is the grass after 8 days than after 3 days?   11.5 inches

Day	Plant Height (inches)	Day	Plant Height (inches)
0	0	6	6
1	0.5	7	9
2	1	8	13
3	1.5	9	18
4	3	10	23
5	4		

Overall Plant Height

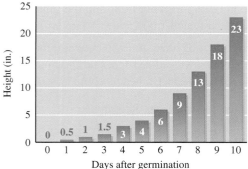

100. **Improving Your Quantitative Literacy** The chart shown here appeared in *USA Today* during the first week of June 2004. Use the chart to answer the following questions.

   a. Do you think the numbers in the chart have been rounded? If so, to which place were they rounded? Yes, nearest hundred

   b. How many more participants were there in 2003 than in 2000? 6,100

   c. If the trend shown in the table continues, estimate how many participants there will be in 2006. 25,200, though other answers are possible.

USA TODAY Snapshots

Triathlon appeals to more women
The 15th annual Danskin Women's Triathlon Series continues this weekend in San Dimas, Calif. The increasing number of overall series participants since 1990:

1990	1,100
1995	4,600
2000	13,000
2003	19,100

Source: Danskin Women's Triathlon Series

By Ellen J. Horrow and Keith Simmons, USA TODAY

1.5 Properties of Real Numbers

OBJECTIVES

A Rewrite expressions using the commutative and associative properties.

B Multiply using the distributive property.

C Identify properties used to rewrite an expression.

In this section we will list all the facts (properties) that you know from past experience are true about numbers in general. We will give each property a name so we can refer to it later in this book. Mathematics is very much like a game. The game involves numbers. The rules of the game are the properties and rules we are developing in this chapter. The goal of the game is to extend the basic rules to as many situations as possible.

You know from past experience with numbers that it makes no difference in which order you add two numbers; that is, $3 + 5$ is the same as $5 + 3$. This fact about numbers is called the *commutative property of addition*. We say addition is a commutative operation. Changing the order of the numbers does not change the answer.

There is one other basic operation that is commutative. Because $3(5)$ is the same as $5(3)$, we say multiplication is a commutative operation. Changing the order of the two numbers you are multiplying does not change the answer.

For all properties listed in this section, a, b, and c represent real numbers.

> **Commutative Property of Addition**
> *In symbols:* $a + b = b + a$
>
> *In words:* Changing the *order* of the numbers in a sum will not change the result.

> **Commutative Property of Multiplication**
>
> *In symbols:* $a \cdot b = b \cdot a$
>
> *In words:*   Changing the *order* of the numbers in a product will not change the result.

 EXAMPLES

1. The statement $5 + 8 = 8 + 5$ is an example of the commutative property of addition.
2. The statement $2 \cdot y = y \cdot 2$ is an example of the commutative property of multiplication.
3. The expression $5 + x + 3$ can be simplified using the commutative property of addition:

$$5 + x + 3 = x + 5 + 3 \qquad \textbf{Commutative property of addition}$$
$$= x + 8 \qquad \textbf{Addition}$$

Note

At this point, some students are confused by the expression $x + 8$; they feel that there is more to do, but they don't know what. At this point, there isn't any more that can be done with $x + 8$ unless we know what x is. So $x + 8$ is as far as we can go with this problem.

The other two basic operations, subtraction and division, are not commutative. The order in which we subtract or divide two numbers makes a difference in the answer.

Another property of numbers that you have used many times has to do with grouping. You know that when we add three numbers it makes no difference which two we add first. When adding $3 + 5 + 7$, we can add the 3 and 5 first and then the 7, or we can add the 5 and 7 first and then the 3. Mathematically, it looks like this: $(3 + 5) + 7 = 3 + (5 + 7)$. This property is true of multiplication as well. Operations that behave in this manner are called *associative* operations. The answer will not change when we change the association (or grouping) of the numbers.

> **Associative Property of Addition**
>
> *In symbols:* $a + (b + c) = (a + b) + c$
>
> *In words:*   Changing the *grouping* of the numbers in a sum will not change the result.

> **Associative Property of Multiplication**
>
> *In symbols:* $a(bc) = (ab)c$
>
> *In words:*   Changing the *grouping* of the numbers in a product will not change the result.

The following examples illustrate how the associative properties can be used to simplify expressions that involve both numbers and variables.

 EXAMPLES   Simplify.

4. $4 + (5 + x) = (4 + 5) + x$ **Associative property of addition**
   $\qquad\qquad = 9 + x$ **Addition**
5. $5(2x) = (5 \cdot 2)x$ **Associative property of multiplication**
   $\qquad = 10x$ **Multiplication**

6. $\frac{1}{5}(5x) = \left(\frac{1}{5} \cdot 5\right)x$ **Associative property of multiplication**

$\quad\quad\quad = 1x$ **Multiplication**

$\quad\quad\quad = x$

7. $3\left(\frac{1}{3}x\right) = \left(3 \cdot \frac{1}{3}\right)x$ **Associative property of multiplication**

$\quad\quad\quad = 1x$ **Multiplication**

$\quad\quad\quad = x$

8. $12\left(\frac{2}{3}x\right) = \left(12 \cdot \frac{2}{3}\right)x$ **Associative property of multiplication**

$\quad\quad\quad = 8x$ **Multiplication**

The associative and commutative properties apply to problems that are either all multiplication or all addition. There is a third basic property that involves both addition and multiplication. It is called the *distributive property* and looks like this.

Distributive Property

In symbols: $a(b + c) = ab + ac$

In words: Multiplication *distributes* over addition.

Note

Because subtraction is defined in terms of addition, it is also true that the distributive property applies to subtraction as well as addition; that is, $a(b - c) = ab - ac$ for any three real numbers a, b, and c.

You will see as we progress through the book that the distributive property is used very frequently in algebra. We can give a visual justification to the distributive property by finding the areas of rectangles. Figure 1 shows a large rectangle that is made up of two smaller rectangles. We can find the area of the large rectangle two different ways.

Method 1 We can calculate the area of the large rectangle directly by finding its length and width. The width is 5 inches, and the length is (3 + 4) inches.

$$\text{Area of large rectangle} = 5(3 + 4)$$
$$= 5(7)$$
$$= 35 \text{ square inches}$$

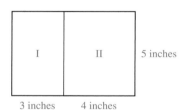

I   II   5 inches

3 inches   4 inches

FIGURE 1

Method 2 Because the area of the large rectangle is the sum of the areas of the two smaller rectangles, we find the area of each small rectangle and then add to find the area of the large rectangle.

$$\text{Area of large rectangle} = \text{Area of rectangle I} + \text{Area of rectangle II}$$
$$= 5(3) + 5(4)$$
$$= 15 + 20$$
$$= 35 \text{ square inches}$$

In both cases the result is 35 square inches. Because the results are the same, the two original expressions must be equal. Stated mathematically, $5(3 + 4) = 5(3) + 5(4)$. We can either add the 3 and 4 first and then multiply that sum by 5, or we can multiply the 3 and the 4 separately by 5 and then add the products. In either case we get the same answer.

Here are some examples that illustrate how we use the distributive property.

 EXAMPLES   Apply the distributive property to each expression, and then simplify the result.

9. $2(x + 3) = 2(x) + 2(3)$ **Distributive property**

$\quad\quad\quad = 2x + 6$ **Multiplication**

10. $5(2x - 8) = 5(2x) - 5(8)$ **Distributive property**

               $= 10x - 40$ **Multiplication**

Notice in this example that multiplication distributes over subtraction as well as addition.

11. $4(x + y) = 4x + 4y$ **Distributive property**

12. $5(2x + 4y) = 5(2x) + 5(4y)$ **Distributive property**

               $= 10x + 20y$ **Multiplication**

13. $\dfrac{1}{2}(3x + 6) = \dfrac{1}{2}(3x) + \dfrac{1}{2}(6)$ **Distributive property**

                 $= \dfrac{3}{2}x + 3$ **Multiplication**

14. $4(2a + 3) + 8 = 4(2a) + 4(3) + 8$ **Distributive property**

                 $= 8a + 12 + 8$ **Multiplication**

                 $= 8a + 20$ **Addition**

15. $a\left(1 + \dfrac{1}{a}\right) = a \cdot 1 + a \cdot \dfrac{1}{a}$ **Distributive property**

               $= a + 1$ **Multiplication**

16. $3\left(\dfrac{1}{3}x + 5\right) = 3 \cdot \dfrac{1}{3}x + 3 \cdot 5$ **Distributive property**

               $= x + 15$ **Multiplication**

17. $12\left(\dfrac{2}{3}x + \dfrac{1}{2}y\right) = 12 \cdot \dfrac{2}{3}x + 12 \cdot \dfrac{1}{2}y$ **Distributive property**

               $= 8x + 6y$ **Multiplication**

Special Numbers

In addition to the three properties mentioned so far, we want to include in our list two special numbers that have unique properties. They are the numbers zero and one.

Additive Identity Property There exists a unique number 0 such that

In symbols: $a + 0 = a$ and $0 + a = a$

In words: Zero preserves identities under addition. (The identity of the number is unchanged after addition with 0.)

Multiplicative Identity Property There exists a unique number 1 such that

In symbols: $a(1) = a$ and $1(a) = a$

In words: The number 1 preserves identities under multiplication. (The identity of the number is unchanged after multiplication by 1.)

Additive Inverse Property For each real number a, there exists a unique number $-a$ such that

In symbols: $a + (-a) = 0$

In words: Opposites add to 0.

> **Multiplicative Inverse Property** For every real number a, except 0, there exists a unique real number $\frac{1}{a}$ such that
>
> *In symbols:* $a\left(\dfrac{1}{a}\right) = 1$
>
> *In words:* Reciprocals multiply to 1.

Of all the basic properties listed, the commutative, associative, and distributive properties are the ones we will use most often. They are important because they will be used as justifications or reasons for many of the things we will do.

The following examples illustrate how we use the preceding properties. Each one contains an algebraic expression that has been changed in some way. The property that justifies the change is written to the right.

EXAMPLES   State the property that justifies the given statement.

18. $x + 5 = 5 + x$ **Commutative property of addition**

19. $(2 + x) + y = 2 + (x + y)$ **Associative property of addition**

20. $6(x + 3) = 6x + 18$ **Distributive property**

21. $2 + (-2) = 0$ **Additive inverse property**

22. $3\left(\dfrac{1}{3}\right) = 1$ **Multiplicative inverse property**

23. $(2 + 0) + 3 = 2 + 3$ **Additive identity property**

24. $(2 + 3) + 4 = 3 + (2 + 4)$ **Commutative and associative properties of addition**

25. $(x + 2) + y = (x + y) + 2$ **Commutative and associative properties of addition**

As a final note on the properties of real numbers, we should mention that although some of the properties are stated for only two or three real numbers, they hold for as many numbers as needed. For example, the distributive property holds for expressions like $3(x + y + z + 5 + 2)$; that is,

$$3(x + y + z + 5 + 2) = 3x + 3y + 3z + 15 + 6$$

It is not important how many numbers are contained in the sum, only that it is a sum. Multiplication, you see, distributes over addition, whether there are two numbers in the sum or 200.

LINKING OBJECTIVES AND EXAMPLES

Next to each **objective** we have listed the examples that are best described by that objective.

A	1–8
B	9–17
C	18–25

GETTING READY FOR CLASS

After reading through the preceding section, respond in your own words and in complete sentences.

1. What is the commutative property of addition?
2. Do you know from your experience with numbers that the commutative property of addition is true? Explain why.
3. Write the commutative property of multiplication in symbols and words.
4. How do you rewrite expressions using the distributive property?

State the property or properties that justify the following.

1. $3 + 2 = 2 + 3$ Commutative

2. $5 + 0 = 5$ Additive identity

3. $4\left(\dfrac{1}{4}\right) = 1$ Multiplicative inverse

4. $10(0.1) = 1$ Multiplicative inverse

5. $4 + x = x + 4$ Commutative

6. $3(x - 10) = 3x - 30$ Distributive

7. $2(y + 8) = 2y + 16$ Distributive

8. $3 + (4 + 5) = (3 + 4) + 5$ Associative

9. $(3 + 1) + 2 = 1 + (3 + 2)$ Commutative, associative

10. $(5 + 2) + 9 = (2 + 5) + 9$ Commutative

11. $(8 + 9) + 10 = (8 + 10) + 9$ Commutative, associative

12. $(7 + 6) + 5 = (5 + 6) + 7$ Commutative, associative

13. $3(x + 2) = 3(2 + x)$ Commutative

14. $2(7y) = (7 \cdot 2)y$ Commutative, associative

15. $x(3y) = 3(xy)$ Commutative, associative

16. $a(5b) = 5(ab)$ Commutative, associative

17. $4(xy) = 4(yx)$ Commutative

18. $3[2 + (-2)] = 3(0)$ Additive inverse

19. $8[7 + (-7)] = 8(0)$ Additive inverse

20. $7(1) = 7$ Multiplicative identity

Each of the following problems has a mistake in it. Correct the right-hand side.

21. $3(x + 2) = 3x + 2$ $3x + 6$

22. $5(4 + x) = 4 + 5x$ $20 + 5x$

23. $9(a + b) = 9a + b$ $9a + 9b$

24. $2(y + 1) = 2y + 1$ $2y + 2$

25. $3(0) = 3$ 0

26. $5\left(\dfrac{1}{5}\right) = 5$ 1

27. $3 + (-3) = 1$ 0

28. $8(0) = 8$ 0

29. $10(1) = 0$ 10

30. $3 \cdot \dfrac{1}{3} = 0$ 1

Use the associative property to rewrite each of the following expressions, and then simplify the result. (See Examples 4, 5, and 6.)

31. $4 + (2 + x)$ $(4 + 2) + x = 6 + x$

32. $5 + (6 + x)$ $(5 + 6) + x = 11 + x$

33. $(x + 2) + 7$ $x + (2 + 7) = x + 9$

34. $(x + 8) + 2$ $x + (8 + 2) = x + 10$

35. $3(5x)$ $(3 \cdot 5)x = 15x$

36. $5(3x)$ $(5 \cdot 3)x = 15x$

37. $9(6y)$ $(9 \cdot 6)y = 54y$

38. $6(9y)$ $(6 \cdot 9)y = 54y$

39. $\dfrac{1}{2}(3a)$ $\left(\dfrac{1}{2} \cdot 3\right)a = \dfrac{3}{2}a$

40. $\dfrac{1}{3}(2a)$ $\left(\dfrac{1}{3} \cdot 2\right)a = \dfrac{2}{3}a$

41. $\dfrac{1}{3}(3x)$ $\left(\dfrac{1}{3} \cdot 3\right)x = x$

42. $\dfrac{1}{4}(4x)$ $\left(\dfrac{1}{4} \cdot 4\right)x = x$

43. $\dfrac{1}{2}(2y)$ $\left(\dfrac{1}{2} \cdot 2\right)y = y$

44. $\dfrac{1}{7}(7y)$ $\left(\dfrac{1}{7} \cdot 7\right)y = y$

45. $\dfrac{3}{4}\left(\dfrac{4}{3}x\right)$ $\left(\dfrac{3}{4} \cdot \dfrac{4}{3}\right)x = x$

46. $\dfrac{3}{2}\left(\dfrac{2}{3}x\right)$ $\left(\dfrac{3}{2} \cdot \dfrac{2}{3}\right)x = x$

47. $\dfrac{6}{5}\left(\dfrac{5}{6}a\right)$ $\left(\dfrac{6}{5} \cdot \dfrac{5}{6}\right)a = a$

48. $\dfrac{2}{5}\left(\dfrac{5}{2}a\right)$ $\left(\dfrac{2}{5} \cdot \dfrac{5}{2}\right)a = a$

Apply the distributive property to each of the following expressions. Simplify when possible.

49. $8(x + 2)$ $8x + 16$

50. $5(x + 3)$ $5x + 15$

51. $8(x - 2)$ $8x - 16$

52. $5(x - 3)$ $5x - 15$

53. $4(y + 1)$ $4y + 4$

54. $4(y - 1)$ $4y - 4$

55. $3(6x + 5)$ $18x + 15$

56. $3(5x + 6)$ $15x + 18$

57. $2(3a + 7)$ $6a + 14$

58. $5(3a + 2)$ $15a + 10$

59. $9(6y - 8)$ $54y - 72$

60. $2(7y - 4)$ $14y - 8$

Apply the distributive property to each of the following expressions. Simplify when possible.

61. $\dfrac{1}{2}(3x - 6)$ $\dfrac{3}{2}x - 3$

62. $\dfrac{1}{3}(2x - 6)$ $\dfrac{2}{3}x - 2$

63. $\dfrac{1}{3}(3x + 6)$ $x + 2$

64. $\dfrac{1}{2}(2x + 4)$ $x + 2$

65. $3(x + y)$ $3x + 3y$

66. $2(x - y)$ $2x - 2y$

67. $8(a - b)$ $8a - 8b$

68. $7(a + b)$ $7a + 7b$

69. $6(2x + 3y)$ $12x + 18y$

70. $8(3x + 2y)$ $24x + 16y$

71. $4(3a - 2b)$ $12a - 8b$

72. $5(4a - 8b)$ $20a - 40b$

73. $\dfrac{1}{2}(6x + 4y)$ $3x + 2y$

74. $\dfrac{1}{3}(6x + 9y)$ $2x + 3y$

75. $4(a + 4) + 9$ $4a + 25$

76. $6(a + 2) + 8$ $6a + 20$

77. $2(3x + 5) + 2$ $6x + 12$

78. $7(2x + 1) + 3$ $14x + 10$

79. $7(2x + 4) + 10$ $14x + 38$

80. $3(5x + 6) + 20$ $15x + 38$

Here are some problems you will see later in the book. Apply the distributive property and simplify, if possible.

▶ **81.** $\frac{1}{2}(4x + 2)$ $2x + 1$ ▶ **82.** $\frac{1}{3}(6x + 3)$ $2x + 1$

▶ **83.** $\frac{3}{4}(8x - 4)$ $6x - 3$ ▶ **84.** $\frac{2}{5}(5x + 10)$ $2x + 4$

▶ **85.** $\frac{5}{6}(6x + 12)$ $5x + 10$ ▶ **86.** $\frac{2}{3}(9x - 3)$ $6x - 2$

▶ **87.** $10\left(\frac{3}{5}x + \frac{1}{2}\right)$ $6x + 5$ ▶ **88.** $8\left(\frac{1}{4}x - \frac{5}{8}\right)$ $2x - 5$

89. $15\left(\frac{1}{3}x + \frac{2}{5}\right)$ $5x + 6$ **90.** $12\left(\frac{1}{12}m + \frac{1}{6}\right)$ $m + 2$

91. $12\left(\frac{1}{2}m - \frac{5}{12}\right)$ $6m - 5$ **92.** $8\left(\frac{1}{8} + \frac{1}{2}m\right)$ $1 + 4m$

93. $21\left(\frac{1}{3} + \frac{1}{7}x\right)$ $7 + 3x$ **94.** $6\left(\frac{3}{2}y + \frac{1}{3}\right)$ $9y + 2$

95. $6\left(\frac{1}{2}x - \frac{1}{3}y\right)$ $3x - 2y$ **96.** $12\left(\frac{1}{4}x + \frac{2}{3}y\right)$ $3x + 8y$

97. $0.09(x + 2,000)$ $0.09x + 180$

98. $0.04(x + 7,000)$ $0.04x + 280$

99. $0.12(x + 500)$ $0.12x + 60$

100. $0.06(x + 800)$ $0.06x + 48$

101. $a\left(1 + \frac{1}{a}\right)$ $a + 1$ **102.** $a\left(1 - \frac{1}{a}\right)$ $a - 1$

103. $a\left(\frac{1}{a} - 1\right)$ $1 - a$ **104.** $a\left(\frac{1}{a} + 1\right)$ $1 + a$

Applying the Concepts

105. Getting Dressed While getting dressed for work, a man puts on his socks and puts on his shoes. Are the two statements "put on your socks" and "put on your shoes" commutative?   No

106. Getting Dressed Are the statements "put on your left shoe" and "put on your right shoe" commutative?   Yes

107. Skydiving A skydiver flying over the jump area is about to do two things: jump out of the plane and pull the rip cord. Are the two events "jump out of the plane" and "pull the rip cord" commutative? That is, will changing the order of the events always produce the same result?   No, not commutative

108. Commutative Property Give an example of two events in your daily life that are commutative.
Brushing your teeth and brushing your hair

109. Division Give an example that shows that division is not a commutative operation; that is, find two numbers for which changing the order of division gives two different answers. $8 \div 4 \neq 4 \div 8$

110. Subtraction Simplify the expression $10 - (5 - 2)$ and the expression $(10 - 5) - 2$ to show that subtraction is not an associative operation.
$10 - (5 - 2) = 10 - 3 = 7; (10 - 5) - 2 = 5 - 2 = 3$

111. Take-Home Pay Jose works at a winery. His monthly salary is $2,400. To cover his taxes and retirement, the winery withholds $480 from each check. Calculate his yearly "take-home" pay using the numbers 2,400, 480, and 12. Do the calculation two different ways so that the results give further justification for the distributive property.
$12(2,400 - 480) = \$23,040;$
$12(2,400) - 12(480) = \$23,040$

112. Hours Worked Carlo works as a waiter. He works double shifts 4 days a week. The lunch shift is 2 hours and the dinner shift is 3 hours. Find the total number of hours he works per week using the numbers 2, 3, and 4. Do the calculation two different ways so that the results give further justification for the distributive property.
$4(2 + 3) = 20$
$(4 \cdot 2) + (4 \cdot 3) = 20$

113. College Expenses Maria is estimating her expenses for attending college for a year. Tuition is $650 per academic quarter. She estimates she will spend $225 on books each quarter. If she plans on attending 3 academic quarters during the year, how much can she expect to spend? Do the calculation two different ways so that the results give further justification for the distributive property.
$3(650 + 225) = \$2,625$
$3(650) + 3(225) = \$2,625$

114. Improving Your Quantitative Literacy Although everything you do in this course will help improve your quantitative literacy, these problems will extend the type of reasoning and thinking you are using in the classroom to situations you will find outside of the classroom. Here is what the *Mathematical*

Foreshadowing Problems
If you have taught this class before, you will notice that Problems 81–104 are the problems that students get stuck on when simplifying complex fractions, or clearing equations of fractions, later in the course.

Association of America has to say about quantitative literacy:

. . . . every college graduate should be able to apply simple mathematical methods to the solution of real-world problems. A quantitatively literate college graduate should be able to:

1. Interpret mathematical models such as formulas, graphs, tables, and schematics, and draw inferences from them.
2. Represent mathematical information symbolically, visually, numerically, and verbally.
3. Estimate and check answers to mathematical problems in order to determine reasonableness, identify alternatives, and select optimal results.
4. Recognize that mathematical and statistical methods have limits.

List two specific skills that you would like to acquire from this course that will improve your quantitative literacy. For example, you may want to better understand the financing options available for purchasing a new car.

1.6 Multiplication of Real Numbers

OBJECTIVES

A Multiply any combination of positive and negative numbers.

B Simplify expressions using the rule for order of operations.

C Multiply fractions.

D Multiply using the associative property.

E Multiply using the distributive property.

F Extend a geometric sequence.

Suppose that you own 5 shares of a stock and the price per share drops $3. How much money have you lost? Intuitively, we know the loss is $15. Because it is a loss, we can express it as −$15. To describe this situation with numbers, we would write

5 shares   each lose $3   for a total of $15

$$5(-3) = -15$$

Reasoning in this manner, we conclude that the product of a positive number with a negative number is a negative number. Let's look at multiplication in more detail.

From our experience with counting numbers, we know that multiplication is simply repeated addition; that is, $3(5) = 5 + 5 + 5$. We will use this fact, along with our knowledge of negative numbers, to develop the rule for multiplication of any two real numbers. The following examples illustrate multiplication with three of the possible combinations of positive and negative numbers.

EXAMPLES   Multiply.

1. Two positives: $3(5) = 5 + 5 + 5$
   $= 15$ **Positive answer**
2. One positive: $3(-5) = -5 + (-5) + (-5)$
   $= -15$ **Negative answer**
3. One negative: $-3(5) = 5(-3)$ **Commutative property**
   $= -3 + (-3) + (-3) + (-3) + (-3)$
   $= -15$ **Negative answer**
4. Two negatives: $-3(-5) = ?$

Note
You may have to read the explanation for Example 4 several times before you understand it completely. The purpose of the explanation in Example 4 is simply to justify the fact that the product of two negative numbers is a positive number. If you have no trouble believing that, then it is not so important that you understand everything in the explanation.

With two negatives, $-3(-5)$, it is not possible to work the problem in terms of repeated addition. (It doesn't "make sense" to write −5 down a −3 number of times.) The answer is probably +15 (that's just a guess), but we need some justification for saying so. We will solve a different problem and in so doing get the answer to the problem $(-3)(-5)$.

Here is a problem to which we know the answer. We will work it two different ways.

$$-3[5 + (-5)] = -3(0) = 0$$

The answer is zero. We also can work the problem using the distributive property.

$$-3[5 + (-5)] = -3(5) + (-3)(-5) \qquad \textbf{Distributive property}$$
$$= -15 + ?$$

Because the answer to the problem is 0, our ? must be $+15$. (What else could we add to -15 to get 0? Only $+15$.)

Here is a summary of the results we have obtained from the first four examples.

Original Numbers Have		The Answer is
the same sign	$3(5) = 15$	positive
different signs	$3(-5) = -15$	negative
different signs	$-3(5) = -15$	negative
the same sign	$-3(-5) = 15$	positive

By examining Examples 1 through 4 and the preceding table, we can use the information there to write the following rule. This rule tells us how to multiply any two real numbers.

> **Rule** To multiply any two real numbers, simply multiply their absolute values. The sign of the answer is
>
> 1. *Positive* if both numbers have the same sign (both + or both −).
>
> 2. *Negative* if the numbers have opposite signs (one +, the other −).

The following examples illustrate how we use the preceding rule to multiply real numbers.

 EXAMPLES   Multiply.

5. $-8(-3) = 24$ ⎫
6. $-10(-5) = 50$ ⎬ If the two numbers in the product have the same
7. $-4(-7) = 28$ ⎭ sign, the answer is positive

8. $5(-7) = -35$ ⎫
9. $-4(8) = -32$ ⎬ If the two numbers in the product have different
10. $-6(10) = -60$ ⎭ signs, the answer is negative

In the following examples, we combine the rule for order of operations with the rule for multiplication to simplify expressions. Remember, the rule for order of operations specifies that we are to work inside the parentheses first and then simplify numbers containing exponents. After this, we multiply and divide, left to right. The last step is to add and subtract, left to right.

Note

Some students have trouble with the expression $-8(-3)$ because they want to subtract rather than multiply. Because we are very precise with the notation we use in algebra, the expression $-8(-3)$ has only one meaning—multiplication. A subtraction problem that uses the same numbers is $-8 - 3$. Compare the two following lists.

All Multiplication	No Multiplication
$5(4)$	$5 + 4$
$-5(4)$	$-5 + 4$
$5(-4)$	$5 - 4$
$-5(-4)$	$-5 - 4$

EXAMPLES   Simplify as much as possible.

11. $-5(-3)(-4) = 15(-4)$
$= -60$

12. $4(-3) + 6(-5) - 10 = -12 + (-30) - 10$ **Multiply**
$= -42 - 10$ **Add**
$= -52$ **Subtract**

13. $(-2)^3 = (-2)(-2)(-2)$ **Definition of exponents**
$= -8$ **Multiply, left to right**

14. $-3(-2)^3 - 5(-4)^2 = -3(-8) - 5(16)$ **Exponents first**
$= 24 - 80$ **Multiply**
$= -56$ **Subtract**

15. $6 - 4(7 - 2) = 6 - 4(5)$ **Inside parentheses first**
$= 6 - 20$ **Multiply**
$= -14$ **Subtract**

Multiplying Fractions

Previously, we mentioned that to multiply two fractions we multiply numerators and multiply denominators. We can apply the rule for multiplication of positive and negative numbers to fractions in the same way we apply it to other numbers. We multiply absolute values: The product is positive if both fractions have the same sign and negative if they have different signs. Here are some examples.

EXAMPLES   Multiply.

16. $-\dfrac{3}{4}\left(\dfrac{5}{7}\right) = -\dfrac{3 \cdot 5}{4 \cdot 7}$ **Different signs give a negative answer**

$= -\dfrac{15}{28}$

17. $-6\left(\dfrac{1}{2}\right) = -\dfrac{6}{1}\left(\dfrac{1}{2}\right)$ **Different signs give a negative answer**

$= -\dfrac{6}{2}$

$= -3$

18. $-\dfrac{2}{3}\left(-\dfrac{3}{2}\right) = \dfrac{2 \cdot 3}{3 \cdot 2}$ **Same signs give a positive answer**

$= \dfrac{6}{6}$

$= 1$

EXAMPLE 19   Figure 1 gives the calories that are burned in 1 hour for a variety of forms of exercise by a person weighing 150 pounds. Figure 2 gives the calories that are consumed by eating some popular fast foods. Find the net change in calories for a 150-pound person playing handball for 2 hours and then eating a Whopper.

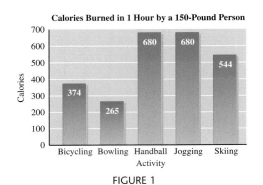

FIGURE 1

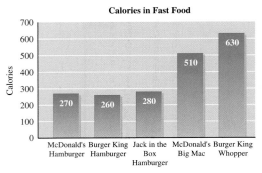

FIGURE 2

SOLUTION The net change in calories will be the difference of the calories gained from eating and the calories lost from exercise.

$$\text{Net change in calories} = 630 - 2(680) = -730 \text{ calories}$$

We can use the rule for multiplication of real numbers, along with the associative property, to multiply expressions that contain numbers and variables.

 EXAMPLES Apply the associative property, and then multiply.

20. $-3(2x) = (-3 \cdot 2)x$ **Associative property**
$= -6x$ **Multiplication**

21. $6(-5y) = [6(-5)]y$ **Associative property**
$= -30y$ **Multiplication**

22. $-2\left(-\dfrac{1}{2}x\right) = \left[(-2)\left(-\dfrac{1}{2}\right)\right]x$ **Associative property**

$= 1x$ **Multiplication**
$= x$ **Multiplication**

The following examples show how we can use both the distributive property and multiplication with real numbers.

 EXAMPLES Apply the distributive property to each expression.

23. $-2(a + 3) = -2a + (-2)(3)$ **Distributive property**
$= -2a + (-6)$ **Multiplication**
$= -2a - 6$

24. $-3(2x + 1) = -3(2x) + (-3)(1)$ **Distributive property**
$= -6x + (-3)$ **Multiplication**
$= -6x - 3$

25. $-\dfrac{1}{3}(2x - 6) = -\dfrac{1}{3}(2x) - \left(-\dfrac{1}{3}\right)(6)$ **Distributive property**

$= -\dfrac{2}{3}x - (-2)$ **Multiplication**

$= -\dfrac{2}{3}x + 2$

26. $-4(3x - 5) - 8 = -4(3x) - (-4)(5) - 8$ **Distributive property**
$= -12x - (-20) - 8$ **Multiplication**
$= -12x + 20 - 8$ **Definition of subtraction**
$= -12x + 12$ **Subtraction**

The next examples continue the work we did previously with finding the value of an algebraic expression for given values of the variable or variables.

 EXAMPLE 27   Find the value of $-\frac{2}{3}x - 4$ when

a. $x = 0$ **b.** $x = 3$ **c.** $x = -\frac{9}{2}$

SOLUTION   Substituting the values of x into our expression one at a time we have

a. $-\frac{2}{3}(0) - 4 = 0 - 4 = -4$

b. $-\frac{2}{3}(3) - 4 = -2 - 4 = -6$

c. $-\frac{2}{3}\left(-\frac{9}{2}\right) - 4 = 3 - 4 = -1$

 EXAMPLE 28   Find the value of $5x - 4y$ when
a. $x = 4$ and $y = 0$
b. $x = 0$ and $y = -8$
c. $x = -2$ and $y = 3$

SOLUTION   We substitute the given values for x and y and then simplify
a. $5(4) - 4(0) = 20 - 0 = 20$
b. $5(0) - 4(-8) = 0 + 32 = 32$
c. $5(-2) - 4(3) = -10 - 12 = -22$

Geometric Sequences

A *geometric sequence* is a sequence of numbers in which each number (after the first number) comes from the number before it by multiplying by the same amount each time. For example, the sequence

$$2, 6, 18, 54, \ldots$$

is a geometric sequence because each number is obtained by multiplying the number before it by 3.

EXAMPLE 29   Each sequence below is a geometric sequence. Find the next number in each sequence.
a. $5, 10, 20, \ldots$ **b.** $3, -15, 75, \ldots$ **c.** $\frac{1}{8}, \frac{1}{4}, \frac{1}{2}, \ldots$

SOLUTION   Because each sequence is a geometric sequence, we know that each term is obtained from the previous term by multiplying by the same number each time.

a. $5, 10, 20, \ldots$: Starting with 5, each number is obtained from the previous number by multiplying by 2 each time. The next number will be $20 \cdot 2 = 40$.

b. $3, -15, 75, \ldots$: The sequence starts with 3. After that, each number is obtained by multiplying by -5 each time. The next number will be $75(-5) = -375$.

LINKING OBJECTIVES AND EXAMPLES

Next to each objective we have listed the examples that are best described by that objective.

A	1–10
B	11–15
C	16–18
D	20–22
E	23–26
F	29

c. $\frac{1}{8}, \frac{1}{4}, \frac{1}{2}, \ldots$: This sequence starts with $\frac{1}{8}$. Multiplying each number in the sequence by 2 produces the next number in the sequence. To extend the sequence, we multiply $\frac{1}{2}$ by 2: $\frac{1}{2} \cdot 2 = 1$ The next number in the sequence is 1.

GETTING READY FOR CLASS

After reading through the preceding section, respond in your own words and in complete sentences.

1. How do you multiply two negative numbers?
2. How do you multiply two numbers with different signs?
3. Explain how some multiplication problems can be thought of as repeated addition.
4. What is a geometric sequence?

Problem Set 1.6

Online support materials can be found at www.thomsonedu.com/login

Use the rule for multiplying two real numbers to find each of the following products.

1. $7(-6)$ -42
2. $8(-4)$ -32
3. $-8(2)$ -16
4. $-16(3)$ -48
5. $-3(-1)$ 3
6. $-7(-1)$ 7
7. $-11(-11)$ 121
8. $-12(-12)$ 144

Use the rule for order of operations to simplify each expression as much as possible.

9. $-3(2)(-1)$ 6
10. $-2(3)(-4)$ 24
11. $-3(-4)(-5)$ -60
12. $-5(-6)(-7)$ -210
13. $-2(-4)(-3)(-1)$ 24
14. $-1(-3)(-2)(-1)$ 6
15. $(-7)^2$ 49
16. $(-8)^2$ 64
17. $(-3)^3$ -27
18. $(-2)^4$ 16
19. $-2(2-5)$ 6
20. $-3(3-7)$ 12
21. $-5(8-10)$ 10
22. $-4(6-12)$ 24
23. $(4-7)(6-9)$ 9
24. $(3-10)(2-6)$ 28
25. $(-3-2)(-5-4)$ 45
26. $(-3-6)(-2-8)$ 90
27. $-3(-6)+4(-1)$ 14
28. $-4(-5)+8(-2)$ 4

29. $2(3)-3(-4)+4(-5)$ -2
30. $5(4)-2(-1)+5(6)$ 52
31. $4(-3)^2+5(-6)^2$ 216
32. $2(-5)^2+4(-3)^2$ 86
33. $7(-2)^3-2(-3)^3$ -2
34. $10(-2)^3-5(-2)^4$ -160
35. $6-4(8-2)$ -18
36. $7-2(6-3)$ 1
37. $9-4(3-8)$ 29
38. $8-5(2-7)$ 33
39. $-4(3-8)-6(2-5)$ 38
40. $-8(2-7)-9(3-5)$ 58
41. $7-2[-6-4(-3)]$ -5
42. $6-3[-5-3(-1)]$ 12
43. $7-3[2(-4-4)-3(-1-1)]$ 37
44. $5-3[7(-2-2)-3(-3+1)]$ 71

45. Simplify each expression.
   a. $5(-4)(-3)$ 60
   b. $5(-4)-3$ -23
   c. $5-4(-3)$ 17
   d. $5-4-3$ -2

46. Simplify each expression.
   a. $-2(-3)(-5)$ -30
   b. $-2(-3)-5$ 1
   c. $-2-3(-5)$ 13
   d. $-2-3-5$ -10

= Videos available by instructor request

▶ = Online student support materials available at www.thomsonedu.com/login

Multiply the following fractions.

47. $-\dfrac{2}{3} \cdot \dfrac{5}{7}$ $-\frac{10}{21}$

48. $-\dfrac{6}{5} \cdot \dfrac{2}{7}$ $-\frac{12}{35}$

49. $-8\left(\dfrac{1}{2}\right)$ -4

50. $-12\left(\dfrac{1}{3}\right)$ -4

51. $\left(-\dfrac{3}{4}\right)^2$ $\frac{9}{16}$

52. $\left(-\dfrac{2}{5}\right)^2$ $\frac{4}{25}$

53. Simplify each expression.

   a. $\dfrac{5}{8}(24) + \dfrac{3}{7}(28)$ 27

   b. $\dfrac{5}{8}(24) - \dfrac{3}{7}(28)$ 3

   c. $\dfrac{5}{8}(-24) + \dfrac{3}{7}(-28)$ -27

   d. $-\dfrac{5}{8}(24) - \dfrac{3}{7}(28)$ -27

54. Simplify each expression.

   a. $\dfrac{5}{6}(18) + \dfrac{3}{5}(15)$ 24

   b. $\dfrac{5}{6}(18) - \dfrac{3}{5}(15)$ 6

   c. $\dfrac{5}{6}(-18) + \dfrac{3}{5}(-15)$ -24

   d. $-\dfrac{5}{6}(18) - \dfrac{3}{5}(15)$ -24

Simplify.

▸ **55.** $\left(\dfrac{1}{2} \cdot 6\right)^2$ 9

▸ **56.** $\left(\dfrac{1}{2} \cdot 10\right)^2$ 25

▸ **57.** $\left(\dfrac{1}{2} \cdot 5\right)^2$ $\frac{25}{4}$

▸ **58.** $\left[\dfrac{1}{2}(0.8)\right]^2$ 0.16

▸ **59.** $\left[\dfrac{1}{2}(-4)\right]^2$ 4

▸ **60.** $\left[\dfrac{1}{2}(-12)\right]^2$ 36

▸ **61.** $\left[\dfrac{1}{2}(-3)\right]^2$ $\frac{9}{4}$

▸ **62.** $\left[\dfrac{1}{2}(-0.8)\right]^2$ 0.16

Find the following products.

63. $-2(4x)$ $-8x$

64. $-8(7x)$ $-56x$

65. $-7(-6x)$ $42x$

66. $-8(-9x)$ $72x$

67. $-\dfrac{1}{3}(-3x)$ x

68. $-\dfrac{1}{5}(-5x)$ x

Apply the distributive property to each expression, and then simplify the result.

69. $-\dfrac{1}{2}(3x - 6)$ $-\frac{3}{2}x + 3$

70. $-\dfrac{1}{4}(2x - 4)$ $-\frac{1}{2}x + 1$

▸ **71.** $-3(2x - 5) - 7$ $-6x + 8$

72. $-4(3x - 1) - 8$ $-12x - 4$

73. $-5(3x + 4) - 10$ $-15x - 30$

74. $-3(4x + 5) - 20$ $-12x - 35$

75. $-4(3x + 5y)$ $-12x - 20y$

76. $5(5x + 4y)$ $25x + 20y$

77. $-2(3x + 5y)$ $-6x - 10y$

78. $-2(2x - y)$ $-4x + 2y$

▸ **79.** $\dfrac{1}{2}(-3x + 6)$ $-\frac{3}{2}x + 3$

▸ **80.** $\dfrac{1}{4}(5x - 20)$ $\frac{5}{4}x - 5$

▸ **81.** $\dfrac{1}{3}(-2x + 6)$ $-\frac{2}{3}x + 2$

▸ **82.** $\dfrac{1}{5}(-4x + 20)$ $-\frac{4}{5}x + 4$

83. $-\dfrac{1}{3}(-2x + 6)$ $\frac{2}{3}x - 2$

84. $-\dfrac{1}{2}(-2x + 6)$ $x - 3$

85. $8\left(-\dfrac{1}{4}x + \dfrac{1}{8}y\right)$ $-2x + y$

86. $9\left(-\dfrac{1}{9}x + \dfrac{1}{3}y\right)$ $-x + 3y$

87. Find the value of $-\dfrac{1}{3}x + 2$ when

   a. $x = 0$ 2

   b. $x = 3$ 1

   c. $x = -3$ 3

88. Find the value of $-\dfrac{2}{3}x + 1$ when

   a. $x = 0$ 1

   b. $x = 3$ -1

   c. $x = -3$ 3

89. Find the value of $2x + y$ when

   a. $x = 2$ and $y = -1$ 3

   b. $x = 0$ and $y = 3$ 3

   c. $x = \dfrac{3}{2}$ and $y = -7$ -4

90. Find the value of $2x - 5y$ when

   a. $x = 2$ and $y = 3$ -11

   b. $x = 0$ and $y = -2$ 10

   c. $x = \dfrac{5}{2}$ and $y = 1$ 0

91. Find the value of $2x^2 - 5x$ when

   a. $x = 4$ 12

   b. $x = -\dfrac{3}{2}$ 12

92. Find the value of $49a^2 - 16$ when

   a. $a = \dfrac{4}{7}$ 0

   b. $a = -\dfrac{4}{7}$ 0

93. Find the value of $y(2y + 3)$ when

   a. $y = 4$ 44

   b. $y = -\dfrac{11}{2}$ 44

Foreshadowing Problems: Problems 79–86 are the problems students need later in the chapter to work with formulas and to clear inequalities of fractions.

94. Find the value of $x(13 - x)$ when
  a. $x = 5$ 40
  b. $x = 8$ 40

95. Five added to the product of 3 and -10 is what number? -25

96. If the product of -8 and -2 is decreased by 4, what number results? 12

97. Write an expression for twice the product of -4 and x, and then simplify it. $2(-4x) = -8x$

98. Write an expression for twice the product of -2 and $3x$, and then simplify it. $2[-2(3x)] = -12x$

99. What number results if 8 is subtracted from the product of -9 and 2? -26

100. What number results if -8 is subtracted from the product of -9 and 2? -10

Each of the following is a geometric sequence. In each case, find the next number in the sequence.

101. $1, 2, 4, \ldots$ 8
102. $1, 5, 25, \ldots$ 125
103. $10, -20, 40, \ldots$ -80
104. $10, -30, 90, \ldots$ -270
105. $1, \dfrac{1}{2}, \dfrac{1}{4}, \ldots$ $\frac{1}{8}$
106. $1, \dfrac{1}{3}, \dfrac{1}{9}, \ldots$ $\frac{1}{27}$
107. $3, -6, 12, \ldots$ -24
108. $-3, 6, -12, \ldots$ 24

Applying the Concepts

109. Stock Value Suppose you own 20 shares of a stock. If the price per share drops $3, how much money have you lost? $60

110. Stock Value Imagine that you purchase 50 shares of a stock at a price of $18 per share. If the stock is selling for $11 a share a week after you purchased it, how much money have you lost? $350

111. Temperature Change The temperature is 25°F at 5:00 in the afternoon. If the temperature drops 6°F every hour after that, what is the temperature at 9:00 in the evening? 1°F

112. Temperature Change The temperature is -5°F at 6:00 in the evening. If the temperature drops 3°F every hour after that, what is the temperature at midnight? -23°F

113. Improving Your Quantitative Literacy This table appeared as part of an article on tuition costs. Although tuition costs rose considerably between 1998 and 2003, the average cost per student declined.
  a. Find the difference of student grants in 2003 and student grants in 1998. $544
  b. Find the difference of actual cost in 2003 and actual cost in 1998. $-$521
  c. How does the information in the table explain how students are paying less to go to college when tuition costs have increased?

Actual tuition cost declines

The actual cost of tuition and fees at four-year public universities has fallen nearly one-third since 1998 because of new tax breaks and an increase in federal and state grants. Average per student:

1998

Actual cost	Student grants	Total cost
$1,636	$1,940	$3,576

2003 Tax credits/deductions

Actual cost		Student grants	Total cost
$1,115	$603	$2,484	$4,202

Sources: USA TODAY research: College Board, Office of Management and Budget

By Karl Geties. USA TODAY

1.7 Division of Real Numbers

OBJECTIVES

A Divide any combination of positive and negative numbers.

B Divide fractions.

C Simplify expressions using the rule for order of operations.

Suppose that you and four friends bought equal shares of an investment for a total of $15,000 and then sold it later for only $13,000. How much did each person lose? Because the total amount of money that was lost can be represented by −$2,000, and there are 5 people with equal shares, we can represent each person's loss with division:

$$\frac{-\$2{,}000}{5} = -\$400$$

From this discussion it seems reasonable to say that a negative number divided by a positive number is a negative number. Here is a more detailed discussion of division with positive and negative numbers.

The last of the four basic operations is division. We will use the same approach to define division as we used for subtraction; that is, we will define division in terms of rules we already know.

Recall that we developed the rule for subtraction of real numbers by defining subtraction in terms of addition. We changed our subtraction problems to addition problems and then added to get our answers. Because we already have a rule for multiplication of real numbers, and division is the inverse operation of multiplication, we will simply define division in terms of multiplication.

We know that division by the number 2 is the same as multiplication by $\frac{1}{2}$; that is, 6 divided by 2 is 3, which is the same as 6 times $\frac{1}{2}$. Similarly, dividing a number by 5 gives the same result as multiplying by $\frac{1}{5}$. We can extend this idea to all real numbers with the following rule.

Note

We are defining division this way simply so that we can use what we already know about multiplication to do division problems. We actually want as few rules as possible. Defining division in terms of multiplication allows us to avoid writing a separate rule for division.

> **Rule** If a and b represent any two real numbers (b cannot be 0), then it is always true that
> $$a \div b = \frac{a}{b} = a\left(\frac{1}{b}\right)$$

Division by a number is the same as multiplication by its reciprocal. Because every division problem can be written as a multiplication problem and because we already know the rule for multiplication of two real numbers, we do not have to write a new rule for division of real numbers. We will simply replace our division problem with multiplication and use the rule we already have.

 EXAMPLES   Write each division problem as an equivalent multiplication problem, and then multiply.

1. $\dfrac{6}{2} = 6\left(\dfrac{1}{2}\right) = 3$ **The product of two positives is positive**

2. $\dfrac{6}{-2} = 6\left(-\dfrac{1}{2}\right) = -3$ ⎫

   The product of a positive and a negative is a negative

3. $\dfrac{-6}{2} = -6\left(\dfrac{1}{2}\right) = -3$ ⎭

4. $\dfrac{-6}{-2} = -6\left(-\dfrac{1}{2}\right) = 3$ **The product of two negatives is positive**

Note

What we are saying here is that the work shown in Examples 1 through 4 is shown simply to justify the answers we obtain. In the future we won't show the middle step in these kinds of problems. Even so, we need to know that division is *defined* to be multiplication by the reciprocal.

The second step in these examples is used only to show that we *can* write division in terms of multiplication. [In actual practice we wouldn't write $\frac{6}{2}$ as $6(\frac{1}{2})$.] The answers, therefore, follow from the rule for multiplication; that is, like signs produce a positive answer, and unlike signs produce a negative answer.

Here are some examples. This time we will not show division as multiplication by the reciprocal. We will simply divide. If the original numbers have the same signs, the answer will be positive. If the original numbers have different signs, the answer will be negative.

EXAMPLES   Divide.

5. $\frac{12}{6} = 2$ **Like signs give a positive answer**

6. $\frac{12}{-6} = -2$ **Unlike signs give a negative answer**

7. $\frac{-12}{6} = -2$ **Unlike signs give a negative answer**

8. $\frac{-12}{-6} = 2$ **Like signs give a positive answer**

9. $\frac{15}{-3} = -5$ **Unlike signs give a negative answer**

10. $\frac{-40}{-5} = 8$ **Like signs give a positive answer**

11. $\frac{-14}{2} = -7$ **Unlike signs give a negative answer**

Division with Fractions

We can apply the definition of division to fractions. Because dividing by a fraction is equivalent to multiplying by its reciprocal, we can divide a number by the fraction $\frac{3}{4}$ by multiplying it by the reciprocal of $\frac{3}{4}$, which is $\frac{4}{3}$. For example,

$$\frac{2}{5} \div \frac{3}{4} = \frac{2}{5} \cdot \frac{4}{3} = \frac{8}{15}$$

You may have learned this rule in previous math classes. In some math classes, multiplication by the reciprocal is referred to as "inverting the divisor and multiplying." No matter how you say it, division by any number (except 0) is always equivalent to multiplication by its reciprocal. Here are additional examples that involve division by fractions.

EXAMPLES   Divide.

12. $\frac{2}{3} \div \frac{5}{7} = \frac{2}{3} \cdot \frac{7}{5}$ **Rewrite as multiplication by the reciprocal**

    $= \frac{14}{15}$ **Multiply**

13. $-\dfrac{3}{4} \div \dfrac{7}{9} = -\dfrac{3}{4} \cdot \dfrac{9}{7}$ **Rewrite as multiplication by the reciprocal**

$= -\dfrac{27}{28}$ **Multiply**

14. $8 \div \left(-\dfrac{4}{5}\right) = \dfrac{8}{1}\left(-\dfrac{5}{4}\right)$ **Rewrite as multiplication by the reciprocal**

$= -\dfrac{40}{4}$ **Multiply**

$= -10$ **Divide 40 by 4**

The last step in each of the following examples involves reducing a fraction to lowest terms. To reduce a fraction to lowest terms, we divide the numerator and denominator by the largest number that divides each of them exactly. For example, to reduce $\dfrac{15}{20}$ to lowest terms, we divide 15 and 20 by 5 to get $\dfrac{3}{4}$.

 EXAMPLES   Simplify as much as possible.

15. $\dfrac{-4(5)}{6} = \dfrac{-20}{6}$ **Simplify numerator**

$= -\dfrac{10}{3}$ **Reduce to lowest terms by dividing numerator and denominator by 2**

16. $\dfrac{30}{-4-5} = \dfrac{30}{-9}$ **Simplify denominator**

$= -\dfrac{10}{3}$ **Reduce to lowest terms by dividing numerator and denominator by 3**

In the examples that follow, the numerators and denominators contain expressions that are somewhat more complicated than those we have seen thus far. To apply the rule for order of operations to these examples, we treat fraction bars the same way we treat grouping symbols; that is, fraction bars separate numerators and denominators so that each will be simplified separately.

EXAMPLES   Simplify.

17. $\dfrac{-8-8}{-5-3} = \dfrac{-16}{-8}$ **Simplify numerator and denominator separately**

$= 2$ **Divide**

18. $\dfrac{2(-3)+4}{12} = \dfrac{-6+4}{12}$ **In the numerator, we multiply before we add**

$= \dfrac{-2}{12}$ **Addition**

$= -\dfrac{1}{6}$ **Reduce to lowest terms by dividing numerator and denominator by 2**

19. $\dfrac{5(-4)+6(-1)}{2(3)-4(1)} = \dfrac{-20+(-6)}{6-4}$ **Multiplication before addition**

$= \dfrac{-26}{2}$ **Simplify numerator and denominator**

$= -13$ **Divide −26 by 2**

We must be careful when we are working with expressions such as $(-5)^2$ and -5^2 that we include the negative sign with the base only when parentheses indicate we are to do so.

Unless there are parentheses to indicate otherwise, we consider the base to be only the number directly below and to the left of the exponent. If we want to include a negative sign with the base, we must use parentheses.

To simplify a more complicated expression, we follow the same rule. For example,

$$7^2 - 3^2 = 49 - 9 \qquad \text{The bases are 7 and 3; the sign between the two terms is a subtraction sign}$$

For another example,

$$5^3 - 3^4 = 125 - 81 \qquad \text{We simplify exponents first, then subtract}$$

 EXAMPLES   Simplify.

20. $\dfrac{5^2 - 3^2}{-5 + 3} = \dfrac{25 - 9}{-2}$ **Simplify numerator and denominator separately**

$\qquad\qquad = \dfrac{16}{-2}$

$\qquad\qquad = -8$

21. $\dfrac{(3 + 2)^2}{-3^2 - 2^2} = \dfrac{5^2}{-9 - 4}$ **Simplify numerator and denominator separately**

$\qquad\qquad = \dfrac{25}{-13}$

$\qquad\qquad = -\dfrac{25}{13}$

We can combine our knowledge of the properties of multiplication with our definition of division to simplify more expressions involving fractions. Here are two examples:

 EXAMPLES   Simplify each expression.

22. $10\left(\dfrac{x}{2}\right) = 10\left(\dfrac{1}{2}x\right)$ **Dividing by 2 is the same as multiplying by $\frac{1}{2}$**

$\qquad\quad = \left(10 \cdot \dfrac{1}{2}\right)x$ **Associative property**

$\qquad\quad = 5x$ **Multiplication**

23. $a\left(\dfrac{3}{a} - 4\right) = a \cdot \dfrac{3}{a} - a \cdot 4$ **Distributive property**

$\qquad\qquad = 3 - 4a$ **Multiplication**

Division with the Number 0

For every division problem there is an associated multiplication problem involving the same numbers. For example, the following two problems say the same thing about the numbers 2, 3, and 6:

Division   *Multiplication*

$\dfrac{6}{3} = 2$ $6 = 2(3)$

We can use this relationship between division and multiplication to clarify division involving the number 0.

First, dividing 0 by a number other than 0 is allowed and always results in 0. To see this, consider dividing 0 by 5. We know the answer is 0 because of the relationship between multiplication and division. This is how we write it:

$$\frac{0}{5} = 0 \qquad \text{because} \qquad 0 = 0(5)$$

However, dividing a nonzero number by 0 is not allowed in the real numbers. Suppose we were attempting to divide 5 by 0. We don't know if there is an answer to this problem, but if there is, let's say the answer is a number that we can represent with the letter n. If 5 divided by 0 is a number n, then

$$\frac{5}{0} = n \qquad \text{and} \qquad 5 = n(0)$$

This is impossible, however, because no matter what number n is, when we multiply it by 0 the answer must be 0. It can never be 5. In algebra, we say expressions like $\frac{5}{0}$ are undefined because there is no answer to them; that is, division by 0 is not allowed in the real numbers.

The only other possibility for division involving the number 0 is 0 divided by 0. We will treat problems like $\frac{0}{0}$ as if they were undefined also.

LINKING OBJECTIVES AND EXAMPLES

Next to each **objective** we have listed the examples that are best described by that objective.

A	1–11
B	12–14
C	15–21

GETTING READY FOR CLASS

After reading through the preceding section, respond in your own words and in complete sentences.

1. Why do we define division in terms of multiplication?
2. What is the reciprocal of a number?
3. How do we divide fractions?
4. Why is division by 0 not allowed with real numbers?

Problem Set 1.7

Online support materials can be found at www.thomsonedu.com/login

Find the following quotients (divide).

▶ **1.** $\dfrac{8}{-4}$ $\quad -2$

2. $\dfrac{10}{-5}$ $\quad -2$

▶ **3.** $\dfrac{-48}{16}$ $\quad -3$

4. $\dfrac{-32}{4}$ $\quad -8$

5. $\dfrac{-7}{21}$ $\quad -\frac{1}{3}$

6. $\dfrac{-25}{100}$ $\quad -\frac{1}{4}$

▶ **7.** $\dfrac{-39}{-13}$ $\quad 3$

8. $\dfrac{-18}{-6}$ $\quad 3$

9. $\dfrac{-6}{-42}$ $\quad \frac{1}{7}$

10. $\dfrac{-4}{-28}$ $\quad \frac{1}{7}$

11. $\dfrac{0}{-32}$ $\quad 0$

12. $\dfrac{0}{17}$ $\quad 0$

The following problems review all four operations with positive and negative numbers. Perform the indicated operations.

13. $-3 + 12$ $\quad 9$

14. $5 + (-10)$ $\quad -5$

15. $-3 - 12$ $\quad -15$

16. $5 - (-10)$ $\quad 15$

▨ = Videos available by instructor request
▶ = Online student support materials available at www.thomsonedu.com/login

17. $-3(12)$ -36

18. $5(-10)$ -50

19. $-3 \div 12$ $-\frac{1}{4}$

20. $5 \div (-10)$ $-\frac{1}{2}$

Divide and reduce all answers to lowest terms.

21. $\frac{4}{5} \div \frac{3}{4}$ $\frac{16}{15}$

22. $\frac{6}{8} \div \frac{3}{4}$ 1

23. $-\frac{5}{6} \div \left(-\frac{5}{8}\right)$ $\frac{4}{3}$

24. $-\frac{7}{9} \div \left(-\frac{1}{6}\right)$ $\frac{14}{3}$

25. $\frac{10}{13} \div \left(-\frac{5}{4}\right)$ $-\frac{8}{13}$

26. $\frac{5}{12} \div \left(-\frac{10}{3}\right)$ $-\frac{1}{8}$

27. $-\frac{5}{6} \div \frac{5}{6}$ -1

28. $-\frac{8}{9} \div \frac{8}{9}$ -1

▶ **29.** $-\frac{3}{4} \div \left(-\frac{3}{4}\right)$ 1

30. $-\frac{6}{7} \div \left(-\frac{6}{7}\right)$ 1

The following problems involve more than one operation. Simplify as much as possible.

31. $\frac{3(-2)}{-10}$ $\frac{3}{5}$

32. $\frac{4(-3)}{24}$ $-\frac{1}{2}$

33. $\frac{-5(-5)}{-15}$ $-\frac{5}{3}$

34. $\frac{-7(-3)}{-35}$ $-\frac{3}{5}$

35. $\frac{-8(-7)}{-28}$ -2

36. $\frac{-3(-9)}{-6}$ $-\frac{9}{2}$

37. $\frac{27}{4-13}$ -3

38. $\frac{27}{13-4}$ 3

39. $\frac{20-6}{5-5}$ Undefined

40. $\frac{10-12}{3-3}$ Undefined

41. $\frac{-3+9}{2 \cdot 5 - 10}$ Undefined

42. $\frac{-4+8}{2 \cdot 4 - 8}$ Undefined

▶ **43.** $\frac{15(-5)-25}{2(-10)}$ 5

44. $\frac{10(-3)-20}{5(-2)}$ 5

45. $\frac{27-2(-4)}{-3(5)}$ $-\frac{7}{3}$

46. $\frac{20-5(-3)}{10(-3)}$ $-\frac{7}{6}$

47. $\frac{12-6(-2)}{12(-2)}$ -1

48. $\frac{3(-4)+5(-6)}{10-6}$ $-\frac{21}{2}$

49. $\frac{5^2-2^2}{-5+2}$ -7

50. $\frac{7^2-4^2}{-7+4}$ -11

51. $\frac{8^2-2^2}{8^2+2^2}$ $\frac{15}{17}$

52. $\frac{4^2-6^2}{4^2+6^2}$ $-\frac{5}{13}$

53. $\frac{(5+3)^2}{-5^2-3^2}$ $-\frac{32}{17}$

54. $\frac{(7+2)^2}{-7^2-2^2}$ $-\frac{81}{53}$

55. $\frac{(8-4)^2}{8^2-4^2}$ $\frac{1}{3}$

56. $\frac{(6-2)^2}{6^2-2^2}$ $\frac{1}{2}$

▶ **57.** $\frac{-4 \cdot 3^2 - 5 \cdot 2^2}{-8(7)}$ 1

58. $\frac{-2 \cdot 5^2 + 3 \cdot 2^3}{-3(13)}$ $\frac{2}{3}$

59. $\frac{3 \cdot 10^2 + 4 \cdot 10 + 5}{345}$ 1

60. $\frac{5 \cdot 10^2 + 6 \cdot 10 + 7}{567}$ 1

61. $\frac{7 - [(2-3) - 4]}{-1-2-3}$ -2

62. $\frac{2 - [(3-5) - 8]}{-3-4-5}$ -1

63. $\frac{6(-4) - 2(5-8)}{-6-3-5}$ $\frac{9}{7}$

64. $\frac{3(-4) - 5(9-11)}{-9-2-3}$ $\frac{1}{7}$

65. $\frac{3(-5-3) + 4(7-9)}{5(-2) + 3(-4)}$ $\frac{16}{11}$

66. $\frac{-2(6-10) - 3(8-5)}{6(-3) - 6(-2)}$ $\frac{1}{6}$

67. $\frac{|3-9|}{3-9}$ -1

68. $\frac{|4-7|}{4-7}$ -1

69. $\frac{2 + 0.15(10)}{10}$ $\frac{7}{20} = 0.35$

70. $\frac{5(5) + 250}{640(5)}$ $\frac{11}{128} \approx 0.086$

▶ **71.** $\frac{1-3}{3-1}$ -1

▶ **72.** $\frac{25-16}{16-25}$ -1

73. Simplify.

   a. $\frac{5-2}{3-1}$ $\frac{3}{2}$

   b. $\frac{2-5}{1-3}$ $\frac{3}{2}$

74. Simplify.

   a. $\frac{6-2}{3-5}$ -2

   b. $\frac{2-6}{5-3}$ -2

75. Simplify.

   a. $\frac{-4-1}{5-(-2)}$ $-\frac{5}{7}$

   b. $\frac{1-(-4)}{-2-5}$ $-\frac{5}{7}$

76. Simplify.

   a. $\frac{-6-1}{4-(-5)}$ $-\frac{7}{9}$

   b. $\frac{1-(-6)}{-5-4}$ $-\frac{7}{9}$

77. Simplify each expression.

   a. $\frac{3 + 2.236}{2}$ 2.618

   b. $\frac{3 - 2.236}{2}$ 0.382

   c. $\frac{3 + 2.236}{2} + \frac{3 - 2.236}{2}$ 3

78. Simplify each expression.

   a. $\frac{1 + 1.732}{2}$ 1.366

   b. $\frac{1 - 1.732}{2}$ -0.366

   c. $\frac{1 + 1.732}{2} + \frac{1 - 1.732}{2}$ 1

79. Simplify each expression.

   a. $20 \div 4 \cdot 5$ 25

   b. $-20 \div 4 \cdot 5$ -25

   c. $20 \div (-4) \cdot 5$ -25

   d. $20 \div 4(-5)$ -25

   e. $-20 \div 4(-5)$ 25

Foreshadowing Problems

Notice that Problems 73–76 will get students ready to work with the slope formula in Chapter 3.

80. Simplify each expression.

   a. $32 \div 8 \cdot 4$ 16

   b. $-32 \div 8 \cdot 4$ -16

   c. $32 \div (-8) \cdot 4$ -16

   d. $32 \div 8(-4)$ -16

   e. $-32 \div 8(-4)$ 16

81. Simplify each expression.

   a. $8 \div \dfrac{4}{5}$ 10

   b. $8 \div \dfrac{4}{5} - 10$ 0

   c. $8 \div \dfrac{4}{5}(-10)$ -100

   d. $8 \div \left(-\dfrac{4}{5}\right) - 10$ -20

82. Simplify each expression.

   a. $10 \div \dfrac{5}{6}$ 12

   b. $10 \div \dfrac{5}{6} - 12$ 0

   c. $10 \div \dfrac{5}{6}(-12)$ -144

   d. $10 \div \left(-\dfrac{5}{6}\right) - 12$ -24

Apply the distributive property.

83. $10\left(\dfrac{x}{2} + \dfrac{3}{5}\right)$ $5x + 6$ **84.** $6\left(\dfrac{x}{3} + \dfrac{5}{2}\right)$ $2x + 15$

85. $15\left(\dfrac{x}{5} + \dfrac{4}{3}\right)$ $3x + 20$ **86.** $6\left(\dfrac{x}{3} + \dfrac{1}{2}\right)$ $2x + 3$

87. $x\left(\dfrac{3}{x} + 1\right)$ $3 + x$ **88.** $x\left(\dfrac{4}{x} + 3\right)$ $4 + 3x$

89. $21\left(\dfrac{x}{7} - \dfrac{y}{3}\right)$ $3x - 7y$ **90.** $36\left(\dfrac{x}{4} - \dfrac{y}{9}\right)$ $9x - 4y$

91. $a\left(\dfrac{3}{a} - \dfrac{2}{a}\right)$ 1   **92.** $a\left(\dfrac{7}{a} + \dfrac{1}{a}\right)$ 8

93. $2y\left(\dfrac{1}{y} - \dfrac{1}{2}\right)$ $2 - y$ **94.** $5y\left(\dfrac{3}{y} - \dfrac{4}{5}\right)$ $15 - 4y$

Answer the following questions.

95. What is the quotient of -12 and -4?   3

96. The quotient of -4 and -12 is what number? $\frac{1}{3}$

97. What number do we divide by -5 to get 2? -10

98. What number do we divide by -3 to get 4? -12

99. Twenty-seven divided by what number is -9? -3

100. Fifteen divided by what number is -3? -5

101. If the quotient of -20 and 4 is decreased by 3, what number results? -8

102. If -4 is added to the quotient of 24 and -8, what number results? -7

Applying the Concepts

103. Investment Suppose that you and 3 friends bought equal shares of an investment for a total of $15,000 and then sold it later for only $13,600. How much did each person lose? $350

104. Investment If 8 people invest $500 each in a stamp collection and after a year the collection is worth $3,800, how much did each person lose? $25

105. Temperature Change Suppose that the temperature outside is dropping at a constant rate. If the temperature is 75°F at noon and drops to 61°F by 4:00 in the afternoon, by how much did the temperature change each hour?   Drops 3.5°F each hour

106. Temperature Change In a chemistry class, a thermometer is placed in a beaker of hot water. The initial temperature of the water is 165°F. After 10 minutes the water has cooled to 72°F. If the water temperature drops at a constant rate, by how much does the water temperature change each minute?   Drops 9.3°F each minute

107. Internet Mailing Lists A company sells products on the Internet through an email list. They predict that they sell one $50 product for every 25 people on their mailing list.

   a. What is their projected revenue if their list contains 10,000 email addresses? $20,000

   b. What is their projected revenue if their list contains 25,000 email addresses? $50,000

   c. They can purchase a list of 5,000 email addresses for $5,000. Is this a wise purchase?   yes

108. Internet Mailing Lists A new band has a following on the Internet. They sell their CDs through an email list. They predict that they sell one $15 CD for every 10 people on their mailing list.

   a. What is their projected revenue if their list contains 5,000 email addresses? $7,500

   b. What is their projected revenue if their list contains 20,000 email addresses? $30,000

   c. If they need to make $45,000, how many people do they need on their email list?   30,000 people

Foreshadowing Problems: Problems 83–94 are more problems that get students ready for simplifying complex fractions, or clearing equations of fractions, later in the course

OBJECTIVES

A Associate numbers with subsets of the real numbers.

B Factor whole numbers into the product of prime factors.

C Reduce fractions to lowest terms.

In Section 1.2 we introduced the real numbers and defined them as the numbers associated with points on the real number line. At that time, we said the real numbers include whole numbers, fractions, and decimals, as well as other numbers that are not as familiar to us as these numbers. In this section we take a more detailed look at the kinds of numbers that make up the set of real numbers.

The numbers that make up the set of real numbers can be classified as *counting numbers, whole numbers, integers, rational numbers,* and *irrational numbers;* each is said to be a *subset* of the real numbers.

> **DEFINITION** Set *A* is called a **subset** of set *B* if set *A* is contained in set *B*; that is, if each and every element in set *A* is also a member of set *B*.

Here is a detailed description of the major subsets of the real numbers.

The counting numbers are the numbers with which we count. They are the numbers 1, 2, 3, and so on. The notation we use to specify a group of numbers like this is *set notation*. We use the symbols { and } to enclose the members of the set.

$$\text{Counting numbers} = \{1, 2, 3, \dots\}$$

EXAMPLE 1 Which of the numbers in the following set are not counting numbers?

$$\left\{-3, 0, \frac{1}{2}, 1, 1.5, 3\right\}$$

SOLUTION The numbers -3, 0, $\frac{1}{2}$, and 1.5 are not counting numbers.

The whole numbers include the counting numbers and the number 0.

$$\text{Whole numbers} = \{0, 1, 2, \dots\}$$

The set of integers includes the whole numbers and the opposites of all the counting numbers.

$$\text{Integers} = \{\dots, -3, -2, -1, 0, 1, 2, 3, \dots\}$$

When we refer to positive integers, we are referring to the numbers 1, 2, 3, ... Likewise, the negative integers are -1, -2, -3, ... The number 0 is neither positive nor negative.

EXAMPLE 2 Which of the numbers in the following set are not integers?

$$\left\{-5, -1.75, 0, \frac{2}{3}, 1, \pi, 3\right\}$$

SOLUTION The only numbers in the set that are not integers are -1.75, $\frac{2}{3}$, and π.

The set of *rational numbers* is the set of numbers commonly called "fractions" together with the integers. The set of rational numbers is difficult to list in the same way we have listed the other sets, so we will use a different kind of notation:

$$\text{Rational numbers} = \left\{ \frac{a}{b} \;\middle|\; a \text{ and } b \text{ are integers } (b \neq 0) \right\}$$

This notation is read "The set of elements $\frac{a}{b}$ such that a and b are integers (and b is not 0)." If a number can be put in the form $\frac{a}{b}$, where a and b are both from the set of integers, then it is called a rational number.

Rational numbers include any number that can be written as the ratio of two integers; that is, rational numbers are numbers that can be put in the form

$$\frac{\text{integer}}{\text{integer}}$$

 EXAMPLE 3   Show why each of the numbers in the following set is a rational number.

$$\left\{ -3, -\frac{2}{3}, 0, 0.333\ldots, 0.75 \right\}$$

SOLUTION   The number -3 is a rational number because it can be written as the ratio of -3 to 1; that is,

$$-3 = \frac{-3}{1}$$

Similarly, the number $-\frac{2}{3}$ can be thought of as the ratio of -2 to 3, whereas the number 0 can be thought of as the ratio of 0 to 1.

Any repeating decimal, such as $0.333\ldots$ (the dots indicate that the 3's repeat forever), can be written as the ratio of two integers. In this case $0.333\ldots$ is the same as the fraction $\frac{1}{3}$.

Finally, any decimal that terminates after a certain number of digits can be written as the ratio of two integers. The number 0.75 is equal to the fraction $\frac{3}{4}$ and is therefore a rational number.

Still other numbers exist, each of which is associated with a point on the real number line, that cannot be written as the ratio of two integers. In decimal form they never terminate and never repeat a sequence of digits indefinitely. They are called *irrational numbers* (because they are not rational):

$$\text{Irrational numbers} = \{\text{nonrational numbers; nonrepeating,} \\ \text{nonterminating decimals}\}$$

We cannot write any irrational number in a form that is familiar to us because they are all nonterminating, nonrepeating decimals. Because they are not rational, they cannot be written as the ratio of two integers. They have to be represented in other ways. One irrational number you have probably seen before is π. It is not 3.14. Rather, 3.14 is an approximation to π. It cannot be written as a terminating decimal number. Other representations for irrational numbers are $\sqrt{2}$, $\sqrt{3}$, $\sqrt{5}$, $\sqrt{6}$, and, in general, the square root of any number that is not itself a perfect square. (If you are not familiar with square roots, you will be after Chapter 8.) Right now it is enough to know that some numbers on the number line cannot be written as the ratio of two integers or in decimal form. We call them irrational numbers.

The set of real numbers is the set of numbers that are either rational or irrational; that is, a real number is either rational or irrational.

Real numbers = {all rational numbers and all irrational numbers}

Prime Numbers and Factoring

The following diagram shows the relationship between multiplication and factoring:

Multiplication

Factors $\longrightarrow$ 3 · 4 = 12 $\longleftarrow$ Product

Factoring

When we read the problem from left to right, we say the product of 3 and 4 is 12. Or we multiply 3 and 4 to get 12. When we read the problem in the other direction, from right to left, we say we have *factored* 12 into 3 times 4, or 3 and 4 are *factors* of 12.

The number 12 can be factored still further:

$$12 = 4 \cdot 3$$
$$= 2 \cdot 2 \cdot 3$$
$$= 2^2 \cdot 3$$

The numbers 2 and 3 are called *prime factors* of 12 because neither of them can be factored any further.

> **DEFINITION** If a and b represent integers, then a is said to be a **factor** (or divisor) of b if a divides b evenly; that is, if a divides b with no remainder.

Note
The number 15 is not a prime number because it has factors of 3 and 5; that is, $15 = 3 \cdot 5$. When a whole number larger than 1 is not prime, it is said to be *composite*.

> **DEFINITION** A **prime number** is any positive integer larger than 1 whose only positive factors (divisors) are itself and 1.

Here is a list of the first few prime numbers.

Prime numbers = {2, 3, 5, 7, 11, 13, 17, 19, 23, 29, 31, 37, 41, . . . }

When a number is not prime, we can factor it into the product of prime numbers. To factor a number into the product of primes, we simply factor it until it cannot be factored further.

EXAMPLE 4   Factor the number 60 into the product of prime numbers.

SOLUTION   We begin by writing 60 as the product of any two positive integers whose product is 60, like 6 and 10:

$$60 = 6 \cdot 10$$

We then factor these numbers:

$$60 = 6 \cdot 10$$
$$= (2 \cdot 3) \cdot (2 \cdot 5)$$
$$= 2 \cdot 2 \cdot 3 \cdot 5$$
$$= 2^2 \cdot 3 \cdot 5$$

Note
It is customary to write the prime factors in order from smallest to largest.

EXAMPLE 5   Factor the number 630 into the product of primes.

SOLUTION   Let's begin by writing 630 as the product of 63 and 10:

$$630 = 63 \cdot 10$$
$$= (7 \cdot 9) \cdot (2 \cdot 5)$$
$$= 7 \cdot 3 \cdot 3 \cdot 2 \cdot 5$$
$$= 2 \cdot 3^2 \cdot 5 \cdot 7$$

It makes no difference which two numbers we start with, as long as their product is 630. We always will get the same result because a number has only one set of prime factors.

$$630 = 18 \cdot 35$$
$$= 3 \cdot 6 \cdot 5 \cdot 7$$
$$= 3 \cdot 2 \cdot 3 \cdot 5 \cdot 7$$
$$= 2 \cdot 3^2 \cdot 5 \cdot 7$$

Note
There are some "tricks" to finding the divisors of a number. For instance, if a number ends in 0 or 5, then it is divisible by 5. If a number ends in an even number (0, 2, 4, 6, or 8), then it is divisible by 2. A number is divisible by 3 if the sum of its digits is divisible by 3. For example, 921 is divisible by 3 because the sum of its digits is $9 + 2 + 1 = 12$, which is divisible by 3.

When we have factored a number into the product of its prime factors, we not only know what prime numbers divide the original number, but we also know all of the other numbers that divide it as well. For instance, if we were to factor 210 into its prime factors, we would have $210 = 2 \cdot 3 \cdot 5 \cdot 7$, which means that 2, 3, 5, and 7 divide 210, as well as any combination of products of 2, 3, 5, and 7; that is, because 3 and 7 divide 210, then so does their product 21. Because 3, 5, and 7 each divide 210, then so does their product 105:

$$\begin{array}{c} \text{21 divides 210} \\ 210 = 2 \cdot 3 \cdot 5 \cdot 7 \\ \text{105 divides 210} \end{array}$$

Although there are many ways in which factoring is used in arithmetic and algebra, one simple application is in reducing fractions to lowest terms.

Recall that we reduce fractions to lowest terms by dividing the numerator and denominator by the same number. We can use the prime factorization of numbers to help us reduce fractions with large numerators and denominators.

EXAMPLE 6   Reduce $\frac{210}{231}$ to lowest terms.

SOLUTION   First we factor 210 and 231 into the product of prime factors. Then we reduce to lowest terms by dividing the numerator and denominator by any factors they have in common.

Note

The small lines we have drawn through the factors that are common to the numerator and denominator are used to indicate that we have divided the numerator and denominator by those factors.

$$\frac{210}{231} = \frac{2 \cdot 3 \cdot 5 \cdot 7}{3 \cdot 7 \cdot 11} \quad \text{Factor the numerator and denominator completely}$$

$$= \frac{2 \cdot \cancel{3} \cdot 5 \cdot \cancel{7}}{\cancel{3} \cdot \cancel{7} \cdot 11} \quad \text{Divide the numerator and denominator by } 3 \cdot 7$$

$$= \frac{2 \cdot 5}{11}$$

$$= \frac{10}{11}$$

When we are working with fractions or with division, some of the instructions we use are equivalent; they mean the same thing. For example, each of the problems below will yield the same result:

Reduce to lowest terms: $\dfrac{50}{-80}$.

Divide: $\dfrac{50}{-80}$.

Simplify: $\dfrac{50}{-80}$.

Whether you think of the problem as a division problem, a simplification, or reducing a fraction to lowest terms, the answer will be $-\frac{5}{8}$, or -0.625, if a decimal is more appropriate for the situation. Sometimes you will see the instruction *simplify*, and sometimes you will see the instruction *reduce*. In either case, you will work the problem in the same way.

LINKING OBJECTIVES AND EXAMPLES

Next to each **objective** we have listed the examples that are best described by that objective.

A	1–3
B	4, 5
C	6

GETTING READY FOR CLASS

After reading through the preceding section, respond in your own words and in complete sentences.

1. What is a whole number?
2. How are factoring and multiplication related?
3. Is every integer also a rational number? Explain.
4. What is a prime number?

Problem Set 1.8

Online support materials can be found at www.thomsonedu.com/login

Given the numbers in the set $\{-3, -2.5, 0, 1, \frac{3}{2}, \sqrt{15}\}$:

▶ **1.** List all the whole numbers.   0, 1

▶ **2.** List all the integers.   −3, 0, 1

▶ **3.** List all the rational numbers. $-3, -2.5, 0, 1, \frac{3}{2}$

▶ **4.** List all the irrational numbers. $\sqrt{15}$

▶ **5.** List all the real numbers.   All

Given the numbers in the set $\{-10, -8, -0.333\ldots, -2, 9, \frac{25}{3}, \pi\}$:

6. List all the whole numbers.   9

7. List all the integers.   −10, −8, −2, 9

8. List all the rational numbers.
$-10, -8, -0.333\ldots, -2, 9, \frac{25}{3}$

9. List all the irrational numbers. π

10. List all the real numbers.   All

Identify the following statements as either true or false.

11. Every whole number is also an integer.   T

12. The set of whole numbers is a subset of the set of integers.   T

13. A number can be both rational and irrational.   F

14. The set of rational numbers and the set of irrational numbers have some elements in common.   F

15. Some whole numbers are also negative integers.   F

16. Every rational number is also a real number.   T

17. All integers are also rational numbers.   T

18. The set of integers is a subset of the set of rational numbers.   T

Label each of the following numbers as *prime* or *composite*. If a number is composite, then factor it completely.

19. 48   Composite, $2^4 \cdot 3$

20. 72   Composite, $2^3 \cdot 3^2$

21. 37   Prime

22. 23   Prime

23. 1,023   Composite, $3 \cdot 11 \cdot 31$

24. 543   Composite, $3 \cdot 181$

Factor the following into the product of primes. When the number has been factored completely, write its prime factors from smallest to largest.

25. 144 $2^4 \cdot 3^2$

26. 288 $2^5 \cdot 3^2$

27. 38 $2 \cdot 19$

28. 63 $3^2 \cdot 7$

29. 105 $3 \cdot 5 \cdot 7$

30. 210 $2 \cdot 3 \cdot 5 \cdot 7$

31. 180 $2^2 \cdot 3^2 \cdot 5$

32. 900 $2^2 \cdot 3^2 \cdot 5^2$

33. 385 $5 \cdot 7 \cdot 11$

34. 1,925 $5^2 \cdot 7 \cdot 11$

35. 121 11^2

36. 546 $2 \cdot 3 \cdot 7 \cdot 13$

37. 420 $2^2 \cdot 3 \cdot 5 \cdot 7$

38. 598 $2 \cdot 13 \cdot 23$

39. 620 $2^2 \cdot 5 \cdot 31$

40. 2,310 $2 \cdot 3 \cdot 5 \cdot 7 \cdot 11$

Reduce each fraction to lowest terms by first factoring the numerator and denominator into the product of prime factors and then dividing out any factors they have in common.

41. $\frac{105}{165}$ $\frac{7}{11}$

42. $\frac{165}{385}$ $\frac{3}{7}$

▶ **43.** $\frac{525}{735}$ $\frac{5}{7}$

44. $\frac{550}{735}$ $\frac{110}{147}$

45. $\frac{385}{455}$ $\frac{11}{13}$

46. $\frac{385}{735}$ $\frac{11}{21}$

47. $\frac{322}{345}$ $\frac{14}{15}$

48. $\frac{266}{285}$ $\frac{14}{15}$

49. $\frac{205}{369}$ $\frac{5}{9}$

50. $\frac{111}{185}$ $\frac{3}{5}$

51. $\frac{215}{344}$ $\frac{5}{8}$

52. $\frac{279}{310}$ $\frac{9}{10}$

The next two problems are intended to give you practice reading, and paying attention to, the instructions that accompany the problems you are working. You will see a number of problems like this throughout the book. Working these problems is an excellent way to get ready for a test or a quiz.

▶ **53.** Work each problem according to the instructions given. (Note that each of these instructions could be replaced with the instruction *Simplify*.)
  a. Add: $50 + (-80)$ −30
  b. Subtract: $50 - (-80)$ 130
  c. Multiply: $50(-80)$ −4,000
  d. Divide: $\frac{50}{-80}$ $-\frac{5}{8}$

▶ **Chalkboard Problems**
Problem 53 is a problem I work in class. I always emphasize to students that they should pay attention to instructions. Notice how the numbers and expressions in Problem 53 are all similar. Although students sometimes complain that these problems are confusing, especially in the early part of the course, I reassure them that it is better to be confused now, doing homework, than when they are taking a test or quiz.

□ = Videos available by instructor request
▶ = Online student support materials available at www.thomsonedu.com/login

1.8 Subsets of the Real Numbers   63

54. Work each problem according to the instructions given.

   a. Add: $-2.5 + 7.5$ 5

   b. Subtract: $-2.5 - 7.5$ -10

   c. Multiply: $-2.5(7.5)$ -18.75

   d. Divide: $\dfrac{-2.5}{7.5}$ $-\dfrac{1}{3}$

Simplify each expression without using a calculator.

55. $\dfrac{6.28}{9(3.14)}$ $\dfrac{2}{9}$

56. $\dfrac{12.56}{4(3.14)}$ 1

57. $\dfrac{9.42}{2(3.14)}$ $\dfrac{3}{2}$

58. $\dfrac{12.56}{2(3.14)}$ 2

▶ **59.** $\dfrac{32}{0.5}$ 64

▶ **60.** $\dfrac{16}{0.5}$ 32

▶ **61.** $\dfrac{5{,}599}{11}$ 509

▶ **62.** $\dfrac{840}{80}$ 10.5

63. Find the value of $\dfrac{2 + 0.15x}{x}$ for each of the values of x given below. Write your answers as decimals, to the nearest hundredth.

   a. $x = 10$ 0.35

   b. $x = 15$ 0.28

   c. $x = 20$ 0.25

64. Find the value of $\dfrac{5x + 250}{640x}$ for each of the values of x given below. Write your answers as decimals, to the nearest thousandth.

   a. $x = 10$ 0.047

   b. $x = 15$ 0.034

   c. $x = 20$ 0.027

65. Factor 6^3 into the product of prime factors by first factoring 6 and then raising each of its factors to the third power. $6^3 = (2 \cdot 3)^3 = 2^3 \cdot 3^3$

66. Factor 12^2 into the product of prime factors by first factoring 12 and then raising each of its factors to the second power. $12^2 = (2^2 \cdot 3)^2 = (2^2)^2(3)^2 = 2^4 \cdot 3^2$

67. Factor $9^4 \cdot 16^2$ into the product of prime factors by first factoring 9 and 16 completely.

$9^4 \cdot 16^2 = (3^2)^4(2^4)^2 = 2^8 \cdot 3^8$

68. Factor $10^2 \cdot 12^3$ into the product of prime factors by first factoring 10 and 12 completely.

$10^2 \cdot 12^3 = (2 \cdot 5)^2(2^2 \cdot 3)^3 = 2^8 \cdot 3^3 \cdot 5^2$

69. Simplify the expression $3 \cdot 8 + 3 \cdot 7 + 3 \cdot 5$, and then factor the result into the product of primes. (Notice one of the factors of the answer is 3.)

$3 \cdot 8 + 3 \cdot 7 + 3 \cdot 5 = 24 + 21 + 15 = 60 = 2^2 \cdot 3 \cdot 5$

70. Simplify the expression $5 \cdot 4 + 5 \cdot 9 + 5 \cdot 3$, and then factor the result into the product of primes.

$5 \cdot 4 + 5 \cdot 9 + 5 \cdot 3 = 20 + 45 + 15 = 80 = 2^4 \cdot 5$

Recall the Fibonacci sequence we introduced earlier in this chapter.

Fibonacci sequence = 1, 1, 2, 3, 5, 8, . . .

Any number in the Fibonacci sequence is a *Fibonacci number*.

71. The Fibonacci numbers are not a subset of which of the following sets: real numbers, rational numbers, irrational numbers, whole numbers?

Irrational numbers

72. Name three Fibonacci numbers that are prime numbers.   2, 3, 5

73. Name three Fibonacci numbers that are composite numbers.   8, 21, 34

74. Is the sequence of odd numbers a subset of the Fibonacci numbers?   No

OBJECTIVES

A Add or subtract two or more fractions with the same denominator.

B Find the least common denominator for a set of fractions.

C Add or subtract fractions with different denominators.

D Extend a sequence of numbers containing fractions.

You may recall from previous math classes that to add two fractions with the same denominator, you simply add their numerators and put the result over the common denominator:

$$\frac{3}{4} + \frac{2}{4} = \frac{3+2}{4} = \frac{5}{4}$$

The reason we add numerators but do not add denominators is that we must follow the distributive property. To see this, you first have to recall that $\frac{3}{4}$ can be written as $3 \cdot \frac{1}{4}$, and $\frac{2}{4}$ can be written as $2 \cdot \frac{1}{4}$ (dividing by 4 is equivalent to multiplying by $\frac{1}{4}$). Here is the addition problem again, this time showing the use of the distributive property:

$$\frac{3}{4} + \frac{2}{4} = 3 \cdot \frac{1}{4} + 2 \cdot \frac{1}{4}$$

$$= (3 + 2) \cdot \frac{1}{4} \qquad \textbf{Distributive property}$$

$$= 5 \cdot \frac{1}{4}$$

$$= \frac{5}{4}$$

Note

Most people who have done any work with adding fractions know that you add fractions that have the same denominator by adding their numerators but not their denominators. However, most people don't know why this works. The reason why we add numerators but not denominators is because of the distributive property. That is what the discussion at the right is all about. If you really want to understand addition of fractions, pay close attention to this discussion.

What we have here is the sum of the numerators placed over the *common denominator.* In symbols we have the following.

Addition and Subtraction of Fractions If a, b, and c are integers and c is not equal to 0, then

$$\frac{a}{c} + \frac{b}{c} = \frac{a+b}{c}$$

This rule holds for subtraction as well; that is,

$$\frac{a}{c} - \frac{b}{c} = \frac{a-b}{c}$$

In Examples 1–4, find the sum or difference. (Add or subtract as indicated.) Reduce all answers to lowest terms. (Assume all variables represent nonzero numbers.)

EXAMPLES

1. $\dfrac{3}{8} + \dfrac{1}{8} = \dfrac{3+1}{8}$ **Add numerators; keep the same denominator**

$\qquad\qquad = \dfrac{4}{8}$ **The sum of 3 and 1 is 4**

$\qquad\qquad = \dfrac{1}{2}$ **Reduce to lowest terms**

2. $\dfrac{a+5}{8} - \dfrac{3}{8} = \dfrac{a+5-3}{8}$ **Combine numerators; keep the same denominator**

$\qquad\qquad\quad = \dfrac{a+2}{8}$

3. $\dfrac{9}{x} - \dfrac{3}{x} = \dfrac{9-3}{x}$ Subtract numerators; keep the same denominator

$\qquad\qquad = \dfrac{6}{x}$ The difference of 9 and 3 is 6

4. $\dfrac{3}{7} + \dfrac{2}{7} - \dfrac{9}{7} = \dfrac{3+2-9}{7}$

$\qquad\qquad\qquad = \dfrac{-4}{7}$

$\qquad\qquad\qquad = -\dfrac{4}{7}$ Unlike signs give a negative answer

As Examples 1–4 indicate, addition and subtraction are simple, straightforward processes when all the fractions have the same denominator. We will now turn our attention to the process of adding fractions that have different denominators. To get started, we need the following definition.

> **DEFINITION** The **least common denominator (LCD)** for a set of denominators is the smallest number that is exactly divisible by each denominator. (Note that in some books the least common denominator is also called the *least common multiple*.)
>
> In other words, all the denominators of the fractions involved in a problem must divide into the least common denominator exactly; that is, they divide it without giving a remainder.

  EXAMPLE 5   Find the LCD for the fractions $\frac{5}{12}$ and $\frac{7}{18}$.

SOLUTION   The least common denominator for the denominators 12 and 18 must be the smallest number divisible by both 12 and 18. We can factor 12 and 18 completely and then build the LCD from these factors. Factoring 12 and 18 completely gives us

$$12 = 2 \cdot 2 \cdot 3 \qquad\qquad 18 = 2 \cdot 3 \cdot 3$$

Now, if 12 is going to divide the LCD exactly, then the LCD must have factors of $2 \cdot 2 \cdot 3$. If 18 is to divide it exactly, it must have factors of $2 \cdot 3 \cdot 3$. We don't need to repeat the factors that 12 and 18 have in common:

12 divides the LCD

$\left.\begin{array}{l}12 = 2 \cdot 2 \cdot 3\\[4pt]18 = 2 \cdot 3 \cdot 3\end{array}\right\}$ LCD $= 2 \cdot 2 \cdot 3 \cdot 3 = 36$

18 divides the LCD

Note

The ability to find least common denominators is very important in mathematics. The discussion here is a detailed explanation of how to do it.

In other words, first we write down the factors of 12, then we attach the factors of 18 that do not already appear as factors of 12. We start with $2 \cdot 2 \cdot 3$ because those are the factors of 12. Then we look at the first factor of 18. It is 2. Because 2 already appears in the expression $2 \cdot 2 \cdot 3$, we don't need to attach another one. Next, we look at the factors $3 \cdot 3$. The expression $2 \cdot 2 \cdot 3$ has one 3. For it to contain the expression $3 \cdot 3$, we attach another 3. The final expression, our LCD, is $2 \cdot 2 \cdot 3 \cdot 3$.

The LCD for 12 and 18 is 36. It is the smallest number that is divisible by both 12 and 18; 12 divides it exactly three times, and 18 divides it exactly two times.

We can use the results of Example 5 to find the sum of the fractions $\frac{5}{12}$ and $\frac{7}{18}$.

EXAMPLE 6   Add $\frac{5}{12} + \frac{7}{18}$.

SOLUTION   We can add fractions only when they have the same denominators. In Example 5 we found the LCD for $\frac{5}{12}$ and $\frac{7}{18}$ to be 36. We change $\frac{5}{12}$ and $\frac{7}{18}$ to equivalent fractions that each have 36 for a denominator by applying Property 1, on page 14, for fractions:

$$\frac{5}{12} = \frac{5 \cdot \mathbf{3}}{12 \cdot \mathbf{3}} = \frac{15}{36}$$

$$\frac{7}{18} = \frac{7 \cdot \mathbf{2}}{18 \cdot \mathbf{2}} = \frac{14}{36}$$

The fraction $\frac{15}{36}$ is equivalent to $\frac{5}{12}$, because it was obtained by multiplying both the numerator and denominator by 3. Likewise, $\frac{14}{36}$ is equivalent to $\frac{7}{18}$ because it was obtained by multiplying the numerator and denominator by 2. All we have left to do is to add numerators:

$$\frac{15}{36} + \frac{14}{36} = \frac{29}{36}$$

The sum of $\frac{5}{12}$ and $\frac{7}{18}$ is the fraction $\frac{29}{36}$. Let's write the complete problem again step-by-step.

$$\frac{5}{12} + \frac{7}{18} = \frac{5 \cdot \mathbf{3}}{12 \cdot \mathbf{3}} + \frac{7 \cdot \mathbf{2}}{18 \cdot \mathbf{2}}$$ **Rewrite each fraction as an equivalent fraction with denominator 36**

$$= \frac{15}{36} + \frac{14}{36}$$

$$= \frac{29}{36}$$ **Add numerators; keep the common denominator**

EXAMPLE 7   Find the LCD for $\frac{3}{4}$ and $\frac{1}{6}$.

SOLUTION   We factor 4 and 6 into products of prime factors and build the LCD from these factors:

$$\left.\begin{array}{l} 4 = 2 \cdot 2 \\ 6 = 2 \cdot 3 \end{array}\right\}   \text{LCD} = 2 \cdot 2 \cdot 3 = 12$$

The LCD is 12. Both denominators divide it exactly; 4 divides 12 exactly three times, and 6 divides 12 exactly two times.

EXAMPLE 8   Add $\frac{3}{4} + \frac{1}{6}$.

SOLUTION   In Example 7 we found that the LCD for these two fractions is 12. We begin by changing $\frac{3}{4}$ and $\frac{1}{6}$ to equivalent fractions with denominator 12:

$$\frac{3}{4} = \frac{3 \cdot \mathbf{3}}{4 \cdot \mathbf{3}} = \frac{9}{12}$$

$$\frac{1}{6} = \frac{1 \cdot \mathbf{2}}{6 \cdot \mathbf{2}} = \frac{2}{12}$$

The fraction $\frac{9}{12}$ is equal to the fraction $\frac{3}{4}$ because it was obtained by multiplying the numerator and denominator of $\frac{3}{4}$ by 3. Likewise, $\frac{2}{12}$ is equivalent to $\frac{1}{6}$ because it was obtained by multiplying the numerator and denominator of $\frac{1}{6}$ by 2. To complete the problem, we add numerators:

$$\frac{9}{12} + \frac{2}{12} = \frac{11}{12}$$

The sum of $\frac{3}{4}$ and $\frac{1}{6}$ is $\frac{11}{12}$. Here is how the complete problem looks:

$$\frac{3}{4} + \frac{1}{6} = \frac{3 \cdot 3}{4 \cdot 3} + \frac{1 \cdot 2}{6 \cdot 2} \qquad \textbf{Rewrite each fraction as an equivalent}$$
$$\textbf{fraction with denominator 12}$$

$$= \frac{9}{12} + \frac{2}{12}$$

$$= \frac{11}{12} \qquad \textbf{Add numerators; keep the same denominator}$$

 EXAMPLE 9   Subtract $\frac{7}{15} - \frac{3}{10}$.

SOLUTION   Let's factor 15 and 10 completely and use these factors to build the LCD:

15 divides the LCD

$$\left.\begin{array}{l} 15 = 3 \cdot 5 \\ 10 = 2 \cdot 5 \end{array}\right\} \text{LCD} = 2 \cdot 3 \cdot 5 = 30$$

10 divides the LCD

Changing to equivalent fractions and subtracting, we have

$$\frac{7}{15} - \frac{3}{10} = \frac{7 \cdot 2}{15 \cdot 2} - \frac{3 \cdot 3}{10 \cdot 3} \qquad \textbf{Rewrite as equivalent fractions with}$$
$$\textbf{the LCD for denominator}$$

$$= \frac{14}{30} - \frac{9}{30}$$

$$= \frac{5}{30} \qquad \textbf{Subtract numerators; keep the LCD}$$

$$= \frac{1}{6} \qquad \textbf{Reduce to lowest terms}$$

As a summary of what we have done so far and as a guide to working other problems, we will now list the steps involved in adding and subtracting fractions with different denominators.

To Add or Subtract Any Two Fractions

Step 1:   Factor each denominator completely and use the factors to build the LCD. (Remember, the LCD is the smallest number divisible by each of the denominators in the problem.)

Step 2:   Rewrite each fraction as an equivalent fraction that has the LCD for its denominator. This is done by multiplying both the numerator and denominator of the fraction in question by the appropriate whole number.

Step 3:   Add or subtract the numerators of the fractions produced in step 2. This is the numerator of the sum or difference. The denominator of the sum or difference is the LCD.

Step 4:   Reduce the fraction produced in step 3 to lowest terms if it is not already in lowest terms.

The idea behind adding or subtracting fractions is really very simple. We can add or subtract only fractions that have the same denominators. If the fractions we are trying to add or subtract do not have the same denominators, we rewrite each of them as an equivalent fraction with the LCD for a denominator.

Here are some further examples of sums and differences of fractions.

EXAMPLE 10   Add $\frac{1}{6} + \frac{1}{8} + \frac{1}{4}$.

SOLUTION   We begin by factoring the denominators completely and building the LCD from the factors that result:

$$
\begin{array}{l}
6 = 2 \cdot 3 \\
8 = 2 \cdot 2 \cdot 2 \\
4 = 2 \cdot 2
\end{array}
\qquad
\text{LCD} = 2 \cdot 2 \cdot 2 \cdot 3 = 24
$$

8 divides the LCD

4 divides the LCD   6 divides the LCD

We then change to equivalent fractions and add as usual:

$$
\begin{aligned}
\frac{1}{6} + \frac{1}{8} + \frac{1}{4} &= \frac{1 \cdot \mathbf{4}}{6 \cdot \mathbf{4}} + \frac{1 \cdot \mathbf{3}}{8 \cdot \mathbf{3}} + \frac{1 \cdot \mathbf{6}}{4 \cdot \mathbf{6}} \\
&= \frac{4}{24} + \frac{3}{24} + \frac{6}{24} \\
&= \frac{13}{24}
\end{aligned}
$$

EXAMPLE 11   Subtract $3 - \frac{5}{6}$.

SOLUTION   The denominators are 1 (because $3 = \frac{3}{1}$) and 6. The smallest number divisible by both 1 and 6 is 6.

$$
\begin{aligned}
3 - \frac{5}{6} &= \frac{3}{1} - \frac{5}{6} \\
&= \frac{3 \cdot \mathbf{6}}{1 \cdot \mathbf{6}} - \frac{5}{6} \\
&= \frac{18}{6} - \frac{5}{6} \\
&= \frac{13}{6}
\end{aligned}
$$

EXAMPLE 12   Find the next number in each sequence.

a. $\frac{1}{2}, 0, -\frac{1}{2}, \ldots$ b. $\frac{1}{2}, 1, \frac{3}{2}, \ldots$ c. $\frac{1}{2}, \frac{1}{4}, \frac{1}{8}, \ldots$

SOLUTION   a. $\frac{1}{2}, 0, -\frac{1}{2}, \ldots$: Adding $-\frac{1}{2}$ to each term produces the next term. The fourth term will be $-\frac{1}{2} + (-\frac{1}{2}) = -1$. This is an arithmetic sequence.

b. $\frac{1}{2}, 1, \frac{3}{2}, \ldots$: Each term comes from the term before it by adding $\frac{1}{2}$. The fourth term will be $\frac{3}{2} + \frac{1}{2} = 2$. This sequence is also an arithmetic sequence.

c. $\frac{1}{2}, \frac{1}{4}, \frac{1}{8}, \dots$: This is a geometric sequence in which each term comes from the term before it by multiplying by $\frac{1}{2}$ each time. The next term will be $\frac{1}{8} \cdot \frac{1}{2} = \frac{1}{16}$.

 EXAMPLE 13   Subtract: $\frac{x}{5} - \frac{1}{6}$

SOLUTION   The LCD for 5 and 6 is their product, 30. We begin by rewriting each fraction with this common denominator:

$$\frac{x}{5} - \frac{1}{6} = \frac{x \cdot 6}{5 \cdot 6} - \frac{1 \cdot 5}{6 \cdot 5}$$

$$= \frac{6x}{30} - \frac{5}{30}$$

$$= \frac{6x - 5}{30}$$

 EXAMPLE 14   Add: $\frac{4}{x} + \frac{2}{3}$

Note

In Example 14, it is understood that x cannot be 0. Do you know why?

SOLUTION   The LCD for x and 3 is $3x$. We multiply the numerator and the denominator of the first fraction by 3 and the numerator and the denominator of the second fraction by x to get two fractions with the same denominator. We then add the numerators:

$$\frac{4}{x} + \frac{2}{3} = \frac{4 \cdot 3}{x \cdot 3} + \frac{2 \cdot x}{3 \cdot x} \qquad \textbf{Change to equivalent fractions}$$

$$= \frac{12}{3x} + \frac{2x}{3x}$$

$$= \frac{12 + 2x}{3x} \qquad \textbf{Add the numerators}$$

When we are working with fractions, we can change the form of a fraction without changing its value. There will be times when one form is easier to work with than another form. Look over the material below and be sure you see that the pairs of expressions are equal.

The expressions $\frac{x}{2}$ and $\frac{1}{2}x$ are equal.

The expressions $\frac{3a}{4}$ and $\frac{3}{4}a$ are equal.

The expressions $\frac{7y}{3}$ and $\frac{7}{3}y$ are equal.

 EXAMPLE 15   Add: $\frac{x}{3} + \frac{5x}{6}$.

SOLUTION   We can do the problem two ways. One way probably seems easier, but both ways are valid methods of finding this sum. You should understand both of them.

Method 1: $\dfrac{x}{3} + \dfrac{5x}{6} = \dfrac{\mathbf{2} \cdot x}{\mathbf{2} \cdot 3} + \dfrac{5x}{6}$ **LCD**

$$= \dfrac{2x + 5x}{6} \qquad \textbf{Add numerators}$$

$$= \dfrac{7x}{6}$$

Method 2: $\dfrac{x}{3} + \dfrac{5x}{6} = \dfrac{1}{3}x + \dfrac{5}{6}x$

$$= \left(\dfrac{1}{3} + \dfrac{5}{6}\right)x \qquad \textbf{Distributive property}$$

$$= \left(\dfrac{\mathbf{2} \cdot 1}{\mathbf{2} \cdot 3} + \dfrac{5}{6}\right)x$$

$$= \left(\dfrac{2}{6} + \dfrac{5}{6}\right)x$$

$$= \dfrac{7}{6}x$$

LINKING OBJECTIVES AND EXAMPLES

Next to each **objective** we have listed the examples that are best described by that objective.

A	1–4
B	5, 7
C	6, 8–11, 13–15
D	12

GETTING READY FOR CLASS

After reading through the preceding section, respond in your own words and in complete sentences.

1. How do we add two fractions that have the same denominators?
2. What is a least common denominator?
3. What is the first step in adding two fractions that have different denominators?
4. What is the last thing you do when adding two fractions?

Problem Set 1.9

Online support materials can be found at www.thomsonedu.com/login

Find the following sums and differences, and reduce to lowest terms. Add and subtract as indicated. Assume all variables represent nonzero numbers.

1. $\dfrac{3}{6} + \dfrac{1}{6}$ $\tfrac{2}{3}$

2. $\dfrac{2}{5} + \dfrac{3}{5}$ 1

▶ 3. $\dfrac{3}{8} - \dfrac{5}{8}$ $-\tfrac{1}{4}$

4. $\dfrac{1}{7} - \dfrac{6}{7}$ $-\tfrac{5}{7}$

5. $-\dfrac{1}{4} + \dfrac{3}{4}$ $\tfrac{1}{2}$

6. $-\dfrac{4}{9} + \dfrac{7}{9}$ $\tfrac{1}{3}$

7. $\dfrac{x}{3} - \dfrac{1}{3}$ $\tfrac{x-1}{3}$

8. $\dfrac{x}{8} - \dfrac{1}{8}$ $\tfrac{x-1}{8}$

9. $\dfrac{1}{4} + \dfrac{2}{4} + \dfrac{3}{4}$ $\tfrac{3}{2}$

10. $\dfrac{2}{5} + \dfrac{3}{5} + \dfrac{4}{5}$ $\tfrac{9}{5}$

11. $\dfrac{x+7}{2} - \dfrac{1}{2}$ $\tfrac{x+6}{2}$

12. $\dfrac{x+5}{4} - \dfrac{3}{4}$ $\tfrac{x+2}{4}$

= Videos available by instructor request

▶ = Online student support materials available at www.thomsonedu.com/login

13. $\dfrac{1}{10} - \dfrac{3}{10} - \dfrac{4}{10}$ $-\dfrac{3}{5}$ **14.** $\dfrac{3}{20} - \dfrac{1}{20} - \dfrac{4}{20}$ $-\dfrac{1}{10}$

15. $\dfrac{1}{a} + \dfrac{4}{a} + \dfrac{5}{a}$ $\dfrac{10}{a}$ **16.** $\dfrac{5}{a} + \dfrac{4}{a} + \dfrac{3}{a}$ $\dfrac{12}{a}$

Find the LCD for each of the following; then use the methods developed in this section to add and subtract as indicated.

17. $\dfrac{1}{8} + \dfrac{3}{4}$ $\dfrac{7}{8}$ **18.** $\dfrac{1}{6} + \dfrac{2}{3}$ $\dfrac{5}{6}$

19. $\dfrac{3}{10} - \dfrac{1}{5}$ $\dfrac{1}{10}$ **20.** $\dfrac{5}{6} - \dfrac{1}{12}$ $\dfrac{3}{4}$

▶ **21.** $\dfrac{4}{9} + \dfrac{1}{3}$ $\dfrac{7}{9}$ **22.** $\dfrac{1}{2} + \dfrac{1}{4}$ $\dfrac{3}{4}$

23. $2 + \dfrac{1}{3}$ $\dfrac{7}{3}$ **24.** $3 + \dfrac{1}{2}$ $\dfrac{7}{2}$

25. $-\dfrac{3}{4} + 1$ $\dfrac{1}{4}$ **26.** $-\dfrac{3}{4} + 2$ $\dfrac{5}{4}$

27. $\dfrac{1}{2} + \dfrac{2}{3}$ $\dfrac{7}{6}$ **28.** $\dfrac{2}{3} + \dfrac{1}{4}$ $\dfrac{11}{12}$

▶ **29.** $\dfrac{5}{12} - \left(-\dfrac{3}{8}\right)$ $\dfrac{19}{24}$ **30.** $\dfrac{9}{16} - \left(-\dfrac{7}{12}\right)$ $\dfrac{55}{48}$

31. $-\dfrac{1}{20} + \dfrac{8}{30}$ $\dfrac{13}{60}$ **32.** $-\dfrac{1}{30} + \dfrac{9}{40}$ $\dfrac{23}{120}$

33. $\dfrac{17}{30} + \dfrac{11}{42}$ $\dfrac{29}{35}$ **34.** $\dfrac{19}{42} + \dfrac{13}{70}$ $\dfrac{67}{105}$

35. $\dfrac{25}{84} + \dfrac{41}{90}$ $\dfrac{949}{1,260}$ **36.** $\dfrac{23}{70} + \dfrac{29}{84}$ $\dfrac{283}{420}$

37. $\dfrac{13}{126} - \dfrac{13}{180}$ $\dfrac{13}{420}$ **38.** $\dfrac{17}{84} - \dfrac{17}{90}$ $\dfrac{17}{1,260}$

39. $\dfrac{3}{4} + \dfrac{1}{8} + \dfrac{5}{6}$ $\dfrac{41}{24}$ **40.** $\dfrac{3}{8} + \dfrac{2}{5} + \dfrac{1}{4}$ $\dfrac{41}{40}$

41. $\dfrac{1}{2} + \dfrac{1}{3} + \dfrac{1}{4} + \dfrac{1}{6}$ $\dfrac{5}{4}$ **42.** $\dfrac{1}{8} + \dfrac{1}{4} + \dfrac{1}{5} + \dfrac{1}{10}$ $\dfrac{27}{40}$

43. $1 - \dfrac{5}{2}$ $-\dfrac{3}{2}$ **44.** $1 - \dfrac{5}{3}$ $-\dfrac{2}{3}$

45. $1 + \dfrac{1}{2}$ $\dfrac{3}{2}$ **46.** $1 + \dfrac{2}{3}$ $\dfrac{5}{3}$

Find the fourth term in each sequence.

47. $\dfrac{1}{3}, 0, -\dfrac{1}{3}, \ldots$ $-\dfrac{2}{3}$ **48.** $\dfrac{2}{3}, 0, -\dfrac{2}{3}, \ldots$ $-\dfrac{4}{3}$

49. $\dfrac{1}{3}, 1, \dfrac{5}{3}, \ldots$ $\dfrac{7}{3}$ **50.** $1, \dfrac{3}{2}, 2, \ldots$ $\dfrac{5}{2}$

51. $1, \dfrac{1}{5}, \dfrac{1}{25}, \ldots$ $\dfrac{1}{125}$ **52.** $1, -\dfrac{1}{2}, \dfrac{1}{4}, \ldots$ $-\dfrac{1}{8}$

Use the rule for order of operations to simplify.

53. $9 - 3\left(\dfrac{5}{3}\right)$ 4 **54.** $6 - 4\left(\dfrac{7}{2}\right)$ -8

55. $-\dfrac{1}{2} + 2\left(-\dfrac{3}{4}\right)$ -2 **56.** $\dfrac{5}{4} - 3\left(\dfrac{7}{12}\right)$ $-\dfrac{1}{2}$

57. $\dfrac{3}{5}(-10) + \dfrac{4}{7}(-21)$ -18 **58.** $-\dfrac{3}{5}(10) - \dfrac{4}{7}(21)$ -18

59. $16\left(-\dfrac{1}{2}\right)^2 - 125\left(-\dfrac{2}{5}\right)^2$ -16

60. $16\left(-\dfrac{1}{2}\right)^3 - 125\left(-\dfrac{2}{5}\right)^3$ 6

61. $-\dfrac{4}{3} \div 2 \cdot 3$ -2 **62.** $-\dfrac{8}{7} \div 4 \cdot 2$ $-\dfrac{4}{7}$

63. $-\dfrac{4}{3} \div 2(-3)$ 2 **64.** $-\dfrac{8}{7} \div 4(-2)$ $\dfrac{4}{7}$

65. $-6 \div \dfrac{1}{2} \cdot 12$ -144 **66.** $-6 \div \left(-\dfrac{1}{2}\right) \cdot 12$ 144

67. $-15 \div \dfrac{5}{3} \cdot 18$ -162 **68.** $-15 \div \left(-\dfrac{5}{3}\right) \cdot 18$ 162

Add or subtract the following fractions. (Assume all variables represent nonzero numbers. See Examples 13–15.)

69. $\dfrac{x}{4} + \dfrac{1}{5}$ $\dfrac{5x+4}{20}$ **70.** $\dfrac{x}{3} + \dfrac{1}{5}$ $\dfrac{5x+3}{15}$

71. $\dfrac{1}{3} + \dfrac{a}{12}$ $\dfrac{a+4}{12}$ **72.** $\dfrac{1}{8} + \dfrac{a}{32}$ $\dfrac{a+4}{32}$

73. $\dfrac{x}{2} + \dfrac{1}{3} + \dfrac{x}{4}$ $\dfrac{9x+4}{12}$ **74.** $\dfrac{x}{3} + \dfrac{1}{4} + \dfrac{x}{5}$ $\dfrac{32x+15}{60}$

75. $\dfrac{2}{x} + \dfrac{3}{5}$ $\dfrac{10+3x}{5x}$ **76.** $\dfrac{3}{x} - \dfrac{2}{5}$ $\dfrac{15-2x}{5x}$

77. $\dfrac{3}{7} + \dfrac{4}{y}$ $\dfrac{3y+28}{7y}$ **78.** $\dfrac{2}{9} + \dfrac{5}{y}$ $\dfrac{2y+45}{9y}$

79. $\dfrac{3}{a} + \dfrac{3}{4} + \dfrac{1}{5}$ $\dfrac{60+19a}{20a}$ **80.** $\dfrac{4}{a} + \dfrac{2}{3} + \dfrac{1}{2}$ $\dfrac{24+7a}{6a}$

81. $\dfrac{1}{2}x + \dfrac{1}{6}x$ $\dfrac{2}{3}x$ **82.** $\dfrac{2}{3}x + \dfrac{5}{6}x$ $\dfrac{3}{2}x$

83. $\dfrac{1}{2}x - \dfrac{3}{4}x$ $-\dfrac{1}{4}x$ **84.** $\dfrac{2}{3}x - \dfrac{5}{6}x$ $-\dfrac{1}{6}x$

85. $\dfrac{1}{3}x + \dfrac{3}{5}x$ $\dfrac{14}{15}x$ **86.** $\dfrac{2}{3}x - \dfrac{3}{5}x$ $\dfrac{1}{15}x$

87. $\dfrac{3x}{4} + \dfrac{x}{6}$ $\dfrac{11}{12}x$ **88.** $\dfrac{3x}{4} - \dfrac{2x}{3}$ $\dfrac{1}{12}x$

89. $\dfrac{2x}{5} + \dfrac{5x}{8}$ $\dfrac{41}{40}x$ **90.** $\dfrac{3x}{5} - \dfrac{3x}{8}$ $\dfrac{9}{40}x$

▶ **91.** $1 - \dfrac{1}{x}$ $\dfrac{x-1}{x}$ ▶ **92.** $1 + \dfrac{1}{x}$ $\dfrac{x+1}{x}$

The next two problems are intended to give you practice reading, and paying attention to, the instructions that accompany the problems you are working. As we mentioned previously, working these problems is an excellent way to get ready for a test or a quiz.

▶ **93.** Work each problem according to the instructions given. (Note that each of these instructions could be replaced with the instruction *Simplify*.)

   a. Add: $\dfrac{3}{4} + \left(-\dfrac{1}{2}\right)$ $\frac{1}{4}$

   b. Subtract: $\dfrac{3}{4} - \left(-\dfrac{1}{2}\right)$ $\frac{5}{4}$

   c. Multiply: $\dfrac{3}{4}\left(-\dfrac{1}{2}\right)$ $-\frac{3}{8}$

   d. Divide: $\dfrac{3}{4} \div \left(-\dfrac{1}{2}\right)$ $-\frac{3}{2}$

94. Work each problem according to the instructions given.

   a. Add: $-\dfrac{5}{8} + \left(-\dfrac{1}{2}\right)$ $-\frac{9}{8}$

   b. Subtract: $-\dfrac{5}{8} - \left(-\dfrac{1}{2}\right)$ $-\frac{1}{8}$

   c. Multiply: $-\dfrac{5}{8}\left(-\dfrac{1}{2}\right)$ $\frac{5}{16}$

   d. Divide: $-\dfrac{5}{8} \div \left(-\dfrac{1}{2}\right)$ $\frac{5}{4}$

Simplify.

95. $\left(1 - \dfrac{1}{2}\right)\left(1 - \dfrac{1}{3}\right)$ $\frac{1}{3}$

96. $\left(1 + \dfrac{1}{2}\right)\left(1 + \dfrac{1}{3}\right)$ 2

97. $\left(1 + \dfrac{1}{2}\right)\left(1 - \dfrac{1}{2}\right)$ $\frac{3}{4}$

98. $\left(1 + \dfrac{1}{3}\right)\left(1 - \dfrac{1}{3}\right)$ $\frac{8}{9}$

99. Find the value of $1 + \dfrac{1}{x}$ when x is

   a. 2 $\frac{3}{2}$

   b. 3 $\frac{4}{3}$

   c. 4 $\frac{5}{4}$

100. Find the value of $1 - \dfrac{1}{x}$ when x is

   a. 2 $\frac{1}{2}$

   b. 3 $\frac{2}{3}$

   c. 4 $\frac{3}{4}$

101. Find the value of $2x + \dfrac{6}{x}$ when x is

   a. 1 8

   b. 2 7

   c. 3 8

102. Find the value of $x + \dfrac{4}{x}$ when x is

   a. 1 5

   b. 2 4

   c. 3 $\frac{13}{3}$

▶ **Chalkboard Problem**
The only thing new in Problem 93 is Part b. The rest is review. I like to work this problem in class because I can review, cover new material, and emphasize how important it is to read the instructions, all at once.

Chapter 1 SUMMARY

NOTE

We will use the margins in the chapter summaries to give examples that correspond to the topic being reviewed whenever it is appropriate

The number(s) in brackets next to each heading indicates the section(s) in which that topic is discussed.

Symbols [1.1]

$a = b$	a is equal to b
$a \neq b$	a is not equal to b
$a < b$	a is less than b
$a \not< b$	a is not less than b
$a > b$	a is greater than b
$a \not> b$	a is not greater than b
$a \geq b$	a is greater than or equal to b
$a \leq b$	a is less than or equal to b

EXAMPLES

1. $2^5 = 2 \cdot 2 \cdot 2 \cdot 2 \cdot 2 = 32$
$5^2 = 5 \cdot 5 = 25$
$10^3 = 10 \cdot 10 \cdot 10 = 1,000$
$1^4 = 1 \cdot 1 \cdot 1 \cdot 1 = 1$

Exponents [1.1]

Exponents are notation used to indicate repeated multiplication. In the expression 3^4, 3 is the *base* and 4 is the *exponent*.

$$3^4 = 3 \cdot 3 \cdot 3 \cdot 3 = 81$$

2. $10 + (2 \cdot 3^2 - 4 \cdot 2)$
$= 10 + (2 \cdot 9 - 4 \cdot 2)$
$= 10 + (18 - 8)$
$= 10 + 10$
$= 20$

Order of Operations [1.1]

When evaluating a mathematical expression, we will perform the operations in the following order, beginning with the expression in the innermost parentheses or brackets and working our way out.

1. Simplify all numbers with exponents, working from left to right if more than one of these numbers is present.

2. Then do all multiplications and divisions left to right.

3. Finally, perform all additions and subtractions left to right.

3. Add all combinations of positive and negative 10 and 13.
$10 + 13 = 23$
$10 + (-13) = -3$
$-10 + 13 = 3$
$-10 + (-13) = -23$

Addition of Real Numbers [1.3]

To add two real numbers with

1. The same sign: Simply add their absolute values and use the common sign.

2. Different signs: Subtract the smaller absolute value from the larger absolute value. The answer has the same sign as the number with the larger absolute value.

Subtraction of Real Numbers [1.4]

4. Subtracting 2 is the same as adding -2:
$$7 - 2 = 7 + (-2) = 5$$

To subtract one number from another, simply add the opposite of the number you are subtracting; that is, if a and b represent real numbers, then

$$a - b = a + (-b)$$

Multiplication of Real Numbers [1.6]

5.
$$3(5) = 15$$
$$3(-5) = -15$$
$$-3(5) = -15$$
$$-3(-5) = 15$$

To multiply two real numbers, simply multiply their absolute values. Like signs give a positive answer. Unlike signs give a negative answer.

Division of Real Numbers [1.7]

6. $\frac{-6}{2} = -6(\frac{1}{2}) = -3$
$\frac{-6}{-2} = -6(-\frac{1}{2}) = 3$

Division by a number is the same as multiplication by its reciprocal. Like signs give a positive answer. Unlike signs give a negative answer.

Absolute Value [1.2]

7. $|5| = 5$
$|-5| = 5$

The absolute value of a real number is its distance from zero on the real number line. Absolute value is never negative.

Opposites [1.2, 1.5]

8. The numbers 3 and -3 are opposites; their sum is 0:
$$3 + (-3) = 0$$

Any two real numbers the same distance from zero on the number line but in opposite directions from zero are called opposites. Opposites always add to zero.

Reciprocals [1.2, 1.5]

9. The numbers 2 and $\frac{1}{2}$ are reciprocals; their product is 1:
$$2(\frac{1}{2}) = 1$$

Any two real numbers whose product is 1 are called reciprocals. Every real number has a reciprocal except 0.

Properties of Real Numbers [1.5]

	For Addition	For Multiplication
Commutative:	$a + b = b + a$	$a \cdot b = b \cdot a$
Associative:	$a + (b + c) = (a + b) + c$	$a \cdot (b \cdot c) = (a \cdot b) \cdot c$
Identity:	$a + 0 = a$	$a \cdot 1 = a$
Inverse:	$a + (-a) = 0$	$a\left(\frac{1}{a}\right) = 1$
Distributive:	$a(b + c) = ab + ac$	

Subsets of the Real Numbers [1.8]

10. a. 7 and 100 are counting numbers, but 0 and -2 are not.
   b. 0 and 241 are whole numbers, but -4 and $\frac{1}{2}$ are not.
   c. -15, 0, and 20 are integers.
   d. -4, $-\frac{1}{2}$, 0.75, and 0.666 . . . are rational numbers.
   e. $-\pi$, $\sqrt{3}$, and π are irrational numbers.
   f. All the numbers listed above are real numbers.

Counting numbers:	$\{1, 2, 3, \ldots\}$
Whole numbers:	$\{0, 1, 2, 3, \ldots\}$
Integers:	$\{\ldots, -3, -2, -1, 0, 1, 2, 3, \ldots\}$
Rational numbers:	{all numbers that can be expressed as the ratio of two integers}
Irrational numbers:	{all numbers on the number line that cannot be expressed as the ratio of two integers}
Real numbers:	{all numbers that are either rational or irrational}

Factoring [1.8]

11. The number 150 can be factored into the product of prime numbers:
$$150 = 15 \cdot 10$$
$$= (3 \cdot 5)(2 \cdot 5)$$
$$= 2 \cdot 3 \cdot 5^2$$

Factoring is the reverse of multiplication.

Multiplication

Factors $\rightarrow 3 \cdot 5 = 15 \leftarrow$ Product

Factoring

Least Common Denominator (LCD) [1.9]

12. The LCD for $\frac{5}{12}$ and $\frac{7}{18}$ is 36.

The *least common denominator* (LCD) for a set of denominators is the smallest number that is exactly divisible by each denominator.

Addition and Subtraction of Fractions [1.9]

13. $\frac{5}{12} + \frac{7}{18} = \frac{5}{12} \cdot \frac{3}{3} + \frac{7}{18} \cdot \frac{2}{2}$
$= \frac{15}{36} + \frac{14}{36}$
$= \frac{29}{36}$

To add (or subtract) two fractions with a common denominator, add (or subtract) numerators and use the common denominator.

$$\frac{a}{c} + \frac{b}{c} = \frac{a + b}{c} \quad \text{and} \quad \frac{a}{c} - \frac{b}{c} = \frac{a - b}{c}$$

> ### ! COMMON MISTAKES
>
> **1.** Interpreting absolute value as changing the sign of the number inside the absolute value symbols. $|-5| = +5$, $|+5| = -5$. (The first expression is correct; the second one is not.) To avoid this mistake, remember: Absolute value is a distance and distance is always measured in positive units.
> **2.** Using the phrase "two negatives make a positive." This works only with multiplication and division. With addition, two negative numbers produce a negative answer. It is best not to use the phrase "two negatives make a positive" at all.

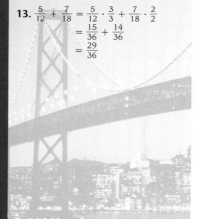

The problems below form a comprehensive review of the material in this chapter. They can be used to study for exams. If you would like to take a practice test on this chapter, you can use the odd-numbered problems. Give yourself an hour and work as many of the odd-numbered problems as possible. When you are finished, or when an hour has passed, check your answers with the answers in the back of the book. You can use the even-numbered problems for a second practice test.

The numbers in brackets refer to the sections of the text in which similar problems can be found.

Write the numerical expression that is equivalent to each phrase, and then simplify. [1.3, 1.4, 1.6, 1.7]

1. The sum of -7 and -10

2. Five added to the sum of -7 and 4

3. The sum of -3 and 12 increased by 5

4. The difference of 4 and 9

5. The difference of 9 and -3

6. The difference of -7 and -9

7. The product of -3 and -7 decreased by 6

8. Ten added to the product of 5 and -6

9. Twice the product of -8 and $3x$

10. The quotient of -25 and -5

Simplify. [1.2]

11. $|-1.8|$

12. $-|-10|$

For each number, give the opposite and the reciprocal. [1.2]

13. 6

14. $-\dfrac{12}{5}$

Multiply. [1.2, 1.6]

15. $\dfrac{1}{2}(-10)$

16. $\left(-\dfrac{4}{5}\right)\left(\dfrac{25}{16}\right)$

Add. [1.3]

17. $-9 + 12$

18. $-18 + (-20)$

19. $-2 + (-8) + [-9 + (-6)]$

20. $(-21) + 40 + (-23) + 5$

Subtract. [1.4]

21. $6 - 9$

22. $14 - (-8)$

23. $-12 - (-8)$

24. $4 - 9 - 15$

Find the products. [1.6]

25. $(-5)(6)$

26. $4(-3)$

27. $-2(3)(4)$

28. $(-1)(-3)(-1)(-4)$

Find the following quotients. [1.7]

29. $\dfrac{12}{-3}$

30. $-\dfrac{8}{9} \div \dfrac{4}{3}$

Simplify. [1.1, 1.6, 1.7]

31. $4 \cdot 5 + 3$

32. $9 \cdot 3 + 4 \cdot 5$

33. $2^3 - 4 \cdot 3^2 + 5^2$

34. $12 - 3(2 \cdot 5 + 7) + 4$

35. $20 + 8 \div 4 + 2 \cdot 5$

36. $2(3 - 5) - (2 - 8)$

37. $30 \div 3 \cdot 2$

38. $(-2)(3) - (4)(-3) - 9$

39. $3(4 - 7)^2 - 5(3 - 8)^2$

40. $(-5 - 2)(-3 - 7)$

41. $\dfrac{4(-3)}{-6}$

42. $\dfrac{3^2 + 5^2}{(3 - 5)^2}$

43. $\dfrac{15 - 10}{6 - 6}$

44. $\dfrac{2(-7) + (-11)(-4)}{7 - (-3)}$

State the property or properties that justify the following. [1.5]

45. $9(3y) = (9 \cdot 3)y$

46. $8(1) = 8$

47. $(4 + y) + 2 = (y + 4) + 2$

48. $5 + (-5) = 0$

49. $(4 + 2) + y = (4 + y) + 2$

50. $5(w - 6) = 5w - 30$

Use the associative property to rewrite each expression, and then simplify the result. [1.5]

51. $7 + (5 + x)$

52. $4(7a)$

53. $\dfrac{1}{9}(9x)$

54. $\dfrac{4}{5}\left(\dfrac{5}{4}y\right)$

Apply the distributive property to each of the following expressions. Simplify when possible. [1.5, 1.6]

55. $7(2x + 3)$

56. $3(2a - 4)$

57. $\dfrac{1}{2}(5x - 6)$

58. $-\dfrac{1}{2}(3x - 6)$

For the set $\{\sqrt{7}, -\frac{1}{3}, 0, 5, -4.5, \frac{2}{5}, \pi, -3\}$ list all the [1.8]

59. rational numbers

60. whole numbers

61. irrational numbers

62. integers

Factor into the product of primes. [1.8]

63. 90   **64.** 840

Combine. [1.9]

65. $\dfrac{18}{35} + \dfrac{13}{42}$ **66.** $\dfrac{x}{6} + \dfrac{7}{12}$

Find the next number in each sequence. [1.1, 1.2, 1.3, 1.6, 1.9]

67. $10, 7, 4, 1, \ldots$ **68.** $10, -30, 90, -270, \ldots$

69. $1, 1, 2, 3, 5, \ldots$ **70.** $4, 6, 8, 10, \ldots$

71. $1, \dfrac{1}{2}, 0, -\dfrac{1}{2}, \ldots$ **72.** $1, -\dfrac{1}{2}, \dfrac{1}{4}, -\dfrac{1}{8}, \ldots$

GROUP PROJECT   Binary Numbers

Students and Instructors: The end of each chapter in this book will contain a section like this one containing two projects. The group project is intended to be done in class. The research projects are to be completed outside of class. They can be done in groups or individually. In my classes, I use the research projects for extra credit. I require all research projects to be done on a word processor and to be free of spelling errors.

Number of People   2 or 3

Time Needed   10 minutes

Equipment   Paper and pencil

Background   Our decimal number system is a base 10 number system. We have 10 digits—0, 1, 2, 3, 4, 5, 6, 7, 8, and 9—which we use to write all the numbers in our number system. The number 10 is the first number that is written with a combination of digits. Although our number system is very useful, there are other number systems that are more appropriate for some disciplines. For example, computers and computer programmers use both the binary number system, which is base 2, and the hexadecimal number system, which is base 16. The binary number system has only digits 0 and 1, which are used to write all the other numbers. Every number in our base 10 number system can be written in the base 2 number system as well.

Procedure   To become familiar with the binary number system, we first learn to count in base 2. Imagine that the odometer on your car had only 0's and 1's. Here is what the odometer would look like for the first 6 miles the car was driven.

ODOMETER READING						MILEAGE
0	0	0	0	0	0	0
0	0	0	0	0	1	1
0	0	0	0	1	0	2
0	0	0	0	1	1	3
0	0	0	1	0	0	4
0	0	0	1	0	1	5
0	0	0	1	1	0	6

Continue the table at left to show the odometer reading for the first 32 miles the car is driven. At 32 miles, the odometer should read

Sophie Germain

Cheryl Slaughter

The photograph at the left shows the street sign in Paris named for the French mathematician Sophie Germain (1776–1831). Among her contributions to mathematics is her work with prime numbers. In this chapter we had an introductory look at some of the classifications for numbers, including the prime numbers. Within the prime numbers themselves, there are still further classifications. In fact, a Sophie Germain prime is a prime number P, for which both P and $2P + 1$ are primes. For example, the prime number 2 is the first Sophie Germain prime because both 2 and $2 \cdot 2 + 1 = 5$ are prime numbers. The next Germain prime is 3 because both 3 and $2 \cdot 3 + 1 = 7$ are primes.

Sophie Germain was born on April 1, 1776, in Paris, France. She taught herself mathematics by reading the books in her father's library at home. Today she is recognized most for her work in number theory, which includes her work with prime numbers. Research the life of Sophie Germain. Write a short essay that includes information on her work with prime numbers and how her results contributed to solving Fermat's Last Theorem almost 200 years later.

Linear Equations and Inequalities

2

J ust before starting work on this edition of your text, I flew to Europe for vacation. From time to time the television screens on the plane displayed statistics about the flight. At one point during the flight the temperature outside the plane was −60°F. When I returned home, I did some research and found that the relationship between temperature T and altitude A can be described with the formula

$$T = -0.0035A + 70$$

when the temperature on the ground is 70°F. The table and the line graph also describe this relationship.

Air Temperature and Altitude	
Altitude (feet)	Temperature (°F)
0	70
10,000	35
20,000	0
30,000	−35
40,000	−70

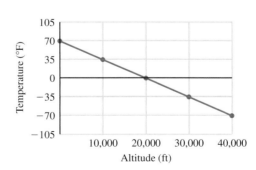

In this chapter we will start our work with formulas, and you will see how we use formulas to produce tables and line graphs like the ones above.

▶ Improve your grade and save time!
Go online to **www.thomsonedu.com/login**
where you can
• Watch videos of instructors working through the in-text examples
• Follow step-by-step online tutorials of in-text examples and review questions
• Work practice problems
• Check your readiness for an exam by taking a pre-test and exploring the modules recommended in your Personalized Study plan
• Receive help from a live tutor online through vMentor™
Try it out! Log in with an access code or purchase access at **www.ichapters.com**.

If you have successfully completed Chapter 1, then you have made a good start at developing the study skills necessary to succeed in all math classes. Here is the list of study skills for this chapter.

1 Imitate Success

Your work should look like the work you see in this book and the work your instructor shows. The steps shown in solving problems in this book were written by someone who has been successful in mathematics. The same is true of your instructor. Your work should imitate the work of people who have been successful in mathematics.

2 List Difficult Problems

Begin to make lists of problems that give you the most difficulty. These are problems in which you are repeatedly making mistakes.

3 Begin to Develop Confidence with Word Problems

It seems that the major difference between those people who are good at working word problems and those who are not is confidence. The people with confidence know that no matter how long it takes them, they eventually will be able to solve the problem. Those without confidence begin by saying to themselves, "I'll never be able to work this problem." Are you like that? If you are, what you need to do is put your old ideas about you and word problems aside for a while and make a decision to be successful. Sometimes that's all it takes. Instead of telling yourself that you can't do word problems, that you don't like them, or that they're not good for anything anyway, decide to do whatever it takes to master them.

2.1 Simplifying Expressions

OBJECTIVES

A Simplify expressions by combining similar terms.

B Simplify expressions by applying the distributive property and then combining similar terms.

C Calculate the value of an expression for a given value of the variable.

If a cellular phone company charges $35 per month plus $0.25 for each minute, or fraction of a minute, that you use one of their cellular phones, then the amount of your monthly bill is given by the expression $35 + 0.25t$. To find the amount you will pay for using that phone 30 minutes in one month, you substitute 30 for t and simplify the resulting expression. This process is one of the topics we will study in this section.

As you will see in the next few sections, the first step in solving an equation is to simplify both sides as much as possible. In the first part of this section, we will practice simplifying expressions by combining what are called *similar* (or *like*) terms.

For our immediate purposes, a term is a number or a number and one or more variables multiplied together. For example, the number 5 is a term, as are the expressions $3x$, $-7y$, and $15xy$.

> **DEFINITION** Two or more terms with the same variable part are called **similar** (or **like**) terms.

The terms $3x$ and $4x$ are similar because their variable parts are identical. Likewise, the terms $18y$, $-10y$, and $6y$ are similar terms.

To simplify an algebraic expression, we simply reduce the number of terms in the expression. We accomplish this by applying the distributive property along with our knowledge of addition and subtraction of positive and negative real numbers. The following examples illustrate the procedure.

 EXAMPLES   Simplify by combining similar terms.

1. $3x + 4x = (3 + 4)x$ **Distributive property**
       $= 7x$ **Addition of 3 and 4**

2. $7a - 10a = (7 - 10)a$ **Distributive property**
       $= -3a$ **Addition of 7 and −10**

3. $18y - 10y + 6y = (18 - 10 + 6)y$ **Distributive property**
       $= 14y$ **Addition of 18, −10, and 6**

When the expression we intend to simplify is more complicated, we use the commutative and associative properties first.

 EXAMPLES   Simplify each expression.

4. $3x + 5 + 2x - 3 = 3x + 2x + 5 - 3$ **Commutative property**
       $= (3x + 2x) + (5 - 3)$ **Associative property**
       $= (3 + 2)x + (5 - 3)$ **Distributive property**
       $= 5x + 2$ **Addition**

5. $4a - 7 - 2a + 3 = (4a - 2a) + (-7 + 3)$ **Commutative and associative properties**

       $= (4 - 2)a + (-7 + 3)$ **Distributive property**
       $= 2a - 4$ **Addition**

6. $5x + 8 - x - 6 = (5x - x) + (8 - 6)$ **Commutative and associative properties**

$$= (5 - 1)x + (8 - 6)$$ **Distributive property**

$$= 4x + 2$$ **Addition**

Notice that in each case the result has fewer terms than the original expression. Because there are fewer terms, the resulting expression is said to be simpler than the original expression.

Simplifying Expressions Containing Parentheses

If an expression contains parentheses, it is often necessary to apply the distributive property to remove the parentheses before combining similar terms.

 EXAMPLE 7   Simplify the expression $5(2x - 8) - 3$.

SOLUTION   We begin by distributing the 5 across $2x - 8$. We then combine similar terms:

$$5(2x - 8) - 3 = 10x - 40 - 3$$ **Distributive property**

$$= 10x - 43$$

 EXAMPLE 8   Simplify $7 - 3(2y + 1)$.

SOLUTION   By the rule for order of operations, we must multiply before we add or subtract. For that reason, it would be incorrect to subtract 3 from 7 first. Instead, we multiply -3 and $2y + 1$ to remove the parentheses and then combine similar terms:

$$7 - 3(2y + 1) = 7 - 6y - 3$$ **Distributive property**

$$= -6y + 4$$

 EXAMPLE 9   Simplify $5(x - 2) - (3x + 4)$.

SOLUTION   We begin by applying the distributive property to remove the parentheses. The expression $-(3x + 4)$ can be thought of as $-1(3x + 4)$. Thinking of it in this way allows us to apply the distributive property:

$$-1(3x + 4) = -1(3x) + (-1)(4)$$

$$= -3x - 4$$

The complete solution looks like this:

$$5(x - 2) - (3x + 4) = 5x - 10 - 3x - 4$$ **Distributive property**

$$= 2x - 14$$ **Combine similar terms**

As you can see from the explanation in Example 9, we use the distributive property to simplify expressions in which parentheses are preceded by a negative sign. In general we can write

$$-(a + b) = -1(a + b)$$
$$= -a + (-b)$$
$$= -a - b$$

The negative sign outside the parentheses ends up changing the sign of each term within the parentheses. In words, we say "the opposite of a sum is the sum of the opposites."

The Value of an Expression

Recall from Chapter 1, an expression like $3x + 2$ has a certain value depending on what number we assign to x. For instance, when x is 4, $3x + 2$ becomes $3(4) + 2$, or 14. When x is -8, $3x + 2$ becomes $3(-8) + 2$, or -22. The value of an expression is found by replacing the variable with a given number.

 EXAMPLES   Find the value of the following expressions by replacing the variable with the given number.

Expression	Value of the Variable	Value of the Expression
10. $3x - 1$	$x = 2$	$3(2) - 1 = 6 - 1 = 5$
11. $7a + 4$	$a = -3$	$7(-3) + 4 = -21 + 4 = -17$
12. $2x - 3 + 4x$	$x = -1$	$2(-1) - 3 + 4(-1)$ $= -2 - 3 + (-4) = -9$
13. $2x - 5 - 8x$	$x = 5$	$2(5) - 5 - 8(5)$ $= 10 - 5 - 40 = -35$
14. $y^2 - 6y + 9$	$y = 4$	$4^2 - 6(4) + 9 = 16 - 24 + 9 = 1$

Simplifying an expression should not change its value; that is, if an expression has a certain value when x is 5, then it will always have that value no matter how much it has been simplified as long as x is 5. If we were to simplify the expression in Example 13 first, it would look like

$$2x - 5 - 8x = -6x - 5$$

When x is 5, the simplified expression $-6x - 5$ is

$$-6(5) - 5 = -30 - 5 = -35$$

It has the same value as the original expression when x is 5.

We also can find the value of an expression that contains two variables if we know the values for both variables.

EXAMPLE 15 Find the value of the expression $2x - 3y + 4$ when x is -5 and y is 6.

SOLUTION Substituting -5 for x and 6 for y, the expression becomes

$$2(-5) - 3(6) + 4 = -10 - 18 + 4$$
$$= -28 + 4$$
$$= -24$$

 EXAMPLE 16 Find the value of the expression $x^2 - 2xy + y^2$ when x is 3 and y is -4.

SOLUTION Replacing each x in the expression with the number 3 and each y in the expression with the number -4 gives us

$$3^2 - 2(3)(-4) + (-4)^2 = 9 - 2(3)(-4) + 16$$
$$= 9 - (-24) + 16$$
$$= 33 + 16$$
$$= 49$$

More About Sequences

As the next example indicates, when we substitute the counting numbers, in order, into algebraic expressions, we form some of the sequences of numbers that we studied in Chapter 1. To review, recall that the sequence of counting numbers (also called the sequence of positive integers) is

Counting numbers = 1, 2, 3, . . .

 EXAMPLE 17 Substitute 1, 2, 3, and 4 for n in the expression $2n - 1$.

SOLUTION Substituting as indicated, we have

When $n = 1$, $2n - 1 = 2 \cdot 1 - 1 = 1$
When $n = 2$, $2n - 1 = 2 \cdot 2 - 1 = 3$
When $n = 3$, $2n - 1 = 2 \cdot 3 - 1 = 5$
When $n = 4$, $2n - 1 = 2 \cdot 4 - 1 = 7$

As you can see, substituting the first four counting numbers into the formula $2n - 1$ produces the first four numbers in the sequence of odd numbers.

The next example is similar to Example 17 but uses tables to display the information.

 EXAMPLE 18 Fill in the tables below to find the sequences formed by substituting the first four counting numbers into the expressions $2n$ and n^2.

a.

n	1	2	3	4
$2n$				

b.

n	1	2	3	4
n^2				

 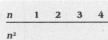

SOLUTION   Proceeding as we did in the previous example, we substitute the numbers 1, 2, 3, and 4 into the given expressions.

a. When $n = 1, 2n = 2 \cdot 1 = 2$
When $n = 2, 2n = 2 \cdot 2 = 4$
When $n = 3, 2n = 2 \cdot 3 = 6$
When $n = 4, 2n = 2 \cdot 4 = 8$

As you can see, the expression $2n$ produces the sequence of even numbers when n is replaced by the counting numbers. Placing these results into our first table gives us

n	1	2	3	4
$2n$	2	4	6	8

b. The expression n^2 produces the sequence of squares when n is replaced by 1, 2, 3, and 4. In table form we have

n	1	2	3	4
n^2	1	4	9	16

LINKING OBJECTIVES AND EXAMPLES

Next to each objective we have listed the examples that are best described by that objective.

A   1–6

B   7–9

C   10–18

GETTING READY FOR CLASS

After reading through the preceding section, respond in your own words and in complete sentences.

1. What are similar terms?
2. Explain how the distributive property is used to combine similar terms.
3. What is wrong with writing $3x + 4x = 7x^2$?
4. Explain how you would find the value of $5x + 3$ when x is 6.

Problem Set 2.1

Online support materials can be found at www.thomsonedu.com/login

Answers appear in the Instructor's Edition only.

Simplify the following expressions.

1. $3x - 6x$ $-3x$
2. $7x - 5x$ $2x$
3. $-2a + a$ $-a$
4. $3a - a$ $2a$
5. $7x + 3x + 2x$ $12x$
6. $8x - 2x - x$ $5x$
7. $3a - 2a + 5a$ $6a$
8. $7a - a + 2a$ $8a$
9. $4x - 3 + 2x$ $6x - 3$
10. $5x + 6 - 3x$ $2x + 6$
11. $3a + 4a + 5$ $7a + 5$
12. $6a + 7a + 8$ $13a + 8$
13. $2x - 3 + 3x - 2$ $5x - 5$
14. $6x + 5 - 2x + 3$ $4x + 8$
15. $3a - 1 + a + 3$ $4a + 2$
16. $-a + 2 + 8a - 7$ $7a - 5$
17. $-4x + 8 - 5x - 10$ $-9x - 2$

18. $-9x - 1 + x - 4$ $-8x - 5$
19. $7a + 3 + 2a + 3a$ $12a + 3$
20. $8a - 2 + a + 5a$ $14a - 2$
21. $5(2x - 1) + 4$ $10x - 1$
22. $2(4x - 3) + 2$ $8x - 4$
23. $7(3y + 2) - 8$ $21y + 6$
24. $6(4y + 2) - 7$ $24y + 5$
25. $-3(2x - 1) + 5$ $-6x + 8$
26. $-4(3x - 2) - 6$ $-12x + 2$
27. $5 - 2(a + 1)$ $-2a + 3$
28. $7 - 8(2a + 3)$ $-16a - 17$
29. $6 - 4(x - 5)$ $-4x + 26$
30. $12 - 3(4x - 2)$ $-12x + 18$
31. $-9 - 4(2 - y) + 1$ $4y - 16$

= Videos available by instructor request

▶ = Online student support materials available at www.thomsonedu.com/login

32. $-10 - 3(2 - y) + 3$ $3y - 13$

33. $-6 + 2(2 - 3x) + 1$ $-6x - 1$

34. $-7 - 4(3 - x) + 1$ $4x - 18$

35. $(4x - 7) - (2x + 5)$ $2x - 12$

36. $(7x - 3) - (4x + 2)$ $3x - 5$

37. $8(2a + 4) - (6a - 1)$ $10a + 33$

38. $9(3a + 5) - (8a - 7)$ $19a + 52$

39. $3(x - 2) + (x - 3)$ $4x - 9$

40. $2(2x + 1) - (x + 4)$ $3x - 2$

41. $4(2y - 8) - (y + 7)$ $7y - 39$

42. $5(y - 3) - (y - 4)$ $4y - 11$

43. $-9(2x + 1) - (x + 5)$ $-19x - 14$

44. $-3(3x - 2) - (2x + 3)$ $-11x + 3$

Evaluate the following expressions when x is 2.

45. $3x - 1$ 5

46. $4x + 3$ 11

47. $-2x - 5$ -9

48. $-3x + 6$ 0

49. $x^2 - 8x + 16$ 4

50. $x^2 - 10x + 25$ 9

51. $(x - 4)^2$ 4

52. $(x - 5)^2$ 9

Find the value of each expression when x is -5. Then simplify the expression, and check to see that it has the same value for $x = -5$.

53. $7x - 4 - x - 3$ -37

54. $3x + 4 + 7x - 6$ -52

55. $5(2x + 1) + 4$ -41

56. $2(3x - 10) + 5$ -45

Find the value of each expression when x is -3 and y is 5.

57. $x^2 - 2xy + y^2$ 64

58. $x^2 + 2xy + y^2$ 4

59. $(x - y)^2$ 64

60. $(x + y)^2$ 4

61. $x^2 + 6xy + 9y^2$ 144

62. $x^2 + 10xy + 25y^2$ 484

63. $(x + 3y)^2$ 144

64. $(x + 5y)^2$ 484

Find the value of $12x - 3$ for each of the following values of x.

65. $\dfrac{1}{2}$ 3

66. $\dfrac{1}{3}$ 1

67. $\dfrac{1}{4}$ 0

68. $\dfrac{1}{6}$ -1

▶ **69.** $\dfrac{3}{2}$ 15

70. $\dfrac{2}{3}$ 5

71. $\dfrac{3}{4}$ 6

72. $\dfrac{5}{6}$ 7

73. Fill in the tables below to find the sequences formed by substituting the first four counting numbers into the expressions $3n$ and n^3.

a.

n	1	2	3	4
$3n$	3	6	9	12

b.

n	1	2	3	4
n^3	1	8	27	64

74. Fill in the tables below to find the sequences formed by substituting the first four counting numbers into the expressions $2n - 1$ and $2n + 1$.

a.

n	1	2	3	4
$2n - 1$	1	3	5	7

b.

n	1	2	3	4
$2n + 1$	3	5	7	9

Find the sequences formed by substituting the counting numbers, in order, into the following expressions.

75. $3n - 2$ 1, 4, 7, 10, . . . an arithmetic sequence

76. $2n - 3$ -1, 1, 3, 5, . . . an arithmetic sequence

77. $n^2 - 2n + 1$ 0, 1, 4, 9, . . . a sequence of squares

78. $(n - 1)^2$ 0, 1, 4, 9, . . . a sequence of squares

Here are some problems you will see later in the book. Simplify.

79. $7 - 3(2y + 1)$ $-6y + 4$

80. $4(3x - 2) - (6x - 5)$ $6x - 3$

81. $0.08x + 0.09x$ $0.17x$

82. $0.04x + 0.05x$ $0.09x$

83. $(x + y) + (x - y)$ $2x$

84. $(-12x - 20y) + (25x + 20y)$ $13x$

Simplify.

▶ **85.** $3x + 2(x - 2)$ $5x - 4$

▶ **86.** $2(x - 2) + 3(5x)$ $17x - 4$

▶ **87.** $4(x + 1) + 3(x - 3)$ $7x - 5$

▶ **88.** $5(x + 2) + 3(x - 1)$ $8x + 7$

▶ **89.** $x + (x + 3)(-3)$ $-2x - 9$

▶ **90.** $x - 2(x + 2)$ $-x - 4$

▶ **91.** $3(4x - 2) - (5x - 8)$ $7x + 2$

▶ **92.** $2(5x - 3) - (2x - 4)$ $8x - 2$

▶ **93.** $-(3x + 1) - (4x - 7)$ $-7x + 6$

▶ **94.** $-(6x + 2) - (8x - 3)$ $-14x + 1$

Foreshadowing Problems
Problems 85–94 are the problems that stdents must work successfully to simplify the equations they will see later in this chapter, and the next. Notice Problems 95–100, 105, and 106. They foreshadow the addition method of solving a system of equations.

95. $(x + 3y) + 3(2x - y)$ $7x$

96. $(2x - y) - 2(x + 3y)$ $-7y$

97. $3(2x + 3y) - 2(3x + 5y)$ $-y$

98. $5(2x + 3y) - 3(3x + 5y)$ x

99. $-6\left(\dfrac{1}{2}x - \dfrac{1}{3}y\right) + 12\left(\dfrac{1}{4}x + \dfrac{2}{3}y\right)$ $10y$

100. $6\left(\dfrac{1}{3}x + \dfrac{1}{2}y\right) - 4\left(x + \dfrac{3}{4}y\right)$ $-2x$

101. $0.08x + 0.09(x + 2{,}000)$ $0.17x + 180$

102. $0.06x + 0.04(x + 7{,}000)$ $0.1x + 280$

103. $0.10x + 0.12(x + 500)$ $0.22x + 60$

104. $0.08x + 0.06(x + 800)$ $0.14x + 48$

105. Find a so the expression $(5x + 4y) + a(2x - y)$ simplifies to an expression that does not contain y. Using that value of a, simplify the expression. $a = 4$, $13x$

106. Find a so the expression $(5x + 4y) - a(x - 2y)$ simplifies to an expression that does not contain x. Using that value of a, simplify the expression. $a = 5$, $14y$

Find the value of $b^2 - 4ac$ for the given values of a, b, and c. (You will see these problems later in the book.)

107. $a = 1$, $b = -5$, $c = -6$ 49

108. $a = 1$, $b = -6$, $c = 7$ 8

109. $a = 2$, $b = 4$, $c = -3$ 40

110. $a = 3$, $b = 4$, $c = -2$ 40

Applying the Concepts

111. **Temperature and Altitude** If the temperature on the ground is 70°F, then the temperature at A feet above the ground can be found from the expression $-0.0035A + 70$. Find the temperature at the following altitudes.
   a. 8,000 feet   b. 12,000 feet   c. 24,000 feet
      42°F   28°F   −14°F

Ed Curry/Corbis

112. **Perimeter of a Rectangle** The expression $2l + 2w$ gives the perimeter of a rectangle with length l and width w. Find the perimeter of the rectangles with the following lengths and widths.
   a. Length = 8 meters, width = 5 meters 26 meters

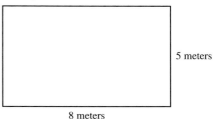

5 meters

8 meters

   b. Length = 10 feet, width = 3 feet 26 feet

3 feet

10 feet

113. **Cellular Phone Rates** A cellular phone company charges $35 per month plus $0.25 for each minute, or fraction of a minute, that you use one of their cellular phones. The expression $35 + 0.25t$ gives the amount of money you will pay for using one of their phones for t minutes a month. Find the monthly bill for using one of their phones.
   a. 10 minutes in a month $37.50
   b. 20 minutes in a month $40.00
   c. 30 minutes in a month $42.50

114. **Cost of Bottled Water** A water bottling company charges $7.00 per month for their water dispenser and $1.10 for each gallon of water delivered. If you have g gallons of water delivered in a month, then the expression $7 + 1.1g$ gives the amount of your bill for that month. Find the monthly bill for each of the following deliveries.

Cool, Refreshing Spring Water

for only **$7.00** per month and **$1.10** per gallon!

   a. 10 gallons $18
   b. 20 gallons $29
   c. 30 gallons $40

115. Taxes We all have to pay taxes. Suppose that 21% of your monthly pay is withheld for federal income taxes and another 8% is withheld for Social Security, state income tax, and other miscellaneous items. If G is your monthly pay before any money is deducted (your gross pay), then the amount of money that you take home each month is given by the expression $G - 0.21G - 0.08G$. Simplify this expression and then find your take-home pay if your gross pay is $1,250 per month. 0.71G; $887.50

116. Improving Your Quantitative Literacy The chart shows how the average cost per minute for using a cellular phone declined over a specific period of years. For each of the following years, use the chart to write an expression for the average cost to talk on a cell phone for x minutes, and then evaluate the expression to find the average cost to talk on a cell phone for an hour.

a. 1997 Approximately $30.00
b. 2001 Approximately $11.40
c. 2004 $6.60

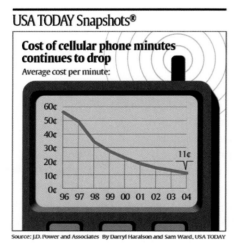

USA TODAY Snapshots®

Cost of cellular phone minutes continues to drop
Average cost per minute:

Source: J.D. Power and Associates By Darryl Haralson and Sam Ward, USA TODAY

From *USA Today*. Copyright 2005. Reprinted with permission.

Maintaining Your Skills

From this point on, each problem set will contain a number of problems under the heading *Maintaining Your Skills*. These problems cover the most important skills you have learned in previous sections and chapters. Hopefully, by working these problems on a regular basis, you will keep yourself current on all the topics we have covered and possibly need less time to study for tests and quizzes.

117. $\dfrac{1}{8} - \dfrac{1}{6} - \dfrac{1}{24}$

118. $\dfrac{x}{8} - \dfrac{x}{6} - \dfrac{x}{24}$

119. $\dfrac{5}{9} - \dfrac{4}{3} - \dfrac{7}{9}$

120. $\dfrac{x}{9} - \dfrac{x}{3} - \dfrac{2x}{9}$

121. $-\dfrac{7}{30} + \dfrac{5}{28} - \dfrac{23}{420}$

122. $-\dfrac{11}{105} + \dfrac{11}{30}\ \dfrac{11}{42}$

Getting Ready for the Next Section

Problems under this heading, *Getting Ready for the Next Section,* are problems that you must be able to work in order to understand the material in the next section. The problems below are exactly the types of problems you will see in the explanations and examples in the next section.

Simplify.

123. $17 - 5$ 12

124. $12 + (-2)$ 10

125. $2 - 5$ -3

126. $25 - 20$ 5

127. $-2.4 + (-7.3)$ -9.7

128. $8.1 + 2.7$ 10.8

129. $-\dfrac{1}{2} + \left(-\dfrac{3}{4}\right)\ -\dfrac{5}{4}$

130. $-\dfrac{1}{6} + \left(-\dfrac{2}{3}\right)\ -\dfrac{5}{6}$

131. $4(2 \cdot 9 - 3) - 7 \cdot 9\ -3$

132. $5(3 \cdot 45 - 4) - 14 \cdot 45$ 25

133. $4(2a - 3) - 7a\ a - 12$

134. $5(3a - 4) - 14a\ a - 20$

135. $-3 - \dfrac{1}{2}\ -\dfrac{7}{2}$

136. $-5 - \dfrac{1}{3}\ -\dfrac{16}{3}$

137. $\dfrac{4}{5} + \dfrac{1}{10} + \dfrac{3}{8}\ \dfrac{51}{40}$

138. $\dfrac{3}{10} + \dfrac{7}{25} + \dfrac{3}{4}\ \dfrac{133}{100}$

139. Find the value of $2x - 3$ when x is 5. 7

140. Find the value of $3x + 4$ when x is -2. -2

Addition Property of Equality

OBJECTIVES

A Check the solution to an equation by substitution.

B Use the addition property of equality to solve an equation.

When light comes into contact with any object, it is reflected, absorbed, and transmitted, as shown in Figure 1.

For a certain type of glass, 88% of the light hitting the glass is transmitted through to the other side, whereas 6% of the light is absorbed into the glass. To find the percent of light that is reflected by the glass, we can solve the equation

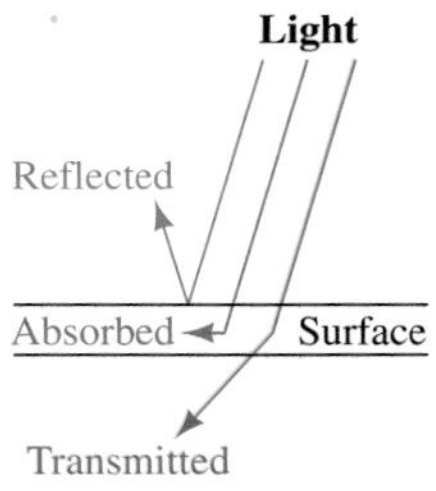

$$88 + R + 6 = 100$$

Light

Reflected

Absorbed ◄ ／ Surface

Transmitted

FIGURE 1

Solving equations of this type is what we study in this section. To solve an equation we must find all replacements for the variable that make the equation a true statement.

> **DEFINITION** The **solution set** for an equation is the set of all numbers that when used in place of the variable make the equation a true statement.

For example, the equation $x + 2 = 5$ has solution set {3} because when x is 3 the equation becomes the true statement $3 + 2 = 5$, or $5 = 5$.

EXAMPLE 1   Is 5 a solution to $2x - 3 = 7$?

SOLUTION   We substitute 5 for x in the equation, and then simplify to see if a true statement results. A true statement means we have a solution; a false statement indicates the number we are using is not a solution.

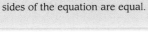

Note

We can use a question mark over the equal signs to show that we don't know yet whether the two sides of the equation are equal.

$$\begin{aligned}
\text{When} \qquad & x = 5 \\
\text{the equation} \qquad & 2x - 3 = 7 \\
\text{becomes} \qquad & 2(5) - 3 \overset{?}{=} 7 \\
& 10 - 3 \overset{?}{=} 7 \\
& 7 = 7 \qquad \textbf{A true statement}
\end{aligned}$$

Because $x = 5$ turns the equation into the true statement $7 = 7$, we know 5 is a solution to the equation.

EXAMPLE 2   Is -2 a solution to $8 = 3x + 4$?

SOLUTION   Substituting -2 for x in the equation, we have

$$\begin{aligned}
8 &\overset{?}{=} 3(-2) + 4 \\
8 &\overset{?}{=} -6 + 4 \\
8 &= -2 \qquad \textbf{A false statement}
\end{aligned}$$

Substituting -2 for x in the equation produces a false statement. Therefore, $x = -2$ is not a solution to the equation.

The important thing about an equation is its solution set. We therefore make the following definition to classify together all equations with the same solution set.

> **DEFINITION** Two or more equations with the same solution set are said to be **equivalent equations.**

Equivalent equations may look different but must have the same solution set.

 EXAMPLE 3

a. $x + 2 = 5$ and $x = 3$ are equivalent equations because both have solution set {3}.

b. $a - 4 = 3$, $a - 2 = 5$, and $a = 7$ are equivalent equations because they all have solution set {7}.

c. $y + 3 = 4$, $y - 8 = -7$, and $y = 1$ are equivalent equations because they all have solution set {1}.

If two numbers are equal and we increase (or decrease) both of them by the same amount, the resulting quantities are also equal. We can apply this concept to equations. Adding the same amount to both sides of an equation always produces an equivalent equation—one with the same solution set. This fact about equations is called the *addition property of equality* and can be stated more formally as follows.

Note

We will use this property many times in the future. Be sure you understand it completely by the time you finish this section.

> **Addition Property of Equality** For any three algebraic expressions A, B, and C,
>
> $$\text{if} \qquad A = B$$
> $$\text{then} \quad A + C = B + C$$
>
> *In words:* Adding the same quantity to both sides of an equation will not change the solution set.

This property is just as simple as it seems. We can add any amount to both sides of an equation and always be sure we have not changed the solution set.

Consider the equation $x + 6 = 5$. We want to solve this equation for the value of x that makes it a true statement. We want to end up with x on one side of the equal sign and a number on the other side. Because we want x by itself, we will add -6 to both sides:

$$x + 6 + (\mathbf{-6}) = 5 + (\mathbf{-6}) \qquad \text{Addition property of equality}$$
$$x + 0 = -1 \qquad\qquad\qquad \text{Addition}$$
$$x = -1$$

All three equations say the same thing about x. They all say that x is -1. All three equations are equivalent. The last one is just easier to read.

Here are some further examples of how the addition property of equality can be used to solve equations.

 EXAMPLE 4   Solve the equation $x - 5 = 12$ for x.

SOLUTION Because we want x alone on the left side, we choose to add $+5$ to both sides:

$$x - 5 + \mathbf{5} = 12 + \mathbf{5} \qquad \text{Addition property of equality}$$
$$x + 0 = 17$$
$$x = 17$$

To check our solution, we substitute 17 for x in the original equation:

When $\qquad\qquad x = 17$

the equation $\qquad x - 5 = 12$

becomes $\qquad 17 - 5 \overset{?}{=} 12$

$\qquad\qquad\qquad 12 = 12 \qquad$ **A true statement**

As you can see, our solution checks. The purpose for checking a solution to an equation is to catch any mistakes we may have made in the process of solving the equation.

EXAMPLE 5 Solve for a: $a + \frac{3}{4} = -\frac{1}{2}$.

SOLUTION Because we want a by itself on the left side of the equal sign, we add the opposite of $\frac{3}{4}$ to each side of the equation.

$$a + \frac{3}{4} + \left(-\frac{3}{4}\right) = -\frac{1}{2} + \left(-\frac{3}{4}\right) \qquad \text{Addition property of equality}$$

$$a + 0 = -\frac{1}{2} \cdot \frac{2}{2} + \left(-\frac{3}{4}\right) \qquad \text{LCD on the right side is 4}$$

$$a = -\frac{2}{4} + \left(-\frac{3}{4}\right) \qquad \frac{2}{4} \text{ is equivalent to } \frac{1}{2}$$

$$a = -\frac{5}{4} \qquad \text{Add fractions}$$

The solution is $a = -\frac{5}{4}$. To check our result, we replace a with $-\frac{5}{4}$ in the original equation. The left side then becomes $-\frac{5}{4} + \frac{3}{4}$, which reduces to $-\frac{1}{2}$, so our solution checks.

EXAMPLE 6 Solve for x: $7.3 + x = -2.4$.

SOLUTION Again, we want to isolate x, so we add the opposite of 7.3 to both sides:

$$7.3 + (-\mathbf{7.3}) + x = -2.4 + (-\mathbf{7.3}) \qquad \text{Addition property of equality}$$

$$0 + x = -9.7$$

$$x = -9.7$$

The addition property of equality also allows us to add variable expressions to each side of an equation.

EXAMPLE 7 Solve for x: $3x - 5 = 4x$.

SOLUTION Adding $-3x$ to each side of the equation gives us our solution.

$$3x - 5 = 4x$$

$$3x + (-\mathbf{3x}) - 5 = 4x + (-\mathbf{3x}) \qquad \text{Distributive property}$$

$$-5 = x$$

Sometimes it is necessary to simplify each side of an equation before using the addition property of equality. The reason we simplify both sides first is that we want as few terms as possible on each side of the equation before we use the addition property of equality. The following examples illustrate this procedure.

EXAMPLE 8   Solve $4(2a - 3) - 7a = 2 - 5$.

SOLUTION   We must begin by applying the distributive property to separate terms on the left side of the equation. Following that, we combine similar terms and then apply the addition property of equality.

$4(2a - 3) - 7a = 2 - 5$	**Original equation**
$8a - 12 - 7a = 2 - 5$	**Distributive property**
$a - 12 = -3$	**Simplify each side**
$a - 12 + \mathbf{12} = -3 + \mathbf{12}$	**Add 12 to each side**
$a = 9$	**Addition**

To check our solution, we replace a with 9 in the original equation.

$$4(2 \cdot 9 - 3) - 7 \cdot 9 \overset{?}{=} 2 - 5$$
$$4(15) - 63 \overset{?}{=} -3$$
$$60 - 63 \overset{?}{=} -3$$
$$-3 = -3 \qquad \textbf{A true statement}$$

Note
Again, we place a question mark over the equal sign because we don't know yet whether the expressions on the left and right side of the equal sign will be equal.

We can also add a term involving a variable to both sides of an equation.

EXAMPLE 9   Solve $3x - 5 = 2x + 7$.

SOLUTION   We can solve this equation in two steps. First, we add $-2x$ to both sides of the equation. When this has been done, x appears on the left side only. Second, we add 5 to both sides:

$3x + (\mathbf{-2x}) - 5 = 2x + (\mathbf{-2x}) + 7$	**Add $-2x$ to both sides**
$x - 5 = 7$	**Simplify each side**
$x - 5 + \mathbf{5} = 7 + \mathbf{5}$	**Add 5 to both sides**
$x = 12$	**Simplify each side**

Note
In my experience teaching algebra, I find that students make fewer mistakes if they think in terms of addition rather than subtraction. So, you are probably better off if you continue to use the addition property just the way we have used it in the examples in this section. But, if you are curious as to whether you can subtract the same number from both sides of an equation, the answer is yes.

A Note on Subtraction Although the addition property of equality is stated for addition only, we can subtract the same number from both sides of an equation as well. Because subtraction is defined as addition of the opposite, subtracting the same quantity from both sides of an equation does not change the solution.

$x + 2 = 12$	**Original equation**
$x + 2 - \mathbf{2} = 12 - \mathbf{2}$	**Subtract 2 from each side**
$x = 10$	**Subtraction**

LINKING OBJECTIVES AND EXAMPLES

Next to each **objective** we have listed the examples that are best described by that objective.

A   1–3

B   4–9

GETTING READY FOR CLASS

After reading through the preceding section, respond in your own words and in complete sentences.

1. What is a solution to an equation?
2. What are equivalent equations?
3. Explain in words the addition property of equality.
4. How do you check a solution to an equation?

Problem Set 2.2

Online support materials can be found at www.thomsonedu.com/login

Find the solution for the following equations. Be sure to show when you have used the addition property of equality.

1. $x - 3 = 8$ 11

2. $x - 2 = 7$ 9

▶ **3.** $x + 2 = 6$ 4

4. $x + 5 = 4$ −1

5. $a + \dfrac{1}{2} = -\dfrac{1}{4}$ $-\frac{3}{4}$

6. $a + \dfrac{1}{3} = -\dfrac{5}{6}$ $-\frac{7}{6}$

7. $x + 2.3 = -3.5$ −5.8

8. $x + 7.9 = -3.4$ −11.3

9. $y + 11 = -6$ −17

10. $y - 3 = -1$ 2

11. $x - \dfrac{5}{8} = -\dfrac{3}{4}$ $-\frac{1}{8}$

12. $x - \dfrac{2}{5} = -\dfrac{1}{10}$ $\frac{3}{10}$

13. $m - 6 = 2m$ −6

14. $3m - 10 = 4m$ −10

15. $6.9 + x = 3.3$ −3.6

16. $7.5 + x = 2.2$ −5.3

17. $5a = 4a - 7$ −7

18. $12a = -3 + 11a$ −3

▶ **19.** $-\dfrac{5}{9} = x - \dfrac{2}{5}$ $-\frac{7}{45}$

20. $-\dfrac{7}{8} = x - \dfrac{4}{5}$ $-\frac{3}{40}$

Simplify both sides of the following equations as much as possible, and then solve.

▶ **21.** $4x + 2 - 3x = 4 + 1$ 3

22. $5x + 2 - 4x = 7 - 3$ 2

23. $8a - \dfrac{1}{2} - 7a = \dfrac{3}{4} + \dfrac{1}{8}$ $\frac{11}{8}$

24. $9a - \dfrac{4}{5} - 8a = \dfrac{3}{10} - \dfrac{1}{5}$ $\frac{9}{10}$

25. $-3 - 4x + 5x = 18$ 21

26. $10 - 3x + 4x = 20$ 10

27. $-11x + 2 + 10x + 2x = 9$ 7

28. $-10x + 5 - 4x + 15x = 0$ −5

29. $-2.5 + 4.8 = 8x - 1.2 - 7x$ 3.5

30. $-4.8 + 6.3 = 7x - 2.7 - 6x$ 4.2

31. $2y - 10 + 3y - 4y = 18 - 6$ 22

32. $15 - 21 = 8x + 3x - 10x$ −6

The following equations contain parentheses. Apply the distributive property to remove the parentheses, then simplify each side before using the addition property of equality.

33. $2(x + 3) - x = 4$ −2

34. $5(x + 1) - 4x = 2$ −3

35. $-3(x - 4) + 4x = 3 - 7$ −16

36. $-2(x - 5) + 3x = 4 - 9$ −15

37. $5(2a + 1) - 9a = 8 - 6$ −3

38. $4(2a - 1) - 7a = 9 - 5$ 8

39. $-(x + 3) + 2x - 1 = 6$ 10

40. $-(x - 7) + 2x - 8 = 4$ 5

▶ **41.** $4y - 3(y - 6) + 2 = 8$ −12

42. $7y - 6(y - 1) + 3 = 9$ 0

43. $-3(2m - 9) + 7(m - 4) = 12 - 9$ 4

44. $-5(m - 3) + 2(3m + 1) = 15 - 8$ −10

▭ = Videos available by instructor request

▶ = Online student support materials available at www.thomsonedu.com/login

Solve the following equations by the method used in Example 9 in this section. Check each solution in the original equation.

45. $4x = 3x + 2$ 2

46. $6x = 5x - 4$ -4

▶ **47.** $8a = 7a - 5$ -5

48. $9a = 8a - 3$ -3

49. $2x = 3x + 1$ -1

50. $4x = 3x + 5$ 5

51. $2y + 1 = 3y + 4$ -3

52. $4y + 2 = 5y + 6$ -4

53. $2m - 3 = m + 5$ 8

54. $8m - 1 = 7m - 3$ -2

55. $4x - 7 = 5x + 1$ -8

56. $3x - 7 = 4x - 6$ -1

57. $4x + \dfrac{4}{3} = 5x - \dfrac{2}{3}$ 2

58. $2x + \dfrac{1}{4} = 3x - \dfrac{5}{4}$ $\dfrac{3}{2}$

59. $8a - 7.1 = 7a + 3.9$ 11

60. $10a - 4.3 = 9a + 4.7$ 9

61. Solve each equation.
  a. $2x = 3$ $\dfrac{3}{2}$
  b. $2 + x = 3$ 1
  c. $2x + 3 = 0$ $-\dfrac{3}{2}$
  d. $2x + 3 = -5$ -4
  e. $2x + 3 = 7x - 5$ $\dfrac{8}{5}$

62. Solve each equation.
  a. $5t = 10$ 2
  b. $5 + t = 10$ 5
  c. $5t + 10 = 0$ -2
  d. $5t + 10 = 12$ $\dfrac{2}{5}$
  e. $5t + 10 = 8t + 12$ $-\dfrac{2}{3}$

Applying the Concepts

63. **Light** When light comes into contact with any object, it is reflected, absorbed, and transmitted, as shown in the following figure. If T represents the percent of light transmitted, R the percent of light reflected, and A the percent of light absorbed by a surface, then the equation $T + R + A = 100$ shows one way these quantities are related.

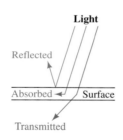

  a. For glass, $T = 88$ and $A = 6$, meaning that 88% of the light hitting the glass is transmitted and 6% is absorbed. Substitute $T = 88$ and $A = 6$ into the equation $T + R + A = 100$ and solve for R to find the percent of light that is reflected. 6%
  b. For flat black paint, $A = 95$ and no light is transmitted, meaning that $T = 0$. What percent of light is reflected by flat black paint? 5%
  c. A pure white surface can reflect 98% of light, so $R = 98$. If no light is transmitted, what percent of light is absorbed by the pure white surface? 2%
  d. Typically, shiny gray metals reflect 70–80% of light. Suppose a thick sheet of aluminum absorbs 25% of light. What percent of light is reflected by this shiny gray metal? (Assume no light is transmitted.) 75%

64. **Improving Your Quantitative Literacy** According to a survey done by *Seventeen* magazine in 2005, the average spending amount for girls going to the prom was $338.

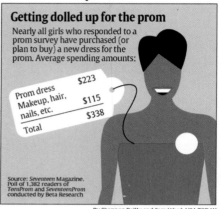

USA TODAY Snapshots®

Getting dolled up for the prom

Nearly all girls who responded to a prom survey have purchased (or plan to buy) a new dress for the prom. Average spending amounts:

Prom dress $223
Makeup, hair, nails, etc. $115
Total $338

Source: *Seventeen* Magazine. Poll of 1,382 readers of *TeenProm* and *SeventeenProm* conducted by Beta Research

By Shannon Reilly and Sam Ward, USA TODAY

From *USA Today*. Copyright 2005. Reprinted with permission.

  a. Suppose Kendra spent $210 for a prom dress, then bought new shoes to go with the dress. If the total bill was $289, does the equation

$$210 + x = 289$$

describe this situation? If so, what does x represent? *Yes, x represents the cost of her shoes.*

  b. Suppose Ava buys a prom dress on sale and then spends $120 on her makeup, nails, and hair. If the total bill is $219, and x represents a positive number, does the equation

$$x + 219 = 120$$

describe the situation?

No, the equation should be x + 120 = 219.

65. **Geometry** The three angles shown in the triangle at the front of the tent in the following figure add up to 180°. Use this fact to write an equation containing x, and then solve the equation to find the number of degrees in the angle at the top of the triangle.

$x + 55 + 55 = 180$; 70°

66. Geometry The figure shows part of a room. From a point on the floor, the angle of elevation to the top of the door is 47°, whereas the angle of elevation to the ceiling above the door is 59°. Use this diagram to write an equation involving x, and then solve the equation to find the number of degrees in the angle that extends from the top of the door to the ceiling. $47 + x = 59; 12°$

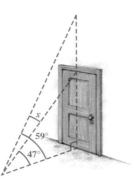

Maintaining Your Skills

The problems that follow review some of the more important skills you have learned in previous sections and chapters. You can consider the time you spend working these problems as time spent studying for exams.

67. $3(6x)$ $18x$

68. $5(4x)$ $20x$

69. $\frac{1}{5}(5x)$ x

70. $\frac{1}{3}(3x)$ x

71. $8\left(\frac{1}{8}y\right)$ y

72. $6\left(\frac{1}{6}y\right)$ y

73. $-2\left(-\frac{1}{2}x\right)$ x

74. $-4\left(-\frac{1}{4}x\right)$ x

75. $-\frac{4}{3}\left(-\frac{3}{4}a\right)$ a

76. $-\frac{5}{2}\left(-\frac{2}{5}a\right)$ a

Getting Ready for the Next Section

To understand all of the explanations and examples in the next section you must be able to work the problems below.

Simplify.

77. $\frac{3}{2}\left(\frac{2}{3}y\right)$ y

78. $\frac{5}{2}\left(\frac{2}{5}y\right)$ y

79. $\frac{1}{5}(5x)$ x

80. $-\frac{1}{4}(-4a)$ a

81. $\frac{1}{5}(30)$ 6

82. $-\frac{1}{4}(24)$ -6

83. $\frac{3}{2}(4)$ 6

84. $\frac{1}{26}(13)$ $\frac{1}{2}$

85. $12\left(-\frac{3}{4}\right)$ -9

86. $12\left(\frac{1}{2}\right)$ 6

87. $\frac{3}{2}\left(-\frac{5}{4}\right)$ $-\frac{15}{8}$

88. $\frac{5}{3}\left(-\frac{6}{5}\right)$ -2

89. $-13 + (-5)$ -18

90. $-14 + (-3)$ -17

91. $-\frac{3}{4} + \left(-\frac{1}{2}\right)$ $-\frac{5}{4}$

92. $-\frac{7}{10} + \left(-\frac{1}{2}\right)$ $-\frac{6}{5}$

93. $7x + (-4x)$ $3x$

94. $5x + (-2x)$ $3x$

2.3   Multiplication Property of Equality

OBJECTIVES

A   Use the multiplication property of equality to solve an equation.

B   Use the addition and multiplication properties of equality together to solve an equation.

As we have mentioned before, we all have to pay taxes. According to Figure 1, people have been paying taxes for quite a long time.

If 21% of your monthly pay is withheld for federal income taxes and another 8% is withheld for Social Security, state income tax, and other miscellaneous items, leaving you with $987.50 a month in take-home pay, then the amount you earned before the deductions were removed from your check is given by the equation

$$G - 0.21G - 0.08G = 987.5$$

In this section we will learn how to solve equations of this type.

FIGURE 1 Collection of taxes, about 3000 B.C. Clerks and scribes appear at the right, with pen and papyrus, and officials and taxpayers appear at the left.

In the previous section, we found that adding the same number to both sides of an equation never changed the solution set. The same idea holds for multiplication by numbers other than zero. We can multiply both sides of an equation by the same nonzero number and always be sure we have not changed the solution set. (The reason we cannot multiply both sides by zero will become apparent later.) This fact about equations is called the *multiplication property of equality,* which can be stated formally as follows.

Note

This property is also used many times throughout the book. Make every effort to understand it completely.

> **Multiplication Property of Equality**   For any three algebraic expressions A, B, and C, where $C \neq 0$,
>
> $$\text{if} \qquad A = B$$
> $$\text{then} \qquad AC = BC$$
>
> *In words:*   Multiplying both sides of an equation by the same nonzero number will not change the solution set.

Suppose we want to solve the equation $5x = 30$. We have $5x$ on the left side but would like to have just x. We choose to multiply both sides by $\frac{1}{5}$ because $(\frac{1}{5})(5) = 1$. Here is the solution:

$$5x = 30$$

$$\frac{1}{5}(5x) = \frac{1}{5}(30) \qquad \textbf{Multiplication property of equality}$$

$$\left(\frac{1}{5} \cdot 5\right)x = \frac{1}{5}(30) \qquad \textbf{Associative property of multiplication}$$

$$1x = 6$$

$$x = 6$$

We chose to multiply by $\frac{1}{5}$ because it is the reciprocal of 5. We can see that multiplication by any number except zero will not change the solution set. If, however, we were to multiply both sides by zero, the result would always be $0 = 0$ because multiplication by zero always results in zero. Although the statement $0 = 0$ is true, we have lost our variable and cannot solve the equation. This is the only restriction of the multiplication property of equality. We are free to multiply both sides of an equation by any number except zero.

Here are some more examples that use the multiplication property of equality.

 EXAMPLE 1   Solve for a: $-4a = 24$.

SOLUTION   Because we want a alone on the left side, we choose to multiply both sides by $-\frac{1}{4}$:

$$-\frac{1}{4}(-4a) = -\frac{1}{4}(24) \qquad \textbf{Multiplication property of equality}$$

$$\left[-\frac{1}{4}(-4)\right]a = -\frac{1}{4}(24) \qquad \textbf{Associative property}$$

$$a = -6$$

EXAMPLE 2   Solve for t: $-\frac{t}{3} = 5$.

SOLUTION   Because division by 3 is the same as multiplication by $\frac{1}{3}$, we can write $-\frac{t}{3}$ as $-\frac{1}{3}t$. To solve the equation, we multiply each side by the reciprocal of $-\frac{1}{3}$, which is -3.

$$-\frac{t}{3} = 5 \qquad \text{\textbf{Original equation}}$$

$$-\frac{1}{3}t = 5 \qquad \text{\textbf{Dividing by 3 is equivalent to multiplying by } } \tfrac{1}{3}$$

$$\mathbf{-3}\left(-\frac{1}{3}t\right) = \mathbf{-3}(5) \qquad \text{\textbf{Multiply each side by } } \mathbf{-3}$$

$$t = -15 \qquad \text{\textbf{Multiplication}}$$

EXAMPLE 3   Solve $\frac{2}{3}y = 4$.

SOLUTION   We can multiply both sides by $\frac{3}{2}$ and have $1y$ on the left side:

$$\frac{\mathbf{3}}{\mathbf{2}}\left(\frac{2}{3}y\right) = \frac{\mathbf{3}}{\mathbf{2}}(4) \qquad \text{\textbf{Multiplication property of equality}}$$

$$\left(\frac{3}{2} \cdot \frac{2}{3}\right)y = \frac{3}{2}(4) \qquad \text{\textbf{Associative property}}$$

$$y = 6 \qquad \text{\textbf{Simplify } } \tfrac{3}{2}(4) = \tfrac{3}{2}\left(\tfrac{4}{1}\right) = \tfrac{12}{2} = 6$$

> *Note*
>
> Notice in Examples 1 through 3 that if the variable is being multiplied by a number like -4 or $\frac{2}{3}$, we always multiply by the number's reciprocal, $-\frac{1}{4}$ or $\frac{3}{2}$, to end up with just the variable on one side of the equation.

EXAMPLE 4   Solve $5 + 8 = 10x + 20x - 4x$.

SOLUTION   Our first step will be to simplify each side of the equation:

$$13 = 26x \qquad \text{\textbf{Simplify both sides first}}$$

$$\frac{\mathbf{1}}{\mathbf{26}}(13) = \frac{\mathbf{1}}{\mathbf{26}}(26x) \qquad \text{\textbf{Multiplication property of equality}}$$

$$\frac{13}{26} = x \qquad \text{\textbf{Multiplication}}$$

$$\frac{1}{2} = x \qquad \text{\textbf{Reduce to lowest terms}}$$

In the next three examples, we will use both the addition property of equality and the multiplication property of equality.

EXAMPLE 5   Solve for x: $6x + 5 = -13$.

SOLUTION   We begin by adding -5 to both sides of the equation:

$$6x + 5 + (\mathbf{-5}) = -13 + (\mathbf{-5}) \qquad \text{\textbf{Add } } -5 \text{ \textbf{to both sides}}$$

$$6x = -18 \qquad \text{\textbf{Simplify}}$$

$$\frac{\mathbf{1}}{\mathbf{6}}(6x) = \frac{\mathbf{1}}{\mathbf{6}}(-18) \qquad \text{\textbf{Multiply both sides by } } \tfrac{1}{6}$$

$$x = -3$$

EXAMPLE 6

Solve for x: $5x = 2x + 12$.

SOLUTION We begin by adding $-2x$ to both sides of the equation:

$$5x + (-\mathbf{2x}) = 2x + (-\mathbf{2x}) + 12 \qquad \text{Add } -2x \text{ to both sides}$$

$$3x = 12 \qquad \text{Simplify}$$

$$\frac{\mathbf{1}}{\mathbf{3}}(3x) = \frac{\mathbf{1}}{\mathbf{3}}(12) \qquad \text{Multiply both sides by } \tfrac{1}{3}$$

$$x = 4 \qquad \text{Simplify}$$

> ### Note
> Notice that in Example 6 we used the addition property of equality first to combine all the terms containing x on the left side of the equation. Once this had been done, we used the multiplication property to isolate x on the left side.

EXAMPLE 7

Solve for x: $3x - 4 = -2x + 6$.

SOLUTION We begin by adding $2x$ to both sides:

$$3x + \mathbf{2x} - 4 = -2x + \mathbf{2x} + 6 \qquad \text{Add } 2x \text{ to both sides}$$

$$5x - 4 = 6 \qquad \text{Simplify}$$

Now we add 4 to both sides:

$$5x - 4 + \mathbf{4} = 6 + \mathbf{4} \qquad \text{Add 4 to both sides}$$

$$5x = 10 \qquad \text{Simplify}$$

$$\frac{\mathbf{1}}{\mathbf{5}}(5x) = \frac{\mathbf{1}}{\mathbf{5}}(10) \qquad \text{Multiply by } \tfrac{1}{5}$$

$$x = 2 \qquad \text{Simplify}$$

The next example involves fractions. You will see that the properties we use to solve equations containing fractions are the same as the properties we used to solve the previous equations. Also, the LCD that we used previously to add fractions can be used with the multiplication property of equality to simplify equations containing fractions.

EXAMPLE 8

Solve $\frac{2}{3}x + \frac{1}{2} = -\frac{3}{4}$.

SOLUTION We can solve this equation by applying our properties and working with the fractions, or we can begin by eliminating the fractions.

Method 1 Working with the fractions.

$$\frac{2}{3}x + \frac{1}{2} + \left(-\frac{\mathbf{1}}{\mathbf{2}}\right) = -\frac{3}{4} + \left(-\frac{\mathbf{1}}{\mathbf{2}}\right) \qquad \text{Add } -\tfrac{1}{2} \text{ to each side}$$

$$\frac{2}{3}x = -\frac{5}{4} \qquad \text{Note that } -\tfrac{3}{4} + \left(-\tfrac{1}{2}\right) = -\tfrac{3}{4} + \left(-\tfrac{2}{4}\right)$$

$$\frac{\mathbf{3}}{\mathbf{2}}\left(\frac{2}{3}x\right) = \frac{\mathbf{3}}{\mathbf{2}}\left(-\frac{5}{4}\right) \qquad \text{Multiply each side by } \tfrac{3}{2}$$

$$x = -\frac{15}{8}$$

> ### Note
> Our original equation has denominators of 3, 2, and 4. The LCD for these three denominators is 12, and it has the property that all three denominators will divide it evenly. Therefore, if we multiply both sides of our equation by 12, each denominator will divide into 12 and we will be left with an equation that does not contain any denominators other than 1.

Method 2 Eliminating the fractions in the beginning.

$$\mathbf{12}\left(\frac{2}{3}x + \frac{1}{2}\right) = \mathbf{12}\left(-\frac{3}{4}\right) \qquad \text{Multiply each side by the LCD 12}$$

$$\mathbf{12}\left(\frac{2}{3}x\right) + \mathbf{12}\left(\frac{1}{2}\right) = \mathbf{12}\left(-\frac{3}{4}\right) \qquad \text{Distributive property on the left side}$$

$$8x + 6 = -9 \qquad \textbf{Multiply}$$

$$8x = -15 \qquad \textbf{Add } -6 \textbf{ to each side}$$

$$x = -\frac{15}{8} \qquad \textbf{Multiply each side by } \tfrac{1}{8}$$

As the third line in Method 2 indicates, multiplying each side of the equation by the LCD eliminates all the fractions from the equation.

As you can see, both methods yield the same solution.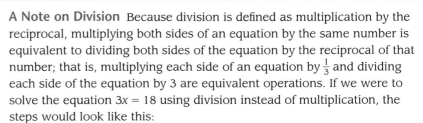

A Note on Division Because division is defined as multiplication by the reciprocal, multiplying both sides of an equation by the same number is equivalent to dividing both sides of the equation by the reciprocal of that number; that is, multiplying each side of an equation by $\frac{1}{3}$ and dividing each side of the equation by 3 are equivalent operations. If we were to solve the equation $3x = 18$ using division instead of multiplication, the steps would look like this:

$$3x = 18 \qquad \textbf{Original equation}$$

$$\frac{3x}{3} = \frac{18}{3} \qquad \textbf{Divide each side by 3}$$

$$x = 6 \qquad \textbf{Division}$$

Using division instead of multiplication on a problem like this may save you some writing. However, with multiplication, it is easier to explain "why" we end up with just one x on the left side of the equation. (The "why" has to do with the associative property of multiplication.) My suggestion is that you continue to use multiplication to solve equations like this one until you understand the process completely. Then, if you find it more convenient, you can use division instead of multiplication.

LINKING OBJECTIVES AND EXAMPLES

Next to each **objective** we have listed the examples that are best described by that objective.

A	1–4
B	5–8

GETTING READY FOR CLASS

After reading through the preceding section, respond in your own words and in complete sentences.

1. Explain in words the multiplication property of equality.
2. If an equation contains fractions, how do you use the multiplication property of equality to clear the equation of fractions?
3. Why is it okay to divide both sides of an equation by the same nonzero number?
4. Explain in words how you would solve the equation $3x = 7$ using the multiplication property of equality.

Solve the following equations. Be sure to show your work.

▶ 1. $5x = 10$ 2

2. $6x = 12$ 2

3. $7a = 28$ 4

4. $4a = 36$ 9

5. $-8x = 4$ $-\frac{1}{2}$

6. $-6x = 2$ $-\frac{1}{3}$

7. $8m = -16$ -2

8. $5m = -25$ -5

9. $-3x = -9$ 3

10. $-9x = -36$ 4

▶ 11. $-7y = -28$ 4

12. $-15y = -30$ 2

13. $2x = 0$ 0

14. $7x = 0$ 0

15. $-5x = 0$ 0

16. $-3x = 0$ 0

▶ 17. $\frac{x}{3} = 2$ 6

18. $\frac{x}{4} = 3$ 12

19. $-\frac{m}{5} = 10$ -50

20. $-\frac{m}{7} = 1$ -7

21. $-\frac{x}{2} = -\frac{3}{4}$ $\frac{3}{2}$

22. $-\frac{x}{3} = \frac{5}{6}$ $-\frac{5}{2}$

23. $\frac{2}{3}a = 8$ 12

24. $\frac{3}{4}a = 6$ 8

25. $-\frac{3}{5}x = \frac{9}{5}$ -3

26. $-\frac{2}{5}x = \frac{6}{15}$ -1

27. $-\frac{5}{8}y = -20$ 32

28. $-\frac{7}{2}y = -14$ 4

Simplify both sides as much as possible, and then solve.

29. $-4x - 2x + 3x = 24$ -8

30. $7x - 5x + 8x = 20$ 2

31. $4x + 8x - 2x = 15 - 10$ $\frac{1}{2}$

32. $5x + 4x + 3x = 4 + 8$ 1

33. $-3 - 5 = 3x + 5x - 10x$ 4

34. $10 - 16 = 12x - 6x - 3x$ -2

35. $18 - 13 = \frac{1}{2}a + \frac{3}{4}a - \frac{5}{8}a$ 8

36. $20 - 14 = \frac{1}{3}a + \frac{5}{6}a - \frac{2}{3}a$ 12

Solve the following equations by multiplying both sides by -1.

37. $-x = 4$ -4

38. $-x = -3$ 3

39. $-x = -4$ 4

40. $-x = 3$ -3

41. $15 = -a$ -15

42. $-15 = -a$ 15

43. $-y = \frac{1}{2}$ $-\frac{1}{2}$

44. $-y = -\frac{3}{4}$ $\frac{3}{4}$

Solve each of the following equations.

▶ 45. $3x - 2 = 7$ 3

46. $2x - 3 = 9$ 6

47. $2a + 1 = 3$ 1

48. $5a - 3 = 7$ 2

49. $\frac{1}{8} + \frac{1}{2}x = \frac{1}{4}$ $\frac{1}{4}$

50. $\frac{1}{3} + \frac{1}{7}x = -\frac{8}{21}$ -5

51. $6x = 2x - 12$ -3

52. $8x = 3x - 10$ -2

53. $2y = -4y + 18$ 3

54. $3y = -2y - 15$ -3

55. $-7x = -3x - 8$ 2

56. $-5x = -2x - 12$ 4

▶ 57. $2x - 5 = 8x + 4$ $-\frac{3}{2}$

58. $3x - 6 = 5x + 6$ -6

59. $x + \frac{1}{2} = \frac{1}{4}x - \frac{5}{8}$ $-\frac{3}{2}$

60. $\frac{1}{3}x + \frac{2}{5} = \frac{1}{5}x - \frac{2}{5}$ -6

61. $m + 2 = 6m - 3$ 1

62. $m + 5 = 6m - 5$ 2

63. $\frac{1}{2}m - \frac{1}{4} = \frac{1}{12}m + \frac{1}{6}$ 1

64. $\frac{1}{2}m - \frac{5}{12} = \frac{1}{12}m + \frac{5}{12}$ 2

65. $6y - 4 = 9y + 2$ -2

66. $2y - 2 = 6y + 14$ -4

67. $\frac{3}{2}y + \frac{1}{3} = y - \frac{2}{3}$ -2

68. $\frac{3}{2}y + \frac{7}{2} = \frac{1}{2}y - \frac{1}{2}$ -4

▶ 69. $5x + 6 = 2$ $-\frac{4}{5}$

▶ 70. $2x + 15 = 3$ -6

71. $\frac{x}{2} = \frac{6}{12}$ 1

72. $\frac{x}{4} = \frac{6}{8}$ 3

73. $\frac{3}{x} = \frac{6}{7}$ $\frac{7}{2}$

74. $\frac{2}{9} = \frac{8}{x}$ 36

75. $\frac{a}{3} = \frac{5}{12}$ $\frac{5}{4}$

76. $\frac{a}{2} = \frac{7}{20}$ $\frac{7}{10}$

77. $\frac{10}{20} = \frac{20}{x}$ 40

78. $\frac{15}{60} = \frac{60}{x}$ 240

79. $\frac{2}{x} = \frac{6}{7}$ $\frac{7}{3}$

80. $\frac{4}{x} = \frac{6}{7}$ $\frac{14}{3}$

Applying the Concepts

81. Break-Even Point Movie theaters pay a certain price for the movies that you and I see. Suppose a theater pays $1,500 for each showing of a popular movie. If they charge $7.50 for each ticket they sell,

Foreshadowing Problems

Problems 71–80 are the equations students will solve when they cover proportions later in the book. With proportions, they will be introduced to the means/extremes property, giving them an alternate way of solving these equations.

▢ = Videos available by instructor request

▶ = Online student support materials available at www.thomsonedu.com/login

then the equation $7.5x = 1,500$ gives the number of tickets they must sell to equal the $1,500 cost of showing the movie. This number is called the break-even point. Solve the equation for x to find the break-even point.   200 tickets

82. **Basketball** Laura plays basketball for her community college. In one game she scored 13 points total, with a combination of free throws, field goals, and three-pointers. Each free throw is worth 1 point, each field goal is 2 points, and each three-pointer is worth 3 points. If she made 1 free throw and 3 field goals, then solving the equation

$$1 + 3(2) + 3x = 13$$

will give us the number of three-pointers she made. Solve the equation to find the number of three-point shots Laura made.   2 three-pointers

83. **Taxes** If 21% of your monthly pay is withheld for federal income taxes and another 8% is withheld for Social Security, state income tax, and other miscellaneous items, leaving you with $987.50 a month in take-home pay, then the amount you earned before the deductions were removed from your check is given by the equation

$$G - 0.21G - 0.08G = 987.5$$

Solve this equation to find your gross income.
$1,391 per month

84. **Rhind Papyrus** The *Rhind Papyrus* is an ancient document that contains mathematical riddles. One problem asks the reader to find a quantity such that when it is added to one-fourth of itself the sum is 15. The equation that describes this situation is

$$x + \frac{1}{4}x = 15$$

Solve this equation.   12

British Museum/Bridgeman Art Library

Maintaining Your Skills

The problems that follow review some of the more important skills you have learned in previous sections and chapters. You can consider the time you spend working these problems as time spent studying for exams.

Apply the distributive property, and then simplify each expression as much as possible.

85. $2(3x - 5)$ $6x - 10$
86. $4(2x - 6)$ $8x - 24$
87. $\frac{1}{2}(3x + 6)$ $\frac{3}{2}x + 3$
88. $\frac{1}{4}(2x + 8)$ $\frac{1}{2}x + 2$
89. $\frac{1}{3}(-3x + 6)$ $-x + 2$
90. $\frac{1}{2}(-2x + 6)$ $-x + 3$

Simplify each expression.

91. $5(2x - 8) - 3$ $10x - 43$
92. $4(3x - 1) + 7$ $12x + 3$
93. $-2(3x + 5) + 3(x - 1)$ $-3x - 13$
94. $6(x + 3) - 2(2x + 4)$ $2x + 10$
95. $7 - 3(2y + 1)$ $-6y + 4$
96. $8 - 5(3y - 4)$ $-15y + 28$

Getting Ready for the Next Section

To understand all of the explanations and examples in the next section you must be able to work the problems below.

Solve each equation.

97. $2x = 4$ 2
98. $3x = 24$ 8
99. $30 = 5x$ 6
100. $0 = 5x$ 0
101. $0.17x = 510$ 3,000
102. $0.1x = 400$ 4,000

Apply the distributive property, then simplify if possible.

103. $3(x - 5) + 4$ $3x - 11$
104. $5(x - 3) + 2$ $5x - 13$
105. $0.09(x + 2,000)$ $0.09x + 180$
106. $0.04(x + 7,000)$ $0.04x + 280$
107. $7 - 3(2y + 1)$ $-6y + 4$
108. $4 - 2(3y + 1)$ $-6y + 2$
109. $3(2x - 5) - (2x - 4)$ $4x - 11$
110. $4(3x - 2) - (6x - 5)$ $6x - 3$

Simplify.

111. $10x + (-5x)$ $5x$
112. $12x + (-7x)$ $5x$
113. $0.08x + 0.09x$ $0.17x$
114. $0.06x + 0.04x$ $0.10x$

2.4 Solving Linear Equations

OBJECTIVES

A Solve a linear equation in one variable.

We will now use the material we have developed in the first three sections of this chapter to build a method for solving any linear equation.

> **DEFINITION** A **linear equation** in one variable is any equation that can be put in the form $ax + b = 0$, where a and b are real numbers and a is not zero.

Each of the equations we will solve in this section is a linear equation in one variable. The steps we use to solve a linear equation in one variable are listed here.

> **Strategy for Solving Linear Equations in One Variable**
>
> ***Step 1a:*** Use the distributive property to separate terms, if necessary.
>
> ***1b:*** If fractions are present, consider multiplying both sides by the LCD to eliminate the fractions. If decimals are present, consider multiplying both sides by a power of 10 to clear the equation of decimals.
>
> ***1c:*** Combine similar terms on each side of the equation.
>
> ***Step 2:*** Use the addition property of equality to get all variable terms on one side of the equation and all constant terms on the other side. A variable term is a term that contains the variable (for example, $5x$). A constant term is a term that does not contain the variable (the number 3, for example).
>
> ***Step 3:*** Use the multiplication property of equality to get x (that is, $1x$) by itself on one side of the equation.
>
> ***Step 4:*** Check your solution in the original equation to be sure that you have not made a mistake in the solution process.

Note

You may have some previous experience solving equations. Even so, you should solve the equations in this section using the method developed here. Your work should look like the examples in the text. If you have learned shortcuts or a different method of solving equations somewhere else, you always can go back to them later. What is important now is that you are able to solve equations by the methods shown here.

As you will see as you work through the examples in this section, it is not always necessary to use all four steps when solving equations. The number of steps used depends on the equation. In Example 1 there are no fractions or decimals in the original equation, so step 1b will not be used. Likewise, after applying the distributive property to the left side of the equation in Example 1, there are no similar terms to combine on either side of the equation, making step 1c also unnecessary.

 EXAMPLE 1 Solve $2(x + 3) = 10$.

SOLUTION To begin, we apply the distributive property to the left side of the equation to separate terms:

Step 1a: $\qquad\qquad 2x + 6 = 10$ $\qquad$ **Distributive property**

Step 2: $\qquad \begin{cases} 2x + 6 + (\mathbf{-6}) = 10 + (\mathbf{-6}) & \textbf{Addition property of equality} \\ \qquad\quad 2x = 4 \end{cases}$

Step 3: $\qquad \begin{cases} \dfrac{\mathbf{1}}{\mathbf{2}}(2x) = \dfrac{\mathbf{1}}{\mathbf{2}}(4) & \textbf{Multiply each side by } \frac{1}{2} \\ \qquad\quad x = 2 & \textbf{The solution is 2} \end{cases}$

The solution to our equation is 2. We check our work (to be sure we have not made either a mistake in applying the properties or an arithmetic mistake) by substituting 2 into our original equation and simplifying each side of the result separately.

Check:

Step 4:

When $x = 2$

the equation $2(x + 3) = 10$

becomes $2(2 + 3) \overset{?}{=} 10$

$2(5) \overset{?}{=} 10$

$10 = 10$ **A true statement**

Our solution checks.

The general method of solving linear equations is actually very simple. It is based on the properties we developed in Chapter 1 and on two very simple new properties. We can add any number to both sides of the equation and multiply both sides by any nonzero number. The equation may change in form, but the solution set will not. If we look back to Example 1, each equation looks a little different from each preceding equation. What is interesting and useful is that each equation says the same thing about x. They all say x is 2. The last equation, of course, is the easiest to read, and that is why our goal is to end up with x by itself.

 EXAMPLE 2   Solve for x: $3(x - 5) + 4 = 13$.

SOLUTION   Our first step will be to apply the distributive property to the left side of the equation:

Step 1a: $3x - 15 + 4 = 13$ **Distributive property**

Step 1c: $3x - 11 = 13$ **Simplify the left side**

Step 2: $3x - 11 + \mathbf{11} = 13 + \mathbf{11}$ **Add 11 to both sides**

$3x = 24$

Step 3: $\dfrac{1}{3}(3x) = \dfrac{1}{3}(24)$ **Multiply both sides by $\frac{1}{3}$**

$x = 8$ **The solution is 8**

Check:

Step 4:

When $x = 8$

the equation $3(x - 5) + 4 = 13$

becomes $3(8 - 5) + 4 \overset{?}{=} 13$

$3(3) + 4 \overset{?}{=} 13$

$9 + 4 \overset{?}{=} 13$

$13 = 13$ **A true statement**

EXAMPLE 3   Solve $5(x - 3) + 2 = 5(2x - 8) - 3$.

SOLUTION   In this case we apply the distributive property on each side of the equation:

Step 1a:	$5x - 15 + 2 = 10x - 40 - 3$	**Distributive property**
Step 1c:	$5x - 13 = 10x - 43$	**Simplify each side**

Step 2:
$$5x + (\mathbf{-5x}) - 13 = 10x + (\mathbf{-5x}) - 43 \qquad \text{Add } -5x \text{ to both sides}$$
$$-13 = 5x - 43$$
$$-13 + \mathbf{43} = 5x - 43 + \mathbf{43} \qquad \text{Add 43 to both sides}$$
$$30 = 5x$$

Step 3:
$$\frac{1}{5}(30) = \frac{1}{5}(5x) \qquad \text{Multiply both sides by } \frac{1}{5}$$
$$6 = x \qquad \text{The solution is 6}$$

Check: Replacing x with 6 in the original equation, we have

Step 4:
$$5(6 - 3) + 2 \stackrel{?}{=} 5(2 \cdot 6 - 8) - 3$$
$$5(3) + 2 \stackrel{?}{=} 5(12 - 8) - 3$$
$$5(3) + 2 \stackrel{?}{=} 5(4) - 3$$
$$15 + 2 \stackrel{?}{=} 20 - 3$$
$$17 = 17 \qquad \text{A true statement}$$

> **Note**
> It makes no difference on which side of the equal sign x ends up. Most people prefer to have x on the left side because we read from left to right, and it seems to sound better to say x is 6 rather than 6 is x. Both expressions, however, have exactly the same meaning.

EXAMPLE 4   Solve the equation $0.08x + 0.09(x + 2{,}000) = 690$.

SOLUTION   We can solve the equation in its original form by working with the decimals, or we can eliminate the decimals first by using the multiplication property of equality and solving the resulting equation. Both methods follow.

Method 1   Working with the decimals.

	$0.08x + 0.09(x + 2{,}000) = 690$	**Original equation**
Step 1a:	$0.08x + 0.09x + 0.09(2{,}000) = 690$	**Distributive property**
Step 1c:	$0.17x + 180 = 690$	**Simplify the left side**

Step 2:
$$0.17x + 180 + (\mathbf{-180}) = 690 + (\mathbf{-180}) \qquad \text{Add } -180 \text{ to each side}$$
$$0.17x = 510$$

Step 3:
$$\frac{0.17x}{\mathbf{0.17}} = \frac{510}{\mathbf{0.17}} \qquad \text{Divide each side by 0.17}$$
$$x = 3{,}000$$

Note that we divided each side of the equation by 0.17 to obtain the solution. This is still an application of the multiplication property of equality because dividing by 0.17 is equivalent to multiplying by $\frac{1}{0.17}$.

Method 2 Eliminating the decimals in the beginning.

	$0.08x + 0.09(x + 2{,}000) = 690$	**Original equation**
Step 1a:	$0.08x + 0.09x + 180 = 690$	**Distributive property**
Step 1b:	$100(0.08x + 0.09x + 180) = 100(690)$	**Multiply both sides by 100**
	$8x + 9x + 18{,}000 = 69{,}000$	
Step 1c:	$17x + 18{,}000 = 69{,}000$	**Simplify the left side**
Step 2:	$17x = 51{,}000$	**Add $-18{,}000$ to each side**
Step 3:	$\dfrac{17x}{17} = \dfrac{51{,}000}{17}$	**Divide each side by 17**
	$x = 3{,}000$	

Check:

Step 4: Substituting 3,000 for x in the original equation, we have

$$0.08(3{,}000) + 0.09(3{,}000 + 2{,}000) \stackrel{?}{=} 690$$

$$0.08(3{,}000) + 0.09(5{,}000) \stackrel{?}{=} 690$$

$$240 + 450 \stackrel{?}{=} 690$$

$$690 = 690 \qquad \textbf{A true statement}$$

 EXAMPLE 5 Solve $7 - 3(2y + 1) = 16$.

SOLUTION We begin by multiplying -3 times the sum of $2y$ and 1:

Step 1a:	$7 - 6y - 3 = 16$	**Distributive property**
Step 1c:	$-6y + 4 = 16$	**Simplify the left side**
Step 2:	$-6y + 4 + (-4) = 16 + (-4)$	**Add -4 to both sides**
	$-6y = 12$	
Step 3:	$-\dfrac{1}{6}(-6y) = -\dfrac{1}{6}(12)$	**Multiply both sides by $-\frac{1}{6}$**
	$y = -2$	

Step 4: Replacing y with -2 in the original equation yields a true statement.

   There are two things to notice about the example that follows: first, the distributive property is used to remove parentheses that are preceded by a negative sign, and, second, the addition property and the multiplication property are not shown in as much detail as in the previous examples.

 EXAMPLE 6 Solve $3(2x - 5) - (2x - 4) = 6 - (4x + 5)$.

SOLUTION When we apply the distributive property to remove the grouping symbols and separate terms, we have to be careful with the signs. Remember, we can think of $-(2x - 4)$ as $-1(2x - 4)$, so that

$$-(2x - 4) = -1(2x - 4) = -2x + 4$$

It is not uncommon for students to make a mistake with this type of simplification and write the result as $-2x - 4$, which is incorrect. Here is the complete solution to our equation:

$$3(2x - 5) - (2x - 4) = 6 - (4x + 5)$$ **Original equation**

$$6x - 15 - 2x + 4 = 6 - 4x - 5$$ **Distributive property**

$$4x - 11 = -4x + 1$$ **Simplify each side**

$$8x - 11 = 1$$ **Add 4x to each side**

$$8x = 12$$ **Add 11 to each side**

$$x = \frac{12}{8}$$ **Multiply each side by $\frac{1}{8}$**

$$x = \frac{3}{2}$$ **Reduce to lowest terms**

The solution, $\frac{3}{2}$, checks when replacing x in the original equation.

LINKING OBJECTIVES AND EXAMPLES

Next to each **objective** we have listed the examples that are best described by that objective.

A   1–6

GETTING READY FOR CLASS

After reading through the preceding section, respond in your own words and in complete sentences.

1. What is the first step in solving a linear equation containing parentheses?
2. What is the last step in solving a linear equation?
3. Explain in words how you would solve the equation $2x - 3 = 8$.
4. If an equation contains decimals, what can you do to eliminate the decimals?

Problem Set 2.4

Online support materials can be found at www.thomsonedu.com/login

Solve each of the following equations using the four steps shown in this section.

1. $2(x + 3) = 12$ 3

2. $3(x - 2) = 6$ 4

3. $6(x - 1) = -18$ −2

4. $4(x + 5) = 16$ −1

5. $2(4a + 1) = -6$ −1

6. $3(2a - 4) = 12$ 4

7. $14 = 2(5x - 3)$ 2

8. $-25 = 5(3x + 4)$ −3

9. $-2(3y + 5) = 14$ −4

10. $-3(2y - 4) = -6$ 3

11. $-5(2a + 4) = 0$ −2

12. $-3(3a - 6) = 0$ 2

13. $1 = \frac{1}{2}(4x + 2)$ 0

14. $1 = \frac{1}{3}(6x + 3)$ 0

15. $3(t - 4) + 5 = -4$ 1

16. $5(t - 1) + 6 = -9$ −2

Solve each equation.

▶ **17.** $4(2y + 1) - 7 = 1$ $\frac{1}{2}$

18. $6(3y + 2) - 8 = -2$ $-\frac{1}{3}$

19. $\frac{1}{2}(x - 3) = \frac{1}{4}(x + 1)$ 7

20. $\frac{1}{3}(x - 4) = \frac{1}{2}(x - 6)$ 10

21. $-0.7(2x - 7) = 0.3(11 - 4x)$ 8

22. $-0.3(2x - 5) = 0.7(3 - x)$ 6

23. $-2(3y + 1) = 3(1 - 6y) - 9$ $-\frac{1}{3}$

24. $-5(4y - 3) = 2(1 - 8y) + 11$ $\frac{1}{2}$

▶ **25.** $\frac{3}{4}(8x - 4) + 3 = \frac{2}{5}(5x + 10) - 1$ $\frac{3}{4}$

26. $\frac{5}{6}(6x + 12) + 1 = \frac{2}{3}(9x - 3) + 5$ 8

= Videos available by instructor request
▶ = Online student support materials available at www.thomsonedu.com/login

27. $0.06x + 0.08(100 - x) = 6.5$ 75

28. $0.05x + 0.07(100 - x) = 6.2$ 40

29. $6 - 5(2a - 3) = 1$ 2   **30.** $-8 - 2(3 - a) = 0$ 7

31. $0.2x - 0.5 = 0.5 - 0.2(2x - 13)$ 6

32. $0.4x - 0.1 = 0.7 - 0.3(6 - 2x)$ 5

33. $2(t - 3) + 3(t - 2) = 28$ 8

34. $-3(t - 5) - 2(2t + 1) = -8$ 3

35. $5(x - 2) - (3x + 4) = 3(6x - 8) + 10$ 0

36. $3(x - 1) - (4x - 5) = 2(5x - 1) - 7$ 1

37. $2(5x - 3) - (2x - 4) = 5 - (6x + 1)$ $\frac{3}{7}$

38. $3(4x - 2) - (5x - 8) = 8 - (2x + 3)$ $\frac{1}{3}$

39. $-(3x + 1) - (4x - 7) = 4 - (3x + 2)$ 1

40. $-(6x + 2) - (8x - 3) = 8 - (5x + 1)$ $-\frac{2}{3}$

▸ **41.** $x + (2x - 1) = 2$ 1   ▸ **42.** $x + (5x + 2) = 20$ 3

▸ **43.** $x - (3x + 5) = -3$ -1 ▸ **44.** $x - (4x - 1) = 7$ -2

▸ **45.** $15 = 3(x - 1)$ 6   ▸ **46.** $12 = 4(x - 5)$ 8

▸ **47.** $4x - (-4x + 1) = 5$ $\frac{3}{4}$ ▸ **48.** $-2x - (4x - 8) = -1$ $\frac{3}{2}$

49. $5x - 8(2x - 5) = 7$ 3   **50.** $3x + 4(8x - 15) = 10$ 2

51. $7(2y - 1) - 6y = -1$ $\frac{3}{4}$ **52.** $4(4y - 3) + 2y = 3$ $\frac{5}{6}$

▸ **53.** $0.2x + 0.5(12 - x) = 3.6$ 8

▸ **54.** $0.3x + 0.6(25 - x) = 12$ 10

▸ **55.** $0.5x + 0.2(18 - x) = 5.4$ 6

▸ **56.** $0.1x + 0.5(40 - x) = 32$ -30

▸ **57.** $x + (x + 3)(-3) = x - 3$ -2

▸ **58.** $x - 2(x + 2) = x - 2$ -1

▸ **59.** $5(x + 2) + 3(x - 1) = -9$ -2

▸ **60.** $4(x + 1) + 3(x - 3) = 2$ 1

▸ **61.** $3(x - 3) + 2(2x) = 5$ 2   **62.** $2(x - 2) + 3(5x) = 30$ 2

63. $5(y + 2) = 4(y + 1)$ -6 **64.** $3(y - 3) = 2(y - 2)$ 5

65. $3x + 2(x - 2) = 6$ 2   **66.** $5x - (x - 5) = 25$ 5

67. $50(x - 5) = 30(x + 5)$ 20   **68.** $34(x - 2) = 26(x + 2)$ 15

▸ **69.** $0.08x + 0.09(x + 2{,}000) = 860$ 4,000

▸ **70.** $0.11x + 0.12(x + 4{,}000) = 940$ 2,000

▸ **71.** $0.10x + 0.12(x + 500) = 214$ 700

72. $0.08x + 0.06(x + 800) = 104$ 400

73. $5x + 10(x + 8) = 245$ 11   **74.** $5x + 10(x + 7) = 175$ 7

75. $5x + 10(x + 3) + 25(x + 5) = 435$ 7

76. $5(x + 3) + 10x + 25(x + 7) = 390$ 5

The next two problems are intended to give you practice reading, and paying attention to, the instructions that accompany the problems you are working. Working these problems is an excellent way to get ready for a test or a quiz.

▸ **77.** Work each problem according to the instructions given.
  a. Solve: $4x - 5 = 0$ $\frac{5}{4} = 1.25$
  b. Solve: $4x - 5 = 25$ $\frac{15}{2} = 7.5$
  c. Add: $(4x - 5) + (2x + 25)$ $6x + 20$
  d. Solve: $4x - 5 = 2x + 25$ 15
  e. Multiply: $4(x - 5)$ $4x - 20$
  f. Solve: $4(x - 5) = 2x + 25$ $\frac{45}{2} = 22.5$

78. Work each problem according to the instructions given.
  a. Solve: $3x + 6 = 0$ -2
  b. Solve: $3x + 6 = 4$ $-\frac{2}{3}$
  c. Add: $(3x + 6) + (7x + 4)$ $10x + 10$
  d. Solve: $3x + 6 = 7x + 4$ $\frac{1}{2}$
  e. Multiply: $3(x + 6)$ $3x + 18$
  f. Solve: $3(x + 6) = 7x + 4$ $\frac{7}{2}$

Maintaining Your Skills

The problems that follow review some of the more important skills you have learned in previous sections and chapters. You can consider the time you spend working these problems as time spent studying for exams.

Multiply.

79. $\frac{1}{2}(3)$ $\frac{3}{2}$ **80.** $\frac{1}{3}(2)$ $\frac{2}{3}$

81. $\frac{2}{3}(6)$ 4   **82.** $\frac{3}{2}(4)$ 6

83. $\frac{5}{9} \cdot \frac{9}{5}$ 1   **84.** $\frac{3}{7} \cdot \frac{7}{3}$ 1

▸ **Chalkboard Problem**
Problem 77 is a Chalkboard Problem. In addition to requiring students to pay attention to the instructions, it reviews combining similar terms, the distributive property, and solving simple equations. I like to work some problems they have already done as I work my way into the new material. You will see this continuous review integrated throughout this book.

Foreshadowing Problems
Problems 41–76 are the equations students need later in this chapter, when they solve applications problem and in later chapters as well.

Fill in the tables by finding the value of each expression for the given values of the variables.

85.

x	$3(x + 2)$	$3x + 2$	$3x + 6$
0	6	2	6
1	9	5	9
2	12	8	12
3	15	11	15

86.

x	$7(x - 5)$	$7x - 5$	$7x - 35$
-3	-56	-26	-56
-2	-49	-19	-49
-1	-42	-12	-42
0	-35	-5	-35

87.

a	$(2a + 1)^2$	$4a^2 + 4a + 1$
1	9	9
2	25	25
3	49	49

88.

a	$(a + 1)^3$	$a^3 + 3a^2 + 3a + 1$
1	8	8
2	27	27
3	64	64

Getting Ready for the Next Section

To understand all of the explanations and examples in the next section you must be able to work the problems below.

Solve each equation.

89. $40 = 2x + 12$ 14

90. $80 = 2x + 12$ 34

91. $12 + 2y = 6$ -3

92. $3x + 18 = 6$ -4

93. $24x = 6$ $\dfrac{1}{4}$

94. $45 = 0.75x$ 60

95. $70 = x \cdot 210$ $\dfrac{1}{3}$

96. $15 = x \cdot 80$ $\dfrac{3}{16}$

Apply the distributive property.

97. $\dfrac{1}{2}(-3x + 6)$ $-\dfrac{3}{2}x + 3$

98. $-\dfrac{1}{4}(-5x + 20)$ $\dfrac{5}{4}x - 5$

2.5 Formulas

OBJECTIVES

A Find the value of a variable in a formula given replacements for the other variables.

B Solve a formula for one of its variables.

C Find the complement and supplement of an angle.

D Solve simple percent problems.

In this section we continue solving equations by working with formulas. To begin, here is the definition of a formula.

> **DEFINITION** In mathematics, a **formula** is an equation that contains more than one variable.

The equation $P = 2l + 2w$, which tells us how to find the perimeter of a rectangle, is an example of a formula.

To begin our work with formulas, we will consider some examples in which we are given numerical replacements for all but one of the variables.

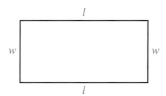

EXAMPLE 1 The perimeter P of a rectangular livestock pen is 40 feet. If the width w is 6 feet, find the length.

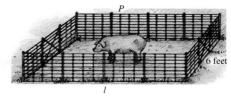

SOLUTION First we substitute 40 for P and 6 for w in the formula $P = 2l + 2w$. Then we solve for l:

When $P = 40$ and $w = 6$

the formula $P = 2l + 2w$

becomes $40 = 2l + 2(6)$

or $40 = 2l + 12$ **Multiply 2 and 6**

$28 = 2l$ **Add -12 to each side**

$14 = l$ **Multiply each side by $\frac{1}{2}$**

To summarize our results, if a rectangular pen has a perimeter of 40 feet and a width of 6 feet, then the length must be 14 feet.

 EXAMPLE 2 Find y when $x = 4$ in the formula $3x + 2y = 6$.

SOLUTION We substitute 4 for x in the formula and then solve for y:

When $x = 4$

the formula $3x + 2y = 6$

becomes $3(4) + 2y = 6$

or $12 + 2y = 6$ **Multiply 3 and 4**

$2y = -6$ **Add -12 to each side**

$y = -3$ **Multiply each side by $\frac{1}{2}$**

In the next examples we will solve a formula for one of its variables without being given numerical replacements for the other variables.

Consider the formula for the area of a triangle:

$A = \frac{1}{2}bh$

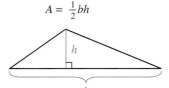

where A = area, b = length of the base, and h = height of the triangle.

Suppose we want to solve this formula for h. What we must do is isolate the variable h on one side of the equal sign. We begin by multiplying both sides by 2:

$$2 \cdot A = 2 \cdot \frac{1}{2}bh$$

$$2A = bh$$

Then we divide both sides by b:

$$\frac{2A}{b} = \frac{bh}{b}$$

$$h = \frac{2A}{b}$$

The original formula $A = \frac{1}{2}bh$ and the final formula $h = \frac{2A}{b}$ both give the same relationship among A, b, and h. The first one has been solved for A and the second one has been solved for h.

> **Rule** To solve a formula for one of its variables, we must isolate that variable on either side of the equal sign. All other variables and constants will appear on the other side.

 EXAMPLE 3 Solve $3x + 2y = 6$ for y.

SOLUTION To solve for y, we must isolate y on the left side of the equation. To begin, we use the addition property of equality to add $-3x$ to each side:

$$3x + 2y = 6 \qquad \text{Original formula}$$
$$3x + (\mathbf{-3x}) + 2y = (\mathbf{-3x}) + 6 \qquad \text{Add } -3x \text{ to each side}$$
$$2y = -3x + 6 \qquad \text{Simplify the left side}$$
$$\frac{\mathbf{1}}{\mathbf{2}}(2y) = \frac{\mathbf{1}}{\mathbf{2}}(-3x + 6) \qquad \text{Multiply each side by } \frac{1}{2}$$
$$y = -\frac{3}{2}x + 3 \qquad \text{Multiplication}$$

 EXAMPLE 4 Solve $h = vt - 16t^2$ for v.

SOLUTION Let's begin by interchanging the left and right sides of the equation. That way, the variable we are solving for, v, will be on the left side.

$$vt - 16t^2 = h \qquad \text{Exchange sides}$$
$$vt - 16t^2 + \mathbf{16t^2} = h + \mathbf{16t^2} \qquad \text{Add } 16t^2 \text{ to each side}$$
$$vt = h + 16t^2$$
$$\frac{vt}{\mathbf{t}} = \frac{h + 16t^2}{\mathbf{t}} \qquad \text{Divide each side by } t$$
$$v = \frac{h + 16t^2}{t}$$

We know we are finished because we have isolated the variable we are solving for on the left side of the equation and it does not appear on the other side.

EXAMPLE 5 Solve for y: $\frac{y-1}{x} = \frac{3}{2}$.

SOLUTION Although we will do more extensive work with formulas of this form later in the book, we need to know how to solve this particular formula for y in order to understand some things in the next chapter. We begin by multiplying each side of the formula by x. Doing so will simplify the left side of the equation and make the rest of the solution process simple.

$$\frac{y-1}{x} = \frac{3}{2} \qquad \text{Original formula}$$

$$x \cdot \frac{y-1}{x} = \frac{3}{2} \cdot x \qquad \textbf{Multiply each side by } x$$

$$y - 1 = \frac{3}{2}x \qquad \textbf{Simplify each side}$$

$$y = \frac{3}{2}x + 1 \qquad \textbf{Add 1 to each side}$$

This is our solution. If we look back to the first step, we can justify our result on the left side of the equation this way: Dividing by x is equivalent to multiplying by its reciprocal $\frac{1}{x}$. Here is what it looks like when written out completely:

$$x \cdot \frac{y-1}{x} = x \cdot \frac{1}{x} \cdot (y-1) = 1(y-1) = y - 1 \qquad$$

FACTS FROM GEOMETRY

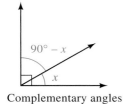

Complementary angles

Supplementary angles

More on Complementary and Supplementary Angles

In Chapter 1 we defined complementary angles as angles that add to 90°; that is, if x and y are complementary angles, then

$$x + y = 90°$$

If we solve this formula for y, we obtain a formula equivalent to our original formula:

$$y = 90° - x$$

Because y is the complement of x, we can generalize by saying that the complement of angle x is the angle $90° - x$. By a similar reasoning process, we can say that the supplement of angle x is the angle $180° - x$. To summarize, if x is an angle, then

The complement of x is $90° - x$, and

The supplement of x is $180° - x$

If you go on to take a trigonometry class, you will see this formula again.

EXAMPLE 6   Find the complement and the supplement of 25°.

SOLUTION   We can use the formulas above with $x = 25°$.

The complement of 25° is $90° - 25° = 65°$.

The supplement of 25° is $180° - 25° = 155°$.

Basic Percent Problems

The last examples in this section show how basic percent problems can be translated directly into equations. To understand these examples, you must recall that percent means "per hundred" that is, 75% is the same as $\frac{75}{100}$, 0.75, and, in reduced fraction form, $\frac{3}{4}$. Likewise, the decimal 0.25 is equivalent to 25%. To change a decimal to a percent, we move the decimal point two places to the right and write the % symbol. To change from a percent to a decimal, we drop the % symbol and move the decimal point two places to the left. The table that follows gives some of the most commonly used fractions and decimals and their equivalent percents.

Fraction	Decimal	Percent
$\frac{1}{2}$	0.5	50%
$\frac{1}{4}$	0.25	25%
$\frac{3}{4}$	0.75	75%
$\frac{1}{3}$	0.333 . . .	$33\frac{1}{3}$%
$\frac{2}{3}$	0.666 . . .	$66\frac{2}{3}$%
$\frac{1}{5}$	0.2	20%
$\frac{2}{5}$	0.4	40%
$\frac{3}{5}$	0.6	60%
$\frac{4}{5}$	0.8	80%

EXAMPLE 7   What number is 25% of 60?

SOLUTION   To solve a problem like this, we let x = the number in question (that is, the number we are looking for). Then, we translate the sentence directly into an equation by using an equal sign for the word "is" and multiplication for the word "of." Here is how it is done:

$$\underbrace{\text{What number}}_{x} \quad \overset{\downarrow}{\underset{=}{\text{is}}} \; \overset{\downarrow}{\underset{0.25}{25\%}} \; \overset{\downarrow}{\underset{\cdot}{\text{of}}} \; \overset{\downarrow}{\underset{60}{60?}}$$

$$x = 15$$

Notice that we must write 25% as a decimal in order to do the arithmetic in the problem.
   The number 15 is 25% of 60.

EXAMPLE 8   What percent of 24 is 6?

SOLUTION   Translating this sentence into an equation, as we did in Example 7, we have:

$$\underbrace{\text{What percent}}_{x} \quad \overset{\downarrow}{\underset{\cdot}{\text{of}}} \; \overset{\downarrow}{\underset{24}{24}} \; \overset{\downarrow}{\underset{=}{\text{is}}} \; \overset{\downarrow}{\underset{6}{6?}}$$

$$\text{or} \qquad 24x = 6$$

Next, we multiply each side by $\frac{1}{24}$. (This is the same as dividing each side by 24.)

$$\frac{1}{24}(24x) = \frac{1}{24}(6)$$

$$x = \frac{6}{24}$$

$$= \frac{1}{4}$$

$$= 0.25, \text{ or } 25\%$$

The number 6 is 25% of 24.

EXAMPLE 9 45 is 75% of what number?

SOLUTION Again, we translate the sentence directly:

$$45 \text{ is } 75\% \text{ of what number?}$$
$$45 = 0.75 \cdot x$$

Next, we multiply each side by $\frac{1}{0.75}$ (which is the same as dividing each side by 0.75):

$$\frac{1}{0.75}(45) = \frac{1}{0.75}(0.75x)$$

$$\frac{45}{0.75} = x$$

$$60 = x$$

The number 45 is 75% of 60.

We can solve application problems involving percent by translating each problem into one of the three basic percent problems shown in Examples 7, 8, and 9.

 EXAMPLE 10 The American Dietetic Association (ADA) recommends eating foods in which the calories from fat are less than 30% of the total calories. The nutrition labels from two kinds of granola bars are shown in Figure 1. For each bar, what percent of the total calories come from fat?

SOLUTION The information needed to solve this problem is located toward the top of each label. Each serving of Bar I contains 210 calories, of which 70 calories come from fat. To find the percent of total calories that come from fat, we must answer this question:

70 is what percent of 210?

Nutrition Facts	
Serving Size 2 bars (47g)	
Servings Per Container 6	
Amount Per Serving	
Calories	210
Calories from Fat	70
	% Daily Value*
Total Fat 8g	12%
Saturated Fat 1g	5%
Cholesterol 0mg	0%
Sodium 150mg	6%
Total Carbohydrate 32g	11%
Dietary Fiber 2g	10%
Sugars 12g	
Protein 4g	
* Percent Daily Values are based on a 2,000 calorie diet. Your daily values may be higher or lower depending on your calorie needs.	

Nutrition Facts	
Serving Size 1 bar (21g)	
Servings Per Container 8	
Amount Per Serving	
Calories	80
Calories from Fat	15
	% Daily Value*
Total Fat 1.5g	2%
Saturated Fat 0g	0%
Cholesterol 0mg	0%
Sodium 60mg	3%
Total Carbohydrate 16g	5%
Dietary Fiber 1g	4%
Sugars 5g	
Protein 2g	
* Percent Daily Values are based on a 2,000 calorie diet. Your daily values may be higher or lower depending on your calorie needs.	

FIGURE 1

For Bar II, one serving contains 80 calories, of which 15 calories come from fat. To find the percent of total calories that come from fat, we must answer this question:

15 is what percent of 80?

Translating each equation into symbols, we have

70 is what percent of 210	15 is what percent of 80
$70 = x \cdot 210$	$15 = x \cdot 80$
$x = \dfrac{70}{210}$	$x = \dfrac{15}{80}$
$x = 0.33$ to the nearest hundredth	$x = 0.19$ to the nearest hundredth
$x = 33\%$	$x = 19\%$

Comparing the two bars, 33% of the calories in Bar I are fat calories, whereas 19% of the calories in Bar II are fat calories. According to the ADA, Bar II is the healthier choice.

LINKING OBJECTIVES AND EXAMPLES

Next to each **objective** we have listed the examples that are best described by that objective.

A	1, 2
B	3–5
C	6
D	7–10

GETTING READY FOR CLASS

After reading through the preceding section, respond in your own words and in complete sentences.

1. What is a formula?

2. How do you solve a formula for one of its variables?

3. What are complementary angles?

4. What does percent mean?

Problem Set 2.5

Online support materials can be found at www.thomsonedu.com/login

Use the formula $P = 2l + 2w$ to find the length l of a rectangular lot if

1. The width w is 50 feet and the perimeter P is 300 feet   100 feet

2. The width w is 75 feet and the perimeter P is 300 feet   75 feet

3. For the equation $2x + 3y = 6$,

  a. Find y when x is 0.   2

  b. Find x when y is 1. $\frac{3}{2}$

▶ **c.** Find y when x is 3.   0

4. For the equation $2x - 5y = 20$,

  a. Find y when x is 0. -4

  b. Find x when y is 0.   10

  c. Find x when y is 2.   15

5. For the equation $y = -\dfrac{1}{3}x + 2$,

  a. Find y when x is 0.   2

  b. Find x when y is 3. -3

  c. Find y when x is 3.   1

6. For the equation $y = -\dfrac{2}{3}x + 1$,

  a. Find y when x is 0.   1

  b. Find x when y is -1.   3

  c. Find y when x is -3.   3

Use the equation $y = (x + 1)^2 - 3$ to find the value of y when

7. $x = -2$ -2 **8.** $x = -1$ -3

9. $x = 1$ 1   **10.** $x = 2$ 6

11. Use the formula $y = \dfrac{20}{x}$ to find y when

  a. $x = 10$ 2   **b.** $x = 5$ 4

12. Use the formula $y = 2x^2$ to find y when

  a. $x = 5$ 50   **b.** $x = -6$ 72

= Videos available by instructor request   ▶ = Chalkboard Problem
▶ = Online student support materials available at www.thomsonedu.com/login

13. Use the formula $y = Kx$ to find K when

   a. $y = 15$ and $x = 3$ 5   **b.** $y = 72$ and $x = 4$ 18

14. Use the formula $y = Kx^2$ to find K when

   a. $y = 32$ and $x = 4$ 2   **b.** $y = 45$ and $x = 3$ 5

15. If $y = \dfrac{K}{x}$, find K if

   a. x is 5 and y is 4.   20
   b. x is 5 and y is 15.   75

16. If $I = \dfrac{K}{d^2}$, find K if

   a. $I = 200$ and $d = 10$.   20,000
   b. $I = 200$ and $d = 5$.   5,000

Solve each of the following for the indicated variable.

▶ **17.** $A = lw$ for l $l = \dfrac{A}{w}$ **18.** $d = rt$ for r $r = \dfrac{d}{t}$

19. $V = lwh$ for h $h = \dfrac{V}{lw}$ **20.** $PV = nRT$ for P $P = \dfrac{nRT}{V}$

▶ **21.** $P = a + b + c$ for a $a = P - b - c$

22. $P = a + b + c$ for b $b = P - a - c$

23. $x - 3y = -1$ for x $x = 3y - 1$

24. $x + 3y = 2$ for x $x = -3y + 2$

25. $-3x + y = 6$ for y $y = 3x + 6$

26. $2x + y = -17$ for y $y = -2x - 17$

27. $2x + 3y = 6$ for y $y = -\frac{2}{3}x + 2$

28. $4x + 5y = 20$ for y $y = -\frac{4}{5}x + 4$

29. $P = 2l + 2w$ for w $w = \dfrac{P - 2l}{2}$

30. $P = 2l + 2w$ for l $l = \dfrac{P - 2w}{2}$

31. $h = vt + 16t^2$ for v $v = \dfrac{h - 16t^2}{t}$

32. $h = vt - 16t^2$ for v $v = \dfrac{h + 16t^2}{t}$

33. $A = \pi r^2 + 2\pi rh$ for h $h = \dfrac{A - \pi r^2}{2\pi r}$

34. $A = 2\pi r^2 + 2\pi rh$ for h $h = \dfrac{A - 2\pi r^2}{2\pi r}$

▶ **35.** Solve for y.

   a. $y - 3 = -2(x + 4)$ $y = -2x - 5$
   b. $y - 5 = 4(x - 3)$ $y = 4x - 7$

36. Solve for y.

   a. $y + 1 = -\dfrac{2}{3}(x - 3)$ $y = -\frac{2}{3}x + 1$

   b. $y - 3 = -\dfrac{2}{3}(x + 3)$ $y = -\frac{2}{3}x + 1$

37. Solve for y.

   a. $y - 1 = \dfrac{3}{4}(x - 1)$ $y = \frac{3}{4}x + \frac{1}{4}$

   b. $y + 2 = \dfrac{3}{4}(x - 4)$ $y = \frac{3}{4}x - 5$

38. Solve for y.

   a. $y + 3 = \dfrac{3}{2}(x - 2)$ $y = \frac{3}{2}x - 6$

   b. $y + 4 = \dfrac{4}{3}(x - 3)$ $y = \frac{4}{3}x - 8$

39. Solve for y.

   a. $\dfrac{y - 1}{x} = \dfrac{3}{5}$ $y = \frac{3}{5}x + 1$

   b. $\dfrac{y - 2}{x} = \dfrac{1}{2}$ $y = \frac{1}{2}x + 2$

   c. $\dfrac{y - 3}{x} = 4$ $y = 4x + 3$

40. Solve for y.

   a. $\dfrac{y + 1}{x} = -\dfrac{3}{5}$ $y = -\frac{3}{5}x - 1$

   b. $\dfrac{y + 2}{x} = -\dfrac{1}{2}$ $y = -\frac{1}{2}x - 2$

   c. $\dfrac{y + 3}{x} = -4$ $y = -4x - 3$

Solve each formula for y.

41. $\dfrac{x}{7} - \dfrac{y}{3} = 1$ $y = \frac{3}{7}x - 3$ **42.** $\dfrac{x}{4} - \dfrac{y}{9} = 1$ $y = \frac{9}{4}x - 9$

43. $-\dfrac{1}{4}x + \dfrac{1}{8}y = 1$ $y = 2x + 8$

44. $-\dfrac{1}{9}x + \dfrac{1}{3}y = 1$ $y = \frac{1}{3}x + 3$

The next two problems are intended to give you practice reading, and paying attention to, the instructions that accompany the problems you are working. As we have mentioned previously, working these problems is an excellent way to get ready for a test or a quiz.

▶ **45.** Work each problem according to the instructions given.

   a. Solve: $4x + 5 = 20$ $\frac{15}{4} = 3.75$

   b. Find the value of $4x + 5$ when x is 3.   17

   c. Solve for y: $4x + 5y = 20$ $y = -\frac{4}{5}x + 4$

   d. Solve for x: $4x + 5y = 20$ $x = -\frac{5}{4}y + 5$

Foreshadowing Problems
The problem set foreshadows many problems students will need later in the course. Problems 37–38 get students ready for the problems in Chapter 3, when they will use the point-slope form of the equation of a line.

46. Work each problem according to the instructions given.

   a. Solve: $-2x + 1 = 4$ $-\frac{3}{2} = -1.5$

   b. Find the value of $-2x + 1$ when x is 8. -15

   c. Solve for y: $-2x + y = 20$ $y = 2x + 20$

   d. Solve for x: $-2x + y = 20$ $x = \frac{1}{2}y - 10$

Find the complement and the supplement of each angle.

47. 30°   60°; 150°   **48.** 60°   30°; 120°

49. 45°   45°; 135°   **50.** 15°   75°; 165°

Translate each of the following into an equation, and then solve that equation.

51. What number is 25% of 40?   10

52. What number is 75% of 40?   30

53. What number is 12% of 2,000?   240

54. What number is 9% of 3,000?   270

55. What percent of 28 is 7?   25%

56. What percent of 28 is 21?   75%

57. What percent of 40 is 14?   35%

58. What percent of 20 is 14?   70%

59. 32 is 50% of what number?   64

60. 16 is 50% of what number?   32

61. 240 is 12% of what number?   2,000

62. 360 is 12% of what number?   3,000

Applying the Concepts

More About Temperatures As we mentioned in Chapter 1, in the U.S. system, temperature is measured on the Fahrenheit scale. In the metric system, temperature is measured on the Celsius scale. On the Celsius scale, water boils at 100 degrees and freezes at 0 degrees. To denote a temperature of 100 degrees on the Celsius scale, we write

      100°C, which is read "100 degrees Celsius"

   Table 1 is intended to give you an intuitive idea of the relationship between the two temperature scales. Table 2 gives the formulas, in both symbols and words, that are used to convert between the two scales.

63. Let F = 212 in the formula $C = \frac{5}{9}(F - 32)$, and solve for C. Does the value of C agree with the information in Table 1?   100°C; yes

TABLE 1

Situation	Temperature	
	Fahrenheit	Celsius
Water freezes	32°F	0°C
Room temperature	68°F	20°C
Normal body temperature	98.6°F	37°C
Water boils	212°F	100°C
Bake cookies	365°F	185°C

TABLE 2

To Convert from	Formula in Symbols	Formula in Words
Fahrenheit to Celsius	$C = \frac{5}{9}(F - 32)$	Subtract 32, multiply by 5, then divide by 9.
Celsius to Fahrenheit	$F = \frac{9}{5}C + 32$	Multiply by $\frac{9}{5}$, then add 32.

64. Let C = 100 in the formula $F = \frac{9}{5}C + 32$, and solve for F. Does the value of F agree with the information in Table 1?   212°F; yes

65. Let F = 68 in the formula $C = \frac{5}{9}(F - 32)$, and solve for C. Does the value of C agree with the information in Table 1?   20°C; yes

66. Let C = 37 in the formula $F = \frac{9}{5}C + 32$, and solve for F. Does the value of F agree with the information in Table 1?   98.6°F; yes

67. Solve the formula $F = \frac{9}{5}C + 32$ for C. $C = \frac{5}{9}(F - 32)$

68. Solve the formula $C = \frac{5}{9}(F - 32)$ for F. $F = \frac{9}{5}C + 32$

Nutrition Labels The nutrition label in Figure 2 is from a quart of vanilla ice cream. The label in Figure 3 is from a pint of vanilla frozen yogurt. Use the information on these labels for problems 69–72. Round your answers to the nearest tenth of a percent.

Nutrition Facts
Serving Size 1/2 cup (65g)
Servings 8

Amount/Serving		
Calories 150		Calories from Fat 90
		% Daily Value*
Total Fat 10g		**16%**
Saturated Fat 6g		**32%**
Cholesterol 35mg		**12%**
Sodium 30mg		**1%**
Total Carbohydrate 14g		**5%**
Dietary Fiber 0g		**0%**
Sugars 11g		
Protein 2g		
Vitamin A 6%	•	Vitamin C 0%
Calcium 6%	•	Iron 0%
* Percent Daily Values are based on a 2,000 calorie diet.		

FIGURE 2   **Vanilla ice cream**

Nutrition Facts
Serving Size 1/2 cup (98g)
Servings Per Container 4

Amount Per Serving

Calories 160   Calories from Fat 25

% Daily Value*

Total Fat 2.5g	4%
Saturated Fat 1.5g	7%
Cholesterol 45mg	15%
Sodium 55mg	2%
Total Carbohydrate 26g	9%
Dietary Fiber 0g	0%
Sugars 19g	
Protein 8g	

Vitamin A 0%   •   Vitamin C 0%

Calcium 25%   •   Iron 0%
* Percent Daily Values are based on a 2,000 calorie diet.

FIGURE 3   **Vanilla frozen yogurt**

69. What percent of the calories in one serving of the vanilla ice cream are fat calories? 60%

70. What percent of the calories in one serving of the frozen yogurt are fat calories? 15.6%

71. One serving of frozen yogurt is 98 grams, of which 26 grams are carbohydrates. What percent of one serving are carbohydrates? 26.5%

72. One serving of vanilla ice cream is 65 grams. What percent of one serving is sugar? 16.9%

Circumference The circumference of a circle is given by the formula $C = 2\pi r$. Find r if

73. The circumference C is 44 meters and π is $\frac{22}{7}$
7 meters

74. The circumference C is 176 meters and π is $\frac{22}{7}$
28 meters

75. The circumference is 9.42 inches and π is 3.14
$\frac{3}{2}$ or 1.5 inches

76. The circumference is 12.56 inches and π is 3.14
2 inches

Volume The volume of a cylinder is given by the formula $V = \pi r^2 h$. Find the height h if

77. The volume V is 42 cubic feet, the radius is $\frac{7}{22}$ feet, and π is $\frac{22}{7}$ 132 feet

78. The volume V is 84 cubic inches, the radius is $\frac{7}{11}$ inches, and π is $\frac{22}{7}$ 66 inches

79. The volume is 6.28 cubic centimeters, the radius is 3 centimeters, and π is 3.14 $\frac{2}{9}$ centimeters

80. The volume is 12.56 cubic centimeters, the radius is 2 centimeters, and π is 3.14   1 centimeter

Maintaining Your Skills

The problems that follow review some of the more important skills you have learned in previous sections and chapters. You can consider the time you spend working these problems as time spent studying for exams.

81. a. $27 - (-68)$ 95
   b. $27 + (-68)$ -41
   c. $-27 - 68$ -95
   d. $-27 + 68$ 41

82. a. $55 - (-29)$ 84
   b. $55 + (-29)$ 26
   c. $-55 - 29$ -84
   d. $-55 + 29$ -26

83. a. $-32 - (-41)$ 9
   b. $-32 + (-41)$ -73
   c. $-32 + 41$ 9
   d. $-32 - 41$ -73

84 a. $-56 - (-35)$ -21
   b. $-56 + (-35)$ -91
   c. $-56 + 35$ -21
   d. $-56 - 35$ -91

Getting Ready for the Next Section

To understand all of the explanations and examples in the next section you must be able to work the problems below.

Write an equivalent expression in English. Include the words *sum* and *difference* when possible.

85. $4 + 1 = 5$ The sum of 4 and 1 is 5.

86. $7 + 3 = 10$ The sum of 7 and 3 is 10.

87. $6 - 2 = 4$ The difference of 6 and 2 is 4.

88. $8 - 1 = 7$ The difference of 8 and 1 is 7.

89. $x - 5 = -12$ The difference of a number and 5 is -12.

90. $2x + 3 = 7$ The sum of twice a number and 3 is 7.

91. $x + 3 = 4(x - 3)$ The sum of a number and 3 is four times the difference of that number and 3.

92. $2(2x - 5) = 2x - 34$ Twice the difference of twice a number and 5 is the difference of twice that number and 34.

For each of the following expressions, write an equivalent equation.

93. Twice the sum of 6 and 3 is 18. $2(6 + 3) = 18$

94. Four added to the product of 3 and -1 is 1.
$3(-1) + 4 = 1$

95. The sum of twice 5 and 3 is 13. $2(5) + 3 = 13$

96. Twice the difference of 8 and 2 is 12. $2(8 - 2) = 12$

97. The sum of a number and five is thirteen. $x + 5 = 13$

98. The difference of ten and a number is negative eight. $10 - x = -8$

99. Five times the sum of a number and seven is thirty.
$5(x + 7) = 30$

100. Five times the difference of twice a number and six is negative twenty. $5(2x - 6) = -20$

2.6 Applications

OBJECTIVES

A Apply the Blueprint for Problem Solving to a variety of application problems.

As you begin reading through the examples in this section, you may find yourself asking why some of these problems seem so contrived. The title of the section is "Applications," but many of the problems here don't seem to have much to do with "real life." You are right about that. Example 3 is what we refer to as an "age problem." But imagine a conversation in which you ask someone how old her children are and she replies, "Bill is 6 years older than Tom. Three years ago the sum of their ages was 21. You figure it out." Although many of the "application" problems in this section are contrived, they are also good for practicing the strategy we will use to solve all application problems.

To begin this section, we list the steps used in solving application problems. We call this strategy the *Blueprint for Problem Solving*. It is an outline that will overlay the solution process we use on all application problems.

BLUEPRINT FOR PROBLEM SOLVING

STEP 1: *Read* the problem, and then mentally *list* the items that are known and the items that are unknown.

STEP 2: *Assign a variable* to one of the unknown items. (In most cases this will amount to letting $x =$ the item that is asked for in the problem.) Then *translate* the other *information* in the problem to expressions involving the variable.

STEP 3: *Reread* the problem, and then *write an equation,* using the items and variables listed in steps 1 and 2, that describes the situation.

STEP 4: *Solve the equation* found in step 3.

STEP 5: *Write* your *answer* using a complete sentence.

STEP 6: *Reread* the problem, and *check* your solution with the original words in the problem.

English	Algebra
The sum of a and b	$a + b$
The difference of a and b	$a - b$
The product of a and b	$a \cdot b$
The quotient of a and b	$\dfrac{a}{b}$
of	$\cdot$ (multiply)
is	$=$ (equals)
A number	x
4 more than x	$x + 4$
4 times x	$4x$
4 less than x	$x - 4$

There are a number of substeps within each of the steps in our blueprint. For instance, with steps 1 and 2 it is always a good idea to draw a diagram or picture if it helps visualize the relationship between the items in the problem. In other cases a table helps organize the information. As you gain more experience using the blueprint to solve application problems, you will find additional techniques that expand the blueprint.

To help with problems of the type shown next in Example 1, in the margin are some common English words and phrases and their mathematical translations.

Number Problems

 EXAMPLE 1   The sum of twice a number and three is seven. Find the number.

SOLUTION   Using the Blueprint for Problem Solving as an outline, we solve the problem as follows:

Step 1: **Read** the problem, and then mentally **list** the items that are known and the items that are unknown.

> *Known items:*   The numbers 3 and 7
>
> *Unknown items:*   The number in question

Step 2: **Assign a variable** to one of the unknown items. Then **translate** the other **information** in the problem to expressions involving the variable.

> Let x = the number asked for in the problem,
> then "The sum of twice a number and
> three" translates to $2x + 3$.

Step 3: **Reread** the problem, and then **write an equation,** using the items and variables listed in steps 1 and 2, that describes the situation.
With all word problems, the word *is* translates to =.

$$\underbrace{\text{The sum of twice } x \text{ and 3}}_{2x + 3} \underset{= 7}{\text{ is 7}}$$

Step 4: **Solve the equation** found in step 3.

$$2x + 3 = 7$$
$$2x + 3 + (\mathbf{-3}) = 7 + (\mathbf{-3})$$
$$2x = 4$$
$$\frac{\mathbf{1}}{\mathbf{2}}(2x) = \frac{\mathbf{1}}{\mathbf{2}}(4)$$
$$x = 2$$

Step 5: **Write** your **answer** using a complete sentence.

> The number is 2.

Step 6: **Reread** the problem, and **check** your solution with the original words in the problem.

> The sum of twice 2 and 3 is 7; a true statement.

You may find some examples and problems in this section that you can solve without using algebra or our blueprint. It is very important that you solve these problems using the methods we are showing here. The purpose behind these problems is to give you experience using the blueprint as a guide to solving problems written in words. Your answers are much less important than the work that you show to obtain your answer. You will be able to condense the steps in the blueprint later in the course. For now, though, you need to show your work in the same detail that we are showing in the examples in this section.

 EXAMPLE 2   One number is three more than twice another; their sum is eighteen. Find the numbers.

SOLUTION

*Step 1: **Read and list.***

> *Known items:*   Two numbers that add to 18. One is 3 more than twice the other.

> *Unknown items:*   The numbers in question.

*Step 2: **Assign a variable, and translate information.***
Let x = the first number. The other is $2x + 3$.

*Step 3: **Reread, and write an equation.***

$$\underbrace{\text{Their sum}}\ \text{is 18}$$
$$x + (2x + 3) = 18$$

*Step 4: **Solve the equation.***

$$x + (2x + 3) = 18$$
$$3x + 3 = 18$$
$$3x + 3 + (-3) = 18 + (-3)$$
$$3x = 15$$
$$x = 5$$

*Step 5: **Write the answer.***
The first number is 5. The other is $2 \cdot 5 + 3 = 13$.

*Step 6: **Reread, and check.***
The sum of 5 and 13 is 18, and 13 is 3 more than twice 5.

Age Problem

Remember as you read through the steps in the solutions to the examples in this section that step 1 is done mentally. Read the problem, and then mentally list the items that you know and the items that you don't know. The purpose of step 1 is to give you direction as you begin to work application problems. Finding the solution to an application problem is a process; it doesn't happen all at once. The first step is to read the problem with a purpose in mind. That purpose is to mentally note the items that are known and the items that are unknown.

EXAMPLE 3   Bill is 6 years older than Tom. Three years ago Bill's age was four times Tom's age. Find the age of each boy now.

SOLUTION   Applying the Blueprint for Problem Solving, we have

Step 1: Read and list.

Known items: Bill is 6 years older than Tom. Three years ago Bill's age was four times Tom's age.

Unknown items: Bill's age and Tom's age

Step 2: Assign a variable, and translate information.
Let x = Tom's age now. That makes Bill $x + 6$ years old now. A table like the one shown here can help organize the information in an age problem. Notice how we placed the x in the box that corresponds to Tom's age now.

	Three Years Ago	Now
Bill		$x + 6$
Tom		x

If Tom is x years old now, 3 years ago he was $x - 3$ years old. If Bill is $x + 6$ years old now, 3 years ago he was $x + 6 - 3 = x + 3$ years old. We use this information to fill in the remaining squares in the table.

	Three Years Ago	Now
Bill	$x + 3$	$x + 6$
Tom	$x - 3$	x

Step 3: Reread, and write an equation.
Reading the problem again, we see that 3 years ago Bill's age was four times Tom's age. Writing this as an equation, we have Bill's age 3 years ago = 4 · (Tom's age 3 years ago):

$$x + 3 = 4(x - 3)$$

Step 4: Solve the equation.

$$x + 3 = 4(x - 3)$$
$$x + 3 = 4x - 12$$
$$x + (-x) + 3 = 4x + (-x) - 12$$
$$3 = 3x - 12$$
$$3 + 12 = 3x - 12 + 12$$
$$15 = 3x$$
$$x = 5$$

Step 5: Write the answer.
Tom is 5 years old. Bill is 11 years old.

Step 6: Reread, and check.
If Tom is 5 and Bill is 11, then Bill is 6 years older than Tom. Three years ago Tom was 2 and Bill was 8. At that time, Bill's age was four times Tom's age. As you can see, the answers check with the original problem.

Geometry Problem

To understand Example 4 completely, you need to recall from Chapter 1 that the perimeter of a rectangle is the sum of the lengths of the sides. The formula for the perimeter is $P = 2l + 2w$.

 EXAMPLE 4   The length of a rectangle is 5 inches more than twice the width. The perimeter is 34 inches. Find the length and width.

SOLUTION   When working problems that involve geometric figures, a sketch of the figure helps organize and visualize the problem.

Step 1: Read and list.

> *Known items:*   The figure is a rectangle. The length is 5 inches more than twice the width. The perimeter is 34 inches.

> *Unknown items:*   The length and the width

Step 2: Assign a variable, and translate information.
Because the length is given in terms of the width (the length is 5 more than twice the width), we let x = the width of the rectangle. The length is 5 more than twice the width, so it must be $2x + 5$. The diagram is a visual description of the relationships we have listed so far.

Step 3: Reread, and write an equation.
The equation that describes the situation is

Twice the length + twice the width is the perimeter

$$2(2x + 5) \quad + \quad 2x \quad = \quad 34$$

Step 4: Solve the equation.

$2(2x + 5) + 2x = 34$	**Original equation**
$4x + 10 + 2x = 34$	**Distributive property**
$6x + 10 = 34$	**Add 4x and 2x**
$6x = 24$	**Add −10 to each side**
$x = 4$	**Divide each side by 6**

Step 5: Write the answer.
The width x is 4 inches. The length is $2x + 5 = 2(4) + 5 = 13$ inches.

Step 6: Reread, and check.
If the length is 13 and the width is 4, then the perimeter must be $2(13) + 2(4) = 26 + 8 = 34$, which checks with the original problem.

Coin Problem

 EXAMPLE 5   Jennifer has $2.45 in dimes and nickels. If she has 8 more dimes than nickels, how many of each coin does she have?

SOLUTION

Step 1: Read and list.

Known items: The type of coins, the total value of the coins, and that there are 8 more dimes than nickels.

Unknown items: The number of nickels and the number of dimes

Step 2: Assign a variable, and translate information.

If we let $x =$ the number of nickels, then $x + 8 =$ the number of dimes. Because the value of each nickel is 5 cents, the amount of money in nickels is $5x$. Similarly, because each dime is worth 10 cents, the amount of money in dimes is $10(x + 8)$. The table summarizes the information we have so far:

	Nickels	Dimes
Number	x	$x + 8$
Value (in cents)	$5x$	$10(x + 8)$

Step 3: Reread, and write an equation.

Because the total value of all the coins is 245 cents, the equation that describes this situation is

Amount of money in nickels		Amount of money in dimes		Total amount of money
$5x$	$+$	$10(x + 8)$	$=$	245

Step 4: Solve the equation.

To solve the equation, we apply the distributive property first.

$5x + 10x + 80 = 245$	**Distributive property**
$15x + 80 = 245$	**Add $5x$ and $10x$**
$15x = 165$	**Add -80 to each side**
$x = 11$	**Divide each side by 15**

Step 5: Write the answer.

The number of nickels is $x = 11$.
The number of dimes is $x + 8 = 11 + 8 = 19$.

Step 6: Reread, and check.

To check our results

11 nickels are worth $5(11) = $ 55 cents
19 dimes are worth $10(19) = 190$ cents
The total value is 245 cents = $2.45

When you begin working the problems in the problem set that follows, there are a couple of things to remember. The first is that you may have to read the problems a number of times before you begin to see how to solve them. The second thing to remember is that word problems are not always solved correctly the first time you try them. Sometimes it takes a couple of attempts and some wrong answers before you can set up and solve these problems correctly.

GETTING READY FOR CLASS

After reading through the preceding section, respond in your own words and in complete sentences.

1. What is the first step in the Blueprint for Problem Solving?
2. What is the last thing you do when solving an application problem?
3. What good does it do you to solve application problems even when they don't have much to do with real life?
4. Write an application problem whose solution depends on solving the equation $2x + 3 = 7$.

LINKING OBJECTIVES AND EXAMPLES

Next to each **objective** we have listed the examples that are best described by that objective.

A   1–5

Problem Set 2.6

Online support materials can be found at www.thomsonedu.com/login

Solve the following word problems. Follow the steps given in the Blueprint for Problem Solving.

Number Problems

1. The sum of a number and five is thirteen. Find the number. 8
2. The difference of ten and a number is negative eight. Find the number. 18
3. The sum of twice a number and four is fourteen. Find the number. 5
4. The difference of four times a number and eight is sixteen. Find the number. 6
5. Five times the sum of a number and seven is thirty. Find the number. −1
6. Five times the difference of twice a number and six is negative twenty. Find the number. 1
7. One number is two more than another. Their sum is eight. Find both numbers. 3 and 5
8. One number is three less than another. Their sum is fifteen. Find the numbers. 9 and 6
9. One number is four less than three times another. If their sum is increased by five, the result is twenty-five. Find the numbers. 6 and 14

10. One number is five more than twice another. If their sum is decreased by ten, the result is twenty-two. Find the numbers. 9 and 23

Age Problems

11. Shelly is 3 years older than Michele. Four years ago the sum of their ages was 67. Find the age of each person now. Shelly is 39; Michele is 36

	Four Years Ago	Now
Shelly	$x - 1$	$x + 3$
Michele	$x - 4$	x

12. Cary is 9 years older than Dan. In 7 years the sum of their ages will be 93. Find the age of each man now. Dan is 35; Cary is 44

	Now	In Seven Years
Cary	$x + 9$	$x + 16$
Dan	x	$x + 7$

= Videos available by instructor request
▶ = Online student support materials available at www.thomsonedu.com/login

13. Cody is twice as old as Evan. Three years ago the sum of their ages was 27. Find the age of each boy now. Evan is 11; Cody is 22

	Three Years Ago	Now
Cody	$2x - 3$	$2x$
Evan	$x - 3$	x

14. Justin is 2 years older than Ethan. In 9 years the sum of their ages will be 30. Find the age of each boy now. Ethan is 5; Justin is 7

	Now	In Nine Years
Justin	$x + 2$	$x + 11$
Ethan	x	$x + 9$

15. Fred is 4 years older than Barney. Five years ago the sum of their ages was 48. How old are they now? Barney is 27; Fred is 31

	Five Years Ago	Now
Fred	$x - 1$	$x + 4$
Barney	$x - 5$	x

16. Tim is 5 years older than JoAnn. Six years from now the sum of their ages will be 79. How old are they now? JoAnn is 31; Tim is 36

	Now	Six Years From Now
Tim	$x + 5$	$x + 11$
JoAnn	x	$x + 6$

17. Jack is twice as old as Lacy. In 3 years the sum of their ages will be 54. How old are they now?
Lacy is 16; Jack is 32

18. John is 4 times as old as Martha. Five years ago the sum of their ages was 50. How old are they now?
Martha is 12; John is 48

19. Pat is 20 years older than his son Patrick. In 2 years Pat will be twice as old as Patrick. How old are they now? Patrick is 18; Pat is 38

20. Diane is 23 years older than her daughter Amy. In 6 years Diane will be twice as old as Amy. How old are they now? Amy is 17; Diane is 40

Perimeter Problems

21. The perimeter of a square is 36 inches. Find the length of one side. $s = 9$ inches

22. The perimeter of a square is 44 centimeters. Find the length of one side. $s = 11$ centimeters

23. The perimeter of a square is 60 feet. Find the length of one side. $s = 15$ feet

24. The perimeter of a square is 84 meters. Find the length of one side. $s = 21$ meters

25. One side of a triangle is three times the shortest side. The third side is 7 feet more than the shortest side. The perimeter is 62 feet. Find all three sides.
11 feet, 18 feet, 33 feet

26. One side of a triangle is half the longest side. The third side is 10 meters less than the longest side. The perimeter is 45 meters. Find all three sides.
22 meters, 11 meters, 12 meters

27. One side of a triangle is half the longest side. The third side is 12 feet less than the longest side. The perimeter is 53 feet. Find all three sides.
26 feet, 13 feet, 14 feet

28. One side of a triangle is 6 meters more than twice the shortest side. The third side is 9 meters more than the shortest side. The perimeter is 75 meters. Find all three sides. 15 meters, 24 meters, 36 meters

29. The length of a rectangle is 5 inches more than the width. The perimeter is 34 inches. Find the length and width. $\ell = 11$ inches; $w = 6$ inches

$x + 5$

30. The width of a rectangle is 3 feet less than the length. The perimeter is 10 feet. Find the length and width. $l = 4$ feet; $w = 1$ foot

31. The length of a rectangle is 7 inches more than twice the width. The perimeter is 68 inches. Find the length and width. $\ell = 25$ inches; $w = 9$ inches

32. The length of a rectangle is 4 inches more than three times the width. The perimeter is 72 inches. Find the length and width. $\ell = 28$ inches; $w = 8$ inches

33. The length of a rectangle is 6 feet more than three times the width. The perimeter is 36 feet. Find the length and width. $\ell = 15$ feet; $w = 3$ feet

34. The length of a rectangle is 3 feet less than twice the width. The perimeter is 54 feet. Find the length and width. $\ell = 17$ feet; $w = 10$ feet

Coin Problems

35. Marissa has $4.40 in quarters and dimes. If she has 5 more quarters than dimes, how many of each coin does she have? 9 dimes; 14 quarters

	Dimes	Quarters
Number	x	$x + 5$
Value (cents)	$10(x)$	$25(x + 5)$

36. Kendra has $2.75 in dimes and nickels. If she has twice as many dimes as nickels, how many of each coin does she have? 11 nickels; 22 dimes

	Nickels	Dimes
Number	x	$2x$
Value (cents)	$5(x)$	$10(2x)$

37. Tanner has $4.35 in nickels and quarters. If he has 15 more nickels than quarters, how many of each coin does he have? 12 quarters; 27 nickels

	Nickels	Quarters
Number	$x + 15$	x
Value (cents)	$5(x + 15)$	$25(x)$

38. Connor has $9.00 in dimes and quarters. If he has twice as many quarters as dimes, how many of each coin does he have? 15 dimes; 30 quarters

	Dimes	Quarters
Number	x	$2x$
Value (cents)	$10(x)$	$25(2x)$

▶ 39. Sue has $2.10 in dimes and nickels. If she has 9 more dimes than nickels, how many of each coin does she have? 8 nickels; 17 dimes

40. Mike has $1.55 in dimes and nickels. If he has 7 more nickels than dimes, how many of each coin does he have? 8 dimes; 15 nickels

41. Katie has a collection of nickels, dimes, and quarters with a total value of $4.35. There are 3 more dimes than nickels and 5 more quarters than nickels. How many of each coin is in her collection? 7 nickels; 10 dimes; 12 quarters

	Nickels	Dimes	Quarters
Number	x	$x + 3$	$x + 5$
Value	$5x$	$10(x + 3)$	$25(x + 5)$

42. Mary Jo has $3.90 worth of nickels, dimes, and quarters. The number of nickels is 3 more than the number of dimes. The number of quarters is 7 more than the number of dimes. How many of each coin is in her collection? 8 nickels; 5 dimes; 12 quarters

	Nickels	Dimes	Quarters
Number	$x + 3$	x	$x + 7$
Value	$5(x + 3)$	$10(x)$	$25(x + 7)$

43. Cory has a collection of nickels, dimes, and quarters with a total value of $2.55. There are 6 more dimes than nickels and twice as many quarters as nickels. How many of each coin is in her collection? 3 nickels; 9 dimes; 6 quarters

	Nickels	Dimes	Quarters
Number	x	$x + 6$	$2x$
Value	$5x$	$10(x + 6)$	$25(2x)$

44. Kelly has a collection of nickels, dimes, and quarters with a total value of $7.40. There are four more nickels than dimes and twice as many quarters as nickels. How many of each coin is in her collection? 12 nickels; 8 dimes; 24 quarters

	Nickels	Dimes	Quarters
Number	$x + 4$	x	$2(x + 4)$
Value	$5(x + 4)$	$10(x)$	$25[2(x + 4)]$

Maintaining Your Skills

Write an equivalent statement in English.

45. $4 < 10$ 4 is less than 10.

46. $4 \le 10$ 4 is less than or equal to 10.

47. $9 \ge -5$ 9 is greater than or equal to -5.

48. $x - 2 > 4$ The difference of x and 2 is greater than 4.

Place the symbol $<$ or the symbol $>$ between the quantities in each expression.

49. 12 20 $<$

50. -12 20 $<$

51. -8 -6 $<$

52. -10 -20 $>$

Simplify.

53. $|8 - 3| - |5 - 2|$ 2

54. $|9 - 2| - |10 - 8|$ 5

55. $15 - |9 - 3(7 - 5)|$ 12

56. $10 - |7 - 2(5 - 3)|$ 7

Getting Ready for the Next Section

To understand all of the explanations and examples in the next section you must be able to work the problems below.

Simplify the following expressions.

57. $x + 2x + 2x$ $5x$

58. $x + 2x + 3x$ $6x$

59. $x + 0.075x$ $1.075x$

60. $x + 0.065x$ $1.065x$

61. $0.09(x + 2,000)$ $0.09x + 180$

62. $0.06(x + 1,500)$ $0.06x + 90$

Solve each of the following equations.

63. $0.05x + 0.06(x - 1,500) = 570$ 6,000

64. $0.08x + 0.09(x + 2,000) = 690$ 3,000

65. $x + 2x + 3x = 180$ 30

66. $2x + 3x + 5x = 180$ 18

2.7 More Applications

OBJECTIVES

A Apply the Blueprint for Problem Solving to a variety of application problems.

Now that you have worked through a number of application problems using our blueprint, you probably have noticed that step 3, in which we write an equation that describes the situation, is the key step. Anyone with experience solving application problems will tell you that there will be times when your first attempt at step 3 results in the wrong equation. Remember, mistakes are part of the process of learning to do things correctly. Many times the correct equation will become obvious after you have written an equation that is partially wrong. In any case it is better to write an equation that is partially wrong and be actively involved with the problem than to write nothing at all. Application problems, like other problems in algebra, are not always solved correctly the first time.

Consecutive Integers

Our first example involves consecutive integers. When we ask for consecutive integers, we mean integers that are next to each other on the number line, like 5 and 6, or 13 and 14, or -4 and -3. In the dictionary, consecutive is defined as following one another in uninterrupted order. If we ask for consecutive *odd* integers, then we mean odd integers that follow one another on the number line. For example, 3 and 5, 11 and 13, and -9 and -7 are consecutive odd integers. As you can see, to get from one odd integer to the next consecutive odd integer we add 2.

If we are asked to find two consecutive integers and we let x equal the first integer, the next one must be $x + 1$, because consecutive integers always differ by 1. Likewise, if we are asked to find two consecutive odd or even integers, and we let x equal the first integer, then the next one will be $x + 2$ because consecutive even or odd integers always differ by 2. Here is a table that summarizes this information.

In Words	Using Algebra	Example
Two consecutive integers	$x, x + 1$	The sum of two consecutive integers is 15. $x + (x + 1) = 15$ or $7 + 8 = 15$
Three consecutive integers	$x, x + 1, x + 2$	The sum of three consecutive integers is 24. $x + (x + 1) + (x + 2) = 24$ or $7 + 8 + 9 = 24$
Two consecutive odd integers	$x, x + 2$	The sum of two consecutive odd integers is 16. $x + (x + 2) = 16$ or $7 + 9 = 16$
Two consecutive even integers	$x, x + 2$	The sum of two consecutive even integers is 18. $x + (x + 2) = 18$ or $8 + 10 = 18$

EXAMPLE 1

The sum of two consecutive odd integers is 28. Find the two integers.

SOLUTION

Step 1: Read and list.

Known items: Two consecutive odd integers. Their sum is equal to 28.
Unknown items: The numbers in question.

Step 2: Assign a variable, and translate information.

If we let x = the first of the two consecutive odd integers, then $x + 2$ is the next consecutive one.

Step 3: Reread, and write an equation.

Their sum is 28.

$$x + (x + 2) = 28$$

Step 4: Solve the equation.

$$2x + 2 = 28 \qquad \text{Simplify the left side}$$

$$2x = 26 \qquad \text{Add } -2 \text{ to each side}$$

$$x = 13 \qquad \text{Multiply each side by } \tfrac{1}{2}$$

Step 5: Write the answer.

The first of the two integers is 13. The second of the two integers will be two more than the first, which is 15.

Step 6: Reread, and check.

Suppose the first integer is 13. The next consecutive odd integer is 15. The sum of 15 and 13 is 28.

Interest

 EXAMPLE 2   Suppose you invest a certain amount of money in an account that earns 8% in annual interest. At the same time, you invest $2,000 more than that in an account that pays 9% in annual interest. If the total interest from both accounts at the end of the year is $690, how much is invested in each account?

SOLUTION

Step 1: ***Read and list.***
Known items: The interest rates, the total interest earned, and how much more is invested at 9%
Unknown items: The amounts invested in each account

Step 2: ***Assign a variable, and translate information.***
Let x = the amount of money invested at 8%. From this, $x + 2,000$ = the amount of money invested at 9%. The interest earned on x dollars invested at 8% is $0.08x$. The interest earned on $x + 2,000$ dollars invested at 9% is $0.09(x + 2,000)$.
Here is a table that summarizes this information:

	Dollars Invested at 8%	Dollars Invested at 9%
Number of	x	$x + 2,000$
Interest on	$0.08x$	$0.09(x + 2,000)$

Step 3: ***Reread, and write an equation.***
Because the total amount of interest earned from both accounts is $690, the equation that describes the situation is

Interest earned at 8%		Interest earned at 9%		Total interest earned
$0.08x$	$+$	$0.09(x + 2,000)$	$=$	690

Step 4: ***Solve the equation.***

$$0.08x + 0.09(x + 2,000) = 690$$

$0.08x + 0.09x + 180 = 690$	**Distributive property**
$0.17x + 180 = 690$	**Add $0.08x$ and $0.09x$**
$0.17x = 510$	**Add -180 to each side**
$x = 3,000$	**Divide each side by 0.17**

Step 5: ***Write the answer.***
The amount of money invested at 8% is $3,000, whereas the amount of money invested at 9% is $x + 2,000 = 3,000 + 2,000 = \$5,000$.

Step 6: ***Reread, and check.***
The interest at 8% is 8% of 3,000 = $0.08(3,000) = \$240$
The interest at 9% is 9% of 5,000 = $0.09(5,000) = \$450$
The total interest is $690

FACTS FROM GEOMETRY

Labeling Triangles and the Sum of the Angles in a Triangle

One way to label the important parts of a triangle is to label the vertices with capital letters and the sides with small letters, as shown in Figure 1.

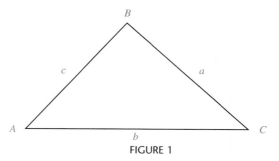

FIGURE 1

In Figure 1, notice that side a is opposite vertex A, side b is opposite vertex B, and side c is opposite vertex C. Also, because each vertex is the vertex of one of the angles of the triangle, we refer to the three interior angles as A, B, and C.

In any triangle, the sum of the interior angles is 180°. For the triangle shown in Figure 1, the relationship is written

$$A + B + C = 180°$$

EXAMPLE 3   The angles in a triangle are such that one angle is twice the smallest angle, whereas the third angle is three times as large as the smallest angle. Find the measure of all three angles.

SOLUTION

Step 1: Read and list.

Known items: The sum of all three angles is 180°, one angle is twice the smallest angle, the largest angle is three times the smallest angle.
Unknown items: The measure of each angle

Step 2: Assign a variable, and translate information.

Let x be the smallest angle, then $2x$ will be the measure of another angle and $3x$ will be the measure of the largest angle.

Step 3: Reread, and write an equation.

When working with geometric objects, drawing a generic diagram sometimes will help us visualize what it is that we are asked to find. In Figure 2, we draw a triangle with angles A, B, and C.

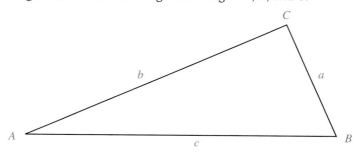

FIGURE 2

We can let the value of $A = x$, the value of $B = 2x$, and the value of $C = 3x$. We know that the sum of angles A, B, and C will be 180°, so our equation becomes

$$x + 2x + 3x = 180°$$

Step 4: Solve the equation.

$$x + 2x + 3x = 180°$$
$$6x = 180°$$
$$x = 30°$$

Step 5: Write the answer.

The smallest angle A measures 30°
Angle B measures $2x$, or $2(30°) = 60°$
Angle C measures $3x$, or $3(30°) = 90°$

Step 6: Reread, and check.

The angles must add to 180°:

$$A + B + C = 180°$$

$$30° + 60° + 90° \stackrel{?}{=} 180°$$

$$180° = 180° \qquad \textbf{Our answers check}$$

GETTING READY FOR CLASS

After reading through the preceding section, respond in your own words and in complete sentences.

1. How do we label triangles?
2. What rule is always true about the three angles in a triangle?
3. Write an application problem whose solution depends on solving the equation $x + 0.075x = 500$.
4. Write an application problem whose solution depends on solving the equation $0.05x + 0.06(x + 200) = 67$.

LINKING OBJECTIVES AND EXAMPLES

Next to each **objective** we have listed the examples that are best described by that objective.

A   1–3

Problem Set 2.7

Online support materials can be found at www.thomsonedu.com/login

Consecutive Integer Problems

1. The sum of two consecutive integers is 11. Find the numbers. 5 and 6

2. The sum of two consecutive integers is 15. Find the numbers. 7 and 8

3. The sum of two consecutive integers is −9. Find the numbers. −4 and −5

4. The sum of two consecutive integers is −21. Find the numbers. −10 and −11

5. The sum of two consecutive odd integers is 28. Find the numbers. 13 and 15

6. The sum of two consecutive odd integers is 44. Find the numbers. 21 and 23

7. The sum of two consecutive even integers is 106. Find the numbers. 52 and 54

= Videos available by instructor request

▶ = Online student support materials available at www.thomsonedu.com/login

8. The sum of two consecutive even integers is 66. Find the numbers.   32 and 34

9. The sum of two consecutive even integers is −30. Find the numbers.   −14 and −16

10. The sum of two consecutive odd integers is −76. Find the numbers.   −37 and −39

11. The sum of three consecutive odd integers is 57. Find the numbers.   17, 19, and 21

12. The sum of three consecutive odd integers is −51. Find the numbers.   −15, −17, and −19

13. The sum of three consecutive even integers is 132. Find the numbers.   42, 44, and 46

14. The sum of three consecutive even integers is −108. Find the numbers.   −34, −36, and −38

Interest Problems

15. Suppose you invest money in two accounts. One of the accounts pays 8% annual interest, whereas the other pays 9% annual interest. If you have $2,000 more invested at 9% than you have invested at 8%, how much do you have invested in each account if the total amount of interest you earn in a year is $860? (Begin by completing the following table.)
$4,000 invested at 8%, $6,000 invested at 9%

	Dollars Invested at 8%	Dollars Invested at 9%
Number of	x	$x + 2,000$
Interest on	$0.08x$	$0.09(x + 2,000)$

16. Suppose you invest a certain amount of money in an account that pays 11% interest annually, and $4,000 more than that in an account that pays 12% annually. How much money do you have in each account if the total interest for a year is $940?
$2,000 invested at 11%, $6,000 invested at 12%

	Dollars Invested at 11%	Dollars Invested at 12%
Number of	x	$x + 4,000$
Interest on	$0.11x$	$0.12(x + 4,000)$

▶ **17.** Tyler has two savings accounts that his grandparents opened for him. The two accounts pay 10%

and 12% in annual interest; there is $500 more in the account that pays 12% than there is in the other account. If the total interest for a year is $214, how much money does he have in each account?
$700 invested at 10%, $1,200 invested at 12%

18. Travis has a savings account that his parents opened for him. It pays 6% annual interest. His uncle also opened an account for him, but it pays 8% annual interest. If there is $800 more in the account that pays 6%, and the total interest from both accounts is $104, how much money is in each of the accounts?
$1,200 at 6%, $400 at 8%

19. A stockbroker has money in three accounts. The interest rates on the three accounts are 8%, 9%, and 10%. If she has twice as much money invested at 9% as she has invested at 8%, three times as much at 10% as she has at 8%, and the total interest for the year is $280, how much is invested at each rate? (*Hint:* Let x = the amount invested at 8%.)
$500 at 8%, $1,000 at 9%, $1,500 at 10%

20. An accountant has money in three accounts that pay 9%, 10%, and 11% in annual interest. He has twice as much invested at 9% as he does at 10% and three times as much invested at 11% as he does at 10%. If the total interest from the three accounts is $610 for the year, how much is invested at each rate? (*Hint:* Let x = the amount invested at 10%.)
$2,000 at 9%, $1,000 at 10%, $3,000 at 11%

Triangle Problems

21. Two angles in a triangle are equal and their sum is equal to the third angle in the triangle. What are the measures of each of the three interior angles?
45°, 45°, 90°

22. One angle in a triangle measures twice the smallest angle, whereas the largest angle is six times the smallest angle. Find the measures of all three angles.   20°, 40°, 120°

23. The smallest angle in a triangle is $\frac{1}{5}$ as large as the largest angle. The third angle is twice the smallest angle. Find the three angles.   22.5°, 45°, 112.5°

24. One angle in a triangle is half the largest angle but three times the smallest. Find all three angles.
18°, 54°, 108°

25. A right triangle has one 37° angle. Find the other two angles. 53°, 90°

26. In a right triangle, one of the acute angles is twice as large as the other acute angle. Find the measure of the two acute angles. 30°, 60°

27. One angle of a triangle measures 20° more than the smallest, while a third angle is twice the smallest. Find the measure of each angle. 80°, 60°, 40°

28. One angle of a triangle measures 50° more than the smallest, while a third angle is three times the smallest. Find the measure of each angle. 78°, 76°, 26°

Miscellaneous Problems

29. **Ticket Prices** Miguel is selling tickets to a barbecue. Adult tickets cost $6.00 and children's tickets cost $4.00. He sells six more children's tickets than adult tickets. The total amount of money he collects is $184. How many adult tickets and how many children's tickets did he sell? 16 adult and 22 children's tickets

	Adult	Child
Number	x	$x + 6$
Income	$6(x)$	$4(x + 6)$

30. **Working Two Jobs** Maggie has a job working in an office for $10 an hour and another job driving a tractor for $12 an hour. One week she works in the office twice as long as she drives the tractor. Her total income for that week is $416. How many hours did she spend at each job? 13 hours driving the tractor and 26 hours working at the office

Job	Office	Tractor
Hours Worked	$2x$	x
Wages Earned	$10(2x)$	$12x$

31. **Phone Bill** The cost of a long-distance phone call is $0.41 for the first minute and $0.32 for each additional minute. If the total charge for a long-distance call is $5.21, how many minutes was the call? 16 minutes

32. **Phone Bill** Danny, who is 1 year old, is playing with the telephone when he accidentally presses one of the buttons his mother has programmed to dial her friend Sue's number. Sue answers the phone and realizes Danny is on the other end. She talks to Danny, trying to get him to hang up. The cost for a call is $0.23 for the first minute and $0.14 for every minute after that. If the total charge for the call is $3.73, how long did it take Sue to convince Danny to hang up the phone? 26 minutes

33. **Hourly Wages** JoAnn works in the publicity office at the state university. She is paid $12 an hour for the first 35 hours she works each week and $18 an hour for every hour after that. If she makes $492 one week, how many hours did she work? 39 hours

34. **Hourly Wages** Diane has a part-time job that pays her $6.50 an hour. During one week she works 26 hours and is paid $178.10. She realizes when she sees her check that she has been given a raise. How much per hour is that raise? $0.35 per hour raise

35. **Office Numbers** Professors Wong and Gil have offices in the mathematics building at Miami Dade College. Their office numbers are consecutive odd integers with a sum of 14,660. What are the office numbers of these two professors? They are in offices 7329 and 7331.

36. **Cell Phone Numbers** Diana and Tom buy two cell phones. The phone numbers assigned to each are consecutive integers with a sum of 11,109,295. If the smaller number is Diana's, what are their phone numbers? Diana's phone number is 555-4647 and Tom's is 555-4648

37. **Age** Marissa and Kendra are 2 years apart in age. Their ages are two consecutive even integers. Kendra is the younger of the two. If Marissa's age is added to twice Kendra's age, the result is 26. How old is each girl? Kendra is 8 years old and Marissa is 10 years old.

38. **Age** Justin's and Ethan's ages form two consecutive odd integers. What is the difference of their ages? The difference is 2 or −2, depending on the order of subtraction.

39. **Arrival Time** Jeff and Carla Cole are driving separately from San Luis Obispo, California, to the north shore of Lake Tahoe, a distance of 425 miles. Jeff leaves San Luis Obispo at 11:00 AM and averages 55 miles per hour on the drive, Carla leaves later, at 1:00 PM but averages 65 miles per hour. Which person arrives in Lake Tahoe first? Jeff

40. **Piano Lessons** Tyler is taking piano lessons. Because he doesn't practice as often as his parents would like him to, he has to pay for part of the lessons himself. His parents pay him $0.50 to do the laun-

dry and $1.25 to mow the lawn. In one month, he does the laundry 6 more times than he mows the lawn. If his parents pay him $13.50 that month, how many times did he mow the lawn?   6 times

At one time, the Texas Junior College Teachers Association annual conference was held in Austin. At that time a taxi ride in Austin was $1.25 for the first $\frac{1}{5}$ of a mile and $0.25 for each additional $\frac{1}{5}$ of a mile. Use this information for Problems 41 and 42.

41. **Cost of a Taxi Ride** If the distance from one of the convention hotels to the airport is 7.5 miles, how much will it cost to take at taxi from that hotel to the airport? $10.38

42. **Cost of a Taxi Ride** Suppose the distance from one of the hotels to one of the western dance clubs in Austin is 12.4 miles. If the fare meter in the taxi gives the charge for that trip as $16.50, is the meter working correctly?   yes

43. **Geometry** The length and width of a rectangle are consecutive even integers. The perimeter is 44 meters. Find the length and width.
    $\ell = 12$ meters; $w = 10$ meters

44. **Geometry** The length and width of a rectangle are consecutive odd integers. The perimeter is 128 meters. Find the length and width.
    $\ell = 33$ meters; $w = 31$ meters

45. **Geometry** The angles of a triangle are three consecutive integers. Find the measure of each angle.
    $59°$, $60°$, $61°$

46. **Geometry** The angles of a triangle are three consecutive even integers. Find the measure of each angle.
    $58°$, $60°$, $62°$

Ike and Nancy Lara give western dance lessons at the Elks Lodge on Sunday nights. The lessons cost $3.00 for members of the lodge and $5.00 for nonmembers. Half of the money collected for the lesson is paid to Ike and Nancy. The Elks Lodge keeps the other half. One Sunday night Ike counts 36 people in the dance lesson. Use this information to work Problems 47 through 50.

47. **Dance Lessons** What is the least amount of money Ike and Nancy will make? $54.00

48. **Dance Lessons** What is the largest amount of money Ike and Nancy will make? $90.00

49. **Dance Lessons** At the end of the evening, the Elks Lodge gives Ike and Nancy a check for $80 to cover half of the receipts. Can this amount be correct?   Yes

50. **Dance Lessons** Besides the number of people in the dance lesson, what additional information does Ike need to know to always be sure he is being paid the correct amount?   How many of the people taking lessons are members, or how many are not members

Maintaining Your Skills

The problems that follow review some of the more important skills you have learned in previous sections and chapters. You can consider the time you spend working these problems as time spent studying for exams.

Simplify the expression $36x - 12$ for each of the following values of x.

51. $\frac{1}{4}$ -3

52. $\frac{1}{6}$ -6

53. $\frac{1}{9}$ -8

54. $\frac{3}{2}$ 42

55. $\frac{1}{3}$ 0

56. $\frac{5}{12}$ 3

57. $\frac{5}{9}$ 8

58. $\frac{2}{3}$ 12

Find the value of each expression when $x = -4$.

59. $3(x - 4)$ -24

60. $-3(x - 4)$ 24

61. $-5x + 8$ 28

62. $5x + 8$ -12

63. $\dfrac{x - 14}{36}$ $-\frac{1}{2}$

64. $\dfrac{x - 12}{36}$ $-\frac{4}{9}$

65. $\dfrac{16}{x} + 3x$ -16

66. $\dfrac{16}{x} - 3x$ 8

67. $7x - \dfrac{12}{x}$ -25

68. $7x + \dfrac{12}{x}$ -31

69. $8\left(\dfrac{x}{2} + 5\right)$ 24

70. $-8\left(\dfrac{x}{2} + 5\right)$ -24

Getting Ready for the Next Section

To understand all of the explanations and examples in the next section you must be able to work the problems below.

Solve the following equations.

71. **a.** $x - 3 = 6$ 9
    b. $x + 3 = 6$ 3
    c. $-x - 3 = 6$ -9
    d. $-x + 3 = 6$ -3

72. a. $x - 7 = 16$ 23

  b. $x + 7 = 16$ 9

  c. $-x - 7 = 16$ -23

  d. $-x + 7 = 16$ -9

73. a. $\dfrac{x}{4} = -2$ -8

  b. $-\dfrac{x}{4} = -2$ 8

  c. $\dfrac{x}{4} = 2$ 8

  d. $-\dfrac{x}{4} = 2$ -8

74. a. $3a = 15$ 5

  b. $3a = -15$ -5

  c. $-3a = 15$ -5

  d. $-3a = -15$ 5

75. $2.5x - 3.48 = 4.9x + 2.07$ -2.3125

76. $2(1 - 3x) + 4 = 4x - 14$ 2

77. $3(x - 4) = -2$ $\dfrac{10}{3}$

78. Solve $2x - 3y = 6$ for y. $y = \dfrac{2}{3}x - 2$

2.8 Linear Inequalities

OBJECTIVES

A Use the addition property for inequalities to solve an inequality.

B Use the multiplication property for inequalities to solve an inequality.

C Use both the addition and multiplication properties to solve an inequality.

D Graph the solution set for an inequality.

E Translate and solve application problems involving inequalities.

Linear inequalities are solved by a method similar to the one used in solving linear equations. The only real differences between the methods are in the multiplication property for inequalities and in graphing the solution set.

An inequality differs from an equation only with respect to the comparison symbol between the two quantities being compared. In place of the equal sign, we use < (less than), ≤ (less than or equal to), > (greater than), or ≥ (greater than or equal to). The addition property for inequalities is almost identical to the addition property for equality.

Addition Property for Inequalities For any three algebraic expressions A, B, and C,

$$\text{if} \qquad A < B$$
$$\text{then} \qquad A + C < B + C$$

In words: Adding the same quantity to both sides of an inequality will not change the solution set.

It makes no difference which inequality symbol we use to state the property. Adding the same amount to both sides always produces an inequality equivalent to the original inequality. Also, because subtraction can be thought of as addition of the opposite, this property holds for subtraction as well as addition.

 EXAMPLE 1   Solve the inequality $x + 5 < 7$.

SOLUTION   To isolate x, we add -5 to both sides of the inequality:

$$x + 5 < 7$$
$$x + 5 + (\mathbf{-5}) < 7 + (\mathbf{-5}) \qquad \textbf{Addition property for inequalities}$$
$$x < 2$$

We can go one step further here and graph the solution set. The solution set is all real numbers less than 2. To graph this set, we simply draw a straight line and label the center 0 (zero) for reference. Then we label the 2 on the right side of zero and extend an arrow beginning at 2 and pointing to the left. We use an open circle at 2 because it is not included in the solution set. Here is the graph.

 EXAMPLE 2   Solve $x - 6 \leq -3$.

SOLUTION   Adding 6 to each side will isolate x on the left side:

$$x - 6 \leq -3$$
$$x - 6 + \mathbf{6} \leq -3 + \mathbf{6} \qquad \text{Add 6 to both sides}$$
$$x \leq 3$$

The graph of the solution set is

Notice that the dot at the 3 is darkened because 3 is included in the solution set. We always will use open circles on the graphs of solution sets with $<$ or $>$ and closed (darkened) circles on the graphs of solution sets with $\leq$ or $\geq$.

To see the idea behind the multiplication property for inequalities, we will consider three true inequality statements and explore what happens when we multiply both sides by a positive number and then what happens when we multiply by a negative number.

Consider the following three true statements:

$$3 < 5 \qquad -3 < 5 \qquad -5 < -3$$

Now multiply both sides by the positive number 4:

$$4(3) < 4(5) \qquad 4(-3) < 4(5) \qquad 4(-5) < 4(-3)$$
$$12 < 20 \qquad\quad -12 < 20 \qquad\quad -20 < -12$$

In each case, the inequality symbol in the result points in the same direction it did in the original inequality. We say the "sense" of the inequality doesn't change when we multiply both sides by a positive quantity.

Notice what happens when we go through the same process but multiply both sides by -4 instead of 4:

$$3 < 5 \qquad\qquad -3 < 5 \qquad\qquad -5 < -3$$

$$-4(3) > -4(5) \qquad -4(-3) > -4(5) \qquad -4(-5) > -4(-3)$$
$$-12 > -20 \qquad\qquad 12 > -20 \qquad\qquad 20 > 12$$

In each case, we have to change the direction in which the inequality symbol points to keep each statement true. Multiplying both sides of an inequality by a negative quantity *always* reverses the sense of the inequality. Our results are summarized in the multiplication property for inequalities.

Note

This discussion is intended to show why the multiplication property for inequalities is written the way it is. You may want to look ahead to the property itself and then come back to this discussion if you are having trouble making sense out of it.

Note

Because division is defined in terms of multiplication, this property is also true for division. We can divide both sides of an inequality by any nonzero number we choose. If that number happens to be negative, we must also reverse the direction of the inequality symbol.

Multiplication Property for Inequalities For any three algebraic expressions A, B, and C,

$$\text{if} \qquad A < B$$
$$\text{then} \qquad AC < BC \qquad \text{when } C \text{ is positive}$$
$$\text{and} \qquad AC > BC \qquad \text{when } C \text{ is negative}$$

In words: Multiplying both sides of an inequality by a positive number does not change the solution set. When multiplying both sides of an inequality by a negative number, it is necessary to reverse the inequality symbol to produce an equivalent inequality.

We can multiply both sides of an inequality by any nonzero number we choose. If that number happens to be negative, we must also reverse the sense of the inequality.

 EXAMPLE 3   Solve $3a < 15$ and graph the solution.

SOLUTION   We begin by multiplying each side by $\frac{1}{3}$. Because $\frac{1}{3}$ is a positive number, we do not reverse the direction of the inequality symbol:

$$3a < 15$$
$$\frac{1}{3}(3a) < \frac{1}{3}(15) \qquad \textbf{Multiply each side by } \frac{1}{3}$$
$$a < 5$$

 EXAMPLE 4   Solve $-3a \le 18$, and graph the solution.

SOLUTION   We begin by multiplying both sides by $-\frac{1}{3}$. Because $-\frac{1}{3}$ is a negative number, we must reverse the direction of the inequality symbol at the same time that we multiply by $-\frac{1}{3}$.

$$-3a \le 18$$
$$-\frac{1}{3}(-3a) \ge -\frac{1}{3}(18) \qquad \textbf{Multiply both sides by } -\frac{1}{3} \textbf{ and reverse}$$
$$\textbf{the direction of the inequality symbol}$$
$$a \ge -6$$

 EXAMPLE 5   Solve $-\frac{x}{4} > 2$ and graph the solution.

SOLUTION   To isolate x, we multiply each side by -4. Because -4 is a negative number, we also must reverse the direction of the inequality symbol:

$$-\frac{x}{4} > 2$$
$$-4\left(-\frac{x}{4}\right) < -4(2) \qquad \textbf{Multiply each side by } -4, \textbf{ and reverse}$$
$$\textbf{the direction of the inequality symbol}$$
$$x < -8$$

To solve more complicated inequalities, we use the following steps.

> ### Strategy for Solving Linear Inequalities in One Variable
> **Step 1a:**   Use the distributive property to separate terms, if necessary.
>
> **1b:**   If fractions are present, consider multiplying both sides by the LCD to eliminate the fractions. If decimals are present, consider multiplying both sides by a power of 10 to clear the inequality of decimals.
>
> **1c:**   Combine similar terms on each side of the inequality.
>
> **Step 2:**   Use the addition property for inequalities to get all variable terms on one side of the inequality and all constant terms on the other side.
>
> **Step 3:**   Use the multiplication property for inequalities to get x by itself on one side of the inequality.
>
> **Step 4:**   Graph the solution set.

   EXAMPLE 6   Solve $2.5x - 3.48 < -4.9x + 2.07$.

SOLUTION   We have two methods we can use to solve this inequality. We can simply apply our properties to the inequality the way it is currently written and work with the decimal numbers, or we can eliminate the decimals to begin with and solve the resulting inequality.

Method 1 Working with the decimals.

$$2.5x - 3.48 < -4.9x + 2.07 \qquad \textbf{Original inequality}$$

$$2.5x + \mathbf{4.9x} - 3.48 < -4.9x + \mathbf{4.9x} + 2.07 \qquad \textbf{Add 4.9x to each side}$$

$$7.4x - 3.48 < 2.07$$

$$7.4x - 3.48 + \mathbf{3.48} < 2.07 + \mathbf{3.48} \qquad \textbf{Add 3.48 to each side}$$

$$7.4x < 5.55$$

$$\frac{7.4x}{\mathbf{7.4}} < \frac{5.55}{\mathbf{7.4}} \qquad \textbf{Divide each side by 7.4}$$

$$x < 0.75$$

Method 2 Eliminating the decimals in the beginning.

Because the greatest number of places to the right of the decimal point in any of the numbers is 2, we can multiply each side of the inequality by 100 and we will be left with an equivalent inequality that contains only whole numbers.

$$2.5x - 3.48 < -4.9x + 2.07 \qquad \textbf{Original inequality}$$

$$\mathbf{100}(2.5x - 3.48) < \mathbf{100}(-4.9x + 2.07) \qquad \textbf{Multiply each side by 100}$$

$$\mathbf{100}(2.5x) - \mathbf{100}(3.48) < \mathbf{100}(-4.9x) + \mathbf{100}(2.07) \qquad \textbf{Distributive property}$$

$$250x - 348 < -490x + 207 \qquad \textbf{Multiplication}$$

$$740x - 348 < 207 \qquad \textbf{Add 490x to each side}$$

$$740x < 555$$ **Add 348 to each side**

$$\frac{740x}{740} < \frac{555}{740}$$ **Divide each side by 740**

$$x < 0.75$$

The solution by either method is $x < 0.75$. Here is the graph:

EXAMPLE 7 Solve $3(x - 4) \geq -2$.

SOLUTION

$$3x - 12 \geq -2$$ **Distributive property**

$$3x - 12 + \mathbf{12} \geq -2 + \mathbf{12}$$ **Add 12 to both sides**

$$3x \geq 10$$

$$\frac{\mathbf{1}}{\mathbf{3}}(3x) \geq \frac{\mathbf{1}}{\mathbf{3}}(10)$$ **Multiply both sides by $\frac{1}{3}$**

$$x \geq \frac{10}{3}$$

EXAMPLE 8 Solve and graph $2(1 - 3x) + 4 < 4x - 14$.

SOLUTION

$$2 - 6x + 4 < 4x - 14$$ **Distributive property**

$$-6x + 6 < 4x - 14$$ **Simplify**

$$-6x + 6 + (\mathbf{-6}) < 4x - 14 + (\mathbf{-6})$$ **Add -6 to both sides**

$$-6x < 4x - 20$$

$$-6x + (\mathbf{-4x}) < 4x + (\mathbf{-4x}) - 20$$ **Add $-4x$ to both sides**

$$-10x < -20$$

$$\left(-\frac{\mathbf{1}}{\mathbf{10}}\right)(-10x) > \left(-\frac{\mathbf{1}}{\mathbf{10}}\right)(-20)$$ **Multiply by $-\frac{1}{10}$, reverse the sense of the inequality**

$$x > 2$$

EXAMPLE 9 Solve $2x - 3y < 6$ for y.

SOLUTION We can solve this inequality for y by first adding $-2x$ to each side and then multiplying each side by $-\frac{1}{3}$. When we multiply by $-\frac{1}{3}$ we must reverse the direction of the inequality symbol. Because this is an inequality, we will not graph the solution.

$$2x - 3y < 6 \qquad \text{Original inequality}$$

$$2x + (-\mathbf{2x}) - 3y < (-\mathbf{2x}) + 6 \qquad \text{Add } -2x \text{ to each side}$$

$$-3y < -2x + 6$$

$$-\frac{1}{\mathbf{3}}(-3y) > -\frac{1}{\mathbf{3}}(-2x + 6) \qquad \text{Multiply each side by } -\tfrac{1}{3}$$

$$y > \frac{2}{3}x - 2 \qquad \text{Distributive property}$$

When working application problems that involve inequalities, the phrases "at least" and "at most" translate as follows:

In Words	In Symbols
x is at least 30	$x \geq 30$
x is at most 20	$x \leq 20$

Applying the Concepts

Our next example is similar to an example done earlier in this chapter. This time it involves an inequality instead of an equation.

We can modify our Blueprint for Problem Solving to solve application problems whose solutions depend on writing and then solving inequalities.

EXAMPLE 10   The sum of two consecutive odd integers is at most 28. What are the possibilities for the first of the two integers?

SOLUTION   When we use the phrase "their sum is at most 28," we mean that their sum is less than or equal to 28.

Step 1: Read and list.
Known items: Two consecutive odd integers. Their sum is less than or equal to 28.
Unknown items: The numbers in question.

Step 2: Assign a variable, and translate information.
If we let $x =$ the first of the two consecutive odd integers, then $x + 2$ is the next consecutive one.

Step 3: Reread, and write an inequality.
Their sum is at most 28.

$$x + (x + 2) \leq 28$$

Step 4: Solve the inequality.

$$2x + 2 \leq 28 \qquad \text{Simplify the left side}$$

$$2x \leq 26 \qquad \text{Add } -2 \text{ to each side}$$

$$x \leq 13 \qquad \text{Multiply each side by } \tfrac{1}{2}$$

Step 5: *Write the answer.*

The first of the two integers must be an odd integer that is less than or equal to 13. The second of the two integers will be two more than whatever the first one is.

Step 6: *Reread, and check.*

Suppose the first integer is 13. The next consecutive odd integer is 15. The sum of 15 and 13 is 28. If the first odd integer is less than 13, the sum of it and the next consecutive odd integer will be less than 28.

LINKING OBJECTIVES AND EXAMPLES

Next to each **objective** we have listed the examples that are best described by that objective.

A	1, 2
B	3–5
C	6–9
D	1–8
E	10

GETTING READY FOR CLASS

After reading through the preceding section, respond in your own words and in complete sentences.

1. State the addition property for inequalities.
2. How is the multiplication property for inequalities different from the multiplication property of equality?
3. When do we reverse the direction of an inequality symbol?
4. Under what conditions do we not change the direction of the inequality symbol when we multiply both sides of an inequality by a number?

Problem Set 2.8

Online support materials can be found at www.thomsonedu.com/login

Solve the following inequalities using the addition property of inequalities. Graph each solution set.

1. $x - 5 < 7$ $x < 12$ **2.** $x + 3 < -5$ $x < -8$

3. $a - 4 \le 8$ $a \le 12$ **4.** $a + 3 \le 10$ $a \le 7$

5. $x - 4.3 > 8.7$ $x > 13$ **6.** $x - 2.6 > 10.4$ $x > 13$

7. $y + 6 \ge 10$ $y \ge 4$ **8.** $y + 3 \ge 12$ $y \ge 9$

9. $2 < x - 7$ $x > 9$ **10.** $3 < x + 8$ $x > -5$

15. $\frac{x}{3} > 5$ $x > 15$ **16.** $\frac{x}{7} > 1$ $x > 7$

17. $-2x > 6$ $x < -3$ **18.** $-3x \ge 9$ $x \le -3$

19. $-3x \ge -18$ $x \le 6$ **20.** $-8x \ge -24$ $x \le 3$

21. $-\frac{x}{5} \le 10$ $x \ge -50$ **22.** $-\frac{x}{9} \ge -1$ $x \le 9$

23. $-\frac{2}{3}y > 4$ $y < -6$ **24.** $-\frac{3}{4}y > 6$ $y < -8$

Solve the following inequalities using the multiplication property of inequalities. If you multiply both sides by a negative number, be sure to reverse the direction of the inequality symbol. Graph the solution set.

11. $3x < 6$ $x < 2$ **12.** $2x < 14$ $x < 7$

13. $5a \le 25$ $a \le 5$ **14.** $4a \le 16$ $a \le 4$

Solve the following inequalities. Graph the solution set in each case.

25. $2x - 3 < 9$ $x < 6$ **26.** $3x - 4 < 17$ $x < 7$

27. $-\frac{1}{5}y - \frac{1}{3} \le \frac{2}{3}$ $y \ge -5$ **28.** $-\frac{1}{6}y - \frac{1}{2} \le \frac{2}{3}$ $y \ge -7$

29. $-7.2x + 1.8 > -19.8$ $x < 3$

= Videos available by instructor request

▶ = Online student support materials available at www.thomsonedu.com/login

30. $-7.8x - 1.3 > 22.1$ $x < -3$

31. $\dfrac{2}{3}x - 5 \le 7$ $x \le 18$ **32.** $\dfrac{3}{4}x - 8 \le 1$ $x \le 12$

33. $-\dfrac{2}{5}a - 3 > 5$ $a < -20$ **34.** $-\dfrac{4}{5}a - 2 > 10$ $a < -15$

35. $5 - \dfrac{3}{5}y > -10$ $y < 25$ **36.** $4 - \dfrac{5}{6}y > -11$ $y < 18$

37. $0.3(a + 1) \le 1.2$ $a \le 3$ **38.** $0.4(a - 2) \le 0.4$ $a \le 3$

▶ **39.** $2(5 - 2x) \le -20$ $x \ge \dfrac{15}{2}$ **40.** $7(8 - 2x) > 28$ $x < 2$

41. $3x - 5 > 8x$ $x < -1$ **42.** $8x - 4 > 6x$ $x > 2$

43. $\dfrac{1}{3}y - \dfrac{1}{2} \le \dfrac{5}{6}y + \dfrac{1}{2}$ $y \ge -2$

44. $\dfrac{7}{6}y + \dfrac{4}{3} \le \dfrac{11}{6}y - \dfrac{7}{6}$ $y \ge \dfrac{15}{4}$

45. $-2.8x + 8.4 < -14x - 2.8$ $x < -1$

46. $-7.2x - 2.4 < -2.4x + 12$ $x > -3$

47. $3(m - 2) - 4 \ge 7m + 14$ $m \le -6$

48. $2(3m - 1) + 5 \ge 8m - 7$ $m \le 5$

49. $3 - 4(x - 2) \le -5x + 6$ $x \le -5$

50. $8 - 6(x - 3) \le -4x + 12$ $x \ge 7$

Solve each of the following inequalities for y.

51. $3x + 2y < 6$ $y < -\dfrac{3}{2}x + 3$ **52.** $-3x + 2y < 6$ $y < \dfrac{3}{2}x + 3$

53. $2x - 5y > 10$ $y < \dfrac{2}{5}x - 2$ **54.** $-2x - 5y > 5$ $y < -\dfrac{2}{5}x - 1$

55. $-3x + 7y \le 21$ $y \le \dfrac{3}{7}x + 3$ **56.** $-7x + 3y \le 21$ $y \le \dfrac{7}{3}x + 7$

57. $2x - 4y \ge -4$ $y \le \dfrac{1}{2}x + 1$ **58.** $4x - 2y \ge -8$ $y \le 2x + 4$

The next two problems are intended to give you practice reading, and paying attention to, the instructions that accompany the problems you are working.

▶ **59.** Work each problem according to the instructions given.
  a. Evaluate when $x = 0$: $-5x + 3$ 3
  b. Solve: $-5x + 3 = -7$ 2
  c. Is 0 a solution to $-5x + 3 < -7$ No
  d. Solve: $-5x + 3 < -7$ $x > 2$

60. Work each problem according to the instructions given.
  a. Evaluate when $x = 0$: $-2x - 5$ -5
  b. Solve: $-2x - 5 = 1$ -3
  c. Is 0 a solution to $-2x - 5 > 1$ No
  d. Solve: $-2x - 5 > 1$ $x < -3$

For each graph below, write an inequality whose solution is the graph.

61.  $x < 3$

62. $x \le 3$

63. $x \ge 3$

64. $x > 3$

Applying the Concepts

65. Consecutive Integers The sum of two consecutive integers is at least 583. What are the possibilities for the first of the two integers? At least 291

66. Consecutive Integers The sum of two consecutive integers is at most 583. What are the possibilities for the first of the two integers? At most 291

67. Number Problems The sum of twice a number and six is less than ten. Find all solutions. $x < 2$

68. Number Problems Twice the difference of a number and three is greater than or equal to the number increased by five. Find all solutions. $x \ge 11$

69. Number Problems The product of a number and four is greater than the number minus eight. Find the solution set. $x > -\dfrac{8}{3}$

70. Number Problems The quotient of a number and five is less than the sum of seven and two. Find the solution set. $x < 45$

71. Geometry Problems The length of a rectangle is 3 times the width. If the perimeter is to be at least 48 meters, what are the possible values for the width? (If the perimeter is at least 48 meters, then it is greater than or equal to 48 meters.) $x \ge 6$; the width is at least 6 meters.

72. Geometry Problems The length of a rectangle is 3 more than twice the width. If the perimeter is to be at least 51 meters, what are the possible values for the width? (If the perimeter is at least 51 meters, then it is greater than or equal to 51 meters.) $x \ge \dfrac{15}{2}$; the width is at least $7\dfrac{1}{2}$ meters.

▶ **Chalkboard Problem**
This chalkboard problem is set up so that students will see the importance of applying the multiplication property for inequalities correctly. If they apply it incorrectly, their solution in part d will contradict their answer to part c.

73. **Geometry Problems** The numerical values of the three sides of a triangle are given by three consecutive even integers. If the perimeter is greater than 24 inches, what are the possibilities for the shortest side? $x > 6$; the shortest side is even and greater than 6 inches.

74. **Geometry Problems** The numerical values of the three sides of a triangle are given by three consecutive odd integers. If the perimeter is greater than 27 inches, what are the possibilities for the shortest side?
$x > 7$; the shortest side is odd and greater than 7 inches.

75. **Car Heaters** If you have ever gotten in a cold car early in the morning you know that the heater does not work until the engine warms up. This is because the heater relies on the heat coming off the engine. Write an equation using an inequality sign to express when the heater will work if the heater works only after the engine is 100°F. $t \geq 100$

76. **Exercise** When Kate exercises, she either swims or runs. She wants to spend a minimum of 8 hours a week exercising, and she wants to swim 3 times the amount she runs. What is the minimum amount of time she must spend doing each exercise? At least 2 hours running and 6 hours swimming.

77. **Profit and Loss** Movie theaters pay a certain price for the movies that you and I see. Suppose a theater pays $1,500 for each showing of a popular movie. If they charge $7.50 for each ticket they sell, then they will lose money if ticket sales are less than $1,500. However, they will make a profit if ticket sales are greater than $1,500. What is the range of tickets they can sell and still lose money? What is the range of tickets they can sell and make a profit?
Lose money if they sell less than 200 tickets. Make a profit if they sell more than 200 tickets.

78. **Stock Sales** Suppose you purchase x shares of a stock at $12 per share. After 6 months you decide to sell all your shares at $20 per share. Your broker charges you $15 for the trade. If your profit is at least $3,985, how many shares did you purchase in the first place? At least 500 shares.

Maintaining Your Skills

The problems that follow review some of the more important skills you have learned in previous sections and chapters. You can consider the time you spend working these problems as time spent studying for exams.

Apply the distributive property, then simplify.

79. $\frac{1}{6}(12x + 6)$ $2x + 1$

80. $\frac{3}{5}(15x - 10)$ $9x - 6$

81. $\frac{2}{3}(-3x - 6)$ $-2x - 4$

82. $\frac{3}{4}(-4x - 12)$ $-3x - 9$

83. $3\left(\frac{5}{6}a + \frac{4}{9}\right)$ $\frac{5}{2}a + \frac{4}{3}$

84. $2\left(\frac{3}{4}a - \frac{5}{6}\right)$ $\frac{3}{2}a - \frac{5}{3}$

85. $-3\left(\frac{2}{3}a + \frac{5}{6}\right)$ $-2a - \frac{5}{2}$

86. $-4\left(\frac{5}{6}a + \frac{4}{9}\right)$ $-\frac{10}{3}a - \frac{16}{9}$

Apply the distributive property, then find the LCD and simplify.

87. $\frac{1}{2}x + \frac{1}{6}x$ $\frac{2}{3}x$

88. $\frac{1}{2}x - \frac{3}{4}x$ $-\frac{1}{4}x$

89. $\frac{2}{3}x - \frac{5}{6}x$ $-\frac{1}{6}x$

90. $\frac{1}{3}x + \frac{3}{5}x$ $\frac{14}{15}x$

91. $\frac{3}{4}x + \frac{1}{6}x$ $\frac{11}{12}x$

92. $\frac{3}{2}x - \frac{2}{3}x$ $\frac{5}{6}x$

93. $\frac{2}{5}x + \frac{5}{8}x$ $\frac{41}{40}x$

94. $\frac{3}{5}x - \frac{3}{8}x$ $\frac{9}{40}x$

Getting Ready for the Next Section

Solve each inequality. Do not graph.

95. $2x - 1 \geq 3$ $x \geq 2$

96. $3x + 1 \geq 7$ $x \geq 2$

97. $-2x > -8$ $x < 4$

98. $-3x > -12$ $x < 4$

99. $-3 \leq 4x + 1$ $x \geq -1$

100. $4x + 1 \leq 9$ $x \leq 2$

OBJECTIVES

A Solve and graph compound inequalities.

The instrument panel on most cars includes a temperature gauge. The one shown below indicates that the normal operating temperature for the engine is from 50°F to 270°F.

We can represent the same situation with an inequality by writing $50 \leq F \leq 270$, where F is the temperature in degrees Fahrenheit. This inequality is a *compound inequality*. In this section we present the notation and definitions associated with compound inequalities.

The *union* of two sets A and B is the set of all elements that are in A or in B. The word *or* is the key word in the definition. The *intersection* of two sets A and B is the set of elements contained in both A and B. The key word in this definition is *and*. We can put the words *and* and *or* together with our methods of graphing inequalities to find the solution sets for compound inequalities.

> **DEFINITION** A **compound inequality** is two or more inequalities connected by the word *and* or *or*.

 EXAMPLE 1 Graph the solution set for the compound inequality

$$x < -1 \quad \text{or} \quad x \geq 3$$

SOLUTION Graphing each inequality separately, we have

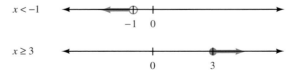

Because the two inequalities are connected by *or*, we want to graph their union; that is, we graph all points that are on either the first graph or the second graph. Essentially, we put the two graphs together on the same number line.

$$x < -1 \quad \text{or} \quad x \geq 3$$

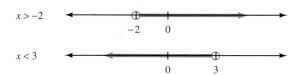 **EXAMPLE 2**   Graph the solution set for the compound inequality

$$x > -2 \quad \text{and} \quad x < 3$$

SOLUTION   Graphing each inequality separately, we have

Because the two inequalities are connected by the word *and,* we will graph their intersection, which consists of all points that are common to both graphs; that is, we graph the region where the two graphs overlap.

EXAMPLE 3   Solve and graph the solution set for

$$2x - 1 \geq 3 \quad \text{and} \quad -3x > -12$$

SOLUTION   Solving the two inequalities separately, we have

$$2x - 1 \geq 3 \quad \text{and} \quad -3x > -12$$
$$2x \geq 4$$
$$x \geq 2 \quad \text{and} \quad -\frac{1}{3}(-3x) < -\frac{1}{3}(-12)$$
$$x < 4$$

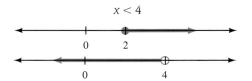

Because the word *and* connects the two graphs, we will graph their intersection—the points they have in common:

Notation Sometimes compound inequalities that use the word *and* can be written in a shorter form. For example, the compound inequality $-2 < x$ and $x < 3$ can be written as $-2 < x < 3$. The word *and* does not appear when an inequality is written in this form. It is implied. The solution set for $-2 < x$ and $x < 3$ is

It is all the numbers between -2 and 3 on the number line. It seems reasonable, then, that this graphs should be the graph of $-2 < x < 3$.

  In both the graph and the inequality, x is said to be between -2 and 3.

EXAMPLE 4 Solve and graph $-3 \leq 2x - 1 \leq 9$.

SOLUTION To solve for x, we must add 1 to the center expression and then divide the result by 2. Whatever we do to the center expression, we also must do to the two expressions on the ends. In this way we can be sure we are producing equivalent inequalities. The solution set will not be affected.

$$-3 \leq 2x - 1 \leq 9$$

$$-2 \leq \quad 2x \quad \leq 10 \qquad \textbf{Add 1 to each expression}$$

$$-1 \leq \quad x \quad \leq 5 \qquad \textbf{Multiply each expression by } \frac{1}{2}$$

-1 5

GETTING READY FOR CLASS

After reading through the preceding section, respond in your own words and in complete sentences.

1. What is a compound inequality?

2. Explain the shorthand notation that can be used to write two inequalities connected by the word *and*.

3. Write two inequalities connected by the word *and* that together are equivalent to $-1 < x < 2$.

4. Explain in words how you would graph the compound inequality

$$x < 2 \text{ or } x > -3$$

LINKING OBJECTIVES AND EXAMPLES

Next to each **objective** we have listed the examples that are best described by that objective.

A   1–4

Problem Set 2.9

Online support materials can be found at www.thomsonedu.com/login

Graph the following compound inequalities.

▶ **1.** $x < -1$ or $x > 5$ **2.** $x \leq -2$ or $x \geq -1$

3. $x < -3$ or $x \geq 0$ **4.** $x < 5$ and $x > 1$

5. $x \leq 6$ and $x > -1$ **6.** $x \leq 7$ and $x > 0$

7. $x > 2$ and $x < 4$ **8.** $x < 2$ or $x > 4$

▶ **9.** $x \geq -2$ and $x \leq 4$ **10.** $x \leq 2$ or $x \geq 4$

11. $x < 5$ and $x > -1$ **12.** $x > 5$ or $x < -1$

13. $-1 < x < 3$ **14.** $-1 \leq x \leq 3$

15. $-3 < x \leq -2$ **16.** $-5 \leq x \leq 0$

Solve the following compound inequalities. Graph the solution set in each case.

17. $3x - 1 < 5$ or $5x - 5 > 10$

18. $x + 1 < -3$ or $x - 2 > 6$

19. $x - 2 > -5$ and $x + 7 < 13$

20. $3x + 2 \leq 11$ and $2x + 2 \geq 0$

21. $11x < 22$ or $12x > 36$

22. $-5x < 25$ and $-2x \geq -12$

23. $3x - 5 < 10$ and $2x + 1 > -5$

☐ = Videos available by instructor request

▶ = Online student support materials available at www.thomsonedu.com/login

24. $5x + 8 < -7$ or $3x - 8 > 10$

▶ **25.** $2x - 3 < 8$ and $3x + 1 > -10$

26. $11x - 8 > 3$ or $12x + 7 < -5$

27. $2x - 1 < 3$ and $3x - 2 > 1$

28. $3x + 9 < 7$ or $2x - 7 > 11$

29. $-1 \le x - 5 \le 2$ **30.** $0 \le x + 2 \le 3$

31. $-4 \le 2x \le 6$ **32.** $-5 < 5x < 10$

▶ **33.** $-3 < 2x + 1 < 5$ **34.** $-7 \le 2x - 3 \le 7$

35. $0 \le 3x + 2 \le 7$ **36.** $2 \le 5x - 3 \le 12$

37. $-7 < 2x + 3 < 11$ **38.** $-5 < 6x - 2 < 8$

39. $-1 \le 4x + 5 \le 9$ **40.** $-8 \le 7x - 1 \le 13$

For each graph below, write an inequality whose solution is the graph.

41.
−2   3
$-2 < x < 3$

42.
−2   3
$x \le -2$ or $x \ge 3$

43.
−2   3
$-2 \le x \le 3$

44.
−2   3
$x < -2$ or $x > 3$

Applying the Concepts

Triangle Inequality The triangle inequality states that the sum of any two sides of a triangle must be greater than the third side.

45. The following triangle RST has sides of length x, $2x$, and 10 as shown.

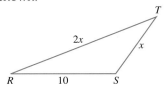

a. Find the three inequalities, which must be true based on the sides of the triangle.
$2x + x > 10$; $x + 10 > 2x$; $2x + 10 > x$

b. Write a compound inequality based on your results above. $\dfrac{10}{3} < x < 10$

46. The following triangle ABC has sides of length x, $3x$, and 16 as shown.

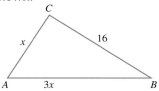

a. Find the three inequalities, which must be true based on the sides of the triangle.
$3x + x > 16$; $3x + 16 > x$; $x + 16 > 3x$

b. Write a compound inequality based on your results above. $4 < x < 8$

47. Engine Temperature The engine in a car gives off a lot of heat due to the combustion in the cylinders. The water used to cool the engine keeps the temperature within the range $50 \le F \le 266$ where F is in degrees Fahrenheit. Graph this inequality.

48. Engine Temperature To find the engine temperature range from Problem 47 in degrees Celsius, we use the fact that $F = \dfrac{9}{5}C + 32$ to rewrite the inequality as

$$50 \le \dfrac{9}{5}C + 32 \le 266$$

Solve this inequality and graph the solution set.
$10 \le C \le 130$

49. Number Problem The difference of twice a number and 3 is between 5 and 7. Find the number. $4 < x < 5$

50. Number Problem The sum of twice a number and 5 is between 7 and 13. Find the number. $1 < x < 4$

51. Perimeter The length of a rectangle is 4 inches longer than the width. If the perimeter is between 20 inches and 30 inches, find all possible values for the width. The width is between 3 inches and $\dfrac{11}{2}$ inches.

52. Perimeter The length of a rectangle is 6 feet longer than the width. If the perimeter is between 24 feet and 36 feet, find all possible values for the width. The width is between 3 feet and 6 feet.

Maintaining Your Skills

The problems that follow review some of the more important skills you have learned in previous sections and chapters. You can consider the time you spend working these problems as time spent studying for exams.

Answer the following percent problems.

53. What number is 25% of 32? 8

54. What number is 15% of 75? 11.25

55. What number is 20% of 120? 24

56. What number is 125% of 300? 375

57. What percent of 36 is 9? 25%

58. What percent of 16 is 9? 56.25%

59. What percent of 50 is 5? 10%

60. What percent of 140 is 35? 25%

61. 16 is 20% of what number? 80

62. 6 is 3% of what number? 200

63. 8 is 2% of what number? 400

64. 70 is 175% of what number? 40

Simplify each expression.

65. $-|-5|$ -5

66. $\left(-\dfrac{2}{3}\right)^3$ $-\dfrac{8}{27}$

67. $-3 - 4(-2)$ 5

68. $2^4 + 3^3 \div 9 - 4^2$ 3

69. $5|3 - 8| - 6|2 - 5|$ 7

70. $7 - 3(2 - 6)$ 19

71. $5 - 2[-3(5 - 7) - 8]$ 9

72. $\dfrac{5 + 3(7 - 2)}{2(-3) - 4}$ -2

73. Find the difference of -3 and -9. 6

74. If you add -4 to the product of -3 and 5, what number results? -19

75. Apply the distributive property to $\frac{1}{2}(4x - 6)$. $2x - 3$

76. Use the associative property to simplify $-6(\frac{1}{3}x)$. $-2x$

Given the numbers in the set $\{-3, -\frac{4}{5}, 0, \frac{5}{8}, 2, \sqrt{5}\}$:

77. List all the integers. $-3, 0, 2$

78. List all the rational numbers. $-3, -\frac{4}{5}, 0, \frac{5}{8}, 2$

EXAMPLES

1. The terms $2x$, $5x$, and $-7x$ are all similar because their variable parts are the same.

Similar Terms [2.1]

A *term* is a number or a number and one or more variables multiplied together. *Similar terms* are terms with the same variable part.

Simplifying Expressions [2.1]

2. Simplify $3x + 4x$.
$$3x + 4x = (3 + 4)x$$
$$= 7x$$

In this chapter we simplified expressions that contained variables by using the distributive property to combine similar terms.

Solution Set [2.2]

3. The solution set for the equation $x + 2 = 5$ is $\{3\}$ because when x is 3 the equation is $3 + 2 = 5$ or $5 = 5$.

The *solution set* for an equation (or inequality) is all the numbers that, when used in place of the variable, make the equation (or inequality) a true statement.

Equivalent Equations [2.2]

4. The equation $a - 4 = 3$ and $a - 2 = 5$ are equivalent because both have solution set $\{7\}$.

Two equations are called *equivalent* if they have the same solution set.

Addition Property of Equality [2.2]

5. Solve $x - 5 = 12$.
$$x - 5 + \mathbf{5} = 12 + \mathbf{5}$$
$$x + 0 = 17$$
$$x = 17$$

When the same quantity is added to both sides of an equation, the solution set for the equation is unchanged. Adding the same amount to both sides of an equation produces an equivalent equation.

Multiplication Property of Equality [2.3]

6. Solve $3x = 18$.
$$\frac{1}{3}(3x) = \frac{1}{3}(18)$$
$$x = 6$$

If both sides of an equation are multiplied by the same nonzero number, the solution set is unchanged. Multiplying both sides of an equation by a nonzero quantity produces an equivalent equation.

Strategy for Solving Linear Equations in One Variable [2.4]

7. Solve $2(x + 3) = 10$.
$$2x + 6 = 10$$
$$2x + 6 + (\mathbf{-6}) = 10 + (\mathbf{-6})$$
$$2x = 4$$
$$\frac{1}{2}(2x) = \frac{1}{2}(4)$$
$$x = 2$$

Step 1a: Use the distributive property to separate terms, if necessary.

    1b: If fractions are present, consider multiplying both sides by the LCD to eliminate the fractions. If decimals are present, consider multiplying both sides by a power of 10 to clear the equation of decimals.

    1c: Combine similar terms on each side of the equation.

Step 2: Use the addition property of equality to get all variable terms on one side of the equation and all constant terms on the other side. A variable term is a term that contains the variable (for example, 5x). A constant term is a term that does not contain the variable (the number 3, for example.)

Step 3: Use the multiplication property of equality to get x (that is, 1x) by itself on one side of the equation.

Step 4: Check your solution in the original equation to be sure that you have not made a mistake in the solution process.

Formulas [2.5]

8. Solving $P = 2l + 2w$ for l, we have

$$P - 2w = 2l$$
$$\frac{P - 2w}{2} = l$$

A formula is an equation with more than one variable. To solve a formula for one of its variables, we use the addition and multiplication properties of equality to move everything except the variable in question to one side of the equal sign so the variable in question is alone on the other side.

Blueprint for Problem Solving [2.6, 2.7]

Step 1: *Read* the problem, and then mentally *list* the items that are known and the items that are unknown.

Step 2: *Assign a variable* to one of the unknown items. (In most cases this will amount to letting x = the item that is asked for in the problem.) Then *translate* the other *information* in the problem to expressions involving the variable.

Step 3: *Reread* the problem, and then *write an equation,* using the items and variables listed in steps 1 and 2, that describes the situation.

Step 4: *Solve the equation* found in step 3.

Step 5: *Write* your *answer* using a complete sentence.

Step 6: *Reread* the problem, and *check* your solution with the original words in the problem.

Addition Property for Inequalities [2.8]

9. Solve $x + 5 < 7$.
$$x + 5 + (-5) < 7 + (-5)$$
$$x < 2$$

Adding the same quantity to both sides of an inequality produces an equivalent inequality, one with the same solution set.

Multiplication Property for Inequalities [2.8]

10. Solve $-3a \leq 18$.
$$-\frac{1}{3}(-3a) \geq -\frac{1}{3}(18)$$
$$a \geq -6$$

Multiplying both sides of an inequality by a positive number never changes the solution set. If both sides are multiplied by a negative number, the sense of the inequality must be reversed to produce an equivalent inequality.

11. Solve $3(x - 4) \geq -2$.

$$3x - 12 \geq -2$$
$$3x - 12 + \mathbf{12} \geq -2 + \mathbf{12}$$
$$3x \geq 10$$
$$\frac{1}{3}(3x) \geq \frac{1}{3}(10)$$
$$x \geq \frac{10}{3}$$

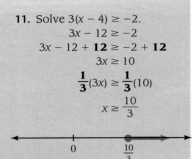

Strategy for Solving Linear Inequalities in One Variable [2.8]

Step 1a: Use the distributive property to separate terms, if necessary.

   1b: If fractions are present, consider multiplying both sides by the LCD to eliminate the fractions. If decimals are present, consider multiplying both sides by a power of 10 to clear the inequality of decimals.

   1c: Combine similar terms on each side of the inequality.

Step 2: Use the addition property for inequalities to get all variable terms on one side of the inequality and all constant terms on the other side.

Step 3: Use the multiplication property for inequalities to get x by itself on one side of the inequality.

Step 4: Graph the solution set.

12. $x < -3$ or $x > 1$

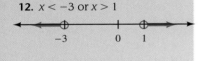

$-2 \leq x \leq 3$

Compound Inequalities [2.9]

Two inequalities connected by the word *and* or *or* form a compound inequality. If the connecting word is *or,* we graph all points that are on either graph. If the connecting word is *and,* we graph only those points that are common to both graphs. The inequality $-2 \leq x \leq 3$ is equivalent to the compound inequality $-2 \leq x$ and $x \leq 3$.

> **! COMMON MISTAKES**
>
> 1. Trying to subtract away coefficients (the number in front of variables) when solving equations. For example:
>
> $$4x = 12$$
> $$4x - \mathbf{4} = 12 - \mathbf{4}$$
> $$x = 8 \leftarrow \text{Mistake}$$
>
>   It is not incorrect to add (-4) to both sides; it's just that $4x - 4$ is not equal to x. Both sides should be multiplied by $\frac{1}{4}$ to solve for x.
>
> 2. Forgetting to reverse the direction of the inequality symbol when multiplying both sides of an inequality by a negative number. For instance:
>
> $$-3x < 12$$
> $$-\frac{1}{3}(-3x) < -\frac{1}{3}(12) \leftarrow \text{Mistake}$$
> $$x < -4$$
>
>   It is not incorrect to multiply both sides by $-\frac{1}{3}$. But if we do, we must also reverse the sense of the inequality.

The problems below form a comprehensive review of the material in this chapter. They can be used to study for exams. If you would like to take a practice test on this chapter, you can use the odd-numbered problems. Give yourself an hour and work as many of the odd-numbered problems as possible. When you are finished, or when an hour has passed, check your answers with the answers in the back of the book. You can use the even-numbered problems for a second practice test.

The numbers in brackets refer to the sections of the text in which similar problems can be found.

Simplify each expression as much as possible. [2.1]

1. $5x - 8x$

2. $6x - 3 - 8x$

3. $-a + 2 + 5a - 9$

4. $5(2a - 1) - 4(3a - 2)$

5. $6 - 2(3y + 1) - 4$

6. $4 - 2(3x - 1) - 5$

Find the value of each expression when x is 3. [2.1]

7. $7x - 2$

8. $-4x - 5 + 2x$

9. $-x - 2x - 3x$

Find the value of each expression when x is -2. [2.1]

10. $5x - 3$

11. $-3x + 2$

12. $7 - x - 3$

Solve each equation. [2.2, 2.3]

13. $x + 2 = -6$

14. $x - \dfrac{1}{2} = \dfrac{4}{7}$

15. $10 - 3y + 4y = 12$

16. $-3 - 4 = -y - 2 + 2y$

17. $2x = -10$

18. $3x = 0$

19. $\dfrac{x}{3} = 4$

20. $-\dfrac{x}{4} = 2$

21. $3a - 2 = 5a$

22. $\dfrac{7}{10}a = \dfrac{1}{5}a + \dfrac{1}{2}$

23. $3x + 2 = 5x - 8$

24. $6x - 3 = x + 7$

25. $0.7x - 0.1 = 0.5x - 0.1$

26. $0.2x - 0.3 = 0.8x - 0.3$

Solve each equation. Be sure to simplify each side first. [2.4]

27. $2(x - 5) = 10$

28. $12 = 2(5x - 4)$

29. $\dfrac{1}{2}(3t - 2) + \dfrac{1}{2} = \dfrac{5}{2}$

30. $\dfrac{3}{5}(5x - 10) = \dfrac{2}{3}(9x + 3)$

31. $2(3x + 7) = 4(5x - 1) + 18$

32. $7 - 3(y + 4) = 10$

Use the formula $4x - 5y = 20$ to find y if [2.5]

33. x is 5

34. x is 0

35. x is -5

36. x is 10

Solve each of the following formulas for the indicated variable. [2.5]

37. $2x - 5y = 10$ for y

38. $5x - 2y = 10$ for y

39. $V = \pi r^2 h$ for h

40. $P = 2l + 2w$ for w

41. What number is 86% of 240? [2.5]

42. What percent of 2,000 is 180? [2.5]

Solve each of the following word problems. In each case, be sure to show the equation that describes the situation. [2.6, 2.7]

43. **Number Problem** The sum of twice a number and 6 is 28. Find the number.

44. **Geometry** The length of a rectangle is 5 times as long as the width. If the perimeter is 60 meters, find the length and the width.

45. **Investing** A man invests a certain amount of money in an account that pays 9% annual interest. He invests $300 more than that in an account that pays 10% annual interest. If his total interest after a year is $125, how much does he have invested in each account?

46. **Coin Problem** A collection of 15 coins is worth $1.00. If the coins are dimes and nickels, how many of each coin are there?

Solve each inequality. [2.8]

47. $-2x < 4$

48. $-5x > -10$

49. $-\dfrac{a}{2} \le -3$

50. $-\dfrac{a}{3} > 5$

Solve each inequality, and graph the solution. [2.8]

51. $-4x + 5 > 37$

52. $2x + 10 < 5x - 11$

53. $2(3t + 1) + 6 \ge 5(2t + 4)$

Graph the solution to each of the following compound inequalities. [2.9]

54. $x < -2$ or $x > 5$

55. $-5x \ge 25$ or $2x - 3 \ge 9$

56. $-1 < 2x + 1 < 9$

GROUP PROJECT Tables and Graphs

Number of People 2-3

Time Needed 5–10 minutes

Equipment Pencil and graph paper

Background Building tables is a method of visualizing information. We can build a table from a situation (as below) or from an equation. In this project, we will first build a table and then write an equation from the information in the table.

Procedure A parking meter, which accepts only dimes and quarters, is emptied at the end of each day. The amount of money in the meter at the end of one particular day is $3.15.

1. Complete the following table so that all possible combinations of dimes and quarters, along with the total number of coins, is shown. Remember, although the number of coins will vary, the value of the dimes and quarters must total $3.15.

2. From the information in the table, answer the following questions.

   a. What is the maximum possible number of coins taken from the meter?

   b. What is the minimum possible number of coins taken from the meter?

   c. When is the number of dimes equal to the number of quarters?

3. Let x = the number of dimes and y = the number of quarters. Write an equation in two variables such that the value of the dimes added to the value of the quarters is $3.15.

Number of Number of Dimes	Quarters	Total Coins	Value
29	1	30	$3.15
24			$3.15
			$3.15
			$3.15
			$3.15
			$3.15

Stand and Deliver

The Kobal Collection

The 1988 film *Stand and Deliver* starring Edward James Olmos and Lou Diamond Phillips is based on a true story. Olmos, in his portrayal of high-school math teacher Jaime Escalante, earned an Academy Award nomination for best actor.

Watch the movie *Stand and Deliver*. After briefly describing the movie, explain how Escalante's students became successful in math. Make a list of specific things you observe that the students had to do to become successful. Indicate which items on this list you think will also help you become successful.

Linear Equations and Inequalities in Two Variables

3

Klaus Hackenberg/zefa/Corbis

W hen light comes into contact with a surface that does not transmit light, then all the light that contacts the surface is either reflected off the surface or absorbed into the surface. If we let R represent the percentage of light reflected and A represent the percentage of light absorbed, then the relationship between these two variables can be written as

$$R + A = 100$$

which is a linear equation in two variables. The following table and graph show the same relationship as that described by the equation. The table is a numerical description; the graph is a visual description.

Reflected and Absorbed Light	
Percent Reflected	Percent Absorbed
0	100
20	80
40	60
60	40
80	20
100	0

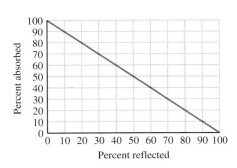

In this chapter we learn how to build tables and draw graphs from linear equations in two variables.

▶ Improve your grade and save time!
Go online to **www.thomsonedu.com/login** where you can
- Watch videos of instructors working through the in-text examples
- Follow step-by-step online tutorials of in-text examples and review questions
- Work practice problems
- Check your readiness for an exam by taking a pre-test and exploring the modules recommended in your Personalized Study plan
- Receive help from a live tutor online through vMentor™

Try it out! Log in with an access code or purchase access at **www.ichapters.com.**

Try to arrange your daily study habits so that you have very little studying to do the night before your next exam. The next two goals will help you achieve this.

1 Review with the Exam in Mind

Each day you should review material that will be covered on the next exam. Your review should consist of working problems. Preferably, the problems you work should be problems from your list of difficult problems.

2 Pay Attention to Instructions

Each of the following is a valid instruction with respect to the equation $y = 3x - 2$, and the result of applying the instructions will be different in each case:

Find x when y is 10.	(Section 2.5)
Solve for x.	(Section 2.5)
Graph the equation.	(Section 3.3)
Find the intercepts.	(Section 3.4)

There are many things to do with the equation $y = 3x - 2$. If you train yourself to pay attention to the instructions that accompany a problem as you work through the assigned problems, you will not find yourself confused about what to do with a problem when you see it on a test.

OBJECTIVES

A Create a bar chart, scatter diagram, or line graph from a table of data.

B Graph ordered pairs on a rectangular coordinate system.

In Chapter 1 we showed the relationship between the table of values for the speed of a race car and the corresponding bar chart. Table 1 and Figure 1 from the introduction of Chapter 1 are reproduced here for reference. In Figure 1, the horizontal line that shows the elapsed time in seconds is called the *horizontal axis,* and the vertical line that shows the speed in miles per hour is called the *vertical axis.*

The data in Table 1 are called *paired data* because the information is organized so that each number in the first column is paired with a specific number in the second column. Each pair of numbers is associated with one of the solid bars in Figure 1. For example, the third bar in the bar chart is associated with the pair of numbers 3 seconds and 162.8 miles per hour. The first number, 3 seconds, is associated with the horizontal axis, and the second number, 162.8 miles per hour, is associated with the vertical axis.

TABLE 1
Speed of a Race Car

Time in Seconds	Speed in Miles per Hour
0	0
1	72.7
2	129.9
3	162.8
4	192.2
5	212.4
6	228.1

Scatter Diagrams and Line Graphs

The information in Table 1 can be visualized with a *scatter diagram* and *line graph* as well. Figure 2 is a scatter diagram of the information in Table 1. We use dots instead of the bars shown in Figure 1 to show the speed of the race car at each second during the race. Figure 3 is called a *line graph.* It is constructed by taking the dots in Figure 2 and connecting each one to the next with a straight line. Notice that we have labeled the axes in these two figures a little differently than we did with the bar chart by making the axes intersect at the number 0.

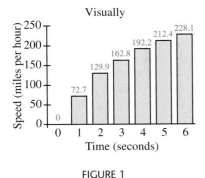

FIGURE 1

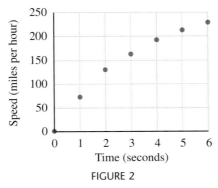

FIGURE 2

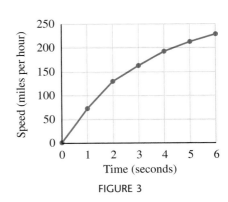

FIGURE 3

The number sequences we have worked with in the past can also be written as paired data by associating each number in the sequence with its position in the sequence. For instance, in the sequence of odd numbers

$$1, 3, 5, 7, 9, \ldots$$

the number 7 is the fourth number in the sequence. Its position is 4, and its value is 7. Here is the sequence of odd numbers written so that the position of each term is noted:

Position   1, 2, 3, 4, 5, . . .

Value   1, 3, 5, 7, 9, . . .

EXAMPLE 1   Tables 2 and 3 give the first five terms of the sequence of odd numbers and the sequence of squares as paired data. In each case construct a scatter diagram.

TABLE 2
Odd Numbers

Position	Value
1	1
2	3
3	5
4	7
5	9

TABLE 3
Squares

Position	Value
1	1
2	4
3	9
4	16
5	25

SOLUTION   The two scatter diagrams are based on the data from Tables 2 and 3 shown here. Notice how the dots in Figure 4 seem to line up in a straight line, whereas the dots in Figure 5 give the impression of a curve. We say the points in Figure 4 suggest a *linear* relationship between the two sets of data, whereas the points in Figure 5 suggest a *nonlinear* relationship.

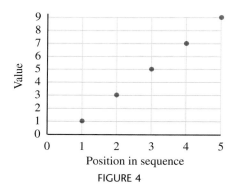

FIGURE 4

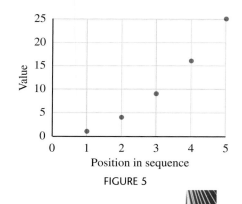

FIGURE 5

As you know, each dot in Figures 4 and 5 corresponds to a pair of numbers, one of which is associated with the horizontal axis and the other with the vertical axis. Paired data play a very important role in the equations we will solve in the next section. To prepare ourselves for those equations, we need to expand the concept of paired data to include negative numbers. At the same time, we want to standardize the position of the axes in the diagrams that we use to visualize paired data.

> **DEFINITION**   A pair of numbers enclosed in parentheses and separated by a comma, such as $(-2, 1)$, is called an **ordered pair** of numbers. The first number in the pair is called the *x*-**coordinate** of the ordered pair; the second number is called the *y*-**coordinate**. For the ordered pair $(-2, 1)$, the *x*-coordinate is -2 and the *y*-coordinate is 1.

Ordered pairs of numbers are important in the study of mathematics because they give us a way to visualize solutions to equations. To see the visual component of ordered pairs, we need the diagram shown in Figure 6. It is called the *rectangular coordinate system*.

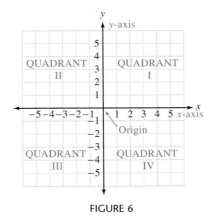

FIGURE 6

The rectangular coordinate system is built from two number lines oriented perpendicular to each other. The horizontal number line is exactly the same as our real number line and is called the *x-axis*. The vertical number line is also the same as our real number line with the positive direction up and the negative direction down. It is called the *y-axis*. The point where the two axes intersect is called the *origin*. As you can see from Figure 6, the axes divide the plane into four quadrants, which are numbered I through IV in a counterclockwise direction.

Graphing Ordered Pairs

To graph the ordered pair (a, b), we start at the origin and move a units forward or back (forward if a is positive and back if a is negative). Then we move b units up or down (up if b is positive, down if b is negative). The point where we end up is the graph of the ordered pair (a, b).

EXAMPLE 2   Graph the ordered pairs $(3, 4)$, $(3, -4)$, $(-3, 4)$, and $(-3, -4)$.

SOLUTION

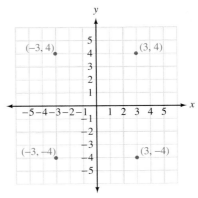

FIGURE 7

Note

It is very important that you graph ordered pairs quickly and accurately. Remember, the first coordinate goes with the horizontal axis and the second coordinate goes with the vertical axis.

We can see in Figure 7 that when we graph ordered pairs, the x-coordinate corresponds to movement parallel to the x-axis (horizontal) and the y-coordinate corresponds to movement parallel to the y-axis (vertical).

 EXAMPLE 3   Graph the ordered pairs $(-1, 3)$, $(2, 5)$, $(0, 0)$, $(0, -3)$, and $(4, 0)$.

SOLUTION   See Figure 8.

> **Note**
> If we do not label the axes of a coordinate system, we assume that each square is one unit long and one unit wide.

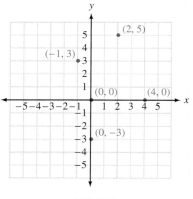

FIGURE 8

LINKING OBJECTIVES AND EXAMPLES

Next to each **objective** we have listed the examples that are best described by that objective.

A   1

B   2, 3

GETTING READY FOR CLASS

After reading through the preceding section, respond in your own words and in complete sentences.

1. What is an ordered pair of numbers?
2. Explain in words how you would graph the ordered pair (3, 4).
3. How do you construct a rectangular coordinate system?
4. Where is the origin on a rectangular coordinate system?

Problem Set 3.1

Online support materials can be found at www.thomsonedu.com/login

Answers appear in the Instructor's Edition only.

Graph the following ordered pairs.

▶ **1.** (3, 2)

▶ **2.** (3, −2)

▶ **3.** (−3, 2)

▶ **4.** (−3, −2)

▶ **5.** (5, 1)

6. (5, −1)

7. (1, 5)

8. (1, −5)

9. (−1, 5)

10. (−1, −5)

11. $(2, \frac{1}{2})$

12. $(3, \frac{3}{2})$

▶ **13.** $(-4, -\frac{5}{2})$

14. $(-5, -\frac{3}{2})$

15. (3, 0)

16. (−2, 0)

▶ **17.** (0, 5)

▶ **18.** (0, 0)

░ = Videos available by instructor request

▶ = Online student support materials available at www.thomsonedu.com/login

Give the coordinates of each numbered point in the figure.

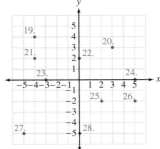

19. (−4, 4)

20. (3, 3)

21. (−4, 2)

22. (0, 2)

23. (−3, 0)

24. (5, 0)

25. (2, −2)

26. (5, −2)

27. (−5, −5)

28. (0, −5)

Graph the points (4, 3) and (−4, −1), and draw a straight line that passes through both of them. Then answer the following questions.

29. Does the graph of (2, 2) lie on the line? Yes

30. Does the graph of (−2, 0) lie on the line? Yes

31. Does the graph of (0, −2) lie on the line? No

32. Does the graph of (−6, 2) lie on the line? No

Graph the points (−2, 4) and (2, −4), and draw a straight line that passes through both of them. Then answer the following questions.

33. Does the graph of (0, 0) lie on the line? Yes

34. Does the graph of (−1, 2) lie on the line? Yes

35. Does the graph of (2, −1) lie on the line? No

36. Does the graph of (1, −2) lie on the line? Yes

Draw a straight line that passes through the points (3, 4) and (3, −4). Then answer the following questions.

37. Is the graph of (3, 0) on this line? Yes

38. Is the graph of (0, 3) on this line? No

39. Is there any point on this line with an *x*-coordinate other than 3? No

40. If you extended the line, would it pass through a point with a *y*-coordinate of 10? Yes

Draw a straight line that passes through the points (3, 4) and (−3, 4). Then answer the following questions.

41. Is the graph of (4, 0) on this line? No

42. Is the graph of (0, 4) on this line? Yes

43. Is there any point on this line with a *y*-coordinate other than 4? No

44. If you extended the line, would it pass through a point with an *x*-coordinate of 10? Yes

Applying the Concepts

45. Hourly Wages Jane takes a job at the local Marcy's department store. Her job pays $8.00 per hour. The graph shows how much Jane earns for working from 0 to 40 hours in a week.

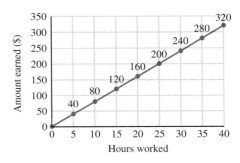

a. List three ordered pairs that lie on the line graph.
(5, 40), (10, 80), (20, 160), answers may vary

b. How much will she earn for working 40 hours? $320

c. If her check for one week is $240, how many hours did she work? 30 hours

d. She works 35 hours one week, but her paycheck before deductions are subtracted out is for $260. Is this correct? Explain.
No, if she works 35 hours, she should be paid $280.

46. Hourly Wages Judy takes a job at Gigi's boutique. Her job pays $6.00 per hour plus $50 per week in commission. The graph shows how much Judy earns for working from 0 to 40 hours in a week.

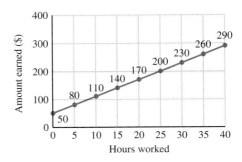

a. List three ordered pairs that lie on the line graph.
(5, 80), (10, 110), (20, 170), answers may vary

b. How much will she earn for working 40 hours? $290

c. If her check for one week is $230, how many hours did she work? 30 hours

d. She works 35 hours one week, but her paycheck before deductions are subtracted out is for $260. Is this correct? Explain. Yes, this is the correct amount.

47. Garbage Production The table and bar chart shown here give the annual production of garbage in the United States for some specific years. Use the information from the table and bar chart to construct a line graph.

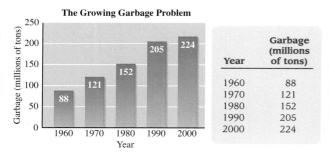

Year	Garbage (millions of tons)
1960	88
1970	121
1980	152
1990	205
2000	224

48. Grass Height The table and bar chart from Problem 99 in Problem Set 1.4 are shown here. Each gives the growth of a certain species of grass over time. Use the information from the table and chart to construct a line graph.

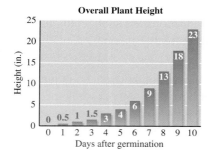

Day	Plant Height (inches)
0	0
1	0.5
2	1
3	1.5
4	3
5	4
6	6
7	9
8	13
9	18
10	23

49. Wireless Phone Costs The table and bar chart shown here provide the projected cost of a wireless phone use. Use the information from the table and chart to construct a line graph.

Year	Cents per Minute
1998	33
1999	28
2000	25
2001	23
2002	22
2003	20

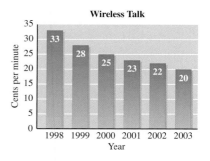

50. Digital Camera Sales The table and bar chart shown here show the sales of digital cameras from 1996 to 1999. Use the information from the table and chart to construct a line graph.

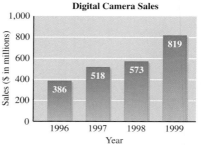

Year	Sales (in millions)
1996	$386
1997	$518
1998	$573
1999	$819

51. Kentucky Derby The bar chart gives the monetary bets placed at the Kentucky Derby for specific years. If x represents the year in question and y represents the total wagering for that year (in millions), write six ordered pairs that describe the information in the table. (1981, 5), (1985, 20.2), (1990, 34.4), (1995, 44.8), (2000, 65.4), (2004, 99.4)

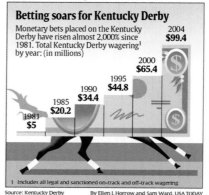

From *USA Today*. Copyright 2005. Reprinted with permission.

52. Improving Your Quantitative Literacy Use the chart to answer the following questions.

   a. Are the expenditures for 2003 more or less than $6,000? *Less*

   b. Are the projected expenditures for 2009 more or less than $8,000? *More*

USA TODAY Snapshots®

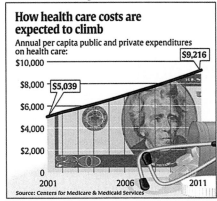

How health care costs are expected to climb

Annual per capita public and private expenditures on health care:

From *USA Today*. Copyright 2004. Reprinted with permission.

53. Right triangle *ABC* has legs of length 5. Point *C* is the ordered pair (6, 2). Find the coordinates of *A* and *B*. *A* = (1, 2), *B* = (6, 7)

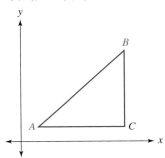

54. Right triangle *ABC* has legs of length 7. Point *C* is the ordered pair (−8, −3). Find the coordinates of *A* and *B*. *A* = (−1, −3), *B* = (−8, −10)

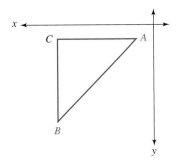

55. Rectangle *ABCD* has a length of 5 and a width of 3. Point *D* is the ordered pair (7, 2). Find points *A*, *B*, and *C*. *A* = (2, 2), *B* = (2, 5), *C* = (7, 5)

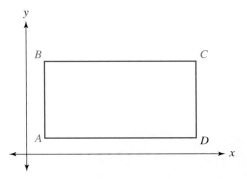

56. Rectangle *ABCD* has a length of 5 and a width of 3. Point *D* is the ordered pair (−1, 1). Find points *A*, *B*, and *C*. *A* = (−6, 1), *B* = (−6, 4), *C* = (−1, 4)

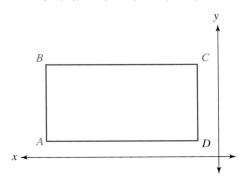

Maintaining Your Skills

Add or subtract as indicated.

57. $\dfrac{x}{5} + \dfrac{3}{5}$ $\dfrac{x+3}{5}$

58. $\dfrac{x}{5} + \dfrac{3}{4}$ $\dfrac{4x+15}{20}$

59. $\dfrac{2}{7} - \dfrac{a}{7}$ $\dfrac{2-a}{7}$

60. $\dfrac{2}{7} - \dfrac{a}{5}$ $\dfrac{10-7a}{35}$

61. $\dfrac{1}{14} - \dfrac{y}{7}$ $\dfrac{1-2y}{14}$

62. $\dfrac{3}{4} + \dfrac{x}{5}$ $\dfrac{15+4x}{20}$

63. $\dfrac{1}{2} + \dfrac{3}{x}$ $\dfrac{x+6}{2x}$

64. $\dfrac{2}{3} - \dfrac{6}{y}$ $\dfrac{2y-18}{3y}$

65. $\dfrac{5+x}{6} - \dfrac{5}{6}$ $\dfrac{x}{6}$

66. $\dfrac{3-x}{3} + \dfrac{2}{3}$ $\dfrac{5-x}{3}$

67. $\dfrac{4}{x} + \dfrac{1}{2}$ $\dfrac{8+x}{2x}$

68. $\dfrac{3}{y} + \dfrac{2}{3}$ $\dfrac{9+2y}{3y}$

Getting Ready for the Next Section

69. Let $2x + 3y = 6$.
   a. Find x if $y = 4$. -3 **b.** Find x if $y = -2$. 6
   c. Find y if $x = 3$. 0 **d.** Find y if $x = 9$. -4

70. Let $2x - 5y = 20$.
   a. Find x if $y = 0$. 10 **b.** Find x if $y = -6$. -5
   c. Find y if $x = 0$. -4 **d.** Find y if $x = 5$. -2

71. Let $y = 2x - 1$.
   a. Find x if $y = 7$. 4 **b.** Find x if $y = 3$. 2
   c. Find y if $x = 0$. -1 **d.** Find y if $x = 5$. 9

72. Let $y = 3x - 2$.
   a. Find x if $y = 4$. 2 **b.** Find x if $y = 7$. 3
   c. Find y if $x = 0$. -2 **d.** Find y if $x = -3$. -11

3.2   Solutions to Linear Equations in Two Variables

OBJECTIVES

A   Find solutions to linear equations in two variables.

B   Decide if an ordered pair is a solution to a linear equation in two variables.

In this section we will begin to investigate equations in two variables. As you will see, equations in two variables have pairs of numbers for solutions. Because we know how to use paired data to construct tables, histograms, and other charts, we can take our work with paired data further by using equations in two variables to construct tables of paired data. Let's begin this section by reviewing the relationship between equations in one variable and their solutions.

If we solve the equation $3x - 2 = 10$, the solution is $x = 4$. If we graph this solution, we simply draw the real number line and place a dot at the point whose coordinate is 4. The relationship between linear equations in one variable, their solutions, and the graphs of those solutions look like this:

Equation	Solution	Graph of Solution Set
$3x - 2 = 10$	$x = 4$	a number line with a dot at 4, 0 marked
$x + 5 = 7$	$x = 2$	a number line with a dot at 2, 0 marked
$2x = -6$	$x = -3$	a number line with a dot at -3, 0 marked

When the equation has one variable, the solution is a single number whose graph is a point on a line.

Now, consider the equation $2x + y = 3$. The first thing we notice is that there are two variables instead of one. Therefore, a solution to the equation $2x + y = 3$ will be not a single number but a pair of numbers, one for x and one for y, that makes the equation a true statement. One pair of numbers that works is $x = 2$, $y = -1$ because when we substitute them for x and y in the equation, we get a true statement.

$$2(2) + (-1) \overset{?}{=} 3$$

$$4 - 1 = 3$$

$$3 = 3 \qquad \textbf{A true statement}$$

The pair of numbers $x = 2, y = -1$ is written as $(2, -1)$. As you know from Section 3.1, $(2, -1)$ is called an *ordered pair* because it is a pair of numbers written in a specific order. The first number is always associated with the variable x,

Note

If this discussion seems a little long and confusing, you may want to look over some of the examples first and then come back and read this. Remember, it isn't always easy to read material in mathematics. What is important is that you understand what you are doing when you work problems. The reading is intended to assist you in understanding what you are doing. It is important to read everything in the book, but you don't always have to read it in the order it is written.

and the second number is always associated with the variable y. We call the first number in the ordered pair the *x-coordinate* (or x component) and the second number the *y-coordinate* (or y component) of the ordered pair.

Let's look back to the equation $2x + y = 3$. The ordered pair $(2, -1)$ is not the only solution. Another solution is $(0, 3)$ because when we substitute 0 for x and 3 for y we get

$$2(0) + 3 \overset{?}{=} 3$$

$$0 + 3 = 3$$

$$3 = 3 \qquad \textbf{A true statement}$$

Still another solution is the ordered pair $(5, -7)$ because

$$2(5) + (-7) \overset{?}{=} 3$$

$$10 - 7 = 3$$

$$3 = 3 \qquad \textbf{A true statement}$$

As a matter of fact, for any number we want to use for x, there is another number we can use for y that will make the equation a true statement. There is an infinite number of ordered pairs that satisfy (are solutions to) the equation $2x + y = 3$; we have listed just a few of them.

 EXAMPLE 1 Given the equation $2x + 3y = 6$, complete the following ordered pairs so they will be solutions to the equation: $(0, \ )$, $(\ , 1)$, $(3, \ )$.

SOLUTION To complete the ordered pair $(0, \ )$, we substitute 0 for x in the equation and then solve for y:

$$2(0) + 3y = 6$$

$$3y = 6$$

$$y = 2$$

The ordered pair is $(0, 2)$.

To complete the ordered pair $(\ , 1)$, we substitute 1 for y in the equation and solve for x:

$$2x + 3(1) = 6$$

$$2x + 3 = 6$$

$$2x = 3$$

$$x = \frac{3}{2}$$

The ordered pair is $(\frac{3}{2}, 1)$.

To complete the ordered pair $(3, \ )$, we substitute 3 for x in the equation and solve for y:

$$2(3) + 3y = 6$$

$$6 + 3y = 6$$

$$3y = 0$$

$$y = 0$$

The ordered pair is $(3, 0)$.

Notice in each case that once we have used a number in place of one of the variables, the equation becomes a linear equation in one variable. We then use the method explained in Chapter 2 to solve for that variable.

 EXAMPLE 2   Complete the following table for the equation $2x - 5y = 20$.

x	y
0	
	2
	0
-5	

SOLUTION   Filling in the table is equivalent to completing the following ordered pairs: $(0, \ ), (\ , 2), (\ , 0), (-5, \ )$. So we proceed as in Example 1.

When $x = 0$, we have	When $y = 2$, we have
$2(0) - 5y = 20$	$2x - 5(2) = 20$
$0 - 5y = 20$	$2x - 10 = 20$
$-5y = 20$	$2x = 30$
$y = -4$	$x = 15$
When $y = 0$, we have	When $x = -5$, we have
$2x - 5(0) = 20$	$2(-5) - 5y = 20$
$2x - 0 = 20$	$-10 - 5y = 20$
$2x = 20$	$-5y = 30$
$x = 10$	$y = -6$

The completed table looks like this:

x	y
0	-4
15	2
10	0
-5	-6

which is equivalent to the ordered pairs $(0, -4)$, $(15, 2)$, $(10, 0)$, and $(-5, -6)$.

 EXAMPLE 3   Complete the following table for the equation $y = 2x - 1$.

x	y
0	
5	
	7
	3

SOLUTION   When $x = 0$, we have   When $x = 5$, we have

$$y = 2(0) - 1 \qquad\qquad y = 2(5) - 1$$
$$y = 0 - 1 \qquad\qquad y = 10 - 1$$
$$y = -1 \qquad\qquad y = 9$$

When $y = 7$, we have   When $y = 3$, we have

$$7 = 2x - 1 \qquad\qquad 3 = 2x - 1$$
$$8 = 2x \qquad\qquad 4 = 2x$$
$$4 = x \qquad\qquad 2 = x$$

The completed table is

x	y
0	−1
5	9
4	7
2	3

which means the ordered pairs $(0, -1)$, $(5, 9)$, $(4, 7)$, and $(2, 3)$ are among the solutions to the equation $y = 2x - 1$.

 EXAMPLE 4   Which of the ordered pairs $(2, 3)$, $(1, 5)$, and $(-2, -4)$ are solutions to the equation $y = 3x + 2$?

SOLUTION   If an ordered pair is a solution to the equation, then it must satisfy the equation; that is, when the coordinates are used in place of the variables in the equation, the equation becomes a true statement.

Try $(2, 3)$ in $y = 3x + 2$:

$$3 \overset{?}{=} 3(2) + 2$$
$$3 = 6 + 2$$
$$3 = 8 \qquad\qquad \textbf{A false statement}$$

Try $(1, 5)$ in $y = 3x + 2$:

$$5 \overset{?}{=} 3(1) + 2$$
$$5 = 3 + 2$$
$$5 = 5 \qquad\qquad \textbf{A true statement}$$

Try $(-2, -4)$ in $y = 3x + 2$:

$$-4 \overset{?}{=} 3(-2) + 2$$
$$-4 = -6 + 2$$
$$-4 = -4 \qquad\qquad \textbf{A true statement}$$

The ordered pairs $(1, 5)$ and $(-2, -4)$ are solutions to the equation $y = 3x + 2$, and $(2, 3)$ is not.

LINKING OBJECTIVES AND EXAMPLES

Next to each **objective** we have listed the examples that are best described by that objective.

| A | 1–3 |
| B | 4 |

GETTING READY FOR CLASS

After reading through the preceding section, respond in your own words and in complete sentences.

1. How can you tell if an ordered pair is a solution to an equation?

2. How would you find a solution to $y = 3x - 5$?

3. Why is $(3, 2)$ not a solution to $y = 3x - 5$?

4. How many solutions are there to an equation that contains two variables?

Problem Set 3.2

Online support materials can be found at www.thomsonedu.com/login

For each equation, complete the given ordered pairs.

▶ **1.** $2x + y = 6$ $(0, 6), (3, 0), (6, -6)$

2. $3x - y = 5$ $(0, -5), (1, -2), (\frac{10}{3}, 5)$

3. $3x + 4y = 12$ $(0, 3), (4, 0), (-4, 6)$

4. $5x - 5y = 20$ $(0, -4), (2, -2), (1, -3)$

5. $y = 4x - 3$ $(1, 1), (\frac{3}{4}, 0), (5, 17)$

6. $y = 3x - 5$ $(6, 13), (0, -5), (-2, -11)$

7. $y = 7x - 1$ $(2, 13), (1, 6), (0, -1)$

8. $y = 8x + 2$ $(3, 26), (-\frac{1}{4}, 0), (-1, -6)$

▶ **9.** $x = -5$ $(-5, 4), (-5, -3), (-5, 0)$

10. $y = 2$ $(5, 2), (-8, 2), \left(\frac{1}{2}, 2\right)$

For each of the following equations, complete the given table.

▶ **11.** $y = 3x$

x	y
1	3
-3	-9
4	12
6	18

12. $y = -2x$

x	y
-4	8
0	0
-5	10
-6	12

13. $y = 4x$

x	y
0	0
$-\frac{1}{2}$	-2
-3	-12
3	12

14. $y = -5x$

x	y
3	-15
0	0
-2	10
4	-20

15. $x + y = 5$

x	y
2	3
3	2
5	0
9	-4

16. $x - y = 8$

x	y
0	-8
4	-4
5	-3
6	-2

17. $2x - y = 4$

x	y
2	0
3	2
1	-2
-3	-10

18. $3x - y = 9$

x	y
3	0
0	-9
5	6
-4	-21

19. $y = 6x - 1$

x	y
0	-1
-1	-7
-3	-19
$\frac{3}{2}$	8

20. $y = 5x + 7$

x	y
0	7
-2	-3
-4	-13
-3	-8

= Videos available by instructor request

▶ = Online student support materials available at www.thomsonedu.com/login

For the following equations, tell which of the given ordered pairs are solutions.

21. $2x - 5y = 10$ $(2, 3)$, $(0, -2)$, $\left(\frac{5}{2}, 1\right)$ $(0, -2)$

22. $3x + 7y = 21$ $(0, 3)$, $(7, 0)$, $(1, 2)$ $(0, 3)$ and $(7, 0)$

23. $y = 7x - 2$ $(1, 5)$, $(0, -2)$, $(-2, -16)$
$(1, 5)$, $(0, -2)$, and $(-2, -16)$

24. $y = 8x - 3$ $(0, 3)$, $(5, 16)$, $(1, 5)$ $(1, 5)$

▶ **25.** $y = 6x$ $(1, 6)$, $(-2, -12)$, $(0, 0)$
$(1, 6)$, $(-2, -12)$, and $(0, 0)$

26. $y = -4x$ $(0, 0)$, $(2, 4)$, $(-3, 12)$ $(0, 0)$ and $(-3, 12)$

27. $x + y = 0$ $(1, 1)$, $(2, -2)$, $(3, 3)$ $(2, -2)$

28. $x - y = 1$ $(0, 1)$, $(0, -1)$, $(1, 2)$ $(0, -1)$

29. $x = 3$ $(3, 0)$, $(3, -3)$, $(5, 3)$ $(3, 0)$ and $(3, -3)$

30. $y = -4$ $(3, -4)$, $(-4, 4)$, $(0, -4)$ $(3, -4)$ and $(0, -4)$

Applying the Concepts

31. Perimeter If the perimeter of a rectangle is 30 inches, then the relationship between the length l and the width w is given by the equation

$$2l + 2w = 30$$

What is the length when the width is 3 inches?
12 inches

32. Perimeter The relationship between the perimeter P of a square and the length of its side s is given by the formula $P = 4s$. If each side of a square is 5 inches, what is the perimeter? If the perimeter of a square is 28 inches, how long is a side?   20 inches, 7 inches

33. Janai earns \$12 per hour working as a math tutor. We can express the amount she earns each week, y, for working x hours with the equation $y = 12x$. Indicate with a *yes* or *no* which of the following could be one of Janai's paychecks. If you answer no, explain your answer.
   a. \$60 for working 5 hours   Yes

   b. \$100 for working nine hours   No, she should earn \$108 for working 9 hours.
   c. \$80 for working 7 hours   No, she should earn \$84 for working 7 hours.
   d. \$168 for working 14 hours   Yes

34. Erin earns \$15 per hour working as a graphic designer. We can express the amount she earns each week, y, for working x hours with the equation $y = 15x$. Indicate with a *yes* or *no* which of the following could be one of Erin's paychecks. If you answer no, explain your answer.
   a. \$75 for working 5 hours   Yes

   b. \$125 for working 9 hours.   No, she should earn \$135 for working 9 hours.
   c. \$90 for working 6 hours   Yes

   d. \$500 for working 35 hours.   No, she should earn \$525 for working 35 hours.

35. The equation $V = -45,000t + 600,000$ can be used to find the value, V, of a small crane at the end of t years.
   a. What is the value of the crane at the end of 5 years?   \$375,000

   b. When is the crane worth \$330,000?   At the end of 6 years.
   c. Is it true that the crane will be worth \$150,000 after 9 years?   No, the crane will be worth \$195,000 after 9 years.
   d. How much did the crane cost?   \$600,000

36. The equation $P = -400t + 2,500$, can be used to find the price, P, of a notebook computer at the end of t years.
   a. What is the value of the notebook computer at the end of 4 years?   \$900

   b. When is the notebook computer worth \$1,700?   After 2 years
   c. Is it true that the notebook computer will be worth \$100 after 5 years?   No, it will be worth \$500 after 5 years.
   d. How much did the notebook computer cost?   \$2,500

Maintaining Your Skills

37. $\dfrac{11(-5) - 17}{2(-6)}$ 6

38. $\dfrac{12(-4) + 15}{3(-11)}$ 1

39. $\dfrac{13(-6) + 18}{4(-5)}$ 3

40. $\dfrac{9^2 - 6^2}{-9 - 6}$ -3

41. $\dfrac{7^2 - 5^2}{(7 - 5)^2}$ 6

42. $\dfrac{7^2 - 2^2}{-7 - 2}$ -5

43. $\dfrac{-3 \cdot 4^2 - 3 \cdot 2^4}{-3(8)}$ 4

44. $\dfrac{-4(8 - 13) - 2(6 - 11)}{-5(3) + 5}$ -3

Getting Ready for the Next Section

45. Find y when x is 4 in the formula $3x + 2y = 6$. -3

46. Find y when x is 0 in the formula $3x + 2y = 6$.   3

47. Find y when x is 0 in $y = -\dfrac{1}{3}x + 2$.   2

48. Find y when x is 3 in $y = -\dfrac{1}{3}x + 2$.   1

49. Find y when x is 2 in $y = \dfrac{3}{2}x - 3$.   0

50. Find y when x is 4 in $y = \dfrac{3}{2}x - 3$.   3

51. Solve $5x + y = 4$ for y. $y = -5x + 4$

52. Solve $-3x + y = 5$ for y. $y = 3x + 5$

53. Solve $3x - 2y = 6$ for y. $y = \frac{3}{2}x - 3$

54. Solve $2x - 3y = 6$ for y. $y = \frac{2}{3}x - 2$

3.3   Graphing Linear Equations in Two Variables

OBJECTIVES

A   Graph a linear equation in two variables.

B   Graph horizontal lines, vertical lines, and lines through the origin.

In this section we will use the rectangular coordinate system introduced in Section 3.1 to obtain a visual picture of *all* solutions to a linear equation in two variables. The process we use to obtain a visual picture of all solutions to an equation is called *graphing*. The picture itself is called the *graph* of the equation. Graphing equations is an important part of algebra. It is a powerful tool that we can use to help apply algebra to the world around us.

 EXAMPLE 1   Graph the solution set for $x + y = 5$.

SOLUTION   We know from the previous section that an infinite number of ordered pairs are solutions to the equation $x + y = 5$. We can't possibly list them all. What we can do is list a few of them and see if there is any pattern to their graphs.

Some ordered pairs that are solutions to $x + y = 5$ are $(0, 5)$, $(2, 3)$, $(3, 2)$, $(5, 0)$. The graph of each is shown in Figure 1.

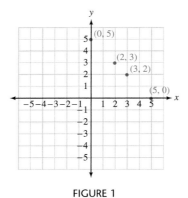

FIGURE 1

Now, by passing a straight line through these points we can graph the solution set for the equation $x + y = 5$. Linear equations in two variables always have graphs that are straight lines. The graph of the solution set for $x + y = 5$ is shown in Figure 2.

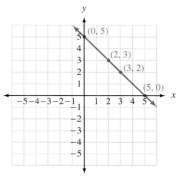

FIGURE 2

Every ordered pair that satisfies $x + y = 5$ has its graph on the line, and any point on the line has coordinates that satisfy the equation. So, there is a one-to-one correspondence between points on the line and solutions to the equation.

Our ability to graph an equation as we have done in Example 1 is due to the invention of the rectangular coordinate system. The French philosopher René Descartes (1596–1650) is the person usually credited with the invention of the rectangular coordinate system. As a philosopher, Descartes is responsible for the statement "I think, therefore I am." Until Descartes invented his coordinate system in 1637, algebra and geometry were treated as separate subjects. The rectangular coordinate system allows us to connect algebra and geometry by associating geometric shapes with algebraic equations.

Here is the precise definition for a linear equation in two variables.

> **DEFINITION** Any equation that can be put in the form $ax + by = c$, where a, b, and c are real numbers and a and b are not both 0, is called a **linear equation in two variables.** The graph of any equation of this form is a straight line (that is why these equations are called "linear"). The form $ax + by = c$ is called **standard form.**

To graph a linear equation in two variables, we simply graph its solution set; that is, we draw a line through all the points whose coordinates satisfy the equation. Here are the steps to follow.

Note

The meaning of the *convenient numbers* referred to in step 1 will become clear as you read the next two examples.

> **To Graph a Linear Equation in Two Variables**
>
> **Step 1:** Find any three ordered pairs that satisfy the equation. This can be done by using a convenient number for one variable and solving for the other variable.
>
> **Step 2:** Graph the three ordered pairs found in step 1. Actually, we need only two points to graph a straight line. The third point serves as a check. If all three points do not line up, there is a mistake in our work.
>
> **Step 3:** Draw a straight line through the three points graphed in step 2.

EXAMPLE 2   Graph the equation $y = 3x - 1$.

SOLUTION   Because $y = 3x - 1$ can be put in the form $ax + by = c$, it is a linear equation in two variables. Hence, the graph of its solution set is a straight line. We can find some specific solutions by substituting numbers for x and then solving for the corresponding values of y. We are free to choose any numbers for x, so let's use 0, 2, and -1.

Let $x = 0$: $y = 3(0) - 1$
$y = 0 - 1$
$y = -1$

In table form

x	y
0	-1
2	5
-1	-4

The ordered pair $(0, -1)$ is one solution.

Let $x = 2$: $y = 3(2) - 1$
$y = 6 - 1$
$y = 5$

The ordered pair $(2, 5)$ is a second solution.

Let $x = -1$: $y = 3(-1) - 1$
$y = -3 - 1$
$y = -4$

The ordered pair $(-1, -4)$ is a third solution.

Next, we graph the ordered pairs $(0, -1)$, $(2, 5)$, $(-1, -4)$ and draw a straight line through them.

The line we have drawn in Figure 3 is the graph of $y = 3x - 1$.

Note
It may seem that we have simply picked the numbers 0, 2, and -1 out of the air and used them for x. In fact we have done just that. Could we have used numbers other than these? The answer is yes, we can substitute any number for x; there will always be a value of y to go with it.

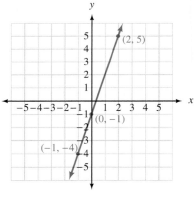

FIGURE 3

Example 2 again illustrates the connection between algebra and geometry that we mentioned previously. Descartes's rectangular coordinate system allows us to associate the equation $y = 3x - 1$ (an algebraic concept) with a specific straight line (a geometric concept). The study of the relationship between equations in algebra and their associated geometric figures is called *analytic geometry*. The rectangular coordinate system often is referred to as the *Cartesian coordinate system* in honor of Descartes.

EXAMPLE 3   Graph the equation $y = -\frac{1}{3}x + 2$.

SOLUTION   We need to find three ordered pairs that satisfy the equation. To do so, we can let x equal any numbers we choose and find corresponding values of

y. But, every value of x we substitute into the equation is going to be multiplied by $-\frac{1}{3}$. Let's use numbers for x that are divisible by 3, like -3, 0, and 3. That way, when we multiply them by $-\frac{1}{3}$, the result will be an integer.

Let $x = -3$: $y = -\dfrac{1}{3}(-3) + 2$ In table form

$y = 1 + 2$

$y = 3$

x	y
-3	3
0	2
3	1

The ordered pair $(-3, 3)$ is one solution.

Let $x = 0$: $y = -\dfrac{1}{3}(0) + 2$

$y = 0 + 2$

$y = 2$

The ordered pair $(0, 2)$ is a second solution.

Let $x = 3$: $y = -\dfrac{1}{3}(3) + 2$

$y = -1 + 2$

$y = 1$

The ordered pair $(3, 1)$ is a third solution.

Graphing the ordered pairs $(-3, 3)$, $(0, 2)$, and $(3, 1)$ and drawing a straight line through their graphs, we have the graph of the equation $y = -\frac{1}{3}x + 2$, as shown in Figure 4.

Note

In Example 3 the values of x we used, -3, 0, and 3, are referred to as convenient values of x because they are easier to work with than some other numbers. For instance, if we let $x = 2$ in our original equation, we would have to add $-\frac{2}{3}$ and 2 to find the corresponding value of y. Not only would the arithmetic be more difficult but also the ordered pair we obtained would have a fraction for its y-coordinate, making it more difficult to graph accurately.

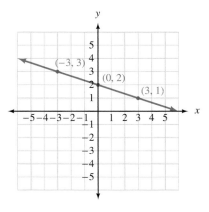

FIGURE 4

 EXAMPLE 4   Graph the solution set for $3x - 2y = 6$.

SOLUTION   It will be easier to find convenient values of x to use in the equation if we first solve the equation for y. To do so, we add $-3x$ to each side, and then we multiply each side by $-\frac{1}{2}$.

$3x - 2y = 6$ **Original equation**

$-2y = -3x + 6$ **Add $-3x$ to each side**

$$-\frac{1}{2}(-2y) = -\frac{1}{2}(-3x + 6) \qquad \textbf{Multiply each side by } -\frac{1}{2}$$

$$y = \frac{3}{2}x - 3 \qquad \textbf{Simplify each side}$$

Now, because each value of x will be multiplied by $\frac{3}{2}$, it will be to our advantage to choose values of x that are divisible by 2. That way, we will obtain values of y that do not contain fractions. This time, let's use 0, 2, and 4 for x.

When $x = 0$: $\qquad y = \frac{3}{2}(0) - 3$

$$y = 0 - 3$$

$$y = -3 \qquad \textbf{(0, −3) is one solution}$$

When $x = 2$: $\qquad y = \frac{3}{2}(2) - 3$

$$y = 3 - 3$$

$$y = 0 \qquad \textbf{(2, 0) is a second solution}$$

When $x = 4$: $\qquad y = \frac{3}{2}(4) - 3$

$$y = 6 - 3$$

$$y = 3 \qquad \textbf{(4, 3) is a third solution}$$

Graphing the ordered pairs $(0, -3)$, $(2, 0)$, and $(4, 3)$ and drawing a line through them, we have the graph shown in Figure 5.

Note

After reading through Example 4, many students ask why we didn't use -2 for x when we were finding ordered pairs that were solutions to the original equation. The answer is, we could have. If we were to let $x = -2$, the corresponding value of y would have been -6. As you can see by looking at the graph in Figure 5, the ordered pair $(-2, -6)$ is on the graph.

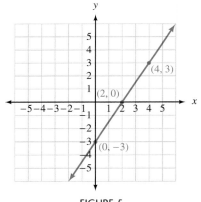

FIGURE 5

EXAMPLE 5 Graph each of the following lines.

a. $y = \frac{1}{2}x$ $\qquad$ b. $x = 3$ $\qquad$ c. $y = -2$

SOLUTION

a. The line $y = \frac{1}{2}x$ passes through the origin because $(0, 0)$ satisfies the equation. To sketch the graph we need at least one more point on the line. When x is 2, we obtain the point $(2, 1)$, and when x is -4, we obtain the point $(-4, -2)$. The graph of $y = \frac{1}{2}x$ is shown in Figure 6a.

b. The line $x = 3$ is the set of all points whose x-coordinate is 3. The variable y does not appear in the equation, so the y-coordinate can be any number. Note that we can write our equation as a linear equation in two variables by writing it as $x + 0y = 3$. Because the product of 0 and y will always be 0, y can be any number. The graph of $x = 3$ is the vertical line shown in Figure 6b.

c. The line $y = -2$ is the set of all points whose y-coordinate is -2. The variable x does not appear in the equation, so the x-coordinate can be any number. Again, we can write our equation as a linear equation in two variables by writing it as $0x + y = -2$. Because the product of 0 and x will always be 0, x can be any number. The graph of $y = -2$ is the horizontal line shown in Figure 6c.

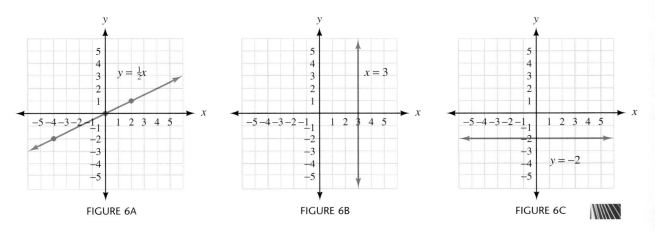

FIGURE 6A   FIGURE 6B   FIGURE 6C

FACTS FROM GEOMETRY

Special Equations and Their Graphs
For the equations below, m, a, and b are real numbers.

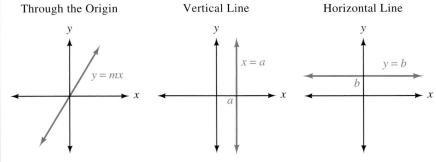

Through the Origin   Vertical Line   Horizontal Line

FIGURE 7A Any equation of the form $y = mx$ has a graph that passes through the origin.

FIGURE 7B Any equation of the form $x = a$ has a vertical line for its graph.

FIGURE 7C Any equation of the form $y = b$ has a horizontal line for its graph.

LINKING OBJECTIVES AND EXAMPLES

Next to each objective we have listed the examples that are best described by that objective.

A   1–4

B   5

GETTING READY FOR CLASS

After reading through the preceding section, respond in your own words and in complete sentences.

1. Explain how you would go about graphing the line $x + y = 5$.

2. When graphing straight lines, why is it a good idea to find three points, when every straight line is determined by only two points?

3. What kind of equations have vertical lines for graphs?

4. What kind of equations have horizontal lines for graphs?

Problem Set 3.3

Online support materials can be found at www.thomsonedu.com/login

For the following equations, complete the given ordered pairs, and use the results to graph the solution set for the equation.

1. $x + y = 4$ $(0, 4), (2, 2), (4, 0)$

2. $x - y = 3$ $(0, -3), (2, -1), (3, 0)$

▶ **3.** $x + y = 3$ $(0, 3), (2, 1), (4, -1)$

4. $x - y = 4$ $(1, -3), (-1, -5), (4, 0)$

5. $y = 2x$ $(0, 0), (-2, -4), (2, 4)$

6. $y = \dfrac{1}{2}x$ $(0, 0), (-2, -1), (2, 1)$

7. $y = \dfrac{1}{3}x$ $(-3, -1), (0, 0), (3, 1)$

8. $y = 3x$ $(-2, -6), (0, 0), (2, 6)$

9. $y = 2x + 1$ $(0, 1), (-1, -1), (1, 3)$

10. $y = -2x + 1$ $(0, 1), (-1, 3), (1, -1)$

11. $y = 4$ $(0, 4), (-1, 4), (2, 4)$

12. $x = 3$ $(3, -2), (3, 0), (3, 5)$

13. $y = \dfrac{1}{2}x + 3$ $(-2, 2), (0, 3), (2, 4)$

14. $y = \dfrac{1}{2}x - 3$ $(-2, -4), (0, -3), (2, -2)$

▶ **15.** $y = -\dfrac{2}{3}x + 1$ $(-3, 3), (0, 1), (3, -1)$

16. $y = -\dfrac{2}{3}x - 1$ $(-3, 1), (0, -1), (3, -3)$

Solve each equation for y. Then, complete the given ordered pairs, and use them to draw the graph.

17. $2x + y = 3$ $(-1, 5), (0, 3), (1, 1)$

18. $3x + y = 2$ $(-1, 5), (0, 2), (1, -1)$

▶ **19.** $3x + 2y = 6$ $(0, 3), (2, 0), (4, -3)$

20. $2x + 3y = 6$ $(0, 2), (3, 0), (6, -2)$

21. $-x + 2y = 6$ $(-2, 2), (0, 3), (2, 4)$

22. $-x + 3y = 6$ $(-3, 1), (0, 2), (3, 3)$

Find three solutions to each of the following equations, and then graph the solution set.

23. $y = -\dfrac{1}{2}x$ **24.** $y = -2x$

25. $y = 3x - 1$ **26.** $y = -3x - 1$

▶ **27.** $-2x + y = 1$ **28.** $-3x + y = 1$

29. $3x + 4y = 8$ **30.** $3x - 4y = 8$

▶ **31.** $x = -2$ **32.** $y = 3$

▶ **33.** $y = 2$ **34.** $x = -3$

Graph each equation.

35. $y = \dfrac{3}{4}x + 1$ **36.** $y = \dfrac{2}{3}x + 1$

37. $y = \dfrac{1}{3}x + \dfrac{2}{3}$ **38.** $y = \dfrac{1}{2}x + \dfrac{1}{2}$

39. $y = \dfrac{2}{3}x + \dfrac{2}{3}$ **40.** $y = -\dfrac{3}{4}x + \dfrac{3}{2}$

For each equation in each table below, indicate whether the graph is horizontal (H), or vertical (V), or whether it passes through the origin (O).

41.

Equation	H, V, and/or O
$x = 3$	V
$y = 3$	H
$y = 3x$	O
$y = 0$	O, H

42.

Equation	H, V, and/or O
$x = \frac{1}{2}$	V
$y = \frac{1}{2}$	H
$y = \frac{1}{2}x$	O
$x = 0$	O, V

43.

Equation	H, V, and/or O
$x = -\frac{3}{5}$	V
$y = -\frac{3}{5}$	H
$y = -\frac{3}{5}x$	O
$x = 0$	O, V

44.

Equation	H, V, and/or O
$x = -4$	V
$y = -4$	H
$y = -4x$	O
$y = 0$	O, H

The next two problems are intended to give you practice reading, and paying attention to, the instructions that accompany the problems you are working. Working these problems is an excellent way to get ready for a test or a quiz.

▶ **45.** Work each problem according to the instructions given.
  a. Solve: $2x + 5 = 10$ $\frac{5}{2}$
  b. Find x when y is 0: $2x + 5y = 10$ 5
  c. Find y when x is 0: $2x + 5y = 10$ 2
  d. Graph: $2x + 5y = 10$
  e. Solve for y: $2x + 5y = 10$ $y = -\frac{2}{5}x + 2$

▶ **Chalkboard Problem**
This chalkboard problem requires students to pay attention to instructions and sets up a discussion on graphing by intercepts for the section that follows this problem set.

46. Work each problem according to the instructions given.
  a. Solve: $x - 2 = 6$ 8
  b. Find x when y is 0: $x - 2y = 6$ 6
  c. Find y when x is 0: $x - 2y = 6$ −3
  d. Graph: $x - 2y = 6$
  e. Solve for y: $x - 2y = 6$ $y = \frac{1}{2}x - 3$

Maintaining Your Skills

Apply the distributive property.

47. $\frac{1}{2}(4x + 10)$ $2x + 5$ **48.** $\frac{1}{2}(6x - 12)$ $3x - 6$

49. $\frac{2}{3}(3x - 9)$ $2x - 6$ **50.** $\frac{1}{3}(2x + 12)$ $\frac{2}{3}x + 4$

51. $\frac{3}{4}(4x + 10)$ $3x + \frac{15}{2}$ **52.** $\frac{3}{4}(8x - 6)$ $6x - \frac{9}{2}$

53. $\frac{3}{5}(10x + 15)$ $6x + 9$ **54.** $\frac{2}{5}(5x - 10)$ $2x - 4$

55. $5\left(\frac{2}{5}x + 10\right)$ $2x + 50$ **56.** $3\left(\frac{2}{3}x + 5\right)$ $2x + 15$

57. $4\left(\frac{3}{2}x - 7\right)$ $6x - 28$ **58.** $4\left(\frac{3}{4}x + 5\right)$ $3x + 20$

59. $\frac{3}{4}(2x + 12y)$ $\frac{3}{2}x + 9y$ **60.** $\frac{3}{4}(8x - 16y)$ $6x - 12y$

61. $\frac{1}{2}(5x - 10y) + 6$ $\frac{5}{2}x - 5y + 6$

62. $\frac{1}{3}(5x - 15y) - 5$ $\frac{5}{3}x - 5y - 5$

Getting Ready for the Next Section

63. Let $3x - 2y = 6$.
  a. Find x when $y = 0$.   2
  b. Find y when $x = 0$.   −3

64. Let $2x - 5y = 10$.
  a. Find x when $y = 0$.   5
  b. Find y when $x = 0$.   −2

65. Let $-x + 2y = 4$.
  a. Find x when $y = 0$.   −4
  b. Find y when $x = 0$.   2

66. Let $3x - y = 6$.
  a. Find x when $y = 0$.   2
  b. Find y when $x = 0$.   −6

67. Let $y = -\dfrac{1}{3}x + 2$.

    a. Find x when $y = 0$. 6

    b. Find y when $x = 0$. 2

68. Let $y = \dfrac{3}{2}x - 3$.

    a. Find x when $y = 0$. 2

    b. Find y when $x = 0$. -3

3.4 More on Graphing: Intercepts

OBJECTIVES

A Find the intercepts of a line from the equation of the line.

B Use intercepts to graph a line.

In this section we continue our work with graphing lines by finding the points where a line crosses the axes of our coordinate system. To do so, we use the fact that any point on the x-axis has a y-coordinate of 0 and any point on the y-axis has an x-coordinate of 0. We begin with the following definition.

> **DEFINITION** The **x-intercept** of a straight line is the x-coordinate of the point where the graph crosses the x-axis. The **y-intercept** is defined similarly. It is the y-coordinate of the point where the graph crosses the y-axis.

If the x-intercept is a, then the point $(a, 0)$ lies on the graph. (This is true because any point on the x-axis has a y-coordinate of 0.)

If the y-intercept is b, then the point $(0, b)$ lies on the graph. (This is true because any point on the y-axis has an x-coordinate of 0.)

Graphically, the relationship is shown in Figure 1.

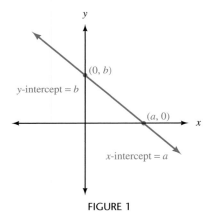

FIGURE 1

 EXAMPLE 1 Find the x- and y-intercepts for $3x - 2y = 6$, and then use them to draw the graph.

SOLUTION To find where the graph crosses the x-axis, we let $y = 0$. (The y-coordinate of any point on the x-axis is 0.)

x-intercept:

$$\text{When} \qquad y = 0$$
$$\text{the equation} \qquad 3x - 2y = 6$$
$$\text{becomes} \qquad 3x - 2(0) = 6$$
$$3x - 0 = 6$$
$$x = 2 \qquad \textbf{Multiply each side by } \tfrac{1}{3}$$

The graph crosses the x-axis at $(2, 0)$, which means the x-intercept is 2.

y-intercept:

When $\qquad x = 0$

the equation $\quad 3x - 2y = 6$

becomes $\qquad 3(0) - 2y = 6$

$\qquad\qquad 0 - 2y = 6$

$\qquad\qquad -2y = 6$

$\qquad\qquad y = -3 \qquad$ **Multiply each side by $-\frac{1}{2}$**

The graph crosses the y-axis at $(0, -3)$, which means the y-intercept is -3.

Plotting the x- and y-intercepts and then drawing a line through them, we have the graph of $3x - 2y = 6$, as shown in Figure 2.

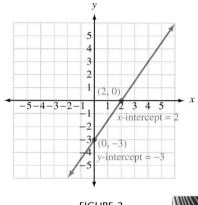

FIGURE 2

EXAMPLE 2 Graph $-x + 2y = 4$ by finding the intercepts and using them to draw the graph.

SOLUTION Again, we find the x-intercept by letting $y = 0$ in the equation and solving for x. Similarly, we find the y-intercept by letting $x = 0$ and solving for y.

x-intercept:

When $\qquad y = 0$

the equation $\quad -x + 2y = 4$

becomes $\qquad -x + 2(0) = 4$

$\qquad\qquad -x + 0 = 4$

$\qquad\qquad -x = 4$

$\qquad\qquad x = -4 \qquad$ **Multiply each side by -1**

The x-intercept is -4, indicating that the point $(-4, 0)$, is on the graph of $-x + 2y = 4$.

y-intercept:

When $\qquad x = 0$

the equation $\quad -x + 2y = 4$

becomes $\qquad -0 + 2y = 4$

$\qquad\qquad 2y = 4$

$\qquad\qquad y = 2 \qquad$ **Multiply each side by $\frac{1}{2}$**

The *y*-intercept is 2, indicating that the point (0, 2) is on the graph of $-x + 2y = 4$.

Plotting the intercepts and drawing a line through them, we have the graph of $-x + 2y = 4$, as shown in Figure 3.

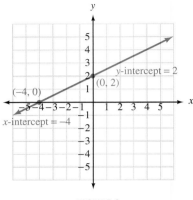

FIGURE 3

Graphing a line by finding the intercepts, as we have done in Examples 1 and 2, is an easy method of graphing if the equation has the form $ax + by = c$ and both the numbers *a* and *b* divide the number *c* evenly.

In our next example we use the intercepts to graph a line in which *y* is given in terms of *x*.

EXAMPLE 3   Use the intercepts for $y = -\frac{1}{3}x + 2$ to draw its graph.

SOLUTION   We graphed this line previously in Example 3 of Section 3.3 by substituting three different values of *x* into the equation and solving for *y*. This time we will graph the line by finding the intercepts.

x-intercept:

When $y = 0$

the equation $y = -\frac{1}{3}x + 2$

becomes $0 = -\frac{1}{3}x + 2$

$-2 = -\frac{1}{3}x$ **Add −2 to each side**

$6 = x$ **Multiply each side by −3**

The *x*-intercept is 6, which means the graph passes through the point (6, 0).

y-intercept:

When $x = 0$

the equation $y = -\frac{1}{3}x + 2$

becomes $y = -\frac{1}{3}(0) + 2$

$y = 2$

The y-intercept is 2, which means the graph passes through the point (0, 2).

The graph of $y = -\frac{1}{3}x + 2$ is shown in Figure 4. Compare this graph, and the method used to obtain it, with Example 3 in Section 3.3.

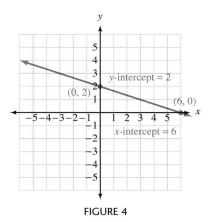

FIGURE 4

LINKING OBJECTIVES AND EXAMPLES

Next to each **objective** we have listed the examples that are best described by that objective.

A   1–3

B   1–3

GETTING READY FOR CLASS

After reading through the preceding section, respond in your own words and in complete sentences.

1. What is the x-intercept for a graph?

2. What is the y-intercept for a graph?

3. How do we find the y-intercept for a line from the equation?

4. How do we graph a line using its intercepts?

Problem Set 3.4

Online support materials can be found at www.thomsonedu.com/login

Find the x- and y-intercepts for the following equations. Then use the intercepts to graph each equation.

1. $2x + y = 4$ **2.** $2x + y = 2$

3. $-x + y = 3$ **4.** $-x + y = 4$

5. $-x + 2y = 2$ **6.** $-x + 2y = 4$

7. $5x + 2y = 10$ **8.** $2x + 5y = 10$

9. $-4x + 5y = 20$ **10.** $-5x + 4y = 20$

11. $3x - 4y = -4$ **12.** $-2x + 3y = 3$

13. $x - 3y = 2$ **14.** $x - 2y = 1$

15. $2x - 3y = -2$ **16.** $3x + 4y = 6$

17. $y = 2x - 6$ **18.** $y = 2x + 6$

19. $y = 2x - 1$ **20.** $y = -2x - 1$

21. $y = \frac{1}{2}x + 3$ **22.** $y = \frac{1}{2}x - 3$

23. $y = -\frac{1}{3}x - 2$ **24.** $y = -\frac{1}{3}x + 2$

For each of the following lines, the x-intercept and the y-intercept are both 0, which means the graph of each will go through the origin, (0, 0). Graph each line by finding a point on each, other than the origin, and then drawing a line through that point and the origin.

25. $y = -2x$ **26.** $y = \frac{1}{2}x$

= Videos available by instructor request

▶ = Online student support materials available at www.thomsonedu.com/login

27. $y = -\dfrac{1}{3}x$

28. $y = -3x$

29. $y = \dfrac{2}{3}x$

30. $y = \dfrac{3}{2}x$

Complete each table.

31.

Equation	x-intercept	y-intercept
$3x + 4y = 12$	4	3
$3x + 4y = 4$	$\dfrac{4}{3}$	1
$3x + 4y = 3$	1	$\dfrac{3}{4}$
$3x + 4y = 2$	$\dfrac{2}{3}$	$\dfrac{1}{2}$

32.

Equation	x-intercept	y-intercept
$-2x + 3y = 6$	-3	2
$-2x + 3y = 3$	$-\dfrac{3}{2}$	1
$-2x + 3y = 2$	-1	$\dfrac{2}{3}$
$-2x + 3y = 1$	$-\dfrac{1}{2}$	$\dfrac{1}{3}$

33.

Equation	x-intercept	y-intercept
$x - 3y = 2$	2	$-\dfrac{2}{3}$
$y = \dfrac{1}{3}x - \dfrac{2}{3}$	2	$-\dfrac{2}{3}$
$x - 3y = 0$	0	0
$y = \dfrac{1}{3}x$	0	0

34.

Equation	x-intercept	y-intercept
$x - 2y = 1$	1	$-\dfrac{1}{2}$
$y = \dfrac{1}{2}x - \dfrac{1}{2}$	1	$-\dfrac{1}{2}$
$x - 2y = 0$	0	0
$y = \dfrac{1}{2}x$	0	0

The next two problems are intended to give you practice reading, and paying attention to, the instructions that accompany the problems you are working. Working these problems is an excellent way to get ready for a test or a quiz.

▶ **35.** Work each problem according to the instructions given.
   a. Solve: $2x - 3 = -3$ 0
   b. Find the x-intercept: $2x - 3y = -3$ $-\dfrac{3}{2}$
   c. Find y when x is 0: $2x - 3y = -3$ 1
   d. Graph: $2x - 3y = -3$
   e. Solve for y: $2x - 3y = -3$ $y = \dfrac{2}{3}x + 1$

36. Work each problem according to the instructions given.
   a. Solve: $3x - 4 = -4$ 0
   b. Find the y-intercept: $3x - 4y = -4$ 1
   c. Find x when y is 0: $3x - 4y = -4$ $-\dfrac{4}{3}$
   d. Graph: $3x - 4y = -4$
   e. Solve for y: $3x - 4y = -4$ $y = \dfrac{3}{4}x + 1$

37. Graph the line that passes through the points $(-2, 5)$ and $(5, -2)$. What are the x- and y-intercepts for this line? x-intercept $= 3$; y-intercept $= 3$

38. Graph the line that passes through the points $(5, 3)$ and $(-3, -5)$. What are the x- and y-intercepts for this line? x-intercept $= 2$; y-intercept $= -2$

From the graphs below, find the x- and y-intercepts for each line.

39.

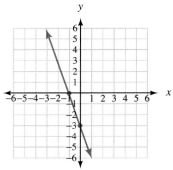

x-intercept $= -1$; y-intercept $= -3$

40.

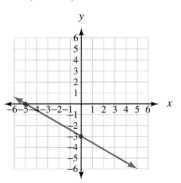

x-intercept $= -5$; y-intercept $= -3$

41. Use the graph at the right to complete the following table.

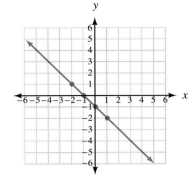

x	y
−2	1
0	−1
−1	0
1	−2

42. Use the graph at the right to complete the following table.

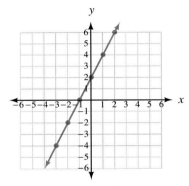

x	y
−2	−2
0	2
−1	0
2	6

43. The vertical line $x = 3$ has only one intercept. Graph $x = 3$, and name its intercept. [Remember, ordered pairs (x, y) that are solutions to the equation $x = 3$ are ordered pairs with an x-coordinate of 3 and any y-coordinate.] x-intercept = 3

44. Graph the vertical line $x = -2$. Then name its intercept. x-intercept = −2

45. The horizontal line $y = 4$ has only one intercept. Graph $y = 4$, and name its intercept. [Ordered pairs (x, y) that are solutions to the equation $y = 4$ are ordered pairs with a y-coordinate of 4 and any x-coordinate.] y-intercept = 4

46. Graph the horizontal line $y = -3$. Then name its intercept. y-intercept = −3

Applying the Concepts

47. **Working Two Jobs** Maggie has a job working in an office for $10 an hour and another job driving a tractor for $12 an hour. Maggie works at both jobs and earns $480 in one week.
   a. Write an equation that gives the number of hours x, she worked for $10 per hour and the number of hours, y, she worked for $12 per hour. 10x + 12y = 480
   b. Find the x- and y-intercepts for this equation. x-intercept = 48; y-intercept = 40
   c. Graph this equation from the intercepts, using only the first quadrant.
   d. From the graph, find how many hours she worked at $12 if she worked 36 hours at $10 per hour. 10 hours
   e. From the graph, find how many hours she worked at $10 if she worked 25 hours at $12 per hour. 18 hours

48. **Improving Your Quantitative Literacy** Give the y-intercept of the graph of poverty rates and tell what it represents. Why is there not an x-intercept?

The y-intercept is 15.1%. In 1993, or the first year represented on the graph, the U.S. poverty rate was 15.1%. There is no x-intercept because the poverty rate has never been 0%.

Maintaining Your Skills

Solve each equation.

49. $-12y - 4 = -148$ 12
50. $-2x - 18 = 4$ −11
51. $-5y - 4 = 51$ −11
52. $-2y + 18 = -14$ 16
53. $11x - 12 = -78$ −6
54. $21 + 9y = -24$ −5
55. $9x + 3 = 66$ 7
56. $-11 - 15a = -71$ 4
57. $-9c - 6 = 12$ −2
58. $-7a + 28 = -84$ 16
59. $4 + 13c = -9$ −1
60. $-3x + 15 = -24$ 13
61. $3y - 12 = 30$ 14
62. $9x + 11 = -16$ −3
63. $-11y + 9 = 75$ −6
64. $9x - 18 = -72$ −6

Getting Ready for the Next Section

65. Evaluate:

a. $\dfrac{5-2}{3-1}$ $\dfrac{3}{2}$ b. $\dfrac{2-5}{1-3}$ $\dfrac{3}{2}$

66. Evaluate.

a. $\dfrac{-4-1}{5-(-2)}$ $-\dfrac{5}{7}$ b. $\dfrac{1+4}{-2-5}$ $-\dfrac{5}{7}$

67. Evaluate the following expressions when $x = 3$ and $y = 5$.

a. $\dfrac{y-2}{x-1}$ $\dfrac{3}{2}$ b. $\dfrac{2-y}{1-x}$ $\dfrac{3}{2}$

68. Evaluate the following expressions when $x = 3$ and $y = 2$.

a. $\dfrac{-4-y}{5-x}$ -3 b. $\dfrac{y+4}{x-5}$ -3

3.5   The Slope of a Line

OBJECTIVES

A   Find the slope of a line from two points on the line.

B   Graph a line given the slope and y-intercept.

In defining the slope of a straight line, we are looking for a number to associate with a straight line that does two things. First of all, we want the slope of a line to measure the "steepness" of the line; that is, in comparing two lines, the slope of the steeper line should have the larger numerical value. Second, we want a line that *rises* going from left to right to have a *positive* slope. We want a line that *falls* going from left to right to have a *negative* slope. (A line that neither rises nor falls going from left to right must, therefore, have 0 slope.) These are illustrated in Figure 1.

Negative slope   Zero slope   Positive slope

FIGURE 1

Note

The 2 in x_2 is called a *subscript*. It is notation that allows us to distinguish between the variables x_1 and x_2, while still showing that they are both x-coordinates.

Suppose we know the coordinates of two points on a line. Because we are trying to develop a general formula for the slope of a line, we will use general points—call the two points $P_1(x_1, y_1)$ and $P_2(x_2, y_2)$. They represent the coordinates of any two different points on our line. We define the *slope* of our line to be the ratio of the vertical change to the horizontal change as we move from point (x_1, y_1) to point (x_2, y_2) on the line. (See Figure 2.)

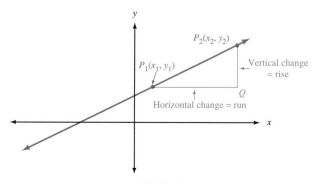

FIGURE 2

We call the vertical change the *rise* in the graph and the horizontal change the *run* in the graph. The slope, then, is

$$\text{Slope} = \frac{\text{vertical change}}{\text{horizontal change}} = \frac{\text{rise}}{\text{run}}$$

We would like to have a numerical value to associate with the rise in the graph and a numerical value to associate with the run in the graph. A quick study of Figure 2 shows that the coordinates of point Q must be (x_2, y_1), because Q is directly below point P_2 and right across from point P_1. We can draw our diagram again in the manner shown in Figure 3. It is apparent from this graph that the rise can be expressed as $(y_2 - y_1)$ and the run as $(x_2 - x_1)$. We usually denote the slope of a line by the letter m. The complete definition of slope follows along with a diagram (Figure 3) that illustrates the definition.

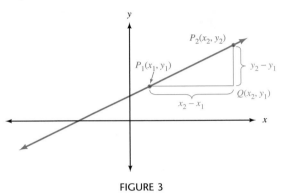

FIGURE 3

DEFINITION If points (x_1, y_1) and (x_2, y_2) are any two different points, then the **slope** of the line on which they lie is

$$\text{Slope} = m = \frac{\text{rise}}{\text{run}} = \frac{y_2 - y_1}{x_2 - x_1}$$

This definition of the slope of a line does just what we want it to do. If the line rises going from left to right, the slope will be positive. If the line falls from left to right, the slope will be negative. Also, the steeper the line, the larger numerical value the slope will have.

EXAMPLE 1   Find the slope of the line through $(1, 2)$ and $(3, 5)$.

SOLUTION We can let

$$(x_1, y_1) = (1, 2)$$

and

$$(x_2, y_2) = (3, 5)$$

then

$$m = \frac{y_2 - y_1}{x_2 - x_1} = \frac{5 - 2}{3 - 1} = \frac{3}{2}$$

The slope is $\frac{3}{2}$. For every vertical change of 3 units, there will be a corresponding horizontal change of 2 units. (See Figure 4.)

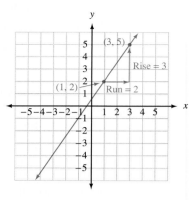

FIGURE 4

EXAMPLE 2 Find the slope of the line through $(-2, 1)$ and $(5, -4)$.

SOLUTION It makes no difference which ordered pair we call (x_1, y_1) and which we call (x_2, y_2).

$$\text{Slope} = m = \frac{y_2 - y_1}{x_2 - x_1} = \frac{-4 - 1}{5 - (-2)} = \frac{-5}{7}$$

The slope is $-\frac{5}{7}$. Every vertical change of -5 units (down 5 units) is accompanied by a horizontal change of 7 units (to the right 7 units). (See Figure 5.)

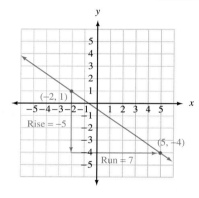

FIGURE 5

EXAMPLE 3 Graph the line with slope $\frac{3}{2}$ and y-intercept 1.

SOLUTION Because the y-intercept is 1, we know that one point on the line is $(0, 1)$. So, we begin by plotting the point $(0, 1)$, as shown in Figure 6.

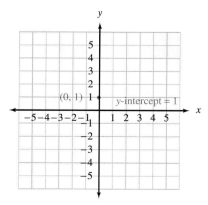

FIGURE 6

There are many lines that pass through the point shown in Figure 6, but only one of those lines has a slope of $\frac{3}{2}$. The slope, $\frac{3}{2}$, can be thought of as the rise in the graph divided by the run in the graph. Therefore, if we start at the point $(0, 1)$ and move 3 units up (that's a rise of 3) and then 2 units to the right (a run of 2), we will be at another point on the graph. Figure 7 shows that the point we reach by doing so is the point $(2, 4)$.

$$\text{Slope} = m = \frac{\text{rise}}{\text{run}} = \frac{3}{2}$$

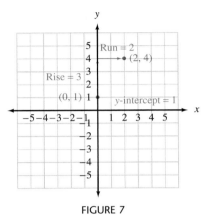

FIGURE 7

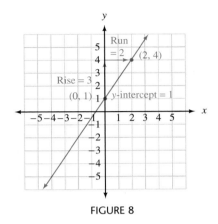

FIGURE 8

To graph the line with slope $\frac{3}{2}$ and y-intercept 1, we simply draw a line through the two points in Figure 7 to obtain the graph shown in Figure 8.

EXAMPLE 4   Find the slope of the line containing $(3, -1)$ and $(3, 4)$.

SOLUTION   Using the definition for slope, we have

$$m = \frac{-1 - 4}{3 - 3} = \frac{-5}{0}$$

The expression $\frac{-5}{0}$ is undefined; that is, there is no real number to associate with it. In this case, we say the line *has no slope.*

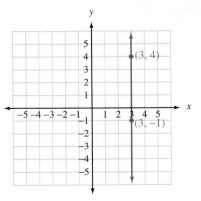

FIGURE 9

The graph of our line is shown in Figure 9. Our line with no slope is a vertical line. All vertical lines have no slope. (And all horizontal lines, as we mentioned earlier, have 0 slope.)

As a final note, the summary reminds us that all horizontal lines have equations of the form $y = b$ and slopes of 0. Because they cross the y-axis at b, the y-intercept is b; there is no x-intercept. Vertical lines have no slope and equations of the form $x = a$. Each will have an x-intercept at a and no y-intercept. Finally, equations of the form $y = mx$ have graphs that pass through the origin. The slope is always m and both the x-intercept and the y-intercept are 0.

FACTS FROM GEOMETRY

Special Equations and Their Graphs, Slopes, and Intercepts
For the equations below, m, a, and b are real numbers.

Through the Origin	Vertical Line	Horizontal Line
Equation: $y = mx$	Equation: $x = a$	Equation: $y = b$
Slope $= m$	No slope	Slope $= 0$
x-intercept $= 0$	x-intercept $= a$	No x-intercept
y-intercept $= 0$	No y-intercept	y-intercept $= b$

Through the Origin

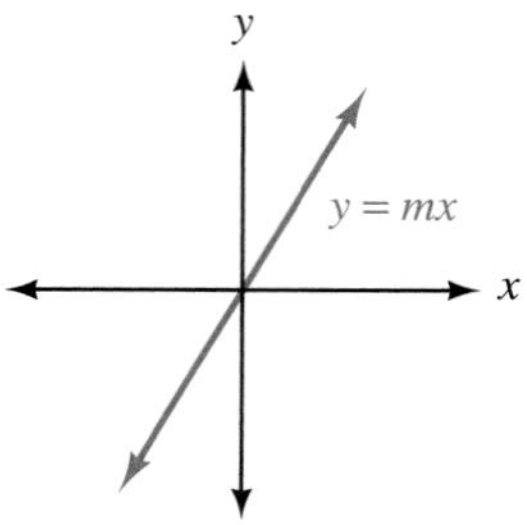

Vertical Line

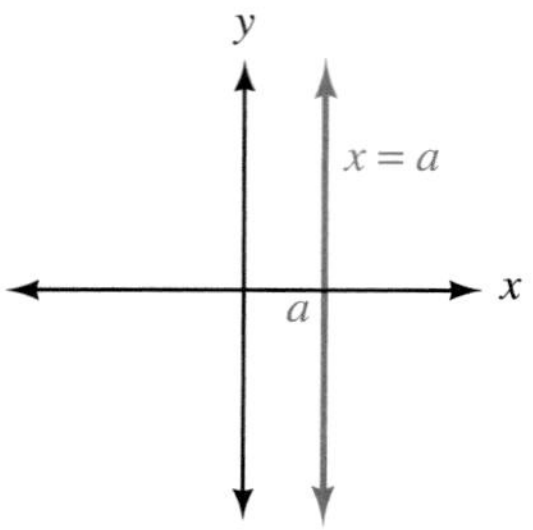

Horizontal Line

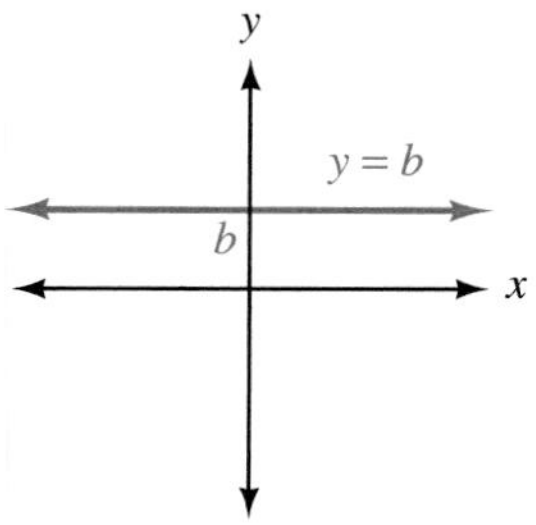

$y = mx$

$x = a$

$y = b$

FIGURE 10A   FIGURE 10B   FIGURE 10C

LINKING OBJECTIVES AND EXAMPLES

Next to each **objective** we have listed the examples that are best described by that objective.

A 1, 2, 4

B 3

GETTING READY FOR CLASS

After reading through the preceding section, respond in your own words and in complete sentences.

1. What is the slope of a line?
2. Would you rather climb a hill with a slope of 1 or a slope of 3? Explain why.
3. Describe how to obtain the slope of a line if you know the coordinates of two points on the line.
4. Describe how you would graph a line from its slope and y-intercept.

Find the slope of the line through the following pairs of points. Then plot each pair of points, draw a line through them, and indicate the rise and run in the graph in the same manner shown in Examples 1 and 2.

1. (2, 1), (4, 4) $\frac{3}{2}$

2. (3, 1), (5, 4) $\frac{3}{2}$

3. (1, 4), (5, 2) $-\frac{1}{2}$

4. (1, 3), (5, 2) $-\frac{1}{4}$

5. (1, −3), (4, 2) $\frac{5}{3}$

6. (2, −3), (5, 2) $\frac{5}{3}$

7. (−3, −2), (1, 3) $\frac{5}{4}$

8. (−3, −1), (1, 4) $\frac{5}{4}$

9. (−3, 2), (3, −2) $-\frac{2}{3}$

10. (−3, 3), (3, −1) $-\frac{2}{3}$

11. (2, −5), (3, −2) 3

12. (2, −4), (3, −1) 3

In each of the following problems, graph the line with the given slope and y-intercept b.

13. $m = \frac{2}{3}$, $b = 1$

14. $m = \frac{3}{4}$, $b = -2$

15. $m = \frac{3}{2}$, $b = -3$

16. $m = \frac{4}{3}$, $b = 2$

17. $m = -\frac{4}{3}$, $b = 5$

18. $m = -\frac{3}{5}$, $b = 4$

19. $m = 2$, $b = 1$

20. $m = -2$, $b = 4$

21. $m = 3$, $b = -1$

22. $m = 3$, $b = -2$

Find the slope and y-intercept for each line.

23.

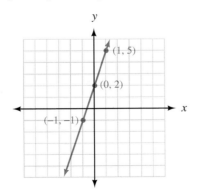

Slope = 3; y-intercept = 2

24.

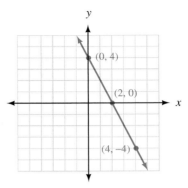

Slope = −2; y-intercept = 4

25.

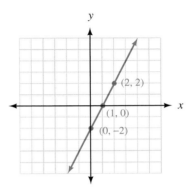

Slope = 2; y-intercept = −2

26.

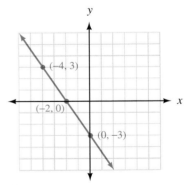

Slope = $-\frac{3}{2}$; y-intercept = −3

27. Graph the line that has an x-intercept of 3 and a y-intercept of −2. What is the slope of this line?

28. Graph the line that has an x-intercept of 2 and a y-intercept of −3. What is the slope of this line?

29. Graph the line with x-intercept 4 and y-intercept 2. What is the slope of this line?

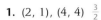

30. Graph the line with x-intercept -4 and y-intercept -2. What is the slope of this line?

31. Graph the line $y = 2x - 3$, then name the slope and y-intercept by looking at the graph.

32. Graph the line $y = -2x + 3$, then name the slope and y-intercept by looking at the graph.

33. Graph the line $y = \frac{1}{2}x + 1$, then name the slope and y-intercept by looking at the graph.

34. Graph the line $y = -\frac{1}{2}x - 2$, then name the slope and y-intercept by looking at the graph.

35. Find y if the line through $(4, 2)$ and $(6, y)$ has a slope of 2. 6

36. Find y if the line through $(1, y)$ and $(7, 3)$ has a slope of 6. -33

For each equation in each table, give the slope of the graph.

37.

Equation	Slope
$x = 3$	undefined
$y = 3$	0
$y = 3x$	3

38.

Equation	Slope
$y = \frac{3}{2}$	0
$x = \frac{3}{2}$	undefined
$y = \frac{3}{2}x$	$\frac{3}{2}$

39.

Equation	Slope
$y = -\frac{2}{3}$	0
$x = -\frac{2}{3}$	undefined
$y = -\frac{2}{3}x$	$-\frac{2}{3}$

40.

Equation	Slope
$x = -2$	undefined
$y = -2$	0
$y = -2x$	-2

Applying the Concepts

41. Plant Height The table and completed line graph give the growth of a certain species of plant over time. Find the slopes of the line segments labeled A, B, and C. Slopes: A, $\frac{1}{2}$; B, 2; C, 5

Day	Plant Height
0	0
1	0.5
2	1
3	1.5
4	3
5	4
6	6
7	9
8	13
9	18
10	23

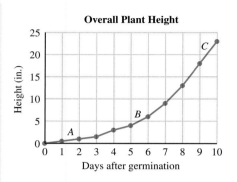

42. Improving Your Quantitative Literacy Let x represent the year in question, and let y represent average taxes paid, then work the following problems.

a. Find the slope of the line that connects the first point and last point on the graph. $m = 4{,}573/24$

b. What does the sign of the slope tell us? The slope is positive, which means the graph is increasing from the first point to the last point.

c. What does the slope represent? The slope is the average annual rate of increase in taxes during this 24-year period.

From *USA Today*. Copyright 2005. Reprinted with permission.

Maintaining Your Skills

Solve each equation.

43. $\frac{1}{2}(4x + 10) = 11$ 3

44. $\frac{1}{2}(6x - 12) = -18$ −4

45. $\frac{2}{3}(3x - 9) = 24$ 15

46. $\frac{1}{3}(2x + 15) = -29$ −51

47. $\frac{3}{4}(4x + 8) = -12$ −6

48. $\frac{3}{4}(8x - 16) = -36$ −4

49. $\frac{3}{5}(10x + 15) = 45$ 6

50. $\frac{2}{5}(5x - 10) = -10$ −3

51. $5\left(\frac{2}{5}x + 10\right) = -28$ −39

52. $3\left(\frac{2}{3}x + 5\right) = -13$ −14

53. $4\left(\frac{3}{2}x - 7\right) = -4$ 4

54. $4\left(\frac{3}{4}x + 5\right) = -40$ −20

55. $\frac{3}{4}(2x + 12) = 24$ 10

56. $-\frac{3}{4}(12x - 16) = -42$ 6

57. $\frac{1}{2}(5x - 10) + 6 = -49$ −20

58. $\frac{1}{3}(5x - 15) - 5 = 20$ 18

Getting Ready for the Next Section

Solve each equation for y.

59. $-2x + y = 4$ $y = 2x + 4$

60. $-4x + y = -2$ $y = 4x - 2$

61. $3x + y = 3$ $y = -3x + 3$

62. $3x + 2y = 6$ $y = -\frac{3}{2}x + 3$

63. $4x - 5y = 20$ $y = \frac{4}{5}x - 4$

64. $-2x - 5y = 10$ $y = -\frac{2}{5}x - 2$

65. $y - 3 = -2(x + 4)$ $y = -2x - 5$

66. $y + 5 = 2(x + 2)$ $y = 2x - 1$

67. $y - 3 = -\frac{2}{3}(x + 3)$ $y = -\frac{2}{3}x + 1$

68. $y - 1 = -\frac{1}{2}(x + 4)$ $y = -\frac{1}{2}x - 1$

69. $\frac{y - 1}{x} = \frac{3}{2}$ $y = \frac{3}{2}x + 1$

70. $\frac{y + 1}{x} = \frac{3}{2}$ $y = \frac{3}{2}x - 1$

3.6 Finding the Equation of a Line

OBJECTIVES

A Find the equation of a line given the slope and y-intercept of the line.

B Find the slope and y-intercept of a line given the equation of the line.

C Find the equation of a line given a point on the line and the slope of the line.

D Find the equation of a line given two points on the line.

To this point in the chapter, most of the problems we have worked have used the equation of a line to find different types of information about the line. For instance, given the equation of a line, we can find points on the line, the graph of the line, the intercepts, and the slope of the line. In this section we reverse things somewhat and move in the other direction; we will use information about a line, such as its slope and y-intercept, to find the equation of a line.

There are three main types of problems to solve in this section.

1. Find the equation of a line from the slope and y-intercept.

2. Find the equation of a line given one point on the line and the slope of the line.

3. Find the equation of a line given two points on the line.

Examples 1 and 2 illustrate the first type of problem. Example 5 solves the second type of problem. The third type of problem is solved in Example 6.

The Slope-Intercept Form of an Equation of a Straight Line

EXAMPLE 1   Find the equation of the line with slope $\frac{3}{2}$ and y-intercept 1.

SOLUTION   We graphed the line with slope $\frac{3}{2}$ and y-intercept 1 in Example 3 of the previous section. Figure 1 shows that graph.

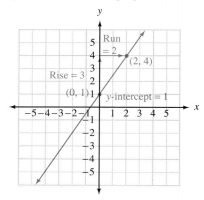

FIGURE 1

What we want to do now is find the equation of the line shown in Figure 1. To do so, we take any other point (x, y) on the line and apply our slope formula to that point and the point $(0, 1)$. We set that result equal to $\frac{3}{2}$, because $\frac{3}{2}$ is the slope of our line. The work is as follows, with a diagram of the situation following.

$$\frac{y-1}{x-0} = \frac{3}{2} \qquad \textbf{Slope} = \frac{\textbf{vertical change}}{\textbf{horizontal change}}$$

$$\frac{y-1}{x} = \frac{3}{2} \qquad \boldsymbol{x - 0 = x}$$

$$y - 1 = \frac{3}{2}x \qquad \textbf{Multiply each side by } \boldsymbol{x}$$

$$y = \frac{3}{2}x + 1 \qquad \textbf{Add 1 to each side}$$

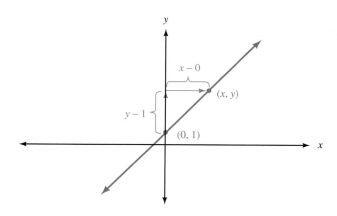

What is interesting and useful about the equation we have just found is that the number in front of x is the slope of the line and the constant term is the y-intercept. It is no coincidence that it turned out this way. Whenever an equation has the form $y = mx + b$, the graph is always a straight line with slope m and y-intercept b. To see that this is true in general, suppose we want the equation of a line with slope m and y-intercept b. Because the y-intercept is b, then the point $(0, b)$ is on the line. If (x, y) is any other point on the line, then we apply our slope formula to get

$$\frac{y - b}{x - 0} = m \qquad \textbf{Slope} = \frac{\textbf{vertical change}}{\textbf{horizontal change}}$$

$$\frac{y - b}{x} = m \qquad x - 0 = x$$

$$y - b = mx \qquad \textbf{Multiply each side by } x$$

$$y = mx + b \qquad \textbf{Add } b \textbf{ to each side}$$

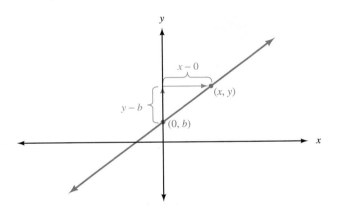

Here is a summary of what we have just found.

Slope-Intercept Form of the Equation of a Line The equation of the line with slope m and y-intercept b is always given by
$$y = mx + b$$

 EXAMPLE 2   Find the equation of the line with slope $-\frac{4}{3}$ and y-intercept 5. Then, graph the line.

SOLUTION   Substituting $m = -\frac{4}{3}$ and $b = 5$ into the equation $y = mx + b$, we have

$$y = -\frac{4}{3}x + 5$$

Finding the equation from the slope and y-intercept is just that easy. If the slope is m and the y-intercept is b, then the equation is always $y = mx + b$.

Because the y-intercept is 5, the graph goes through the point $(0, 5)$. To find a second point on the graph, we start at $(0, 5)$ and move 4 units down (that's a rise of -4) and 3 units to the right (a run of 3). The point we reach is $(3, 1)$. Drawing a line that passes through (0, 5) and (3, 1), we have the graph of our equation. (Note that we could also let the rise = 4 and the run = −3 and obtain the same graph.) The graph is shown in Figure 2.

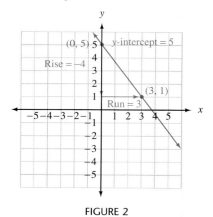

FIGURE 2

 EXAMPLE 3   Find the slope and y-intercept for $-2x + y = -4$. Then, use them to draw the graph.

SOLUTION   To identify the slope and y-intercept from the equation, the equation must be in the form $y = mx + b$ (slope-intercept form). To write our equation in this form, we must solve the equation for y. To do so, we simply add 2x to each side of the equation.

$$-2x + y = -4 \qquad \textbf{Original equation}$$
$$y = 2x - 4 \qquad \textbf{Add 2x to each side}$$

The equation is now in slope-intercept form, so the slope must be 2 and the y-intercept must be −4. The graph, therefore, crosses the y-axis at (0, −4). Because the slope is 2, we can let the rise = 2 and the run = 1 and find a second point on the graph. The graph is shown in Figure 3.

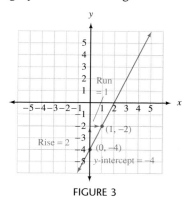

FIGURE 3

 EXAMPLE 4   Find the slope and y-intercept for $3x - 2y = 6$.

SOLUTION   To find the slope and y-intercept from the equation, we must write the equation in the form $y = mx + b$. This means we must solve the equation $3x - 2y = 6$ for y.

$$3x - 2y = 6 \qquad\qquad \textbf{Original equation}$$

$$-2y = -3x + 6 \qquad\qquad \textbf{Add } -3x \textbf{ to each side}$$

$$-\frac{1}{2}(-2y) = -\frac{1}{2}(-3x + 6) \qquad \textbf{Multiply each side by } -\frac{1}{2}$$

$$y = \frac{3}{2}x - 3 \qquad\qquad \textbf{Simplify each side}$$

Now that the equation is written in slope-intercept form, we can identify the slope as $\frac{3}{2}$ and the y-intercept as -3. The graph is shown in Figure 4.

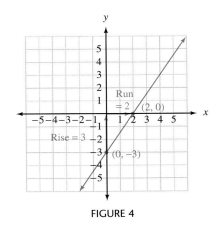

FIGURE 4

The Point-Slope Form of an Equation of a Straight Line

A second useful form of the equation of a straight line is the point-slope form.

Let line l contain the point (x_1, y_1) and have slope m. If (x, y) is any other point on l, then by the definition of slope we have

$$\frac{y - y_1}{x - x_1} = m$$

Multiplying both sides by $(x - x_1)$ gives us

$$(x - x_1) \cdot \frac{y - y_1}{x - x_1} = m(x - x_1)$$

$$y - y_1 = m(x - x_1)$$

This last equation is known as the *point-slope form* of the equation of a straight line.

> **Point-Slope Form of the Equation of a Line** The equation of the line through (x_1, y_1) with slope m is given by
>
> $$y - y_1 = m(x - x_1)$$

This form is used to find the equation of a line, either given one point on the line and the slope, or given two points on the line.

EXAMPLE 5   Find the equation of the line with slope -2 that contains the point $(-4, 3)$. Write the answer in slope-intercept form.

SOLUTION

Using $(x_1, y_1) = (-4, 3)$ and $m = -2$

in $\quad y - y_1 = m(x - x_1)$ **Point-slope form**

gives us $\quad y - 3 = -2(x + 4)$ **Note: $x - (-4) = x + 4$**

$\quad\quad\quad y - 3 = -2x - 8$ **Multiply out right side**

$\quad\quad\quad y = -2x - 5$ **Add 3 to each side**

Figure 5 is the graph of the line that contains $(-4, 3)$ and has a slope of -2. Notice that the y-intercept on the graph matches that of the equation we found.

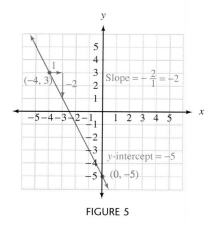

FIGURE 5

 EXAMPLE 6   Find the equation of the line that passes through the points $(-3, 3)$ and $(3, -1)$.

SOLUTION   We begin by finding the slope of the line:

$$m = \frac{3 - (-1)}{-3 - 3} = \frac{4}{-6} = -\frac{2}{3}$$

Using $(x_1, y_1) = (3, -1)$ and $m = -\frac{2}{3}$ in $y - y_1 = m(x - x_1)$ yields

$$y + 1 = -\frac{2}{3}(x - 3)$$

$$y + 1 = -\frac{2}{3}x + 2 \quad \text{\textbf{Multiply out right side}}$$

$$y = -\frac{2}{3}x + 1 \quad \text{\textbf{Add −1 to each side}}$$

Figure 6 shows the graph of the line that passes through the points $(-3, 3)$ and $(3, -1)$. As you can see, the slope and y-intercept are $-\frac{2}{3}$ and 1, respectively.

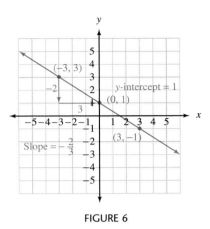

FIGURE 6

Note In Example 6 we could have used the point $(-3, 3)$ instead of $(3, -1)$ and obtained the same equation; that is, using $(x_1, y_1) = (-3, 3)$ and $m = -\frac{2}{3}$ in $y - y_1 = m(x - x_1)$ gives us

$$y - 3 = -\frac{2}{3}(x + 3)$$

$$y - 3 = -\frac{2}{3}x - 2$$

$$y = -\frac{2}{3}x + 1$$

which is the same result we obtained using $(3, -1)$.

Methods of Graphing Lines

1. Substitute convenient values of x into the equation, and find the corresponding values of y. We used this method first for equations like $y = 2x - 3$. To use this method for equations that looked like $2x - 3y = 6$, we first solved them for y.

2. Find the x- and y-intercepts. This method works best for equations of the form $3x + 2y = 6$ where the numbers in front of x and y divide the constant term evenly.

3. Find the slope and y-intercept. This method works best when the equation has the form $y = mx + b$ and b is an integer.

LINKING OBJECTIVES AND EXAMPLES

Next to each **objective** we have listed the examples that are best described by that objective.

A	1, 2
B	3, 4
C	5
D	6

GETTING READY FOR CLASS

After reading through the preceding section, respond in your own words and in complete sentences.

1. What are m and b in the equation $y = mx + b$?
2. How would you find the slope and y-intercept for the line $3x - 2y = 6$?
3. What is the point-slope form of the equation of a line?
4. How would you find the equation of a line from two points on the line?

In each of the following problems, give the equation of the line with the given slope and y-intercept.

1. $m = \frac{2}{3}$, $b = 1$ $y = \frac{2}{3}x + 1$

2. $m = \frac{3}{4}$, $b = -2$ $y = \frac{3}{4}x - 2$

▶ **3.** $m = \frac{3}{2}$, $b = -1$ $y = \frac{3}{2}x - 1$

4. $m = \frac{4}{3}$, $b = 2$ $y = \frac{4}{3}x + 2$

5. $m = -\frac{2}{5}$, $b = 3$ $y = -\frac{2}{5}x + 3$

6. $m = -\frac{3}{5}$, $b = 4$ $y = -\frac{3}{5}x + 4$

7. $m = 2$, $b = -4$ $y = 2x - 4$

8. $m = -2$, $b = 4$ $y = -2x + 4$

Find the slope and y-intercept for each of the following equations by writing them in the form $y = mx + b$. Then, graph each equation.

9. $-2x + y = 4$ $m = 2; b = 4$ **10.** $-2x + y = 2$ $m = 2; b = 2$

11. $3x + y = 3$ $m = -3; b = 3$ **12.** $3x + y = 6$ $m = -3; b = 6$

13. $3x + 2y = 6$ $m = -\frac{3}{2}; b = 3$

14. $2x + 3y = 6$ $m = -\frac{2}{3}; b = 2$

▶ **15.** $4x - 5y = 20$ $m = \frac{4}{5}; b = -4$

16. $2x - 5y = 10$ $m = \frac{2}{5}; b = -2$

17. $-2x - 5y = 10$ $m = -\frac{2}{5}; b = -2$

18. $-4x + 5y = 20$ $m = \frac{4}{5}; b = 4$

For each of the following problems, the slope and one point on a line are given. In each case use the point-slope form to find the equation of that line. (Write your answers in slope-intercept form.)

▶ **19.** $(-2, -5)$, $m = 2$ $y = 2x - 1$

20. $(-1, -5)$, $m = 2$ $y = 2x - 3$

21. $(-4, 1)$, $m = -\frac{1}{2}$ $y = -\frac{1}{2}x - 1$

22. $(-2, 1)$, $m = -\frac{1}{2}$ $y = -\frac{1}{2}x$

23. $(2, -3)$, $m = \frac{3}{2}$ $y = \frac{3}{2}x - 6$

24. $(3, -4)$, $m = \frac{4}{3}$ $y = \frac{4}{3}x - 8$

25. $(-1, 4)$, $m = -3$ $y = -3x + 1$

26. $(-2, 5)$, $m = -3$ $y = -3x - 1$

Find the equation of the line that passes through each pair of points. Write your answers in slope-intercept form.

27. $(-2, -4)$, $(1, -1)$ $y = x - 2$

28. $(2, 4)$, $(-3, -1)$ $y = x + 2$

29. $(-1, -5)$, $(2, 1)$ $y = 2x - 3$

30. $(-1, 6)$, $(1, 2)$ $y = -2x + 4$

31. $(-3, -2)$, $(3, 6)$ $y = \frac{4}{3}x + 2$

32. $(-3, 6)$, $(3, -2)$ $y = -\frac{4}{3}x + 2$

33. $(-3, -1)$, $(3, -5)$ $y = -\frac{2}{3}x - 3$

34. $(-3, -5)$, $(3, 1)$ $y = x - 2$

Find the slope and y-intercept for each line. Then write the equation of each line in slope-intercept form.

35.

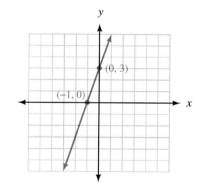

$m = 3$, $b = 3$; $y = 3x + 3$

36.

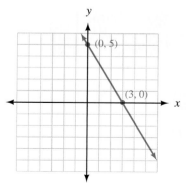

$m = -\frac{5}{3}$, $b = 5$; $y = -\frac{5}{3}x + 5$

37.

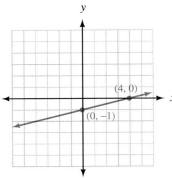

$m = \frac{1}{4}$, $b = -1$; $y = \frac{1}{4}x - 1$

38.

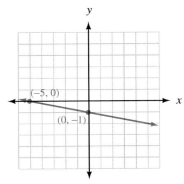

$m = -\frac{1}{5}$, $b = -1$; $y = -\frac{1}{5}x - 1$

The next two problems are intended to give you practice reading, and paying attention to, the instructions that accompany the problems you are working. Working these problems is a excellent way to get ready for a test or a quiz.

▶ **39.** Work each problem according to the instructions given.
  a. Solve: $-2x + 1 = 6$. $-\frac{5}{2}$
  b. Write in slope-intercept form:
     $-2x + y = 6$ $y = 2x + 6$
  c. Find the y-intercept: $-2x + y = 6$ 6
  d. Find the slope: $-2x + y = 6$ 2
  e. Graph: $-2x + y = 6$

40. Work each problem according to the instructions given.
  a. Solve: $x + 3 = -6$. -9
  b. Write in slope-intercept form:
     $x + 3y = -6$ $y = -\frac{1}{3}x - 2$
  c. Find the y-intercept: $x + 3y = -6$ -2
  d. Find the slope: $x + 3y = -6$ $-\frac{1}{3}$
  e. Graph: $x + 3y = -6$

▶ = Chalkboard Problem

41. Find the equation of the line with x-intercept 3 and y-intercept 2. $y = -\frac{2}{3}x + 2$

42. Find the equation of the line with x-intercept 2 and y-intercept 3. $y = -\frac{3}{2}x + 3$

43. Find the equation of the line with x-intercept -2 and y-intercept -5. $y = -\frac{5}{2}x - 5$

44. Find the equation of the line with x-intercept -3 and y-intercept -5. $y = -\frac{5}{3}x - 5$

45. The equation of the vertical line that passes through the points $(3, -2)$ and $(3, 4)$ is either $x = 3$ or $y = 3$. Which one is it? $x = 3$

46. The equation of the horizontal line that passes through the points $(2, 3)$ and $(-1, 3)$ is either $x = 3$ or $y = 3$. Which one is it? $y = 3$

Applying the Concepts

47. **Value of a Copy Machine** Cassandra buys a new color copier for her small business. It will cost $21,000 and will decrease in value each year. The graph below shows the value of the copier after the first 5 years of ownership.

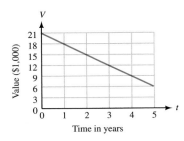

  a. How much is the copier worth after 5 years? $6,000
  b. After how many years is the copier worth $12,000? 3 years
  c. Find the slope of this line. slope = $-3,000$
  d. By how many dollars per year is the copier decreasing in value? $3,000
  e. Find the equation of this line where V is the value after t years. $V = -3,000t + 21,000$

48. **Value of a Forklift** Elliot buys a new forklift for his business. It will cost $140,000 and will decrease in value each year. The graph below shows the value of the forklift after the first 6 years of ownership.

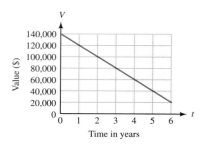

a. How much is the forklift worth after 6 years?
$20,000

b. After how many years is the forklift worth
$80,000? 3 years

c. Find the slope of this line. slope = −20,000

d. By how many dollars per year is the forklift
decreasing in value? $20,000

e. Find the equation of this line where V is the
value after t years. $V = -20,000t + 140,000$

49. Salesperson's Income Kevin starts a new job in sales
next month. He will earn $1,000 per month plus a
certain amount for each shirt he sells. The graph
below shows the amount Kevin will earn per
month based on how many shirts he sells.

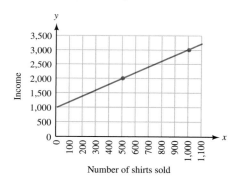

a. How much will he earn for selling 1,000 shirts?
$3,000

b. How many shirts must he sell to earn $2,000 for
a month? 500 shirts

c. Find the slope of this line. slope = 2

d. How much money does Kevin earn for each
shirt he sells? $2

e. Find the equation of this line where y is the
amount he earns for selling x number of shirts.
$y = 2x + 1,000$

50. Improving Your Quantitative Literacy Let x represent the
number of years past 2001 and y represent health
care costs as shown on the graph. (This means that

when $x = 0$, $y = 5,039$.) Use the first and last points
on the graph to find the equation of the line that
connects them. Then use the equation to estimate
health care costs in 2006. $y = 417.7x + 5,039$

In 2006, health care costs were $7,127.50.

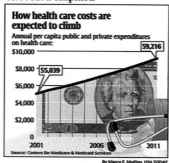

From *USA Today.* Copyright 2004.
Reprinted with permission.

Maintaining Your Skills

Use the equation $y = 3x - 5$ to find x if:

51. $y = 13$ 6

52. $y = -20$ −5

53. $y = -11$ −2

54. $y = 16$ 7

Use the equation $y = \dfrac{3}{7}x + 4$ to find y if:

55. $x = 0$ 4

56. $x = 28$ 16

57. $x = -35$ −11

58. $x = -21$ −5

Use the equation $y = 3x - 5$ to find y if:

59. $x = 5$ 10

60. $x = -5$ −20

61. $x = -11$ −38

62. $x = 11$ 28

Use the equation $y = \dfrac{x - 6}{2}$ to find x if:

63. $y = -9$ −12

64. $y = 9$ 24

65. $y = -12$ −18

66. $y = 16$ 38

Getting Ready for the Next Section

Graph each of the following lines.

67. $x + y = 4$

68. $x - y = -2$

69. $y = 2x - 3$

70. $y = 2x + 3$

71. $y = 2x$

72. $y = -2x$

Linear Inequalities in Two Variables

A linear inequality in two variables is any expression that can be put in the form

$$ax + by < c$$

where a, b, and c are real numbers (a and b not both 0). The inequality symbol can be any of the following four: $<, \leq, >, \geq$.

Some examples of linear inequalities are

$$2x + 3y < 6 \qquad y \geq 2x + 1 \qquad x - y \leq 0$$

Although not all of these inequalities have the form $ax + by < c$, each one can be put in that form.

The solution set for a linear inequality is a section of the coordinate plane. The boundary for the section is found by replacing the inequality symbol with an equal sign and graphing the resulting equation. The boundary is included in the solution set (and represented with a solid line) if the inequality symbol used originally is $\leq$ or $\geq$. The boundary is not included (and is represented with a broken line) if the original symbol is $<$ or $>$.

Let's look at some examples.

EXAMPLE 1   Graph the solution set for $x + y \leq 4$.

SOLUTION   The boundary for the graph is the graph of $x + y = 4$. The boundary is included in the solution set because the inequality symbol is $\leq$.

The graph of the boundary is shown in Figure 1.

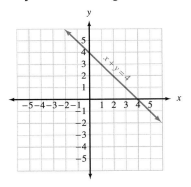

FIGURE 1

The boundary separates the coordinate plane into two sections, or regions: the region above the boundary and the region below the boundary. The solution set for $x + y \leq 4$ is one of these two regions along with the boundary. To find the correct region, we simply choose any convenient point that is *not* on the boundary. We then substitute the coordinates of the point into the original inequality $x + y \leq 4$. If the point we choose satisfies the inequality, then it is a member of the solution set, and we can assume that all points on the same side of the boundary as the chosen point are also in the solution set. If the coordinates of our point do not satisfy the original inequality, then the solution set lies on the other side of the boundary.

In this example a convenient point not on the boundary is the origin. Substituting (0, 0) into $x + y \leq 4$ gives us

$$0 + 0 \overset{?}{\leq} 4$$

$$0 \leq 4 \qquad \textbf{A true statement}$$

Because the origin is a solution to the inequality $x + y \leq 4$, and the origin is below the boundary, all other points below the boundary are also solutions. The graph of $x + y \leq 4$ is shown in Figure 2.

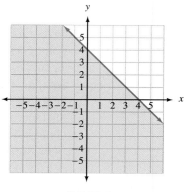

FIGURE 2

The region above the boundary is described by the inequality $x + y > 4$.

Here is a list of steps to follow when graphing the solution set for linear inequalities in two variables.

To Graph the Solution Set for Linear Inequalities in Two Variables

Step 1: Replace the inequality symbol with an equal sign. The resulting equation represents the boundary for the solution set.

Step 2: Graph the boundary found in step 1 using a *solid line* if the boundary is included in the solution set (that is, if the original inequality symbol was either $\leq$ or $\geq$). Use a *broken line* to graph the boundary if it is *not* included in the solution set. (It is not included if the original inequality was either $<$ or $>$.)

Step 3: Choose any convenient point not on the boundary and substitute the coordinates into the *original* inequality. If the resulting statement is *true,* the graph lies on the *same* side of the boundary as the chosen point. If the resulting statement is *false,* the solution set lies on the *opposite* side of the boundary.

 EXAMPLE 2   Graph the solution set for $y < 2x - 3$.

SOLUTION   The boundary is the graph of $y = 2x - 3$. The boundary is not included because the original inequality symbol is $<$. We therefore use a broken line to represent the boundary, as shown in Figure 3.

A convenient test point is again the origin. Using $(0, 0)$ in $y < 2x - 3$, we have

$$0 \overset{?}{<} 2(0) - 3$$

$$0 < -3 \qquad \textbf{A false statement}$$

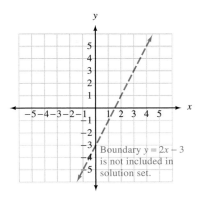

FIGURE 3

Because our test point gives us a false statement and it lies above the boundary, the solution set must lie on the other side of the boundary, as shown in Figure 4.

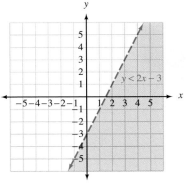

FIGURE 4

EXAMPLE 3   Graph the inequality $2x + 3y \le 6$.

SOLUTION We begin by graphing the boundary $2x + 3y = 6$. The boundary is included in the solution because the inequality symbol is $\le$.

If we use $(0, 0)$ as our test point, we see that it yields a true statement when its coordinates are substituted into $2x + 3y \le 6$. The graph, therefore, lies below the boundary, as shown in Figure 5.

The ordered pair $(0, 0)$ is a solution to $2x + 3y \le 6$; all points on the same side of the boundary as $(0, 0)$ also must be solutions to the inequality $2x + 3y \le 6$.

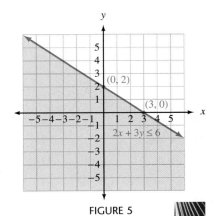

FIGURE 5

EXAMPLE 4 Graph the solution set for $x \le 5$.

SOLUTION The boundary is $x = 5$, which is a vertical line. All points to the left have x-coordinates less than 5, and all points to the right have x-coordinates greater than 5, as shown in Figure 6.

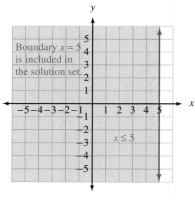

FIGURE 6

GETTING READY FOR CLASS

After reading through the preceding section, respond in your own words and in complete sentences.

1. When graphing a linear inequality in two variables, how do you find the equation of the boundary line?
2. What is the significance of a broken line in the graph of an inequality?
3. When graphing a linear inequality in two variables, how do you know which side of the boundary line to shade?
4. Describe the set of ordered pairs that are solutions to $x + y < 6$.

LINKING OBJECTIVES AND EXAMPLES

Next to each **objective** we have listed the examples that are best described by that objective.

A   1–4

Problem Set 3.7

Online support materials can be found at www.thomsonedu.com/login

Graph the following linear inequalities.

▶ **1.** $2x - 3y < 6$

2. $3x + 2y \ge 6$

3. $x - 2y \le 4$

4. $2x + y > 4$

5. $x - y \le 2$

6. $x - y \le 1$

7. $3x - 4y \ge 12$

8. $4x + 3y < 12$

9. $5x - y \le 5$

10. $4x + y > 4$

11. $2x + 6y \le 12$

12. $x - 5y > 5$

13. $x \ge 1$

14. $x < 5$

15. $x \ge -3$

16. $y \le -4$

▶ **17.** $y < 2$

18. $3x - y > 1$

19. $2x + y > 3$

20. $5x + 2y < 2$

21. $y \le 3x - 1$

22. $y \ge 3x + 2$

23. $y \le -\frac{1}{2}x + 2$

24. $y < \frac{1}{3}x + 3$

The next two problems are intended to give you practice reading, and paying attention to, the instructions that accompany the problems you are working.

▶ **25.** Work each problem according to the instructions given.

  a. Solve: $4 + 3y < 12$ $y < \frac{8}{3}$

  b. Solve: $4 - 3y < 12$ $y > -\frac{8}{3}$

= Videos available by instructor request   ▶ = Chalkboard Problem
▶ = Online student support materials available at www.thomsonedu.com/login

c. Solve for y: $4x + 3y = 12$ $y = -\frac{4}{3}x + 4$

d. Graph: $y < -\frac{4}{3}x + 4$

26. Work each problem according to the instructions given.

a. Solve: $3x + 2 \geq 6$ $x \geq \frac{4}{3}$

b. Solve: $-3x + 2 \geq 6$ $x \leq -\frac{4}{3}$

c. Solve for y: $3x + 2y = 6$ $y = -\frac{3}{2}x + 3$

d. Graph: $y \geq -\frac{3}{2}x + 3$

27. Find the equation of the line in part a, then use this information to find the inequalities for the graphs on parts b and c.

a. $y = \frac{2}{5}x + 2$

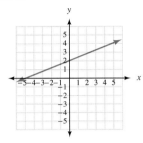

b. $y < \frac{2}{5}x + 2$

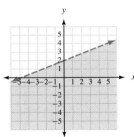

c. $y > \frac{2}{5}x + 2$

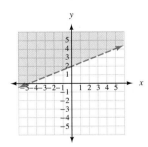

28. Find the equation of the line in part a, then use this information to find the inequalities for the graphs on parts b and c.

a. $y = -\frac{3}{2}x + 3$

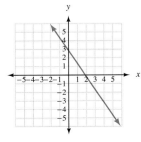

b.

$y \geq -\frac{3}{2}x + 3$

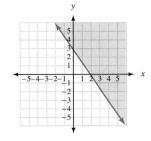

c.

$y \leq -\frac{3}{2}x + 3$

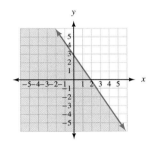

Maintaining Your Skills

29. Simplify the expression $7 - 3(2x - 4) - 8$. $-6x + 11$

30. Find the value of $x^2 - 2xy + y^2$ when $x = 3$ and $y = -4$.   49

Solve each equation.

31. $-\frac{3}{2}x = 12$ -8

32. $2x - 4 = 5x + 2$ -2

33. $8 - 2(x + 7) = 2$ -4

34. $3(2x - 5) - (2x - 4) = 6 - (4x + 5)$ $\frac{3}{2}$

35. Solve the formula $P = 2l + 2w$ for w. $w = \frac{P - 2l}{2}$

Solve each inequality, and graph the solution.

36. $-4x < 20$

37. $3 - 2x > 5$

38. $3 - 4(x - 2) \geq -5x + 6$

39. Solve the formula $3x - 2y \leq 12$ for y. $y \geq \frac{3}{2}x - 6$

40. What number is 12% of 2,000?   240

41. **Geometry** The length of a rectangle is 5 inches more than 3 times the width. If the perimeter is 26 inches, find the length and width.

EXAMPLES

Linear Equation in Two Variables [3.3]

1. The equation $3x + 2y = 6$ is an example of a linear equation in two variables.

Width 2 inches, length 11 inches

A linear equation in two variables is any equation that can be put in the form $ax + by = c$. The graph of every linear equation is a straight line.

2. The graph of $y = -\frac{2}{3}x - 1$ is shown below.

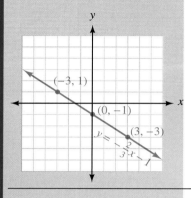

Strategy for Graphing Linear Equations in Two Variables [3.3]

Step 1 Find any three ordered pairs that satisfy the equation. This can be done by using a convenient number for one variable and solving for the other variable.

Step 2: Graph the three ordered pairs found in step 1. Actually, we need only two points to graph a straight line. The third point serves as a check. If all three points do not line up, there is a mistake in our work.

Step 3: Draw a straight line through the three points graphed in step 2.

Intercepts [3.4]

3. To find the x-intercept for $3x + 2y = 6$, we let $y = 0$ and get
$$3x = 6$$
$$x = 2$$
In this case the x-intercept is 2, and the graph crosses the x-axis at (2, 0).

The x-intercept of an equation is the x-coordinate of the point where the graph crosses the x-axis. The y-intercept is the y-coordinate of the point where the graph crosses the y-axis. We find the y-intercept by substituting $x = 0$ into the equation and solving for y. The x-intercept is found by letting $y = 0$ and solving for x.

Slope of a Line [3.5]

4. The slope of the line through $(3, -5)$ and $(-2, 1)$ is

$$m = \frac{-5 - 1}{3 - (-2)} = \frac{-6}{5} = -\frac{6}{5}$$

The *slope* of the line containing the points (x_1, y_1) and (x_2, y_2) is given by

$$\text{Slope} = m = \frac{y_2 - y_1}{x_2 - x_1} = \frac{\text{rise}}{\text{run}}$$

Point-Slope Form of a Straight Line [3.6]

5. The equation of the line through $(1, 2)$ with a slope of 3 is
$$y - 2 = 3(x - 1)$$
$$y - 2 = 3x - 3$$
$$y = 3x - 1$$

If a line has a slope of m and contains the point (x_1, y_1), the equation can be written as

$$y - y_1 = m(x - x_1)$$

Slope-Intercept Form of a Straight Line [3.6]

6. The equation of the line with a slope of 2 and a y-intercept 5 is
$$y = 2x + 5$$

The equation of the line with a slope of m and a y-intercept b is

$$y = mx + b$$

To Graph a Linear Inequality in Two Variables [3.7]

7. Graph $x - y \geq 3$.

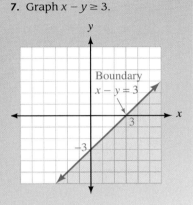

Step 1: Replace the inequality symbol with an equal sign. The resulting equation represents the boundary for the solution set.

Step 2: Graph the boundary found in step 1, using a *solid line* if the original inequality symbol was either $\leq$, or $\geq$. Use a *broken line* otherwise.

Step 3: Choose any convenient point not on the boundary and substitute the coordinates into the *original* inequality. If the resulting statement is *true,* the graph lies on the *same* side of the boundary as the chosen point. If the resulting statement is *false,* the solution set lies on the *opposite* side of the boundary.

Chapter 3 Review Test

The problems below form a comprehensive review of the material in this chapter. They can be used to study for exams. If you would like to take a practice test on this chapter, you can use the odd-numbered problems. Give yourself an hour and work as many of the odd-numbered problems as possible. When you are finished, or when an hour has passed, check your answers with the answers in the back of the book. You can use the even-numbered problems for a second practice test.

The numbers in brackets refer to the sections of the text in which similar problems can be found.

For each equation, complete the given ordered pairs. [3.2]

1. $3x + y = 6$ (4,), (0,), (, 3), (, 0)

2. $2x - 5y = 20$ (5,), (0,), (, 2), (, 0)

3. $y = 2x - 6$ (4,), (, −2), (, 3)

4. $y = 5x + 3$ (2,), (, 0), (, −3)

5. $y = -3$ (2,), (−1,), (−3,)

6. $x = 6$ (, 5), (, 0), (, −1)

For the following equations, tell which of the given ordered pairs are solutions. [3.2]

7. $3x - 4y = 12$ $\left(-2, \frac{9}{2}\right)$, (0, 3), $\left(2, -\frac{3}{2}\right)$

8. $y = 3x + 7$ $\left(-\frac{8}{3}, -1\right)$, $\left(\frac{7}{3}, 0\right)$, (−3, −2)

Graph the following ordered pairs. [3.1]

9. (4, 2)

10. (−3, 1)

11. (0, 5)

12. (−2, −3)

13. (−3, 0)

14. $\left(5, -\frac{3}{2}\right)$

For the following equations, complete the given ordered pairs, and use the results to graph the solution set for the equations. [3.3]

15. $x + y = -2$ (, 0), (0,), (1,)

16. $y = 3x$ (−1,), (1,), (, 0)

17. $y = 2x - 1$ (1,), (0,), (, −3)

18. $x = -3$ (, 0), (, 5), (, −5)

Graph the following equations. [3.3]

19. $3x - y = 3$

20. $x - 2y = 2$

21. $y = -\frac{1}{3}x$

22. $y = \frac{3}{4}x$

23. $y = 2x + 1$

24. $y = -\frac{1}{2}x + 2$

25. $x = 5$

26. $y = -3$

27. $2x - 3y = 3$

28. $5x - 2y = 5$

Find the x- and y-intercepts for each equation. [3.4]

29. $3x - y = 6$

30. $2x - 6y = 24$

31. $y = x - 3$

32. $y = 3x - 6$

33. $y = -5$

34. $x = 4$

Find the slope of the line through the given pair of points. [3.5]

35. (2, 3), (3, 5)

36. (−2, 3), (6, −5)

37. (−1, −4), (−3, −8)

38. $\left(\frac{1}{2}, 4\right), \left(-\frac{1}{2}, 2\right)$

39. Find x if the line through (3, 3) and (x, 9) has slope 2.

40. Find x if the line through (5, −5) and (−5, x) has slope 2.

Find the equation of the line that contains the given point and has the given slope. Write answers in slope-intercept form. [3.6]

41. (−1, 4); $m = -2$

42. (4, 3); $m = \frac{1}{2}$

43. (3, −2); $m = -\frac{3}{4}$

44. (3, 5); $m = 0$

Find the equation of the line with the given slope and y-intercept. [3.6]

45. $m = 3, b = 2$

46. $m = -1, b = 6$

47. $m = -\frac{1}{3}, b = \frac{3}{4}$

48. $m = 0, b = 0$

For each of the following equations, determine the slope and y-intercept. [3.6]

49. $y = 4x - 1$

50. $2x + y = -5$

51. $6x + 3y = 9$

52. $5x + 2y = 8$

Graph the following linear inequalities. [3.7]

53. $x - y < 3$

54. $x \geq -3$

55. $y \leq -4$

56. $y \leq -2x + 3$

GROUP PROJECT Reading Graphs

Number of People 2-3

Time Needed 5–10 minutes

Equipment Pencil and paper

Background Although most of the graphs we have encountered in this chapter have been straight lines, many of the graphs that describe the world around us are not straight lines. In this group project we gain experience working with graphs that are not straight lines.

Procedure Read the introduction to each problem below. Then use the graphs to answer the questions.

1. A patient is taking a prescribed dose of a medication every 4 hours during the day to relieve the symptoms of a cold. Figure 1 shows how the concentration of that medication in the patient's system changes over time. The 0 on the horizontal axis corresponds to the time the patient takes the first dose of medication. (The units of concentration on the vertical axis are nanograms per milliliter.)
    a. Explain what the steep vertical line segments show with regard to the patient and his medication.
    b. What has happened to make the graph fall off on the right?
    c. What is the maximum concentration of the medication in the patient's system during the time period shown in Figure 1?
    d. Find the values of A, B, and C.

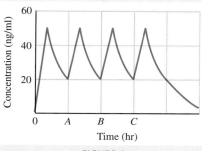

FIGURE 1

2. **Reading Graphs.** Figure 2 shows the number of people in line at a theater box office to buy tickets for a movie that starts at 7:30. The box office opens at 6:45.
    a. How many people are in line at 6:30?
    b. How many people are in line when the box office opens?
    c. How many people are in line when the show starts?
    d. At what times are there 60 people in line?
    e. How long after the show starts is there no one left in line?

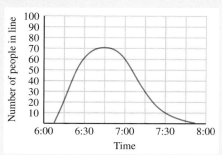

FIGURE 2

Least Squares Curve Fitting

Bettman/Corbis

FIGURE 1

In 1929, the astronomer Edwin Hubble (shown in Figure 1) announced his discovery that the other galaxies in the universe are moving away from us at velocities that increase with distance. The relationship between velocity and distance is described by the linear equation:

$$v = Hr$$

where r is the distance of the galaxy from us, v is its velocity away from us, and H is "Hubble's constant." Figure 2 shows a plot of velocity versus distance, where each point represents a galaxy. The fact that the dots all lie approximately on a straight line is the basis of "Hubble's law."

As you can imagine, there are many lines that could be drawn through the dots in Figure 2. The line shown in Figure 2 is called the *line of best fit* for the points shown in the figure. The method used most often in mathematics to find the line of best fit is called the *least squares method.* Research the least squares method of finding the line of best fit and write an essay that describes the method. Your essay should answer the question: "Why is this method of curve fitting called the *least squares* method?"

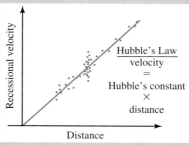

FIGURE 2

Systems of Linear Equations

4

Getty Images

Two companies offer Internet access to their customers. Company A charges $10 a month plus $3 for every hour of Internet connection. Company B charges $18 a month plus $1 for every hour of Internet connection. To compare the monthly charges of the two companies we form what is called a system of equations. Here is that system.

$$y = 3x + 10$$
$$y = x + 18$$

The top equation gives us information on company A; the bottom equation gives information on company B. Tables 1 and 2 and the graphs in Figure 1 give us additional information about this system of equations.

TABLE 1 Company A	
Hours	Cost
0	$10
1	$13
2	$16
3	$19
4	$22
5	$25
6	$28
7	$31
8	$34
9	$37
10	$40

TABLE 2 Company B	
Hours	Cost
0	$18
1	$19
2	$20
3	$21
4	$22
5	$23
6	$24
7	$25
8	$26
9	$27
10	$28

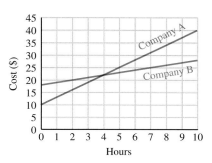

As you can see from looking at the tables and at the graphs in Figure 1, the monthly charges for the two companies will be equal if Internet use is exactly 4 hours. In this chapter we work with systems of linear equations.

▶ Improve your grade and save time! Go online to **www.thomsonedu.com/login** where you can

- Watch videos of instructors working through the in-text examples
- Follow step-by-step online tutorials of in-text examples and review questions
- Work practice problems
- Check your readiness for an exam by taking a pre-test and exploring the modules recommended in your Personalized Study plan
- Receive help from a live tutor online through vMentor™

Try it out! Log in with an access code or purchase access at **www.ichapters.com.**

213

The study skills for this chapter concern the way you approach new situations in mathematics. The first study skill applies to your natural instincts for what does and doesn't work in mathematics. The second study skill gives you a way of testing your instincts.

1 Don't Let Your Intuition Fool You

As you become more experienced and more successful in mathematics, you will be able to trust your mathematical intuition. For now, though, it can get in the way of success. For example, if you ask a beginning algebra student to "subtract 3 from -5" many will answer -2 or 2. Both answers are incorrect, even though they may seem intuitively true.

2 Test Properties About Which You Are Unsure

From time to time you will be in a situation in which you would like to apply a property or rule, but you are not sure if it is true. You can always test a property or statement by substituting numbers for variables. For instance, I always have students that rewrite $(x + 3)^2$ as $x^2 + 9$, thinking that the two expressions are equivalent. The fact that the two expressions are not equivalent becomes obvious when we substitute 10 for x in each one.

$$\text{When } x = 10, \text{ the expression } (x + 3)^2 \text{ is } (10 + 3)^2 = 13^2 = 169$$

$$\text{When } x = 10, \text{ the expression } x^2 + 9 = 10^2 + 9 = 100 + 9 = 109$$

It is not unusual, nor is it wrong, to try occasionally to apply a property that doesn't exist. If you have any doubt about generalizations you are making, test them by replacing variables with numbers and simplifying.

4.1 Solving Linear Systems by Graphing

OBJECTIVES

A Solve a system of linear equations in two variables by graphing.

Two linear equations considered at the same time make up what is called a *system of linear equations.* Both equations contain two variables and, of course, have graphs that are straight lines. The following are systems of linear equations:

$$x + y = 3 \qquad y = 2x + 1 \qquad 2x - y = 1$$
$$3x + 4y = 2 \qquad y = 3x + 2 \qquad 3x - 2y = 6$$

The solution set for a system of linear equations is all ordered pairs that are solutions to both equations. Because each linear equation has a graph that is a straight line, we can expect the intersection of the graphs to be a point whose coordinates are solutions to the system; that is, if we graph both equations on the same coordinate system, we can read the coordinates of the point of intersection and have the solution to our system. Here is an example.

EXAMPLE 1   Solve the following system by graphing.

$$x + y = 4$$
$$x - y = -2$$

SOLUTION   On the same set of coordinate axes we graph each equation separately. Figure 1 shows both graphs, without showing the work necessary to get them. We can see from the graphs that they intersect at the point (1, 3). The point (1, 3) therefore must be the solution to our system because it is the only ordered pair whose graph lies on both lines. Its coordinates satisfy both equations.

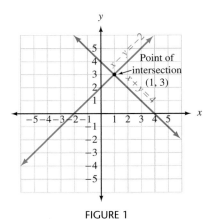

FIGURE 1

We can check our results by substituting the coordinates $x = 1, y = 3$ into both equations to see if they work.

When	$x = 1$	When	$x = 1$
and	$y = 3$	and	$y = 3$
the equation	$x + y = 4$	the equation	$x - y = -2$
becomes	$1 + 3 \overset{?}{=} 4$	becomes	$1 - 3 \overset{?}{=} -2$
or	$4 = 4$	or	$-2 = -2$

The point (1, 3) satisfies both equations.

Here are some steps to follow in solving linear systems by graphing.

Strategy for Solving a Linear System by Graphing

Step 1: Graph the first equation by the methods described in Section 3.3 or 3.4.

Step 2: Graph the second equation on the same set of axes used for the first equation.

Step 3: Read the coordinates of the point of intersection of the two graphs.

Step 4: Check the solution in both equations.

 EXAMPLE 2   Solve the following system by graphing.

$$x + 2y = 8$$
$$2x - 3y = 2$$

SOLUTION   Graphing each equation on the same coordinate system, we have the lines shown in Figure 2.

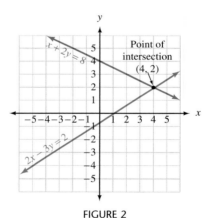

FIGURE 2

From Figure 2, we can see the solution for our system is (4, 2). We check this solution as follows.

When	$x = 4$	When	$x = 4$
and	$y = 2$	and	$y = 2$
the equation	$x + 2y = 8$	the equation	$2x - 3y = 2$
becomes	$4 + 2(2) \overset{?}{=} 8$	becomes	$2(4) - 3(2) \overset{?}{=} 2$
	$4 + 4 = 8$		$8 - 6 = 2$
	$8 = 8$		$2 = 2$

The point (4, 2) satisfies both equations and, therefore, must be the solution to our system.

EXAMPLE 3

Solve this system by graphing.

$$y = 2x - 3$$
$$x = 3$$

SOLUTION Graphing both equations on the same set of axes, we have Figure 3.

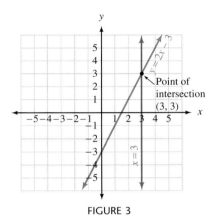

FIGURE 3

The solution to the system is the point (3, 3).

EXAMPLE 4

Solve by graphing.

$$y = x - 2$$
$$y = x + 1$$

SOLUTION Graphing both equations produces the lines shown in Figure 4. We can see in Figure 4 that the lines are parallel and therefore do not intersect. Our system has no ordered pair as a solution because there is no ordered pair that satisfies both equations. We say the solution set is the empty set and write Ø.

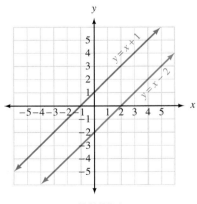

FIGURE 4

Example 4 is one example of two special cases associated with linear systems. The other special case happens when the two graphs coincide. Here is an example.

EXAMPLE 5 Graph the system.

$$2x + y = 4$$
$$4x + 2y = 8$$

SOLUTION Both graphs are shown in Figure 5. The two graphs coincide. The reason becomes apparent when we multiply both sides of the first equation by 2:

$$2x + y = 4$$

$$\mathbf{2}(2x + y) = \mathbf{2}(4) \qquad \textbf{Multiply both sides by 2}$$

$$4x + 2y = 8$$

The equations have the same solution set. Any ordered pair that is a solution to one is a solution to the system. The system has an infinite number of solutions. (Any point on the line is a solution to the system.)

Note

We sometimes use special vocabulary to describe the special cases shown in Examples 4 and 5. When a system of equations has no solution because the lines are parallel (as in Example 4), we say the system is *inconsistent.* When the lines coincide (as in Example 5), we say the system is *dependent.*

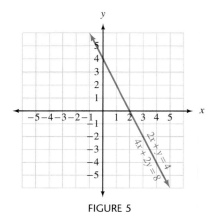

FIGURE 5

The two special cases illustrated in the previous two examples do not happen often. Usually, a system has a single ordered pair as a solution. Solving a system of linear equations by graphing is useful only when the ordered pair in the solution set has integers for coordinates. Two other solution methods work well in all cases. We will develop the other two methods in the next two sections.

GETTING READY FOR CLASS

LINKING OBJECTIVES AND EXAMPLES

Next to each **objective** we have listed the examples that are best described by that objective.

A 1–5

After reading through the preceding section, respond in your own words and in complete sentences.

1. What is a system of two linear equations in two variables?
2. What is a solution to a system of linear equations?
3. How do we solve a system of linear equations by graphing?
4. Under what conditions will a system of linear equations not have a solution?

Answers appear in the Instructor's Edition only.

Solve the following systems of linear equations by graphing.

▶ 1. $x + y = 3$
$x - y = 1$ (2, 1)

2. $x + y = 2$
$x - y = 4$ (3, −1)

3. $x + y = 1$
$-x + y = 3$ (−1, 2)

4. $x + y = 1$
$x - y = -5$ (−2, 3)

5. $x + y = 8$
$-x + y = 2$ (3, 5)

6. $x + y = 6$
$-x + y = -2$ (4, 2)

▶ 7. $3x - 2y = 6$
$x - y = 1$ (4, 3)

8. $5x - 2y = 10$
$x - y = -1$ (4, 5)

9. $6x - 2y = 12$
$3x + y = -6$ (0, −6)

10. $4x - 2y = 8$
$2x + y = -4$ (0, −4)

11. $4x + y = 4$
$3x - y = 3$ (1, 0)

12. $5x - y = 10$
$2x + y = 4$ (2, 0)

13. $x + 2y = 0$
$2x - y = 0$ (0, 0)

14. $3x + y = 0$
$5x - y = 0$ (0, 0)

15. $3x - 5y = 15$
$-2x + y = 4$ (−5, −6)

16. $2x - 4y = 8$
$2x - y = -1$ (−2, −3)

17. $y = 2x + 1$
$y = -2x - 3$ (−1, −1)

18. $y = 3x - 4$
$y = -2x + 1$ (1, −1)

19. $x + 3y = 3$
$y = x + 5$ (−3, 2)

20. $2x + y = -2$
$y = x + 4$ (−2, 2)

▶ 21. $x + y = 2$
$x = -3$ (−3, 5)

22. $x + y = 6$
$y = 2$ (4, 2)

23. $x = -4$
$y = 6$ (−4, 6)

24. $x = 5$
$y = -1$ (5, −1)

25. $x + y = 4$
$2x + 2y = -6$
∅

26. $x - y = 3$
$2x - 2y = 6$
Any point on the line

27. $4x - 2y = 8$
$2x - y = 4$
Any point on the line

28. $3x - 6y = 6$
$x - 2y = 4$
∅

The next two problems are intended to give you practice reading, and paying attention to, the instructions that accompany the problems you are working.

▶ 29. Work each problem according to the instructions given.
   a. Simplify: $(3x - 4y) + (x - y)$ $4x - 5y$
   b. Find y when x is 4 in $3x - 4y = 8$ 1

   c. Find the y-intercept: $3x - 4y = 8$ −2
   d. Graph: $3x - 4y = 8$
   e. Find the point where the graphs of $3x - 4y = 8$ and $x - y = 2$ cross.   (0, −2)

30. Work each problem according to the instructions given.
   a. Simplify: $(x + 4y) + (-2x + 3y)$ $-x + 7y$
   b. Find y when x is 3 in $-2x + 3y = 3$ 3
   c. Find the y-intercept: $-2x + 3y = 3$ 1
   d. Graph: $-2x + 3y = 3$
   e. Find the point where the graphs of $-2x + 3y = 3$ and $x + 4y = 4$ cross.   (0, 1)

31. As you probably have guessed by now, it can be difficult to solve a system of equations by graphing if the solution to the system contains a fraction. The solution to the following system is $(\frac{1}{2}, 1)$. Solve the system by graphing.

$$y = -2x + 2$$
$$y = 4x - 1$$

32. The solution to the following system is $(\frac{1}{3}, -2)$. Solve the system by graphing.

$$y = 3x - 3$$
$$y = -3x - 1$$

33. A second difficulty can arise in solving a system of equations by graphing if one or both of the equations is difficult to graph. The solution to the following system is (2, 1). Solve the system by graphing.

$$3x - 8y = -2$$
$$x - y = 1$$

34. The solution to the following system is (−3, 2). Solve the system by graphing.

$$2x + 5y = 4$$
$$x - y = -5$$

Applying the Concepts

35. **Job Comparison** Jane is deciding between two sales positions. She can work for Marcy's and receive $8.00 per hour or for Gigi's, where she earns $6.00

▶ **Chalkboard Problem**
Problem 29 is nice way to lead into solving systems of equations by graphing. The only thing left to do is graph the second equation. Work it before going to class. I think you will like it.

  = Videos available by instructor request
▶ = Online student support materials available at www.thomsonedu.com/login

per hour but also receives a $50 commission per week. The two lines in the following figure represent the money Jane will make for working at each of the jobs.

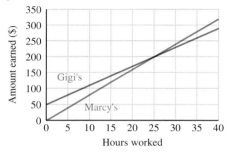

a. From the figure, how many hours would Jane have to work to earn the same amount at each of the positions? 25 hours

b. If Jane expects to work less than 20 hours a week, which job should she choose? Gigi's

c. If Jane expects to work more than 30 hours a week, which job should she choose? Marcy's

36. Improving Your Quantitative Literacy The graph here shows the percentage of alcohol- and nonalcohol-related fatal car crashes for the years 1982 through 2004. Although these graphs are not linear, they do intersect.

a. What is the point of intersection of these two lines? (1989, 50%)

b. Explain the significance of the point of intersection in terms of car crash fatalities.

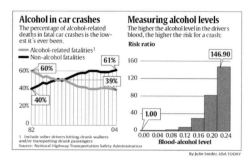

From *USA Today*. Copyright 2005. Reprinted with permission.

The point of intersection is the year in which the number of alcohol- and nonalcohol-related traffic fatalities were the closest.

Maintaining Your Skills

Simplify each expression.

37. $6x + 100(0.04x + 0.75)$ $10x + 75$

38. $5x + 100(0.03x + 0.65)$ $8x + 65$

39. $13x - 1{,}000(0.002x + 0.035)$ $11x - 35$

40. $9x - 1{,}000(0.023x + 0.015)$ $-14x - 15$

41. $16x - 10(1.7x - 5.8)$ $-x + 58$

42. $43x - 10(3.1x - 2.7)$ $12x + 27$

43. $0.04x + 0.06(100 - x)$ $-0.02x + 6$

44. $0.07x + 0.03(100 - x)$ $0.04x + 3$

45. $0.025x - 0.028(1{,}000 + x)$ $-0.003x - 28$

46. $0.065x - 0.037(1{,}000 + x)$ $0.028x - 37$

47. $2.56x - 1.25(100 + x)$ $1.31x - 125$

48. $8.42x - 6.68(100 + x)$ $1.74x - 668$

Getting Ready for the Next Section

Simplify each of the following.

49. $(x + y) + (x - y)$ $2x$

50. $(x + 2y) + (-x + y)$ $3y$

51. $(6x - 3y) + (x + 3y)$ $7x$

52. $(6x + 9y) + (-6x - 10y)$ $-y$

53. $(-12x - 20y) + (25x + 20y)$ $13x$

54. $(-3x + 2y) + (3x + 8y)$ $10y$

55. $-4(3x + 5y)$ $-12x - 20y$

56. $6\left(\dfrac{1}{2}x - \dfrac{1}{3}y\right)$ $3x - 2y$

57. $12\left(\dfrac{1}{4}x + \dfrac{2}{3}y\right)$ $3x + 8y$

58. $5(5x + 4y)$ $25x + 20y$

59. $-2(2x - y)$ $-4x + 2y$

60. $-2(4x - 3y)$ $-8x + 6y$

61. Let $x + y = 4$. If $x = 3$, find y.   1

62. Let $x + 2y = 4$. If $x = 3$, find y.   0.5

63. Let $x + 3y = 3$. If $x = 3$, find y.   0

64. Let $2x + 3y = -1$. If $y = -1$, find x.   1

65. Let $3x + 5y = -7$. If $x = 6$, find y. -5

66. Let $3x - 2y = 12$. If $y = 6$, find x.   8

4.2 The Elimination Method

OBJECTIVES

A Use the elimination method to solve a system of linear equations in two variables.

The addition property states that if equal quantities are added to both sides of an equation, the solution set is unchanged. In the past we have used this property to help solve equations in one variable. We will now use it to solve systems of linear equations. Here is another way to state the addition property of equality.

Let A, B, C, and D represent algebraic expressions.

$$\begin{array}{ll} \text{If} & A = B \\ \text{and} & \underline{C = D} \\ \text{then} & A + C = B + D \end{array}$$

Because C and D are equal (that is, they represent the same number), what we have done is added the same amount to both sides of the equation $A = B$. Let's see how we can use this form of the addition property of equality to solve a system of linear equations.

EXAMPLE 1 Solve the following system.

$$x + y = 4$$
$$x - y = 2$$

SOLUTION The system is written in the form of the addition property of equality as written in this section. It looks like this:

$$A = B$$
$$C = D$$

where A is $x + y$, B is 4, C is $x - y$, and D is 2.

We use the addition property of equality to add the left sides together and the right sides together.

$$\begin{array}{r} x + y = 4 \\ \underline{x - y = 2} \\ 2x + 0 = 6 \end{array}$$

We now solve the resulting equation for x.

$$2x + 0 = 6$$
$$2x = 6$$
$$x = 3$$

Note

The graphs shown in the margin next to our first three examples are not part of the solution shown in each example. The graphs are there simply to show you that the results we obtain by the elimination method are consistent with the results we would obtain by graphing.

The value we get for x is the value of the x-coordinate of the point of intersection of the two lines $x + y = 4$ and $x - y = 2$. To find the y-coordinate, we simply substitute $x = 3$ into either of the two original equations. Using the first equation, we get

$$3 + y = 4$$
$$y = 1$$

The solution to our system is the ordered pair (3, 1). It satisfies both equations.

When	$x = 3$	When	$x = 3$
and	$y = 1$	and	$y = 1$
the equation	$x + y = 4$	the equation	$x - y = 2$
becomes	$3 + 1 \overset{?}{=} 4$	becomes	$3 - 1 \overset{?}{=} 2$
or	$4 = 4$	or	$2 = 2$

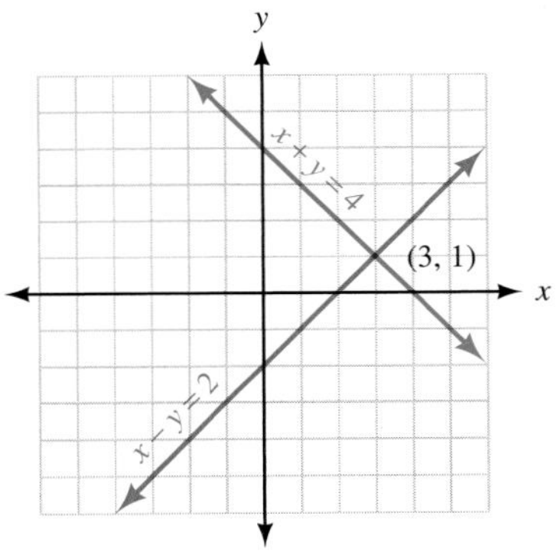

Figure 1 is visual evidence that the solution to our system is (3, 1).

FIGURE 1

The most important part of this method of solving linear systems is eliminating one of the variables when we add the left and right sides together. In our first example, the equations were written so that the y variable was eliminated when we added the left and right sides together. If the equations are not set up this way to begin with, we have to work on one or both of them separately before we can add them together to eliminate one variable.

 EXAMPLE 2   Solve the following system.

$$x + 2y = 4$$

$$x - y = -5$$

SOLUTION   Notice that if we were to add the equations together as they are, the resulting equation would have terms in both x and y. Let's eliminate the variable x by multiplying both sides of the second equation by -1 before we add the equations together. (As you will see, we can choose to eliminate either the x or the y variable.) Multiplying both sides of the second equation by -1 will not change its solution, so we do not need to be concerned that we have altered the system.

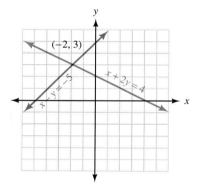

$$
\begin{array}{l}
x + 2y = 4 \xrightarrow{\text{No change}} x + 2y = 4 \quad \textbf{Add left and right sides to get} \\[4pt]
x - y = -5 \xrightarrow[\text{Multiply by } -1]{} \underline{-x + y = 5} \\[4pt]
\hphantom{x - y = -5 \longrightarrow} 0 + 3y = 9 \\[4pt]
\hphantom{x - y = -5 \longrightarrow} 3y = 9 \\[4pt]
\hphantom{x - y = -5 \longrightarrow} y = 3 \quad \begin{cases} \textbf{\textit{y}-Coordinate of the} \\ \textbf{point of intersection} \end{cases}
\end{array}
$$

Substituting $y = 3$ into either of the two original equations, we get $x = -2$. The solution to the system is $(-2, 3)$. It satisfies both equations. Figure 2 shows the solution to the system as the point where the two lines cross.

FIGURE 2

 EXAMPLE 3   Solve the following system.

$$2x - y = 6$$

$$x + 3y = 3$$

SOLUTION   Let's eliminate the y variable from the two equations. We can do this by multiplying the first equation by 3 and leaving the second equation unchanged.

$$
\begin{array}{l}
2x - y = 6 \xrightarrow{\text{3 times both sides}} 6x - 3y = 18 \\[6pt]
x + 3y = 3 \xrightarrow[\text{No change}]{} x + 3y = 3
\end{array}
$$

The important thing about our system now is that the coefficients (the numbers in front) of the y variables are opposites. When we add the terms on each side of the equal sign, then the terms in y will add to zero and be eliminated.

$$6x - 3y = 18$$
$$\underline{x + 3y = 3}$$
$$7x \quad\;\; = 21 \qquad \textbf{Add corresponding terms}$$

This gives us $x = 3$. Using this value of x in the second equation of our original system, we have

$$3 + 3y = 3$$
$$3y = 0$$
$$y = 0$$

We could substitute $x = 3$ into any of the equations with both x and y variables and also get $y = 0$. The solution to our system is the ordered pair $(3, 0)$. Figure 3 is a picture of the system of equations showing the solution $(3, 0)$.

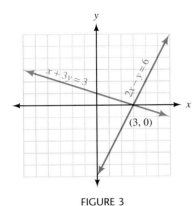

FIGURE 3

 EXAMPLE 4　　Solve the system.

$$2x + 3y = -1$$
$$3x + 5y = -2$$

Note

If you are having trouble understanding this method of solution, it is probably because you can't see why we chose to multiply by 3 and −2 in the first step of Example 4. Look at the result of doing so: the $6x$ and $-6x$ will add to 0. We chose to multiply by 3 and −2 because they produce $6x$ and $-6x$, which will add to 0.

SOLUTION　Let's eliminate x from the two equations. If we multiply the first equation by 3 and the second by −2, the coefficients of x will be 6 and −6, respectively. The x terms in the two equations will then add to zero.

$$2x + 3y = -1 \xrightarrow{\text{Multiply by 3}} 6x + 9y = -3$$
$$3x + 5y = -2 \xrightarrow[\text{Multiply by } -2]{} -6x - 10y = 4$$

We now add the left and right sides of our new system together.

$$6x + 9y = -3$$
$$\underline{-6x - 10y = 4}$$
$$-y = 1$$
$$y = -1$$

Substituting $y = -1$ into the first equation in our original system, we have

$$2x + 3(-1) = -1$$
$$2x - 3 = -1$$
$$2x = 2$$
$$x = 1$$

The solution to our system is $(1, -1)$. It is the only ordered pair that satisfies both equations.

EXAMPLE 5　　Solve the system.

$$3x + 5y = -7$$
$$5x + 4y = 10$$

SOLUTION   Let's eliminate y by multiplying the first equation by -4 and the second equation by 5.

$$3x + 5y = -7 \xrightarrow{\text{Multiply by } -4} -12x - 20y = 28$$

$$5x + 4y = 10 \xrightarrow[\text{Multiply by } 5]{} \underline{\quad 25x + 20y = 50}$$

$$13x \qquad = 78$$

$$x = 6$$

Substitute $x = 6$ into either equation in our original system, and the result will be $y = -5$. The solution is therefore $(6, -5)$.

EXAMPLE 6   Solve the system.

$$\frac{1}{2}x - \frac{1}{3}y = 2$$

$$\frac{1}{4}x + \frac{2}{3}y = 6$$

SOLUTION   Although we could solve this system without clearing the equations of fractions, there is probably less chance for error if we have only integer coefficients to work with. So let's begin by multiplying both sides of the top equation by 6 and both sides of the bottom equation by 12, to clear each equation of fractions.

$$\frac{1}{2}x - \frac{1}{3}y = 2 \xrightarrow{\text{Multiply by } 6} 3x - 2y = 12$$

$$\frac{1}{4}x + \frac{2}{3}y = 6 \xrightarrow[\text{Multiply by } 12]{} 3x + 8y = 72$$

Now we can eliminate x by multiplying the top equation by -1 and leaving the bottom equation unchanged.

$$3x - 2y = 12 \xrightarrow{\text{Multiply by } -1} -3x + 2y = -12$$

$$3x + 8y = 72 \xrightarrow[\text{No change}]{} \underline{\quad 3x + 8y = \quad 72}$$

$$10y = 60$$

$$y = 6$$

We can substitute $y = 6$ into any equation that contains both x and y. Let's use $3x - 2y = 12$.

$$3x - 2(6) = 12$$

$$3x - 12 = 12$$

$$3x = 24$$

$$x = 8$$

The solution to the system is $(8, 6)$.

Our next two examples will show what happens when we apply the elimination method to a system of equations consisting of parallel lines and to a system in which the lines coincide.

EXAMPLE 7   Solve the system.

$$2x - y = 2$$

$$4x - 2y = 12$$

SOLUTION   Let us choose to eliminate y from the system. We can do this by multiplying the first equation by -2 and leaving the second equation unchanged.

$$2x - y = 2 \xrightarrow{\text{Multiply by } -2} -4x + 2y = -4$$

$$4x - 2y = 12 \xrightarrow[\text{No change}]{} 4x - 2y = 12$$

If we add both sides of the resulting system, we have

$$-4x + 2y = -4$$
$$\underline{4x - 2y = 12}$$
$$0 + 0 = 8$$

or $0 = 8$ **A false statement**

Both variables have been eliminated and we end up with the false statement $0 = 8$. We have tried to solve a system that consists of two parallel lines. There is no solution, and that is the reason we end up with a false statement. Figure 4 is a visual representation of the situation and is conclusive evidence that there is no solution to our system.

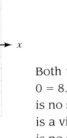

FIGURE 4

EXAMPLE 8   Solve the system.

$$4x - 3y = 2$$

$$8x - 6y = 4$$

SOLUTION   Multiplying the top equation by -2 and adding, we can eliminate the variable x.

$$4x - 3y = 2 \xrightarrow{\text{Multiply by } -2} -8x + 6y = -4$$

$$8x - 6y = 4 \xrightarrow[\text{No change}]{} \underline{8x - 6y = 4}$$
$$0 = 0$$

Both variables have been eliminated, and the resulting statement $0 = 0$ is true. In this case the lines coincide because the equations are equivalent. The solution set consists of all ordered pairs that satisfy either equation.

The preceding two examples illustrate the two special cases in which the graphs of the equations in the system either coincide or are parallel.

Here is a summary of our results from these two examples:

Both variables are eliminated and the resulting statement is false.	$\leftrightarrow$	The lines are parallel and there is no solution to the system.
Both variables are eliminated and the resulting statement is true.	$\leftrightarrow$	The lines coincide and there is an infinite number of solutions to the system.

The main idea in solving a system of linear equations by the elimination method is to use the multiplication property of equality on one or both of the original equations, if necessary, to make the coefficients of either variable opposites. The following box shows some steps to follow when solving a system of linear equations by the elimination method.

Strategy for Solving a System of Linear Equations by the Elimination Method

Step 1: Decide which variable to eliminate. (In some cases one variable will be easier to eliminate than the other. With some practice you will notice which one it is.)

Step 2: Use the multiplication property of equality on each equation separately to make the coefficients of the variable that is to be eliminated opposites.

Step 3: Add the respective left and right sides of the system together.

Step 4: Solve for the variable remaining.

Step 5: Substitute the value of the variable from step 4 into an equation containing both variables and solve for the other variable.

Step 6: Check your solution in both equations, if necessary.

GETTING READY FOR CLASS

After reading through the preceding section, respond in your own words and in complete sentences.

1. How is the addition property of equality used in the elimination method of solving a system of linear equations?
2. What happens when we use the elimination method to solve a system of linear equations consisting of two parallel lines?
3. What does it mean when we solve a system of linear equations by the elimination method and we end up with the statement $0 = 8$?
4. What is the first step in solving a system of linear equations that contains fractions?

LINKING OBJECTIVES AND EXAMPLES

Next to each objective we have listed the examples that are best described by that objective.

A   1–8

Solve the following systems of linear equations by elimination.

▶ **1.** $x + y = 3$
   $x - y = 1$ $(2, 1)$

2. $x + y = -2$
   $x - y = 6$ $(2, -4)$

3. $x + y = 10$
   $-x + y = 4$ $(3, 7)$

4. $x - y = 1$
   $-x - y = -7$ $(4, 3)$

5. $x - y = 7$
   $-x - y = 3$ $(2, -5)$

6. $x - y = 4$
   $2x + y = 8$ $(4, 0)$

7. $x + y = -1$
   $3x - y = -3$ $(-1, 0)$

8. $2x - y = -2$
   $-2x - y = 2$ $(-1, 0)$

9. $3x + 2y = 1$
   $-3x - 2y = -1$
   Lines coincide.

10. $-2x - 4y = 1$
   $2x + 4y = -1$
   Lines coincide.

Solve each of the following systems by eliminating the y variable.

▶ **11.** $3x - y = 4$
   $2x + 2y = 24$ $(4, 8)$

12. $2x + y = 3$
   $3x + 2y = 1$ $(5, -7)$

13. $5x - 3y = -2$
   $10x - y = 1$ $(\frac{1}{5}, 1)$

14. $4x - y = -1$
   $2x + 4y = 13$ $(\frac{1}{2}, 3)$

15. $11x - 4y = 11$
   $5x + y = 5$ $(1, 0)$

16. $3x - y = 7$
   $10x - 5y = 25$ $(2, -1)$

Solve each of the following systems by eliminating the x variable.

17. $3x - 5y = 7$
   $-x + y = -1$ $(-1, -2)$

18. $4x + 2y = 32$
   $x + y = -2$ $(18, -20)$

19. $-x - 8y = -1$
   $-2x + 4y = 13$ $(-5, \frac{3}{4})$

20. $-x + 10y = 1$
   $-5x + 15y = -9$ $(3, \frac{2}{5})$

21. $-3x - y = 7$
   $6x + 7y = 11$ $(-4, 5)$

22. $-5x + 2y = -6$
   $10x + 7y = 34$ $(2, 2)$

Solve each of the following systems of linear equations by the elimination method.

23. $6x - y = -8$
   $2x + y = -16$ $(-3, -10)$

24. $5x - 3y = -3$
   $3x + 3y = -21$ $(-3, -4)$

25. $x + 3y = 9$
   $2x - y = 4$ $(3, 2)$

26. $x + 2y = 0$
   $2x - y = 0$ $(0, 0)$

27. $x - 6y = 3$
   $4x + 3y = 21$ $(5, \frac{1}{3})$

28. $8x + y = -1$
   $4x - 5y = 16$ $(\frac{1}{4}, -3)$

▶ **29.** $2x + 9y = 2$
   $5x + 3y = -8$ $(-2, \frac{2}{3})$

30. $5x + 2y = 11$
   $7x + 8y = 7$ $(\frac{37}{13}, -\frac{21}{13})$

31. $\frac{1}{3}x + \frac{1}{4}y = \frac{7}{6}$
   $\frac{3}{2}x - \frac{1}{3}y = \frac{7}{3}$ $(2, 2)$

32. $\frac{7}{12}x - \frac{1}{2}y = \frac{1}{6}$
   $\frac{2}{5}x - \frac{1}{3}y = \frac{11}{15}$ $(56, 65)$

33. $3x + 2y = -1$
   $6x + 4y = 0$
   Lines are parallel; $\varnothing$

34. $8x - 2y = 2$
   $4x - y = 2$
   Lines are parallel; $\varnothing$

35. $11x + 6y = 17$
   $5x - 4y = 1$ $(1, 1)$

36. $3x - 8y = 7$
   $10x - 5y = 45$ $(5, 1)$

37. $\frac{1}{2}x + \frac{1}{6}y = \frac{1}{3}$
   $-x - \frac{1}{3}y = -\frac{1}{6}$
   Lines are parallel.; $\varnothing$

38. $-\frac{1}{3}x - \frac{1}{2}y = -\frac{2}{3}$
   $-\frac{2}{3}x - y = -\frac{4}{3}$
   Lines coincide.

Solve each system.

▶ **39.** $x + y = 22$
   $5x + 10y = 170$ $(10, 12)$

▶ **40.** $x + y = 14$
   $10x + 25y = 185$ $(11, 3)$

▶ **41.** $x + y = 14$
   $5x + 25y = 230$ $(6, 8)$

▶ **42.** $x + y = 11$
   $5x + 10y = 95$ $(3, 8)$

▶ **43.** $x + y = 15,000$
   $6x + 7y = 98,000$
   $(7,000, 8,000)$

▶ **44.** $x + y = 10,000$
   $6x + 7y = 63,000$
   $(7,000, 3,000)$

▶ **45.** $x + y = 11,000$
   $4x + 7y = 68,000$
   $(3,000, 8,000)$

▶ **46.** $x + y = 20,000$
   $8x + 6y = 138,000$
   $(9,000, 11,000)$

▶ **47.** $x + y = 23$
   $5x + 10y = 175$ $(11, 12)$

▶ **48.** $x + y = 45$
   $25x + 5y = 465$ $(12, 33)$

49. Multiply both sides of the second equation in the following system by 100, and then solve as usual.

$$x + y = 22$$
$$0.05x + 0.10y = 1.70 \quad (10, 12)$$

50. Multiply both sides of the second equation in the following system by 100, and then solve as usual.

$$x + y = 15,000$$
$$0.06x + 0.07y = 980 \quad (7,000, 8,000)$$

Foreshadowing Problems
Problems 39–48 are problems students must work successfully to solve the application problems they will see later in this chapter.

☐ = Videos available by instructor request
▶ = Online student support materials available at www.thomsonedu.com/login

Maintaining Your Skills

For each of the equations, determine the slope and y-intercept.

51. $3x - y = 3$ slope = 3; y-int = −3

52. $2x + y = -2$ slope = −2; y-int = −2

53. $2x - 5y = 25$ slope = $\frac{2}{5}$; y-int = −5

54. $-3x + 4y = -12$ slope = $\frac{3}{4}$; y-int = −3

Find the slope of the line through the given points.

55. $(-2, 3)$ and $(6, -5)$ slope = −1

56. $(2, -4)$ and $(8, -2)$ slope = $\frac{1}{3}$

57. $(5, 3)$ and $(2, -3)$ slope = 2

58. $(-1, -4)$ and $(-4, -1)$ slope = −1

59. Find y if the line through $(-2, 5)$ and $(-4, y)$ has a slope of -3. 11

60. Find y if the line through $(-2, 4)$ and $(6, y)$ has a slope of -2. −12

61. Find y if the line through $(3, -6)$ and $(6, y)$ has a slope of 5. 9

62. Find y if the line through $(3, 4)$ and $(-2, y)$ has a slope of -4. 24

For each of the following problems, the slope and one point on a line are given. Find the equation.

63. $(-2, -6)$, $m = 3$ $y = 3x$ **64.** $(4, 2)$, $m = \frac{1}{2}$ $y = \frac{1}{2}x$

Find the equation of the line that passes through each pair of points.

65. $(-3, -5)$, $(3, 1)$ $y = x - 2$

66. $(-1, -5)$, $(2, 1)$ $y = 2x - 3$

Getting Ready for the Next Section

Solve.

67. $x + (2x - 1) = 2$ 1 **68.** $2x - 3(2x - 8) = 12$ 3

69. $2(3y - 1) - 3y = 4$ 2 **70.** $-2x + 4(3x + 6) = 14$ −1

71. $-2x + 3(5x - 1) = 10$ 1

72. $1.5x + 15 = 0.75x + 24.95$ 13.267

Solve each equation for the indicated variable.

73. $x - 3y = -1$ for x $x = 3y - 1$

74. $-3x + y = 6$ for y $y = 3x + 6$

75. Let $y = 2x - 1$. If $x = 1$, find y. 1

76. Let $y = 2x - 8$. If $x = 5$, find y. 2

77. Let $x = 3y - 1$. If $y = 2$, find x. 5

78. Let $x = 4y - 5$. If $y = 2$, find x. 3

Let $y = 1.5x + 15$.

79. If $x = 13$, find y. 34.5 **80.** If $x = 14$, find y. 36

Let $y = 0.75x + 24.95$.

81. If $x = 12$, find y. 33.95 **82.** If $x = 16$, find y. 36.95

4.3 The Substitution Method

OBJECTIVES

A Use the substitution method to solve a system of linear equations in two variables.

There is a third method of solving systems of equations. It is the substitution method, and, like the elimination method, it can be used on any system of linear equations. Some systems, however, lend themselves more to the substitution method than others do.

 EXAMPLE 1 Solve the following system.

$$x + y = 2$$
$$y = 2x - 1$$

SOLUTION If we were to solve this system by the methods used in the previous section, we would have to rearrange the terms of the second equation so that similar terms would be in the same column. There is no need to do this, however, because the second equation tells us that y is $2x - 1$. We can replace the y variable in the first equation with the expression $2x - 1$ from the second equation; that is, we *substitute* $2x - 1$ from the second equation for y in the first equation. Here is what it looks like:

$$x + (2x - 1) = 2$$

The equation we end up with contains only the variable x. The y variable has been eliminated by substitution.

Solving the resulting equation, we have

$$x + (2x - 1) = 2$$
$$3x - 1 = 2$$
$$3x = 3$$
$$x = 1$$

This is the x-coordinate of the solution to our system. To find the y-coordinate, we substitute $x = 1$ into the second equation of our system. (We could substitute $x = 1$ into the first equation also and have the same result.)

$$y = 2(1) - 1$$
$$y = 2 - 1$$
$$y = 1$$

The solution to our system is the ordered pair $(1, 1)$. It satisfies both of the original equations. Figure 1 provides visual evidence that the substitution method yields the correct solution.

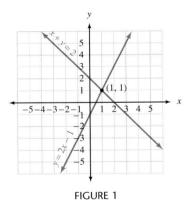

FIGURE 1

EXAMPLE 2   Solve the following system by the substitution method.

$$2x - 3y = 12$$
$$y = 2x - 8$$

SOLUTION   Again, the second equation says y is $2x - 8$. Because we are looking for the ordered pair that satisfies both equations, the y in the first equation must also be $2x - 8$. Substituting $2x - 8$ from the second equation for y in the first equation, we have

$$2x - 3(2x - 8) = 12$$

Note
Sometimes this method of solving systems of equations is confusing the first time you see it. If you are confused, you may want to read through this first example more than once and try it on your own.

This equation can still be read as $2x - 3y = 12$ because $2x - 8$ is the same as y. Solving the equation, we have

$$2x - 3(2x - 8) = 12$$

$$2x - 6x + 24 = 12$$

$$-4x + 24 = 12$$

$$-4x = -12$$

$$x = 3$$

To find the y-coordinate of our solution, we substitute $x = 3$ into the second equation in the original system.

When $\qquad x = 3$

the equation $\qquad y = 2x - 8$

becomes $\qquad y = 2(3) - 8$

$\qquad\qquad\qquad y = 6 - 8 = -2$

The solution to our system is $(3, -2)$.

 EXAMPLE 3 Solve the following system by solving the first equation for x and then using the substitution method:

$$x - 3y = -1$$

$$2x - 3y = 4$$

SOLUTION We solve the first equation for x by adding $3y$ to both sides to get

$$x = 3y - 1$$

Using this value of x in the second equation, we have

$$2(3y - 1) - 3y = 4$$

$$6y - 2 - 3y = 4$$

$$3y - 2 = 4$$

$$3y = 6$$

$$y = 2$$

Next, we find x.

When $\qquad y = 2$

the equation $\qquad x = 3y - 1$

becomes $\qquad x = 3(2) - 1$

$\qquad\qquad\qquad x = 6 - 1$

$\qquad\qquad\qquad x = 5$

The solution to our system is $(5, 2)$.

Here are the steps to use in solving a system of equations by the substitution method.

Strategy for Solving a System of Equations by the Substitution Method

Step 1: Solve either one of the equations for x or y. (This step is not necessary if one of the equations is already in the correct form, as in Examples 1 and 2.)

Step 2: Substitute the expression for the variable obtained in step 1 into the other equation and solve it.

Step 3: Substitute the solution from step 2 into any equation in the system that contains both variables and solve it.

Step 4: Check your results, if necessary.

 EXAMPLE 4   Solve by substitution.

$$-2x + 4y = 14$$

$$-3x + y = 6$$

SOLUTION   We can solve either equation for either variable. If we look at the system closely, it becomes apparent that solving the second equation for y is the easiest way to go. If we add $3x$ to both sides of the second equation, we have

$$y = 3x + 6$$

Substituting the expression $3x + 6$ back into the first equation in place of y yields the following result.

$$-2x + 4(3x + 6) = 14$$

$$-2x + 12x + 24 = 14$$

$$10x + 24 = 14$$

$$10x = -10$$

$$x = -1$$

Substituting $x = -1$ into the equation $y = 3x + 6$ leaves us with

$$y = 3(-1) + 6$$

$$y = -3 + 6$$

$$y = 3$$

The solution to our system is $(-1, 3)$.

 EXAMPLE 5   Solve by substitution.

$$4x + 2y = 8$$

$$y = -2x + 4$$

SOLUTION Substituting the expression $-2x + 4$ for y from the second equation into the first equation, we have

$$4x + 2(-2x + 4) = 8$$

$$4x - 4x + 8 = 8$$

$$8 = 8 \qquad \textbf{A true statement}$$

Both variables have been eliminated, and we are left with a true statement. Recall from the last section that a true statement in this situation tells us the lines coincide; that is, the equations $4x + 2y = 8$ and $y = -2x + 4$ have exactly the same graph. Any point on that graph has coordinates that satisfy both equations and is a solution to the system.

EXAMPLE 6

The following table shows two contract rates charged by GTE Wireless for cellular phone use. At how many minutes will the two rates cost the same amount?

	Flat Rate	Plus	Per Minute Charge
Plan 1	$15		$1.50
Plan 2	$24.95		$0.75

SOLUTION If we let y = the monthly charge for x minutes of phone use, then the equations for each plan are

$$\text{Plan 1:} \quad y = 1.5x + 15$$

$$\text{Plan 2:} \quad y = 0.75x + 24.95$$

We can solve this system by substitution by replacing the variable y in Plan 2 with the expression $1.5x + 15$ from Plan 1. If we do so, we have

$$1.5x + 15 = 0.75x + 24.95$$

$$0.75x + 15 = 24.95$$

$$0.75x = 9.95$$

$$x = 13.27 \qquad \textbf{to the nearest hundredth}$$

The monthly bill is based on the number of minutes you use the phone, with any fraction of a minute moving you up to the next minute. If you talk for a total of 13 minutes, you are billed for 13 minutes. If you talk for 13 minutes, 10 seconds, you are billed for 14 minutes. The number of minutes on your bill always will be a whole number. So, to calculate the cost for talking 13.27 minutes, we would replace x with 14 and find y. Let's compare the two plans at $x = 13$ minutes and at $x = 14$ minutes.

$$\text{Plan 1:} \quad y = 1.5x + 15$$

$$\text{When} \quad x = 13, y = \$34.50$$

$$\text{When} \quad x = 14, y = \$36.00$$

Plan 2: $y = 0.75x + 24.95$

When $x = 13, y = \$34.70$

When $x = 14, y = \$35.45$

The two plans will never give the same cost for talking x minutes. If you talk 13 or less minutes, Plan 1 will cost less. If you talk for more than 13 minutes, you will be billed for 14 minutes, and Plan 2 will cost less than Plan 1.

GETTING READY FOR CLASS

After reading through the preceding section, respond in your own words and in complete sentences.

1. What is the first step in solving a system of linear equations by substitution?
2. When would substitution be more efficient than the elimination method in solving two linear equations?
3. What does it mean when we solve a system of linear equations by the substitution method and we end up with the statement $8 = 8$?
4. How would you begin solving the following system using the substitution method?

$$x + y = 2$$

$$y = 2x - 1$$

LINKING OBJECTIVES AND EXAMPLES

Next to each **objective** we have listed the examples that are best described by that objective.

A   1–6

Problem Set 4.3

Online support materials can be found at www.thomsonedu.com/login

Solve the following systems by substitution. Substitute the expression in the second equation into the first equation and solve.

▶ **1.** $x + y = 11$
     $y = 2x - 1$ $(4, 7)$

2. $x - y = -3$
     $y = 3x + 5$ $(-1, 2)$

3. $x + y = 20$
     $y = 5x + 2$ $(3, 17)$

4. $3x - y = -1$
     $x = 2y - 7$ $(1, 4)$

5. $-2x + y = -1$
     $y = -4x + 8$ $\left(\frac{3}{2}, 2\right)$

6. $4x - y = 5$
     $y = -4x + 1$ $\left(\frac{3}{4}, -2\right)$

7. $3x - 2y = -2$
     $x = -y + 6$ $(2, 4)$

8. $2x - 3y = 17$
     $x = -y + 6$ $(7, -1)$

9. $5x - 4y = -16$
     $y = 4$ $(0, 4)$

10. $6x + 2y = 18$
     $x = 3$ $(3, 0)$

11. $5x + 4y = 7$
     $y = -3x$ $(-1, 3)$

12. $10x + 2y = -6$
     $y = -5x$
     Lines are parallel;.$\varnothing$

Solve the following systems by solving one of the equations for x or y and then using the substitution method.

13. $x + 3y = 4$
     $x - 2y = -1$ $(1, 1)$

14. $x - y = 5$
     $x + 2y = -1$ $(3, -2)$

▶ **15.** $2x + y = 1$
     $x - 5y = 17$ $(2, -3)$

16. $2x - 2y = 2$
     $x - 3y = -7$ $(5, 4)$

17. $3x + 5y = -3$
     $x - 5y = -5$ $\left(-2, \frac{3}{5}\right)$

18. $2x - 4y = -4$
     $x + 2y = 8$ $\left(3, \frac{5}{2}\right)$

19. $5x + 3y = 0$
     $x - 3y = -18$ $(-3, 5)$

20. $x - 3y = -5$
     $x - 2y = 0$ $(10, 5)$

21. $-3x - 9y = 7$
     $x + 3y = 12$
     Lines are parallel;.$\varnothing$

22. $2x + 6y = -18$
     $x + 3y = -9$
     Lines coincide.

     = Videos available by instructor request

▶ = Online student support materials available at www.thomsonedu.com/login

Solve the following systems using the substitution method.

23. $5x - 8y = 7$
    $y = 2x - 5$ (3, 1)

24. $3x + 4y = 10$
    $y = 8x - 15$ (2, 1)

▶ **25.** $7x - 6y = -1$
    $x = 2y - 1$ $(\frac{1}{2}, \frac{3}{4})$

26. $4x + 2y = 3$
    $x = 4y - 3$ $(\frac{1}{3}, \frac{5}{6})$

27. $-3x + 2y = 6$
    $y = 3x$ (2, 6)

28. $-2x - y = -3$
    $y = -3x$ (-3, 9)

29. $5x - 6y = -4$
    $x = y$ (4, 4)

30. $2x - 4y = 0$
    $y = x$ (0, 0)

31. $3x + 3y = 9$
    $y = 2x - 12$ (5, -2)

32. $7x + 6y = -9$
    $y = -2x + 1$ (3, -5)

33. $7x - 11y = 16$
    $y = 10$ (18, 10)

34. $9x - 7y = -14$
    $x = 7$ (7, 11)

35. $-4x + 4y = -8$
    $y = x - 2$
    Lines coincide.

36. $-4x + 2y = -10$
    $y = 2x - 5$
    Lines coincide.

Solve each system.

▶ **37.** $2x + 5y = 36$
    $y = 12 - x$ (8, 4)

▶ **38.** $3x + 6y = 120$
    $y = 25 - x$ (10, 15)

▶ **39.** $5x + 2y = 54$
    $y = 18 - x$ (6, 12)

▶ **40.** $10x + 5y = 320$
    $y = 40 - x$ (24, 16)

▶ **41.** $2x + 2y = 96$
    $y = 2x$ (16, 32)

▶ **42.** $x + y = 22$
    $y = x + 9$ (6.5, 15.5)

Solve each system by substitution. You can eliminate the decimals if you like, but you don't have to. The solution will be the same in either case.

43. $0.05x + 0.10y = 1.70$
    $y = 22 - x$
    (10, 12)

44. $0.20x + 0.50y = 3.60$
    $y = 12 - x$
    (8, 4)

The next two problems are intended to give you practice reading, and paying attention to, the instructions that accompany the problems you are working. Working these problems is an excellent way to get ready for a test or quiz.

▶ **45.** Work each problem according to the instructions given.
   a. Solve: $4y - 5 = 20$ $\frac{25}{4}$
   b. Solve for y: $4x - 5y = 20$ $y = \frac{4}{5}x - 4$

c. Solve for x: $x - y = 5$ $x = y + 5$
d. Solve the system: $4x - 5y = 20$
        $x - y = 5$ (5, 0)

46. Work each problem according to the instructions given.
   a. Solve: $2x - 1 = 4$ $\frac{5}{2}$
   b. Solve for y: $2x - y = 4$ $y = 2x - 4$
   c. Solve for x: $x + 3y = 9$ $x = -3y + 9$
   d. Solve the system: $2x - y = 4$
        $x + 3y = 9$ (3, 2)

Applying the Concepts

47. Gas Mileage Daniel is trying to decide whether to buy a car or a truck. The truck he is considering will cost him $150 a month in loan payments, and it gets 20 miles per gallon in gas mileage. The car will cost $180 a month in loan payments, but it gets 35 miles per gallon in gas mileage. Daniel estimates that he will pay $1.40 per gallon for gas. This means that the monthly cost to drive the truck x miles will be $y = \frac{1.40}{20}x + 150$. The total monthly cost to drive the car x miles will be $y = \frac{1.40}{35}x + 180$. The following figure shows the graph of each equation.

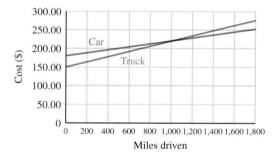

a. At how many miles do the car and the truck cost the same to operate? 1,000 miles
b. If Daniel drives more than 1,200 miles, which will be cheaper? Car
c. If Daniel drives fewer than 800 miles, which will be cheaper? Truck
d. Why do the graphs appear in the first quadrant only? We are only working with positive numbers.

48. Video Production Pat runs a small company that duplicates videotapes. The daily cost and daily revenue for a company duplicating videos are shown

▶ **Chalkboard Problem**
Problem 45 will show students that solving one of the equations for a variable is easier than solving the other equation. A good problem to do for your second example in class.

in the following figure. The daily cost for duplicating x videos is $y = \frac{6}{5}x + 20$; the daily revenue (the amount of money he brings in each day) for duplicating x videos is $y = 1.7x$. The graphs of the two lines are shown in the following figure.

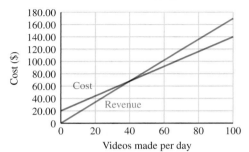

a. Pat will "break even" when his cost and his revenue are equal. How many videos does he need to duplicate to break even? 40
b. Pat will incur a loss when his revenue is less than his cost. If he duplicates 30 videos in one day, will he incur a loss? Yes
c. Pat will make a profit when his revenue is larger than his costs. For what values of x will Pat make a profit? $x > 40$
d. Why does the graph appear in the first quadrant only? We are only working with positive numbers.

Maintaining Your Skills

49. $6(3 + 4) + 5$ 47
50. $[(1 + 2)(2 + 3)] + (4 \div 2)$ 17
51. $1^2 + 2^2 + 3^2$ 14   52. $(1 + 2 + 3)^2$ 36
53. $5(6 + 3 \cdot 2) + 4 + 3 \cdot 2$ 70
54. $(1 + 2)^3 + [(2 \cdot 3) + (4 \cdot 5)]$ 53
55. $(1^3 + 2^3) + [(2 \cdot 3) + (4 \cdot 5)]$ 35
56. $[2(3 + 4 + 5)] \div 3$ 8
57. $(2 \cdot 3 + 4 + 5) \div 3$ 5
58. $10^4 + 10^3 + 10^2 + 10^1$ 11,110
59. $6 \cdot 10^3 + 5 \cdot 10^2 + 4 \cdot 10^1$ 6,540
60. $5 \cdot 10^3 + 2 \cdot 10^2 + 8 \cdot 10^1$ 5,280

61. $1 \cdot 10^3 + 7 \cdot 10^2 + 6 \cdot 10^1 + 0$ 1,760
62. $4(2 - 1) + 5(3 - 2)$ 9
63. $4 \cdot 2 - 1 + 5 \cdot 3 - 2$ 20   64. $2^3 + 3^2 \cdot 4 - 5$ 39
65. $(2^3 + 3^2) \cdot 4 - 5$ 63   66. $4^2 - 2^4 + (2 \cdot 2)^2$ 16
67. $2(2^2 + 3^2) + 3(3^2)$ 53   68. $2 \cdot 2^2 + 3^2 + 3 \cdot 3^2$ 44

Getting Ready for the Next Section

69. One number is eight more than five times another, their sum is 26. Find the numbers. 3 and 23
70. One number is three less than four times another; their sum is 27. Find the numbers. 6 and 21
71. The difference of two positive numbers is nine. The larger number is six less than twice the smaller number. Find the numbers. 15 and 24
72. The difference of two positive numbers is 17. The larger number is one more than twice the smaller number. Find the numbers. 16 and 33
73. The length of a rectangle is 5 inches more than three times the width. The perimeter is 58 inches. Find the length and width. Length = 23 in.; width = 6 in.
74. The length of a rectangle is 3 inches less than twice the width. The perimeter is 36 inches. Find the length and width. Length = 11 in.; width = 7 in.
75. John has $1.70 in nickels and dimes in his pocket. He has four more nickels than he does dimes. How many of each does he have? 14 nickels and 10 dimes
76. Jamie has $2.65 in dimes and quarters in her pocket. She has two more dimes than she does quarters. How many of each does she have?
7 quarters and 9 dimes

Solve the systems by any method.

77. $y = 5x + 2$ 78. $y = 2x + 3$
      $x + y = 20$ (3, 17) $x + y = 9$ (2, 7)
79. $x + y = 15,000$ 80. $x + y = 22$
      $0.06x + 0.07y = 980$ $0.05x + 0.1y = 1.70$
      (7,000, 8,000)   (10, 12)

Applications

OBJECTIVES

A Apply the Blueprint for Problem Solving to a variety of application problems involving systems of equations.

I often have heard students remark about the word problems in elementary algebra: "What does this have to do with real life?" Most of the word problems we will encounter don't have much to do with "real life." We are actually just practicing. Ultimately, all problems requiring the use of algebra are word problems; that is, they are stated in words first, then translated to symbols. The problem then is solved by some system of mathematics, like algebra. Most real applications involve calculus or higher levels of mathematics. So, if the problems we solve are upsetting or frustrating to you, then you probably are taking them too seriously.

The word problems in this section have two unknown quantities. We will write two equations in two variables (each of which represents one of the unknown quantities), which of course is a system of equations. We then solve the system by one of the methods developed in the previous sections of this chapter. Here are the steps to follow in solving these word problems.

Blueprint for Problem Solving Using a System of Equations

Step 1: **Read** the problem, and then mentally *list* the items that are known and the items that are unknown.

Step 2: **Assign variables** to each of the unknown items; that is, let x = one of the unknown items and y = the other unknown item. Then **translate** the other **information** in the problem to expressions involving the two variables.

Step 3: **Reread** the problem, and then **write a system of equations,** using the items and variables listed in steps 1 and 2, that describes the situation.

Step 4: **Solve the system** found in step 3.

Step 5: **Write** your **answers** using complete sentences.

Step 6: **Reread** the problem, and **check** your solution with the original words in the problem.

Remember, the more problems you work, the more problems you will be able to work. If you have trouble getting started on the problem set, come back to the examples and work through them yourself. The examples are similar to the problems found in the problem set.

Number Problem

 EXAMPLE 1   One number is 2 more than 5 times another number. Their sum is 20. Find the two numbers.

SOLUTION   Applying the steps in our blueprint, we have

Step 1: We know that the two numbers have a sum of 20 and that one of them is 2 more than 5 times the other. We don't know what the numbers themselves are.

Step 2: Let x represent one of the numbers and y represent the other. "One number is 2 more than 5 times another" translates to

$$y = 5x + 2$$

"Their sum is 20" translates to

$$x + y = 20$$

Step 3: The system that describes the situation must be

$$x + y = 20$$
$$y = 5x + 2$$

Step 4: We can solve this system by substituting the expression $5x + 2$ in the second equation for y in the first equation:

$$x + (5x + 2) = 20$$
$$6x + 2 = 20$$
$$6x = 18$$
$$x = 3$$

Using $x = 3$ in either of the first two equations and then solving for y, we get $y = 17$.

Step 5: So 17 and 3 are the numbers we are looking for.

Step 6: The number 17 is 2 more than 5 times 3, and the sum of 17 and 3 is 20.

> **Note**
> We are using the substitution method here because the system we are solving is one in which the substitution method is the more convenient method.

Interest Problem

 EXAMPLE 2   Mr. Hicks had $15,000 to invest. He invested part at 6% and the rest at 7%. If he earns $980 in interest, how much did he invest at each rate?

SOLUTION   Remember, step 1 is done mentally.

Step 1: We do not know the specific amounts invested in the two accounts. We do know that their sum is $15,000 and that the interest rates on the two accounts are 6% and 7%.

Step 2: Let x = the amount invested at 6% and y = the amount invested at 7%. Because Mr. Hicks invested a total of $15,000, we have

$$x + y = 15{,}000$$

The interest he earns comes from 6% of the amount invested at 6% and 7% of the amount invested at 7%. To find 6% of x, we multiply x by 0.06, which gives us $0.06x$. To find 7% of y, we multiply 0.07 times y and get $0.07y$.

$$\begin{array}{ccccc} \text{Interest} & + & \text{interest} & = & \text{total} \\ \text{at 6\%} & & \text{at 7\%} & & \text{interest} \\ 0.06x & + & 0.07y & = & 980 \end{array}$$

Step 3: The system is

$$x + y = 15{,}000$$
$$0.06x + 0.07y = 980$$

Step 4:   We multiply the first equation by −6 and the second by 100 to eliminate x:

$$x + y = 15,000 \xrightarrow{\text{Multiply by } -6} -6x - 6y = -90,000$$

$$0.06x + 0.07y = 980 \xrightarrow{\text{Multiply by } 100} \frac{6x + 7y = 98,000}{y = 8,000}$$

Substituting $y = 8,000$ into the first equation and solving for x, we get $x = 7,000$.

Step 5:   He invested $7,000 at 6% and $8,000 at 7%.

Step 6:   Checking our solutions in the original problem, we have: The sum of $7,000 and $8,000 is $15,000, the total amount he invested. To complete our check, we find the total interest earned from the two accounts:

The interest on $7,000 at 6% is $0.06(7,000) = 420$
The interest on $8,000 at 7% is $0.07(8,000) = 560$
The total interest is $980

> **Note**
> In this case we are using the elimination method. Notice also that multiplying the second equation by 100 clears it of decimals.

Coin Problem

EXAMPLE 3   John has $1.70 all in dimes and nickels. He has a total of 22 coins. How many of each kind does he have?

SOLUTION

Step 1:   We know that John has 22 coins that are dimes and nickels. We know that a dime is worth 10 cents and a nickel is worth 5 cents. We do not know the specific number of dimes and nickels he has.

Step 2:   Let x = the number of nickels and y = the number of dimes. The total number of coins is 22, so

$$x + y = 22$$

The total amount of money he has is $1.70, which comes from nickels and dimes:

$$\underset{\text{in nickels}}{\text{Amount of money}} + \underset{\text{in dimes}}{\text{amount of money}} = \underset{\text{of money}}{\text{total amount}}$$

$$0.05x \quad + \quad 0.10y \quad = \quad 1.70$$

Step 3:   The system that represents the situation is

$$x + y = 22 \qquad \text{The number of coins}$$

$$0.05x + 0.10y = 1.70 \qquad \text{The value of the coins}$$

Step 4:   We multiply the first equation by −5 and the second by 100 to eliminate the variable x:

$$x + y = 22 \xrightarrow{\text{Multiply by } -5} -5x - 5y = -110$$

$$0.05x + 0.10y = 1.70 \xrightarrow{\text{Multiply by } 100} \frac{5x + 10y = 170}{5y = 60}$$

$$y = 12$$

Substituting $y = 12$ into our first equation, we get $x = 10$.

Step 5: John has 12 dimes and 10 nickels.

Step 6: Twelve dimes and 10 nickels total 22 coins.

12 dimes are worth $12(0.10) = 1.20$

10 nickels are worth $10(0.05) = 0.50$

The total value is $1.70

Mixture Problem

EXAMPLE 4 How much 20% alcohol solution and 50% alcohol solution must be mixed to get 12 gallons of 30% alcohol solution?

SOLUTION To solve this problem we must first understand that a 20% alcohol solution is 20% alcohol and 80% water.

Step 1: We know there are two solutions that together must total 12 gallons. 20% of one of the solutions is alcohol and the rest is water, whereas the other solution is 50% alcohol and 50% water. We do not know how many gallons of each individual solution we need.

Step 2: Let x = the number of gallons of 20% alcohol solution needed and y = the number of gallons of 50% alcohol solution needed. Because the total number of gallons we will end up with is 12, and this 12 gallons must come from the two solutions we are mixing, our first equation is

$$x + y = 12$$

To obtain our second equation, we look at the amount of alcohol in our two original solutions and our final solution. The amount of alcohol in the x gallons of 20% solution is $0.20x$, and the amount of alcohol in y gallons of 50% solution is $0.50y$. The amount of alcohol in the 12 gallons of 30% solution is $0.30(12)$. Because the amount of alcohol we start with must equal the amount of alcohol we end up with, our second equation is

$$0.20x + 0.50y = 0.30(12)$$

The information we have so far can also be summarized with a table. Sometimes by looking at a table like the one that follows it is easier to see where the equations come from.

	20% Solution	50% Solution	Final Solution
Number of Gallons	x	y	12
Gallons of Alcohol	$0.20x$	$0.50y$	$0.30(12)$

Step 3: Our system of equations is

$$x + y = 12$$

$$0.20x + 0.50y = 0.30(12)$$

Step 4:   We can solve this system by substitution. Solving the first equation for y and substituting the result into the second equation, we have

$$0.20x + 0.50(12 - x) = 0.30(12)$$

Multiplying each side by 10 gives us an equivalent equation that is a little easier to work with.

$$2x + 5(12 - x) = 3(12)$$

$$2x + 60 - 5x = 36$$

$$-3x + 60 = 36$$

$$-3x = -24$$

$$x = 8$$

If x is 8, then y must be 4 because $x + y = 12$.

Step 5:   It takes 8 gallons of 20% alcohol solution and 4 gallons of 50% alcohol solution to produce 12 gallons of 30% alcohol solution.

Step 6:   Try it and see.

GETTING READY FOR CLASS

After reading through the preceding section, respond in your own words and in complete sentences.

1. If you were to apply the Blueprint for Problem Solving from Section 2.6 to the examples in this section, what would be the first step?
2. If you were to apply the Blueprint for Problem Solving from Section 2.6 to the examples in this section, what would be the last step?
3. Which method of solving these systems do you prefer? Why?
4. Write an application problem for which the solution depends on solving a system of equations.

LINKING OBJECTIVES AND EXAMPLES

Next to each **objective** we have listed the examples that are best described by that objective.

A   1–4

Problem Set 4.4

Solve the following word problems. Be sure to show the equations used.

Number Problems

1. Two numbers have a sum of 25. One number is 5 more than the other. Find the numbers. 10 and 15

2. The difference of two numbers is 6. Their sum is 30. Find the two numbers. 12 and 18

3. The sum of two numbers is 15. One number is 4 times the other. Find the numbers. 3 and 12

4. The difference of two positive numbers is 28. One number is 3 times the other. Find the two numbers. 14 and 42

▶ 5. Two positive numbers have a difference of 5. The larger number is one more than twice the smaller. Find the two numbers. 4 and 9

6. One number is 2 more than 3 times another. Their sum is 26. Find the two numbers. 6 and 20

= Videos available by instructor request

▶ = Online student support materials available at www.thomsonedu.com/login

7. One number is 5 more than 4 times another. Their sum is 35. Find the two numbers. 6 and 29

8. The difference of two positive numbers is 8. The larger is twice the smaller decreased by 7. Find the two numbers. 15 and 23

Interest Problems

9. Mr. Wilson invested money in two accounts. His total investment was $20,000. If one account pays 6% in interest and the other pays 8% in interest, how much does he have in each account if he earned a total of $1,380 in interest in 1 year? $9,000 at 8%, $11,000 at 6%

10. A total of $11,000 was invested. Part of the $11,000 was invested at 4%, and the rest was invested at 7%. If the investments earn $680 per year, how much was invested at each rate? $3,000 at 4%, $8,000 at 7%

▸ 11. A woman invested 4 times as much at 5% as she did at 6%. The total amount of interest she earns in 1 year from both accounts is $520. How much did she invest at each rate? $2,000 at 6%, $8,000 at 5%

12. Ms. Hagan invested twice as much money in an account that pays 7% interest as she did in an account that pays 6% in interest. Her total investment pays her $1,000 a year in interest. How much did she invest at each rate? $5,000 at 6%, $10,000 at 7%

Coin Problems

13. Ron has 14 coins with a total value of $2.30. The coins are nickels and quarters. How many of each coin does he have? 6 nickels, 8 quarters

14. Diane has $0.95 in dimes and nickels. She has a total of 11 coins. How many of each kind does she have? 3 nickels, 8 dimes

▸ 15. Suppose Tom has 21 coins totaling $3.45. If he has only dimes and quarters, how many of each type does he have? 12 dimes, 9 quarters

16. A coin collector has 31 dimes and nickels with a total face value of $2.40. (They are actually worth a lot more.) How many of each coin does she have? 14 nickels, 17 dimes

Mixture Problems

▸ 17. How many liters of 50% alcohol solution and 20% alcohol solution must be mixed to obtain 18 liters of 30% alcohol solution?

6 liters of 50% solution, 12 liters of 20% solution

	50% Solution	20% Solution	Final Solution
Number of Liters	x	y	18
Liters of Alcohol	$0.50x$	$0.20y$	$0.30(18)$

18. How many liters of 10% alcohol solution and 5% alcohol solution must be mixed to obtain 40 liters of 8% alcohol solution?

24 liters of 10% solution, 16 liters of 5% solution

	10% Solution	5% Solution	Final Solution
Number of Liters	x	y	40
Liters of Alcohol	$0.10x$	$0.05y$	$0.08(40)$

19. A mixture of 8% disinfectant solution is to be made from 10% and 7% disinfectant solutions. How much of each solution should be used if 30 gallons of 8% solution are needed?

10 gallons of 10% solution, 20 gallons of 7% solution

20. How much 50% antifreeze solution and 40% antifreeze solution should be combined to give 50 gallons of 46% antifreeze solution?

30 gallons of 50% solution, 20 gallons of 40% solution

Miscellaneous Problems

21. For a Saturday matinee, adult tickets cost $5.50 and kids under 12 pay only $4.00. If 70 tickets are sold for a total of $310, how many of the tickets were adult tickets and how many were sold to kids under 12? 20 adults, 50 kids

22. The Bishop's Peak 4-H club is having its annual fundraising dinner. Adults pay $15 apiece and children pay $10 apiece. If the number of adult tickets sold is twice the number of children's tickets sold, and the total income for the dinner is $1,600, how many of each kind of ticket did the 4-H club sell?

80 adults, 40 children

23. A farmer has 96 feet of fence with which to make a corral. If he arranges it into a rectangle that is twice as long as it is wide, what are the dimensions? 16 feet wide, 32 feet long

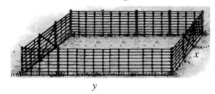

24. If a 22-inch rope is to be cut into two pieces so that one piece is 3 inches longer than twice the other, how long is each piece? $6\frac{1}{3}$ inches, $15\frac{2}{3}$ inches

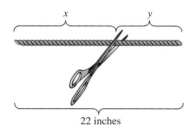

22 inches

25. A gambler finishes a session of blackjack with $5 chips and $25 chips. If he has 45 chips in all, with a total value of $465, how many of each kind of chip does the gambler have? 33 $5 chips, 12 $25 chips

26. Tyler has been saving his winning lottery tickets. He has 23 tickets that are worth a total of $175. If each ticket is worth either $5 or $10, how many of each does he have? 11 $5 tickets, 12 $10 tickets

27. Mary Jo spends $2,550 to buy stock in two companies. She pays $11 a share to one of the companies and $20 a share to the other. If she ends up with a total of 150 shares, how many shares did she buy at $11 a share and how many did she buy at $20 a share? 50 at $11, 100 at $20

28. Kelly sells 62 shares of stock she owns for a total of $433. If the stock was in two different companies, one selling at $6.50 a share and the other at $7.25 a share, how many of each did she sell? 22 at $6.50, 40 at $7.25

Maintaining Your Skills

29. Fill in each ordered pair so that it is a solution to $y = \frac{1}{2}x + 3$.　(−2, 2), (0, 3), (2, 4)

30. Graph the line $y = \frac{1}{2}x + 3$.

31. Graph the line $x = -2$.

32. Graph $3x - 2y = 6$.

33. Find the slope of the line through (2, 5) and (0, 1).　2

34. Find the slope and y-intercept for the line $2x - 5y = 10$.　$m = \frac{2}{5}$, $b = -2$

35. Find the equation of the line through (−2, 1) with slope $\frac{1}{2}$.　$y = \frac{1}{2}x + 2$

36. Write the equation of the line with slope −2 and y-intercept $\frac{3}{2}$.　$y = -2x + \frac{3}{2}$

37. Find the equation of the line through (2, 5) and (0, 1).　$y = 2x + 1$

38. Graph the solution set for $2x - y < 4$.

EXAMPLES

Definitions [4.1]

1. The solution to the system
$$x + 2y = 4$$
$$x - y = 1$$
is the ordered pair (2, 1). It is the only ordered pair that satisfies both equations.

1. A *system of linear equations,* as the term is used in this book, is two linear equations that each contain the same two variables.

2. The *solution set* for a system of equations is the set of all ordered pairs that satisfy *both* equations. The solution set to a system of linear equations will contain:

    Case I   One ordered pair when the graphs of the two equations intersect at only one point (this is the most common situation)

    Case II   No ordered pairs when the graphs of the two equations are parallel lines

    Case III   An infinite number of ordered pairs when the graphs of the two equations coincide (are the same line)

Strategy for Solving a System by Graphing [4.1]

2. Solving the system in Example 1 by graphing looks like

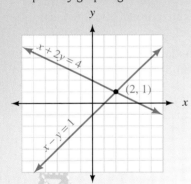

Step 1: Graph the first equation.

Step 2: Graph the second equation on the same set of axes.

Step 3: Read the coordinates of the point where the graphs cross each other (the coordinates of the point of intersection).

Step 4: Check the solution to see that it satisfies *both* equations.

Strategy for Solving a System by the Elimination Method [4.2]

3. We can eliminate the *y* variable from the system in Example 1 by multiplying both sides of the second equation by 2 and adding the result to the first equation

$$x + 2y = 4 \qquad\qquad x + 2y = 4$$
$$x - y = 1 \xrightarrow[\text{Multiply by 2}]{} 2x - 2y = 2$$
$$3x \qquad = 6$$
$$x \qquad = 2$$

Substituting $x = 2$ into either of the original two equations gives $y = 1$. The solution is (2, 1).

Step 1: Look the system over to decide which variable will be easier to eliminate.

Step 2: Use the multiplication property of equality on each equation separately to ensure that the coefficients of the variable to be eliminated are opposites.

Step 3: Add the left and right sides of the system produced in step 2, and solve the resulting equation.

Step 4: Substitute the solution from step 3 back into any equation with both x and y variables, and solve.

Step 5: Check your solution in both equations, if necessary.

Strategy for Solving a System by the Substitution Method [4.3]

4. We can apply the substitution method to the system in Example 1 by first solving the second equation for x to get $x = y + 1$. Substituting this expression for x into the first equation, we have

$$(y + 1) + 2y = 4$$
$$3y + 1 = 4$$
$$3y = 3$$
$$y = 1$$

Using $y = 1$ in either of the original equations gives $x = 2$.

Step 1: Solve either of the equations for one of the variables (this step is not necessary if one of the equations has the correct form already).

Step 2: Substitute the results of step 1 into the other equation, and solve.

Step 3: Substitute the results of step 2 into an equation with both x and y variables, and solve. (The equation produced in step 1 is usually a good one to use.)

Step 4: Check your solution, if necessary.

Special Cases [4.1, 4.2, 4.3]

In some cases, using the elimination or substitution method eliminates both variables. The situation is interpreted as follows.

1. If the resulting statement is *false,* then the lines are parallel and there is no solution to the system.

2. If the resulting statement is *true,* then the equations represent the same line (the lines coincide). In this case any ordered pair that satisfies either equation is a solution to the system.

! COMMON MISTAKES

The most common mistake encountered in solving linear systems is the failure to complete the problem. Here is an example.

$$x + y = 8$$
$$x - y = 4$$
$$2x = 12$$
$$x = 6$$

This is only half the solution. To find the other half, we must substitute the 6 back into one of the original equations and then solve for y.

Remember, solutions to systems of linear equations always consist of ordered pairs. We need an x-coordinate and a y-coordinate; $x = 6$ can never be a solution to a system of linear equations.

The problems below form a comprehensive review of the material in this chapter. They can be used to study for exams. If you would like to take a practice test on this chapter, you can use the odd-numbered problems. Give yourself an hour and work as many of the odd-numbered problems as possible. When you are finished, or when an hour has passed, check your answers with the answers in the back of the book. You can use the even-numbered problems for a second practice test.

Solve the following systems by graphing. [4.1]

1. $x + y = 2$
$x - y = 6$

2. $x + y = -1$
$-x + y = 5$

3. $2x - 3y = 12$
$-2x + y = -8$

4. $4x - 2y = 8$
$3x + y = 6$

5. $y = 2x - 3$
$y = -2x + 5$

6. $y = -x - 3$
$y = 3x + 1$

Solve the following systems by the elimination method. [4.2]

7. $x - y = 4$
$x + y = -2$

8. $-x - y = -3$
$2x + y = 1$

9. $5x - 3y = 2$
$-10x + 6y = -4$

10. $2x + 3y = -2$
$3x - 2y = 10$

11. $-3x + 4y = 1$
$-4x + y = -3$

12. $-4x - 2y = 3$
$2x + y = 1$

13. $-2x + 5y = -11$
$7x - 3y = -5$

14. $-2x + 5y = -15$
$3x - 4y = 19$

Solve the following systems by substitution. [4.3]

15. $x + y = 5$
$y = -3x + 1$

16. $x - y = -2$
$y = -2x - 10$

17. $4x - 3y = -16$
$y = 3x + 7$

18. $5x + 2y = -2$
$y = -8x + 10$

19. $x - 4y = 2$
$-3x + 12y = -8$

20. $4x - 2y = 8$
$3x + y = -19$

21. $10x - 5y = 20$
$x + 6y = -11$

22. $3x - y = 2$
$-6x + 2y = -4$

Solve the following word problems. Be sure to show the equations used. [4.4]

23. Number Problem The sum of two numbers is 18. If twice the smaller number is 6 more than the larger, find the two numbers.

24. Number Problem The difference of two positive numbers is 16. One number is 3 times the other. Find the two numbers.

25. Investing A total of $12,000 was invested. Part of the $12,000 was invested at 4%, and the rest was invested at 5%. If the interest for one year is $560, how much was invested at each rate?

26. Investing A total of $14,000 was invested. Part of the $14,000 was invested at 6%, and the rest was invested at 8%. If the interest for one year is $1,060, how much was invested at each rate?

27. Coin Problem Barbara has $1.35 in dimes and nickels. She has a total of 17 coins. How many of each does she have?

28. Coin Problem Tom has $2.40 in dimes and quarters. He has a total of 15 coins. How many of each does he have?

29. Mixture Problem How many liters of 20% alcohol solution and 10% alcohol solution must be mixed to obtain 50 liters of a 12% alcohol solution?

30. Mixture Problem How many liters of 25% alcohol solution and 15% alcohol solution must be mixed to obtain 40 liters of a 20% alcohol solution?

Chapter 4 Projects

Systems of Linear Equations

GROUP PROJECT   Tables and Graphs

Number of People   2–3

Time Needed   8–12 minutes

Equipment   Pencil, ruler, and graph paper

Background   In the Chapter 2 project, we built the table below. As you may recall, we built the table based on the different combinations of coins from a parking meter that accepts only dimes and quarters. The amount of money in the meter at the end of one particular day was $3.15.

Number of Dimes	Number of Quarters	Total Coins	Value
29	1	30	$3.15
24	3	27	$3.15
19	5	24	$3.15
14	7	21	$3.15
9	9	18	$3.15
4	11	15	$3.15
x	y	$x + y$	

Procedure

1. From each row in the table, plot the points (x, y) where x is the number of dimes and y is the number of quarters. For example, the first point to plot is (29, 1). Then connect the points by drawing a straight line through them. The line should pass through all the points.

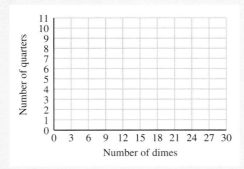

2. What is the equation of the line that passes through all the points?

3. Draw six additional lines on the same grid using the following equations:

   $x + y = 30, x + y = 27, x + y = 24, x + y = 21,$
   $x + y = 18,$ and $x + y = 15.$

4. Write a parking meter application problem, the solution to which is the pair of lines $10x + 25y = 315$ and $x + y = 24$.

Cartesian Coordinate System

The stamp shown here was issued by Albania in 1966. It shows the French philosopher and mathematician René Descartes. As we mentioned earlier in this chapter, Descartes is credited with the discovery of the rectangular coordinate system. (Notice the coordinate system in the background of the stamp.)

Descartes insisted that his best theories came to him while resting in bed. He once said "the only way to do good work in mathematics and preserve one's health is never to get up in the morning before one feels inclined to do so." One story of how he came to develop the Cartesian (from Des*cartes*) coordinate system is as follows: One morning, while lying in bed, he noticed a fly

crawling on the ceiling. After studying it, he realized he could state the fly's position on the ceiling by giving its distance from each of the edges of the ceiling. Research this story and then put your results into an essay that shows the connection between the position of the fly on the ceiling and the coordinates of points in our rectangular coordinate system.

Exponents and Polynomials

5

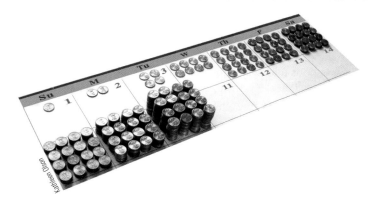

Kathleen Olson

I f you were given a penny on the first day of September, and then each day after that you were given twice the amount of money you received the day before, how much money would you receive on September 30? To begin, Table 1 and Figure 1 show the amount of money you would receive on each of the first 10 days of the month. As you can see, on the tenth day of the month you would receive $5.12.

TABLE 1

Money That Doubles Each Day

Day	Money (in cents)
1	$1 = 2^0$
2	$2 = 2^1$
3	$4 = 2^2$
4	$8 = 2^3$
5	$16 = 2^4$
6	$32 = 2^5$
7	$64 = 2^6$
8	$128 = 2^7$
9	$256 = 2^8$
10	$512 = 2^9$

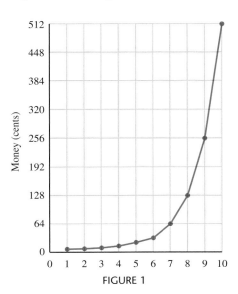

FIGURE 1

To find the amount of money on day 30, we could continue to double the amount on each of the next 20 days. Or, we could notice the pattern of exponents in the second column of the table and reason that the amount of money on day 30 would be 2^{29} cents, which is a very large number. In fact, 2^{29} cents is $5,368,709.12—a little less than $5.4 million. When you are finished with this chapter, you will have a good working knowledge of exponents.

▶ Improve your grade and save time!
Go online to **www.thomsonedu.com/login**
where you can
• Watch videos of instructors working through the in-text examples
• Follow step-by-step online tutorials of in-text examples and review questions
• Work practice problems
• Check your readiness for an exam by taking a pre-test and exploring the modules recommended in your Personalized Study plan
• Receive help from a live tutor online through vMentor™
Try it out! Log in with an access code or purchase access at **www.ichapters.com**.

The study skills for this chapter are about attitude. They are points of view that point toward success.

1   Be Focused, Not Distracted

I have students who begin their assignments by asking themselves, "Why am I taking this class?" Or, "When am I ever going to use this stuff?" If you are asking yourself similar questions, you may be distracting yourself away from doing the things that will produce the results you want in this course. Don't dwell on questions and evaluations of the class that can be used as excuses for not doing well. If you want to succeed in this course, focus your energy and efforts toward success, rather than distracting yourself away from your goals.

2   Be Resilient

Don't let setbacks keep you from your goals. You want to put yourself on the road to becoming a person who can succeed in this class or any class in college. Failing a test or quiz, or having a difficult time on some topics, is normal. No one goes through college without some setbacks. Don't let a temporary disappointment keep you from succeeding in this course. A low grade on a test or quiz is simply a signal that some reevaluation of your study habits needs to take place.

3   Intend to Succeed

I always have a few students who simply go through the motions of studying without intending to master the material. It is more important to them to look like they are studying than to actually study. You need to study with the intention of being successful in the course. Intend to master the material, no matter what it takes.

Multiplication with Exponents

OBJECTIVES

A Use the definition of integer exponents to evaluate expressions containing exponents.

B Use Property 1 for exponents.

C Use Property 2 for exponents.

D Use Property 3 for exponents.

E Write numbers in scientific notation and expanded form.

Recall that an *exponent* is a number written just above and to the right of another number, which is called the *base*. In the expression 5^2, for example, the exponent is 2 and the base is 5. The expression 5^2 is read "5 to the second power" or "5 squared." The meaning of the expression is

$$5^2 = 5 \cdot 5 = 25$$

In the expression 5^3, the exponent is 3 and the base is 5. The expression 5^3 is read "5 to the third power" or "5 cubed." The meaning of the expression is

$$5^3 = 5 \cdot 5 \cdot 5 = 125$$

Here are some further examples.

EXAMPLES

1. $4^3 = 4 \cdot 4 \cdot 4 = 16 \cdot 4 = 64$ Exponent 3, base 4

2. $-3^4 = -3 \cdot 3 \cdot 3 \cdot 3 = -81$ Exponent 4, base 3

3. $(-2)^5 = (-2)(-2)(-2)(-2)(-2) = -32$ Exponent 5, base -2

4. $\left(-\dfrac{3}{4}\right)^2 = \left(-\dfrac{3}{4}\right)\left(-\dfrac{3}{4}\right) = \dfrac{9}{16}$ Exponent 2, base $-\dfrac{3}{4}$

QUESTION: In what way are $(-5)^2$ and -5^2 different?

ANSWER: In the first case, the base is -5. In the second case, the base is 5. The answer to the first is 25. The answer to the second is -25. Can you tell why? Would there be a difference in the answers if the exponent in each case were changed to 3?

We can simplify our work with exponents by developing some properties of exponents. We want to list the things we know are true about exponents and then use these properties to simplify expressions that contain exponents.

The first property of exponents applies to products with the same base. We can use the definition of exponents, as indicating repeated multiplication, to simplify expressions like $7^4 \cdot 7^2$.

$$7^4 \cdot 7^2 = (7 \cdot 7 \cdot 7 \cdot 7)(7 \cdot 7)$$
$$= (7 \cdot 7 \cdot 7 \cdot 7 \cdot 7 \cdot 7)$$
$$= 7^6 \qquad \textit{Notice: } \mathbf{4 + 2 = 6}$$

As you can see, multiplication with the same base resulted in addition of exponents. We can summarize this result with the following property.

> **Property 1 for Exponents** If a is any real number and r and s are integers, then
> $$a^r \cdot a^s = a^{r+s}$$
> *In words:* To multiply two expressions with the same base, add exponents and use the common base.

Here are some examples using Property 1.

 EXAMPLES Use Property 1 to simplify the following expressions. Leave your answers in terms of exponents:

5. $5^3 \cdot 5^6 = 5^{3+6} = 5^9$

6. $x^7 \cdot x^8 = x^{7+8} = x^{15}$

7. $3^4 \cdot 3^8 \cdot 3^5 = 3^{4+8+5} = 3^{17}$

Note

In Examples 5, 6, and 7, notice that in each case the base in the original problem is the same base that appears in the answer and that it is written only once in the answer. A very common mistake that people make when they first begin to use Property 1 is to write a 2 in front of the base in the answer. For example, people making this mistake would get $2x^{15}$ or $(2x)^{15}$ as the result in Example 6. To avoid this mistake, you must be sure you understand the meaning of Property 1 exactly as it is written.

Another common type of expression involving exponents is one in which an expression containing an exponent is raised to another power. The expression $(5^3)^2$ is an example:

$$(5^3)^2 = (5^3)(5^3)$$

$$= 5^{3+3}$$

$$= 5^6 \quad \textbf{Notice: } 3 \cdot 2 = 6$$

This result offers justification for the second property of exponents.

> **Property 2 for Exponents** If a is any real number and r and s are integers, then
> $$(a^r)^s = a^{r \cdot s}$$
> *In words:* A power raised to another power is the base raised to the product of the powers.

 EXAMPLES Simplify the following expressions:

8. $(4^5)^6 = 4^{5 \cdot 6} = 4^{30}$

9. $(x^3)^5 = x^{3 \cdot 5} = x^{15}$

The third property of exponents applies to expressions in which the product of two or more numbers or variables is raised to a power. Let's look at how the expression $(2x)^3$ can be simplified:

$$(2x)^3 = (2x)(2x)(2x)$$

$$= (2 \cdot 2 \cdot 2)(x \cdot x \cdot x)$$

$$= 2^3 \cdot x^3 \quad \textbf{Notice: The exponent 3 distributes over the product 2x}$$

$$= 8x^3$$

We can generalize this result into a third property of exponents.

> **Property 3 for Exponents** If a and b are any two real numbers and r is an integer, then
> $$(ab)^r = a^r b^r$$
> *In words:* The power of a product is the product of the powers.

Here are some examples using Properties 1–3 to simplify expressions.

EXAMPLES   Simplify the following expressions:

10. $(5x)^2 = 5^2 \cdot x^2$ — Property 3
$= 25x^2$

11. $(2xy)^3 = 2^3 \cdot x^3 \cdot y^3$ — Property 3
$= 8x^3y^3$

12. $(3x^2)^3 = 3^3(x^2)^3$ — Property 3
$= 27x^6$ — Property 2

13. $\left(-\dfrac{1}{4}x^2y^3\right)^2 = \left(-\dfrac{1}{4}\right)^2(x^2)^2(y^3)^2$ — Property 3
$= \dfrac{1}{16}x^4y^6$ — Property 2

14. $(x^4)^3(x^2)^5 = x^{12} \cdot x^{10}$ — Property 2
$= x^{22}$ — Property 1

15. $(2y)^3(3y^2) = 2^3y^3(3y^2)$ — Property 3
$= 8 \cdot 3(y^3 \cdot y^2)$ — Commutative and associative properties
$= 24y^5$ — Property 1

16. $(2x^2y^5)^3(3x^4y)^2 = 2^3(x^2)^3(y^5)^3 \cdot 3^2(x^4)^2y^2$ — Property 3
$= 8x^6y^{15} \cdot 9x^8y^2$ — Property 2
$= (8 \cdot 9)(x^6x^8)(y^{15}y^2)$ — Commutative and associative properties
$= 72x^{14}y^{17}$ — Property 1

Note

If we include units with the dimensions of the diagrams, then the units for the area will be square units and the units for volume will be cubic units. More specifically:

If a square has a side 5 inches long, then its area will be

$A = (5 \text{ inches})^2 = 25 \text{ inches}^2$

where the unit inches² stands for square inches.

If a cube has a single side 5 inches long, then its volume will be

$V = (5 \text{ inches})^3 = 125 \text{ inches}^3$

where the unit inches³ stands for cubic inches.

If a rectangular solid has a length of 5 inches, a width of 4 inches, and a height of 3 inches, then its volume is

$V = (5 \text{ inches})(4 \text{ inches})(3 \text{ inches})$
$= 60 \text{ inches}^3$

FACTS FROM GEOMETRY

Volume of a Rectangular Solid

It is easy to see why the phrase "five squared" is associated with the expression 5^2. Simply find the area of the square shown in Figure 1 with a side of 5.

FIGURE 1

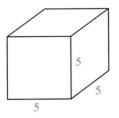

FIGURE 2

To see why the phrase "five cubed" is associated with the expression 5^3, we have to find the *volume* of a cube for which all three dimensions are 5 units long. The volume of a cube is a measure of the space occupied by the cube. To calculate the volume of the cube shown in Figure 2, we multiply the three dimensions together to get $5 \cdot 5 \cdot 5 = 5^3$.

The cube shown in Figure 2 is a special case of a general category of three-dimensional geometric figures called *rectangular solids*. Rectangular solids have rectangles for sides, and all connecting sides meet at right angles. The three dimensions are length, width, and height. To find the volume of a rectangular solid, we find the product of the three dimensions, as shown in Figure 3.

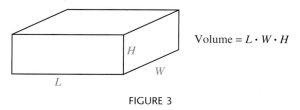

$$\text{Volume} = L \cdot W \cdot H$$

FIGURE 3

Scientific Notation

Many branches of science require working with very large numbers. In astronomy, for example, distances commonly are given in light-years. A light-year is the distance light travels in a year. It is approximately

$$5{,}880{,}000{,}000{,}000 \text{ miles}$$

This number is difficult to use in calculations because of the number of zeros it contains. Scientific notation provides a way of writing very large numbers in a more manageable form.

> **DEFINITION** A number is in **scientific notation** when it is written as the product of a number between 1 and 10 and an integer power of 10. A number written in scientific notation has the form
>
> $$n \times 10^r$$
>
> where $1 \leq n < 10$ and $r =$ an integer.

 EXAMPLE 17   Write 376,000 in scientific notation.

SOLUTION   We must rewrite 376,000 as the product of a number between 1 and 10 and a power of 10. To do so, we move the decimal point 5 places to the left so that it appears between the 3 and the 7. Then we multiply this number by 10^5. The number that results has the same value as our original number and is written in scientific notation:

$$376{,}000 = 3.76 \times 10^5$$

Moved 5 places.

Decimal point originally here.

Keeps track of the 5 places we moved the decimal point.

 EXAMPLE 18   Write 4.52×10^3 in expanded form.

SOLUTION   Since 10^3 is 1,000, we can think of this as simply a multiplication problem; that is,

$$4.52 \times 10^3 = 4.52 \times 1{,}000 = 4{,}520$$

LINKING OBJECTIVES AND EXAMPLES

Next to each **objective** we have listed the examples that are best described by that objective.

A	1–4
B	5–7, 14–16
C	8, 9, 12–14, 16
D	10–13, 15, 16
E	17, 18

On the other hand, we can think of the exponent 3 as indicating the number of places we need to move the decimal point to write our number in expanded form. Since our exponent is positive 3, we move the decimal point three places to the right:

$$4.52 \times 10^3 = 4,520$$

GETTING READY FOR CLASS

After reading through the preceding section, respond in your own words and in complete sentences.

1. Explain the difference between -5^2 and $(-5)^2$.
2. How do you multiply two expressions containing exponents when they each have the same base?
3. What is Property 2 for exponents?
4. When is a number written in scientific notation?

Problem Set 5.1

Online support materials can be found at www.thomsonedu.com/login

Answers appear in the Instructor's Edition only.

Name the base and exponent in each of the following expressions. Then use the definition of exponents as repeated multiplication to simplify.

1. 4^2 16

2. 6^2 36

3. $(0.3)^2$ 0.09

4. $(0.03)^2$ 0.0009

5. 4^3 64

6. 10^3 1,000

7. $(-5)^2$ 25

8. -5^2 -25

9. -2^3 -8

10. $(-2)^3$ -8

11. 3^4 81

12. $(-3)^4$ 81

13. $\left(\dfrac{2}{3}\right)^2$ $\dfrac{4}{9}$

14. $\left(\dfrac{2}{3}\right)^3$ $\dfrac{8}{27}$

▶ 15. $\left(\dfrac{1}{2}\right)^4$ $\dfrac{1}{16}$

16. $\left(\dfrac{4}{5}\right)^2$ $\dfrac{16}{25}$

17. **a.** Complete the following table.

Number x	Square x^2
1	1
2	4
3	9
4	16
5	25
6	36
7	49

b. Using the results of part a, fill in the blank in the following statement: For numbers larger than 1, the square of the number is _____ than the number. *Either larger or greater will work.*

= Videos available by instructor request

▶ = Online student support materials available at www.thomsonedu.com/login

18. a. Complete the following table.

Number x	Square x^2
$\frac{1}{2}$	$\frac{1}{4}$
$\frac{1}{3}$	$\frac{1}{9}$
$\frac{1}{4}$	$\frac{1}{16}$
$\frac{1}{5}$	$\frac{1}{25}$
$\frac{1}{6}$	$\frac{1}{36}$
$\frac{1}{7}$	$\frac{1}{49}$
$\frac{1}{8}$	$\frac{1}{64}$

b. Using the results of part a, fill in the blank in the following statement: For numbers between 0 and 1, the square of the number is _____ than the number. Either *smaller* or *less* will work.

Use Property 1 for exponents to simplify each expression. Leave all answers in terms of exponents.

▶ **19.** $x^4 \cdot x^5$ x^9
20. $x^7 \cdot x^3$ x^{10}

21. $y^{10} \cdot y^{20}$ y^{30}
22. $y^{30} \cdot y^{30}$ y^{60}

23. $2^5 \cdot 2^4 \cdot 2^3$ 2^{12}
24. $4^2 \cdot 4^3 \cdot 4^4$ 4^9

25. $x^4 \cdot x^6 \cdot x^8 \cdot x^{10}$ x^{28}
26. $x^{20} \cdot x^{18} \cdot x^{16} \cdot x^{14}$ x^{68}

Use Property 2 for exponents to write each of the following problems with a single exponent. (Assume all variables are positive numbers.)

▶ **27.** $(x^2)^5$ x^{10}
28. $(x^5)^2$ x^{10}

29. $(5^4)^3$ 5^{12}
30. $(5^3)^4$ 5^{12}

31. $(y^3)^3$ y^9
32. $(y^2)^2$ y^4

33. $(2^5)^{10}$ 2^{50}
34. $(10^5)^2$ 10^{10}

35. $(a^3)^x$ a^{3x}
36. $(a^5)^x$ a^{5x}

37. $(b^x)^y$ b^{xy}
38. $(b^r)^s$ b^{rs}

Use Property 3 for exponents to simplify each of the following expressions.

39. $(4x)^2$ $16x^2$
40. $(2x)^4$ $16x^4$

41. $(2y)^5$ $32y^5$
42. $(5y)^2$ $25y^2$

43. $(-3x)^4$ $81x^4$
44. $(-3x)^3$ $-27x^3$

45. $(0.5ab)^2$ $0.25a^2b^2$
46. $(0.4ab)^2$ $0.16a^2b^2$

▶ **47.** $(4xyz)^3$ $64x^3y^3z^3$
48. $(5xyz)^3$ $125x^3y^3z^3$

▶ **Chalkboard Problems**
Problems 73 and 74 lead right into the properties of exponents.

Simplify the following expressions by using the properties of exponents.

49. $(2x^4)^3$ $8x^{12}$
50. $(3x^5)^2$ $9x^{10}$

51. $(4a^3)^2$ $16a^6$
52. $(5a^2)^2$ $25a^4$

53. $(x^2)^3(x^4)^2$ x^{14}
54. $(x^5)^2(x^3)^5$ x^{25}

55. $(a^3)^1(a^2)^4$ a^{11}
56. $(a^4)^1(a^1)^3$ a^7

57. $(2x)^3(2x)^4$ $128x^7$
58. $(3x)^2(3x)^3$ $243x^5$

59. $(3x^2)^3(2x)^4$ $432x^{10}$
60. $(3x)^3(2x^3)^2$ $108x^9$

61. $(4x^2y^3)^2$ $16x^4y^6$
62. $(9x^3y^5)^2$ $81x^6y^{10}$

63. $\left(\frac{2}{3}a^4b^5\right)^3$ $\frac{8}{27}a^{12}b^{15}$
64. $\left(\frac{3}{4}ab^7\right)^3$ $\frac{27}{64}a^3b^{21}$

Write each expression as a perfect square.

▶ **65.** $x^4 = (\ x^2\ )^2$
▶ **66.** $x^6 = (\ x^3\ )^2$

▶ **67.** $16x^2 = (\ 4x\ )^2$
▶ **68.** $256x^4 = (\ 16x^2\ )^2$

Write each expression as a perfect cube.

▶ **69.** $8 = (\ 2\ )^3$
▶ **70.** $27 = (\ 3\ )^3$

▶ **71.** $64x^3 = (\ 4x\ )^3$
▶ **72.** $27x^6 = (\ 3x^2\ )^3$

▶ **73.** Let $x = 2$ in each of the following expressions and simplify.
a. x^3x^2 32
b. $(x^3)^2$ 64
c. x^5 32
d. x^6 64

▶ **74.** Let $x = -1$ in each of the following expressions and simplify.
a. x^3x^4 -1
b. $(x^3)^4$ 1
c. x^7 -1
d. x^{12} 1

75. Complete the following table, and then construct a line graph of the information in the table.

Number x	Square x^2
-3	9
-2	4
-1	1
0	0
1	1
2	4
3	9

76. Complete the table, and then use the template to construct a line graph of the information in the table.

Number x	Cube x^3
−3	−27
−2	−8
−1	−1
0	0
1	1
2	8
3	27

77. Complete the table. When you are finished, notice how the points in this table could be used to refine the line graph you created in Problem 75.

Number x	Square x^2
−2.5	6.25
−1.5	2.25
−0.5	0.25
0	0
0.5	0.25
1.5	2.25
2.5	6.25

78. Complete the following table. When you are finished, notice that this table contains exactly the same entries as the table from Problem 77. This table uses fractions, whereas the table from Problem 77 uses decimals.

Number x	Square x^2
$-\frac{5}{2}$	$\frac{25}{4}$
$-\frac{3}{2}$	$\frac{9}{4}$
$-\frac{1}{2}$	$\frac{1}{4}$
0	0
$\frac{1}{2}$	$\frac{1}{4}$
$\frac{3}{2}$	$\frac{9}{4}$
$\frac{5}{2}$	$\frac{25}{4}$

Write each number in scientific notation.

▶ **79.** 43,200 4.32×10^4

80. 432,000 4.32×10^5

81. 570 5.7×10^2

82. 5,700 5.7×10^3

83. 238,000 2.38×10^5

84. 2,380,000 2.38×10^6

Write each number in expanded form.

▶ **85.** 2.49×10^3 2,490

86. 2.49×10^4 24,900

87. 3.52×10^2 352

88. 3.52×10^5 352,000

89. 2.8×10^4 28,000

90. 2.8×10^3 2,800

Applying the Concepts

91. Volume of a Cube Find the volume of a cube if each side is 3 inches long. 27 inches³

92. Volume of a Cube Find the volume of a cube if each side is 3 feet long. 27 feet³

▶ **93. Volume of a Cube** A bottle of perfume is packaged in a box that is in the shape of a cube. Find the volume of the box if each side is 2.5 inches long. Round to the nearest tenth. 15.6 inches³

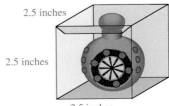

2.5 inches

2.5 inches

2.5 inches

94. Volume of a Cube A television set is packaged in a box that is in the shape of a cube. Find the volume of the box if each side is 18 inches long. 5,832 inches³

95. Volume of a Box A rented videotape is in a plastic container that has the shape of a rectangular solid. Find the volume of the container if the length is 8 inches, the width is 4.5 inches, and the height is 1 inch. 36 inches³

96. Volume of a Box Your textbook is in the shape of a rectangular solid. Find the volume in cubic inches. Around 100 inches³

97. Volume of a Box If a box has a volume of 42 cubic feet, is it possible for you to fit inside the box? Explain your answer. Answers will vary.

98. Volume of a Box A box has a volume of 45 cubic inches. Will a can of soup fit inside the box? Explain your answer. Answers will vary.

99. Age in Seconds If you are 21 years old, you have been alive for more than 650,000,000 seconds. Write this last number in scientific notation. 6.5×10^8

100. Distance Around the Earth The distance around the Earth at the equator is more than 130,000,000 feet. Write this number in scientific notation. 1.3×10^8

101. Lifetime Earnings If you earn at least $12 an hour and work full-time for 30 years, you will make at least 7.4×10^5 dollars. Write this last number in expanded form. $740,000

102. Heart Beats per Year If your pulse is 72, then in one year your heart will beat at least 3.78×10^7 times. Write this last number in expanded form. 37,800,000

103. Investing If you put $1,000 into a savings account every year from the time you are 25 years old until you are 55 years old, you will have more than 1.8×10^5 dollars in the account when you reach 55 years of age (assuming 10% annual interest). Write 1.8×10^5 in expanded form. $180,000

104. Investing If you put $20 into a savings account every month from the time you are 20 years old until you are 30 years old, you will have more than 3.27×10^3 dollars in the account when you reach 30 years of age (assuming 6% annual interest compounded monthly). Write 3.27×10^3 in expanded form. $3,270

Displacement The displacement, in cubic inches, of a car engine is given by the formula

$$d = \pi \cdot s \cdot c \cdot \left(\frac{1}{2} \cdot b\right)^2$$

where s is the stroke and b is the bore, as shown in the figure, and c is the number of cylinders.

Calculate the engine displacement for each of the following cars. Use 3.14 to approximate π. Round your answers to the nearest cubic inch.

105. Ferrari Modena 8 cylinders, 3.35 inches of bore, 3.11 inches of stroke 219 inches³

106. Audi A8 8 cylinders, 3.32 inches of bore, 3.66 inches of stroke 253 inches³

107. Mitsubishi Eclipse 6 cylinders, 3.59 inches of bore, 2.99 inches of stroke 182 inches³

108. Porsche 911 GT3 6 cylinders, 3.94 inches of bore, 3.01 inches of stroke 220 inches³

Maintaining Your Skills

Factor each of the following into its product of prime factors.

109. 128 2^7

110. 200 $2^3 \cdot 5^2$

111. 250 $2 \cdot 5^3$

112. 512 2^9

113. 720 $2^4 \cdot 3^2 \cdot 5$

114. 555 $3 \cdot 5 \cdot 37$

115. 820 $2^2 \cdot 5 \cdot 41$

116. 1,024 2^{10}

Factor the following by first factoring the base and then raising each of its factors to the third power.

117. 6^3 $2^3 \cdot 3^3$

118. 10^3 $2^3 \cdot 5^3$

119. 30^3 $2^3 \cdot 3^3 \cdot 5^3$

120. 42^3 $2^3 \cdot 3^3 \cdot 7^3$

121. 25^3 5^6

122. 8^3 2^9

123. 12^3 $2^6 \cdot 3^3$

124. 36^3 $2^6 \cdot 3^6$

Getting Ready for the Next Section

Subtract.

125. $4 - 7$ -3

126. $-4 - 7$ -11

127. $4 - (-7)$ 11

128. $-4 - (-7)$ 3

129. $15 - 20$ -5

130. $15 - (-20)$ 35

131. $-15 - (-20)$ 5

132. $-15 - 20$ -35

Simplify.

133. $2(3) - 4$ 2

134. $5(3) - 10$ 5

135. $4(3) - 3(2)$ 6

136. $-8 - 2(3)$ -14

137. $2(5 - 3)$ 4

138. $2(3) - 4 - 3(-4)$ 14

139. $5 + 4(-2) - 2(-3)$ 3

140. $2(3) + 4(5) - 5(2)$ 16

5.2 Division with Exponents

OBJECTIVES

A Apply the definition for negative exponents.

B Use Property 4 for exponents.

C Use Property 5 for exponents.

D Simplify expressions involving exponents of 0 and 1.

E Simplify expressions using combinations of the properties of exponents.

F Write numbers in scientific notation and expanded form.

In Section 5.1 we found that multiplication with the same base results in addition of exponents; that is, $a^r \cdot a^s = a^{r+s}$. Since division is the inverse operation of multiplication, we can expect division with the same base to result in subtraction of exponents.

To develop the properties for exponents under division, we again apply the definition of exponents:

$$\frac{x^5}{x^3} = \frac{x \cdot x \cdot x \cdot x \cdot x}{x \cdot x \cdot x}$$

$$\frac{2^4}{2^7} = \frac{2 \cdot 2 \cdot 2 \cdot 2}{2 \cdot 2 \cdot 2 \cdot 2 \cdot 2 \cdot 2 \cdot 2}$$

$$= \frac{x \cdot x \cdot x}{x \cdot x \cdot x}(x \cdot x)$$

$$= \frac{2 \cdot 2 \cdot 2 \cdot 2}{2 \cdot 2 \cdot 2 \cdot 2} \cdot \frac{1}{2 \cdot 2 \cdot 2}$$

$$= 1(x \cdot x)$$

$$= \frac{1}{2 \cdot 2 \cdot 2}$$

$$= x^2 \quad \textit{Notice: } 5 - 3 = 2$$

$$= \frac{1}{2^3} \quad \textit{Notice: } 7 - 4 = 3$$

In both cases division with the same base resulted in subtraction of the smaller exponent from the larger. The problem is deciding whether the answer is a fraction. The problem is resolved easily by the following definition.

DEFINITION If r is a positive integer, then $a^{-r} = \dfrac{1}{a^r} = \left(\dfrac{1}{a}\right)^r$ $\quad (a \neq 0)$

The following examples illustrate how we use this definition to simplify expressions that contain negative exponents.

 EXAMPLES Write each expression with a positive exponent and then simplify:

1. $2^{-3} = \dfrac{1}{2^3} = \dfrac{1}{8}$ $\qquad$ *Notice:* **Negative exponents do not indicate negative numbers. They indicate reciprocals**

2. $5^{-2} = \dfrac{1}{5^2} = \dfrac{1}{25}$

3. $3x^{-6} = 3 \cdot \dfrac{1}{x^6} = \dfrac{3}{x^6}$

Now let us look back to our original problem and try to work it again with the help of a negative exponent. We know that $\frac{2^4}{2^7} = \frac{1}{2^3}$. Let us decide now that with division of the same base, we will always subtract the exponent in the denominator from the exponent in the numerator and see if this conflicts with what we know is true.

$$\frac{2^4}{2^7} = 2^{4-7} \qquad \text{Subtracting the bottom exponent from the top exponent}$$

$$= 2^{-3} \qquad \text{Subtraction}$$

$$= \frac{1}{2^3} \qquad \text{Definition of negative exponents}$$

Subtracting the exponent in the denominator from the exponent in the numerator and then using the definition of negative exponents gives us the same result we obtained previously. We can now continue the list of properties of exponents we started in Section 5.1.

Property 4 for Exponents If a is any real number and r and s are integers, then

$$\frac{a^r}{a^s} = a^{r-s} \qquad (a \neq 0)$$

In words: To divide with the same base, subtract the exponent in the denominator from the exponent in the numerator and raise the base to the exponent that results.

The following examples show how we use Property 4 and the definition for negative exponents to simplify expressions involving division.

 EXAMPLES Simplify the following expressions:

4. $\dfrac{x^9}{x^6} = x^{9-6} = x^3$

5. $\dfrac{x^4}{x^{10}} = x^{4-10} = x^{-6} = \dfrac{1}{x^6}$

6. $\dfrac{2^{15}}{2^{20}} = 2^{15-20} = 2^{-5} = \dfrac{1}{2^5} = \dfrac{1}{32}$

Our final property of exponents is similar to Property 3 from Section 5.1, but it involves division instead of multiplication. After we have stated the property, we will give a proof of it. The proof shows why this property is true.

Property 5 for Exponents If a and b are any two real numbers ($b \neq 0$) and r is an integer, then

$$\left(\frac{a}{b}\right)^r = \frac{a^r}{b^r}$$

In words: A quotient raised to a power is the quotient of the powers.

Proof

$$\left(\frac{a}{b}\right)^r = \left(a \cdot \frac{1}{b}\right)^r \qquad \textbf{By the definition of division}$$

$$= a^r \cdot \left(\frac{1}{b}\right)^r \qquad \textbf{By Property 3}$$

$$= a^r \cdot b^{-r} \qquad \textbf{By the definition of negative exponents}$$

$$= a^r \cdot \frac{1}{b^r} \qquad \textbf{By the definition of negative exponents}$$

$$= \frac{a^r}{b^r} \qquad \textbf{By the definition of division}$$

EXAMPLES   Simplify the following expressions.

7. $\left(\dfrac{x}{2}\right)^3 = \dfrac{x^3}{2^3} = \dfrac{x^3}{8}$

8. $\left(\dfrac{5}{y}\right)^2 = \dfrac{5^2}{y^2} = \dfrac{25}{y^2}$

9. $\left(\dfrac{2}{3}\right)^4 = \dfrac{2^4}{3^4} = \dfrac{16}{81}$

Zero and One as Exponents

We have two special exponents left to deal with before our rules for exponents are complete: 0 and 1. To obtain an expression for x^1, we will solve a problem two different ways:

$$\left.\begin{array}{c}\dfrac{x^3}{x^2} = \dfrac{x \cdot x \cdot x}{x \cdot x} = x \\[2mm] \dfrac{x^3}{x^2} = x^{3-2} = x^1\end{array}\right\}\quad \textbf{Hence } \boldsymbol{x^1 = x}$$

Stated generally, this rule says that $a^1 = a$. This seems reasonable and we will use it since it is consistent with our property of division using the same base.

We use the same procedure to obtain an expression for x^0:

$$\left.\begin{array}{c}\dfrac{5^2}{5^2} = \dfrac{25}{25} = 1 \\[2mm] \dfrac{5^2}{5^2} = 5^{2-2} = 5^0\end{array}\right\}\quad \textbf{Hence } \boldsymbol{5^0 = 1}$$

It seems, therefore, that the best definition of x^0 is 1 for all x except $x = 0$. In the case of $x = 0$, we have 0^0, which we will not define. This definition will probably seem awkward at first. Most people would like to define x^0 as 0 when they first encounter it. Remember, the zero in this expression is an exponent, so x^0 does not mean to multiply by zero. Thus, we can make the general statement that $a^0 = 1$ for all real numbers except $a = 0$.

Here are some examples involving the exponents 0 and 1.

EXAMPLES   Simplify the following expressions:

10. $8^0 = 1$

11. $8^1 = 8$

12. $4^0 + 4^1 = 1 + 4 = 5$

13. $(2x^2y)^0 = 1$

Here is a summary of the definitions and properties of exponents we have developed so far. For each definition or property in the list, a and b are real numbers, and r and s are integers.

Definitions	Properties
$a^{-r} = \dfrac{1}{a^r} = \left(\dfrac{1}{a}\right)^r \qquad a \neq 0$	**1.** $a^r \cdot a^s = a^{r+s}$
$a^1 = a$	**2.** $(a^r)^s = a^{rs}$
$a^0 = 1 \qquad a \neq 0$	**3.** $(ab)^r = a^r b^r$
	4. $\dfrac{a^r}{a^s} = a^{r-s} \qquad a \neq 0$
	5. $\left(\dfrac{a}{b}\right)^r = \dfrac{a^r}{b^r} \qquad b \neq 0$

Here are some additional examples. These examples use a combination of the preceding properties and definitions.

EXAMPLES   Simplify each expression. Write all answers with positive exponents only:

14. $\dfrac{(5x^3)^2}{x^4} = \dfrac{25x^6}{x^4}$ Properties 2 and 3

$\qquad\qquad = 25x^2$ Property 4

15. $\dfrac{x^{-8}}{(x^2)^3} = \dfrac{x^{-8}}{x^6}$ Property 2

$\qquad\qquad = x^{-8-6}$ Property 4

$\qquad\qquad = x^{-14}$ Subtraction

$\qquad\qquad = \dfrac{1}{x^{14}}$ Definition of negative exponents

16. $\left(\dfrac{y^5}{y^3}\right)^2 = \dfrac{(y^5)^2}{(y^3)^2}$ Property 5

$\qquad\qquad = \dfrac{y^{10}}{y^6}$ Property 2

$\qquad\qquad = y^4$ Property 4

Notice in Example 16 that we could have simplified inside the parentheses first and then raised the result to the second power:

$$\left(\frac{y^5}{y^3}\right)^2 = (y^2)^2 = y^4$$

17. $(3x^5)^{-2} = \dfrac{1}{(3x^5)^2}$ Definition of negative exponents

$\qquad\qquad = \dfrac{1}{9x^{10}}$ Properties 2 and 3

18. $x^{-8} \cdot x^5 = x^{-8+5}$ Property 1

$\qquad\qquad = x^{-3}$ Addition

$\qquad\qquad = \dfrac{1}{x^3}$ Definition of negative exponents

19. $\dfrac{(a^3)^2 a^{-4}}{(a^{-4})^3} = \dfrac{a^6 a^{-4}}{a^{-12}}$ Property 2

$\qquad\qquad = \dfrac{a^2}{a^{-12}}$ Property 1

$\qquad\qquad = a^{14}$ Property 4

In the next two examples we use division to compare the area and volume of geometric figures.

EXAMPLE 20   Suppose you have two squares, one of which is larger than the other. If the length of a side of the larger square is 3 times as long as the length of a side of the smaller square, how many of the smaller squares will it take to cover up the larger square?

SOLUTION   If we let x represent the length of a side of the smaller square, then the length of a side of the larger square is $3x$. The area of each square, along with a diagram of the situation, is given in Figure 1.

Square 1: $A = x^2$ Square 2: $A = (3x)^2 = 9x^2$

FIGURE 1

To find out how many smaller squares it will take to cover up the larger square, we divide the area of the larger square by the area of the smaller square.

$$\frac{\text{Area of square 2}}{\text{Area of square 1}} = \frac{9x^2}{x^2} = 9$$

It will take 9 of the smaller squares to cover the larger square.

EXAMPLE 21   Suppose you have two boxes, each of which is a cube. If the length of a side in the second box is 3 times as long as the length of a side of the first box, how many of the smaller boxes will fit inside the larger box?

SOLUTION   If we let x represent the length of a side of the smaller box, then the length of a side of the larger box is $3x$. The volume of each box, along with a diagram of the situation, is given in Figure 2.

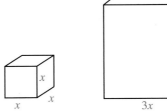

Box 1: $V = x^3$ Box 2: $V = (3x)^3 = 27x^3$

FIGURE 2

To find out how many smaller boxes will fit inside the larger box, we divide the volume of the larger box by the volume of the smaller box.

$$\frac{\text{Volume of box 2}}{\text{Volume of box 1}} = \frac{27x^3}{x^3} = 27$$

We can fit 27 of the smaller boxes inside the larger box.

More on Scientific Notation

Now that we have completed our list of definitions and properties of exponents, we can expand the work we did previously with scientific notation.

Recall that a number is in scientific notation when it is written in the form

$$n \times 10^r$$

where $1 \le n < 10$ and r is an integer.

Since negative exponents give us reciprocals, we can use negative exponents to write very small numbers in scientific notation. For example, the number 0.00057, when written in scientific notation, is equivalent to 5.7×10^{-4}. Here's why:

$$5.7 \times 10^{-4} = 5.7 \times \frac{1}{10^4} = 5.7 \times \frac{1}{10,000} = \frac{5.7}{10,000} = 0.00057$$

The table below lists some other numbers in both scientific notation and expanded form.

 EXAMPLE 22

Number Written the Long Way		Number Written Again in Scientific Notation
376,000	=	3.76×10^5
49,500	=	4.95×10^4
3,200	=	3.2×10^3
591	=	5.91×10^2
46	=	4.6×10^1
8	=	8×10^0
0.47	=	4.7×10^{-1}
0.093	=	9.3×10^{-2}
0.00688	=	6.88×10^{-3}
0.0002	=	2×10^{-4}
0.000098	=	9.8×10^{-5}

Notice that in each case, when the number is written in scientific notation, the decimal point in the first number is placed so that the number is between 1 and 10. The exponent on 10 in the second number keeps track of the number of places we moved the decimal point in the original number to get a number between 1 and 10:

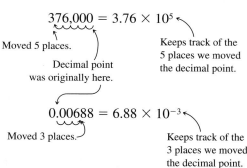

$$376,000 = 3.76 \times 10^5$$

Moved 5 places.

Decimal point was originally here.

Keeps track of the 5 places we moved the decimal point.

$$0.00688 = 6.88 \times 10^{-3}$$

Moved 3 places.

Keeps track of the 3 places we moved the decimal point.

LINKING OBJECTIVES AND EXAMPLES

Next to each **objective** we have listed the examples that are best described by that objective.

A	1–3
B	4–6
C	7–9
D	10–13
E	14–19
F	22

GETTING READY FOR CLASS

After reading through the preceding section, respond in your own words and in complete sentences.

1. How do you divide two expressions containing exponents when they each have the same base?
2. Explain the difference between 3^2 and 3^{-2}.
3. If a positive base is raised to a negative exponent, can the result be a negative number?
4. Explain what happens when we use 0 as an exponent.

Problem Set 5.2

Online support materials can be found at www.thomsonedu.com/login

Write each of the following with positive exponents, and then simplify, when possible.

▶ **1.** 3^{-2} $\frac{1}{9}$ **2.** 3^{-3} $\frac{1}{27}$

3. 6^{-2} $\frac{1}{36}$ **4.** 2^{-6} $\frac{1}{64}$

5. 8^{-2} $\frac{1}{64}$ **6.** 3^{-4} $\frac{1}{81}$

▶ **7.** 5^{-3} $\frac{1}{125}$ **8.** 9^{-2} $\frac{1}{81}$

9. $2x^{-3}$ $\frac{2}{x^3}$ **10.** $5x^{-1}$ $\frac{5}{x}$

11. $(2x)^{-3}$ $\frac{1}{8x^3}$ **12.** $(5x)^{-1}$ $\frac{1}{5x}$

13. $(5y)^{-2}$ $\frac{1}{25y^2}$ **14.** $5y^{-2}$ $\frac{5}{y^2}$

15. 10^{-2} $\frac{1}{100}$ **16.** 10^{-3} $\frac{1}{1,000}$

17. Complete the following table.

Number x	Square x^2	Power of 2 2^x
−3	9	$\frac{1}{8}$
−2	4	$\frac{1}{4}$
−1	1	$\frac{1}{2}$
0	0	1
1	1	2
2	4	4
3	9	8

18. Complete the following table.

Number x	Cube x^3	Power of 3 3^x
−3	−27	$\frac{1}{27}$
−2	−8	$\frac{1}{9}$
−1	−1	$\frac{1}{3}$
0	0	1
1	1	3
2	8	9
3	27	27

Use Property 4 to simplify each of the following expressions. Write all answers that contain exponents with positive exponents only.

▶ **19.** $\dfrac{5^1}{5^3}$ $\frac{1}{25}$ **20.** $\dfrac{7^6}{7^8}$ $\frac{1}{49}$

▶ **21.** $\dfrac{x^{10}}{x^4}$ x^6 **22.** $\dfrac{x^4}{x^{10}}$ $\frac{1}{x^6}$

    = Videos available by instructor request
▶ = Online student support materials available at www.thomsonedu.com/login

23. $\dfrac{4^3}{4^0}$ 64

24. $\dfrac{4^0}{4^3}$ $\dfrac{1}{64}$

25. $\dfrac{(2x)^7}{(2x)^4}$ $8x^3$

26. $\dfrac{(2x)^4}{(2x)^7}$ $\dfrac{1}{8x^3}$

27. $\dfrac{6^{11}}{6}$ 6^{10}

28. $\dfrac{8^7}{8}$ 8^6

29. $\dfrac{6}{6^{11}}$ $\dfrac{1}{6^{10}}$

30. $\dfrac{8}{8^7}$ $\dfrac{1}{8^6}$

31. $\dfrac{2^{-5}}{2^3}$ $\dfrac{1}{2^8}$

32. $\dfrac{2^{-5}}{2^{-3}}$ $\dfrac{1}{4}$

▶ **33.** $\dfrac{2^5}{2^{-3}}$ 2^8

34. $\dfrac{2^{-3}}{2^{-5}}$ 4

35. $\dfrac{(3x)^{-5}}{(3x)^{-8}}$ $27x^3$

36. $\dfrac{(2x)^{-10}}{(2x)^{-15}}$ $32x^5$

Simplify the following expressions. Any answers that contain exponents should contain positive exponents only.

37. $(3xy)^4$ $81x^4y^4$

38. $(4xy)^3$ $64x^3y^3$

39. 10^0 1

40. 10^1 10

41. $(2a^2b)^1$ $2a^2b$

42. $(2a^2b)^0$ 1

43. $(7y^3)^{-2}$ $\dfrac{1}{49y^6}$

44. $(5y^4)^{-2}$ $\dfrac{1}{25y^8}$

45. $x^{-3}x^{-5}$ $\dfrac{1}{x^8}$

46. $x^{-6}\cdot x^8$ x^2

47. $y^7\cdot y^{-10}$ $\dfrac{1}{y^3}$

48. $y^{-4}\cdot y^{-6}$ $\dfrac{1}{y^{10}}$

49. $\dfrac{(x^2)^3}{x^4}$ x^2

50. $\dfrac{(x^5)^3}{x^{10}}$ x^5

51. $\dfrac{(a^4)^3}{(a^3)^2}$ a^6

52. $\dfrac{(a^5)^3}{(a^5)^2}$ a^5

53. $\dfrac{y^7}{(y^2)^8}$ $\dfrac{1}{y^9}$

54. $\dfrac{y^2}{(y^3)^4}$ $\dfrac{1}{y^{10}}$

55. $\left(\dfrac{y^7}{y^2}\right)^8$ y^{40}

56. $\left(\dfrac{y^2}{y^3}\right)^4$ $\dfrac{1}{y^4}$

57. $\dfrac{(x^{-2})^3}{x^{-5}}$ $\dfrac{1}{x}$

58. $\dfrac{(x^2)^{-3}}{x^{-5}}$ $\dfrac{1}{x}$

59. $\left(\dfrac{x^{-2}}{x^{-5}}\right)^3$ x^9

60. $\left(\dfrac{x^2}{x^{-5}}\right)^{-3}$ $\dfrac{1}{x^{21}}$

▶ **61.** $\dfrac{(a^3)^2(a^4)^5}{(a^5)^2}$ a^{16}

62. $\dfrac{(a^4)^8(a^2)^5}{(a^3)^4}$ a^{30}

63. $\dfrac{(a^{-2})^3(a^4)^2}{(a^{-3})^{-2}}$ $\dfrac{1}{a^4}$

64. $\dfrac{(a^{-5})^{-3}(a^7)^{-1}}{(a^{-3})^5}$ a^{23}

▶ **65.** Let $x = 2$ in each of the following expressions and simplify.

    a. $\dfrac{x^7}{x^2}$ 32   **b.** x^5 32

    c. $\dfrac{x^2}{x^7}$ $\dfrac{1}{32}$ **d.** x^{-5} $\dfrac{1}{32}$

66. Let $x = -1$ in each of the following expressions and simplify.

    a. $\dfrac{x^{12}}{x^9}$ -1 **b.** x^3 -1

    c. $\dfrac{x^{11}}{x^9}$ 1   **d.** x^2 1

67. Write each expression as a perfect square.

    a. $\dfrac{1}{25} = \left(\dfrac{1}{5}\right)^2$ **b.** $\dfrac{1}{64} = \left(\dfrac{1}{8}\right)^2$

    c. $\dfrac{1}{x^2} = \left(\dfrac{1}{x}\right)^2$ **d.** $\dfrac{1}{x^4} = \left(\dfrac{1}{x^2}\right)^2$

68. Write each expression as a perfect cube.

    a. $\dfrac{1}{125} = \left(\dfrac{1}{5}\right)^3$ **b.** $\dfrac{1}{27} = \left(\dfrac{1}{3}\right)^3$

    c. $\dfrac{x^6}{125} = \left(\dfrac{x^2}{5}\right)^3$ **d.** $\dfrac{x^3}{27} = \left(\dfrac{x}{3}\right)^3$

69. Complete the following table, and then construct a line graph of the information in the table.

Number x	Power of 2 2^x
−3	$\dfrac{1}{8}$
−2	$\dfrac{1}{4}$
−1	$\dfrac{1}{2}$
0	1
1	2
2	4
3	8

▶ = Chalkboard Problem

70. Complete the following table, and then construct a line graph of the information in the table.

Number x	Power of 3 3^x
-3	$\frac{1}{27}$
-2	$\frac{1}{9}$
-1	$\frac{1}{3}$
0	1
1	3
2	9
3	27

Write each of the following numbers in scientific notation.

71. 0.0048 4.8×10^{-3}

72. 0.000048 4.8×10^{-5}

73. 25 2.5×10^{1}

74. 35 3.5×10^{1}

75. 0.000009 9×10^{-6}

76. 0.0009 9×10^{-4}

77. Complete the following table.

Expanded Form	Scientific Notation $n \times 10^r$
▶ 0.000357	3.57×10^{-4}
0.00357	3.57×10^{-3}
0.0357	3.57×10^{-2}
0.357	3.57×10^{-1}
3.57	3.57×10^{0}
35.7	3.57×10^{1}
357	3.57×10^{2}
$3,570$	3.57×10^{3}
$35,700$	3.57×10^{4}

78. Complete the following table.

Expanded Form	Scientific Notation $n \times 10^r$
0.000123	1.23×10^{-4}
0.00123	1.23×10^{-3}
0.0123	1.23×10^{-2}
0.123	1.23×10^{-1}
1.23	1.23×10^{0}
12.3	1.23×10^{1}
123	1.23×10^{2}
$1,230$	1.23×10^{3}
$12,300$	1.23×10^{4}

Write each of the following numbers in expanded form.

79. 4.23×10^{-3} 0.00423

80. 4.23×10^{3} $4,230$

81. 8×10^{-5} 0.00008

82. 8×10^{5} $800,000$

83. 4.2×10^{0} 4.2

84. 4.2×10^{1} 42

Applying the Concepts

85. Some home computers can do a calculation in 2×10^{-3} seconds. Write this number in expanded form. 0.002

86. Some of the cells in the human body have a radius of 3×10^{-5} inches. Write this number in expanded form. 0.00003

87. Margin of Victory In this graph are the five closest margins of victory of the IRL IndyCar Series. Write each number in scientific notation.

$0.0024 = 2.4 \times 10^{-3}$
$0.0051 = 5.1 \times 10^{-3}$
$0.0096 = 9.6 \times 10^{-3}$
$0.0099 = 9.9 \times 10^{-3}$
$0.0111 = 1.11 \times 10^{-2}$

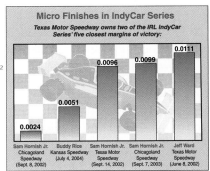

88. Some cameras used in scientific research can take one picture every 0.000000167 second. Write this number in scientific notation. 1.67×10^{-7}

89. The number 25×10^{3} is not in scientific notation because 25 is larger than 10. Write 25×10^{3} in scientific notation. 2.5×10^{4}

90. The number 0.25×10^{3} is not in scientific notation because 0.25 is less than 1. Write 0.25×10^{3} in scientific notation. 2.5×10^{2}

91. The number 23.5×10^{4} is not in scientific notation because 23.5 is not between 1 and 10. Rewrite 23.5×10^{4} in scientific notation. 2.35×10^{5}

92. The number 375×10^{3} is not in scientific notation because 375 is not between 1 and 10. Rewrite 375×10^{3} in scientific notation. 3.75×10^{5}

93. The number 0.82×10^{-3} is not in scientific notation because 0.82 is not between 1 and 10. Rewrite 0.82×10^{-3} in scientific notation. 8.2×10^{-4}

94. The number 0.93×10^{-2} is not in scientific notation because 0.93 is not between 1 and 10. Rewrite 0.93×10^{-2} in scientific notation. 9.3×10^{-3}

Comparing Areas Suppose you have two squares, one of which is larger than the other. Suppose further that the side of the larger square is twice as long as the side of the smaller square.

95. If the length of the side of the smaller square is 10 inches, give the area of each square. Then find the number of smaller squares it will take to cover the larger square. 100 inches², 400 inches²; 4

96. How many smaller squares will it take to cover the larger square if the length of the side of the smaller square is 1 foot? 4

97. If the length of the side of the smaller square is x, find the area of each square. Then find the number of smaller squares it will take to cover the larger square. x^2; $4x^2$; 4

98. Suppose the length of the side of the larger square is 1 foot. How many smaller squares will it take to cover the larger square? 4

Comparing Volumes Suppose you have two boxes, each of which is a cube. Suppose further that the length of a side of the second box is twice as long as the length of a side of the first box.

99. If the length of a side of the first box is 6 inches, give the volume of each box. Then find the number of smaller boxes that will fit inside the larger box. 216 inches³; 1,728 inches³; 8

100. How many smaller boxes can be placed inside the larger box if the length of a side of the second box is 1 foot? 8

101. If the length of a side of the first box is x, find the volume of each box. Then find the number of smaller boxes that will fit inside the larger box. x^3; $8x^3$; 8

102. Suppose the length of a side of the larger box is 12 inches. How many smaller boxes will fit inside the larger box? 8

Maintaining Your Skills

Simplify the following expressions.

103. $4x + 3x$ $7x$

104. $9x + 7x$ $16x$

105. $5a - 3a$ $2a$

106. $10a - 2a$ $8a$

107. $4y + 5y + y$ $10y$

108. $6y - y + 2y$ $7y$

Getting Ready for the Next Section

Simplify.

109. $3(4.5)$ 13.5

110. $\dfrac{1}{2} \cdot \dfrac{5}{7}$ $\dfrac{5}{14}$

111. $\dfrac{4}{5}(10)$ 8

112. $\dfrac{9.6}{3}$ 3.2

113. $6.8(3.9)$ 26.52

114. $9 - 20$ -11

115. $-3 + 15$ 12

116. $2x \cdot x \cdot \dfrac{1}{2}x$ x^3

117. $x^5 \cdot x^3$ x^8

118. $y^2 \cdot y$ y^3

119. $\dfrac{x^3}{x^2}$ x

120. $\dfrac{x^2}{x}$ x

121. $\dfrac{y^3}{y^5}$ $\dfrac{1}{y^2}$

122. $\dfrac{x^2}{x^5}$ $\dfrac{1}{x^3}$

Write in expanded form.

123. 3.4×10^2 340

124. 6.0×10^{-4} 0.0006

5.3 Operations with Monomials

OBJECTIVES

A Multiply monomials.

B Divide monomials.

C Multiply and divide numbers written in scientific notation.

D Add and subtract monomials.

We have developed all the tools necessary to perform the four basic operations on the simplest of polynomials: monomials.

> **DEFINITION** A **monomial** is a one-term expression that is either a constant (number) or the product of a constant and one or more variables raised to whole number exponents.

The following are examples of monomials:

$$-3 \qquad 15x \qquad -23x^2y \qquad 49x^4y^2z^4 \qquad \frac{3}{4}a^2b^3$$

The numerical part of each monomial is called the *numerical coefficient,* or just *coefficient.* Monomials are also called *terms.*

Multiplication and Division of Monomials

There are two basic steps involved in the multiplication of monomials. First, we rewrite the products using the commutative and associative properties. Then, we simplify by multiplying coefficients and adding exponents of like bases.

 EXAMPLES Multiply:

1. $(-3x^2)(4x^3) = (-3 \cdot 4)(x^2 \cdot x^3)$ **Commutative and associative properties**

$$= -12x^5$$ **Multiply coefficients, add exponents**

2. $\left(\frac{4}{5}x^5y^2\right)(10x^3y) = \left(\frac{4}{5}10\right)(x^5x^3)(y^2y)$ **Commutative and associative properties**

$$= 8x^8y^3$$ **Multiply coefficients, add exponents**

You can see that in each case the work was the same—multiply coefficients and add exponents of the same base. We can expect division of monomials to proceed in a similar way. Since our properties are consistent, division of monomials will result in division of coefficients and subtraction of exponents of like bases.

 EXAMPLES Divide:

3. $\dfrac{15x^3}{3x^2} = \dfrac{15}{3} \cdot \dfrac{x^3}{x^2}$ **Write as separate fractions**

$$= 5x$$ **Divide coefficients, subtract exponents**

4. $\dfrac{39x^2y^3}{3xy^5} = \dfrac{39}{3} \cdot \dfrac{x^2}{x} \cdot \dfrac{y^3}{y^5}$ **Write as separate fractions**

$$= 13x \cdot \dfrac{1}{y^2}$$ **Divide coefficients, subtract exponents**

$$= \dfrac{13x}{y^2}$$ **Write answer as a single fraction**

In Example 4, the expression $\frac{y^3}{y^5}$ simplifies to $\frac{1}{y^2}$ because of Property 4 for exponents and the definition of negative exponents. If we were to show all the work in this simplification process, it would look like this:

$$\frac{y^3}{y^5} = y^{3-5} \qquad \text{Property 4 for exponents}$$

$$= y^{-2} \qquad \text{Subtraction}$$

$$= \frac{1}{y^2} \qquad \text{Definition of negative exponents}$$

The point of this explanation is this: Even though we may not show all the steps when simplifying an expression involving exponents, the result we obtain still can be justified using the properties of exponents. We have not introduced any new properties in Example 4; we have just not shown the details of each simplification.

EXAMPLE 5   Divide:

$$\frac{25a^5b^3}{50a^2b^7} = \frac{25}{50} \cdot \frac{a^5}{a^2} \cdot \frac{b^3}{b^7} \qquad \text{Write as separate fractions}$$

$$= \frac{1}{2} \cdot a^3 \cdot \frac{1}{b^4} \qquad \text{Divide coefficients, subtract exponents}$$

$$= \frac{a^3}{2b^4} \qquad \text{Write answer as a single fraction} \quad \text{}$$

Notice in Example 5 that dividing 25 by 50 results in $\frac{1}{2}$. This is the same result we would obtain if we reduced the fraction $\frac{25}{50}$ to lowest terms, and there is no harm in thinking of it that way. Also, notice that the expression $\frac{b^3}{b^7}$ simplifies to $\frac{1}{b^4}$ by Property 4 for exponents and the definition of negative exponents, even though we have not shown the steps involved in doing so.

Multiplication and Division of Numbers Written in Scientific Notation

We multiply and divide numbers written in scientific notation using the same steps we used to multiply and divide monomials.

EXAMPLE 6   Multiply $(4 \times 10^7)(2 \times 10^{-4})$.

SOLUTION   Since multiplication is commutative and associative, we can rearrange the order of these numbers and group them as follows:

$$(4 \times 10^7)(2 \times 10^{-4}) = (4 \times 2)(10^7 \times 10^{-4})$$

$$= 8 \times 10^3$$

Notice that we add exponents, $7 + (-4) = 3$, when we multiply with the same base.

EXAMPLE 7   Divide $\frac{9.6 \times 10^{12}}{3 \times 10^4}$.

SOLUTION   We group the numbers between 1 and 10 separately from the powers of 10 and proceed as we did in Example 6:

$$\frac{9.6 \times 10^{12}}{3 \times 10^4} = \frac{9.6}{3} \times \frac{10^{12}}{10^4}$$

$$= 3.2 \times 10^8$$

Notice that the procedure we used in both of these examples is very similar to multiplication and division of monomials, for which we multiplied or divided coefficients and added or subtracted exponents.

Addition and Subtraction of Monomials

Addition and subtraction of monomials will be almost identical since subtraction is defined as addition of the opposite. With multiplication and division of monomials, the key was rearranging the numbers and variables using the commutative and associative properties. With addition, the key is application of the distributive property. We sometimes use the phrase *combine monomials* to describe addition and subtraction of monomials.

> **DEFINITION** Two terms (monomials) with the same variable part (same variables raised to the same powers) are called **similar** (or *like*) **terms.**

You can add only similar terms. This is because the distributive property (which is the key to addition of monomials) cannot be applied to terms that are not similar.

EXAMPLES   Combine the following monomials.

8. $-3x^2 + 15x^2 = (-3 + 15)x^2$ **Distributive property**
$$= 12x^2 \qquad \textbf{Add coefficients}$$

9. $9x^2y - 20x^2y = (9 - 20)x^2y$ **Distributive property**
$$= -11x^2y \qquad \textbf{Add coefficients}$$

10. $5x^2 + 8y^2$ **In this case we cannot apply the distributive property, so we cannot add the monomials**

The next examples show how we simplify expressions containing monomials when more than one operation is involved.

EXAMPLE 11   Simplify $\dfrac{(6x^4y)(3x^7y^5)}{9x^5y^2}$.

SOLUTION   We begin by multiplying the two monomials in the numerator:

$$\frac{(6x^4y)(3x^7y^5)}{9x^5y^2} = \frac{18x^{11}y^6}{9x^5y^2} \qquad \textbf{Simplify numerator}$$

$$= 2x^6y^4 \qquad \textbf{Divide}$$

EXAMPLE 12   Simplify $\dfrac{(6.8 \times 10^5)(3.9 \times 10^{-7})}{7.8 \times 10^{-4}}$.

SOLUTION   We group the numbers between 1 and 10 separately from the powers of 10:

$$\frac{(6.8)(3.9)}{7.8} \times \frac{(10^5)(10^{-7})}{10^{-4}} = 3.4 \times 10^{5+(-7)-(-4)}$$

$$= 3.4 \times 10^2$$

EXAMPLE 13   Simplify $\dfrac{14x^5}{2x^2} + \dfrac{15x^8}{3x^5}$.

SOLUTION   Simplifying each expression separately and then combining similar terms gives

$$\frac{14x^5}{2x^2} + \frac{15x^8}{3x^5} = 7x^3 + 5x^3 \qquad \textbf{Divide}$$

$$= 12x^3 \qquad \textbf{Add}$$

Our work with exponents and division allows us to multiply fractions and other expressions involving exponents. For example,

$$x^5 \cdot \frac{6}{x^2} = \frac{x^5}{1} \cdot \frac{6}{x^2} = \frac{6x^5}{x^2} = 6x^3$$

It is not necessary to show the intermediate steps in a problem like this. We are showing them here just so you can see why we subtract exponents to get the x^3 in the answer.

EXAMPLES   Apply the distributive property, then simplify, if possible.

14. $x^2\left(1 - \dfrac{6}{x}\right) = x^2 \cdot 1 - x^2 \cdot \dfrac{6}{x} = x^2 - \dfrac{6x^2}{x} = x^2 - 6x$

15. $ab\left(\dfrac{1}{b} - \dfrac{1}{a}\right) = ab \cdot \dfrac{1}{b} - ab \cdot \dfrac{1}{a} = \dfrac{ab}{b} - \dfrac{ab}{a} = a - b$

EXAMPLE 16   A rectangular solid is twice as long as it is wide and one-half as high as it is wide. Write an expression for the volume.

SOLUTION   We begin by making a diagram of the object (Figure 1) with the dimensions labeled as given in the problem.

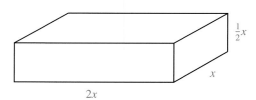

FIGURE 1

The volume is the product of the three dimensions:

$$V = 2x \cdot x \cdot \frac{1}{2}x = x^3$$

The box has the same volume as a cube with side x, as shown in Figure 2.

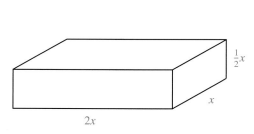

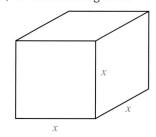

Equal Volumes

FIGURE 2

LINKING OBJECTIVES
AND EXAMPLES

Next to each **objective** we
have listed the examples
that are best described by
that objective.

A	1, 2, 11
B	3–5, 11
C	6, 7, 12
D	8–10

GETTING READY FOR CLASS

*After reading through the preceding section, respond in your own words
and in complete sentences.*

1. What is a monomial?
2. Describe how you would multiply $3x^2$ and $5x^2$.
3. Describe how you would add $3x^2$ and $5x^2$.
4. Describe how you would multiply two numbers written in scientific
   notation.

Problem Set 5.3

Online support materials can be found at www.thomsonedu.com/login

Multiply.

1. $(3x^4)(4x^3)$ $12x^7$

2. $(6x^5)(-2x^2)$ $-12x^7$

3. $(-2y^4)(8y^7)$ $-16y^{11}$

4. $(5y^{10})(2y^5)$ $10y^{15}$

5. $(8x)(4x)$ $32x^2$

6. $(7x)(5x)$ $35x^2$

7. $(10a^3)(10a)(2a^2)$ $200a^6$

8. $(5a^4)(10a)(10a^4)$ $500a^9$

9. $(6ab^2)(-4a^2b)$ $-24a^3b^3$

10. $(-5a^3b)(4ab^4)$ $-20a^4b^5$

▶ 11. $(4x^2y)(3x^3y^3)(2xy^4)$ $24x^6y^8$

12. $(5x^6)(-10xy^4)(-2x^2y^6)$ $100x^9y^{10}$

Divide. Write all answers with positive exponents only.

13. $\dfrac{15x^3}{5x^2}$ $3x$

14. $\dfrac{25x^5}{5x^4}$ $5x$

15. $\dfrac{18y^9}{3y^{12}}$ $\dfrac{6}{y^3}$

16. $\dfrac{24y^4}{-8y^7}$ $-\dfrac{3}{y^3}$

17. $\dfrac{32a^3}{64a^4}$ $\dfrac{1}{2a}$

18. $\dfrac{25a^5}{75a^6}$ $\dfrac{1}{3a}$

▶ 19. $\dfrac{21a^2b^3}{-7ab^5}$ $-\dfrac{3a}{b^2}$

20. $\dfrac{32a^5b^6}{8ab^5}$ $4a^4b$

21. $\dfrac{3x^3y^2z}{27xy^2z^3}$ $\dfrac{x^2}{9z^2}$

22. $\dfrac{5x^5y^4z}{30x^3yz^2}$ $\dfrac{x^2y^3}{6z}$

23. Fill in the table.

a	b	ab	$\dfrac{a}{b}$	$\dfrac{b}{a}$
10	$5x$	$50x$	$\dfrac{2}{x}$	$\dfrac{x}{2}$
$20x^3$	$6x^2$	$120x^5$	$\dfrac{10x}{3}$	$\dfrac{3}{10x}$
$25x^5$	$5x^4$	$125x^9$	$5x$	$\dfrac{1}{5x}$
$3x^{-2}$	$3x^2$	9	$\dfrac{1}{x^4}$	x^4
$-2y^4$	$8y^7$	$-16y^{11}$	$-\dfrac{1}{4y^3}$	$-4y^3$

24. Fill in the table.

a	b	ab	$\dfrac{a}{b}$	$\dfrac{b}{a}$
$10y$	$2y^2$	$20y^3$	$\dfrac{5}{y}$	$\dfrac{y}{5}$
$10y^2$	$2y$	$20y^3$	$5y$	$\dfrac{1}{5y}$
$5y^3$	15	$75y^3$	$\dfrac{y^3}{3}$	$\dfrac{3}{y^3}$
5	$15y^3$	$75y^3$	$\dfrac{1}{3y^3}$	$3y^3$
$4y^{-3}$	$4y^3$	16	$\dfrac{1}{y^6}$	y^6

▢ = Videos available by instructor request
▶ = Online student support materials available at www.thomsonedu.com/login

Find each product. Write all answers in scientific notation.

25. $(3 \times 10^3)(2 \times 10^5)$ 6×10^8

26. $(4 \times 10^8)(1 \times 10^6)$ 4×10^{14}

27. $(3.5 \times 10^4)(5 \times 10^{-6})$ 1.75×10^{-1}

28. $(7.1 \times 10^5)(2 \times 10^{-8})$ 1.42×10^{-2}

29. $(5.5 \times 10^{-3})(2.2 \times 10^{-4})$ 1.21×10^{-6}

30. $(3.4 \times 10^{-2})(4.5 \times 10^{-6})$ 1.53×10^{-7}

Find each quotient. Write all answers in scientific notation.

▶ 31. $\dfrac{8.4 \times 10^5}{2 \times 10^2}$ 4.2×10^3 **32.** $\dfrac{9.6 \times 10^{20}}{3 \times 10^6}$ 3.2×10^{14}

33. $\dfrac{6 \times 10^8}{2 \times 10^{-2}}$ 3×10^{10} **34.** $\dfrac{8 \times 10^{12}}{4 \times 10^{-3}}$ 2×10^{15}

35. $\dfrac{2.5 \times 10^{-6}}{5 \times 10^{-4}}$ 5×10^{-3} **36.** $\dfrac{4.5 \times 10^{-8}}{9 \times 10^{-4}}$ 5×10^{-5}

Combine by adding or subtracting as indicated.

37. $3x^2 + 5x^2$ $8x^2$ **38.** $4x^3 + 8x^3$ $12x^3$

39. $8x^5 - 19x^5$ $-11x^5$ **40.** $75x^6 - 50x^6$ $25x^6$

41. $2a + a - 3a$ 0 **42.** $5a + a - 6a$ 0

▶ 43. $10x^3 - 8x^3 + 2x^3$ $4x^3$ **44.** $7x^5 + 8x^5 - 12x^5$ $3x^5$

45. $20ab^2 - 19ab^2 + 30ab^2$ $31ab^2$

46. $18a^3b^2 - 20a^3b^2 + 10a^3b^2$ $8a^3b^2$

47. Fill in the table.

a	b	ab	a + b
$5x$	$3x$	$15x^2$	$8x$
$4x^2$	$2x^2$	$8x^4$	$6x^2$
$3x^3$	$6x^3$	$18x^6$	$9x^3$
$2x^4$	$-3x^4$	$-6x^8$	$-x^4$
x^5	$7x^5$	$7x^{10}$	$8x^5$

48. Fill in the table.

a	b	ab	a − b
$2y$	$3y$	$6y^2$	$-y$
$-2y$	$3y$	$-6y^2$	$-5y$
$4y^2$	$5y^2$	$20y^4$	$-y^2$
y^3	$-3y^3$	$-3y^6$	$4y^3$
$5y^4$	$7y^4$	$35y^8$	$-2y^4$

Simplify. Write all answers with positive exponents only.

49. $\dfrac{(3x^2)(8x^5)}{6x^4}$ $4x^3$ **50.** $\dfrac{(7x^3)(6x^8)}{14x^5}$ $3x^6$

51. $\dfrac{(9a^2b)(2a^3b^4)}{18a^5b^7}$ $\dfrac{1}{b^2}$ **52.** $\dfrac{(21a^5b)(2a^8b^4)}{14ab}$ $3a^{12}b^4$

53. $\dfrac{(4x^3y^2)(9x^4y^{10})}{(3x^5y)(2x^6y)}$ $\dfrac{6y^{10}}{x^4}$ **54.** $\dfrac{(5x^4y^4)(10x^3y^3)}{(25xy^5)(2xy^7)}$ $\dfrac{x^5}{y^5}$

Simplify each expression, and write all answers in scientific notation.

55. $\dfrac{(6 \times 10^8)(3 \times 10^5)}{9 \times 10^7}$ 2×10^6

56. $\dfrac{(8 \times 10^4)(5 \times 10^{10})}{2 \times 10^7}$ 2×10^8

57. $\dfrac{(5 \times 10^3)(4 \times 10^{-5})}{2 \times 10^{-2}}$ 1×10^1

58. $\dfrac{(7 \times 10^6)(4 \times 10^{-4})}{1.4 \times 10^{-3}}$ 2×10^6

59. $\dfrac{(2.8 \times 10^{-7})(3.6 \times 10^4)}{2.4 \times 10^3}$ 4.2×10^{-6}

60. $\dfrac{(5.4 \times 10^2)(3.5 \times 10^{-9})}{4.5 \times 10^6}$ 4.2×10^{-13}

Simplify.

61. $\dfrac{18x^4}{3x} + \dfrac{21x^7}{7x^4}$ $9x^3$ **62.** $\dfrac{24x^{10}}{6x^4} + \dfrac{32x^7}{8x}$ $8x^6$

63. $\dfrac{45a^6}{9a^4} - \dfrac{50a^8}{2a^6}$ $-20a^2$ **64.** $\dfrac{16a^9}{4a} - \dfrac{28a^{12}}{4a^4}$ $-3a^8$

65. $\dfrac{6x^7y^4}{3x^2y^2} + \dfrac{8x^5y^8}{2y^6}$ $6x^5y^2$ **66.** $\dfrac{40x^{10}y^{10}}{8x^2y^5} + \dfrac{10x^8y^8}{5y^3}$ $7x^8y^5$

Apply the distributive property.

▶ 67. $xy\left(x + \dfrac{1}{y}\right)$ $x^2y + x$ **▶ 68.** $xy\left(y + \dfrac{1}{x}\right)$ $xy^2 + y$

▶ 69. $xy\left(\dfrac{1}{y} + \dfrac{1}{x}\right)$ $x + y$ **▶ 70.** $xy\left(\dfrac{1}{x} - \dfrac{1}{y}\right)$ $y - x$

71. $x^2\left(1 - \dfrac{4}{x^2}\right)$ $x^2 - 4$ **72.** $x^2\left(1 - \dfrac{9}{x^2}\right)$ $x^2 - 9$

73. $x^2\left(1 - \dfrac{1}{x} - \dfrac{6}{x^2}\right)$ $x^2 - x - 6$

74. $x^2\left(1 - \dfrac{5}{x} + \dfrac{6}{x^2}\right)$ $x^2 - 5x + 6$

75. $x^2\left(1 - \dfrac{5}{x}\right)$ $x^2 - 5x$

76. $x^2\left(1 - \dfrac{3}{x}\right)$ $x^2 - 3x$

Foreshadowing Problems
Problems 67–78 get students ready for the equations containing rational expressions, and the complex fractions, they will see later in the book.

77. $x^2\left(1 - \dfrac{8}{x}\right)$ $x^2 - 8x$ **78.** $x^2\left(1 - \dfrac{6}{x}\right)$ $x^2 - 6x$

79. Divide each monomial by $5a^2$.
   a. $10a^2$ 2
   b. $-15a^2b$ $-3b$
   c. $25a^2b^2$ $5b^2$

80. Divide each monomial by $36x^2$.
   a. $6x^2a$ $\dfrac{a}{6}$
   b. $12x^2a$ $\dfrac{a}{3}$
   c. $-6x^2a$ $-\dfrac{a}{6}$

81. Divide each monomial by $8x^2y$.
   a. $24x^3y^2$ $3xy$
   b. $16x^2y^2$ $2y$
   c. $-4x^2y^3$ $-\dfrac{y^2}{2}$

82. Divide each monomial by $7x^2y$.
   a. $21x^3y^2$ $3xy$
   b. $14x^2y^2$ $2y$
   c. $-7x^2y^3$ $-y^2$

Maintaining Your Skills

Find the value of each expression when $x = -2$.

83. $4x$ -8 **84.** $-3x$ 6

85. $-2x + 5$ 9   **86.** $-4x - 1$ 7

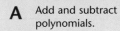

Foreshadowing Problems
Problems 79–82 get students ready for division by monomial later in the chapter.

87. $x^2 + 5x + 6$ 0   **88.** $x^2 - 5x + 6$ 20

For each of the following equations complete the given ordered pairs so each is a solution to the equation, and then use the ordered pairs to graph the equation.

89. $y = 2x + 2$ $(-2, -2)$, $(0, 2)$, $(2, 6)$

90. $y = 2x - 3$ $(-1, -5)$, $(0, -3)$, $(2, 1)$

91. $y = \dfrac{1}{3}x + 1$ $(-3, 0)$, $(0, 1)$, $(3, 2)$

92. $y = \dfrac{1}{2}x - 2$ $(-2, -3)$, $(0, -2)$, $(2, -1)$

Getting Ready for the Next Section

Simplify.

93. $3 - 8$ -5 **94.** $-5 + 7$ 2

95. $-1 + 7$ 6   **96.** $1 - 8$ -7

97. $3(5)^2 + 1$ 76   **98.** $3(-2)^2 - 5(-2) + 4$ 26

99. $2x^2 + 4x^2$ $6x^2$ **100.** $3x^2 - x^2$ $2x^2$

101. $-5x + 7x$ $2x$ **102.** $x - 2x$ $-x$

103. $-(2x + 9)$ $-2x - 9$

104. $-(4x^2 - 2x - 6)$ $-4x^2 + 2x + 6$

105. Find the value of $2x + 3$ when $x = 4$. 11

106. Find the value of $(3x)^2$ when $x = 3$. 81

5.4   Addition and Subtraction of Polynomials

OBJECTIVES

A Add and subtract polynomials.

B Find the value of a polynomial for a given value of the variable.

In this section we will extend what we learned in Section 5.3 to expressions called polynomials. We begin this section with the definition of a polynomial.

> **DEFINITION** A **polynomial** is a finite sum of monomials (terms).

Examples The following are polynomials:
$$3x^2 + 2x + 1 \qquad 15x^2y + 21xy^2 - y^2 \qquad 3a - 2b + 4c - 5d$$

Polynomials can be further classified by the number of terms they contain. A polynomial with two terms is called a binomial. If it has three terms, it is a trinomial. As stated before, a monomial has only one term.

> **DEFINITION** The **degree** of a polynomial in one variable is the highest power to which the variable is raised.

Examples

$3x^5 + 2x^3 + 1$ **A trinomial of degree 5**

$2x + 1$ **A binomial of degree 1**

$3x^2 + 2x + 1$ **A trinomial of degree 2**

$3x^5$ **A monomial of degree 5**

-9 **A monomial of degree 0**

There are no new rules for adding one or more polynomials. We rely only on our previous knowledge. Here are some examples.

 EXAMPLE 1 Add $(2x^2 - 5x + 3) + (4x^2 + 7x - 8)$.

SOLUTION We use the commutative and associative properties to group similar terms together and then apply the distributive property to add:

$$(2x^2 - 5x + 3) + (4x^2 + 7x - 8)$$

$$= (2x^2 + 4x^2) + (-5x + 7x) + (3 - 8) \qquad \textbf{Commutative and associative properties}$$

$$= (2 + 4)x^2 + (-5 + 7)x + (3 - 8) \qquad \textbf{Distributive property}$$

$$= 6x^2 + 2x - 5 \qquad \textbf{Addition}$$

The results here indicate that to add two polynomials, we add coefficients of similar terms.

 EXAMPLE 2 Add $x^2 + 3x + 2x + 6$.

SOLUTION The only similar terms here are the two middle terms. We combine them as usual to get

$$x^2 + 3x + 2x + 6 = x^2 + 5x + 6$$

You will recall from Chapter 1 the definition of subtraction: $a - b = a + (-b)$. To subtract one expression from another, we simply add its opposite. The letters a and b in the definition can each represent polynomials. The opposite of a polynomial is the polynomial with opposite terms. When you subtract one polynomial from another you subtract each of its terms.

 EXAMPLE 3 Subtract $(3x^2 + x + 4) - (x^2 + 2x + 3)$.

SOLUTION To subtract $x^2 + 2x + 3$, we change the sign of each of its terms and add. If you are having trouble remembering why we do this, remember that we can think of $-(x^2 + 2x + 3)$ as $-1(x^2 + 2x + 3)$. If we distribute the -1 across $x^2 + 2x + 3$, we get $-x^2 - 2x - 3$:

$$(3x^2 + x + 4) - (x^2 + 2x + 3)$$

$$= 3x^2 + x + 4 - x^2 - 2x - 3 \qquad \textbf{Take the opposite of each term in the second polynomial}$$

$$= (3x^2 - x^2) + (x - 2x) + (4 - 3)$$

$$= 2x^2 - x + 1$$

EXAMPLE 4   Subtract $-4x^2 + 5x - 7$ from $x^2 - x - 1$.

SOLUTION   The polynomial $x^2 - x - 1$ comes first, then the subtraction sign, and finally the polynomial $-4x^2 + 5x - 7$ in parentheses.

$$(x^2 - x - 1) - (-4x^2 + 5x - 7)$$

$$= x^2 - x - 1 + 4x^2 - 5x + 7 \qquad \textbf{Take the opposite of each term in}$$
$$\textbf{the second polynomial}$$

$$= (x^2 + 4x^2) + (-x - 5x) + (-1 + 7)$$

$$= 5x^2 - 6x + 6$$

The last topic we want to consider in this section is finding the value of a polynomial for a given value of the variable.

To find the value of the polynomial $3x^2 + 1$ when x is 5, we replace x with 5 and simplify the result:

When $x = 5$

the polynomial $3x^2 + 1$

becomes $3(5)^2 + 1 = 3(25) + 1$

$$= 75 + 1 = 76$$

There are two important points to remember when adding or subtracting polynomials. First, to add or subtract two polynomials, you always add or subtract *coefficients* of similar terms. Second, the exponents never increase in value when you are adding or subtracting similar terms.

 EXAMPLE 5   Find the value of $3x^2 - 5x + 4$ when $x = -2$.

SOLUTION

When $x = -2$

the polynomial $3x^2 - 5x + 4$

becomes $3(-2)^2 - 5(-2) + 4 = 3(4) + 10 + 4$

$$= 12 + 10 + 4 = 26$$

LINKING OBJECTIVES AND EXAMPLES

Next to each **objective** we have listed the examples that are best described by that objective.

A   1–4

B   5

GETTING READY FOR CLASS

After reading through the preceding section, respond in your own words and in complete sentences.

1. What are similar terms?
2. What is the degree of a polynomial?
3. Describe how you would subtract one polynomial from another.
4. How would you find the value of $3x^2 - 5x + 4$ when x is -2?

Identify each of the following polynomials as a trinomial, binomial, or monomial, and give the degree in each case.

1. $2x^3 - 3x^2 + 1$ Trinomial, 3

2. $4x^2 - 4x + 1$ Trinomial, 2

3. $5 + 8a - 9a^3$ Trinomial, 3

4. $6 + 12x^3 + x^4$ Trinomial, 4

5. $2x - 1$ Binomial, 1

6. $4 + 7x$ Binomial, 1

7. $45x^2 - 1$ Binomial, 2

8. $3a^3 + 8$ Binomial, 3

9. $7a^2$ Monomial, 2

10. $90x$ Monomial, 1

11. -4 Monomial, 0

12. 56 Monomial, 0

Perform the following additions and subtractions.

13. $(2x^2 + 3x + 4) + (3x^2 + 2x + 5)$ $5x^2 + 5x + 9$

14. $(x^2 + 5x + 6) + (x^2 + 3x + 4)$ $2x^2 + 8x + 10$

15. $(3a^2 - 4a + 1) + (2a^2 - 5a + 6)$ $5a^2 - 9a + 7$

16. $(5a^2 - 2a + 7) + (4a^2 - 3a + 2)$ $9a^2 - 5a + 9$

17. $x^2 + 4x + 2x + 8$ $x^2 + 6x + 8$

18. $x^2 + 5x - 3x - 15$ $x^2 + 2x - 15$

19. $6x^2 - 3x - 10x + 5$ $6x^2 - 13x + 5$

20. $10x^2 + 30x - 2x - 6$ $10x^2 + 28x - 6$

21. $x^2 - 3x + 3x - 9$ $x^2 - 9$

22. $x^2 - 5x + 5x - 25$ $x^2 - 25$

23. $3y^2 - 5y - 6y + 10$ $3y^2 - 11y + 10$

24. $y^2 - 18y + 2y - 12$ $y^2 - 16y - 12$

25. $(6x^3 - 4x^2 + 2x) + (9x^2 - 6x + 3)$ $6x^3 + 5x^2 - 4x + 3$

26. $(5x^3 + 2x^2 + 3x) + (2x^2 + 5x + 1)$ $5x^3 + 4x^2 + 8x + 1$

27. $\left(\frac{2}{3}x^2 - \frac{1}{5}x - \frac{3}{4}\right) + \left(\frac{4}{3}x^2 - \frac{4}{5}x + \frac{7}{4}\right)$ $2x^2 - x + 1$

28. $\left(\frac{3}{8}x^3 - \frac{5}{7}x^2 - \frac{2}{5}\right) + \left(\frac{5}{8}x^3 - \frac{2}{7}x^2 + \frac{7}{5}\right)$ $x^3 - x^2 + 1$

29. $(a^2 - a - 1) - (-a^2 + a + 1)$ $2a^2 - 2a - 2$

30. $(5a^2 - a - 6) - (-3a^2 - 2a + 4)$ $8a^2 + a - 10$

31. $\left(\frac{5}{9}x^3 + \frac{1}{3}x^2 - 2x + 1\right) - \left(\frac{2}{3}x^3 + x^2 + \frac{1}{2}x - \frac{3}{4}\right)$

$-\frac{1}{9}x^3 - \frac{2}{3}x^2 - \frac{5}{2}x + \frac{7}{4}$

32. $\left(4x^3 - \frac{2}{5}x^2 + \frac{3}{8}x - 1\right) - \left(\frac{9}{2}x^3 + \frac{1}{4}x^2 - x + \frac{5}{6}\right)$

$-\frac{1}{2}x^3 - \frac{13}{20}x^2 + \frac{11}{8}x - \frac{11}{6}$

33. $(4y^2 - 3y + 2) + (5y^2 + 12y - 4) - (13y^2 - 6y + 20)$

$-4y^2 + 15y - 22$

34. $(2y^2 - 7y - 8) - (6y^2 + 6y - 8) + (4y^2 - 2y + 3)$

$-15y + 3$

Simplify.

35. $(x^2 - 5x) - (x^2 - 3x)$ $-2x$

36. $(-2x + 8) - (-2x + 6)$ 2

37. $(6x^2 - 11x) - (6x^2 - 15x)$ $4x$

38. $(10x^2 - 3x) - (10x^2 - 50x)$ $47x$

39. $(x^3 + 3x^2 + 9x) - (3x^2 + 9x + 27)$ $x^3 - 27$

40. $(x^3 + 2x^2 + 4x) - (2x^2 + 4x - 8)$ $x^3 + 8$

41. $(x^3 + 4x^2 + 4x) + (2x^2 + 8x + 8)$ $x^3 + 6x^2 + 12x + 8$

42. $(x^3 + 2x^2 + x) + (x^2 + 2x + 1)$ $x^3 + 3x^2 + 3x + 1$

43. $(x^2 - 4) - (x^2 - 4x + 4)$ $4x - 8$

44. $(4x^2 - 9) - (4x^2 - 12x + 9)$ $12x - 18$

45. Subtract $10x^2 + 23x - 50$ from $11x^2 - 10x + 13$.
$x^2 - 33x + 63$

46. Subtract $2x^2 - 3x + 5$ from $4x^2 - 5x + 10$. $2x^2 - 2x + 5$

47. Subtract $3y^2 + 7y - 15$ from $11y^2 + 11y + 11$.
$8y^2 + 4y + 26$

48. Subtract $15y^2 - 8y - 2$ from $3y^2 - 3y + 2$.
$-12y^2 + 5y + 4$

49. Add $50x^2 - 100x - 150$ to $25x^2 - 50x + 75$.
$75x^2 - 150x - 75$

50. Add $7x^2 - 8x + 10$ to $-8x^2 + 2x - 12$. $-x^2 - 6x - 2$

51. Subtract $2x + 1$ from the sum of $3x - 2$ and $11x + 5$.
$12x + 2$

52. Subtract $3x - 5$ from the sum of $5x + 2$ and $9x - 1$.
$11x + 6$

53. Find the value of the polynomial $x^2 - 2x + 1$ when x is 3.   4

54. Find the value of the polynomial $(x - 1)^2$ when x is 3.   4

55. Find the value of $100p^2 - 1,300p + 4,000$ when
 a. $p = 5$ 0
 b. $p = 8$ 0

56. Find the value of $100p^2 - 800p + 1,200$ when
 a. $p = 2$ 0
 b. $p = 6$ 0

= Videos available by instructor request

▶ = Online student support materials available at www.thomsonedu.com/login

57. Find the value of $600 + 1,000x - 100x^2$ when
  a. $x = 8$ 2,200
  b. $x = -2$ −1,800

58. Find the value of $500 + 800x - 100x^2$ when
  a. $x = 6$ 1,700
  b. $x = -1$ −400

Applying the Concepts

59. Packaging A crystal ball with a diameter of 6 inches is being packaged for shipment. If the crystal ball is placed inside a circular cylinder with radius 3 inches and height 6 inches, how much volume will need to be filled with padding? (The volume of a sphere with radius r is $\frac{4}{3}\pi r^3$, and the volume of a right circular cylinder with radius r and height h is $\pi r^2 h$.)   18 π inches³

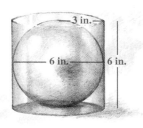

60. Packaging Suppose the circular cylinder of Problem 45 has a radius of 4 inches and a height of 7 inches. How much volume will need to be filled with padding?   76 π inches³

Maintaining Your Skills

61. $3x(-5x)$ $-15x^2$
62. $-3x(-7x)$ $21x^2$
63. $2x(3x^2)$ $6x^3$
64. $x^2(3x)$ $3x^3$
65. $3x^2(2x^2)$ $6x^4$
66. $4x^2(2x^2)$ $8x^4$

Getting Ready for the Next Section

Simplify.

67. $(-5)(-1)$ 5
68. $3(-4)$ -12
69. $(-1)(6)$ -6
70. $(-7)8$ -56
71. $(5x)(-4x)$ $-20x^2$
72. $(3x)(2x)$ $6x^2$
73. $3x(-7)$ $-21x$
74. $3x(-1)$ $-3x$
75. $5x + (-3x)$ $2x$
76. $-3x - 10x$ $-13x$

Multiply.

77. $3(2x - 6)$ $6x - 18$
78. $-4x(x + 5)$ $-4x^2 - 20x$

5.5 Multiplication with Polynomials

OBJECTIVES

A Multiply a monomial with a polynomial.

B Multiply two binomials.

C Multiply two polynomials.

We begin our discussion of multiplication of polynomials by finding the product of a monomial and a trinomial.

 EXAMPLE 1   Multiply $3x^2(2x^2 + 4x + 5)$.

SOLUTION   Applying the distributive property gives us

$$3x^2(2x^2 + 4x + 5) = 3x^2(2x^2) + 3x^2(4x) + 3x^2(5) \quad \textbf{Distributive property}$$

$$= 6x^4 + 12x^3 + 15x^2 \quad \textbf{Multiplication}$$

The distributive property is the key to multiplication of polynomials. We can use it to find the product of any two polynomials. There are some shortcuts we can use in certain situations, however. Let's look at an example that involves the product of two binomials.

 **EXAMPLE 2** Multiply $(3x - 5)(2x - 1)$.

SOLUTION

$$(3x - 5)(2x - 1) = 3x(2x - 1) - 5(2x - 1)$$
$$= 3x(2x) + 3x(-1) + (-5)(2x) + (-5)(-1)$$
$$= 6x^2 - 3x - 10x + 5$$
$$= 6x^2 - 13x + 5$$

If we look closely at the second and third lines of work in this example, we can see that the terms in the answer come from all possible products of terms in the first binomial with terms in the second binomial. This result is generalized as follows.

> **Rule** To multiply any two polynomials, multiply each term in the first with each term in the second.

There are two ways we can put this rule to work.

FOIL Method

If we look at the original problem in Example 2 and then at the answer, we see that the first term in the answer came from multiplying the first terms in each binomial:

$$3x \cdot 2x = 6x^2 \qquad \text{FIRST}$$

The middle term in the answer came from adding the products of the two outside terms and the two inside terms in each binomial:

$$3x(-1) = -3x \qquad \text{OUTSIDE}$$
$$-5(2x) = \underline{-10x} \qquad \text{INSIDE}$$
$$-13x$$

The last term in the answer came from multiplying the two last terms:

$$-5(-1) = 5 \qquad \text{LAST}$$

To summarize the FOIL method, we will multiply another two binomials.

 EXAMPLE 3 Multiply $(2x + 3)(5x - 4)$.

SOLUTION

$$(2x + 3)(5x - 4) = \underbrace{2x(5x)}_{\text{FIRST}} + \underbrace{2x(-4)}_{\text{OUTSIDE}} + \underbrace{3(5x)}_{\text{INSIDE}} + \underbrace{3(-4)}_{\text{LAST}}$$

$$= 10x^2 - 8x + 15x - 12$$
$$= 10x^2 + 7x - 12$$

With practice $-8x + 15x = 7x$ can be done mentally.

COLUMN Method

The FOIL method can be applied only when multiplying two binomials. To find products of polynomials with more than two terms, we use what is called the COLUMN method.

The COLUMN method of multiplying two polynomials is very similar to long multiplication with whole numbers. It is just another way of finding all possible products of terms in one polynomial with terms in another polynomial.

 EXAMPLE 4   Multiply $(2x + 3)(3x^2 - 2x + 1)$.

SOLUTION

$$
\begin{array}{r}
3x^2 - 2x + 1 \\
2x + 3 \\
\hline
6x^3 - 4x^2 + 2x \\
9x^2 - 6x + 3 \\
\hline
6x^3 + 5x^2 - 4x + 3
\end{array}
$$

$\leftarrow 2x(3x^2 - 2x + 1)$
$\leftarrow 3(3x^2 - 2x + 1)$
$\leftarrow$ **Add similar terms**

It will be to your advantage to become very fast and accurate at multiplying polynomials. You should be comfortable using either method. The following examples illustrate the three types of multiplication.

 EXAMPLES   Multiply:

5. $4a^2(2a^2 - 3a + 5) = 4a^2(2a^2) + 4a^2(-3a) + 4a^2(5)$
$\qquad\qquad\qquad\qquad = 8a^4 - 12a^3 + 20a^2$

6. $(x - 2)(y + 3) = x(y) + x(3) + (-2)(y) + (-2)(3)$
$\qquad\qquad\qquad\quad$ F $\qquad$ O $\qquad$ I $\qquad$ L
$\qquad\qquad\qquad = xy + 3x - 2y - 6$

7. $(x + y)(a - b) = x(a) + x(-b) + y(a) + y(-b)$
$\qquad\qquad\qquad\quad$ F $\qquad$ O $\qquad$ I $\qquad$ L
$\qquad\qquad\qquad = xa - xb + ya - yb$

8. $(5x - 1)(2x + 6) = 5x(2x) + 5x(6) + (-1)(2x) + (-1)(6)$
$\qquad\qquad\qquad\qquad$ F $\qquad$ O $\qquad$ I $\qquad$ L
$\qquad\qquad\qquad\quad = 10x^2 + 30x + (-2x) + (-6)$
$\qquad\qquad\qquad\quad = 10x^2 + 28x - 6$

 EXAMPLE 9   The length of a rectangle is 3 more than twice the width. Write an expression for the area of the rectangle.

SOLUTION We begin by drawing a rectangle and labeling the width with x. Since the length is 3 more than twice the width, we label the length with $2x + 3$.

$2x + 3$

x

Since the area A of a rectangle is the product of the length and width, we write our formula for the area of this rectangle as

$$A = x(2x + 3)$$
$$A = 2x^2 + 3x \qquad \textbf{Multiply}$$

Revenue

Suppose that a store sells x items at p dollars per item. The total amount of money obtained by selling the items is called the *revenue*. It can be found by multiplying the number of items sold, x, by the price per item, p. For example, if 100 items are sold for $6 each, the revenue is $100(6) = \$600$. Similarly, if 500 items are sold for $8 each, the total revenue is $500(8) = \$4,000$. If we denote the revenue with the letter R, then the formula that relates R, x, and p is

Revenue = (number of items sold)(price of each item)

In symbols: $R = xp$.

EXAMPLE 10 A store selling diskettes for home computers knows from past experience that it can sell x diskettes each day at a price of p dollars per diskette, according to the equation $x = 800 - 100p$. Write a formula for the daily revenue that involves only the variables R and p.

SOLUTION From our previous discussion we know that the revenue R is given by the formula

$$R = xp$$

But, since $x = 800 - 100p$, we can substitute $800 - 100p$ for x in the revenue equation to obtain

$$R = (800 - 100p)p$$
$$R = 800p - 100p^2$$

This last formula gives the revenue, R, in terms of the price, p.

LINKING OBJECTIVES AND EXAMPLES

Next to each **objective** we have listed the examples that are best described by that objective.

A	1, 5
B	2, 3, 6–8
C	1–10

GETTING READY FOR CLASS

After reading through the preceding section, respond in your own words and in complete sentences.

1. How do we multiply two polynomials?
2. Describe how the distributive property is used to multiply a monomial and a polynomial.
3. Describe how you would use the FOIL method to multiply two binomials.
4. Show how the product of two binomials can be a trinomial.

Multiply the following by applying the distributive property.

1. $2x(3x + 1)$ $6x^2 + 2x$

2. $4x(2x - 3)$ $8x^2 - 12x$

▶ **3.** $2x^2(3x^2 - 2x + 1)$ $6x^4 - 4x^3 + 2x^2$

4. $5x(4x^3 - 5x^2 + x)$ $20x^4 - 25x^3 + 5x^2$

5. $2ab(a^2 - ab + 1)$ $2a^3b - 2a^2b^2 + 2ab$

6. $3a^2b(a^3 + a^2b^2 + b^3)$ $3a^5b + 3a^4b^3 + 3a^2b^4$

7. $y^2(3y^2 + 9y + 12)$ $3y^4 + 9y^3 + 12y^2$

8. $5y(2y^2 - 3y + 5)$ $10y^3 - 15y^2 + 25y$

9. $4x^2y(2x^3y + 3x^2y^2 + 8y^3)$ $8x^5y^2 + 12x^4y^3 + 32x^2y^4$

10. $6xy^3(2x^2 + 5xy + 12y^2)$ $12x^3y^3 + 30x^2y^4 + 72xy^5$

Multiply the following binomials. You should do about half the problems using the FOIL method and the other half using the COLUMN method. Remember, you want to be comfortable using both methods.

▶ **11.** $(x + 3)(x + 4)$ $x^2 + 7x + 12$

12. $(x + 2)(x + 5)$ $x^2 + 7x + 10$

13. $(x + 6)(x + 1)$ $x^2 + 7x + 6$ **14.** $(x + 1)(x + 4)$ $x^2 + 5x + 4$

15. $\left(x + \dfrac{1}{2}\right)\left(x + \dfrac{3}{2}\right)$ $x^2 + 2x + \frac{3}{4}$

16. $\left(x + \dfrac{3}{5}\right)\left(x + \dfrac{2}{5}\right)$ $x^2 + x + \frac{6}{25}$

17. $(a + 5)(a - 3)$ $a^2 + 2a - 15$

18. $(a - 8)(a + 2)$ $a^2 - 6a - 16$

19. $(x - a)(y + b)$ $xy + bx - ay - ab$

20. $(x + a)(y - b)$ $xy - xb + ay - ab$

21. $(x + 6)(x - 6)$ $x^2 - 36$ **22.** $(x + 3)(x - 3)$ $x^2 - 9$

23. $\left(y + \dfrac{5}{6}\right)\left(y - \dfrac{5}{6}\right)$ $y^2 - \frac{25}{36}$ **24.** $\left(y - \dfrac{4}{7}\right)\left(y + \dfrac{4}{7}\right)$ $y^2 - \frac{16}{49}$

25. $(2x - 3)(x - 4)$ $2x^2 - 11x + 12$

26. $(3x - 5)(x - 2)$ $3x^2 - 11x + 10$

27. $(a + 2)(2a - 1)$ $2a^2 + 3a - 2$

28. $(a - 6)(3a + 2)$ $3a^2 - 16a - 12$

▶ **29.** $(2x - 5)(3x - 2)$ $6x^2 - 19x + 10$

30. $(3x + 6)(2x - 1)$ $6x^2 + 9x - 6$

31. $(2x + 3)(a + 4)$ $2ax + 8x + 3a + 12$

32. $(2x - 3)(a - 4)$ $2ax - 8x - 3a + 12$

33. $(5x - 4)(5x + 4)$ $25x^2 - 16$

34. $(6x + 5)(6x - 5)$ $36x^2 - 25$

35. $\left(2x - \dfrac{1}{2}\right)\left(x + \dfrac{3}{2}\right)$ $2x^2 + \frac{5}{2}x - \frac{3}{4}$

36. $\left(4x - \dfrac{3}{2}\right)\left(x + \dfrac{1}{2}\right)$ $4x^2 + \frac{1}{2}x - \frac{3}{4}$

37. $(1 - 2a)(3 - 4a)$ $3 - 10a + 8a^2$

38. $(1 - 3a)(3 + 2a)$ $3 - 7a - 6a^2$

For each of the following problems, fill in the area of each small rectangle and square, and then add the results together to find the indicated product.

39. $(x + 2)(x + 3)$ $(x + 2)(x + 3) = x^2 + 2x + 3x + 6$
$$= x^2 + 5x + 6$$

40. $(x + 4)(x + 5)$ $(x + 4)(x + 5) = x^2 + 4x + 5x + 20$
$$= x^2 + 9x + 20$$

41. $(x + 1)(2x + 2)$ $(x + 1)(2x + 2) = 2x^2 + 4x + 2$

42. $(2x + 1)(2x + 2)$ $(2x + 1)(2x + 2) = 4x^2 + 6x + 2$

= Videos available by instructor request

▶ = Online student support materials available at www.thomsonedu.com/login

Multiply the following.

43. $(a - 3)(a^2 - 3a + 2)$ $a^3 - 6a^2 + 11a - 6$

44. $(a + 5)(a^2 + 2a + 3)$ $a^3 + 7a^2 + 13a + 15$

▶ **45.** $(x + 2)(x^2 - 2x + 4)$ $x^3 + 8$

46. $(x + 3)(x^2 - 3x + 9)$ $x^3 + 27$

47. $(2x + 1)(x^2 + 8x + 9)$ $2x^3 + 17x^2 + 26x + 9$

48. $(3x - 2)(x^2 - 7x + 8)$ $3x^3 - 23x^2 + 38x - 16$

49. $(5x^2 + 2x + 1)(x^2 - 3x + 5)$ $5x^4 - 13x^3 + 20x^2 + 7x + 5$

50. $(2x^2 + x + 1)(x^2 - 4x + 3)$ $2x^4 - 7x^3 + 3x^2 - x + 3$

Multiply.

51. $(x^2 + 3)(2x^2 - 5)$ $2x^4 + x^2 - 15$

52. $(4x^3 - 8)(5x^3 + 4)$ $20x^6 - 24x^3 - 32$

53. $(3a^4 + 2)(2a^2 + 5)$ $6a^6 + 15a^4 + 4a^2 + 10$

54. $(7a^4 - 8)(4a^3 - 6)$ $28a^7 - 42a^4 - 32a^3 + 48$

55. $(x + 3)(x + 4)(x + 5)$ $x^3 + 12x^2 + 47x + 60$

56. $(x - 3)(x - 4)(x - 5)$ $x^3 - 12x^2 + 47x - 60$

Simplify.

▶ **57.** $(x - 3)(x - 2) + 2$ $x^2 - 5x + 8$

▶ **58.** $(2x - 5)(3x + 2) - 4$ $6x^2 - 11x - 14$

▶ **59.** $(2x - 3)(4x + 3) + 4$ $8x^2 - 6x - 5$

▶ **60.** $(3x + 8)(5x - 7) + 52$ $15x^2 + 19x - 4$

▶ **61.** $(x + 4)(x - 5) + (-5)(2)$ $x^2 - x - 30$

▶ **62.** $(x + 3)(x - 4) + (-4)(2)$ $x^2 - x - 20$

▶ **63.** $2(x - 3) + x(x + 2)$ $x^2 + 4x - 6$

▶ **64.** $5(x + 3) + 1(x + 4)$ $6x + 19$

▶ **65.** $3x(x + 1) - 2x(x - 5)$ $x^2 + 13x$

▶ **66.** $4x(x - 2) - 3x(x - 4)$ $x^2 + 4x$

▶ **67.** $x(x + 2) - 3$ $x^2 + 2x - 3$

▶ **68.** $2x(x - 4) + 6$ $2x^2 - 8x + 6$

▶ **69.** $a(a - 3) + 6$ $a^2 - 3a + 6$

▶ **70.** $a(a - 4) + 8$ $a^2 - 4a + 8$

▶ **71.** Find each product.
   a. $(x + 1)(x - 1)$ $x^2 - 1$
   b. $(x + 1)(x + 1)$ $x^2 + 2x + 1$
   c. $(x + 1)(x^2 + 2x + 1)$ $x^3 + 3x^2 + 3x + 1$
   d. $(x + 1)(x^3 + 3x^2 + 3x + 1)$ $x^4 + 4x^3 + 6x^2 + 4x + 1$

Foreshadowing Problems
Problems 61–70 get students ready for work they will do with rational expressions later in the book.

72. Find each product.
   a. $(x + 1)(x^2 - x + 1)$ $x^3 + 1$
   b. $(x + 2)(x^2 - 2x + 4)$ $x^3 + 8$
   c. $(x + 3)(x^2 - 3x + 9)$ $x^3 + 27$
   d. $(x + 4)(x^2 - 4x + 16)$ $x^3 + 64$

73. Find each product.
   a. $(x + 1)(x - 1)$ $x^2 - 1$
   b. $(x + 1)(x - 2)$ $x^2 - x - 2$
   c. $(x + 1)(x - 3)$ $x^2 - 2x - 3$
   d. $(x + 1)(x - 4)$ $x^2 - 3x - 4$

74. Find each product.
   a. $(x + 2)(x - 2)$ $x^2 - 4$
   b. $(x - 2)(x^2 + 2x + 4)$ $x^3 - 8$
   c. $(x^2 + 4)(x^2 - 4)$ $x^4 - 16$
   d. $(x^3 + 8)(x^3 - 8)$ $x^6 - 64$

If the product of two expressions is 0, then one or both of the expressions must be zero. That is, the only way to multiply and get 0, is to multiply by 0. For each expression below, find all values of x that make the expression 0. (If the expression cannot be 0, say so.)

▶ **75.** $5x$ 0

▶ **77.** $x + 5$ -5

▶ **79.** $(x - 3)(x + 2)$ $-2, 3$

81. $x^2 + 16$ Never 0

▶ **83.** $(x - a)(x - b)$ a, b

76. $3x^2$ 0

78. $x^2 + 5$ Never 0

80. $x(x - 5)$ 0, 5

82. $x^2 - 100$ $-10, 10$

▶ **84.** $x(x + a)$ 0, $-a$

Applying the Concepts

85. **Area** The length of a rectangle is 5 units more than twice the width. Write an expression for the area of the rectangle. $A = x(2x + 5) = 2x^2 + 5x$

$2x + 5$

x

86. **Area** The length of a rectangle is 2 more than three times the width. Write an expression for the area of the rectangle. $A = x(3x + 2) = 3x^2 + 2x$

87. **Area** The width and length of a rectangle are given by two consecutive integers. Write an expression for the area of the rectangle. $A = x(x + 1) = x^2 + x$

▶ **Chalkboard Problems**
Here are a variety of Chalkboard Problems. Do Problem 71 if you want to bring up Pascal's triangle. Problem 72 gets students ready for the next section and also for factoring the sum and difference of two cubes in the next chapter.

88. Area The width and length of a rectangle are given by two consecutive even integers. Write an expression for the area of the rectangle. $A = x(x + 2) = x^2 + 2x$

89. Revenue A store selling typewriter ribbons knows that the number of ribbons it can sell each week, x, is related to the price per ribbon, p, by the equation $x = 1,200 - 100p$. Write an expression for the weekly revenue that involves only the variables R and p. (*Remember:* The equation for revenue is $R = xp$.) $R = (1,200 - 100p)p = 1,200p - 100p^2$

90. Revenue A store selling small portable radios knows from past experience that the number of radios it can sell each week, x, is related to the price per radio, p, by the equation $x = 1,300 - 100p$. Write an expression for the weekly revenue that involves only the variables R and p. $R = (1,300 - 100p)p = 1,300p - 100p^2$

Maintaining Your Skills

Solve the following systems by the graphing method.

91. $x + y = 4$
$x - y = 2$ (3, 1)

92. $x - y = 4$
$x + y = 2$ (3, -1)

Solve each system by the elimination method.

93. $3x + 2y = 1$
$2x + y = 3$ (5, -7)

94. $2x + 3y = -1$
$3x + 5y = -2$ (1, -1)

Solve each system by the substitution method.

95. $x + y = 20$
$y = 5x + 2$ (3, 17)

96. $2x - 6y = 2$
$y = 3x + 1$ $(-\frac{1}{2}, -\frac{1}{2})$

97. Investing A total of $1,200 is invested in two accounts. One of the accounts pays 8% interest annually and the other pays 10% interest annually. If the total amount of interest earned from both accounts for the year is $104, how much is invested in each account? $800 at 8%, $400 at 10%

98. Coin Problem Amy has $1.85 in dimes and quarters. If she has a total of 11 coins, how many of each coin does she have? 6 dimes, 5 quarters

Getting Ready for the Next Section

Simplify.

99. $13 \cdot 13$ 169

100. $3x \cdot 3x$ $9x^2$

101. $2(x)(-5)$ $-10x$

102. $2(2x)(-3)$ $-12x$

103. $6x + (-6x)$ 0

104. $3x + (-3x)$ 0

105. $(2x)(-3) + (2x)(3)$ 0

106. $(2x)(-5y) + (2x)(5y)$ 0

Multiply.

107. $-4(3x - 4)$ $-12x + 16$

108. $-2x(2x + 7)$ $-4x^2 - 14x$

109. $(x - 1)(x + 2)$ $x^2 + x - 2$

110. $(x + 5)(x - 6)$ $x^2 - x - 30$

111. $(x + 3)(x + 3)$ $x^2 + 6x + 9$

112. $(3x - 2)(3x - 2)$ $9x^2 - 12x + 4$

5.6 Binomial Squares and Other Special Products

OBJECTIVES

A Find the square of a binomial.

B Multiply expressions of the form $(a + b)(a - b)$.

In this section we will combine the results of the last section with our definition of exponents to find some special products.

EXAMPLE 1 Find the square of $(3x - 2)$.

SOLUTION To square $(3x - 2)$, we multiply it by itself:

$$(3x - 2)^2 = (3x - 2)(3x - 2) \qquad \text{Definition of exponents}$$
$$= 9x^2 - 6x - 6x + 4 \qquad \text{FOIL method}$$
$$= 9x^2 - 12x + 4 \qquad \text{Combine similar terms}$$

Notice that the first and last terms in the answer are the squares of the first and last terms in the original binomial and that the middle term is twice the product of the two terms in the original binomial.

EXAMPLES

2. $(a + b)^2 = (a + b)(a + b)$
$= a^2 + 2ab + b^2$
3. $(a - b)^2 = (a - b)(a - b)$
$= a^2 - 2ab + b^2$

Note

A very common mistake when squaring binomials is to write
$(a + b)^2 = a^2 + b^2$
which just isn't true. The mistake becomes obvious when we substitute 2 for a and 3 for b:
$(2 + 3)^2 \neq 2^2 + 3^2$
$25 \neq 13$
Exponents do not distribute over addition or subtraction.

Binomial squares having the form of Examples 2 and 3 occur very frequently in algebra. It will be to your advantage to memorize the following rule for squaring a binomial.

> **Rule** The square of a binomial is the sum of the square of the first term, the square of the last term, and twice the product of the two original terms. In symbols this rule is written as follows:
>
> $(x + y)^2 = \qquad x^2 \qquad + \qquad 2xy \qquad + \qquad y^2$
>
> Square of first term Twice product of the two terms Square of last term

EXAMPLES

Multiply using the preceding rule:

			First term squared		*Twice their product*		*Last term squared*		*Answer*
4.	$(x - 5)^2$	$=$	x^2	$+$	$2(x)(-5)$	$+$	25	$=$	$x^2 - 10x + 25$
5.	$(x + 2)^2$	$=$	x^2	$+$	$2(x)(2)$	$+$	4	$=$	$x^2 + 4x + 4$
6.	$(2x - 3)^2$	$=$	$4x^2$	$+$	$2(2x)(-3)$	$+$	9	$=$	$4x^2 - 12x + 9$
7.	$(5x - 4)^2$	$=$	$25x^2$	$+$	$2(5x)(-4)$	$+$	16	$=$	$25x^2 - 40x + 16$

Another special product that occurs frequently is $(a + b)(a - b)$. The only difference in the two binomials is the sign between the two terms. The interesting thing about this type of product is that the middle term is always zero. Here are some examples.

EXAMPLES

Multiply using the FOIL method:

8. $(2x - 3)(2x + 3) = 4x^2 + 6x - 6x - 9$ FOIL method
$= 4x^2 - 9$
9. $(x - 5)(x + 5) = x^2 + 5x - 5x - 25$ FOIL method
$= x^2 - 25$
10. $(3x - 1)(3x + 1) = 9x^2 + 3x - 3x - 1$ FOIL method
$= 9x^2 - 1$

Notice that in each case the middle term is zero and therefore doesn't appear in the answer. The answers all turn out to be the difference of two squares. Here is a rule to help you memorize the result.

> **Rule** When multiplying two binomials that differ only in the sign between their terms, subtract the square of the last term from the square of the first term.
>
> $$(a - b)(a + b) = a^2 - b^2$$

Here are some problems that result in the difference of two squares.

 EXAMPLES Multiply using the preceding rule:

11. $(x + 3)(x - 3) = x^2 - 9$
12. $(a + 2)(a - 2) = a^2 - 4$
13. $(9a + 1)(9a - 1) = 81a^2 - 1$
14. $(2x - 5y)(2x + 5y) = 4x^2 - 25y^2$
15. $(3a - 7b)(3a + 7b) = 9a^2 - 49b^2$

Although all the problems in this section can be worked correctly using the methods in the previous section, they can be done much faster if the two rules are *memorized*. Here is a summary of the two rules:

$$(a + b)^2 = (a + b)(a + b) = a^2 + 2ab + b^2$$
$$(a - b)^2 = (a - b)(a - b) = a^2 - 2ab + b^2$$
$$(a - b)(a + b) = a^2 - b^2$$

 EXAMPLE 16 Write an expression in symbols for the sum of the squares of three consecutive even integers. Then, simplify that expression.

SOLUTION If we let x = the first of the even integers, then $x + 2$ is the next consecutive even integer, and $x + 4$ is the one after that. An expression for the sum of their squares is

$$x^2 + (x + 2)^2 + (x + 4)^2 \qquad \textbf{Sum of squares}$$
$$= x^2 + (x^2 + 4x + 4) + (x^2 + 8x + 16) \qquad \textbf{Expand squares}$$
$$= 3x^2 + 12x + 20 \qquad \textbf{Add similar terms}$$

LINKING OBJECTIVES AND EXAMPLES

Next to each **objective** we have listed the examples that are best described by that objective.

A	1–7
B	8–15

GETTING READY FOR CLASS

After reading through the preceding section, respond in your own words and in complete sentences.

1. Describe how you would square the binomial $a + b$.
2. Explain why $(x + 3)^2$ is not equal to $x^2 + 9$.
3. What kind of products result in the difference of two squares?
4. When multiplied out, how will $(x + 3)^2$ and $(x - 3)^2$ differ?

Perform the indicated operations.

1. $(x - 2)^2$ $x^2 - 4x + 4$ **2.** $(x + 2)^2$ $x^2 + 4x + 4$

▶ **3.** $(a + 3)^2$ $a^2 + 6a + 9$ **4.** $(a - 3)^2$ $a^2 - 6a + 9$

5. $(x - 5)^2$ $x^2 - 10x + 25$ **6.** $(x - 4)^2$ $x^2 - 8x + 16$

7. $\left(a - \dfrac{1}{2}\right)^2$ $a^2 - a + \dfrac{1}{4}$ **8.** $\left(a + \dfrac{1}{2}\right)^2$ $a^2 + a + \dfrac{1}{4}$

9. $(x + 10)^2$ $x^2 + 20x + 100$ **10.** $(x - 10)^2$ $x^2 - 20x + 100$

11. $(a + 0.8)^2$ $a^2 + 1.6a + 0.64$ **12.** $(a - 0.4)^2$ $a^2 - 0.8a + 0.16$

13. $(2x - 1)^2$ $4x^2 - 4x + 1$ **14.** $(3x + 2)^2$ $9x^2 + 12x + 4$

15. $(4a + 5)^2$ $16a^2 + 40a + 25$ **16.** $(4a - 5)^2$ $16a^2 - 40a + 25$

17. $(3x - 2)^2$ $9x^2 - 12x + 4$ **18.** $(2x - 3)^2$ $4x^2 - 12x + 9$

▶ **19.** $(3a + 5b)^2$ $9a^2 + 30ab + 25b^2$

20. $(5a - 3b)^2$ $25a^2 - 30ab + 9b^2$

21. $(4x - 5y)^2$ $16x^2 - 40xy + 25y^2$

22. $(5x + 4y)^2$ $25x^2 + 40xy + 16y^2$

23. $(7m + 2n)^2$ $49m^2 + 28mn + 4n^2$

24. $(2m - 7n)^2$ $4m^2 - 28mn + 49n^2$

25. $(6x - 10y)^2$ $36x^2 - 120xy + 100y^2$

26. $(10x + 6y)^2$ $100x^2 + 120xy + 36y^2$

27. $(x^2 + 5)^2$ $x^4 + 10x^2 + 25$ **28.** $(x^2 + 3)^2$ $x^4 + 6x^2 + 9$

29. $(a^2 + 1)^2$ $a^4 + 2a^2 + 1$ **30.** $(a^2 - 2)^2$ $a^4 - 4a^2 + 4$

▶ **31.** $\left(y + \dfrac{3}{2}\right)^2$ $y^2 + 3y + \dfrac{9}{4}$ ▶ **32.** $\left(y - \dfrac{3}{2}\right)^2$ $y^2 - 3y + \dfrac{9}{4}$

▶ **33.** $\left(a + \dfrac{1}{2}\right)^2$ $a^2 + a + \dfrac{1}{4}$ ▶ **34.** $\left(a - \dfrac{5}{2}\right)^2$ $a^2 - 5a + \dfrac{25}{4}$

▶ **35.** $\left(x + \dfrac{3}{4}\right)^2$ $x^2 + \dfrac{3}{2}x + \dfrac{9}{16}$ ▶ **36.** $\left(x - \dfrac{3}{8}\right)^2$ $x^2 - \dfrac{3}{4}x + \dfrac{9}{64}$

▶ **37.** $\left(t + \dfrac{1}{5}\right)^2$ $t^2 + \dfrac{2}{5}t + \dfrac{1}{25}$ ▶ **38.** $\left(t - \dfrac{3}{5}\right)^2$ $t^2 - \dfrac{6}{5}t + \dfrac{9}{25}$

Comparing Expressions Fill in each table.

39.

x	$(x + 3)^2$	$x^2 + 9$	$x^2 + 6x + 9$
1	16	10	16
2	25	13	25
3	36	18	36
4	49	25	49

Foreshadowing Problems
Problems 31–38 get students ready for the completing the square problems later in the book.

40.

x	$(x - 5)^2$	$x^2 + 25$	$x^2 - 10x + 25$
1	16	26	16
2	9	29	9
3	4	34	4
4	1	41	1

41.

a	b	$(a + b)^2$	$a^2 + b^2$	$a^2 + ab + b^2$	$a^2 + 2ab + b^2$
1	1	4	2	3	4
3	5	64	34	49	64
3	4	49	25	37	49
4	5	81	41	61	81

42.

a	b	$(a - b)^2$	$a^2 - b^2$	$a^2 - 2ab + b^2$
2	1	1	3	1
5	2	9	21	9
2	5	9	−21	9
4	3	1	7	1

Multiply.

43. $(a + 5)(a - 5)$ $a^2 - 25$ **44.** $(a - 6)(a + 6)$ $a^2 - 36$

45. $(y - 1)(y + 1)$ $y^2 - 1$ **46.** $(y - 2)(y + 2)$ $y^2 - 4$

47. $(9 + x)(9 - x)$ $81 - x^2$ **48.** $(10 - x)(10 + x)$ $100 - x^2$

49. $(2x + 5)(2x - 5)$ $4x^2 - 25$ **50.** $(3x + 5)(3x - 5)$ $9x^2 - 25$

51. $\left(4x + \dfrac{1}{3}\right)\left(4x - \dfrac{1}{3}\right)$ $16x^2 - \dfrac{1}{9}$

52. $\left(6x + \dfrac{1}{4}\right)\left(6x - \dfrac{1}{4}\right)$ $36x^2 - \dfrac{1}{16}$

53. $(2a + 7)(2a - 7)$ $4a^2 - 49$

54. $(3a + 10)(3a - 10)$ $9a^2 - 100$

▶ **55.** $(6 - 7x)(6 + 7x)$ $36 - 49x^2$

56. $(7 - 6x)(7 + 6x)$ $49 - 36x^2$

57. $(x^2 + 3)(x^2 - 3)$ $x^4 - 9$ **58.** $(x^2 + 2)(x^2 - 2)$ $x^4 - 4$

59. $(a^2 + 4)(a^2 - 4)$ $a^4 - 16$ **60.** $(a^2 + 9)(a^2 - 9)$ $a^4 - 81$

61. $(5y^4 - 8)(5y^4 + 8)$ $25y^8 - 64$

62. $(7y^5 + 6)(7y^5 - 6)$ $49y^{10} - 36$

= Videos available by instructor request

▶ = Online student support materials available at www.thomsonedu.com/login

Multiply and simplify.

63. $(x + 3)(x - 3) + (x - 5)(x + 5)$ $2x^2 - 34$

64. $(x - 7)(x + 7) + (x - 4)(x + 4)$ $2x^2 - 65$

65. $(2x + 3)^2 - (4x - 1)^2$ $-12x^2 + 20x + 8$

66. $(3x - 5)^2 - (2x + 3)^2$ $5x^2 - 42x + 16$

67. $(a + 1)^2 - (a + 2)^2 + (a + 3)^2$ $a^2 + 4a + 6$

68. $(a - 1)^2 + (a - 2)^2 - (a - 3)^2$ $a^2 - 4$

69. $(2x + 3)^3$ $8x^3 + 36x^2 + 54x + 27$

70. $(3x - 2)^3$ $27x^3 - 54x^2 + 36x - 8$

71. Find the value of each expression when x is 6.
    a. $x^2 - 25$ 11
    b. $(x - 5)^2$ 1
    c. $(x + 5)(x - 5)$ 11

72. Find the value of each expression when x is 5.
    a. $x^2 - 9$ 16
    b. $(x - 3)^2$ 4
    c. $(x + 3)(x - 3)$ 16

73. Evaluate each expression when x is -2.
    a. $(x + 3)^2$ 1
    b. $x^2 + 9$ 13
    c. $x^2 + 6x + 9$ 1

74. Evaluate each expression when x is -3.
    a. $(x + 2)^2$ 1
    b. $x^2 + 4$ 13
    c. $x^2 + 4x + 4$ 1

Applying the Concepts

75. **Number Problem** Write an expression for the sum of the squares of two consecutive integers. Then, simplify that expression. $x^2 + (x + 1)^2 = 2x^2 + 2x + 1$

76. **Number Problem** Write an expression for the sum of the squares of two consecutive odd integers. Then, simplify that expression. $x^2 + (x + 2)^2 = 2x^2 + 4x + 4$

77. **Number Problem** Write an expression for the sum of the squares of three consecutive integers. Then, simplify that expression.
    $x^2 + (x + 1)^2 + (x + 2)^2 = 3x^2 + 6x + 5$

78. **Number Problem** Write an expression for the sum of the squares of three consecutive odd integers. Then, simplify that expression.
    $x^2 + (x + 2)^2 + (x + 4)^2 = 3x^2 + 12x + 20$

79. **Area** We can use the concept of area to further justify our rule for squaring a binomial. The length of each side of the square shown in the figure is $a + b$. (The longer line segment has length a and the shorter line segment has length b.) The area of the whole square is $(a + b)^2$. However, the whole area is the sum of the areas of the two smaller squares and the two smaller rectangles that make it up. Write the area of the two smaller squares and the two smaller rectangles and then add them together to verify the formula $(a + b)^2 = a^2 + 2ab + b^2$. $a^2 + ab + ba + b^2 = a^2 + 2ab + b^2$

80. **Area** The length of each side of the large square shown in the figure is $x + 5$. Therefore, its area is $(x + 5)^2$. Find the area of the two smaller squares and the two smaller rectangles that make up the large square, then add them together to verify the formula $(x + 5)^2 = x^2 + 10x + 25$. $x^2 + 5x + 5x + 25 = x^2 + 10x + 25$

Chalkboard Problems
Problems 71–74 can be used to reinforce the formulas from this section. They also give you a chance to talk about the common mistakes students make with these problems.

Maintaining Your Skills

Solve each system by graphing.

81. $x + y = 2$
$x - y = 4$ (3, −1)

82. $x + y = 1$
$x - y = -3$ (−1, 2)

83. $y = 2x + 3$
$y = -2x - 1$ (−1, 1)

84. $y = 2x - 1$
$y = -2x + 3$ (1, 1)

Getting Ready for the Next Section

Simplify each expression (divide).

85. $\dfrac{10x^3}{5x}$ $2x^2$

86. $\dfrac{15x^2}{5x}$ $3x$

87. $\dfrac{3x^2}{3}$ x^2

88. $\dfrac{4x^2}{2}$ $2x^2$

89. $\dfrac{9x^2}{3x}$ $3x$

90. $\dfrac{3x^4}{9x^2}$ $\dfrac{x^2}{3}$

91. $\dfrac{24x^3y^2}{8x^2y}$ $3xy$

92. $-\dfrac{4x^2y^3}{8x^2y}$ $-\dfrac{y^2}{2}$

93. $\dfrac{15x^2y}{3xy}$ $5x$

94. $\dfrac{21xy^2}{3xy}$ $7y$

95. $\dfrac{35a^6b^8}{70a^2b^{10}}$ $\dfrac{a^4}{2b^2}$

96. $\dfrac{75a^2b^6}{25a^4b^3}$ $\dfrac{3b^3}{a^2}$

5.7 Dividing a Polynomial by a Monomial

OBJECTIVES

A Divide a polynomial by a monomial.

To divide a polynomial by a monomial, we will use the definition of division and apply the distributive property. Follow the steps in this example closely.

EXAMPLE 1   Divide $10x^3 - 15x^2$ by $5x$.

SOLUTION

$$\frac{10x^3 - 15x^2}{5x} = (10x^3 - 15x^2)\frac{1}{5x}$$

Division by $5x$ is the same as multiplication by $\frac{1}{5x}$

$$= 10x^3\left(\frac{1}{5x}\right) - 15x^2\left(\frac{1}{5x}\right)$$

Distribute $\frac{1}{5x}$ to both terms

$$= \frac{10x^3}{5x} - \frac{15x^2}{5x}$$

Multiplication by $\frac{1}{5x}$ is the same as division by $5x$

$$= 2x^2 - 3x$$

Division of monomials as done in Section 5.3

If we were to leave out the first steps, the problem would look like this:

$$\frac{10x^3 - 15x^2}{5x} = \frac{10x^3}{5x} - \frac{15x^2}{5x}$$

$$= 2x^2 - 3x$$

The problem is much shorter and clearer this way. You may leave out the first two steps from Example 1 when working problems in this section. They are part of Example 1 only to help show you why the following rule is true.

> **Rule** To divide a polynomial by a monomial, simply divide each term in the polynomial by the monomial.

Here are some further examples using our rule for division of a polynomial by a monomial.

 EXAMPLE 2   Divide $\dfrac{3x^2 - 6}{3}$.

SOLUTION   We begin by writing the 3 in the denominator under each term in the numerator. Then we simplify the result:

$$\frac{3x^2 - 6}{3} = \frac{3x^2}{3} - \frac{6}{3} \qquad \textbf{Divide each term in the numerator by 3}$$

$$= x^2 - 2 \qquad \textbf{Simplify} \qquad$$

 EXAMPLE 3   Divide $\dfrac{4x^2 - 2}{2}$.

SOLUTION   Dividing each term in the numerator by 2, we have

$$\frac{4x^2 - 2}{2} = \frac{4x^2}{2} - \frac{2}{2} \qquad \textbf{Divide each term in the numerator by 2}$$

$$= 2x^2 - 1 \qquad \textbf{Simplify} \qquad$$

 EXAMPLE 4   Find the quotient of $27x^3 - 9x^2$ and $3x$.

SOLUTION   We again are asked to divide the first polynomial by the second one:

$$\frac{27x^3 - 9x^2}{3x} = \frac{27x^3}{3x} - \frac{9x^2}{3x} \qquad \textbf{Divide each term by 3x}$$

$$= 9x^2 - 3x \qquad \textbf{Simplify} \qquad$$

 EXAMPLE 5   Divide $(15x^2y - 21xy^2) \div (-3xy)$.

SOLUTION   This is the same type of problem we have shown in the first four examples; it is just worded a little differently. Note that when we divide each term in the first polynomial by $-3xy$, the negative sign must be taken into account:

$$\frac{15x^2y - 21xy^2}{-3xy} = \frac{15x^2y}{-3xy} - \frac{21xy^2}{-3xy} \qquad \textbf{Divide each term by } -3xy$$

$$= -5x - (-7y) \qquad \textbf{Simplify}$$

$$= -5x + 7y \qquad \textbf{Simplify} \qquad$$

 EXAMPLE 6   Divide $\dfrac{24x^3y^2 + 16x^2y^2 - 4x^2y^3}{8x^2y}$.

SOLUTION   Writing $8x^2y$ under each term in the numerator and then simplifying, we have

$$\frac{24x^3y^2 + 16x^2y^2 - 4x^2y^3}{8x^2y} = \frac{24x^3y^2}{8x^2y} + \frac{16x^2y^2}{8x^2y} - \frac{4x^2y^3}{8x^2y}$$

$$= 3xy + 2y - \frac{y^2}{2} \qquad$$

From the examples in this section, it is clear that to divide a polynomial by a monomial, we must divide each term in the polynomial by the monomial. Often, students taking algebra for the first time will make the following mistake:

$$\frac{x + \cancel{2}}{\cancel{2}} = x + 1 \qquad \textbf{Mistake}$$

The mistake here is in not dividing both terms in the numerator by 2. The correct way to divide $x + 2$ by 2 looks like this:

$$\frac{x + 2}{2} = \frac{x}{2} + \frac{2}{2} = \frac{x}{2} + 1 \qquad \textbf{Correct}$$

LINKING OBJECTIVES AND EXAMPLES

Next to each **objective** we have listed the examples that are best described by that objective.

A   1–6

GETTING READY FOR CLASS

After reading through the preceding section, respond in your own words and in complete sentences.

1. What property of real numbers is the key to dividing a polynomial by a monomial?

2. Describe how you would divide a polynomial by 5x.

3. Is our answer in Example 6 a polynomial?

4. Explain the mistake in the problem $\frac{x + 2}{2} = x + 1$.

Problem Set 5.7

Online support materials can be found at www.thomsonedu.com/login

Divide the following polynomials by $5x$.

1. $5x^2 - 10x$ $x - 2$

2. $10x^3 - 15x$ $2x^2 - 3$

3. $15x - 10x^3$ $3 - 2x^2$

4. $50x^3 - 20x^2$ $10x^2 - 4x$

5. $25x^2y - 10xy$ $5xy - 2y$

6. $15xy^2 + 20x^2y$ $3y^2 + 4xy$

7. $35x^5 - 30x^4 + 25x^3$ $7x^4 - 6x^3 + 5x^2$

8. $40x^4 - 30x^3 + 20x^2$ $8x^3 - 6x^2 + 4x$

9. $50x^5 - 25x^3 + 5x$ $10x^4 - 5x^2 + 1$

10. $75x^6 + 50x^3 - 25x$ $15x^5 + 10x^2 - 5$

Divide the following by $-2a$.

11. $8a^2 - 4a$ $-4a + 2$

12. $a^3 - 6a^2$ $-\frac{a^2}{2} + 3a$

13. $16a^5 + 24a^4$ $-8a^4 - 12a^3$

14. $30a^6 + 20a^3$ $-15a^5 - 10a^2$

15. $8ab + 10a^2$ $-4b - 5a$

16. $6a^2b - 10ab^2$ $-3ab + 5b^2$

17. $12a^3b - 6a^2b^2 + 14ab^3$ $-6a^2b + 3ab^2 - 7b^3$

18. $4ab^3 - 16a^2b^2 - 22a^3b$ $-2b^3 + 8ab^2 + 11a^2b$

19. $a^2 + 2ab + b^2$ $-\frac{a}{2} - b - \frac{b^2}{2a}$

20. $a^2b - 2ab^2 + b^3$ $-\frac{ab}{2} + b^2 - \frac{b^3}{2a}$

Perform the following divisions (find the following quotients).

21. $\frac{6x + 8y}{2}$ $3x + 4y$

22. $\frac{9x - 3y}{3}$ $3x - y$

23. $\frac{7y - 21}{-7}$ $-y + 3$

24. $\frac{14y - 12}{2}$ $7y - 6$

25. $\frac{2x^2 + 16x - 18}{2}$ $x^2 + 8x - 9$

26. $\frac{3x^2 - 3x - 18}{3}$ $x^2 - x - 6$

27. $\frac{3y^2 - 9y + 3}{3}$ $y^2 - 3y + 1$

28. $\frac{2y^2 - 8y + 2}{2}$ $y^2 - 4y + 1$

29. $\frac{x^2y - x^3y^2}{-x^2y}$ $-1 + xy$

30. $\frac{ab + a^2b^2}{ab}$ $1 + ab$

= Videos available by instructor request

▶ = Online student support materials available at www.thomsonedu.com/login

31. $\dfrac{a^2b^2 - ab^2}{-ab^2}$ $-a + 1$ **32.** $\dfrac{a^2b^2c + ab^2c^2}{abc}$ $ab + bc$

53. $\dfrac{(x + 5)^2 + (x + 5)(x - 5)}{2x}$ $x + 5$

33. $\dfrac{x^3 - 3x^2y + xy^2}{x}$ $x^2 - 3xy + y^2$

54. $\dfrac{(x - 4)^2 + (x + 4)(x - 4)}{2x}$ $x - 4$

34. $\dfrac{x^2 - 3xy^2 + xy^3}{x}$ $x - 3y^2 + y^3$

▶ **55.** Find the value of each expression when x is 2.

    a. $2x + 3$ 7

35. $\dfrac{10a^2 - 15a^2b + 25a^2b^2}{5a^2}$ $2 - 3b + 5b^2$

    b. $\dfrac{10x + 15}{5}$ 7

36. $\dfrac{11a^2b^2 - 33ab}{-11ab}$ $-ab + 3$

    c. $10x + 3$ 23

37. $\dfrac{26x^2y^2 - 13xy}{-13xy}$ $-2xy + 1$

▶ **56.** Find the value of each expression when x is 5.

    a. $3x + 2$ 17

38. $\dfrac{6x^2y^2 - 3xy}{6xy}$ $xy - \dfrac{1}{2}$

    b. $\dfrac{6x^2 + 4x}{2x}$ 17

39. $\dfrac{4x^2y^2 - 2xy}{4xy}$ $xy - \dfrac{1}{2}$

    c. $6x^2 + 2$ 152

40. $\dfrac{6x^2a + 12x^2b - 6x^2c}{36x^2}$ $\dfrac{a}{6} + \dfrac{b}{3} - \dfrac{c}{6}$

▶ **57.** Evaluate each expression for $x = 10$.

    a. $\dfrac{3x + 8}{2}$ 19

41. $\dfrac{5a^2x - 10ax^2 + 15a^2x^2}{20a^2x^2}$ $\dfrac{1}{4x} - \dfrac{1}{2a} + \dfrac{3}{4}$

    b. $3x + 4$ 34

42. $\dfrac{12ax - 9bx + 18cx}{6x^2}$ $\dfrac{2a}{x} - \dfrac{3b}{2x} + \dfrac{3c}{x}$

    c. $\dfrac{3}{2}x + 4$ 19

▶ **43.** $\dfrac{16x^5 + 8x^2 + 12x}{12x^3}$ $\dfrac{4x^2}{3} + \dfrac{2}{3x} + \dfrac{1}{x^2}$

▶ **58.** Evaluate each expression for $x = 10$.

    a. $\dfrac{5x - 6}{2}$ 22

44. $\dfrac{27x^2 - 9x^3 - 18x^4}{-18x^3}$ $-\dfrac{3}{2x} + \dfrac{1}{2} + x$

    b. $5x - 3$ 47

Divide. Assume all variables represent positive numbers.

    c. $\dfrac{5}{2}x - 3$ 22

45. $\dfrac{9a^{5m} - 27a^{3m}}{3a^{2m}}$ $3a^{3m} - 9a^m$

59. Find the value of each expression when x is 2.

    a. $2x^2 - 3x$ 2

46. $\dfrac{26a^{3m} - 39a^{5m}}{13a^{3m}}$ $2 - 3a^{2m}$

    b. $\dfrac{10x^3 - 15x^2}{5x}$ 2

47. $\dfrac{10x^{5m} - 25x^{3m} + 35x^m}{5x^m}$ $2x^{4m} - 5x^{2m} + 7$

60. Find the value of each expression when a is 3.

    a. $-4a + 2$ -10

48. $\dfrac{18x^{2m} + 24x^{4m} - 30x^{6m}}{6x^{2m}}$ $3 + 4x^{2m} - 5x^{4m}$

    b. $\dfrac{8a^2 - 4a}{-2a}$ -10

Simplify each numerator, and then divide.

Maintaining Your Skills

49. $\dfrac{2x^3(3x + 2) - 3x^2(2x - 4)}{2x^2}$ $3x^2 - x + 6$

Solve each system of equations by the elimination method.

50. $\dfrac{5x^2(6x - 3) + 6x^3(3x - 1)}{3x}$ $6x^3 + 8x^2 - 5x$

61. $x + y = 6$
     $x - y = 8$ $(7, -1)$

62. $2x + y = 5$
     $-x + y = -4$ $(3, -1)$

51. $\dfrac{(x + 2)^2 - (x - 2)^2}{2x}$ 4

63. $2x - 3y = -5$
     $x + y = 5$ $(2, 3)$

64. $2x - 4y = 10$
     $3x - 2y = -1$ $(-3, -4)$

52. $\dfrac{(x - 3)^2 - (x + 3)^2}{3x}$ -4

▶ **Chalkboard Problems**

Problems 55–58 reinforce the correct way to divide by a monomial. They also give you another opportunity to show that only common factors can be eliminated from a quotient.

Solve each system by the substitution method.

65. $x + y = 2$
$y = 2x - 1$ (1, 1)

66. $2x - 3y = 4$
$x = 3y - 1$ (5, 2)

67. $4x + 2y = 8$
$y = -2x + 4$
Lines coincide.

68. $4x + 2y = 8$
$y = -2x + 5$
Lines are parallel.

Multiply.

73. $(x - 3)x$ $x^2 - 3x$

74. $(x - 3)(-2)$ $-2x + 6$

75. $2x^2(x - 5)$ $2x^3 - 10x^2$

76. $10x(x - 5)$ $10x^2 - 50x$

Getting Ready for the Next Section

Divide.

69. $27\overline{)3,962}$ $146\frac{20}{27}$

70. $13\overline{)18,780}$ $1,444\frac{8}{13}$

71. $\dfrac{2x^2 + 5x}{x}$ $2x + 5$

72. $\dfrac{7x^2 + 9x^3 + 3x^7}{x^3}$ $\dfrac{7}{x} + 9 + 3x^4$

Subtract.

77. $(x^2 - 5x) - (x^2 - 3x)$ $-2x$

78. $(2x^3 + 0x^2) - (2x^3 - 10x^2)$ $10x^2$

79. $(-2x + 8) - (-2x + 6)$ 2

80. $(4x - 14) - (4x - 10)$ -4

5.8 Dividing a Polynomial by a Polynomial

OBJECTIVES

A Divide a polynomial by a polynomial.

Since long division for polynomials is very similar to long division with whole numbers, we will begin by reviewing a division problem with whole numbers. You may realize when looking at Example 1 that you don't have a very good idea why you proceed as you do with long division. What you do know is that the process always works. We are going to approach the explanations in this section in much the same manner; that is, we won't always be sure why the steps we will use are important, only that they always produce the correct result.

 EXAMPLE 1   Divide $27\overline{)3,962}$.

SOLUTION

$$
\begin{array}{r}
1 \quad\leftarrow \textbf{Estimate 27 into 39} \\
27\overline{)3,962} \\
2\,7 \quad\leftarrow \textbf{Multiply } 1 \times 27 = 27 \\
\hline
1\,2 \quad\leftarrow \textbf{Subtract } 39 - 27 = 12
\end{array}
$$

$$
\begin{array}{r}
1 \\
27\overline{)3,962} \\
2\,7{\downarrow} \\
\hline
1\,26 \quad\leftarrow \textbf{Bring down the 6}
\end{array}
$$

These are the four basic steps in long division. Estimate, multiply, subtract, and bring down the next term. To finish the problem, we simply perform the same four steps again:

$$
\begin{array}{r}
14 \quad \leftarrow \textbf{4 is the estimate} \\
27\overline{)3{,}962} \\
\underline{2\ 7\downarrow} \\
1\ 26 \\
\underline{1\ 08} \quad \leftarrow \textbf{Multiply to get 108} \\
182 \quad \leftarrow \textbf{Subtract to get 18, then bring down the 2}
\end{array}
$$

One more time.

$$
\begin{array}{r}
146 \quad \leftarrow \textbf{6 is the estimate} \\
27\overline{)3{,}962} \\
\underline{2\ 7} \\
1\ 26 \\
\underline{1\ 08} \\
182 \\
\underline{162} \quad \leftarrow \textbf{Multiply to get 162} \\
20 \quad \leftarrow \textbf{Subtract to get 20}
\end{array}
$$

Since there is nothing left to bring down, we have our answer.

$$
\frac{3{,}962}{27} = 146 + \frac{20}{27} \qquad \text{or} \qquad 146\frac{20}{27}
$$

Here is how it works with polynomials.

EXAMPLE 2

Divide $\dfrac{x^2 - 5x + 8}{x - 3}$.

SOLUTION

$$
\begin{array}{r}
x \qquad\qquad \leftarrow \textbf{Estimate } x^2 \div x = x \\
x - 3\overline{)\ x^2 - 5x + 8} \\
\underline{\overset{-}{\cancel{+}}\,x^2 \overset{+}{\cancel{/}}\, 3x} \quad \leftarrow \textbf{Multiply } x(x - 3) = x^2 - 3x \\
- 2x \quad \leftarrow \textbf{Subtract } (x^2 - 5x) - (x^2 - 3x) = -2x
\end{array}
$$

$$
\begin{array}{r}
x \qquad\qquad \\
x - 3\overline{)\ x^2 - 5x + 8} \\
\underline{\overset{-}{\cancel{+}}\,x^2 \overset{+}{\cancel{/}}\, 3x} \\
- 2x + 8 \quad \leftarrow \textbf{Bring down the 8}
\end{array}
$$

Notice that to subtract one polynomial from another, we add its opposite. That is why we change the signs on $x^2 - 3x$ and add what we get to $x^2 - 5x$. (To subtract the second polynomial, simply change the signs and add.)

We perform the same four steps again:

$$
\begin{array}{r}
x - 2 \quad \leftarrow \textbf{-2 is the estimate } (-2x \div x = -2) \\
x - 3\overline{)\ x^2 - 5x + 8} \\
\underline{\overset{-}{\cancel{+}}\,x^2 \overset{+}{\cancel{/}}\, 3x}\ \ \downarrow \\
- 2x + 8 \\
\underline{\overset{+}{\cancel{/}}\,2x \overset{-}{\cancel{/}}\, 6} \quad \leftarrow \textbf{Multiply } -2(x - 3) = -2x + 6 \\
2 \quad \leftarrow \textbf{Subtract } (-2x + 8) - (-2x + 6) = 2
\end{array}
$$

Since there is nothing left to bring down, we have our answer:

$$\frac{x^2 - 5x + 8}{x - 3} = x - 2 + \frac{2}{x - 3}$$

To check our answer, we multiply $(x - 3)(x - 2)$ to get $x^2 - 5x + 6$. Then, adding on the remainder, 2, we have $x^2 - 5x + 8$.

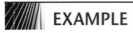

EXAMPLE 3   Divide $\dfrac{6x^2 - 11x - 14}{2x - 5}$.

SOLUTION

$$
\begin{array}{r}
3x + 2 \\
2x - 5{\overline{\smash{\big)}\,6x^2 - 11x - 14}} \\
\underline{\mp 6x^2 \mp 15x} \\
+\ 4x - 14 \\
\underline{\mp\ 4x \mp 10} \\
-\ 4
\end{array}
$$

$$\frac{6x^2 - 11x - 14}{2x - 5} = 3x + 2 + \frac{-4}{2x - 5}$$

One last step is sometimes necessary. The two polynomials in a division problem must both be in descending powers of the variable and cannot skip any powers from the highest power down to the constant term.

EXAMPLE 4   Divide $\dfrac{2x^3 - 3x + 2}{x - 5}$.

SOLUTION   The problem will be much less confusing if we write $2x^3 - 3x + 2$ as $2x^3 + 0x^2 - 3x + 2$. Adding $0x^2$ does not change our original problem.

$$
\begin{array}{r}
2x^2 \quad \leftarrow \textbf{Estimate } 2x^3 \div x = 2x^2 \\
x - 5{\overline{\smash{\big)}\,2x^3 + \ 0x^2 - 3x + 2}} \\
\underline{\mp 2x^3 \mp 10x^2} \quad \leftarrow \textbf{Multiply } 2x^2(x - 5) = 2x^3 - 10x^2 \\
+\ 10x^2 - 3x \quad \leftarrow \textbf{Subtract:}
\end{array}
$$

$(2x^3 + 0x^2) - (2x^3 - 10x^2) = 10x^2$
— **Bring down the next term**

Adding the term $0x^2$ gives us a column in which to write $10x^2$. (Remember, you can add and subtract only similar terms.)

Here is the completed problem:

$$
\begin{array}{r}
2x^2 + 10x + 47 \\
x - 5{\overline{\smash{\big)}\,2x^3 + \ 0x^2 - \ 3x + \ \ 2}} \\
\underline{\mp 2x^3 \mp 10x^2} \\
+\ 10x^2 - \ 3x \\
\underline{\mp 10x^2 \mp 50x} \\
+\ 47x + \ \ 2 \\
\underline{\mp 47x \mp 235} \\
237
\end{array}
$$

Our answer is $\dfrac{2x^3 - 3x + 2}{x - 5} = 2x^2 + 10x + 47 + \dfrac{237}{x - 5}$.

As you can see, long division with polynomials is a mechanical process. Once you have done it correctly a couple of times, it becomes very easy to produce the correct answer.

GETTING READY FOR CLASS

After reading through the preceding section, respond in your own words and in complete sentences.

1. What are the four steps used in long division with whole numbers?
2. How is division of two polynomials similar to long division with whole numbers?
3. What are the four steps used in long division with polynomials?
4. How do we use 0 when dividing the polynomial $2x^3 - 3x + 2$ by $x - 5$?

LINKING OBJECTIVES AND EXAMPLES

Next to each **objective** we have listed the examples that are best described by that objective.

A 2–4

Problem Set 5.8

Online support materials can be found at www.thomsonedu.com/login

Divide.

1. $\dfrac{x^2 - 5x + 6}{x - 3}$ $x - 2$

2. $\dfrac{x^2 - 5x + 6}{x - 2}$ $x - 3$

3. $\dfrac{a^2 + 9a + 20}{a + 5}$ $a + 4$

4. $\dfrac{a^2 + 9a + 20}{a + 4}$ $a + 5$

5. $\dfrac{x^2 - 6x + 9}{x - 3}$ $x - 3$

6. $\dfrac{x^2 + 10x + 25}{x + 5}$ $x + 5$

7. $\dfrac{2x^2 + 5x - 3}{2x - 1}$ $x + 3$

8. $\dfrac{4x^2 + 4x - 3}{2x - 1}$ $2x + 3$

9. $\dfrac{2a^2 - 9a - 5}{2a + 1}$ $a - 5$

10. $\dfrac{4a^2 - 8a - 5}{2a + 1}$ $2a - 5$

11. $\dfrac{x^2 + 5x + 8}{x + 3}$ $x + 2 + \dfrac{2}{x + 3}$

12. $\dfrac{x^2 + 5x + 4}{x + 3}$ $x + 2 + \dfrac{-2}{x + 3}$

13. $\dfrac{a^2 + 3a + 2}{a + 5}$ $a - 2 + \dfrac{12}{a + 5}$

14. $\dfrac{a^2 + 4a + 3}{a + 5}$ $a - 1 + \dfrac{8}{a + 5}$

15. $\dfrac{x^2 + 2x + 1}{x - 2}$ $x + 4 + \dfrac{9}{x - 2}$

16. $\dfrac{x^2 + 6x + 9}{x - 3}$ $x + 9 + \dfrac{36}{x - 3}$

17. $\dfrac{x^2 + 5x - 6}{x + 1}$ $x + 4 + \dfrac{-10}{x + 1}$

18. $\dfrac{x^2 - x - 6}{x + 1}$ $x - 2 + \dfrac{-4}{x + 1}$

19. $\dfrac{a^2 + 3a + 1}{a + 2}$ $a + 1 + \dfrac{-1}{a + 2}$

20. $\dfrac{a^2 - a + 3}{a + 1}$ $a - 2 + \dfrac{5}{a + 1}$

21. $\dfrac{2x^2 - 2x + 5}{2x + 4}$ $x - 3 + \dfrac{17}{2x + 4}$

22. $\dfrac{15x^2 + 19x - 4}{3x + 8}$ $5x - 7 + \dfrac{52}{3x + 8}$

▶ 23. $\dfrac{6a^2 + 5a + 1}{2a + 3}$ $3a - 2 + \dfrac{7}{2a + 3}$

24. $\dfrac{4a^2 + 4a + 3}{2a + 1}$ $2a + 1 + \dfrac{2}{2a + 1}$

25. $\dfrac{6a^3 - 13a^2 - 4a + 15}{3a - 5}$ $2a^2 - a - 3$

26. $\dfrac{2a^3 - a^2 + 3a + 2}{2a + 1}$ $a^2 - a + 2$

Fill in the missing terms in the numerator, and then use long division to find the quotients (see Example 4).

27. $\dfrac{x^3 + 4x + 5}{x + 1}$ $x^2 - x + 5$

28. $\dfrac{x^3 + 4x^2 - 8}{x + 2}$ $x^2 + 2x - 4$

29. $\dfrac{x^3 - 1}{x - 1}$ $x^2 + x + 1$

30. $\dfrac{x^3 + 1}{x + 1}$ $x^2 - x + 1$

31. $\dfrac{x^3 - 8}{x - 2}$ $x^2 + 2x + 4$

32. $\dfrac{x^3 + 27}{x + 3}$ $x^2 - 3x + 9$

= Videos available by instructor request
▶ = Online student support materials available at www.thomsonedu.com/login

▶ **33.** Find the value of each expression when x is 3.

   a. $x^2 + 2x + 4$ 19

   b. $\dfrac{x^3 - 8}{x - 2}$ 19

   c. $x^2 - 4$ 5

▶ **34.** Find the value of each expression when x is 2.

   a. $x^2 - 3x + 9$ 7

   b. $\dfrac{x^3 + 27}{x + 3}$ 7

   c. $x^2 + 9$ 13

▶ **35.** Find the value of each expression when x is 4.

   a. $x + 3$ 7

   b. $\dfrac{x^2 + 9}{x + 3}$ $\frac{25}{7}$

   c. $x - 3 + \dfrac{18}{x + 3}$ $\frac{25}{7}$

▶ **36.** Find the value of each expression when x is 2.

   a. $x + 1$ 3

   b. $\dfrac{x^2 + 1}{x + 1}$ $\frac{5}{3}$

   c. $x - 1 + \dfrac{2}{x + 1}$ $\frac{5}{3}$

Long Division Use the information in the table to find the monthly payment for auto insurance for the cities below. Round to the nearest cent.

USA TODAY Snapshots®

Cities with the most expensive auto insurance annual premiums

Detroit	$5,162
Philadelphia	$4,142
Newark, N.J.	$3,482
Los Angeles	$3,225
New York City	$3,127

Note: Based on drivers over a minimum age with clean driving records and including comprehensive, collision, bodily injury, property damage and uninsured motorist coverage.

Source: Runzheimer International

By Darryl Haralson and Marcy E. Mullins, USA TODAY

From *USA Today.* Copyright 2005. Reprinted with permission.

37. Detroit $430.17   **38.** Philadelphia $345.17

39. Newark, N.J. $290.17   **40.** Los Angeles $268.75

Maintaining Your Skills

Use systems of equations to solve the following word problems.

41. Number Problem The sum of two numbers is 25. One of the numbers is 4 times the other. Find the numbers.   5, 20

42. Number Problem The sum of two numbers is 24. One of the numbers is 3 more than twice the other. Find the numbers.   7, 17

43. Investing Suppose you have a total of $1,200 invested in two accounts. One of the accounts pays 8% annual interest and the other pays 9% annual interest. If your total interest for the year is $100, how much money did you invest in each of the accounts? $800 at 8%, $400 at 9%

44. Investing If you invest twice as much money in an account that pays 12% annual interest as you do in an account that pays 11% annual interest, how much do you have in each account if your total interest for a year is $210? $600 at 11%, $1,200 at 12%

45. Money Problem If you have a total of $160 in $5 bills and $10 bills, how many of each type of bill do you have if you have 4 more $10 bills than $5 bills?   8 $5 bills, 12 $10 bills

46. Coin Problem Suppose you have 20 coins worth a total of $2.80. If the coins are all nickels and quarters, how many of each type do you have?   11 nickels, 9 quarters

47. Mixture Problem How many gallons of 20% antifreeze solution and 60% antifreeze solution must be mixed to get 16 gallons of 35% antifreeze solution?   10 gallons of 20%, 6 gallons of 60%

48. Mixture Problem A chemist wants to obtain 80 liters of a solution that is 12% hydrochloric acid. How many liters of 10% hydrochloric acid solution and 20% hydrochloric acid solution should he mix to do so?   64 liters of 10%, 16 liters of 20%

▶ **Chalkboard Problems**

Problems 33–36 reinforce the correct way to do division with polynomials. They also give you another opportunity to show that only common factors can be eliminated from a quotient.

Exponents: Definition and Properties [5.1, 5.2]

EXAMPLES

1. a. $2^3 = 2 \cdot 2 \cdot 2 = 8$
b. $x^5 \cdot x^3 = x^{5+3} = x^8$

c. $\dfrac{x^5}{x^3} = x^{5-3} = x^2$

d. $(3x)^2 = 3^2 \cdot x^2 = 9x^2$

e. $\left(\dfrac{2}{3}\right)^3 = \dfrac{2^3}{3^3} = \dfrac{8}{27}$

f. $(x^5)^3 = x^{5 \cdot 3} = x^{15}$

g. $3^{-2} = \dfrac{1}{3^2} = \dfrac{1}{9}$

Integer exponents indicate repeated multiplications.

$$a^r \cdot a^s = a^{r+s} \qquad \text{To multiply with the same base, you add exponents}$$

$$\dfrac{a^r}{a^s} = a^{r-s} \qquad \text{To divide with the same base, you subtract exponents}$$

$$(ab)^r = a^r \cdot b^r \qquad \text{Exponents distribute over multiplication}$$

$$\left(\dfrac{a}{b}\right)^r = \dfrac{a^r}{b^r} \qquad \text{Exponents distribute over division}$$

$$(a^r)^s = a^{r \cdot s} \qquad \text{A power of a power is the product of the powers}$$

$$a^{-r} = \dfrac{1}{a^r} \qquad \text{Negative exponents imply reciprocals}$$

Multiplication of Monomials [5.3]

2. $(5x^2)(3x^4) = 15x^6$

To multiply two monomials, multiply coefficients and add exponents.

Division of Monomials [5.3]

3. $\dfrac{12x^9}{4x^5} = 3x^4$

To divide two monomials, divide coefficients and subtract exponents.

Scientific Notation [5.1, 5.2, 5.3]

4. $768,000 = 7.68 \times 10^5$
$0.00039 = 3.9 \times 10^{-4}$

A number is in scientific notation when it is written as the product of a number between 1 and 10 and an integer power of 10.

Addition of Polynomials [5.4]

5. $(3x^2 - 2x + 1) + (2x^2 + 7x - 3)$
$\quad = 5x^2 + 5x - 2$

To add two polynomials, add coefficients of similar terms.

Subtraction of Polynomials [5.4]

6. $(3x + 5) - (4x - 3)$
$\quad = 3x + 5 - 4x + 3$
$\quad = -x + 8$

To subtract one polynomial from another, add the opposite of the second to the first.

7. a. $2a^2(5a^2 + 3a - 2)$
   $= 10a^4 + 6a^3 - 4a^2$

  b. $(x + 2)(3x - 1)$
   $= 3x^2 - x + 6x - 2$
   $= 3x^2 + 5x - 2$

  c. $2x^2 - 3x + 4$
   $\underline{3x - 2}$
   $6x^3 - 9x^2 + 12x$
   $\underline{- 4x^2 + 6x - 8}$
   $6x^3 - 13x^2 + 18x - 8$

Multiplication of Polynomials [5.5]

To multiply a polynomial by a monomial, we apply the distributive property. To multiply two binomials we use the FOIL method. In other situations we use the COLUMN method. Each method achieves the same result: To multiply any two polynomials, we multiply each term in the first polynomial by each term in the second polynomial.

Special Products [5.6]

8.
$(x + 3)^2 = x^2 + 6x + 9$
$(x - 3)^2 = x^2 - 6x + 9$
$(x + 3)(x - 3) = x^2 - 9$

$\left.\begin{array}{l} (a + b)^2 = a^2 + 2ab + b^2 \\[6pt] (a - b)^2 = a^2 - 2ab + b^2 \end{array}\right\}$ **Binomial squares**

$(a + b)(a - b) = a^2 - b^2$ **Difference of two squares**

Dividing a Polynomial by a Monomial [5.7]

9. $\dfrac{12x^3 - 18x^2}{6x} = 2x^2 - 3x$

To divide a polynomial by a monomial, divide each term in the polynomial by the monomial.

Long Division with Polynomials [5.8]

10.
$$\begin{array}{r} x - 2 \\ x - 3 \overline{)\ x^2 - 5x + 8} \\ \end{array}$$

Division with polynomials is similar to long division with whole numbers. The steps in the process are estimate, multiply, subtract, and bring down the next term. The divisors in all the long-division problems in this chapter were binomials.

COMMON MISTAKES

1. If a term contains a variable that is raised to a power, then the exponent on the variable is associated only with that variable, unless there are parentheses; that is, the expression $3x^2$ means $3 \cdot x \cdot x$, not $3x \cdot 3x$. It is a mistake to write $3x^2$ as $9x^2$. The only way to end up with $9x^2$ is to start with $(3x)^2$.

2. It is a mistake to add nonsimilar terms. For example, $2x$ and $3x^2$ are nonsimilar terms and therefore cannot be combined; that is, $2x + 3x^2 \neq 5x^3$. If you were to substitute 10 for x in the preceding expression, you would see that the two sides are not equal.

3. It is a mistake to distribute exponents over sums and differences; that is, $(a + b)^2$ is not $a^2 + b^2$. Convince yourself of this by letting $a = 2$ and $b = 3$ and then simplifying both sides.

4. Another common mistake can occur when dividing a polynomial by a monomial. Here is an example:

$$\frac{x + \cancel{2}}{\cancel{2}} = x + 1 \qquad \textbf{Mistake}$$

The mistake here is in not dividing both terms in the numerator by 2. The correct way to divide $x + 2$ by 2 looks like this:

$$\frac{x + 2}{2} = \frac{x}{2} + \frac{2}{2} \qquad \textbf{Correct}$$

$$= \frac{x}{2} + 1$$

Chapter 5 Review Test

The problems below form a comprehensive review of the material in this chapter. They can be used to study for exams. If you would like to take a practice test on this chapter, you can use the odd-numbered problems. Give yourself an hour and work as many of the odd-numbered problems as possible. When you are finished, or when an hour has passed, check your answers with the answers in the back of the book. You can use the even-numbered problems for a second practice test.

The numbers in brackets refer to the sections of the text in which similar problems can be found.

Simplify. [5.1]

1. $(-1)^3$

2. -8^2

3. $\left(\dfrac{3}{7}\right)^2$

4. $y^3 \cdot y^9$

5. $x^{15} \cdot x^7 \cdot x^5 \cdot x^3$

6. $(x^7)^5$

7. $(2^6)^4$

8. $(3y)^3$

9. $(-2xyz)^3$

Simplify each expression. Any answers that contain exponents should contain positive exponents only. [5.2]

10. 7^{-2}

11. $4x^{-5}$

12. $(3y)^{-3}$

13. $\dfrac{a^9}{a^3}$

14. $\left(\dfrac{x^3}{x^5}\right)^2$

15. $\dfrac{x^9}{x^{-6}}$

16. $\dfrac{x^{-7}}{x^{-2}}$

17. $(-3xy)^0$

18. $3^0 - 5^1 + 5^0$

Simplify. Any answers that contain exponents should contain positive exponents only. [5.1, 5.2]

19. $(3x^3y^2)^2$

20. $(2a^3b^2)^4(2a^5b^6)^2$

21. $(-3xy^2)^{-3}$

22. $\dfrac{(b^3)^4(b^2)^5}{(b^7)^3}$

23. $\dfrac{(x^{-3})^3(x^6)^{-1}}{(x^{-5})^{-4}}$

Simplify. Write all answers with positive exponents only. [5.3]

24. $\dfrac{(2x^4)(15x^9)}{6x^6}$

25. $\dfrac{(10x^3y^5)(21x^2y^6)}{(7xy^3)(5x^9y)}$

26. $\dfrac{21a^{10}}{3a^4} - \dfrac{18a^{17}}{6a^{11}}$

27. $\dfrac{8x^8y^3}{2x^3y} - \dfrac{10x^6y^9}{5xy^7}$

Simplify, and write all answers in scientific notation. [5.3]

28. $(3.2 \times 10^3)(2 \times 10^4)$

29. $\dfrac{4.6 \times 10^5}{2 \times 10^{-3}}$

30. $\dfrac{(4 \times 10^6)(6 \times 10^5)}{3 \times 10^8}$

Perform the following additions and subtractions. [5.4]

31. $(3a^2 - 5a + 5) + (5a^2 - 7a - 8)$

32. $(-7x^2 + 3x - 6) - (8x^2 - 4x + 7) + (3x^2 - 2x - 1)$

33. Subtract $8x^2 + 3x - 2$ from $4x^2 - 3x - 2$.

34. Find the value of $2x^2 - 3x + 5$ when $x = 3$.

Multiply. [5.5]

35. $3x(4x - 7)$

36. $8x^3y(3x^2y - 5xy^2 + 4y^3)$

37. $(a + 1)(a^2 + 5a - 4)$

38. $(x + 5)(x^2 - 5x + 25)$

39. $(3x - 7)(2x - 5)$

40. $\left(5y + \dfrac{1}{5}\right)\left(5y - \dfrac{1}{5}\right)$

41. $(a^2 - 3)(a^2 + 3)$

Perform the indicated operations. [5.6]

42. $(a - 5)^2$

43. $(3x + 4)^2$

44. $(y^2 + 3)^2$

45. Divide $10ab + 20a^2$ by $-5a$. [5.7]

46. Divide $40x^5y^4 - 32x^3y^3 - 16x^2y$ by $-8xy$. [5.7]

Divide using long division. [5.8]

47. $\dfrac{x^2 + 15x + 54}{x + 6}$

48. $\dfrac{6x^2 + 13x - 5}{3x - 1}$

49. $\dfrac{x^3 + 64}{x + 4}$

50. $\dfrac{3x^2 - 7x + 10}{3x + 2}$

51. $\dfrac{2x^3 - 7x^2 + 6x + 10}{2x + 1}$

GROUP PROJECT   Discovering Pascal's Triangle

Number of People 3

Time Needed 20 minutes

Equipment Paper and pencils

Background The triangular array of numbers shown here is known as Pascal's triangle, after the French philosopher Blaise Pascal (1623–1662).

```
            1
          1   1
        1   2   1
      1   3   3   1
    1   4   6   4   1
  1   5  10  10   5   1
```

Procedure Look at Pascal's triangle and discover how the numbers in each row of the triangle are obtained from the numbers in the row above it.

1. Once you have discovered how to extend the triangle, write the next two rows.

2. Pascal's triangle can be linked to the Fibonacci sequence by rewriting Pascal's triangle so that the 1s on the left side of the triangle line up under one another, and the other columns are equally spaced to the right of the first column. Rewrite Pascal's triangle as indicated and then look along the diagonals of the new array until you discover how the Fibonacci sequence can be obtained from it.

3. The diagram below shows Pascal's triangle as written in Japanese in 1781. Use your knowledge of Pascal's triangle to translate the numbers written in Japanese into our number system. Then write down the Japanese numbers from 1 to 20.

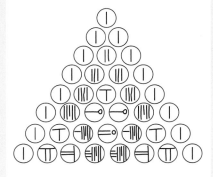

Pascal's triangle in Japanese (1781)

Working Mathematicians

It may seem at times as if all the mathematicians of note lived 100 or more years ago. However, that is not the case. There are mathematicians doing research today who are discovering new mathematical ideas and extending what is known about mathematics.

Use the Internet to find a mathematician working in the field today. Find out what drew them to mathematics in the first place, what it took for them to be successful, and what they like about their career in mathematics. Then summarize your results into an essay that gives anyone reading it a profile of a working mathematician.

Factoring

Matt May/Icon SMI/Corbis

If you watch professional football on television, you will hear the announcers refer to "hang time" when the punter punts the ball. Hang time is the amount of time the ball is in the air, and it depends on only one thing—the initial vertical velocity imparted to the ball by the kicker's foot. We can find the hang time of a football by solving equations. Table 1 shows the equations to solve for hang time, given various initial vertical velocities. Figure 1 is a visual representation of some equations associated with the ones in Table 1. In Figure 1, you can find hang time on the horizontal axis.

TABLE 1
Hang Time for a Football

Initial Vertical Velocity (feet/second)	Equation in Factored Form	Hang Time (seconds)
16	$16t(1 - t) = 0$	1
32	$16t(2 - t) = 0$	2
48	$16t(3 - t) = 0$	3
64	$16t(4 - t) = 0$	4
80	$16t(5 - t) = 0$	5

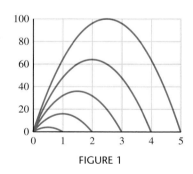

FIGURE 1

▶ Improve your grade and save time!
Go online to **www.thomsonedu.com/login** where you can
• Watch videos of instructors working through the in-text examples
• Follow step-by-step online tutorials of in-text examples and review questions
• Work practice problems
• Check your readiness for an exam by taking a pre-test and exploring the modules recommended in your Personalized Study plan
• Receive help from a live tutor online through vMentor™
Try it out! Log in with an access code or purchase access at **www.ichapters.com**.

The equations in the second column of the table are in what is called "factored form." Once the equation is in factored form, hang time can be read from the second factor. In this chapter we develop techniques that allow us to factor a variety of polynomials. Factoring is the key to solving equations like the ones in Table 1.

The study skills for this chapter are a continuation of the skills from the previous chapter.

1 Continue to Set and Keep a Schedule

Sometimes I find students do well at first and then become overconfident. They begin to put in less time with their homework. Don't do that. Keep to the same schedule that you started with.

2 Increase Effectiveness

You want to become more and more effective with the time you spend on your homework. You want to increase the amount of learning you obtain in the time you have set aside. Increase those activities that you feel are the most beneficial and decrease those that have not given you the results you want.

3 Continue to List Difficult Problems

This study skill was started in Chapter 2. You should continue to list and rework the problems that give you the most difficulty. It is this list that you will use to study for the next exam. Your goal is to go into the next exam knowing that you successfully can work any problem from your list of hard problems.

6.1 The Greatest Common Factor and Factoring by Grouping

OBJECTIVES

A Factor the greatest common factor from a polynomial.

B Factor by grouping.

In Chapter 1 we used the following diagram to illustrate the relationship between multiplication and factoring.

$$\text{Factors} \rightarrow 3 \cdot 5 = 15 \leftarrow \text{Product}$$

Multiplication

Factoring

A similar relationship holds for multiplication of polynomials. Reading the following diagram from left to right, we say the product of the binomials $x + 2$ and $x + 3$ is the trinomial $x^2 + 5x + 6$. However, if we read in the other direction, we can say that $x^2 + 5x + 6$ factors into the product of $x + 2$ and $x + 3$.

$$\text{Factors} \rightarrow (x + 2)(x + 3) = x^2 + 5x + 6 \leftarrow \text{Product}$$

Multiplication

Factoring

In this chapter we develop a systematic method of factoring polynomials.

In this section we will apply the distributive property to polynomials to factor from them what is called the greatest common factor.

> **DEFINITION** The **greatest common factor** for a polynomial is the largest monomial that divides (is a factor of) each term of the polynomial.

We use the term *largest monomial* to mean the monomial with the greatest coefficient and highest power of the variable.

 EXAMPLE 1 Find the greatest common factor for the polynomial:

$$3x^5 + 12x^2$$

SOLUTION The terms of the polynomial are $3x^5$ and $12x^2$. The largest number that divides the coefficients is 3, and the highest power of x that is a factor of x^5 and x^2 is x^2. Therefore, the greatest common factor for $3x^5 + 12x^2$ is $3x^2$; that is, $3x^2$ is the largest monomial that divides each term of $3x^5 + 12x^2$.

 EXAMPLE 2 Find the greatest common factor for:

$$8a^3b^2 + 16a^2b^3 + 20a^3b^3$$

SOLUTION The largest number that divides each of the coefficients is 4. The highest power of the variable that is a factor of a^3b^2, a^2b^3, and a^3b^3 is a^2b^2. The greatest common factor for $8a^3b^2 + 16a^2b^3 + 20a^3b^3$ is $4a^2b^2$. It is the largest monomial that is a factor of each term.

Once we have recognized the greatest common factor of a polynomial, we can apply the distributive property and factor it out of each term. We rewrite the polynomial as the product of its greatest common factor with the polynomial that remains after the greatest common factor has been factored from each term in the original polynomial.

EXAMPLE 3   Factor the greatest common factor from $3x - 15$.

SOLUTION   The greatest common factor for the terms $3x$ and 15 is 3. We can rewrite both $3x$ and 15 so that the greatest common factor 3 is showing in each term. It is important to realize that $3x$ means $3 \cdot x$. The 3 and the x are not "stuck" together:

$$3x - 15 = 3 \cdot x - 3 \cdot 5$$

Now, applying the distributive property, we have:

$$3 \cdot x - 3 \cdot 5 = 3(x - 5)$$

To check a factoring problem like this, we can multiply 3 and $x - 5$ to get $3x - 15$, which is what we started with. Factoring is simply a procedure by which we change sums and differences into products. In this case we changed the difference $3x - 15$ into the product $3(x - 5)$. Note, however, that we have not changed the meaning or value of the expression. The expression we end up with is equivalent to the expression we started with.

 EXAMPLE 4   Factor the greatest common factor from:

$$5x^3 - 15x^2$$

SOLUTION   The greatest common factor is $5x^2$. We rewrite the polynomial as:

$$5x^3 - 15x^2 = 5x^2 \cdot x - 5x^2 \cdot 3$$

Then we apply the distributive property to get:

$$5x^2 \cdot x - 5x^2 \cdot 3 = 5x^2(x - 3)$$

To check our work, we simply multiply $5x^2$ and $(x - 3)$ to get $5x^3 - 15x^2$, which is our original polynomial.

 EXAMPLE 5   Factor the greatest common factor from:

$$16x^5 - 20x^4 - 8x^3$$

SOLUTION   The greatest common factor is $4x^3$. We rewrite the polynomial so we can see the greatest common factor $4x^3$ in each term; then we apply the distributive property to factor it out.

$$16x^5 - 20x^4 - 8x^3 = 4x^3 \cdot 4x^2 - 4x^3 \cdot 5x - 4x^3 \cdot 2$$

$$= 4x^3(4x^2 - 5x - 2)$$

EXAMPLE 6   Factor the greatest common factor from:

$$6x^3y - 18x^2y^2 - 12xy^3$$

SOLUTION   The greatest common factor is $6xy$. We rewrite the polynomial in terms of $6xy$ and then apply the distributive property as follows:

$$6x^3y - 18x^2y^2 - 12xy^3 = 6xy \cdot x^2 - 6xy \cdot 3xy - 6xy \cdot 2y^2$$

$$= 6xy(x^2 - 3xy - 2y^2)$$

EXAMPLE 7   Factor the greatest common factor from:

$$3a^2b - 6a^3b^2 + 9a^3b^3$$

SOLUTION   The greatest common factor is $3a^2b$:

$$3a^2b - 6a^3b^2 + 9a^3b^3 = 3a^2b(1) - 3a^2b(2ab) + 3a^2b(3ab^2)$$

$$= 3a^2b(1 - 2ab + 3ab^2)$$

Factoring by Grouping

To develop our next method of factoring, called *factoring by grouping,* we start by examining the polynomial $xc + yc$. The greatest common factor for the two terms is c. Factoring c from each term we have:

$$xc + yc = c(x + y)$$

> ### Note
> The phrase *greatest common factor* doesn't fit what we are doing here as well as it fit the examples in the beginning of this section. Certainly our original definition of the greatest common factor can be applied to the expression $xc + yc$ to obtain a greatest common factor of c. However, for the expression
> $$x(a + b) + y(a + b)$$
> the common factor $(a + b)$ is not itself a monomial, and so it doesn't fit our definition of greatest common factor. However, the idea of factoring out as much as possible from each term still applies to this expression. The expression $(a + b)$ is common to both terms, and so we factor it from each term using the distributive property.

But suppose that c itself was a more complicated expression, such as $a + b$, so that the expression we were trying to factor was $x(a + b) + y(a + b)$, instead of $xc + yc$. The expression $x(a + b) + y(a + b)$ has $(a + b)$ common to each term. Factoring this common factor from each term looks like this:

$$x(a + b) + y(a + b) = (a + b)(x + y)$$

To see how all of this applies to factoring polynomials, consider the polynomial

$$xy + 3x + 2y + 6$$

There is no greatest common factor other than the number 1. However, if we group the terms together two at a time, we can factor an x from the first two terms and a 2 from the last two terms:

$$xy + 3x + 2y + 6 = x(y + 3) + 2(y + 3)$$

The expression on the right can be thought of as having two terms: $x(y + 3)$ and $2(y + 3)$. Each of these expressions contains the common factor $y + 3$, which can be factored out using the distributive property:

$$x(y + 3) + 2(y + 3) = (y + 3)(x + 2)$$

This last expression is in factored form. The process we used to obtain it is called factoring by grouping. Here are some additional examples.

EXAMPLE 8   Factor $ax + bx + ay + by$.

SOLUTION   We begin by factoring x from the first two terms and y from the last two terms:

$$ax + bx + ay + by = x(a + b) + y(a + b)$$

$$= (a + b)(x + y)$$

To convince yourself that this is factored correctly, multiply the two factors $(a + b)$ and $(x + y)$.

EXAMPLE 9 Factor by grouping: $3ax - 2a + 15x - 10$.

SOLUTION First, we factor a from the first two terms and 5 from the last two terms. Then, we factor $3x - 2$ from the remaining two expressions:

$$3ax - 2a + 15x - 10 = a(3x - 2) + 5(3x - 2)$$
$$= (3x - 2)(a + 5)$$

Again, multiplying $(3x - 2)$ and $(a + 5)$ will convince you that these are the correct factors.

 EXAMPLE 10 Factor $2x^2 + 5ax - 2xy - 5ay$.

SOLUTION From the first two terms we factor x. From the second two terms we must factor $-y$ so that the binomial that remains after we do so matches the binomial produced by the first two terms:

$$2x^2 + 5ax - 2xy - 5ay = x(2x + 5a) - y(2x + 5a)$$
$$= (2x + 5a)(x - y)$$

Another way to accomplish the same result is to use the commutative property to interchange the middle two terms, and then factor by grouping:

$$2x^2 + 5ax - 2xy - 5ay = 2x^2 - 2xy + 5ax - 5ay \quad \textbf{Commutative}$$
$$\textbf{property}$$
$$= 2x(x - y) + 5a(x - y)$$
$$= (x - y)(2x + 5a)$$

This is the same result we obtained previously.

EXAMPLE 11 Factor $6x^2 - 3x - 4x + 2$ by grouping.

SOLUTION The first two terms have $3x$ in common, and the last two terms have either a 2 or a -2 in common. Suppose we factor $3x$ from the first two terms and 2 from the last two terms. We get:

$$6x^2 - 3x - 4x + 2 = 3x(2x - 1) + 2(-2x + 1)$$

We can't go any further because there is no common factor that will allow us to factor further. However, if we factor -2, instead of 2, from the last two terms, our problem is solved:

$$6x^2 - 3x - 4x + 2 = 3x(2x - 1) - 2(2x - 1)$$
$$= (2x - 1)(3x - 2)$$

In this case, factoring -2 from the last two terms gives us an expression that can be factored further.

LINKING OBJECTIVES AND EXAMPLES

Next to each **objective** we have listed the examples that are best described by that objective.

A 1–7

B 8–11

GETTING READY FOR CLASS

After reading through the preceding section, respond in your own words and in complete sentences.

1. What is the greatest common factor for a polynomial?
2. After factoring a polynomial, how can you check your result?
3. When would you try to factor by grouping?
4. What is the relationship between multiplication and factoring?

Problem Set 6.1

Online support materials can be found at www.thomsonedu.com/login

Answers appear in the Instructor's Edition only.

Factor the following by taking out the greatest common factor.

1. $15x + 25$ $5(3x + 5)$
2. $14x + 21$ $7(2x + 3)$
▶ **3.** $6a + 9$ $3(2a + 3)$
4. $8a + 10$ $2(4a + 5)$
5. $4x - 8y$ $4(x - 2y)$
6. $9x - 12y$ $3(3x - 4y)$
7. $3x^2 - 6x - 9$ $3(x^2 - 2x - 3)$
8. $2x^2 + 6x + 4$ $2(x^2 + 3x + 2)$
9. $3a^2 - 3a - 60$ $3(a^2 - a - 20)$
10. $2a^2 - 18a + 28$ $2(a^2 - 9a + 14)$
11. $24y^2 - 52y + 24$ $4(6y^2 - 13y + 6)$
12. $18y^2 + 48y + 32$ $2(9y^2 + 24y + 16)$
13. $9x^2 - 8x^3$ $x^2(9 - 8x)$ **14.** $7x^3 - 4x^2$ $x^2(7x - 4)$
15. $13a^2 - 26a^3$ $13a^2(1 - 2a)$ **16.** $5a^2 - 10a^3$ $5a^2(1 - 2a)$
17. $21x^2y - 28xy^2$ $7xy(3x - 4y)$
18. $30xy^2 - 25x^2y$ $5xy(6y - 5x)$
19. $22a^2b^2 - 11ab^2$ $11ab^2(2a - 1)$
▶ **20.** $15x^3 - 25x^2 + 30x$ $5x(3x^2 - 5x + 6)$
21. $7x^3 + 21x^2 - 28x$ $7x(x^2 + 3x - 4)$
22. $16x^4 - 20x^2 - 16x$ $4x(4x^3 - 5x - 4)$
23. $121y^4 - 11x^4$ $11(11y^4 - x^4)$
24. $25a^4 - 5b^4$ $5(5a^4 - b^4)$
25. $100x^4 - 50x^3 + 25x^2$ $25x^2(4x^2 - 2x + 1)$
26. $36x^5 + 72x^3 - 81x^2$ $9x^2(4x^3 + 8x - 9)$

27. $8a^2 + 16b^2 + 32c^2$ $8(a^2 + 2b^2 + 4c^2)$
28. $9a^2 - 18b^2 - 27c^2$ $9(a^2 - 2b^2 - 3c^2)$
29. $4a^2b - 16ab^2 + 32a^2b^2$ $4ab(a - 4b + 8ab)$
30. $5ab^2 + 10a^2b^2 + 15a^2b$ $5ab(b + 2ab + 3a)$

Factor the following by taking out the greatest common factor.

31. $121a^3b^2 - 22a^2b^3 + 33a^3b^3$ $11a^2b^2(11a - 2b + 3ab)$
32. $20a^4b^3 - 18a^3b^4 + 22a^4b^4$ $2a^3b^3(10a - 9b + 11ab)$
33. $12x^2y^3 - 72x^5y^3 - 36x^4y^4$ $12x^2y^3(1 - 6x^3 - 3x^2y)$
34. $49xy - 21x^2y^2 + 35x^3y^3$ $7xy(7 - 3xy + 5x^2y^2)$

Factor by grouping.

▶ **35.** $xy + 5x + 3y + 15$ $(x + 3)(y + 5)$
36. $xy + 2x + 4y + 8$ $(x + 4)(y + 2)$
37. $xy + 6x + 2y + 12$ $(x + 2)(y + 6)$
38. $xy + 2y + 6x + 12$ $(x + 2)(y + 6)$
▶ **39.** $ab + 7a - 3b - 21$ $(a - 3)(b + 7)$
40. $ab + 3b - 7a - 21$ $(a + 3)(b - 7)$
41. $ax - bx + ay - by$ $(a - b)(x + y)$
42. $ax - ay + bx - by$ $(a + b)(x - y)$
43. $2ax + 6x - 5a - 15$ $(2x - 5)(a + 3)$
44. $3ax + 21x - a - 7$ $(3x - 1)(a + 7)$
45. $3xb - 4b - 6x + 8$ $(b - 2)(3x - 4)$

= Videos available by instructor request
▶ = Online student support materials available at www.thomsonedu.com/login

46. $3xb - 4b - 15x + 20$ $(b - 5)(3x - 4)$

47. $x^2 + ax + 2x + 2a$ $(x + 2)(x + a)$

48. $x^2 + ax + 3x + 3a$ $(x + 3)(x + a)$

49. $x^2 - ax - bx + ab$ $(x - b)(x - a)$

50. $x^2 + ax - bx - ab$ $(x - b)(x + a)$

Factor by grouping. You can group the terms together two at a time or three at a time. Either way will produce the same result.

51. $ax + ay + bx + by + cx + cy$ $(x + y)(a + b + c)$

52. $ax + bx + cx + ay + by + cy$ $(x + y)(a + b + c)$

Factor the following polynomials by grouping the terms together two at a time.

53. $6x^2 + 9x + 4x + 6$ $(3x + 2)(2x + 3)$

54. $6x^2 - 9x - 4x + 6$ $(3x - 2)(2x - 3)$

55. $20x^2 - 2x + 50x - 5$ $(10x - 1)(2x + 5)$

56. $20x^2 + 25x + 4x + 5$ $(4x + 5)(5x + 1)$

57. $20x^2 + 4x + 25x + 5$ $(4x + 5)(5x + 1)$

58. $20x^2 + 4x - 25x - 5$ $(4x - 5)(5x + 1)$

59. $x^3 + 2x^2 + 3x + 6$ $(x + 2)(x^2 + 3)$

60. $x^3 - 5x^2 - 4x + 20$ $(x - 5)(x^2 - 4)$

61. $6x^3 - 4x^2 + 15x - 10$ $(3x - 2)(2x^2 + 5)$

62. $8x^3 - 12x^2 + 14x - 21$ $(2x - 3)(4x^2 + 7)$

63. The greatest common factor of the binomial $3x + 6$ is 3. The greatest common factor of the binomial $2x + 4$ is 2. What is the greatest common factor of their product $(3x + 6)(2x + 4)$ when it has been multiplied out? 6

64. The greatest common factors of the binomials $4x + 2$ and $5x + 10$ are 2 and 5, respectively. What is the greatest common factor of their product $(4x + 2)(5x + 10)$ when it has been multiplied out? 10

65. The following factorization is incorrect. Find the mistake, and correct the right-hand side:

$$12x^2 + 6x + 3 = 3(4x^2 + 2x) \quad 3(4x^2 + 2x + 1)$$

66. Find the mistake in the followzing factorization, and then rewrite the right-hand side correctly:

$$10x^2 + 2x + 6 = 2(5x^2 + 3) \quad 2(5x^2 + x + 3)$$

Maintaining Your Skills

Divide.

67. $\dfrac{y^3 - 16y^2 + 64y}{y}$ $y^2 - 16y + 64$

68. $\dfrac{5x^3 + 35x^2 + 60x}{5x}$ $x^2 + 7x + 12$

69. $\dfrac{-12x^4 + 48x^3 + 144x^2}{-12x^2}$ $x^2 - 4x - 12$

70. $\dfrac{16x^5 + 20x^4 + 60x^3}{4x^3}$ $4x^2 + 5x + 15$

71. $\dfrac{-18y^5 + 63y^4 - 108y^3}{-9y^3}$ $2y^2 - 7y + 12$

72. $\dfrac{36y^6 - 66y^5 + 54y^4}{6y^4}$ $6y^2 - 11y + 9$

Subtract.

73. $(5x^2 + 5x - 4) - (3x^2 - 2x + 7)$ $2x^2 + 7x - 11$

74. $(7x^4 - 4x^2 - 5) - (2x^4 - 4x^2 + 5)$ $5x^4 - 10$

75. Subtract $4x - 5$ from $7x + 3$. $3x + 8$

76. Subtract $3x + 2$ from $-6x + 1$. $-9x - 1$

77. Subtract $2x^2 - 4x$ from $5x^2 - 5$. $3x^2 + 4x - 5$

78. Subtract $6x^2 + 3$ from $2x^2 - 4x$. $-4x^2 - 4x - 3$

Getting Ready for the Next Section

Multiply each of the following.

79. $(x - 7)(x + 2)$ $x^2 - 5x - 14$

80. $(x - 7)(x - 2)$ $x^2 - 9x + 14$

81. $(x - 3)(x + 2)$ $x^2 - x - 6$

82. $(x + 3)(x - 2)$ $x^2 + x - 6$

83. $(x + 3)(x^2 - 3x + 9)$ $x^3 + 27$

84. $(x - 2)(x^2 + 2x + 4)$ $x^3 - 8$

85. $(2x + 1)(x^2 + 4x - 3)$ $2x^3 + 9x^2 - 2x - 3$

86. $(3x + 2)(x^2 - 2x - 4)$ $3x^3 - 4x^2 - 16x - 8$

87. $3x^4(6x^3 - 4x^2 + 2x)$ $18x^7 - 12x^6 + 6x^5$

88. $2x^4(5x^3 + 4x^2 - 3x)$ $10x^7 + 8x^6 - 6x^5$

89. $\left(x + \dfrac{1}{3}\right)\left(x + \dfrac{2}{3}\right)$ $x^2 + x + \dfrac{2}{9}$

90. $\left(x + \dfrac{1}{4}\right)\left(x + \dfrac{3}{4}\right)$ $x^2 + x + \dfrac{3}{16}$

91. $(6x + 4y)(2x - 3y)$ $12x^2 - 10xy - 12y^2$

92. $(8a - 3b)(4a - 5b)$ $32a^2 - 52ab + 15b^2$

93. $(9a + 1)(9a - 1)$ $81a^2 - 1$

94. $(7b + 1)(7b + 1)$ $49b^2 + 14b + 1$

95. $(x - 9)(x - 9)$ $x^2 - 18x + 81$

96. $(x - 8)(x - 8)$ $x^2 - 16x + 64$

97. $(x + 2)(x^2 - 2x + 4)$ $x^3 + 8$

98. $(x - 3)(x^2 + 3x + 9)$ $x^3 - 27$

6.2 Factoring Trinomials

OBJECTIVES

A Factor a trinomial whose leading coefficient is the number 1.

B Factor a polynomial by first factoring out the greatest common factor and then factoring the polynomial that remains.

In this section we will factor trinomials in which the coefficient of the squared term is 1. The more familiar we are with multiplication of binomials the easier factoring trinomials will be.

Recall multiplication of binomials from Chapter 5:

$$(x + 3)(x + 4) = x^2 + 7x + 12$$

$$(x - 5)(x + 2) = x^2 - 3x - 10$$

The first term in the answer is the product of the first terms in each binomial. The last term in the answer is the product of the last terms in each binomial. The middle term in the answer comes from adding the product of the outside terms to the product of the inside terms.

Let's have a and b represent real numbers and look at the product of $(x + a)$ and $(x + b)$:

$$(x + a)(x + b) = x^2 + ax + bx + ab$$

$$= x^2 + (a + b)x + ab$$

The coefficient of the middle term is the sum of a and b. The last term is the product of a and b. Writing this as a factoring problem, we have:

$$x^2 + \underset{\text{Sum}}{(a + b)}x + \underset{\text{Product}}{ab} = (x + a)(x + b)$$

To factor a trinomial in which the coefficient of x^2 is 1, we need only find the numbers a and b whose sum is the coefficient of the middle term and whose product is the constant term (last term).

Note

As you will see as we progress through the book, factoring is a tool that is used in solving a number of problems. Before seeing how it is used, however, we first must learn how to do it. So, in this section and the two sections that follow, we will be developing our factoring skills.

 EXAMPLE 1 Factor $x^2 + 8x + 12$.

SOLUTION The coefficient of x^2 is 1. We need two numbers whose sum is 8 and whose product is 12. The numbers are 6 and 2:

$$x^2 + 8x + 12 = (x + 6)(x + 2)$$

We can easily check our work by multiplying $(x + 6)$ and $(x + 2)$

$$\text{Check:} \quad (x + 6)(x + 2) = x^2 + 6x + 2x + 12$$

$$= x^2 + 8x + 12$$

EXAMPLE 2 Factor $x^2 - 2x - 15$.

SOLUTION The coefficient of x^2 is again 1. We need to find a pair of numbers whose sum is -2 and whose product is -15. Here are all the possibilities for products that are -15.

Products	Sums
$-1(15) = -15$	$-1 + 15 = 14$
$1(-15) = -15$	$1 + (-15) = -14$
$-5(3) = -15$	$-5 + 3 = -2$
$5(-3) = -15$	$5 + (-3) = 2$

Note
Again, we can check our results by multiplying our factors to see if their product is the original polynomial.

The third line gives us what we want. The factors of $x^2 - 2x - 15$ are $(x - 5)$ and $(x + 3)$:

$$x^2 - 2x - 15 = (x - 5)(x + 3)$$

EXAMPLE 3 Factor $2x^2 + 10x - 28$.

SOLUTION The coefficient of x^2 is 2. We begin by factoring out the greatest common factor, which is 2:

$$2x^2 + 10x - 28 = 2(x^2 + 5x - 14)$$

Now, we factor the remaining trinomial by finding a pair of numbers whose sum is 5 and whose product is -14. Here are the possibilities:

Products	Sums
$-1(14) = -14$	$-1 + 14 = 13$
$1(-14) = -14$	$1 + (-14) = -13$
$-7(2) = -14$	$-7 + 2 = -5$
$7(-2) = -14$	$7 + (-2) = 5$

Note
In Example 3 we began by factoring out the greatest common factor. The first step in factoring any trinomial is to look for the greatest common factor. If the trinomial in question has a greatest common factor other than 1, we factor it out first and then try to factor the trinomial that remains.

From the last line we see that the factors of $x^2 + 5x - 14$ are $(x + 7)$ and $(x - 2)$. Here is the complete problem:

$$2x^2 + 10x - 28 = 2(x^2 + 5x - 14)$$
$$= 2(x + 7)(x - 2)$$

EXAMPLE 4 Factor $3x^3 - 3x^2 - 18x$.

SOLUTION We begin by factoring out the greatest common factor, which is $3x$. Then we factor the remaining trinomial. Without showing the table of products and sums as we did in Examples 2 and 3, here is the complete solution:

$$3x^3 - 3x^2 - 18x = 3x(x^2 - x - 6)$$
$$= 3x(x - 3)(x + 2)$$

Note
Trinomials in which the coefficient of the second-degree term is 1 are the easiest to factor. Success in factoring any type of polynomial is directly related to the amount of time spent working the problems. The more we practice, the more accomplished we become at factoring.

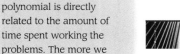

EXAMPLE 5 Factor $x^2 + 8xy + 12y^2$.

SOLUTION This time we need two expressions whose product is $12y^2$ and whose sum is $8y$. The two expressions are $6y$ and $2y$ (see Example 1 in this section):

$$x^2 + 8xy + 12y^2 = (x + 6y)(x + 2y)$$

You should convince yourself that these factors are correct by finding their product.

LINKING OBJECTIVES AND EXAMPLES

Next to each **objective** we have listed the examples that are best described by that objective.

A   1, 2, 5

B   3, 4

GETTING READY FOR CLASS

After reading through the preceding section, respond in your own words and in complete sentences.

1. When the leading coefficient of a trinomial is 1, what is the relationship between the other two coefficients and the factors of the trinomial?
2. When factoring polynomials, what should you look for first?
3. How can you check to see that you have factored a trinomial correctly?
4. Describe how you would find the factors of $x^2 + 8x + 12$.

Problem Set 6.2

Online support materials can be found at www.thomsonedu.com/login

Factor the following trinomials.

1. $x^2 + 7x + 12$ $(x + 3)(x + 4)$
2. $x^2 + 7x + 10$ $(x + 5)(x + 2)$
3. $x^2 + 3x + 2$ $(x + 1)(x + 2)$
4. $x^2 + 7x + 6$ $(x + 6)(x + 1)$
5. $a^2 + 10a + 21$ $(a + 3)(a + 7)$
6. $a^2 - 7a + 12$ $(a - 3)(a - 4)$
7. $x^2 - 7x + 10$ $(x - 2)(x - 5)$
8. $x^2 - 3x + 2$ $(x - 1)(x - 2)$
9. $y^2 - 10y + 21$ $(y - 3)(y - 7)$
10. $y^2 - 7y + 6$ $(y - 1)(y - 6)$
11. $x^2 - x - 12$ $(x - 4)(x + 3)$
12. $x^2 - 4x - 5$ $(x - 5)(x + 1)$
13. $y^2 + y - 12$ $(y + 4)(y - 3)$
14. $y^2 + 3y - 18$ $(y + 6)(y - 3)$
15. $x^2 + 5x - 14$ $(x + 7)(x - 2)$
16. $x^2 - 5x - 24$ $(x - 8)(x + 3)$
17. $r^2 - 8r - 9$ $(r - 9)(r + 1)$
18. $r^2 - r - 2$ $(r - 2)(r + 1)$
19. $x^2 - x - 30$ $(x - 6)(x + 5)$

20. $x^2 + 8x + 12$ $(x + 2)(x + 6)$
21. $a^2 + 15a + 56$ $(a + 7)(a + 8)$
22. $a^2 - 9a + 20$ $(a - 4)(a - 5)$
23. $y^2 - y - 42$ $(y + 6)(y - 7)$
24. $y^2 + y - 42$ $(y + 7)(y - 6)$
25. $x^2 + 13x + 42$ $(x + 6)(x + 7)$
26. $x^2 - 13x + 42$ $(x - 6)(x - 7)$

Factor the following problems completely. First, factor out the greatest common factor, and then factor the remaining trinomial.

27. $2x^2 + 6x + 4$ $2(x + 1)(x + 2)$
28. $3x^2 - 6x - 9$ $3(x + 1)(x - 3)$
29. $3a^2 - 3a - 60$ $3(a + 4)(a - 5)$
30. $2a^2 - 18a + 28$ $2(a - 2)(a - 7)$
31. $100x^2 - 500x + 600$ $100(x - 2)(x - 3)$
32. $100x^2 - 900x + 2,000$ $100(x - 4)(x - 5)$
33. $100p^2 - 1,300p + 4,000$ $100(p - 5)(p - 8)$
34. $100p^2 - 1,200p + 3,200$ $100(p - 4)(p - 8)$
35. $x^4 - x^3 - 12x^2$ $x^2(x + 3)(x - 4)$
36. $x^4 - 11x^3 + 24x^2$ $x^2(x - 3)(x - 8)$

= Videos available by instructor request
▶ = Online student support materials available at www.thomsonedu.com/login

37. $2r^3 + 4r^2 - 30r$ $2r(r + 5)(r - 3)$

38. $5r^3 + 45r^2 + 100r$ $5r(r + 4)(r + 5)$

39. $2y^4 - 6y^3 - 8y^2$ $2y^2(y + 1)(y - 4)$

40. $3r^3 - 3r^2 - 6r$ $3r(r + 1)(r - 2)$

41. $x^5 + 4x^4 + 4x^3$ $x^3(x + 2)(x + 2)$

42. $x^5 + 13x^4 + 42x^3$ $x^3(x + 6)(x + 7)$

43. $3y^4 - 12y^3 - 15y^2$ $3y^2(y + 1)(y - 5)$

44. $5y^4 - 10y^3 + 5y^2$ $5y^2(y - 1)(y - 1)$

45. $4x^4 - 52x^3 + 144x^2$ $4x^2(x - 4)(x - 9)$

46. $3x^3 - 3x^2 - 18x$ $3x(x + 2)(x - 3)$

Factor the following trinomials.

47. $x^2 + 5xy + 6y^2$ $(x + 2y)(x + 3y)$

48. $x^2 - 5xy + 6y^2$ $(x - 2y)(x - 3y)$

49. $x^2 - 9xy + 20y^2$ $(x - 4y)(x - 5y)$

50. $x^2 + 9xy + 20y^2$ $(x + 4y)(x + 5y)$

51. $a^2 + 2ab - 8b^2$ $(a + 4b)(a - 2b)$

52. $a^2 - 2ab - 8b^2$ $(a - 4b)(a + 2b)$

53. $a^2 - 10ab + 25b^2$ $(a - 5b)(a - 5b)$

54. $a^2 + 6ab + 9b^2$ $(a + 3b)(a + 3b)$

55. $a^2 + 10ab + 25b^2$ $(a + 5b)(a + 5b)$

56. $a^2 - 6ab + 9b^2$ $(a - 3b)(a - 3b)$

57. $x^2 + 2xa - 48a^2$ $(x - 6a)(x + 8a)$

58. $x^2 - 3xa - 10a^2$ $(x + 2a)(x - 5a)$

59. $x^2 - 5xb - 36b^2$ $(x + 4b)(x - 9b)$

60. $x^2 - 13xb + 36b^2$ $(x - 4b)(x - 9b)$

Factor completely.

61. $x^4 - 5x^2 + 6$ $(x^2 - 3)(x^2 - 2)$

62. $x^6 - 2x^3 - 15$ $(x^3 - 5)(x^3 + 3)$

63. $x^2 - 80x - 2,000$ $(x - 100)(x + 20)$

64. $x^2 - 190x - 2,000$ $(x - 200)(x + 10)$

65. $x^2 - x + \dfrac{1}{4}$ $\left(x - \dfrac{1}{2}\right)\left(x - \dfrac{1}{2}\right)$

66. $x^2 - \dfrac{2}{3}x + \dfrac{1}{9}$ $\left(x - \dfrac{1}{3}\right)\left(x - \dfrac{1}{3}\right)$

67. $x^2 + 0.6x + 0.08$ $(x + 0.2)(x + 0.4)$

68. $x^2 + 0.8x + 0.15$ $(x + 0.5)(x + 0.3)$

69. If one of the factors of $x^2 + 24x + 128$ is $x + 8$, what is the other factor? $x + 16$

70. If one factor of $x^2 + 260x + 2,500$ is $x + 10$, what is the other factor? $x + 250$

71. What polynomial, when factored, gives $(4x + 3)$ $(x - 1)$? $4x^2 - x - 3$

72. What polynomial factors to $(4x - 3)(x + 1)$? $4x^2 + x - 3$

Maintaining Your Skills

Simplify each expression. Write using only positive exponents.

73. $\left(-\dfrac{2}{5}\right)^2$ $\dfrac{4}{25}$

74. $\left(-\dfrac{3}{8}\right)^2$ $\dfrac{9}{64}$

75. $(3a^3)^2(2a^2)^3$ $72a^{12}$

76. $(-4x^4)^2(2x^5)^4$ $256x^{28}$

77. $\dfrac{(4x)^{-7}}{(4x)^{-5}}$ $\dfrac{1}{16x^2}$

78. $\dfrac{(2x)^{-3}}{(2x)^{-5}}$ $4x^2$

79. $\dfrac{12a^5b^3}{72a^2b^5}$ $\dfrac{a^3}{6b^2}$

80. $\dfrac{25x^5y^3}{50x^2y^7}$ $\dfrac{x^3}{2y^4}$

81. $\dfrac{15x^{-5}y^3}{45x^2y^5}$ $\dfrac{1}{3x^7y^2}$

82. $\dfrac{25a^2b^7}{75a^5b^3}$ $\dfrac{b^4}{3a^3}$

83. $(-7x^3y)(3xy^4)$ $-21x^4y^5$

84. $(9a^6b^4)(6a^4b^3)$ $54a^{10}b^7$

85. $(-5a^3b^{-1})(4a^{-2}b^4)$ $-20ab^3$

86. $(-3a^2b^{-4})(6a^5b^{-2})$ $-\dfrac{18a^7}{b^6}$

87. $(9a^2b^3)(-3a^3b^5)$ $-27a^5b^8$

88. $(-7a^5b^8)(6a^7b^4)$ $-42a^{12}b^{12}$

Getting Ready for the Next Section

Multiply using the FOIL method.

89. $(6a + 1)(a + 2)$ $6a^2 + 13a + 2$

90. $(6a - 1)(a - 2)$ $6a^2 - 13a + 2$

91. $(3a + 2)(2a + 1)$ $6a^2 + 7a + 2$

92. $(3a - 2)(2a - 1)$ $6a^2 - 7a + 2$

93. $(6a + 2)(a + 1)$ $6a^2 + 8a + 2$

94. $(3a + 1)(2a + 2)$ $6a^2 + 8a + 2$

6.3 More Trinomials to Factor

A Factor a trinomial whose leading coefficient is other than 1.

B Factor a polynomial by first factoring out the greatest common factor and then factoring the polynomial that remains.

We now will consider trinomials whose greatest common factor is 1 and whose leading coefficient (the coefficient of the squared term) is a number other than 1. We present two methods for factoring trinomials of this type. The first method involves listing possible factors until the correct pair of factors is found. This requires a certain amount of trial and error. The second method is based on the factoring by grouping process that we covered previously. Either method can be used to factor trinomials whose leading coefficient is a number other than 1.

Method 1: Factoring $ax^2 + bx + c$ by trial and error

Suppose we want to factor the trinomial $2x^2 - 5x - 3$. We know the factors (if they exist) will be a pair of binomials. The product of their first terms is $2x^2$ and the product of their last term is -3. Let us list all the possible factors along with the trinomial that would result if we were to multiply them together. Remember, the middle term comes from the product of the inside terms plus the product of the outside terms.

Binomial Factors	First Term	Middle Term	Last Term
$(2x - 3)(x + 1)$	$2x^2$	$-x$	-3
$(2x + 3)(x - 1)$	$2x^2$	$+x$	-3
$(2x - 1)(x + 3)$	$2x^2$	$+5x$	-3
$(2x + 1)(x - 3)$	$2x^2$	$-5x$	-3

We can see from the last line that the factors of $2x^2 - 5x - 3$ are $(2x + 1)$ and $(x - 3)$. There is no straightforward way, as there was in the previous section, to find the factors, other than by trial and error or by simply listing all the possibilities. We look for possible factors that, when multiplied, will give the correct first and last terms, and then we see if we can adjust them to give the correct middle term.

EXAMPLE 1 Factor $6a^2 + 7a + 2$.

SOLUTION We list all the possible pairs of factors that, when multiplied together, give a trinomial whose first term is $6a^2$ and whose last term is $+2$.

Binomial Factors	First Term	Middle Term	Last Term
$(6a + 1)(a + 2)$	$6a^2$	$+13a$	$+2$
$(6a - 1)(a - 2)$	$6a^2$	$-13a$	$+2$
$(3a + 2)(2a + 1)$	$6a^2$	$+7a$	$+2$
$(3a - 2)(2a - 1)$	$6a^2$	$-7a$	$+2$

Note
Remember, we can always check our results by multiplying the factors we have and comparing that product with our original polynomial.

The factors of $6a^2 + 7a + 2$ are $(3a + 2)$ and $(2a + 1)$.

Check: $(3a + 2)(2a + 1) = 6a^2 + 7a + 2$

Notice that in the preceding list we did not include the factors $(6a + 2)$ and $(a + 1)$. We do not need to try these since the first factor has a 2 common to each term and so could be factored again, giving $2(3a + 1)(a + 1)$. Since our original trinomial, $6a^2 + 7a + 2$, did *not* have a greatest common factor of 2, neither of its factors will.

EXAMPLE 2   Factor $4x^2 - x - 3$.

SOLUTION   We list all the possible factors that, when multiplied, give a trinomial whose first term is $4x^2$ and whose last term is -3.

Binomial Factors	First Term	Middle Term	Last Term
$(4x + 1)(x - 3)$	$4x^2$	$-11x$	-3
$(4x - 1)(x + 3)$	$4x^2$	$+11x$	-3
$(4x + 3)(x - 1)$	$4x^2$	$-x$	-3
$(4x - 3)(x + 1)$	$4x^2$	$+x$	-3
$(2x + 1)(2x - 3)$	$4x^2$	$-4x$	-3
$(2x - 1)(2x + 3)$	$4x^2$	$+4x$	-3

The third line shows that the factors are $(4x + 3)$ and $(x - 1)$.

Check: $(4x + 3)(x - 1) = 4x^2 - x - 3$

You will find that the more practice you have at factoring this type of trinomial, the faster you will get the correct factors. You will pick up some shortcuts along the way, or you may come across a system of eliminating some factors as possibilities. Whatever works best for you is the method you should use. Factoring is a very important tool, and you must be good at it.

EXAMPLE 3   Factor $12y^3 + 10y^2 - 12y$.

SOLUTION   We begin by factoring out the greatest common factor, $2y$:

$$12y^3 + 10y^2 - 12y = 2y(6y^2 + 5y - 6)$$

We now list all possible factors of a trinomial with the first term $6y^2$ and last term -6, along with the associated middle terms.

Note
Once again, the first step in any factoring problem is to factor out the greatest common factor if it is other than 1.

Possible Factors	Middle Term When Multiplied
$(3y + 2)(2y - 3)$	$-5y$
$(3y - 2)(2y + 3)$	$+5y$
$(6y + 1)(y - 6)$	$-35y$
$(6y - 1)(y + 6)$	$+35y$

The second line gives the correct factors. The complete problem is:

$$12y^3 + 10y^2 - 12y = 2y(6y^2 + 5y - 6)$$
$$= 2y(3y - 2)(2y + 3)$$

EXAMPLE 4   Factor $30x^2y - 5xy^2 - 10y^3$.

SOLUTION   The greatest common factor is $5y$:

$$30x^2y - 5xy^2 - 10y^3 = 5y(6x^2 - xy - 2y^2)$$
$$= 5y(2x + y)(3x - 2y)$$

Method 2: Factoring $ax^2 + bx + c$ by grouping

Recall previously that we can use factoring by grouping to factor the polynomial $6x^2 - 3x - 4x + 2$. We begin by factoring $3x$ from the first two terms and -2 from the last two terms. For review, here is the complete problem:

$$6x^2 - 3x - 4x + 2 = 3x(2x - 1) - 2(2x - 1)$$
$$= (2x - 1)(3x - 2)$$

Now, let's back up a little and notice that our original polynomial $6x^2 - 3x - 4x + 2$ can be simplified to $6x^2 - 7x + 2$ by adding $-3x$ and $-4x$. This means that $6x^2 - 7x + 2$ can be factored to $(2x - 1)(3x - 2)$ by the grouping method shown previously. The key to using this process is to rewrite the middle term $-7x$ as $-3x - 4x$.

To generalize this discussion, here are the steps we use to factor trinomials by grouping.

Strategy for Factoring $ax^2 + bx + c$ by Grouping

Step 1: Form the product ac.

Step 2: Find a pair of numbers whose product is ac and whose sum is b.

Step 3: Rewrite the polynomial to be factored so that the middle term bx is written as the sum of two terms whose coefficients are the two numbers found in step 2.

Step 4: Factor by grouping.

 EXAMPLE 5   Factor $3x^2 - 10x - 8$ using these steps.

SOLUTION   The trinomial $3x^2 - 10x - 8$ has the form $ax^2 + bx + c$, where $a = 3$, $b = -10$, and $c = -8$.

Step 1: The product ac is $3(-8) = -24$.

Step 2: We need to find two numbers whose product is -24 and whose sum is -10. Let's systematically begin to list all the pairs of numbers whose product is -24 to find the pair whose sum is -10.

Product	Sum
$-24(1) = -24$	$-24 + 1 = -23$
$-12(2) = -24$	$-12 + 2 = -10$

We stop here because we have found the pair of numbers whose product is -24 and whose sum is -10. The numbers are -12 and 2.

Step 3: We now rewrite our original trinomial so the middle term $-10x$ is written as the sum of $-12x$ and $2x$:

$$3x^2 - 10x - 8 = 3x^2 - 12x + 2x - 8$$

Step 4: Factoring by grouping, we have:

$$3x^2 - 12x + 2x - 8 = 3x(x - 4) + 2(x - 4)$$
$$= (x - 4)(3x + 2)$$

We can check our work by multiplying $x - 4$ and $3x + 2$ to get $3x^2 - 10x - 8$.

EXAMPLE 6 Factor $4x^2 - x - 3$.

SOLUTION In this case, $a = 4$, $b = -1$, and $c = -3$. The product ac is $4(-3) = -12$. We need a pair of numbers whose product is -12 and whose sum is -1. We begin listing pairs of numbers whose product is -12 and whose sum is -1.

Product	Sum
$-12(1) = -12$	$-12 + 1 = -11$
$-6(2) = -12$	$-6 + 2 = -4$
$-4(3) = -12$	$-4 + 3 = -1$

We stop here because we have found the pair of numbers for which we are looking. They are -4 and 3. Next, we rewrite the middle term $-x$ as the sum $-4x + 3x$ and proceed to factor by grouping.

$$4x^2 - x - 3 = 4x^2 - 4x + 3x - 3$$
$$= 4x(x - 1) + 3(x - 1)$$
$$= (x - 1)(4x + 3)$$

Compare this procedure and the result with those shown in Example 2 of this section.

 EXAMPLE 7 Factor $8x^2 - 2x - 15$.

SOLUTION The product ac is $8(-15) = -120$. There are many pairs of numbers whose product is -120. We are looking for the pair whose sum is also -2. The numbers are -12 and 10. Writing $-2x$ as $-12x + 10x$ and then factoring by grouping, we have:

$$8x^2 - 2x - 15 = 8x^2 - 12x + 10x - 15$$
$$= 4x(2x - 3) + 5(2x - 3)$$
$$= (2x - 3)(4x + 5)$$

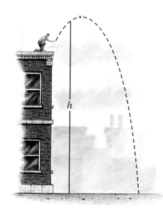

 EXAMPLE 8 A ball is tossed into the air with an upward velocity of 16 feet per second from the top of a building 32 feet high. The equation that gives the height of the ball above the ground at any time t is

$$h = 32 + 16t - 16t^2$$

Factor the right side of this equation and then find h when t is 2.

SOLUTION We begin by factoring out the greatest common factor, 16. Then, we factor the trinomial that remains:

$$h = 32 + 16t - 16t^2$$
$$h = 16(2 + t - t^2)$$
$$h = 16(2 - t)(1 + t)$$

Letting $t = 2$ in the equation, we have

$$h = 16(0)(3) = 0$$

When t is 2, h is 0.

LINKING OBJECTIVES AND EXAMPLES

Next to each **objective** we have listed the examples that are best described by that objective.

A 1–8

B 3, 4, 8

GETTING READY FOR CLASS

After reading through the preceding section, respond in your own words and in complete sentences.

1. What is the first step in factoring a trinomial?
2. Describe the criteria you would use to set up a table of possible factors of a trinomial.
3. What does it mean if you factor a trinomial and one of your factors has a greatest common factor of 3?
4. Describe how you would look for possible factors of $6a^2 + 7a + 2$.

Problem Set 6.3

Online support materials can be found at www.thomsonedu.com/login

Factor the following trinomials.

▶ 1. $2x^2 + 7x + 3$ $(2x + 1)(x + 3)$

2. $2x^2 + 5x + 3$ $(2x + 3)(x + 1)$

3. $2a^2 - a - 3$ $(2a - 3)(a + 1)$

4. $2a^2 + a - 3$ $(2a + 3)(a - 1)$

5. $3x^2 + 2x - 5$ $(3x + 5)(x - 1)$

6. $3x^2 - 2x - 5$ $(3x - 5)(x + 1)$

7. $3y^2 - 14y - 5$ $(3y + 1)(y - 5)$

8. $3y^2 + 14y - 5$ $(3y - 1)(y + 5)$

9. $6x^2 + 13x + 6$ $(2x + 3)(3x + 2)$

10. $6x^2 - 13x + 6$ $(2x - 3)(3x - 2)$

11. $4x^2 - 12xy + 9y^2$ $(2x - 3y)(2x - 3y)$

12. $4x^2 + 12xy + 9y^2$ $(2x + 3y)(2x + 3y)$

▶ 13. $4y^2 - 11y - 3$ $(4y + 1)(y - 3)$

14. $4y^2 + y - 3$ $(4y - 3)(y + 1)$

15. $20x^2 - 41x + 20$ $(4x - 5)(5x - 4)$

16. $20x^2 + 9x - 20$ $(4x + 5)(5x - 4)$

17. $20a^2 + 48ab - 5b^2$ $(10a - b)(2a + 5b)$

18. $20a^2 + 29ab + 5b^2$ $(4a + 5b)(5a + b)$

19. $20x^2 - 21x - 5$ $(4x - 5)(5x + 1)$

20. $20x^2 - 48x - 5$ $(10x + 1)(2x - 5)$

21. $12m^2 + 16m - 3$ $(6m - 1)(2m + 3)$

22. $12m^2 + 20m + 3$ $(6m + 1)(2m + 3)$

23. $20x^2 + 37x + 15$ $(4x + 5)(5x + 3)$

24. $20x^2 + 13x - 15$ $(4x + 5)(5x - 3)$

25. $12a^2 - 25ab + 12b^2$ $(3a - 4b)(4a - 3b)$

26. $12a^2 + 7ab - 12b^2$ $(3a + 4b)(4a - 3b)$

▶ 27. $3x^2 - xy - 14y^2$ $(3x - 7y)(x + 2y)$

28. $3x^2 + 19xy - 14y^2$ $(3x - 2y)(x + 7y)$

29. $14x^2 + 29x - 15$ $(2x + 5)(7x - 3)$

30. $14x^2 + 11x - 15$ $(2x + 3)(7x - 5)$

31. $6x^2 - 43x + 55$ $(3x - 5)(2x - 11)$

32. $6x^2 - 7x - 55$ $(3x - 11)(2x + 5)$

33. $15t^2 - 67t + 38$ $(5t - 19)(3t - 2)$

34. $15t^2 - 79t - 34$ $(5t + 2)(3t - 17)$

Factor each of the following completely. Look first for the greatest common factor.

35. $4x^2 + 2x - 6$ $2(2x + 3)(x - 1)$

36. $6x^2 - 51x + 63$ $3(2x - 3)(x - 7)$

37. $24a^2 - 50a + 24$ $2(4a - 3)(3a - 4)$

38. $18a^2 + 48a + 32$ $2(3a + 4)(3a + 4)$

39. $10x^3 - 23x^2 + 12x$ $x(5x - 4)(2x - 3)$

40. $10x^4 + 7x^3 - 12x^2$ $x^2(5x - 4)(2x + 3)$

41. $6x^4 - 11x^3 - 10x^2$ $x^2(3x + 2)(2x - 5)$

42. $6x^3 + 19x^2 + 10x$ $x(3x + 2)(2x + 5)$

= Videos available by instructor request

▶ = Online student support materials available at www.thomsonedu.com/login

▶ **43.** $10a^3 - 6a^2 - 4a$ $2a(5a + 2)(a - 1)$

44. $6a^3 + 15a^2 + 9a$ $3a(2a + 3)(a + 1)$

45. $15x^3 - 102x^2 - 21x$ $3x(5x + 1)(x - 7)$

46. $2x^4 - 24x^3 + 64x^2$ $2x^2(x - 4)(x - 8)$

47. $35y^3 - 60y^2 - 20y$ $5y(7y + 2)(y - 2)$

48. $14y^4 - 32y^3 + 8y^2$ $2y^2(7y - 2)(y - 2)$

49. $15a^4 - 2a^3 - a^2$ $a^2(5a + 1)(3a - 1)$

50. $10a^5 - 17a^4 + 3a^3$ $a^3(5a - 1)(2a - 3)$

51. $24x^2y - 6xy - 45y$ $3y(4x + 5)(2x - 3)$

52. $8x^2y^2 + 26xy^2 + 15y^2$ $y^2(4x + 3)(2x + 5)$

53. $12x^2y - 34xy^2 + 14y^3$ $2y(2x - y)(3x - 7y)$

54. $12x^2y - 46xy^2 + 14y^3$ $2y(2x - 7y)(3x - y)$

55. Evaluate the expression $2x^2 + 7x + 3$ and the expression $(2x + 1)(x + 3)$ for $x = 2$. Both equal 25.

56. Evaluate the expression $2a^2 - a - 3$ and the expression $(2a - 3)(a + 1)$ for $a = 5$. Both equal 42.

57. What polynomial factors to $(2x + 3)(2x - 3)$? $4x^2 - 9$

58. What polynomial factors to $(5x + 4)(5x - 4)$? $25x^2 - 16$

59. What polynomial factors to $(x + 3)(x - 3)(x^2 + 9)$? $x^4 - 81$

60. What polynomial factors to $(x + 2)(x - 2)(x^2 + 4)$? $x^4 - 16$

▶ **61.** One factor of $12x^2 - 71x + 105$ is $x - 3$. Find the other factor. $12x - 35$

▶ **62.** One factor of $18x^2 + 121x - 35$ is $x + 7$. Find the other factor. $18x - 5$

▶ **63.** One factor of $54x^2 + 111x + 56$ is $6x + 7$. Find the other factor. $9x + 8$

▶ **64.** One factor of $63x^2 + 110x + 48$ is $7x + 6$. Find the other factor. $9x + 8$

▶ **65.** One factor of $16t^2 - 64t + 48$ is $t - 1$. Find the other factors, then write the polynomial in factored form. $16(t - 1)(t - 3)$

▶ **66.** One factor of $16t^2 - 32t + 12$ is $2t - 1$. Find the other factors, then write the polynomial in factored form. $4(2t - 1)(2t - 3)$

Applying the Concepts

67. **Archery** Margaret shoots an arrow into the air. The equation for the height (in feet) of the tip of the arrow is:

$$h = 8 + 62t - 16t^2$$

Factor the right side of this equation. Then fill in the table for various heights of the arrow, using the factored form of the equation. $h = 2(4 - t)(1 + 8t)$

Time t (seconds)	Height h (feet)
0	8
1	54
2	68
3	50
4	0

68. **Coin Toss** At the beginning of every football game, the referee flips a coin to see who will kick off. The equation that gives the height (in feet) of the coin tossed in the air is:

$$h = 6 + 29t - 16t^2$$

a. Factor this equation. $h = (3 + 16t)(2 - t)$

b. Use the factored form of the equation to find the height of the quarter after 0 seconds, 1 second, and 2 seconds. 6 feet, 19 feet, 0 feet

Maintaining Your Skills

Perform the following additions and subtractions.

69. $(6x^3 - 4x^2 + 2x) + (9x^2 - 6x + 3)$ $6x^3 + 5x^2 - 4x + 3$

70. $(6x^3 - 4x^2 + 2x) - (9x^2 - 6x + 3)$ $6x^3 - 13x^2 + 8x - 3$

71. $(-7x^4 + 4x^3 - 6x) + (8x^4 + 7x^3 - 9)$ $x^4 + 11x^3 - 6x - 9$

72. $(-7x^4 + 4x^3 - 6x) - (8x^4 + 7x^3 - 9)$ $-15x^4 - 3x^3 - 6x + 9$

73. $(2x^5 + 3x^3 + 4x) + (5x^3 - 6x - 7)$ $2x^5 + 8x^3 - 2x - 7$

74. $(2x^5 + 3x^3 + 4x) - (5x^3 - 6x - 7)$ $2x^5 - 2x^3 + 10x + 7$

75. $(-8x^5 - 5x^4 + 7) + (7x^4 + 2x^2 + 5)$ $-8x^5 + 2x^4 + 2x^2 + 12$

76. $(-8x^5 - 5x^4 + 7) - (7x^4 + 2x^2 + 5)$ $-8x^5 - 12x^4 - 2x^2 + 2$

77. $\dfrac{24x^3y^7}{6x^{-2}y^4} + \dfrac{27x^{-2}y^{10}}{9x^{-7}y^7}$ $7x^5y^3$

78. $\dfrac{15x^8y^4}{5x^2y^2} - \dfrac{4x^7y^5}{2xy^3}$ x^6y^2

79. $\dfrac{18a^5b^9}{3a^3b^6} - \dfrac{48a^{-3}b^{-1}}{16a^{-5}b^{-4}}$ $3a^2b^3$

80. $\dfrac{54a^{-3}b^5}{6a^{-7}b^{-2}} - \dfrac{32a^6b^5}{8a^2b^{-2}}$ $5a^4b^7$

Getting Ready for the Next Section

Multiply each of the following.

81. $(x + 3)(x - 3)$ $x^2 - 9$
82. $(x - 4)(x + 4)$ $x^2 - 16$

83. $(x + 5)(x - 5)$ $x^2 - 25$
84. $(x - 6)(x + 6)$ $x^2 - 36$

85. $(x + 7)(x - 7)$ $x^2 - 49$
86. $(x - 8)(x + 8)$ $x^2 - 64$

87. $(x + 9)(x - 9)$ $x^2 - 81$
88. $(x - 10)(x + 10)$ $x^2 - 100$

89. $(2x - 3y)(2x + 3y)$ $4x^2 - 9y^2$

90. $(5x - 6y)(5x + 6y)$ $25x^2 - 36y^2$

91. $(x^2 + 4)(x + 2)(x - 2)$ $x^4 - 16$

92. $(x^2 + 9)(x + 3)(x - 3)$ $x^4 - 81$

93. $(x + 3)^2$ $x^2 + 6x + 9$
94. $(x - 4)^2$ $x^2 - 8x + 16$

95. $(x + 5)^2$ $x^2 + 10x + 25$
96. $(x - 6)^2$ $x^2 - 12x + 36$

97. $(x + 7)^2$ $x^2 + 14x + 49$
98. $(x - 8)^2$ $x^2 - 16x + 64$

99. $(x + 9)^2$ $x^2 + 18x + 81$
100. $(x - 10)^2$ $x^2 - 20x + 100$

101. $(2x + 3)^2$ $4x^2 + 12x + 9$ **102.** $(3x - y)^2$ $9x^2 - 6xy + y^2$

103. $(4x - 2y)^2$ $16x^2 - 16xy + 4y^2$

104. $(5x - 6y)^2$ $25x^2 - 60xy + 36y^2$

6.4 The Difference of Two Squares

OBJECTIVES

A Factor the difference of two squares.

B Factor a perfect square trinomial.

C Factor a polynomial by first factoring out the greatest common factor and then factoring the polynomial that remains.

In Chapter 5 we listed the following three special products:

$$(a + b)^2 = (a + b)(a + b) = a^2 + 2ab + b^2$$
$$(a - b)^2 = (a - b)(a - b) = a^2 - 2ab + b^2$$
$$(a + b)(a - b) = a^2 - b^2$$

Since factoring is the reverse of multiplication, we can also consider the three special products as three special factorings:

$$a^2 + 2ab + b^2 = (a + b)^2$$
$$a^2 - 2ab + b^2 = (a - b)^2$$
$$a^2 - b^2 = (a + b)(a - b)$$

Any trinomial of the form $a^2 + 2ab + b^2$ or $a^2 - 2ab + b^2$ can be factored by the methods of Section 6.3. The last line is the factoring to obtain the difference of two squares. The difference of two squares always factors in this way. Again, these are patterns you must be able to recognize on sight.

EXAMPLE 1   Factor $16x^2 - 25$.

SOLUTION   We can see that the first term is a perfect square, and the last term is also. This fact becomes even more obvious if we rewrite the problem as:

$$16x^2 - 25 = (4x)^2 - (5)^2$$

The first term is the square of the quantity $4x$, and the last term is the square of 5. The completed problem looks like this:

$$16x^2 - 25 = (4x)^2 - (5)^2$$
$$= (4x + 5)(4x - 5)$$

To check our results, we multiply:

$$(4x + 5)(4x - 5) = 16x^2 + 20x - 20x - 25$$
$$= 16x^2 - 25$$

EXAMPLE 2   Factor $36a^2 - 1$.

SOLUTION   We rewrite the two terms to show they are perfect squares and then factor. Remember, 1 is its own square, $1^2 = 1$.

$$36a^2 - 1 = (6a)^2 - (1)^2$$
$$= (6a + 1)(6a - 1)$$

To check our results, we multiply:

$$(6a + 1)(6a - 1) = 36a^2 + 6a - 6a - 1$$
$$= 36a^2 - 1$$

EXAMPLE 3   Factor $x^4 - y^4$.

SOLUTION x^4 is the perfect square $(x^2)^2$, and y^4 is $(y^2)^2$:

$$x^4 - y^4 = (x^2)^2 - (y^2)^2$$
$$= (x^2 - y^2)(x^2 + y^2)$$

Note

If you think the sum of two squares $x^2 + y^2$ factors, you should try it. Write down the factors you think it has, and then multiply them using the FOIL method. You won't get $x^2 + y^2$.

The factor $(x^2 - y^2)$ is itself the difference of two squares and therefore can be factored again. The factor $(x^2 + y^2)$ is the *sum* of two squares and cannot be factored again. The complete solution is this:

$$x^4 - y^4 = (x^2)^2 - (y^2)^2$$
$$= (x^2 - y^2)(x^2 + y^2)$$
$$= (x + y)(x - y)(x^2 + y^2)$$

EXAMPLE 4   Factor $25x^2 - 60x + 36$.

SOLUTION   Although this trinomial can be factored by the method we used in Section 6.3, we notice that the first and last terms are the perfect squares $(5x)^2$ and $(6)^2$. Before going through the method for factoring trinomials by listing all possible factors, we can check to see if $25x^2 - 60x + 36$ factors to $(5x - 6)^2$. We need only multiply to check:

$$(5x - 6)^2 = (5x - 6)(5x - 6)$$
$$= 25x^2 - 30x - 30x + 36$$
$$= 25x^2 - 60x + 36$$

The trinomial $25x^2 - 60x + 36$ factors to $(5x - 6)(5x - 6) = (5x - 6)^2$.

Note

As we have indicated before, perfect square trinomials like the ones in Examples 4 and 5 can be factored by the methods developed in previous sections. Recognizing that they factor to binomial squares simply saves time in factoring.

 EXAMPLE 5   Factor $5x^2 + 30x + 45$.

SOLUTION   We begin by factoring out the greatest common factor, which is 5. Then we notice that the trinomial that remains is a perfect square trinomial:

$$5x^2 + 30x + 45 = 5(x^2 + 6x + 9)$$
$$= 5(x + 3)^2$$

 EXAMPLE 6   Factor $(x - 3)^2 - 25$.

SOLUTION   This example has the form $a^2 - b^2$, where a is $x - 3$ and b is 5. We factor it according to the formula for the difference of two squares:

$$(x - 3)^2 - 25 = (x - 3)^2 - 5^2 \qquad \textbf{Write 25 as } \mathbf{5^2}$$
$$= [(x - 3) - 5][(x - 3) + 5] \qquad \textbf{Factor}$$
$$= (x - 8)(x + 2) \qquad \textbf{Simplify}$$

Notice in this example we could have expanded $(x - 3)^2$, subtracted 25, and then factored to obtain the same result:

$$(x - 3)^2 - 25 = x^2 - 6x + 9 - 25 \qquad \textbf{Expand } \mathbf{(x - 3)^2}$$
$$= x^2 - 6x - 16 \qquad \textbf{Simplify}$$
$$= (x - 8)(x + 2) \qquad \textbf{Factor}$$

LINKING OBJECTIVES AND EXAMPLES

Next to each **objective** we have listed the examples that are best described by that objective.

A	1–3, 6
B	4, 5
C	5

GETTING READY FOR CLASS

After reading through the preceding section, respond in your own words and in complete sentences.

1. Describe how you factor the difference of two squares.
2. What is a perfect square trinomial?
3. How do you know when you've factored completely?
4. Describe how you would factor $25x^2 - 60x + 36$.

Factor the following.

▶ **1.** $x^2 - 9$ $(x + 3)(x - 3)$

2. $x^2 - 25$ $(x + 5)(x - 5)$

3. $a^2 - 36$ $(a + 6)(a - 6)$

4. $a^2 - 64$ $(a + 8)(a - 8)$

5. $x^2 - 49$ $(x + 7)(x - 7)$

6. $x^2 - 121$ $(x + 11)(x - 11)$

7. $4a^2 - 16$ $4(a + 2)(a - 2)$

8. $4a^2 + 16$ $4(a^2 + 4)$

9. $9x^2 + 25$ Cannot be factored.

10. $16x^2 - 36$ $4(2x + 3)(2x - 3)$

11. $25x^2 - 169$ $(5x + 13)(5x - 13)$

12. $x^2 - y^2$ $(x + y)(x - y)$

▶ **13.** $9a^2 - 16b^2$ $(3a + 4b)(3a - 4b)$

14. $49a^2 - 25b^2$ $(7a + 5b)(7a - 5b)$

15. $9 - m^2$ $(3 + m)(3 - m)$

16. $16 - m^2$ $(4 + m)(4 - m)$

17. $25 - 4x^2$ $(5 + 2x)(5 - 2x)$

18. $36 - 49y^2$ $(6 + 7y)(6 - 7y)$

19. $2x^2 - 18$ $2(x + 3)(x - 3)$

20. $3x^2 - 27$ $3(x + 3)(x - 3)$

21. $32a^2 - 128$ $32(a + 2)(a - 2)$

22. $3a^3 - 48a$ $3a(a + 4)(a - 4)$

23. $8x^2y - 18y$ $2y(2x + 3)(2x - 3)$

24. $50a^2b - 72b$ $2b(5a + 6)(5a - 6)$

25. $a^4 - b^4$ $(a^2 + b^2)(a + b)(a - b)$

26. $a^4 - 16$ $(a^2 + 4)(a + 2)(a - 2)$

27. $16m^4 - 81$ $(4m^2 + 9)(2m + 3)(2m - 3)$

28. $81 - m^4$ $(9 + m^2)(3 + m)(3 - m)$

29. $3x^3y - 75xy^3$ $3xy(x + 5y)(x - 5y)$

30. $2xy^3 - 8x^3y$ $2xy(y + 2x)(y - 2x)$

Factor the following.

31. $x^2 - 2x + 1$ $(x - 1)^2$

32. $x^2 - 6x + 9$ $(x - 3)^2$

33. $x^2 + 2x + 1$ $(x + 1)^2$

34. $x^2 + 6x + 9$ $(x + 3)^2$

▶ **35.** $a^2 - 10a + 25$ $(a - 5)^2$

36. $a^2 + 10a + 25$ $(a + 5)^2$

37. $y^2 + 4y + 4$ $(y + 2)^2$

38. $y^2 - 8y + 16$ $(y - 4)^2$

39. $x^2 - 4x + 4$ $(x - 2)^2$

40. $x^2 + 8x + 16$ $(x + 4)^2$

41. $m^2 - 12m + 36$ $(m - 6)^2$

42. $m^2 + 12m + 36$ $(m + 6)^2$

▶ **43.** $4a^2 + 12a + 9$ $(2a + 3)^2$

44. $9a^2 - 12a + 4$ $(3a - 2)^2$

45. $49x^2 - 14x + 1$ $(7x - 1)^2$

46. $64x^2 - 16x + 1$ $(8x - 1)^2$

47. $9y^2 - 30y + 25$ $(3y - 5)^2$

48. $25y^2 + 30y + 9$ $(5y + 3)^2$

49. $x^2 + 10xy + 25y^2$ $(x + 5y)^2$

50. $25x^2 + 10xy + y^2$ $(5x + y)^2$

51. $9a^2 + 6ab + b^2$ $(3a + b)^2$

52. $9a^2 - 6ab + b^2$ $(3a - b)^2$

Factor.

▶ **53.** $y^2 - 3y + \dfrac{9}{4}$ $\left(y - \dfrac{3}{2}\right)^2$

▶ **54.** $y^2 + 3y + \dfrac{9}{4}$ $\left(y + \dfrac{3}{2}\right)^2$

▶ **55.** $a^2 + a + \dfrac{1}{4}$ $\left(a + \dfrac{1}{2}\right)^2$

▶ **56.** $a^2 - 5a + \dfrac{25}{4}$ $\left(a - \dfrac{5}{2}\right)^2$

▶ **57.** $x^2 - 7x + \dfrac{49}{4}$ $\left(x - \dfrac{7}{2}\right)^2$

▶ **58.** $x^2 + 9x + \dfrac{81}{4}$ $\left(x + \dfrac{9}{2}\right)^2$

▶ **59.** $x^2 - \dfrac{3}{4}x + \dfrac{9}{64}$ $\left(x - \dfrac{3}{8}\right)^2$

▶ **60.** $x^2 - \dfrac{3}{2}x + \dfrac{9}{16}$ $\left(x - \dfrac{3}{4}\right)^2$

▶ **61.** $t^2 - \dfrac{2}{5}t + \dfrac{1}{25}$ $\left(t - \dfrac{1}{5}\right)^2$

▶ **62.** $t^2 + \dfrac{6}{5}t + \dfrac{9}{25}$ $\left(t + \dfrac{3}{5}\right)^2$

Factor the following by first factoring out the greatest common factor.

63. $3a^2 + 18a + 27$ $3(a + 3)^2$

64. $4a^2 - 16a + 16$ $4(a - 2)^2$

65. $2x^2 + 20xy + 50y^2$ $2(x + 5y)^2$

66. $3x^2 + 30xy + 75y^2$ $3(x + 5y)^2$

67. $5x^3 + 30x^2y + 45xy^2$ $5x(x + 3y)^2$

68. $12x^2y - 36xy^2 + 27y^3$ $3y(2x - 3y)^2$

Factor by grouping the first three terms together.

69. $x^2 + 6x + 9 - y^2$ $(x + 3 + y)(x + 3 - y)$

70. $x^2 + 10x + 25 - y^2$ $(x + 5 + y)(x + 5 - y)$

71. $x^2 + 2xy + y^2 - 9$ $(x + y + 3)(x + y - 3)$

72. $a^2 + 2ab + b^2 - 25$ $(a + b + 5)(a + b - 5)$

73. Find a value for b so that the polynomial $x^2 + bx + 49$ factors to $(x + 7)^2$. 14

74. Find a value of b so that the polynomial $x^2 + bx + 81$ factors to $(x + 9)^2$. 18

75. Find the value of c for which the polynomial $x^2 + 10x + c$ factors to $(x + 5)^2$. 25

76. Find the value of a for which the polynomial $ax^2 + 12x + 9$ factors to $(2x + 3)^2$. 4

Foreshadowing Problems

Problems 53–62 will help students get ready for some of the more difficult problems on completing the square that they will see later in the book.

= Videos available by instructor request

▶ = Online student support materials available at www.thomsonedu.com/login

Maintaining Your Skills

Divide.

77. $\dfrac{24y^3 - 36y^2 - 18y}{6y}$ $4y^2 - 6y - 3$

78. $\dfrac{77y^3 + 35y^2 + 14y}{-7y}$ $-11y^2 - 5y - 2$

79. $\dfrac{48x^7 - 36x^5 + 12x^2}{4x^2}$ $12x^5 - 9x^3 + 3$

80. $\dfrac{-50x^5 + 15x^4 + 10x^2}{5x^2}$ $-10x^3 + 3x^2 + 2$

81. $\dfrac{18x^7 + 12x^6 - 6x^5}{-3x^4}$ $-6x^3 - 4x^2 + 2x$

82. $\dfrac{-64x^5 - 18x^4 - 56x^3}{2x^3}$ $-32x^2 - 9x - 28$

83. $\dfrac{-42x^5 + 24x^4 - 66x^2}{6x^2}$ $-7x^3 + 4x^2 - 11$

84. $\dfrac{63x^7 - 27x^6 - 99x^5}{-9x^4}$ $-7x^3 + 3x^2 + 11x$

Use long division to divide.

85. $\dfrac{x^2 - 5x + 8}{x - 3}$ $x - 2 + \dfrac{2}{x - 3}$

86. $\dfrac{x^2 + 7x + 12}{x + 4}$ $x + 3$

87. $\dfrac{6x^2 + 5x + 3}{2x + 3}$ $3x - 2 + \dfrac{9}{2x + 3}$

88. $\dfrac{x^3 + 27}{x + 3}$ $x^2 - 3x + 9$

Getting Ready for the Next Section

Multiply each of the following.

89. **a.** 1^3 1
    b. 2^3 8
    c. 3^3 27
    d. 4^3 64
    e. 5^3 125

90. **a.** $(-1)^3$ -1
    b. $(-2)^3$ -8
    c. $(-3)^3$ -27
    d. $(-4)^3$ -64
    e. $(-5)^3$ -125

91. **a.** $x(x^2 - x + 1)$ $x^3 - x^2 + x$
    b. $1(x^2 - x + 1)$ $x^2 - x + 1$
    c. $(x + 1)(x^2 - x + 1)$ $x^3 + 1$

92. **a.** $x(x^2 + x + 1)$ $x^3 + x^2 + x$
    b. $-1(x^2 + x + 1)$ $-x^2 - x - 1$
    c. $(x - 1)(x^2 + x + 1)$ $x^3 - 1$

93. **a.** $x(x^2 - 2x + 4)$ $x^3 - 2x^2 + 4x$
    b. $2(x^2 - 2x + 4)$ $2x^2 - 4x + 8$
    c. $(x + 2)(x^2 - 2x + 4)$ $x^3 + 8$

94. **a.** $x(x^2 + 2x + 4)$ $x^3 + 2x^2 + 4x$
    b. $-2(x^2 + 2x + 4)$ $-2x^2 - 4x - 8$
    c. $(x - 2)(x^2 + 2x + 4)$ $x^3 - 8$

95. **a.** $x(x^2 - 3x + 9)$ $x^3 - 3x^2 + 9x$
    b. $3(x^2 - 3x + 9)$ $3x^2 - 9x + 27$
    c. $(x + 3)(x^2 - 3x + 9)$ $x^3 + 27$

96. **a.** $x(x^2 + 3x + 9)$ $x^3 + 3x^2 + 9x$
    b. $-3(x^2 + 3x + 9)$ $-3x^2 - 9x - 27$
    c. $(x - 3)(x^2 + 3x + 9)$ $x^3 - 27$

97. **a.** $x(x^2 - 4x + 16)$ $x^3 - 4x^2 + 16x$
    b. $4(x^2 - 4x + 16)$ $4x^2 - 16x + 64$
    c. $(x + 4)(x^2 - 4x + 16)$ $x^3 + 64$

98. **a.** $x(x^2 + 4x + 16)$ $x^3 + 4x^2 + 16x$
    b. $-4(x^2 + 4x + 16)$ $-4x^2 - 16x - 64$
    c. $(x - 4)(x^2 + 4x + 16)$ $x^3 - 64$

99. **a.** $x(x^2 - 5x + 25)$ $x^3 - 5x^2 + 25x$
    b. $5(x^2 - 5x + 25)$ $5x^2 - 25x + 125$
    c. $(x + 5)(x^2 - 5x + 25)$ $x^3 + 125$

100. **a.** $x(x^2 + 5x + 25)$ $x^3 + 5x^2 + 25x$
    b. $-5(x^2 + 5x + 25)$ $-5x^2 - 25x - 125$
    c. $(x - 5)(x^2 + 5x + 25)$ $x^3 - 125$

The Sum and Difference of Two Cubes

Previously, we factored a variety of polynomials. Among the polynomials we factored were polynomials that were the difference of two squares. The formula we used to factor the difference of two squares looks like this:

$$a^2 - b^2 = (a + b)(a - b)$$

If we ran across a binomial that had the form of the difference of two squares, we factored it by applying this formula. For example, to factor $x^2 - 25$, we simply notice that it can be written in the form $x^2 - 5^2$, which looks like the difference of two squares. According to the formula above, this binomial factors into $(x + 5)(x - 5)$.

In this section we want to use two new formulas that will allow us to factor the sum and difference of two cubes. For example, we want to factor the binomial $x^3 - 8$, which is the difference of two cubes. (To see that it is the differrence of two cubes, notice that it can be written $x^3 - 2^3$.) We also want to factor $y^3 + 27$, which is the sum of two cubes. (To see this, notice that $y^3 + 27$ can be written as $y^3 + 3^3$.)

The formulas that allow us to factor the sum of two cubes and the difference of two cubes are not as simple as the formula for factoring the difference of two squares. Here is what they look like:

$$a^3 + b^3 = (a + b)(a^2 - ab + b^2)$$
$$a^3 - b^3 = (a - b)(a^2 + ab + b^2)$$

Let's begin our work with these two formulas by showing that they are true. To do so, we multiply out the right side of each formula.

 EXAMPLE 1   Verify the two formulas.

SOLUTION   We verify the formulas by multiplying the right sides and comparing the results with the left sides:

$$
\begin{array}{r}
a^2 - ab + b^2 \\
a + b \\
\hline
a^3 - a^2b + ab^2 \\
a^2b - ab^2 + b^3 \\
\hline
a^3 \qquad\qquad\quad + b^3
\end{array}
\qquad\qquad
\begin{array}{r}
a^2 + ab + b^2 \\
a - b \\
\hline
a^3 + a^2b + ab^2 \\
- a^2b - ab^2 - b^3 \\
\hline
a^3 \qquad\qquad\quad - b^3
\end{array}
$$

The first formula is correct.   The second formula is correct.

Here are some examples that use the formulas for factoring the sum and difference of two cubes.

 EXAMPLE 2   Factor $x^3 - 8$.

SOLUTION   Since the two terms are perfect cubes, we write them as such and apply the formula:

$$
\begin{aligned}
x^3 - 8 &= x^3 - 2^3 \\
&= (x - 2)(x^2 + 2x + 2^2) \\
&= (x - 2)(x^2 + 2x + 4)
\end{aligned}
$$

EXAMPLE 3   Factor $y^3 + 27$.

SOLUTION   Proceeding as we did in Example 2, we first write 27 as 3^3. Then, we apply the formula for factoring the sum of two cubes, which is $a^3 + b^3 = (a + b)(a^2 - ab + b^2)$:

$$y^3 + 27 = y^3 + 3^3$$

$$= (y + 3)(y^2 - 3y + 3^2)$$

$$= (y + 3)(y^2 - 3y + 9)$$

EXAMPLE 4   Factor $64 + t^3$.

SOLUTION   The first term is the cube of 4 and the second term is the cube of t. Therefore,

$$64 + t^3 = 4^3 + t^3$$

$$= (4 + t)(16 - 4t + t^2)$$

EXAMPLE 5   Factor $27x^3 + 125y^3$.

SOLUTION   Writing both terms as perfect cubes, we have

$$27x^3 + 125y^3 = (3x)^3 + (5y)^3$$

$$= (3x + 5y)(9x^2 - 15xy + 25y^2)$$

EXAMPLE 6   Factor $a^3 - \frac{1}{8}$.

SOLUTION   The first term is the cube of a, whereas the second term is the cube of $\frac{1}{2}$:

$$a^3 - \frac{1}{8} = a^3 - \left(\frac{1}{2}\right)^3$$

$$= \left(a - \frac{1}{2}\right)\left(a^2 + \frac{1}{2}a + \frac{1}{4}\right)$$

EXAMPLE 7   Factor $x^6 - y^6$.

SOLUTION   We have a choice of how we want to write the two terms to begin. We can write the expression as the difference of two squares, $(x^3)^2 - (y^3)^2$, or as the difference of two cubes, $(x^2)^3 - (y^2)^3$. It is better to use the difference of two squares if we have a choice:

$$x^6 - y^6 = (x^3)^2 - (y^3)^2$$

$$= (x^3 - y^3)(x^3 + y^3)$$

$$= (x - y)(x^2 + xy + y^2)(x + y)(x^2 - xy + y^2)$$

LINKING OBJECTIVES AND EXAMPLES

Next to each objective we have listed the examples that are best described by that objective.

A   2–7

GETTING READY FOR CLASS

After reading through the preceding section, respond in your own words and in complete sentences.

1. How can you check your work when factoring?
2. Why are the numbers 8, 27, 64, and 125 used so frequently in this section?
3. List the cubes of the numbers 1 through 10.
4. How are you going to remember that the sum of two cubes factors, while the sum of two squares is prime?

Problem Set 6.5

Online support materials can be found at www.thomsonedu.com/login

Factor each of the following as the sum or difference of two cubes.

▶ **1.** $x^3 - y^3$ $(x - y)(x^2 + xy + y^2)$

2. $x^3 + y^3$ $(x + y)(x^2 - xy + y^2)$

▶ **3.** $a^3 + 8$ $(a + 2)(a^2 - 2a + 4)$

4. $a^3 - 8$ $(a - 2)(a^2 + 2a + 4)$

5. $27 + x^3$ $(3 + x)(9 - 3x + x^2)$

6. $27 - x^3$ $(3 - x)(9 + 3x + x^2)$

7. $y^3 - 1$ $(y - 1)(y^2 + y + 1)$

8. $y^3 + 1$ $(y + 1)(y^2 - y + 1)$

▶ **9.** $y^3 - 64$ $(y - 4)(y^2 + 4y + 16)$

10. $y^3 + 64$ $(y + 4)(y^2 - 4y + 16)$

11. $125h^3 - t^3$ $(5h - t)(25h^2 + 5ht + t^2)$

12. $t^3 + 125h^3$ $(t + 5h)(t^2 - 5ht + 25h^2)$

13. $x^3 - 216$ $(x - 6)(x^2 + 6x + 36)$

14. $216 + x^3$ $(6 + x)(36 - 6x + x^2)$

15. $2y^3 - 54$ $2(y - 3)(y^2 + 3y + 9)$

16. $81 + 3y^3$ $3(3 + y)(9 - 3y + y^2)$

17. $2a^3 - 128b^3$ $2(a - 4b)(a^2 + 4ab + 16b^2)$

18. $128a^3 + 2b^3$ $2(4a + b)(16a^2 - 4ab + b^2)$

19. $2x^3 + 432y^3$ $2(x + 6y)(x^2 - 6xy + 36y^2)$

20. $432x^3 - 2y^3$ $2(6x - y)(36x^2 + 6xy + y^2)$

21. $10a^3 - 640b^3$ $10(a - 4b)(a^2 + 4ab + 16b^2)$

22. $640a^3 + 10b^3$ $10(4a + b)(16a^2 - 4ab + b^2)$

23. $10r^3 - 1{,}250$ $10(r - 5)(r^2 + 5r + 25)$

24. $10r^3 + 1{,}250$ $10(r + 5)(r^2 - 5r + 25)$

25. $64 + 27a^3$ $(4 + 3a)(16 - 12a + 9a^2)$

26. $27 - 64a^3$ $(3 - 4a)(9 + 12a + 16a^2)$

▶ **27.** $8x^3 - 27y^3$ $(2x - 3y)(4x^2 + 6xy + 9y^2)$

28. $27x^3 - 8y^3$ $(3x - 2y)(9x^2 + 6xy + 4y^2)$

29. $t^3 + \frac{1}{27}$ $(t + \frac{1}{3})(t^2 - \frac{1}{3}t + \frac{1}{9})$

30. $t^3 - \frac{1}{27}$ $(t - \frac{1}{3})(t^2 + \frac{1}{3}t + \frac{1}{9})$

▶ **31.** $27x^3 - \frac{1}{27}$ $(3x - \frac{1}{3})(9x^2 + x + \frac{1}{9})$

32. $8x^3 + \frac{1}{8}$ $(2x + \frac{1}{2})(4x^2 - x + \frac{1}{4})$

33. $64a^3 + 125b^3$ $(4a + 5b)(16a^2 - 20ab + 25b^2)$

34. $125a^3 - 27b^3$ $(5a - 3b)(25a^2 + 15ab + 9b^2)$

35. $\frac{1}{8}x^3 - \frac{1}{27}y^3$ $(\frac{1}{2}x - \frac{1}{3}y)(\frac{1}{4}x^2 + \frac{1}{6}xy + \frac{1}{9}y^2)$

36. $\frac{1}{27}x^3 + \frac{1}{8}y^3$ $(\frac{1}{3}x + \frac{1}{2}y)(\frac{1}{9}x^2 - \frac{1}{6}xy + \frac{1}{4}y^2)$

37. $a^6 - b^6$ $(a - b)(a^2 + ab + b^2)(a + b)(a^2 - ab + b^2)$

38. $x^6 - 64y^6$ $(x - 2y)(x^2 + 2xy + 4y^2)(x + 2y)(x^2 - 2xy + 4y^2)$

39. $64x^6 - y^6$ $(2x - y)(4x^2 + 2xy + y^2)(2x + y)(4x^2 - 2xy + y^2)$

▶ **40.** $x^6 - (3y)^6$ $(x - 3y)(x^2 + 3xy + 9y^2)(x + 3y)(x^2 - 3xy + 9y^2)$

41. $x^6 - (5y)^6$ $(x - 5y)(x^2 + 5xy + 25y^2)(x + 5y)(x^2 - 5xy + 25y^2)$

42. $(4x)^6 - (7y)^6$
$(4x - 7y)(16x^2 + 28xy + 49y^2)(4x + 7y)(16x^2 - 28xy + 49y^2)$

= Videos available by instructor request

▶ = Online student support materials available at www.thomsonedu.com/login

Maintaining Your Skills

Solve each equation for x.

43. $2x - 6y = 8$ $\quad x = 3y + 4$

44. $-3x + 9y = 12$ $\quad x = 3y - 4$

45. $4x - 6y = 8$ $\quad x = \dfrac{3}{2}y + 2$

46. $-20x + 15y = -10$ $\quad x = \dfrac{3}{4}y + \dfrac{1}{2}$

Solve each equation for y.

47. $3x - 6y = -18$ $\quad y = \dfrac{1}{2}x + 3$

48. $-3x + 9y = -18$ $\quad y = \dfrac{1}{3}x - 2$

49. $4x - 6y = 24$ $\quad y = \dfrac{2}{3}x - 4$

50. $-20x + 5y = -10$ $\quad y = 4x - 2$

Getting Ready for the Next Section

Multiply each of the following.

51. $2x^3(x + 2)(x - 2)$ $\quad 2x^5 - 8x^3$

52. $3x^2(x + 3)(x - 3)$ $\quad 3x^4 - 27x^2$

53. $3x^2(x - 3)^2$ $\quad 3x^4 - 18x^3 + 27x^2$

54. $2x^3(x + 5)^2$ $\quad 2x^5 + 20x^4 + 50x^3$

55. $y(y^2 + 25)$ $\quad y^3 + 25y$

56. $y^3(y^2 + 36)$ $\quad y^5 + 36y^3$

57. $(5a - 2)(3a + 1)$ $\quad 15a^2 - a - 2$

58. $(3a - 4)(2a - 1)$ $\quad 6a^2 - 11a + 4$

59. $4x^2(x - 5)(x + 2)$ $\quad 4x^4 - 12x^3 - 40x^2$

60. $6x(x - 4)(x + 2)$ $\quad 6x^3 - 12x^2 - 48x$

61. $2ab^3(b^2 - 4b + 1)$ $\quad 2ab^5 - 8ab^4 + 2ab^3$

62. $2a^3b(a^2 + 3a + 1)$ $\quad 2a^5b + 6a^4b + 2a^3b$

6.6 Factoring: A General Review

OBJECTIVES

A Factor a variety of polynomials.

In this section we will review the different methods of factoring that we presented in the previous sections of the chapter. This section is important because it will give you an opportunity to factor a variety of polynomials. Prior to this section, the polynomials you worked with were grouped together according to the method used to factor them; that is, in Section 6.4 all the polynomials you factored were either the difference of two squares or perfect square trinomials. What usually happens in a situation like this is that you become proficient at factoring the kind of polynomial you are working with at the time but have trouble when given a variety of polynomials to factor.

We begin this section with a checklist that can be used in factoring polynomials of any type. When you have finished this section and the problem set that follows, you want to be proficient enough at factoring that the checklist is second nature to you.

> **Strategy for Factoring a Polynomial**
> **Step 1:** If the polynomial has a greatest common factor other than 1, then factor out the greatest common factor.
>
> **Step 2:** If the polynomial has two terms (it is a binomial), then see if it is the difference of two squares or the sum or difference of two cubes, and then factor accordingly. Remember, if it is the sum of two squares, it will not factor.
>
> **Step 3:** If the polynomial has three terms (a trinomial), then either it is a perfect square trinomial, which will factor into the square of a binomial, or it is not a perfect square trinomial, in which case you use the trial and error method developed in Section 6.3.
>
> **Step 4:** If the polynomial has more than three terms, try to factor it by grouping.
>
> **Step 5:** As a final check, see if any of the factors you have written can be factored further. If you have overlooked a common factor, you can catch it here.

Here are some examples illustrating how we use the checklist.

 EXAMPLE 1   Factor $2x^5 - 8x^3$.

SOLUTION   First, we check to see if the greatest common factor is other than 1. Since the greatest common factor is $2x^3$, we begin by factoring it out. Once we have done so, we notice that the binomial that remains is the difference of two squares:

$$2x^5 - 8x^3 = 2x^3(x^2 - 4)$$ **Factor out the greatest common factor, $2x^3$**

$$= 2x^3(x + 2)(x - 2)$$ **Factor the difference of two squares**

Note that the greatest common factor $2x^3$ that we factored from each term in the first step of Example 1 remains as part of the answer to the problem; that is because it is one of the factors of the original binomial. Remember, the expression we end up with when factoring must be equal to the expression we start with. We can't just drop a factor and expect the resulting expression to equal the original expression.

 EXAMPLE 2   Factor $3x^4 - 18x^3 + 27x^2$.

SOLUTION   Step 1 is to factor out the greatest common factor, $3x^2$. After we have done so, we notice that the trinomial that remains is a perfect square trinomial, which will factor as the square of a binomial:

$$3x^4 - 18x^3 + 27x^2 = 3x^2(x^2 - 6x + 9)$$ **Factor out $3x^2$**

$$= 3x^2(x - 3)^2$$ **$x^2 - 6x + 9$ is the square of $x - 3$**

EXAMPLE 3   Factor $y^3 + 25y$.

SOLUTION   We begin by factoring out the y that is common to both terms. The binomial that remains after we have done so is the sum of two squares, which does not factor, so after the first step we are finished:

$$y^3 + 25y = y(y^2 + 25)$$ **Factor out the greatest common factor, y; then notice that $y^2 + 25$ cannot be factored further**

EXAMPLE 4   Factor $6a^2 - 11a + 4$.

SOLUTION   Here we have a trinomial that does not have a greatest common factor other than 1. Since it is not a perfect square trinomial, we factor it by trial and error; that is, we look for binomial factors the product of whose first terms is $6a^2$ and the product of whose last terms is 4. Then we look for the combination of these types of binomials whose product gives us a middle term of $-11a$. Without showing all the different possibilities, here is the answer:

$$6a^2 - 11a + 4 = (3a - 4)(2a - 1)$$

EXAMPLE 5   Factor $6x^3 - 12x^2 - 48x$.

SOLUTION   This trinomial has a greatest common factor of $6x$. The trinomial that remains after the $6x$ has been factored from each term must be factored by trial and error:

$$6x^3 - 12x^2 - 48x = 6x(x^2 - 2x - 8)$$
$$= 6x(x - 4)(x + 2)$$

EXAMPLE 6   Factor $2ab^5 + 8ab^4 + 2ab^3$.

SOLUTION   The greatest common factor is $2ab^3$. We begin by factoring it from each term. After that we find the trinomial that remains cannot be factored further:

$$2ab^5 + 8ab^4 + 2ab^3 = 2ab^3(b^2 + 4b + 1)$$

EXAMPLE 7   Factor $xy + 8x + 3y + 24$.

SOLUTION   Since our polynomial has four terms, we try factoring by grouping:

$$xy + 8x + 3y + 24 = x(y + 8) + 3(y + 8)$$
$$= (y + 8)(x + 3)$$

LINKING OBJECTIVES AND EXAMPLES

Next to each **objective** we have listed the examples that are best described by that objective.

A   1–7

GETTING READY FOR CLASS

After reading through the preceding section, respond in your own words and in complete sentences.

1. What is the first step in factoring any polynomial?
2. If a polynomial has four terms, what method of factoring should you try?
3. If a polynomial has two terms, what method of factoring should you try?
4. What is the last step in factoring any polynomial?

Factor each of the following polynomials completely; that is, once you are finished factoring, none of the factors you obtain should be factorable. Also, note that the even-numbered problems are not necessarily similar to the odd-numbered problems that precede them in this problem set.

▶ **1.** $x^2 - 81$ $(x + 9)(x - 9)$

2. $x^2 - 18x + 81$ $(x - 9)^2$

3. $x^2 + 2x - 15$ $(x + 5)(x - 3)$

4. $15x^2 + 11x - 6$ Cannot be factored.

5. $x^2 + 6x + 9$ $(x + 3)^2$

6. $12x^2 - 11x + 2$ $(4x - 1)(3x - 2)$

▶ **7.** $y^2 - 10y + 25$ $(y - 5)^2$

8. $21y^2 - 25y - 4$ $(7y + 1)(3y - 4)$

9. $2a^3b + 6a^2b + 2ab$ $2ab(a^2 + 3a + 1)$

▶ **10.** $6a^2 - ab - 15b^2$ $(2a + 3b)(3a - 5b)$

11. $x^2 + x + 1$ Cannot be factored.

12. $2x^2 - 4x + 2$ $2(x - 1)^2$

13. $12a^2 - 75$ $3(2a + 5)(2a - 5)$

14. $18a^2 - 50$ $2(3a + 5)(3a - 5)$

15. $9x^2 - 12xy + 4y^2$ $(3x - 2y)^2$

16. $x^3 - x^2$ $x^2(x - 1)$

17. $4x^3 + 16xy^2$ $4x(x^2 + 4y^2)$

18. $16x^2 + 49y^2$ Cannot be factored.

19. $2y^3 + 20y^2 + 50y$ $2y(y + 5)^2$

20. $3y^2 - 9y - 30$ $3(y - 5)(y + 2)$

21. $a^6 + 4a^4b^2$ $a^4(a^2 + 4b^2)$

22. $5a^2 - 45b^2$ $5(a + 3b)(a - 3b)$

23. $xy + 3x + 4y + 12$ $(x + 4)(y + 3)$

24. $xy + 7x + 6y + 42$ $(x + 6)(y + 7)$

25. $x^3 - 27$ $(x - 3)(x^2 + 3x + 9)$

26. $x^4 - 81$ $(x^2 + 9)(x + 3)(x - 3)$

▶ **27.** $xy - 5x + 2y - 10$ $(x + 2)(y - 5)$

28. $xy - 7x + 3y - 21$ $(x + 3)(y - 7)$

29. $5a^2 + 10ab + 5b^2$ $5(a + b)^2$

30. $3a^3b^2 + 15a^2b^2 + 3ab^2$ $3ab^2(a^2 + 5a + 1)$

31. $x^2 + 49$ Cannot be factored.

32. $16 - x^4$ $(4 + x^2)(2 + x)(2 - x)$

▶ **33.** $3x^2 + 15xy + 18y^2$ $3(x + 2y)(x + 3y)$

34. $3x^2 + 27xy + 54y^2$ $3(x + 3y)(x + 6y)$

35. $2x^2 + 15x - 38$ $(2x + 19)(x - 2)$

36. $2x^2 + 7x - 85$ $(2x + 17)(x - 5)$

37. $100x^2 - 300x + 200$ $100(x - 2)(x - 1)$

38. $100x^2 - 400x + 300$ $100(x - 3)(x - 1)$

39. $x^2 - 64$ $(x + 8)(x - 8)$

40. $9x^2 - 4$ $(3x + 2)(3x - 2)$

41. $x^2 + 3x + ax + 3a$ $(x + a)(x + 3)$

42. $x^2 + 4x + bx + 4b$ $(x + b)(x + 4)$

43. $49a^7 - 9a^5$ $a^5(7a + 3)(7a - 3)$

44. $8a^3 + 1$ $(2a + 1)(4a^2 - 2a + 1)$

▶ **45.** $49x^2 + 9y^2$ Cannot be factored.

46. $12x^4 - 62x^3 + 70x^2$ $2x^2(2x - 7)(3x - 5)$

47. $25a^3 + 20a^2 + 3a$ $a(5a + 1)(5a + 3)$

48. $36a^4 - 100a^2$ $4a^2(3a + 5)(3a - 5)$

49. $xa - xb + ay - by$ $(x + y)(a - b)$

50. $xy - bx + ay - ab$ $(x + a)(y - b)$

51. $48a^4b - 3a^2b$ $3a^2b(4a + 1)(4a - 1)$

52. $18a^4b^2 - 12a^3b^3 + 8a^2b^4$ $2a^2b^2(9a^2 - 6ab + 4b^2)$

53. $20x^4 - 45x^2$ $5x^2(2x + 3)(2x - 3)$

54. $16x^3 + 16x^2 + 3x$ $x(4x + 1)(4x + 3)$

55. $3x^2 + 35xy - 82y^2$ $(3x + 41y)(x - 2y)$

56. $3x^2 + 37xy - 86y^2$ $(3x + 43y)(x - 2y)$

57. $16x^5 - 44x^4 + 30x^3$ $2x^3(2x - 3)(4x - 5)$

58. $16x^2 + 16x - 1$ Cannot be factored.

59. $2x^2 + 2ax + 3x + 3a$ $(2x + 3)(x + a)$

60. $2x^2 + 2ax + 5x + 5a$ $(2x + 5)(x + a)$

61. $y^4 - 1$ $(y^2 + 1)(y + 1)(y - 1)$

62. $a^7 + 8a^4b^3$ $a^4(a + 2b)(a^2 - 2ab + 4b^2)$

63. $12x^4y^2 + 36x^3y^3 + 27x^2y^4$ $3x^2y^2(2x + 3y)^2$

64. $16x^3y^2 - 4xy^2$ $4xy^2(2x + 1)(2x - 1)$

▶ **65.** $16t^2 - 64t + 48$ $16(t - 1)(t - 3)$

▶ **66.** $16t^2 - 32t + 12$ $4(2t - 1)(2t - 3)$

= Videos available by instructor request

▶ = Online student support materials available at www.thomsonedu.com/login

▶ **67.** $54x^2 + 111x + 56$ $(9x + 8)(6x + 7)$

▶ **68.** $63x^2 + 110x + 48$ $(9x + 8)(7x + 6)$

Maintaining Your Skills

Solve each equation.

69. $-2(x + 4) = -10$ 1

70. $\frac{3}{4}(-4x - 8) = 21$ -9

71. $\frac{3}{5}x + 4 = 22$ 30

72. $-10 = 4 - \frac{7}{4}x$ 8

73. $6x - 4(9 - x) = -96$ -6

74. $-2(x - 5) + 5x = 4 - 9$ -5

75. $2x - 3(4x - 7) = -3x$ 3

76. $\frac{3}{4}(8 + x) = \frac{1}{5}(5x - 15)$ 36

77. $\frac{1}{2}x - \frac{5}{12} = \frac{1}{12}x + \frac{5}{12}$ 2

78. $\frac{3}{10}x + \frac{5}{2} = \frac{3}{5}x - \frac{1}{2}$ 10

Getting Ready for the Next Section

Solve each equation.

79. $3x - 6 = 9$ 5

80. $5x - 1 = 14$ 3

81. $2x + 3 = 0$ $-\frac{3}{2}$

82. $4x - 5 = 0$ $\frac{5}{4}$

83. $4x + 3 = 0$ $-\frac{3}{4}$

84. $3x - 1 = 0$ $\frac{1}{3}$

6.7 Solving Equations by Factoring

OBJECTIVES

A Solve an equation by writing it in standard form and then factoring.

In this section we will use the methods of factoring developed in previous sections, along with a special property of 0, to solve quadratic equations.

> **DEFINITION** Any equation that can be put in the form $ax^2 + bx + c = 0$, where a, b, and c are real numbers ($a \neq 0$), is called a **quadratic equation.** The equation $ax^2 + bx + c = 0$ is called **standard form** for a quadratic equation:
>
> an x^2 term   an x term   and a constant term
> $a(\text{variable})^2 + b(\text{variable}) + (\text{absence of the variable}) = 0$

The number 0 has a special property. If we multiply two numbers and the product is 0, then one or both of the original two numbers must be 0. In symbols, this property looks like this.

> **Zero-Factor Property** Let a and b represent real numbers. If $a \cdot b = 0$, then $a = 0$ or $b = 0$.

Suppose we want to solve the quadratic equation $x^2 + 5x + 6 = 0$. We can factor the left side into $(x + 2)(x + 3)$. Then we have:

$$x^2 + 5x + 6 = 0$$
$$(x + 2)(x + 3) = 0$$

Now, $(x + 2)$ and $(x + 3)$ both represent real numbers. Their product is 0; therefore, either $(x + 3)$ is 0 or $(x + 2)$ is 0. Either way we have a solution to our equation. We use the property of 0 stated to finish the problem:

$$x^2 + 5x + 6 = 0$$
$$(x + 2)(x + 3) = 0$$
$$x + 2 = 0 \quad \text{or} \quad x + 3 = 0$$
$$x = -2 \quad \text{or} \quad x = -3$$

Our solution set is $\{-2, -3\}$. Our equation has two solutions. To check our solutions we have to check each one separately to see that they both produce a true statement when used in place of the variable:

When $x = -3$
the equation $x^2 + 5x + 6 = 0$
becomes $(-3)^2 + 5(-3) + 6 \overset{?}{=} 0$
$9 + (-15) + 6 = 0$
$0 = 0$

When $x = -2$
the equation $x^2 + 5x + 6 = 0$
becomes $(-2)^2 + 5(-2) + 6 \overset{?}{=} 0$
$4 + (-10) + 6 = 0$
$0 = 0$

We have solved a quadratic equation by replacing it with two linear equations in one variable.

Strategy for Solving a Quadratic Equation by Factoring

Step 1: Put the equation in standard form; that is, 0 on one side and decreasing powers of the variable on the other.

Step 2: Factor completely.

Step 3: Use the zero-factor property to set each variable factor from step 2 to 0.

Step 4: Solve each equation produced in step 3.

Step 5: Check each solution, if necessary.

Note
Notice that to solve a quadratic equation by this method, it must be possible to factor it. If we can't factor it, we can't solve it by this method. We will learn how to solve quadratic equations that do not factor when we get to Chapter 9.

EXAMPLE 1   Solve the equation $2x^2 - 5x = 12$.

SOLUTION

Step 1: Begin by adding -12 to both sides, so the equation is in standard form:

$$2x^2 - 5x = 12$$
$$2x^2 - 5x - 12 = 0$$

Step 2: Factor the left side completely:

$$(2x + 3)(x - 4) = 0$$

Step 3: Set each factor to 0:

$$2x + 3 = 0 \quad \text{or} \quad x - 4 = 0$$

Step 4: Solve each of the equations from step 3:

$$2x + 3 = 0 \qquad x - 4 = 0$$
$$2x = -3 \qquad x = 4$$
$$x = -\frac{3}{2}$$

Step 5: Substitute each solution into $2x^2 - 5x = 12$ to check:

$$\text{Check: } -\frac{3}{2} \qquad\qquad \text{Check: } 4$$

$$2\left(-\frac{3}{2}\right)^2 - 5\left(-\frac{3}{2}\right) \overset{?}{=} 12 \qquad 2(4)^2 - 5(4) \overset{?}{=} 12$$

$$2\left(\frac{9}{4}\right) + 5\left(\frac{3}{2}\right) = 12 \qquad 2(16) - 20 = 12$$

$$\frac{9}{2} + \frac{15}{2} = 12 \qquad 32 - 20 = 12$$

$$\frac{24}{2} = 12 \qquad\qquad 12 = 12$$

$$12 = 12$$

EXAMPLE 2 Solve for a: $16a^2 - 25 = 0$.

SOLUTION The equation is already in standard form:

$$16a^2 - 25 = 0$$
$$(4a - 5)(4a + 5) = 0 \qquad \textbf{Factor left side}$$
$$4a - 5 = 0 \quad \text{or} \quad 4a + 5 = 0 \qquad \textbf{Set each factor to 0}$$
$$4a = 5 \qquad\qquad 4a = -5 \qquad \textbf{Solve the resulting equations}$$
$$a = \frac{5}{4} \qquad\qquad a = -\frac{5}{4}$$

EXAMPLE 3 Solve $4x^2 = 8x$.

SOLUTION We begin by adding $-8x$ to each side of the equation to put it in standard form. Then we factor the left side of the equation by factoring out the greatest common factor.

$$4x^2 = 8x$$
$$4x^2 - 8x = 0 \qquad \textbf{Add } -8x \textbf{ to each side}$$
$$4x(x - 2) = 0 \qquad \textbf{Factor the left side}$$
$$4x = 0 \quad \text{or} \quad x - 2 = 0 \qquad \textbf{Set each factor to 0}$$
$$x = 0 \quad \text{or} \quad x = 2 \qquad \textbf{Solve the resulting equations}$$

The solutions are 0 and 2.

EXAMPLE 4   Solve $x(2x + 3) = 44$.

SOLUTION   We must multiply out the left side first and then put the equation in standard form:

$$x(2x + 3) = 44$$
$$2x^2 + 3x = 44 \qquad \text{Multiply out the left side}$$
$$2x^2 + 3x - 44 = 0 \qquad \text{Add } -44 \text{ to each side}$$
$$(2x + 11)(x - 4) = 0 \qquad \text{Factor the left side}$$
$$2x + 11 = 0 \quad \text{or} \quad x - 4 = 0 \qquad \text{Set each factor to 0}$$
$$2x = -11 \quad \text{or} \quad x = 4 \qquad \text{Solve the resulting equations}$$
$$x = -\frac{11}{2}$$

The two solutions are $-\frac{11}{2}$ and 4.

EXAMPLE 5   Solve for x: $5^2 = x^2 + (x + 1)^2$.

SOLUTION   Before we can put this equation in standard form we must square the binomial. Remember, to square a binomial, we use the formula $(a + b)^2 = a^2 + 2ab + b^2$:

$$5^2 = x^2 + (x + 1)^2$$
$$25 = x^2 + x^2 + 2x + 1 \qquad \textbf{Expand } 5^2 \textbf{ and } (x + 1)^2$$
$$25 = 2x^2 + 2x + 1 \qquad \textbf{Simplify the right side}$$
$$0 = 2x^2 + 2x - 24 \qquad \textbf{Add } -25 \textbf{ to each side}$$
$$0 = 2(x^2 + x - 12) \qquad \textbf{Begin factoring}$$
$$0 = 2(x + 4)(x - 3) \qquad \textbf{Factor completely}$$
$$x + 4 = 0 \quad \text{or} \quad x - 3 = 0 \qquad \textbf{Set each variable factor to 0}$$
$$x = -4 \quad \text{or} \quad x = 3$$

Note, in the second to the last line, that we do not set 2 equal to 0. That is because 2 can never be 0. It is always 2. We only use the zero-factor property to set variable factors to 0 because they are the only factors that can possibly be 0.

Also notice that it makes no difference which side of the equation is 0 when we write the equation in standard form.

Although the equation in the next example is not a quadratic equation, it can be solved by the method shown in the first five examples.

EXAMPLE 6   Solve $24x^3 = -10x^2 + 6x$ for x.

SOLUTION   First, we write the equation in standard form:

$$24x^3 + 10x^2 - 6x = 0 \qquad \textbf{Standard form}$$
$$2x(12x^2 + 5x - 3) = 0 \qquad \textbf{Factor out } 2x$$
$$2x(3x - 1)(4x + 3) = 0 \qquad \textbf{Factor remaining trinomial}$$
$$2x = 0 \quad \text{or} \quad 3x - 1 = 0 \quad \text{or} \quad 4x + 3 = 0 \qquad \textbf{Set factors to 0}$$
$$x = 0 \quad \text{or} \quad x = \frac{1}{3} \quad \text{or} \quad x = -\frac{3}{4} \qquad \textbf{Solutions}$$

LINKING OBJECTIVES AND EXAMPLES

Next to each **objective** we have listed the examples that are best described by that objective.

A 1–6

GETTING READY FOR CLASS

After reading through the preceding section, respond in your own words and in complete sentences.

1. When is an equation in standard form?
2. What is the first step in solving an equation by factoring?
3. Describe the zero-factor property in your own words.
4. Describe how you would solve the equation $2x^2 - 5x = 12$.

Problem Set 6.7

Online support materials can be found at www.thomsonedu.com/login

The following equations are already in factored form. Use the special zero factor property to set the factors to 0 and solve.

1. $(x + 2)(x - 1) = 0$ $-2, 1$
2. $(x + 3)(x + 2) = 0$ $-3, -2$
3. $(a - 4)(a - 5) = 0$ $4, 5$
4. $(a + 6)(a - 1) = 0$ $-6, 1$
5. $x(x + 1)(x - 3) = 0$ $0, -1, 3$
6. $x(2x + 1)(x - 5) = 0$ $0, -\frac{1}{2}, 5$
7. $(3x + 2)(2x + 3) = 0$ $-\frac{2}{3}, -\frac{3}{2}$
8. $(4x - 5)(x - 6) = 0$ $\frac{5}{4}, 6$
9. $m(3m + 4)(3m - 4) = 0$ $0, -\frac{4}{3}, \frac{4}{3}$
10. $m(2m - 5)(3m - 1) = 0$ $0, \frac{5}{2}, \frac{1}{3}$
11. $2y(3y + 1)(5y + 3) = 0$ $0, -\frac{1}{3}, -\frac{3}{5}$
12. $3y(2y - 3)(3y - 4) = 0$ $0, \frac{3}{2}, \frac{4}{3}$

Solve the following equations

13. $x^2 + 3x + 2 = 0$ $-1, -2$
14. $x^2 - x - 6 = 0$ $3, -2$
15. $x^2 - 9x + 20 = 0$ $4, 5$
16. $x^2 + 2x - 3 = 0$ $1, -3$
17. $a^2 - 2a - 24 = 0$ $6, -4$
18. $a^2 - 11a + 30 = 0$ $5, 6$
19. $100x^2 - 500x + 600 = 0$ $2, 3$
20. $100x^2 - 300x + 200 = 0$ $1, 2$
21. $x^2 = -6x - 9$ -3
22. $x^2 = 10x - 25$ 5
23. $a^2 - 16 = 0$ $4, -4$
24. $a^2 - 36 = 0$ $6, -6$
25. $2x^2 + 5x - 12 = 0$ $\frac{3}{2}, -4$
26. $3x^2 + 14x - 5 = 0$ $\frac{1}{3}, -5$

27. $9x^2 + 12x + 4 = 0$ $-\frac{2}{3}$
28. $12x^2 - 24x + 9 = 0$ $\frac{1}{2}, \frac{3}{2}$
29. $a^2 + 25 = 10a$ 5
30. $a^2 + 16 = 8a$ 4
31. $2x^2 = 3x + 20$ $4, -\frac{5}{2}$
32. $6x^2 = x + 2$ $\frac{2}{3}, -\frac{1}{2}$
33. $3m^2 = 20 - 7m$ $\frac{5}{3}, -4$
34. $2m^2 = -18 + 15m$ $\frac{3}{2}, 6$
35. $4x^2 - 49 = 0$ $\frac{7}{2}, -\frac{7}{2}$
36. $16x^2 - 25 = 0$ $\frac{5}{4}, -\frac{5}{4}$
37. $x^2 + 6x = 0$ $0, -6$
38. $x^2 - 8x = 0$ $0, 8$
39. $x^2 - 3x = 0$ $0, 3$
40. $x^2 + 5x = 0$ $0, -5$
41. $2x^2 = 8x$ $0, 4$
42. $2x^2 = 10x$ $0, 5$
43. $3x^2 = 15x$ $0, 5$
44. $5x^2 = 15x$ $0, 3$
45. $1,400 = 400 + 700x - 100x^2$ $2, 5$
46. $2,700 = 700 + 900x - 100x^2$ $4, 5$
47. $6x^2 = -5x + 4$ $\frac{1}{2}, -\frac{4}{3}$
48. $9x^2 = 12x - 4$ $\frac{2}{3}$
49. $x(2x - 3) = 20$ $4, -\frac{5}{2}$
50. $x(3x - 5) = 12$ $3, -\frac{4}{3}$
51. $t(t + 2) = 80$ $8, -10$
52. $t(t + 2) = 99$ $9, -11$
53. $4,000 = (1,300 - 100p)p$ $5, 8$
54. $3,200 = (1,200 - 100p)p$ $4, 8$
55. $x(14 - x) = 48$ $6, 8$
56. $x(12 - x) = 32$ $4, 8$
57. $(x + 5)^2 = 2x + 9$ -4
58. $(x + 7)^2 = 2x + 13$ -6
59. $(y - 6)^2 = y - 4$ $5, 8$
60. $(y + 4)^2 = y + 6$ $-2, -5$
61. $10^2 = (x + 2)^2 + x^2$ $6, -8$
62. $15^2 = (x + 3)^2 + x^2$ $9, -12$
63. $2x^3 + 11x^2 + 12x = 0$ $0, -\frac{3}{2}, -4$
64. $3x^3 + 17x^2 + 10x = 0$ $0, -\frac{2}{3}, -5$
65. $4y^3 - 2y^2 - 30y = 0$ $0, 3, -\frac{5}{2}$

= Videos available by instructor request
▶ = Online student support materials available at www.thomsonedu.com/login

66. $9y^3 + 6y^2 - 24y = 0$ $0, \frac{4}{3}, -2$

67. $8x^3 + 16x^2 = 10x$ $0, \frac{1}{2}, -\frac{5}{2}$

68. $24x^3 - 22x^2 = -4x$ $0, \frac{1}{4}, \frac{2}{3}$

69. $20a^3 = -18a^2 + 18a$ $0, \frac{3}{5}, -\frac{3}{2}$

70. $12a^3 = -2a^2 + 10a$ $0, \frac{5}{6}, -1$

▶ **71.** $16t^2 - 32t + 12 = 0$ $\frac{1}{2}, \frac{3}{2}$

▶ **72.** $16t^2 - 64t + 48 = 0$ $1, 3$

Simplify each side as much as possible, then solve the equation.

▶ **73.** $(a - 5)(a + 4) = -2a$ $-5, 4$

▶ **74.** $(a + 2)(a - 3) = -2a$ $-3, 2$

▶ **75.** $3x(x + 1) - 2x(x - 5) = -42$ $-7, -6$

▶ **76.** $4x(x - 2) - 3x(x - 4) = -3$ $-3, -1$

▶ **77.** $2x(x + 3) = x(x + 2) - 3$ $-3, -1$

▶ **78.** $3x(x - 3) = 2x(x - 4) + 6$ $-2, 3$

▶ **79.** $a(a - 3) + 6 = 2a$ $2, 3$

▶ **80.** $a(a - 4) + 8 = 2a$ $2, 4$

▶ **81.** $15(x + 20) + 15x = 2x(x + 20)$ $-15, 10$

▶ **82.** $15(x + 8) + 15x = 2x(x + 8)$ $-5, 12$

▶ **83.** $15 = a(a + 2)$ $-5, 3$

▶ **84.** $6 = a(a - 5)$ $-1, 6$

Use factoring by grouping to solve the following equations.

85. $x^3 + 3x^2 - 4x - 12 = 0$ $-3, -2, 2$

86. $x^3 + 5x^2 - 9x - 45 = 0$ $3, -3, -5$

87. $x^3 + x^2 - 16x - 16 = 0$ $-4, -1, 4$

88. $4x^3 + 12x^2 - 9x - 27 = 0$ $\frac{3}{2}, -\frac{3}{2}, -3$

89. Find a quadratic equation that has two solutions: $x = 3$ and $x = 5$. Write your answer in standard form. $x^2 - 8x + 15 = 0$

90. Find a quadratic equation that has two solutions: $x = 9$ and $x = 1$. Write your answer in standard form. $x^2 - 10x + 9 = 0$

91. Find a quadratic equation that has the two given solutions.
a. $x = 3$ and $x = 2$. $x^2 - 5x + 6 = 0$
b. $x = 1$ and $x = 6$. $x^2 - 7x + 6 = 0$
c. $x = 3$ and $x = -2$. $x^2 - x - 6 = 0$

92. Find a quadratic equation that has the two given solutions.
a. $x = 4$ and $x = 5$. $x^2 - 9x + 20 = 0$
b. $x = 2$ and $x = 10$. $x^2 - 12x + 20 = 0$
c. $x = -4$ and $x = 5$. $x^2 - x - 20 = 0$

Maintaining Your Skills

Use the properties of exponents to simplify each expression.

93. 2^{-3} $\frac{1}{8}$

94. 5^{-2} $\frac{1}{25}$

95. $\frac{x^5}{x^{-3}}$ x^8

96. $\frac{x^{-2}}{x^{-5}}$ x^3

97. $\frac{(x^2)^3}{(x^{-3})^4}$ x^{18}

98. $\frac{(x^2)^{-4}(x^{-2})^3}{(x^{-3})^{-5}}$ $\frac{1}{x^{29}}$

99. Write the number 0.0056 in scientific notation. 5.6×10^{-3}

100. Write the number 2.34×10^{-4} in expanded form. 0.000234

101. Write the number 5,670,000,000 in scientific notation. 5.67×10^9

102. Write the number 0.00000567 in scientific notation. 5.67×10^{-6}

Getting Ready for the Next Section

Write each sentence as an algebraic equation.

103. The product of two consecutive integers is 72. $x(x + 1) = 72$

104. The product of two consecutive even integers is 80. $x(x + 2) = 80$

105. The product of two consecutive odd integers is 99. $x(x + 2) = 99$

106. The product of two consecutive odd integers is 63. $x(x + 2) = 63$

107. The product of two consecutive even integers is 10 less than 5 times their sum. $x(x + 2) = 5[x + (x + 2)] - 10$

108. The product of two consecutive odd integers is 1 less than 4 times their sum. $x(x + 2) = 4[x + (x + 2)] - 1$

Foreshadowing Problems
Problems 73–84 help students get ready for some of the equations they will solve in the next chapter.

The following word problems are taken from the book *Academic Algebra,* written by William J. Milne and published by the American Book Company in 1901. Solve each problem.

109. Cost of a Bicycle and a Suit A bicycle and a suit cost $90. How much did each cost, if the bicycle cost 5 times as much as the suit? Bicycle $75, suit $15

110. Cost of a Cow and a Calf A man bought a cow and a calf for $36, paying 8 times as much for the cow as for the calf. What was the cost of each?
Cow $32, calf $4

111. Cost of a House and a Lot A house and a lot cost $3,000. If the house cost 4 times as much as the lot, what was the cost of each? House $2,400, lot $600

112. Daily Wages A plumber and two helpers together earned $7.50 per day. How much did each earn per day, if the plumber earned 4 times as much as each helper? Plumber $5, helper $1.25

6.8 Applications

OBJECTIVES

A Solve a variety of application problems

In this section we will look at some application problems, the solutions to which require solving a quadratic equation. We will also introduce the Pythagorean theorem, one of the oldest theorems in the history of mathematics. The person whose name we associate with the theorem, Pythagoras (of Samos), was a Greek philosopher and mathematician who lived from about 560 B.C. to 480 B.C. According to the British philosopher Bertrand Russell, Pythagoras was "intellectually one of the most important men that ever lived."

Also in this section, the solutions to the examples show only the essential steps from our Blueprint for Problem Solving. Recall that step 1 is done mentally; we read the problem and mentally list the items that are known and the items that are unknown. This is an essential part of problem solving. However, now that you have had experience with application problems, you are doing step 1 automatically.

Number Problems

 EXAMPLE 1 The product of two consecutive odd integers is 63. Find the integers.

SOLUTION Let x = the first odd integer; then $x + 2$ = the second odd integer. An equation that describes the situation is:

$$x(x + 2) = 63 \quad \textbf{Their product is 63}$$

We solve the equation:

$$x(x + 2) = 63$$
$$x^2 + 2x = 63$$
$$x^2 + 2x - 63 = 0$$
$$(x - 7)(x + 9) = 0$$
$$x - 7 = 0 \quad \text{or} \quad x + 9 = 0$$
$$x = 7 \quad \text{or} \quad x = -9$$

If the first odd integer is 7, the next odd integer is $7 + 2 = 9$. If the first odd integer is -9, the next consecutive odd integer is $-9 + 2 = -7$. We have two pairs of consecutive odd integers that are solutions. They are 7, 9 and -9, -7.

We check to see that their products are 63:

$$7(9) = 63$$

$$-7(-9) = 63$$

Suppose we know that the sum of two numbers is 50. We want to find a way to represent each number using only one variable. If we let x represent one of the two numbers, how can we represent the other? Let's suppose for a moment that x turns out to be 30. Then the other number will be 20, because their sum is 50; that is, if two numbers add up to 50 and one of them is 30, then the other must be $50 - 30 = 20$. Generalizing this to any number x, we see that if two numbers have a sum of 50 and one of the numbers is x, then the other must be $50 - x$. The table that follows shows some additional examples.

If Two Numbers Have a Sum of	And One of Them Is	Then the Other Must Be
50	x	$50 - x$
100	x	$100 - x$
10	y	$10 - y$
12	n	$12 - n$

Now, let's look at an example that uses this idea.

 EXAMPLE 2   The sum of two numbers is 13. Their product is 40. Find the numbers.

SOLUTION   If we let x represent one of the numbers, then $13 - x$ must be the other number because their sum is 13. Since their product is 40, we can write:

$x(13 - x) = 40$	**The product of the two numbers is 40**
$13x - x^2 = 40$	**Multiply the left side**
$x^2 - 13x = -40$	**Multiply both sides by -1 and reverse the order of the terms on the left side**
$x^2 - 13x + 40 = 0$	**Add 40 to each side**
$(x - 8)(x - 5) = 0$	**Factor the left side**

$$x - 8 = 0 \quad \text{or} \quad x - 5 = 0$$

$$x = 8 \qquad\qquad x = 5$$

The two solutions are 8 and 5. If x is 8, then the other number is $13 - x = 13 - 8 = 5$. Likewise, if x is 5, the other number is $13 - x = 13 - 5 = 8$. Therefore, the two numbers we are looking for are 8 and 5. Their sum is 13 and their product is 40.

Geometry Problems

Many word problems dealing with area can best be described algebraically by quadratic equations.

 EXAMPLE 3   The length of a rectangle is 3 more than twice the width. The area is 44 square inches. Find the dimensions (find the length and width).

SOLUTION As shown in Figure 1, let x = the width of the rectangle. Then $2x + 3$ = the length of the rectangle because the length is three more than twice the width.

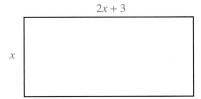

FIGURE 1

Since the area is 44 square inches, an equation that describes the situation is

$$x(2x + 3) = 44 \qquad \textbf{Length} \cdot \textbf{width} = \textbf{area}$$

We now solve the equation:

$$x(2x + 3) = 44$$
$$2x^2 + 3x = 44$$
$$2x^2 + 3x - 44 = 0$$
$$(2x + 11)(x - 4) = 0$$
$$2x + 11 = 0 \qquad \text{or} \qquad x - 4 = 0$$
$$x = -\frac{11}{2} \qquad \text{or} \qquad x = 4$$

The solution $x = -\frac{11}{2}$ cannot be used since length and width are always given in positive units. The width is 4. The length is 3 more than twice the width or $2(4) + 3 = 11$.

$$\text{Width} = 4 \text{ inches}$$

$$\text{Length} = 11 \text{ inches}$$

The solutions check in the original problem since $4(11) = 44$.

 EXAMPLE 4   The numerical value of the area of a square is twice its perimeter. What is the length of its side?

SOLUTION As shown in Figure 2, let x = the length of its side. Then x^2 = the area of the square and $4x$ = the perimeter of the square:

FIGURE 2

An equation that describes the situation is

$$x^2 = 2(4x) \qquad \textbf{The area is 2 times the perimeter}$$

$$x^2 = 8x$$

$$x^2 - 8x = 0$$

$$x(x - 8) = 0$$

$$x(x - 8) = 0$$

$$x = 0 \qquad \text{or} \qquad x - 8 = 0$$

$$x = 8$$

Since $x = 0$ does not make sense in our original problem, we use $x = 8$. If the side has length 8, then the perimeter is $4(8) = 32$ and the area is $8^2 = 64$. Since 64 is twice 32, our solution is correct.

FACTS FROM GEOMETRY

The Pythagorean Theorem

Next, we will work some problems involving the Pythagorean theorem, which we mentioned in the introduction to this section. It may interest you to know that Pythagoras formed a secret society around the year 540 B.C. Known as the Pythagoreans, members kept no written record of their work; everything was handed down by spoken word. They influenced not only mathematics, but religion, science, medicine, and music as well. Among other things, they discovered the correlation between musical notes and the reciprocals of counting numbers, $\frac{1}{2}, \frac{1}{3}, \frac{1}{4}$, and so on. In their daily lives, they followed strict dietary and moral rules to achieve a higher rank in future lives.

Pythagorean Theorem In any right triangle (Figure 3), the square of the longer side (called the hypotenuse) is equal to the sum of the squares of the other two sides (called legs).

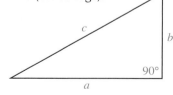

$$c^2 = a^2 + b^2$$

FIGURE 3

EXAMPLE 5   The three sides of a right triangle are three consecutive integers. Find the lengths of the three sides.

SOLUTION   Let x = the first integer (shortest side)

then $x + 1$ = the next consecutive integer

and $x + 2$ = the third consecutive integer (longest side)

A diagram of the triangle is shown in Figure 4.

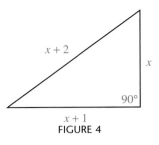

$x + 2$

x

90°

$x + 1$

FIGURE 4

The Pythagorean theorem tells us that the square of the longest side $(x + 2)^2$ is equal to the sum of the squares of the two shorter sides, $(x + 1)^2 + x^2$. Here is the equation:

$$(x + 2)^2 = (x + 1)^2 + x^2$$

$$x^2 + 4x + 4 = x^2 + 2x + 1 + x^2 \qquad \textbf{Expand squares}$$

$$x^2 - 2x - 3 = 0 \qquad \textbf{Standard form}$$

$$(x - 3)(x + 1) = 0 \qquad \textbf{Factor}$$

$$x - 3 = 0 \quad \text{or} \quad x + 1 = 0 \qquad \textbf{Set factors to 0}$$

$$x = 3 \quad \text{or} \quad x = -1$$

Since a triangle cannot have a side with a negative number for its length, we must not use -1 for a solution to our original problem; therefore, the shortest side is 3. The other two sides are the next two consecutive integers, 4 and 5.

 EXAMPLE 6   The hypotenuse of a right triangle is 5 inches, and the lengths of the two legs (the other two sides) are given by two consecutive integers. Find the lengths of the two legs.

SOLUTION   If we let $x =$ the length of the shortest side, then the other side must be $x + 1$. A diagram of the triangle is shown in Figure 5.

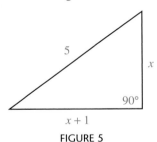

5

x

90°

$x + 1$

FIGURE 5

The Pythagorean theorem tells us that the square of the longest side, 5^2, is equal to the sum of the squares of the two shorter sides, $x^2 + (x + 1)^2$. Here is the equation:

$$5^2 = x^2 + (x + 1)^2 \qquad \textbf{Pythagorean theorem}$$

$$25 = x^2 + x^2 + 2x + 1 \qquad \textbf{Expand } 5^2 \text{ and } (x + 1)^2$$

$$25 = 2x^2 + 2x + 1 \qquad \textbf{Simplify the right side}$$

$$0 = 2x^2 + 2x - 24 \qquad \textbf{Add } -25 \text{ to each side}$$

$$0 = 2(x^2 + x - 12) \qquad \textbf{Begin factoring}$$

$$0 = 2(x + 4)(x - 3) \qquad \textbf{Factor completely}$$

$$x + 4 = 0 \quad \text{or} \quad x - 3 = 0 \qquad \textbf{Set variable factors to 0}$$

$$x = -4 \quad \text{or} \quad x = 3$$

Since a triangle cannot have a side with a negative number for its length, we cannot use -4; therefore, the shortest side must be 3 inches. The next side is

$x + 1 = 3 + 1 = 4$ inches. Since the hypotenuse is 5, we can check our solutions with the Pythagorean theorem as shown in Figure 6.

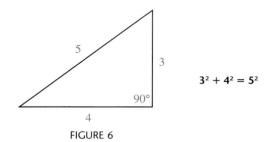

3² + 4² = 5²

FIGURE 6

EXAMPLE 7   A company can manufacture x hundred items for a total cost of $C = 300 + 500x - 100x^2$. How many items were manufactured if the total cost is $900?

SOLUTION   We are looking for x when C is 900. We begin by substituting 900 for C in the cost equation. Then we solve for x:

When $C = 900$

the equation $C = 300 + 500x - 100x^2$

becomes $900 = 300 + 500x - 100x^2$

> **Note**
>
> If you are planning on taking finite mathematics, statistics, or business calculus in the future, Examples 7 and 8 will give you a head start on some of the problems you will see in those classes.

We can write this equation in standard form by adding -300, $-500x$, and $100x^2$ to each side. The result looks like this:

$$100x^2 - 500x + 600 = 0$$
$$100(x^2 - 5x + 6) = 0 \qquad \textbf{Begin factoring}$$
$$100(x - 2)(x - 3) = 0 \qquad \textbf{Factor completely}$$
$$x - 2 = 0 \quad \text{or} \quad x - 3 = 0 \qquad \textbf{Set variable factors to 0}$$
$$x = 2 \quad \text{or} \qquad x = 3$$

Our solutions are 2 and 3, which means that the company can manufacture 2 hundred items or 3 hundred items for a total cost of $900.

EXAMPLE 8   A manufacturer of small portable radios knows that the number of radios she can sell each week is related to the price of the radios by the equation $x = 1{,}300 - 100p$ (x is the number of radios and p is the price per radio). What price should she charge for the radios to have a weekly revenue of $4,000?

SOLUTION   First, we must find the revenue equation. The equation for total revenue is $R = xp$, where x is the number of units sold and p is the price per unit. Since we want R in terms of p, we substitute $1{,}300 - 100p$ for x in the equation $R = xp$:

If $R = xp$

and $x = 1{,}300 - 100p$

then $R = (1{,}300 - 100p)p$

We want to find p when R is 4,000. Substituting 4,000 for R in the equation gives us:

$$4,000 = (1,300 - 100p)p$$

If we multiply out the right side, we have:

$$4,000 = 1,300p - 100p^2$$

To write this equation in standard form, we add $100p^2$ and $-1,300p$ to each side:

$$100p^2 - 1,300p + 4,000 = 0 \quad \text{Add } 100p^2 \text{ and } -1,300p \text{ to each side}$$
$$100(p^2 - 13p + 40) = 0 \quad \text{Begin factoring}$$
$$100(p - 5)(p - 8) = 0 \quad \text{Factor completely}$$
$$p - 5 = 0 \quad \text{or} \quad p - 8 = 0 \quad \text{Set variable factors to 0}$$
$$p = 5 \quad \text{or} \quad p = 8$$

If she sells the radios for \$5 each or for \$8 each, she will have a weekly revenue of \$4,000.

GETTING READY FOR CLASS

After reading through the preceding section, respond in your own words and in complete sentences.

1. What are consecutive integers?
2. Explain the Pythagorean theorem in words.
3. Write an application problem for which the solution depends on solving the equation $x(x + 1) = 12$.
4. Write an application problem for which the solution depends on solving the equation $x(2x - 3) = 40$.

LINKING OBJECTIVES AND EXAMPLES

Next to each **objective** we have listed the examples that are best described by that objective.

A   1–8

Problem Set 6.8

Online support materials can be found at www.thomsonedu.com/login

Solve the following word problems. Be sure to show the equation used.

Number Problems

▶ **1.** The product of two consecutive even integers is 80. Find the two integers. 8, 10 and −10, −8

2. The product of two consecutive integers is 72. Find the two integers. 8, 9 and −9, −8

3. The product of two consecutive odd integers is 99. Find the two integers. 9, 11 and −11, −9

4. The product of two consecutive integers is 132. Find the two integers. 11, 12 and −12, −11

5. The product of two consecutive even integers is 10 less than 5 times their sum. Find the two integers. 8, 10 and 0, 2

6. The product of two consecutive odd integers is 1 less than 4 times their sum. Find the two integers. 7, 9 and −1, 1

7. The sum of two numbers is 14. Their product is 48. Find the numbers. 8, 6

8. The sum of two numbers is 12. Their product is 32. Find the numbers. 4, 8

= Videos available by instructor request
▶ = Online student support materials available at www.thomsonedu.com/login

▶ **9.** One number is 2 more than 5 times another. Their product is 24. Find the numbers. 2, 12 and $-\frac{12}{5}$, -10

10. One number is 1 more than twice another. Their product is 55. Find the numbers. 5, 11 and $-\frac{11}{2}$, -10

11. One number is 4 times another. Their product is 4 times their sum. Find the numbers. 5, 20 and 0, 0

12. One number is 2 more than twice another. Their product is 2 more than twice their sum. Find the numbers. 3, 8 and -1, 0

Geometry Problems

13. The length of a rectangle is 1 more than the width. The area is 12 square inches. Find the dimensions. Width 3 inches, length 4 inches

14. The length of a rectangle is 3 more than twice the width. The area is 44 square inches. Find the dimensions. Width 4 inches, length 11 inches

15. The height of a triangle is twice the base. The area is 9 square inches. Find the base. Base 3 inches

16. The height of a triangle is 2 more than twice the base. The area is 20 square feet. Find the base. Base 4 feet

17. The hypotenuse of a right triangle is 10 inches. The lengths of the two legs are given by two consecutive even integers. Find the lengths of the two legs. 6 inches and 8 inches

18. The hypotenuse of a right triangle is 15 inches. One of the legs is 3 inches more than the other. Find the lengths of the two legs. 9 inches and 12 inches

19. The shorter leg of a right triangle is 5 meters. The hypotenuse is 1 meter longer than the longer leg. Find the length of the longer leg. 12 meters

20. The shorter leg of a right triangle is 12 yards. If the hypotenuse is 20 yards, how long is the other leg? 16 yards

Business Problems

21. A company can manufacture x hundred items for a total cost of $C = 400 + 700x - 100x^2$. Find x if the total cost is \$1,400. 2 hundred items or 5 hundred items

22. If the total cost C of manufacturing x hundred items is given by the equation $C = 700 + 900x - 100x^2$, find x when C is \$2,700. 4 hundred items or 5 hundred items

23. The total cost C of manufacturing x hundred video-tapes is given by the equation

$$C = 600 + 1{,}000x - 100x^2$$

Find x if the total cost is \$2,200.

2 hundred videotapes or 8 hundred videotapes

24. The total cost C of manufacturing x hundred pen and pencil sets is given by the equation

$$C = 500 + 800x - 100x^2.$$

Find x when C is \$1,700. 2 hundred sets or 6 hundred sets

25. A company that manufactures typewriter ribbons knows that the number of ribbons it can sell each week, x, is related to the price p per ribbon by the equation $x = 1{,}200 - 100p$. At what price should the company sell the ribbons if it wants the weekly revenue to be \$3,200? (*Remember:* The equation for revenue is $R = xp$.) \$4 or \$8

26. A company manufactures diskettes for home computers. It knows from experience that the number of diskettes it can sell each day, x, is related to the price p per diskette by the equation $x = 800 - 100p$. At what price should the company sell the diskettes if it wants the daily revenue to be \$1,200? \$2 or \$6

27. The relationship between the number of calculators a company sells per week, x, and the price p of each calculator is given by the equation $x = 1{,}700 - 100p$. At what price should the calculators be sold if the weekly revenue is to be \$7,000? \$7 or \$10

28. The relationship between the number of pencil sharpeners a company can sell each week, x, and the price p of each sharpener is given by the equation $x = 1{,}800 - 100p$. At what price should the sharpeners be sold if the weekly revenue is to be \$7,200? \$6 or \$12

29. Pythagorean Theorem A 13-foot ladder is placed so that it reaches to a point on the wall that is 2 feet higher than twice the distance from the base of the wall to the base of the ladder.

Wall   13 ft

   a. How far from the wall is the base of the ladder?
   5 feet

   b. How high does the ladder reach? 12 feet

Ground

30. Height of a Projectile If a rocket is fired vertically into the air with a speed of 240 feet per second, its height at time t seconds is given by $h(t) = -16t^2 + 240t$. Here is a graph of its height at various times, with the details left out:

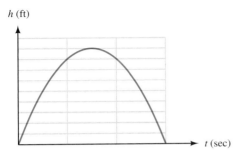

At what time(s) will the rocket be the following number of feet above the ground?

a. 704 feet   4 seconds and 11 seconds
b. 896 feet   7 seconds and 8 seconds
c. Why do parts (a) and (b) each have two answers?
   Once on the way up and again on the way down
d. How long will the rocket be in the air? (*Hint:* How high is it when it hits the ground?)   15 seconds
e. When the equation for part (d) is solved, one of the answers is $t = 0$ second. What does this represent?   When the rocket was fired

31. Projectile Motion A gun fires a bullet almost straight up from the edge of a 100-foot cliff. If the bullet leaves the gun with a speed of 396 feet per second, its height at time t is given by $h(t) = -16t^2 + 396t + 100$, measured from the ground below the cliff.

a. When will the bullet land on the ground below the cliff? (*Hint:* What is its height when it lands? Remember that we are measuring from the ground below, not from the cliff.)   25 seconds later
b. Make a table showing the bullet's height every five seconds, from the time it is fired ($t = 0$) to the time it lands. (*Note:* It is faster to substitute into the factored form.)

t (sec)	h (feet)
0	100
5	1,680
10	2,460
15	2,440
20	1,620
25	0

32. Constructing a Box I have a piece of cardboard that is twice as long as it is wide. If I cut a 2-inch by 2-inch square from each corner and fold up the resulting flaps, I get a box with a volume of 32 cubic inches. What are the dimensions of the cardboard?

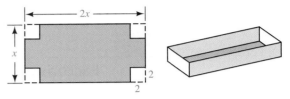

6 inches by 12 inches

Maintaining Your Skills

Simplify each expression. (Write all answers with positive exponents only.)

33. $(5x^3)^2(2x^6)^3$ $200x^{24}$

34. 2^{-3} $\frac{1}{8}$

35. $\dfrac{x^4}{x^{-3}}$ x^7

36. $\dfrac{(20x^2y^3)(5x^4y)}{(2xy^5)(10x^2y^3)}$ $\dfrac{5x^3}{y^4}$

37. $(2 \times 10^{-4})(4 \times 10^5)$ 8×10^1

38. $\dfrac{9 \times 10^{-3}}{3 \times 10^{-2}}$ 3×10^{-1}

39. $20ab^2 - 16ab^2 + 6ab^2$ $10ab^2$

40. Subtract $6x^2 - 5x - 7$ from $9x^2 + 3x - 2$. $3x^2 + 8x + 5$

Multiply.

41. $2x^2(3x^2 + 3x - 1)$ $6x^4 + 6x^3 - 2x^2$

42. $(2x + 3)(5x - 2)$ $10x^2 + 11x - 6$

43. $(3y - 5)^2$ $9y^2 - 30y + 25$

44. $(a - 4)(a^2 + 4a + 16)$ $a^3 - 64$

45. $(2a^2 + 7)(2a^2 - 7)$ $4a^4 - 49$

46. Divide $15x^{10} - 10x^8 + 25x^6$ by $5x^6$. $3x^4 - 2x^2 + 5$

Chapter 6 SUMMARY

Greatest Common Factor [6.1]

EXAMPLES

1. $8x^4 - 10x^3 + 6x^2$
$= 2x^2 \cdot 4x^2 - 2x^2 \cdot 5x + 2x^2 \cdot 3$
$= 2x^2(4x^2 - 5x + 3)$

The largest monomial that divides each term of a polynomial is called the greatest common factor for that polynomial. We begin all factoring by factoring out the greatest common factor.

Factoring Trinomials [6.2, 6.3]

2. $x^2 + 5x + 6 = (x + 2)(x + 3)$
$x^2 - 5x + 6 = (x - 2)(x - 3)$
$6x^2 - x - 2 = (2x + 1)(3x - 2)$
$6x^2 + 7x + 2 = (2x + 1)(3x + 2)$

One method of factoring a trinomial is to list all pairs of binomials the product of whose first terms gives the first term of the trinomial and the product of whose last terms gives the last term of the trinomial. We then choose the pair that gives the correct middle term for the original trinomial.

Special Factoring [6.4]

3. $x^2 + 10x + 25 = (x + 5)^2$
$x^2 - 10x + 25 = (x - 5)^2$
$x^2 - 25 = (x + 5)(x - 5)$

$$a^2 + 2ab + b^2 = (a + b)^2$$

$$a^2 - 2ab + b^2 = (a - b)^2$$

$$a^2 - b^2 = (a + b)(a - b)$$

Sum and Difference of Two Cubes [6.5]

4. $x^3 - 27 = (x - 3)(x^2 + 3x + 9)$
$x^3 + 27 = (x + 3)(x^2 - 3x + 9)$

$$a^3 - b^3 = (a - b)(a^2 + ab + b^2) \quad \text{Difference of two cubes}$$

$$a^3 + b^3 = (a + b)(a^2 - ab + b^2) \quad \text{Sum of two cubes}$$

Strategy for Factoring a Polynomial [6.6]

5. a. $2x^5 - 8x^3 = 2x^3(x^2 - 4)$
$= 2x^3(x + 2)(x - 2)$
b. $3x^4 - 18x^3 + 27x^2$
$= 3x^2(x^2 - 6x + 9)$
$= 3x^2(x - 3)^2$
c. $6x^3 - 12x^2 - 48x$
$= 6x(x^2 - 2x - 8)$
$= 6x(x - 4)(x + 2)$
d. $x^2 + ax + bx + ab$
$= x(x + a) + b(x + a)$
$= (x + a)(x + b)$

Step 1: If the polynomial has a greatest common factor other than 1, then factor out the greatest common factor.

Step 2: If the polynomial has two terms (it is a binomial), then see if it is the difference of two squares or the sum or difference of two cubes, and then factor accordingly. Remember, if it is the sum of two squares, it will not factor.

Step 3: If the polynomial has three terms (a trinomial), then it is either a perfect square trinomial that will factor into the square of a binomial, or it is not a perfect square trinomial, in which case you use the trial and error method developed in Section 6.3.

Step 4: If the polynomial has more than three terms, then try to factor it by grouping.

Step 5: As a final check, see if any of the factors you have written can be factored further. If you have overlooked a common factor, you can catch it here.

Strategy for Solving a Quadratic Equation [6.7]

6. Solve $x^2 - 6x = -8$.
$$x^2 - 6x + 8 = 0$$
$$(x - 4)(x - 2) = 0$$
$$x - 4 = 0 \quad \text{or} \quad x - 2 = 0$$
$$x = 4 \quad \text{or} \quad x = 2$$
Both solutions check.

Step 1: Write the equation in standard form:
$$ax^2 + bx + c = 0$$

Step 2: Factor completely.

Step 3: Set each variable factor equal to 0.

Step 4: Solve the equations found in step 3.

Step 5: Check solutions, if necessary.

The Pythagorean Theorem [6.8]

7. The hypotenuse of a right triangle is 5 inches, and the lengths of the two legs (the other two sides) are given by two consecutive integers. Find the lengths of the two legs.

If we let x = the length of the shortest side, then the other side must be $x + 1$. The Pythagorean theorem tells us that
$$5^2 = x^2 + (x + 1)^2$$
$$25 = x^2 + x^2 + 2x + 1$$
$$25 = 2x^2 + 2x + 1$$
$$0 = 2x^2 + 2x - 24$$
$$0 = 2(x^2 + x - 12)$$
$$0 = 2(x + 4)(x - 3)$$
$$x + 4 = 0 \quad \text{or} \quad x - 3 = 0$$
$$x = -4 \quad \text{or} \quad x = 3$$
Since a triangle cannot have a side with a negative number for its length, we cannot use -4. One leg is $x = 3$ and the other leg is $x + 1 = 3 + 1 = 4$.

In any right triangle, the square of the longest side (called the hypotenuse) is equal to the sum of the squares of the other two sides (called legs).

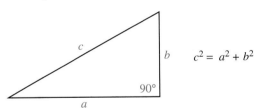

$$c^2 = a^2 + b^2$$

! COMMON MISTAKES

It is a mistake to apply the zero-factor property to numbers other than zero. For example, consider the equation $(x - 3)(x + 4) = 18$. A fairly common mistake is to attempt to solve it with the following steps:

$$(x - 3)(x + 4) = 18$$
$$x - 3 = 18 \quad \text{or} \quad x + 4 = 18 \leftarrow \textbf{Mistake}$$
$$x = 21 \quad \text{or} \quad x = 14$$

These are obviously not solutions, as a quick check will verify:

Check: $x = 21$

$$(21 - 3)(21 + 4) \stackrel{?}{=} 18$$
$$18 \cdot 25 = 18$$
$$450 = 18$$

Check: $x = 14$

$$(14 - 3)(14 + 4) \stackrel{?}{=} 18$$
$$11 \cdot 18 = 18$$
$$198 = 18$$

$\xleftarrow{\text{false statements}}\rightarrow$

The mistake is in setting each factor equal to 18. It is not necessarily true that when the product of two numbers is 18, either one of them is itself 18. The correct solution looks like this:

$$(x - 3)(x + 4) = 18$$
$$x^2 + x - 12 = 18$$
$$x^2 + x - 30 = 0$$
$$(x + 6)(x - 5) = 0$$

$$x + 6 = 0 \quad \text{or} \quad x - 5 = 0$$
$$x = -6 \quad \text{or} \quad x = 5$$

To avoid this mistake, remember that before you factor a quadratic equation, you must write it in standard form. It is in standard form only when 0 is on one side and decreasing powers of the variable are on the other.

The problems below form a comprehensive review of the material in this chapter. They can be used to study for exams. If you would like to take a practice test on this chapter, you can use the odd-numbered problems. Give yourself an hour and work as many of the odd-numbered problems as possible. When you are finished, or when an hour has passed, check your answers with the answers in the back of the book. You can use the even-numbered problems for a second practice test.

The numbers in brackets refer to the sections of the text in which similar problems can be found.

Factor the following by factoring out the greatest common factor. [6.1]

1. $10x - 20$ **2.** $4x^3 - 9x^2$

3. $5x - 5y$ **4.** $7x^3 + 2x$

5. $8x + 4$ **6.** $2x^2 + 14x + 6$

7. $24y^2 - 40y + 48$ **8.** $30xy^3 - 45x^3y^2$

9. $49a^3 - 14b^3$ **10.** $6ab^2 + 18a^3b^3 - 24a^2b$

Factor by grouping. [6.1]

11. $xy + bx + ay + ab$ **12.** $xy + 4x - 5y - 20$

13. $2xy + 10x - 3y - 15$ **14.** $5x^2 - 4ax - 10bx + 8ab$

Factor the following trinomials. [6.2]

15. $y^2 + 9y + 14$ **16.** $w^2 + 15w + 50$

17. $a^2 - 14a + 48$ **18.** $r^2 - 18r + 72$

19. $y^2 + 20y + 99$ **20.** $y^2 + 8y + 12$

Factor the following trinomials. [6.3]

21. $2x^2 + 13x + 15$ **22.** $4y^2 - 12y + 5$

23. $5y^2 + 11y + 6$ **24.** $20a^2 - 27a + 9$

25. $6r^2 + 5rt - 6t^2$ **26.** $10x^2 - 29x - 21$

Factor the following if possible. [6.4]

27. $n^2 - 81$ **28.** $4y^2 - 9$

29. $x^2 + 49$ **30.** $36y^2 - 121x^2$

31. $64a^2 - 121b^2$ **32.** $64 - 9m^2$

Factor the following. [6.4]

33. $y^2 + 20y + 100$ **34.** $m^2 - 16m + 64$

35. $64t^2 + 16t + 1$ **36.** $16n^2 - 24n + 9$

37. $4r^2 - 12rt + 9t^2$ **38.** $9m^2 + 30mn + 25n^2$

Factor the following. [6.2]

39. $2x^2 + 20x + 48$ **40.** $a^3 - 10a^2 + 21a$

41. $3m^3 - 18m^2 - 21m$ **42.** $5y^4 + 10y^3 - 40y^2$

Factor the following trinomials. [6.3]

43. $8x^2 + 16x + 6$ **44.** $3a^3 - 14a^2 - 5a$

45. $20m^3 - 34m^2 + 6m$ **46.** $30x^2y - 55xy^2 + 15y^3$

Factor the following. [6.4]

47. $4x^2 + 40x + 100$ **48.** $4x^3 + 12x^2 + 9x$

49. $5x^2 - 45$ **50.** $12x^3 - 27xy^2$

Factor the following. [6.5]

51. $8a^3 - b^3$ **52.** $27x^3 + 8y^3$

53. $125x^3 - 64y^3$

Factor the following polynomials completely. [6.6]

54. $6a^3b + 33a^2b^2 + 15ab^3$ **55.** $x^5 - x^3$

56. $4y^6 + 9y^4$ **57.** $12x^5 + 20x^4y - 8x^3y^2$

58. $30a^4b + 35a^3b^2 - 15a^2b^3$

59. $18a^3b^2 + 3a^2b^3 - 6ab^4$

Solve. [6.7]

60. $(x - 5)(x + 2) = 0$ **61.** $3(2y + 5)(2y - 5) = 0$

62. $m^2 + 3m = 10$ **63.** $a^2 - 49 = 0$

64. $m^2 - 9m = 0$ **65.** $6y^2 = -13y - 6$

66. $9x^4 + 9x^3 = 10x^2$

Solve the following word problems. [6.8]

67. Number Problem The product of two consecutive even integers is 120. Find the two integers.

68. Number Problem The product of two consecutive integers is 110. Find the two integers.

69. Number Problem The product of two consecutive odd integers is 1 less than 3 times their sum. Find the integers.

70. Number Problem The sum of two numbers is 20. Their product is 75. Find the numbers.

71. Number Problem One number is 1 less than twice another. Their product is 66. Find the numbers.

72. Geometry The height of a triangle is 8 times the base. The area is 16 square inches. Find the base.

GROUP PROJECT   Visual Factoring

Number of People 2 or 3

Time Needed 10–15 minutes

Equipment Pencil, graph paper, and scissors

Background When a geometric figure is divided into smaller figures, the area of the original figure and the area of any rearrangement of the smaller figures must be the same. We can use this fact to help visualize some factoring problems.

Procedure Use the diagram below to work the following problems.

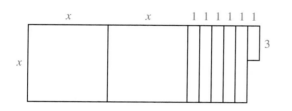

1. Write a polynomial involving x that gives the area of the diagram.

2. Factor the polynomial found in Part 1.

3. Copy the figure onto graph paper, then cut along the lines so that you end up with 2 squares and 6 rectangles.

4. Rearrange the pieces from Part 3 to show that the factorization you did in Part 2 is correct.

Factoring and Internet Security

The security of the information on computers is directly related to factoring of whole numbers. The key lies in the fact that multiplying whole numbers is a straightforward, simple task, whereas factoring can be very time-consuming. For example, multiplying the numbers 1,234 and 3,433 to obtain 4,236,322 takes very little time, even if done by hand. But given the number 4,236,322, finding its factors, even with a calculator or computer, is more than you want to try. The discipline that studies how to make and break codes is cryptography. The current Web browsers, such as Internet Explorer and Netscape, use a system called RSA public-key cryptosystem invented by Adi Shamir of Israel's Weizmann Institute of Science. In 1999 Shamir announced that he had found a method of factoring large numbers quickly that will put the current Internet security system at risk.

Research the connection between computer security and factoring, the RSA cryptosystem, and the current state of security on the Internet, and then write an essay summarizing your results.

Rational Expressions

7

F irst Bank of San Luis Obispo charges $2.00 per month and $0.15 per check for a regular checking account. If we write x checks in one month, the total monthly cost of the checking account will be $C = 2.00 + 0.15x$. From this formula we see that the more checks we write in a month, the more we pay for the account. But, it is also true that the more checks we write in a month, the lower the cost per check. To find the average cost per check, we divide the total cost by the number of checks written:

$$\text{Average cost} = A = \frac{C}{X} = \frac{2.00 + 0.15x}{X}$$

We can use this formula to create Table 1 and Figure 1, giving us a visual interpretation of the relationship between the number of checks written and the average cost per check.

TABLE 1 Average Cost	
Number of Checks	Average Cost Per Check
1	2.15
2	1.15
5	0.55
10	0.35
15	0.28
20	0.25

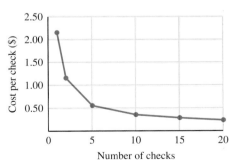

As you can see, if we write one check per month, the cost per check is relatively high, $2.15. However, if we write 20 checks per month, each check costs us only $0.25. Using average cost per check is a good way to compare different checking accounts. The expression $\frac{2.00 + 0.15x}{X}$ in the average cost formula is a rational expression. When you have finished this chapter you will have a good working knowledge of rational expressions.

▶ Improve your grade and save time!
Go online to **www.thomsonedu.com/login** where you can
• Watch videos of instructors working through the in-text examples
• Follow step-by-step online tutorials of in-text examples and review questions
• Work practice problems
• Check your readiness for an exam by taking a pre-test and exploring the modules recommended in your Personalized Study plan
• Receive help from a live tutor online through vMentor™
Try it out! Log in with an access code or purchase access at **www.ichapters.com**.

This is the last chapter in which we will mention study skills. You know by now what works best for you and what you have to do to achieve your goals for this course. From now on, it is simply a matter of sticking with the things that work for you, and avoiding the things that do not work. It seems simple, but as with anything that takes effort, it is up to you to see that you maintain the skills that get you where you want to be in the course.

If you intend to take more classes in mathematics, and you want to en-sure your success in those classes, then you can work toward this goal: *Become a student who can learn mathematics on your own.* Most people who have degrees in mathematics were students who could learn mathe-matics on their own. This doesn't mean that you have to learn it all on your own; it simply means that if you have to, you can learn it on your own. Attaining this goal gives you independence and puts you in control of your success in any math class you take.

7.1 Reducing Rational Expressions to Lowest Terms

OBJECTIVES

A Find the restrictions on the variable in a rational expression.

B Reduce a rational expression to lowest terms.

C Work problems involving ratios.

In Chapter 1 we defined the set of rational numbers to be the set of all numbers that could be put in the form $\dfrac{a}{b}$, where a and b are integers ($b \neq 0$):

$$\text{Rational numbers} = \left\{ \frac{a}{b} \,\middle|\, a \text{ and } b \text{ are integers, } b \neq 0 \right\}$$

A *rational expression* is any expression that can be put in the form $\dfrac{P}{Q}$, where P and Q are polynomials and $Q \neq 0$:

$$\text{Rational expressions} = \left\{ \frac{P}{Q} \,\middle|\, P \text{ and } Q \text{ are polynomials, } Q \neq 0 \right\}$$

Each of the following is an example of a rational expression:

$$\frac{2x + 3}{x} \qquad \frac{x^2 - 6x + 9}{x^2 - 4} \qquad \frac{5}{x^2 + 6} \qquad \frac{2x^2 + 3x + 4}{2}$$

For the rational expression

$$\frac{x^2 - 6x + 9}{x^2 - 4}$$

the polynomial on top, $x^2 - 6x + 9$, is called the numerator, and the polynomial on the bottom, $x^2 - 4$, is called the denominator. The same is true of the other rational expressions.

We must be careful that we do not use a value of the variable that will give us a denominator of zero. Remember, division by zero is not defined.

EXAMPLES

State the restrictions on the variable in the following rational expressions:

1. $\dfrac{x + 2}{x - 3}$

SOLUTION The variable x can be any real number except $x = 3$ since, when $x = 3$, the denominator is $3 - 3 = 0$. We state this restriction by writing $x \neq 3$.

2. $\dfrac{5}{x^2 - x - 6}$

SOLUTION If we factor the denominator, we have $x^2 - x - 6 = (x - 3)(x + 2)$. If either of the factors is zero, the whole denominator is zero. Our restrictions are $x \neq 3$ and $x \neq -2$ since either one makes $x^2 - x - 6 = 0$.

We will not always list each restriction on a rational expression, but we should be aware of them and keep in mind that no rational expression can have a denominator of zero.

The two fundamental properties of rational expressions are listed next. We will use these two properties many times in this chapter.

Properties of Rational Expressions

Property 1

Multiplying the numerator and denominator of a rational expression by the same nonzero quantity will not change the value of the rational expression.

Property 2

Dividing the numerator and denominator of a rational expression by the same nonzero quantity will not change the value of the rational expression.

We can use Property 2 to reduce rational expressions to lowest terms. Since this process is almost identical to the process of reducing fractions to lowest terms, let's recall how the fraction $\frac{6}{15}$ is reduced to lowest terms:

$$\frac{6}{15} = \frac{2 \cdot 3}{5 \cdot 3} \qquad \textbf{Factor numerator and denominator}$$

$$= \frac{2 \cdot \cancel{3}}{5 \cdot \cancel{3}} \qquad \textbf{Divide out the common factor, 3}$$

$$= \frac{2}{5} \qquad \textbf{Reduce to lowest terms}$$

The same procedure applies to reducing rational expressions to lowest terms. The process is summarized in the following rule.

Rule To reduce a rational expression to lowest terms, first factor the numerator and denominator completely and then divide both the numerator and denominator by any factors they have in common.

EXAMPLE 3 Reduce $\dfrac{x^2 - 9}{x^2 + 5x + 6}$ to lowest terms.

SOLUTION We begin by factoring:

$$\frac{x^2 - 9}{x^2 + 5x + 6} = \frac{(x - 3)(x + 3)}{(x + 2)(x + 3)}$$

Notice that both polynomials contain the factor $(x + 3)$. If we divide the numerator by $(x + 3)$, we are left with $(x - 3)$. If we divide the denominator by $(x + 3)$, we are left with $(x + 2)$. The complete solution looks like this:

$$\frac{x^2 - 9}{x^2 + 5x + 6} = \frac{(x - 3)\cancel{(x + 3)}}{(x + 2)\cancel{(x + 3)}} \qquad \begin{array}{l}\textbf{Factor the numerator and}\\\textbf{denominator completely}\end{array}$$

$$= \frac{x - 3}{x + 2} \qquad \begin{array}{l}\textbf{Divide out the common}\\\textbf{factor, } x + 3\end{array}$$

It is convenient to draw a line through the factors as we divide them out. It is especially helpful when the problems become longer.

EXAMPLE 4   Reduce to lowest terms $\dfrac{10a + 20}{5a^2 - 20}$.

SOLUTION   We begin by factoring out the greatest common factor from the numerator and denominator:

$$\frac{10a + 20}{5a^2 - 20} = \frac{10(a + 2)}{5(a^2 - 4)}$$ Factor the greatest common factor from the numerator and denominator

$$= \frac{10\cancel{(a + 2)}}{5\cancel{(a + 2)}(a - 2)}$$ Factor the denominator as the difference of two squares

$$= \frac{2}{a - 2}$$ Divide out the common factors 5 and $a + 2$

 EXAMPLE 5   Reduce $\dfrac{2x^3 + 2x^2 - 24x}{x^3 + 2x^2 - 8x}$ to lowest terms.

SOLUTION   We begin by factoring the numerator and denominator completely. Then we divide out all factors common to the numerator and denominator. Here is what it looks like:

$$\frac{2x^3 + 2x^2 - 24x}{x^3 + 2x^2 - 8x} = \frac{2x(x^2 + x - 12)}{x(x^2 + 2x - 8)}$$ Factor out the greatest common factor first

$$= \frac{2\cancel{x}(x - 3)\cancel{(x + 4)}}{\cancel{x}(x - 2)\cancel{(x + 4)}}$$ Factor the remaining trinomials

$$= \frac{2(x - 3)}{x - 2}$$ Divide out the factors common to the numerator and denominator

 EXAMPLE 6   Reduce $\dfrac{x - 5}{x^2 - 25}$ to lowest terms.

SOLUTION

$$\frac{x - 5}{x^2 - 25} = \frac{\cancel{x - 5}}{\cancel{(x - 5)}(x + 5)}$$ Factor numerator and denominator completely

$$= \frac{1}{x + 5}$$ Divide out the common factor, $x - 5$

 EXAMPLE 7   Reduce $\dfrac{x^3 + y^3}{x^2 - y^2}$ to lowest terms.

SOLUTION   We begin by factoring the numerator and denominator completely. (Remember, we can only reduce to lowest terms when the numerator and denominator are in factored form. Trying to reduce before factoring will only lead to mistakes.)

$$\frac{x^3 + y^3}{x^2 - y^2} = \frac{\cancel{(x + y)}(x^2 - xy + y^2)}{\cancel{(x + y)}(x - y)}$$ Factor

$$= \frac{x^2 - xy + y^2}{x - y}$$ Divide out the common factor

Ratios

For the rest of this section we will concern ourselves with ratios, a topic closely related to reducing fractions and rational expressions to lowest terms. Let's start with a definition.

> **DEFINITION** If a and b are any two numbers, $b \neq 0$, then the **ratio** of a and b is
>
> $$\frac{a}{b}$$

As you can see, ratios are another name for fractions or rational numbers. They are a way of comparing quantities. Since we also can think of $\frac{a}{b}$ as the quotient of a and b, ratios are also quotients. The following table gives some ratios in words and as fractions.

Note

With ratios it is common to leave the 1 in the denominator.

Ratio	As a Fraction	In Lowest Terms
25 to 75	$\frac{25}{75}$	$\frac{1}{3}$
8 to 2	$\frac{8}{2}$	$\frac{4}{1}$
20 to 16	$\frac{20}{16}$	$\frac{5}{4}$

 EXAMPLE 8   A solution of hydrochloric acid (HCl) and water contains 49 milliliters of water and 21 milliliters of HCl. Find the ratio of HCl to water and of HCl to the total volume of the solution.

SOLUTION   The ratio of HCl to water is 21 to 49, or

$$\frac{21}{49} = \frac{3}{7}$$

The amount of total solution volume is $49 + 21 = 70$ milliliters. Therefore, the ratio of HCl to total solution is 21 to 70, or

$$\frac{21}{70} = \frac{3}{10}$$

Rate Equation

Many of the problems in this chapter will use what is called the *rate equation*. You use this equation on an intuitive level when you are estimating how long it will take you to drive long distances. For example, if you drive at 50 miles per hour for 2 hours, you will travel 100 miles. Here is the rate equation:

$$\text{Distance} = \text{rate} \cdot \text{time}$$

$$d = r \cdot t$$

The rate equation has two equivalent forms, the most common of which is obtained by solving for r. Here it is:

$$r = \frac{d}{t}$$

The rate *r* in the rate equation is the ratio of distance to time and also is referred to as *average speed*. The units for rate are miles per hour, feet per second, kilometers per hour, and so on.

The Forest Chair Lift

L = 5,603 feet

EXAMPLE 9   The Forest chair lift at the Northstar ski resort in Lake Tahoe is 5,603 feet long. If a ride on this chair lift takes 11 minutes, what is the average speed of the lift in feet per minute?

SOLUTION   To find the speed of the lift, we find the ratio of distance covered to time. (Our answer is rounded to the nearest whole number.)

$$\text{Rate} = \frac{\text{distance}}{\text{time}} = \frac{5,603 \text{ feet}}{11 \text{ minutes}} = \frac{5,603}{11} \text{ feet/minute} = 509 \text{ feet/minute}$$

Note how we separate the numerical part of the problem from the units. In the next section, we will convert this rate to miles per hour.

EXAMPLE 10   A Ferris wheel was built in St. Louis in 1986. It is named *Colossus*. The circumference of the wheel is 518 feet. It has 40 cars, each of which holds 6 passengers. A trip around the wheel takes 40 seconds. Find the average speed of a rider on *Colossus*.

Circumference

Solution   To find the average speed, we divide the distance traveled, which in this case is the circumference, by the time it takes to travel once around the wheel.

$$r = \frac{d}{t} = \frac{518 \text{ feet}}{40 \text{ seconds}} = 13.0 \text{ (rounded)}$$

The average speed of a rider on the *Colossus* is 13.0 feet per second.

In the next section, you will convert the ratio into an equivalent ratio that gives the speed of the rider in miles per hour.

LINKING OBJECTIVES AND EXAMPLES

Next to each **objective** we have listed the examples that are best described by that objective.

A   1, 2

B   3–7

C   8–10

GETTING READY FOR CLASS

After reading through the preceding section, respond in your own words and in complete sentences.

1. How do you reduce a rational expression to lowest terms?
2. What are the properties we use to manipulate rational expressions?
3. For what values of the variable is a rational expression undefined?
4. What is a ratio?

Answers appear in the Instructor's Edition only.

▶ **1.** Simplify each expression.

a. $\dfrac{5 + 1}{25 - 1}$ $\dfrac{1}{4}$

b. $\dfrac{x + 1}{x^2 - 1}$ $\dfrac{1}{x - 1}$, $x \neq -1, 1$

c. $\dfrac{x^2 - x}{x^2 - 1}$ $\dfrac{x}{x + 1}$, $x \neq -1, 1$

d. $\dfrac{x^3 - 1}{x^2 - 1}$ $\dfrac{x^2 + x + 1}{x + 1}$, $x \neq -1, 1$

e. $\dfrac{x^3 - 1}{x^3 - x^2}$ $\dfrac{x^2 + x + 1}{x^2}$, $x \neq 0, 1$

▶ **2.** Simplify each expression.

a. $\dfrac{25 - 30 + 9}{25 - 9}$ $\dfrac{1}{4}$

b. $\dfrac{x^2 - 6x + 9}{x^2 - 9}$ $\dfrac{x - 3}{x + 3}$, $x \neq -3, 3$

c. $\dfrac{x^2 - 10x + 9}{x^2 - 9x}$ $\dfrac{x - 1}{x}$, $x \neq 0, 9$

d. $\dfrac{x^2 + 3x + ax + 3a}{x^2 - 9}$ $\dfrac{x + a}{x - 3}$, $x \neq -3, 3$

e. $\dfrac{x^3 + 27}{x^3 - 9x}$ $\dfrac{x^2 - 3x + 9}{x(x - 3)}$, $x \neq -3, 0, 3$

Reduce the following rational expressions to lowest terms, if possible. Also, specify any restrictions on the variable in Problems 1 through 10.

3. $\dfrac{a - 3}{a^2 - 9}$ $\dfrac{1}{a + 3}$, $a \neq -3, 3$

4. $\dfrac{a + 4}{a^2 - 16}$ $\dfrac{1}{a - 4}$, $a \neq -4, 4$

5. $\dfrac{x + 5}{x^2 - 25}$ $\dfrac{1}{x - 5}$, $x \neq -5, 5$

6. $\dfrac{x - 2}{x^2 - 4}$ $\dfrac{1}{x + 2}$, $x \neq -2, 2$

7. $\dfrac{2x^2 - 8}{4}$ $\dfrac{(x + 2)(x - 2)}{2}$

8. $\dfrac{5x - 10}{x - 2}$ 5, $x \neq 2$

9. $\dfrac{2x - 10}{3x - 6}$ $\dfrac{2(x - 5)}{3(x - 2)}$, $x \neq 2$

10. $\dfrac{4x - 8}{x - 2}$ 4, $x \neq 2$

▶ **11.** $\dfrac{10a + 20}{5a + 10}$ 2

12. $\dfrac{11a + 33}{6a + 18}$ $\dfrac{11}{6}$

13. $\dfrac{5x^2 - 5}{4x + 4}$ $\dfrac{5(x - 1)}{4}$

14. $\dfrac{7x^2 - 28}{2x + 4}$ $\dfrac{7(x - 2)}{2}$

15. $\dfrac{x - 3}{x^2 - 6x + 9}$ $\dfrac{1}{x - 3}$

16. $\dfrac{x^2 - 10x + 25}{x - 5}$ $x - 5$

17. $\dfrac{3x + 15}{3x^2 + 24x + 45}$ $\dfrac{1}{x + 3}$

18. $\dfrac{5x + 15}{5x^2 + 40x + 75}$ $\dfrac{1}{x + 5}$

19. $\dfrac{a^2 - 3a}{a^3 - 8a^2 + 15a}$ $\dfrac{1}{a - 5}$

20. $\dfrac{a^2 + 3a}{a^3 - 2a^2 - 15a}$ $\dfrac{1}{a - 5}$

21. $\dfrac{3x - 2}{9x^2 - 4}$ $\dfrac{1}{3x + 2}$

22. $\dfrac{2x - 3}{4x^2 - 9}$ $\dfrac{1}{2x + 3}$

23. $\dfrac{x^2 + 8x + 15}{x^2 + 5x + 6}$ $\dfrac{x + 5}{x + 2}$

24. $\dfrac{x^2 - 8x + 15}{x^2 - x - 6}$ $\dfrac{x - 5}{x + 2}$

25. $\dfrac{2m^3 - 2m^2 - 12m}{m^2 - 5m + 6}$ $\dfrac{2m(m + 2)}{m - 2}$

26. $\dfrac{2m^3 + 4m^2 - 6m}{m^2 - m - 12}$ $\dfrac{2m(m - 1)}{m - 4}$

▶ **27.** $\dfrac{x^3 + 3x^2 - 4x}{x^3 - 16x}$ $\dfrac{x - 1}{x - 4}$

28. $\dfrac{3a^2 - 8a + 4}{9a^3 - 4a}$ $\dfrac{a - 2}{a(3a + 2)}$

29. $\dfrac{4x^3 - 10x^2 + 6x}{2x^3 + x^2 - 3x}$ $\dfrac{2(2x - 3)}{2x + 3}$

30. $\dfrac{3a^3 - 8a^2 + 5a}{4a^3 - 5a^2 + 1a}$ $\dfrac{3a - 5}{4a - 1}$

▶ **31.** $\dfrac{4x^2 - 12x + 9}{4x^2 - 9}$ $\dfrac{2x - 3}{2x + 3}$

32. $\dfrac{5x^2 + 18x - 8}{5x^2 + 13x - 6}$ $\dfrac{x + 4}{x + 3}$

33. $\dfrac{x + 3}{x^4 - 81}$ $\dfrac{1}{(x^2 + 9)(x - 3)}$

34. $\dfrac{x^2 + 9}{x^4 - 81}$ $\dfrac{1}{(x + 3)(x - 3)}$

35. $\dfrac{3x^2 + x - 10}{x^4 - 16}$ $\dfrac{3x - 5}{(x^2 + 4)(x - 2)}$

36. $\dfrac{5x^2 - 26x + 24}{x^4 - 64}$ $\dfrac{5x^2 - 26x + 24}{x^4 - 64}$

37. $\dfrac{42x^3 - 20x^2 - 48x}{6x^2 - 5x - 4}$ $\dfrac{2x(7x + 6)}{2x + 1}$

38. $\dfrac{36x^3 + 132x^2 - 135x}{6x^2 + 25x - 9}$ $\dfrac{3x(6x - 5)}{3x - 1}$

39. $\dfrac{x^3 - y^3}{x^2 - y^2}$ $\dfrac{x^2 + xy + y^2}{x + y}$

40. $\dfrac{x^3 + y^3}{x^2 - y^2}$ $\dfrac{x^2 - xy + y^2}{x - y}$

41. $\dfrac{x^3 + 8}{x^2 - 4}$ $\dfrac{x^2 - 2x + 4}{x - 2}$

42. $\dfrac{x^3 - 125}{x^2 - 25}$ $\dfrac{x^2 + 5x + 25}{x + 5}$

43. $\dfrac{x^3 + 8}{x^2 + x - 2}$ $\dfrac{x^2 - 2x + 4}{x - 1}$

44. $\dfrac{x^2 - 2x - 3}{x^3 - 27}$ $\dfrac{x + 1}{x^2 + 3x + 9}$

To reduce each of the following rational expressions to lowest terms, you will have to use factoring by grouping. Be sure to factor each numerator and denominator completely before dividing out any common factors. (Remember, factoring by grouping takes two steps.)

45. $\dfrac{xy + 3x + 2y + 6}{xy + 3x + 5y + 15}$ $\dfrac{x + 2}{x + 5}$

46. $\dfrac{xy + 7x + 4y + 28}{xy + 3x + 4y + 12}$ $\dfrac{y + 7}{y + 3}$

▶ **Chalkboard Problems**

Problems 1 and 2 are a good way to start the lecture on this section. They each contain a variety of factoring problems that together form a comprehensive review of factoring

▢ = Videos available by instructor request

▶ = Online student support materials available at www.thomsonedu.com/login

47. $\dfrac{x^2 - 3x + ax - 3a}{x^2 - 3x + bx - 3b}$ $\dfrac{x + a}{x + b}$

48. $\dfrac{x^2 - 6x + ax - 6a}{x^2 - 7x + ax - 7a}$ $\dfrac{x - 6}{x - 7}$

The next two problems are intended to give you practice reading, and paying attention to, the instructions that accompany the problems you are working. Working these problems is an excellent way to get ready for a test or quiz

▶ **49.** Work each problem according to the instructions given.

    a. Add: $(x^2 - 4x) + (4x - 16)$ $x^2 - 16$

    b. Subtract: $(x^2 - 4x) - (4x - 16)$ $x^2 - 8x + 16$

    c. Multiply: $(x^2 - 4x)(4x - 16)$ $4x^3 - 32x^2 + 64x$

    d. Reduce: $\dfrac{x^2 - 4x}{4x - 16}$ $\dfrac{x}{4}$

50. Work each problem according to the instructions given.

    a. Add: $(9x^2 - 3x) + (6x - 2)$ $9x^2 + 3x - 2$

    b. Subtract: $(9x^2 - 3x) - (6x - 2)$ $9x^2 - 9x + 2$

    c. Multiply: $(9x^2 - 3x)(6x - 2)$ $54x^3 - 36x^2 + 6x$

    d. Reduce: $\dfrac{9x^2 - 3x}{6x - 2}$ $\dfrac{3x}{2}$

Write each ratio as a fraction in lowest terms.

51. 8 to 6 $\dfrac{4}{3}$ **52.** 6 to 8 $\dfrac{3}{4}$

53. 200 to 250 $\dfrac{4}{5}$ **54.** 250 to 200 $\dfrac{5}{4}$

55. 32 to 4 $\dfrac{8}{1}$ **56.** 4 to 32 $\dfrac{1}{8}$

Applying the Concepts

57. Cost and Average Cost As we mentioned in the introduction to this chapter, if a bank charges $2.00 per month and $0.15 per check for one of its checking accounts, then the total monthly cost to write x checks is $C = 2.00 + 0.15x$, and the average cost of each of the x checks written is $A = \dfrac{2.00 + 0.15x}{x}$.

Compare these two formulas by filling in the following table. Round to the nearest cent.

Checks Written x	Total Cost $2.00 + 0.15x$	Cost per Check $\dfrac{2.00 + 0.15x}{x}$
0	$2.00	Undefined
5	$2.75	$0.55
10	$3.50	$0.35
15	$4.25	$0.28
20	$5.00	$0.25

58. Cost and Average Cost A rewritable CD-ROM drive for a computer costs $250. An individual CD for the drive costs $5.00 and can store 640 megabytes of information. The total cost of filling x CDs with information is $C = 250 + 5x$ dollars. The average cost per megabyte of information is given by $A = \dfrac{5x + 250}{640x}$. Compare the total cost and average cost per megabyte of storage by completing the following table. Round all answers to the nearest tenth of a cent.

CDs Purchased x	Total Cost $250 + 5x$	Cost per Megabyte $\dfrac{5x + 250}{640x}$
0	$250	Undefined
5	$275	$0.086
10	$300	$0.047
15	$325	$0.034
20	$350	$0.027

59. Speed of a Car A car travels 122 miles in 3 hours. Find the average speed of the car in miles per hour. Round to the nearest tenth. 40.7 miles/hour

60. Speed of a Bullet A bullet fired from a gun travels a distance of 4,500 feet in 3 seconds. Find the average speed of the bullet in feet per second. 1,500 feet/second

61. Ferris Wheel The first Ferris wheel was designed and built by George Ferris in 1893. It was a large wheel with a circumference of 785 feet. If one trip around the circumference of the wheel took 20 minutes, find the average speed of a rider in feet per minute. 39.25 feet/minute

▶ **Chalkboard Problems**

Problems 49 and 50 are Chalkboard problems. In addition to requiring students to pay attention to the instructions, they review addition, subtraction, and multiplication with polynomials. I like to work some problems they have already done as I work my way into the new material as a way of integrating continuous review into my course.

62. Ferris Wheel In 1897 a large Ferris wheel was built in Vienna; it is still in operation today. Known as *The Great Wheel,* it has a circumference of 618 feet. If one trip around the wheel takes 15 minutes, find the average speed of a rider on this wheel in feet per minute.   41.2 feet/minute

63. Baseball The chart shown here appeared in *USA Today* in 2005. For the four pitchers mentioned in the chart, calculate the number of strikeouts per inning. Round to the nearest hundredth.

Johnson	1.23
Wood	1.16
Martinez	1.16
Nomo	0.98

USA TODAY Snapshots®

Mowing down the opposition
Strikeout king Nolan Ryan struck out an average of 9.55 batters per nine innings. Active major league pitchers with the most strikeouts per nine innings:

Randy Johnson, N.Y. Yankees   11.04

Kerry Wood, Chicago Cubs   10.45

Pedro Martinez, N.Y. Mets   10.40

Hideo Nomo, Tampa Bay   8.81

Through Sunday
Source: MLB

By Quang Lam and Keith Simmons, USA TODAY

From *USA Today.* Copyright 2005. Reprinted with permission.

64. Ferris Wheel A person riding a Ferris wheel travels once around the wheel, a distance of 188 feet, in 30 seconds. What is the average speed of the rider in feet per second? Round to the nearest tenth.
6.3 feet/second

65. Average Speed Tina is training for a biathlon. As part of her training, she runs an 8-mile course, 2 miles of which is on level ground and 6 miles of which is downhill. It takes her 20 minutes to run the level part of the course and 40 minutes to run the downhill part of the course. Find her average speed in minutes per mile and in miles per minute for each part of the course. Round to the nearest hundredth, if rounding is necessary.   Level ground: 10 minutes/mile or 0.1 miles/minute; Downhill: $\frac{20}{3}$ minutes/mile or $\frac{3}{20}$ miles/minute

66. Jogging A jogger covers a distance of 3 miles in 24 minutes. Find the average speed of the jogger in miles per minute.   0.125 miles/minute

67. Fuel Consumption An economy car travels 168 miles on 3.5 gallons of gas. Give the average fuel consumption of the car in miles per gallon.   48 miles/gallon

68. Fuel Consumption A luxury car travels 100 miles on 8 gallons of gas. Give the average fuel consumption of the car in miles per gallon.   12.5 miles/gallon

69. Comparing Expressions Replace x with 5 and y with 4 in the expression
$$\frac{x^2 - y^2}{x - y}$$
and simplify the result. Is the result equal to $5 - 4$ or $5 + 4$? $5 + 4 = 9$

70. Comparing Expressions Replace x with 2 in the expression
$$\frac{x^3 - 1}{x - 1}$$
and simplify the result. Your answer should be equal to what you would get if you replaced x with 2 in $x^2 + x + 1$. $\frac{2^3 - 1}{2 - 1} = 7$; $2^2 + 2 + 1 = 7$

71. Comparing Expressions Complete the following table; then show why the table turns out as it does.
$\frac{x - 3}{3 - x} = -1$

x	$\dfrac{x-3}{3-x}$
-2	-1
-1	-1
0	-1
1	-1
2	-1

72. Comparing Expressions Complete the following table; then show why the table turns out as it does.

$$\frac{25 - x^2}{x^2 - 25} = -1$$

x	$\dfrac{25 - x^2}{x^2 - 25}$
-4	-1
-2	-1
0	-1
2	-1
4	-1

73. Comparing Expressions You know from reading through Example 6 in this section that $\dfrac{x - 5}{x^2 - 25} = \dfrac{1}{x + 5}$.

Compare these expressions by completing the following table. (Be careful—not all the rows have equal entries.)

x	$\dfrac{x - 5}{x^2 - 25}$	$\dfrac{1}{x + 5}$
0	$\frac{1}{5}$	$\frac{1}{5}$
2	$\frac{1}{7}$	$\frac{1}{7}$
-2	$\frac{1}{3}$	$\frac{1}{3}$
5	Undefined	$\frac{1}{10}$
-5	Undefined	Undefined

74. Comparing Expressions You know from your work in this section that $\dfrac{x^2 - 6x + 9}{x^2 - 9} = \dfrac{x - 3}{x + 3}$. Compare these expressions by completing the following table. (Be careful—not all the rows have equal entries.)

x	$\dfrac{x^2 - 6x + 9}{x^2 - 9}$	$\dfrac{x - 3}{x + 3}$
-3	Undefined	Undefined
-2	-5	-5
-1	-2	-2
0	-1	-1
1	$-\frac{1}{2}$	$-\frac{1}{2}$
2	$-\frac{1}{5}$	$-\frac{1}{5}$
3	Undefined	0

75. Stock Market One method of comparing stocks on the stock market is the price to earnings ratio, or P/E.

$$P/E = \frac{\text{Current Stock Price}}{\text{Earnings per Share}}$$

Most stocks have a P/E between 25 and 40. A stock with a P/E of less than 25 may be undervalued, while a stock with a P/E greater than 40 may be overvalued. Fill in the P/E for each stock listed in the table below. Based on your results, are any of the stocks undervalued? Disney, Nike

Stock	Price	Earnings per Share	P/E
Yahoo	37.80	1.07	35.33
Google	381.24	4.51	84.53
Disney	24.96	1.34	18.63
Nike	85.46	4.88	17.51
Ebay	40.96	0.73	56.11

76. Improving Your Quantitative Literacy Each of the numbers in the chart is a rate with units of students per computer (students/computer). Use the information in the chart to answer the following questions.

a. For each of the categories in the chart, find the rate of computers per student. Round to the nearest hundredth.

b. If a senior high actually had 3.2 students per computer, and students had access to 550 computers, how many students would be enrolled in the school? 1,760 students

c. If an elementary school actually had 4.2 students per computer, and the total enrollment in the school was 651 students, how many computers would the students have access to? 155 computers

76a Elementary: 0.24 computers per student
Middle/junior high: 0.26 computers per student
Senior high: 0.31 computers per student

USA TODAY Snapshots®

Older pupils have greater PC access

In U.S. public schools, the number of students who must share a computer decreases as they get older. How many students per computer in 2004:

Elementary 4.2

Middle/Junior high 3.8

Senior high 3.2

Source: 2004 Technology in Education, Market Data Retrieval

By Ashley Burrell and Robert W. Ahrens, USA TODAY

From *USA Today*. Copyright 2005. Reprinted with permission.

Maintaining Your Skills

Simplify.

77. $\dfrac{27x^5}{9x^2} - \dfrac{45x^8}{15x^5}$ 0

78. $\dfrac{36x^9}{4x} - \dfrac{45x^3}{5x^{-5}}$ 0

79. $\dfrac{72a^3b^7}{9ab^5} + \dfrac{64a^5b^3}{8a^3b}$ $16a^2b^2$

80. $\dfrac{80a^5b^{11}}{10a^2b} + \dfrac{33a^6b^{12}}{11a^3b^2}$ $11a^3b^{10}$

Divide.

81. $\dfrac{38x^7 + 42x^5 - 84x^3}{2x^3}$ $19x^4 + 21x^2 - 42$

82. $\dfrac{49x^6 - 63x^4 - 35x^2}{7x^2}$ $7x^4 - 9x^2 - 5$

83. $\dfrac{28a^5b^5 + 36ab^4 - 44a^4b}{4ab}$ $7a^4b^4 + 9b^3 - 11a^3$

84. $\dfrac{30a^3b - 12a^2b^2 + 6ab^3}{6ab}$ $5a^2 - 2ab + b^2$

Getting Ready for the Next Section

Perform the indicated operation.

85. $\dfrac{3}{4} \cdot \dfrac{10}{21}$ $\dfrac{5}{14}$

86. $\dfrac{2}{9} \cdot \dfrac{15}{22}$ $\dfrac{5}{33}$

87. $\dfrac{4}{5} \div \dfrac{8}{9}$ $\dfrac{9}{10}$

88. $\dfrac{3}{5} \div \dfrac{15}{7}$ $\dfrac{7}{25}$

Factor completely.

89. $x^2 - 9$ $(x + 3)(x - 3)$

90. $x^2 - 25$ $(x + 5)(x - 5)$

91. $3x - 9$ $3(x - 3)$

92. $2x - 4$ $2(x - 2)$

93. $x^2 - x - 20$ $(x - 5)(x + 4)$

94. $x^2 + 7x + 12$ $(x + 3)(x + 4)$

95. $a^2 + 5a$ $a(a + 5)$

96. $a^2 - 4a$ $a(a - 4)$

Reduce to lowest terms.

97. $\dfrac{a(a + 5)(a - 5)(a + 4)}{a^2 + 5a}$ $(a - 5)(a + 4)$

98. $\dfrac{a(a + 2)(a - 4)(a + 5)}{a^2 - 4a}$ $(a + 2)(a + 5)$

Multiply. Give the answers as decimals rounded to the nearest tenth.

99. $\dfrac{5,603}{11} \cdot \dfrac{1}{5,280} \cdot \dfrac{60}{1}$ 5.8

100. $\dfrac{772}{2.2} \cdot \dfrac{1}{5,280} \cdot \dfrac{60}{1}$ 4.0

7.2 Multiplication and Division of Rational Expressions

OBJECTIVES

A Multiply and divide rational expressions by factoring and then dividing out common factors.

B Convert between units using unit analysis.

Recall that to multiply two fractions we simply multiply numerators and multiply denominators and then reduce to lowest terms, if possible:

$$\dfrac{3}{4} \cdot \dfrac{10}{21} = \dfrac{30}{84} \qquad \begin{array}{l}\leftarrow \text{ Multiply numerators} \\ \leftarrow \text{ Multiply denominators}\end{array}$$

$$= \dfrac{5}{14} \qquad \leftarrow \text{ Reduce to lowest terms}$$

Recall also that the same result can be achieved by factoring numerators and denominators first and then dividing out the factors they have in common:

$$\dfrac{3}{4} \cdot \dfrac{10}{21} = \dfrac{3}{2 \cdot 2} \cdot \dfrac{2 \cdot 5}{3 \cdot 7} \qquad \text{Factor}$$

$$= \dfrac{3 \cdot 2 \cdot 5}{2 \cdot 2 \cdot 3 \cdot 7} \qquad \begin{array}{l}\text{Multiply numerators} \\ \text{Multiply denominators}\end{array}$$

$$= \dfrac{5}{14} \qquad \text{Divide out common factors}$$

We can apply the second process to the product of two rational expressions, as the following example illustrates.

EXAMPLE 1   Multiply $\dfrac{x-2}{x+3} \cdot \dfrac{x^2-9}{2x-4}$.

SOLUTION   We begin by factoring numerators and denominators as much as possible. Then we multiply the numerators and denominators. The last step consists of dividing out all factors common to the numerator and denominator:

$$\frac{x-2}{x+3} \cdot \frac{x^2-9}{2x-4} = \frac{x-2}{x+3} \cdot \frac{(x-3)(x+3)}{2(x-2)} \qquad \textbf{Factor completely}$$

$$= \frac{(x-2)(x-3)(x+3)}{(x+3)(2)(x-2)} \qquad \begin{array}{l}\textbf{Multiply numerators}\\\textbf{and denominators}\end{array}$$

$$= \frac{x-3}{2} \qquad \textbf{Divide out common factors}$$

In Chapter 1 we defined division as the equivalent of multiplication by the reciprocal. This is how it looks with fractions:

$$\frac{4}{5} \div \frac{8}{9} = \frac{4}{5} \cdot \frac{9}{8} \qquad \textbf{Division as multiplication by the reciprocal}$$

$$\left.\begin{array}{l}= \dfrac{2 \cdot 2 \cdot 3 \cdot 3}{5 \cdot 2 \cdot 2 \cdot 2} \\[1.5em] = \dfrac{9}{10}\end{array}\right\} \textbf{Factor and divide out common factors}$$

The same idea holds for division with rational expressions. The rational expression that follows the division symbol is called the *divisor;* to divide, we multiply by the reciprocal of the divisor.

EXAMPLE 2   Divide $\dfrac{3x-9}{x^2-x-20} \div \dfrac{x^2+2x-15}{x^2-25}$.

SOLUTION   We begin by taking the reciprocal of the divisor and writing the problem again in terms of multiplication. We then factor, multiply, and, finally, divide out all factors common to the numerator and denominator of the resulting expression. The complete solution looks like this:

$$\frac{3x-9}{x^2-x-20} \div \frac{x^2+2x-15}{x^2-25}$$

$$= \frac{3x-9}{x^2-x-20} \cdot \frac{x^2-25}{x^2+2x-15} \qquad \begin{array}{l}\textbf{Multiply by the reciprocal of}\\\textbf{the divisor}\end{array}$$

$$= \frac{3(x-3)}{(x+4)(x-5)} \cdot \frac{(x-5)(x+5)}{(x+5)(x-3)} \qquad \textbf{Factor}$$

$$= \frac{3(x-3)(x-5)(x+5)}{(x+4)(x-5)(x+5)(x-3)} \qquad \textbf{Multiply}$$

$$= \frac{3}{x+4} \qquad \textbf{Divide out common factors}$$

As you can see, factoring is the single most important tool we use in working with rational expressions. Most of the work we have done or will do with rational expressions is accomplished most easily if the rational expressions are in factored form. Here are some more examples of multiplication and division with rational expressions.

EXAMPLES

3. Multiply $\dfrac{3a + 6}{a^2} \cdot \dfrac{a}{2a + 4}$.

SOLUTION

$$\frac{3a + 6}{a^2} \cdot \frac{a}{2a + 4}$$

$$= \frac{3(a + 2)}{a^2} \cdot \frac{a}{2(a + 2)} \qquad \text{Factor completely}$$

$$= \frac{3(a + 2)a}{a^2(2)(a + 2)} \qquad \text{Multiply}$$

$$= \frac{3}{2a} \qquad \begin{array}{l}\text{Divide numerator and denominator} \\ \text{by common factors } a(a + 2)\end{array}$$

4. Divide $\dfrac{x^2 + 7x + 12}{x^2 - 16} \div \dfrac{x^2 + 6x + 9}{2x - 8}$.

SOLUTION

$$\frac{x^2 + 7x + 12}{x^2 - 16} \div \frac{x^2 + 6x + 9}{2x - 8}$$

$$= \frac{x^2 + 7x + 12}{x^2 - 16} \cdot \frac{2x - 8}{x^2 + 6x + 9} \qquad \begin{array}{l}\text{Division is multiplication} \\ \text{by the reciprocal}\end{array}$$

$$= \frac{(x + 3)(x + 4)(2)(x - 4)}{(x - 4)(x + 4)(x + 3)(x + 3)} \qquad \text{Factor and multiply}$$

$$= \frac{2}{x + 3} \qquad \text{Divide out common factors}$$

In Example 4 we factored and multiplied the two expressions in a single step. This saves writing the problem one extra time.

EXAMPLE 5   Multiply $(x^2 - 49)\left(\dfrac{x + 4}{x + 7}\right)$.

SOLUTION   We can think of the polynomial $x^2 - 49$ as having a denominator of 1. Thinking of $x^2 - 49$ in this way allows us to proceed as we did in previous examples:

$$(x^2 - 49)\left(\frac{x + 4}{x + 7}\right) = \frac{x^2 - 49}{1} \cdot \frac{x + 4}{x + 7} \qquad \begin{array}{l}\text{Write } x^2 - 49 \text{ with} \\ \text{denominator 1}\end{array}$$

$$= \frac{(x + 7)(x - 7)(x + 4)}{x + 7} \qquad \text{Factor and multiply}$$

$$= (x - 7)(x + 4) \qquad \text{Divide out common factors}$$

We can leave the answer in this form or multiply to get $x^2 - 3x - 28$. In this section let's agree to leave our answers in factored form.

EXAMPLE 6 Multiply $a(a + 5)(a - 5)\left(\dfrac{a + 4}{a^2 + 5a}\right)$.

SOLUTION We can think of the expression $a(a + 5)(a - 5)$ as having a denominator of 1:

$$a(a + 5)(a - 5)\left(\frac{a + 4}{a^2 + 5a}\right)$$

$$= \frac{a(a + 5)(a - 5)}{1} \cdot \frac{a + 4}{a^2 + 5a}$$

$$= \frac{\cancel{a}(\cancel{a + 5})(a - 5)(a + 4)}{\cancel{a}(\cancel{a + 5})} \qquad \textbf{Factor and multiply}$$

$$= (a - 5)(a + 4) \qquad \textbf{Divide out common factors}$$

Unit Analysis

Unit analysis is a method of converting between units of measure by multiplying by the number 1. Here is our first illustration: Suppose you are flying in a commercial airliner and the pilot tells you the plane has reached its cruising altitude of 35,000 feet. How many miles is the plane above the ground?

If you know that 1 mile is 5,280 feet, then it is simply a matter of deciding what to do with the two numbers, 5,280 and 35,000. By using unit analysis, this decision is unnecessary:

$$35,000 \text{ feet} = \frac{35,000 \text{ feet}}{1} \cdot \frac{1 \text{ mile}}{5,280 \text{ feet}}$$

We treat the units common to the numerator and denominator in the same way we treat factors common to the numerator and denominator; common units can be divided out, just as common factors are. In the previous expression, we have feet common to the numerator and denominator. Dividing them out leaves us with miles only. Here is the complete solution:

$$35,000 \text{ feet} = \frac{35,000 \text{ \cancel{feet}}}{1} \cdot \frac{1 \text{ mile}}{5,280 \text{ \cancel{feet}}}$$

$$= \frac{35,000}{5,280} \text{ miles}$$

$$= 6.6 \text{ miles to the nearest tenth of a mile}$$

The expression $\dfrac{1 \text{ mile}}{5,280 \text{ feet}}$ is called a *conversion factor*. It is simply the number 1 written in a convenient form. Because it is the number 1, we can multiply any other number by it and always be sure we have not changed that number. The key to unit analysis is choosing the right conversion factors.

EXAMPLE 7 The Mall of America in the Twin Cities covers 78 acres of land. If 1 square mile = 640 acres, how many square miles does the Mall of America cover? Round your answer to the nearest hundredth of a square mile.

SOLUTION We are starting with acres and want to end up with square miles. We need to multiply by a conversion factor that will allow acres to divide out and leave us with square miles:

$$78 \text{ acres} = \frac{78 \text{ acres}}{1} \cdot \frac{1 \text{ square mile}}{640 \text{ acres}}$$

$$= \frac{78}{640} \text{ square miles}$$

$$= 0.12 \text{ square miles to the nearest hundredth}$$

The next example is a continuation of Example 9 from Section 7.1.

The Forest Chair Lift

EXAMPLE 8 The Forest chair lift at the Northstar ski resort in Lake Tahoe is 5,603 feet long. If a ride on this chair lift takes 11 minutes, what is the average speed of the lift in miles per hour?

SOLUTION First, we find the speed of the lift in feet per minute, as we did in Example 9 of Section 7.1, by taking the ratio of distance to time.

$$\text{Rate} = \frac{\text{distance}}{\text{time}} = \frac{5{,}603 \text{ feet}}{11 \text{ minutes}} = \frac{5{,}603}{11} \text{ feet per minute}$$

$$= 509 \text{ feet per minute}$$

Next, we convert feet per minute to miles per hour. To do this, we need to know that

$$1 \text{ mile} = 5{,}280 \text{ feet}$$
$$1 \text{ hour} = 60 \text{ minutes}$$

$$\text{Speed} = 509 \text{ feet per minute} = \frac{509 \text{ feet}}{1 \text{ minute}} \cdot \frac{1 \text{ mile}}{5{,}280 \text{ feet}} \cdot \frac{60 \text{ minutes}}{1 \text{ hour}}$$

$$= \frac{509 \cdot 60}{5{,}280} \text{ miles per hour}$$

$$= 5.8 \text{ miles per hour to the nearest tenth}$$

LINKING OBJECTIVES AND EXAMPLES

Next to each **objective** we have listed the examples that are best described by that objective.

A   1–6

B   7, 8

GETTING READY FOR CLASS

After reading through the preceding section, respond in your own words and in complete sentences.

1. How do we multiply rational expressions?
2. Explain the steps used to divide rational expressions.
3. What part does factoring play in multiplying and dividing rational expressions?
4. Why are all conversion factors the same as the number 1?

Multiply or divide as indicated. Be sure to reduce all answers to lowest terms. (The numerator and denominator of the answer should not have any factors in common.)

1. $\dfrac{x+y}{3} \cdot \dfrac{6}{x+y}$ 2

2. $\dfrac{x-1}{x+1} \cdot \dfrac{5}{x-1}$ $\dfrac{5}{x+1}$

▶ 3. $\dfrac{2x+10}{x^2} \cdot \dfrac{x^3}{4x+20}$ $\dfrac{x}{2}$

4. $\dfrac{3x^4}{3x-6} \cdot \dfrac{x-2}{x^2}$ x^2

5. $\dfrac{9}{2a-8} \div \dfrac{3}{a-4}$ $\dfrac{3}{2}$

6. $\dfrac{8}{a^2-25} \div \dfrac{16}{a+5}$ $\dfrac{1}{2(a-5)}$

7. $\dfrac{x+1}{x^2-9} \div \dfrac{2x+2}{x+3}$ $\dfrac{1}{2(x-3)}$

8. $\dfrac{11}{x-2} \div \dfrac{22}{2x^2-8}$ $x+2$

9. $\dfrac{a^2+5a}{7a} \cdot \dfrac{4a^2}{a^2+4a}$ $\dfrac{4a(a+5)}{7(a+4)}$

10. $\dfrac{4a^2+4a}{a^2-25} \cdot \dfrac{a^2-5a}{8a}$ $\dfrac{a(a+1)}{2(a+5)}$

▶ 11. $\dfrac{y^2-5y+6}{2y+4} \div \dfrac{2y-6}{y+2}$ $\dfrac{y-2}{4}$

12. $\dfrac{y^2-7y}{3y^2-48} \div \dfrac{y^2-9}{y^2-7y+12}$ $\dfrac{y(y-7)}{3(y+4)(y+3)}$

13. $\dfrac{2x-8}{x^2-4} \cdot \dfrac{x^2+6x+8}{x-4}$ $\dfrac{2(x+4)}{x-2}$

14. $\dfrac{x^2+5x+1}{7x-7} \cdot \dfrac{x-1}{x^2+5x+1}$ $\dfrac{1}{7}$

15. $\dfrac{x-1}{x^2-x-6} \cdot \dfrac{x^2+5x+6}{x^2-1}$ $\dfrac{x+3}{(x-3)(x+1)}$

16. $\dfrac{x^2-3x-10}{x^2-4x+3} \cdot \dfrac{x^2-5x+6}{x^2-3x-10}$ $\dfrac{x-2}{x-1}$

17. $\dfrac{a^2+10a+25}{a+5} \div \dfrac{a^2-25}{a-5}$ 1

18. $\dfrac{a^2+a-2}{a^2+5a+6} \div \dfrac{a-1}{a}$ $\dfrac{a}{a+3}$

19. $\dfrac{y^3-5y^2}{y^4+3y^3+2y^2} \div \dfrac{y^2-5y+6}{y^2-2y-3}$ $\dfrac{y-5}{(y+2)(y-2)}$

20. $\dfrac{y^2-5y}{y^2+7y+12} \div \dfrac{y^3-7y^2+10y}{y^2+9y+18}$ $\dfrac{y+6}{(y+4)(y-2)}$

21. $\dfrac{2x^2+17x+21}{x^2+2x-35} \cdot \dfrac{x^3-125}{2x^2-7x-15}$ $\dfrac{x^2+5x+25}{x-5}$

22. $\dfrac{x^2+x-42}{4x^2+31x+21} \cdot \dfrac{4x^2-5x-6}{x^3-8}$ $\dfrac{x-6}{x^2+2x+4}$

23. $\dfrac{2x^2+10x+12}{4x^2+24x+32} \cdot \dfrac{2x^2+18x+40}{x^2+8x+15}$ 1

24. $\dfrac{3x^2-3}{6x^2+18x+12} \cdot \dfrac{2x^2-8}{x^2-3x+2}$ 1

25. $\dfrac{2a^2+7a+3}{a^2-16} \div \dfrac{4a^2+8a+3}{2a^2-5a-12}$ $\dfrac{a+3}{a+4}$

26. $\dfrac{3a^2+7a-20}{a^2+3a-4} \div \dfrac{3a^2-2a-5}{a^2-2a+1}$ $\dfrac{a-1}{a+1}$

▶ 27. $\dfrac{4y^2-12y+9}{y^2-36} \div \dfrac{2y^2-5y+3}{y^2+5y-6}$ $\dfrac{2y-3}{y-6}$

28. $\dfrac{5y^2-6y+1}{y^2-1} \div \dfrac{16y^2-9}{4y^2+7y+3}$ $\dfrac{5y-1}{4y-3}$

29. $\dfrac{x^2-1}{6x^2+42x+60} \cdot \dfrac{7x^2+17x+6}{x^3+1} \cdot \dfrac{6x+30}{7x^2-11x-6}$ $\dfrac{x-1}{(x-2)(x^2-x+1)}$

30. $\dfrac{4x^2-1}{3x-15} \cdot \dfrac{4x^2-17x-15}{4x^2-9x-9} \cdot \dfrac{3x-9}{8x^3-1}$ $\dfrac{2x+1}{4x^2+2x+1}$

31. $\dfrac{18x^3+21x^2-60x}{21x^2-25x-4} \cdot \dfrac{28x^2-17x-3}{16x^3+28x^2-30x}$ $\dfrac{3}{2}$

32. $\dfrac{56x^3+54x^2-20x}{8x^2-2x-15} \cdot \dfrac{6x^2+5x-21}{63x^3+129x^2-42x}$ $\dfrac{2}{3}$

The next two problems are intended to give you practice reading, and paying attention to, the instructions that accompany the problems you are working. Working these problems is an excellent way to get ready for a test or quiz.

▶ 33. Work each problem according to the instructions given.

   a. Simplify: $\dfrac{9-1}{27-1}$ $\dfrac{4}{13}$

   b. Reduce: $\dfrac{x^2-1}{x^3-1}$ $\dfrac{x+1}{x^2+x+1}$

   c. Multiply: $\dfrac{x^2-1}{x^3-1} \cdot \dfrac{x-1}{x+1}$ $\dfrac{x-1}{x^2+x+1}$

   d. Divide: $\dfrac{x^2-1}{x^3-1} \div \dfrac{x-1}{x^2+x+1}$ $\dfrac{x+1}{x-1}$

34. Work each problem according to the instructions given.

   a. Simplify: $\dfrac{16-9}{16+24+9}$ $\dfrac{1}{7}$

   b. Reduce: $\dfrac{4x^2-9}{4x^2+12x+9}$ $\dfrac{2x-3}{2x+3}$

   c. Multiply: $\dfrac{4x^2-9}{4x^2+12x+9} \cdot \dfrac{2x+3}{2x-3}$ 1

   d. Divide: $\dfrac{4x^2-9}{4x^2+12x+9} \div \dfrac{2x+3}{2x-3}$ $\dfrac{(2x-3)^2}{(2x+3)^2}$

Multiply the following expressions using the method shown in Examples 5 and 6 in this section.

▶ 35. $(x^2-9)\left(\dfrac{2}{x+3}\right)$ $2(x-3)$

▶ = Chalkboard Problem

   = Videos available by instructor request

▶ = Online student support materials available at www.thomsonedu.com/login

36. $(x^2 - 9)\left(\dfrac{-3}{x - 3}\right)$　$-3(x + 3)$

37. $(x^2 - x - 6)\left(\dfrac{x + 1}{x - 3}\right)$　$(x + 2)(x + 1)$

38. $(x^2 - 2x - 8)\left(\dfrac{x + 3}{x - 4}\right)$　$(x + 2)(x + 3)$

39. $(x^2 - 4x - 5)\left(\dfrac{-2x}{x + 1}\right)$　$-2x(x - 5)$

40. $(x^2 - 6x + 8)\left(\dfrac{4x}{x - 2}\right)$　$4x(x - 4)$

Each of the following problems involves some factoring by grouping. Remember, before you can divide out factors common to the numerators and denominators of a product, you must factor completely.

41. $\dfrac{x^2 - 9}{x^2 - 3x} \cdot \dfrac{2x + 10}{xy + 5x + 3y + 15}$　$\dfrac{2(x + 5)}{x(y + 5)}$

42. $\dfrac{x^2 - 16}{x^2 - 4x} \cdot \dfrac{3x + 18}{xy + 6x + 4y + 24}$　$\dfrac{3(x + 6)}{x(y + 6)}$

43. $\dfrac{2x^2 + 4x}{x^2 - y^2} \cdot \dfrac{x^2 + 3x + xy + 3y}{x^2 + 5x + 6}$　$\dfrac{2x}{x - y}$

44. $\dfrac{x^2 - 25}{3x^2 + 3xy} \cdot \dfrac{x^2 + 4x + xy + 4y}{x^2 + 9x + 20}$　$\dfrac{x - 5}{3x}$

45. $\dfrac{x^3 - 3x^2 + 4x - 12}{x^4 - 16} \cdot \dfrac{3x^2 + 5x - 2}{3x^2 - 10x + 3}$　$\dfrac{1}{x - 2}$

46. $\dfrac{x^3 - 5x^2 + 9x - 45}{x^4 - 81} \cdot \dfrac{5x^2 + 18x + 9}{5x^2 - 22x - 15}$　$\dfrac{1}{x - 3}$

Simplify each expression. Work inside parentheses first, and then divide out common factors.

47. $\left(1 - \dfrac{1}{2}\right)\left(1 - \dfrac{1}{3}\right)\left(1 - \dfrac{1}{4}\right)\left(1 - \dfrac{1}{5}\right)$　$\dfrac{1}{5}$

48. $\left(1 + \dfrac{1}{2}\right)\left(1 + \dfrac{1}{3}\right)\left(1 + \dfrac{1}{4}\right)\left(1 + \dfrac{1}{5}\right)$　3

The dots in the following problems represent factors not written that are in the same pattern as the surrounding factors. Simplify.

49. $\left(1 - \dfrac{1}{2}\right)\left(1 - \dfrac{1}{3}\right)\left(1 - \dfrac{1}{4}\right) \cdots \left(1 - \dfrac{1}{99}\right)\left(1 - \dfrac{1}{100}\right)$　$\dfrac{1}{100}$

50. $\left(1 - \dfrac{1}{3}\right)\left(1 - \dfrac{1}{4}\right)\left(1 - \dfrac{1}{5}\right) \cdots \left(1 - \dfrac{1}{98}\right)\left(1 - \dfrac{1}{99}\right)$　$\dfrac{2}{99}$

Applying the Concepts

51. Mount Whitney The top of Mount Whitney, the highest point in California, is 14,494 feet above sea level. Give this height in miles to the nearest tenth of a mile.　2.7 miles

52. Motor Displacement The relationship between liters and cubic inches, both of which are measures of volume, is 0.0164 liters = 1 cubic inch. If a Ford Mustang has a motor with a displacement of 4.9 liters, what is the displacement in cubic inches? Round your answer to the nearest cubic inch.　299 inches³

53. Speed of Sound The speed of sound is 1,088 feet per second. Convert the speed of sound to miles per hour. Round your answer to the nearest whole number.　742 miles per hour

54. Average Speed A car travels 122 miles in 3 hours. Find the average speed of the car in feet per second. Round to the nearest whole number.　60 feet per second

55. Ferris Wheel As we mentioned in Problem Set 7.1, the first Ferris wheel was built in 1893. It was a large wheel with a circumference of 785 feet. If one trip around the circumference of the wheel took 20 minutes, find the average speed of a rider in miles per hour. Round to the nearest hundredth.

0.45 miles per hour

56. Unit Analysis The photograph shows the Cone Nebula as seen by the Hubble telescope in April 2002. The distance across the photograph is about 2.5 light-years. If we assume light travels 186,000 miles in 1 second, we can find the number of miles in 1 light-year by converting 186,000 miles/second to miles/year. Find the number of miles in 1 light-year. Write your answer in expanded form and in scientific notation.　5,865,696,000,000 miles/year; 5.865696×10^{12}

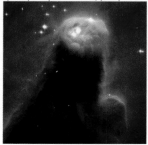

NASA

57. Ferris Wheel A Ferris wheel called *Colossus* has a circumference of 518 feet. If a trip around the circumference of *Colossus* takes 40 seconds, find the average speed of a rider in miles per hour. Round to the nearest tenth. 8.8 miles per hour

58. Average Speed Tina is training for a biathlon. As part of her training, she runs an 8-mile course, 2 miles of which is on level ground and 6 miles of which is downhill. It takes her 20 minutes to run the level part of the course and 40 minutes to run the downhill part of the course. Find her average speed in miles per hour for each part of the course.
Level ground: 6 miles per hour; downhill: 9 miles per hour

59. Running Speed Have you ever wondered about the miles per hour speed of a fast runner? This problem will give you some insight into this question.

USA TODAY Snapshots®

Fast times at Central Park
ING New York City Marathon champion Hendrik Ramaala takes aim at another New York victory at the Healthy Kidney 10K road race Sunday in Central Park. Best times in the 10K at Central Park:

Paul Koech, Kenya (1997)	28 minutes, 10 seconds
Geoff Smith, England (1982)	28:41
Geoff Smith, England (1983)	28:59
Girma Tola, Ethiopia (1997)	29:01
Peter Githuka, Kenya (1997)	29:03

Source: New York Road Runners
By Ellen J. Horrow and Keith Simmons, USA TODAY

From *USA Today*. Copyright 2004. Reprinted with permission.

a. Calculate the average speed in kilometers per hour for Paul Koech. 21.3 kilometers/hour

b. If a 10K race is approximately 6.2 miles, convert the answer from part a to miles per hour. 13.2 miles/hour

60. Improving Your Quantitative Literacy The guidelines for fitness now indicate that a person who walks 10,000 steps daily is physically fit.
  According to *The Walking Site* on the Internet, "The average person's stride length is approximately 2.5 feet long. That means it

Jim Cummings/Getty Images

takes just over 2,000 steps to walk one mile, and 10,000 steps is close to 5 miles." Use your knowledge of unit analysis to determine if these facts are correct.

Maintaining Your Skills

Add the following fractions.

61. $\dfrac{1}{2} + \dfrac{5}{2}$ 3

62. $\dfrac{2}{3} + \dfrac{8}{3}$ $\dfrac{10}{3}$

63. $2 + \dfrac{3}{4}$ $\dfrac{11}{4}$

64. $1 + \dfrac{4}{7}$ $\dfrac{11}{7}$

Simplify each term, then add.

65. $\dfrac{10x^4}{2x^2} + \dfrac{12x^6}{3x^4}$ $9x^2$

66. $\dfrac{32x^8}{8x^3} + \dfrac{27x^7}{3x^2}$ $13x^5$

67. $\dfrac{12a^2b^5}{3ab^3} + \dfrac{14a^4b^7}{7a^3b^5}$ $6ab^2$

68. $\dfrac{16a^3b^2}{4ab} + \dfrac{25a^6b^5}{5a^4b^4}$ $9a^2b$

Getting Ready for the Next Section

Perform the indicated operation.

69. $\dfrac{1}{5} + \dfrac{3}{5}$ $\dfrac{4}{5}$

70. $\dfrac{1}{7} + \dfrac{5}{7}$ $\dfrac{6}{7}$

71. $\dfrac{1}{10} + \dfrac{3}{14}$ $\dfrac{11}{35}$

72. $\dfrac{1}{21} + \dfrac{4}{15}$ $\dfrac{11}{35}$

73. $\dfrac{1}{10} - \dfrac{3}{14}$ $-\dfrac{4}{35}$

74. $\dfrac{1}{21} - \dfrac{4}{15}$ $-\dfrac{23}{105}$

Multiply.

75. $2(x - 3)$ $2x - 6$

76. $x(x + 2)$ $x^2 + 2x$

77. $(x + 4)(x - 5)$ $x^2 - x - 20$

78. $(x + 3)(x - 4)$ $x^2 - x - 12$

Reduce to lowest terms.

79. $\dfrac{x + 3}{x^2 - 9}$ $\dfrac{1}{x - 3}$

80. $\dfrac{x + 7}{x^2 - 49}$ $\dfrac{1}{x - 7}$

81. $\dfrac{x^2 - x - 30}{2(x + 5)(x - 5)}$ $\dfrac{x - 6}{2(x - 5)}$

82. $\dfrac{x^2 - x - 20}{2(x + 4)(x - 4)}$ $\dfrac{x - 5}{2(x - 4)}$

Simplify.

83. $(x + 4)(x - 5) - 10$ $x^2 - x - 30$

84. $(x + 3)(x - 4) - 8$ $x^2 - x - 20$

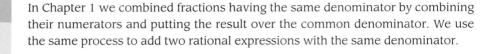
Addition and Subtraction of Rational Expressions

OBJECTIVES

A Add and subtract rational expressions that have the same denominators.

B Add and subtract rational expressions that have different denominators.

In Chapter 1 we combined fractions having the same denominator by combining their numerators and putting the result over the common denominator. We use the same process to add two rational expressions with the same denominator.

EXAMPLES

1. Add $\dfrac{5}{x} + \dfrac{3}{x}$.

SOLUTION Adding numerators, we have:

$$\frac{5}{x} + \frac{3}{x} = \frac{8}{x}$$

2. Add $\dfrac{x}{x^2 - 9} + \dfrac{3}{x^2 - 9}$.

SOLUTION Since both expressions have the same denominator, we add numerators and reduce to lowest terms:

$$\frac{x}{x^2 - 9} + \frac{3}{x^2 - 9} = \frac{x + 3}{x^2 - 9}$$

$$= \frac{\cancel{x + 3}}{(\cancel{x + 3})(x - 3)}$$

$$= \frac{1}{x - 3}$$

Reduce to lowest terms by factoring the denominator and then dividing out the common factor $x + 3$

Remember, it is the distributive property that allows us to add rational expressions by simply adding numerators. Because of this, we must begin all addition problems involving rational expressions by first making sure all the expressions have the same denominator.

> **DEFINITION** The **least common denominator** (LCD) for a set of denominators is the simplest quantity that is exactly divisible by all the denominators.

 EXAMPLE 3 Add $\dfrac{1}{10} + \dfrac{3}{14}$.

SOLUTION

Step 1: Find the LCD for 10 and 14. To do so, we factor each denominator and build the LCD from the factors:

$$\left. \begin{array}{l} 10 = 2 \cdot 5 \\ 14 = 2 \cdot 7 \end{array} \right\} \quad \text{LCD} = 2 \cdot 5 \cdot 7 = 70$$

We know the LCD is divisible by 10 because it contains the factors 2 and 5. It is also divisible by 14 because it contains the factors 2 and 7.

Step 2: Change to equivalent fractions that each have denominator 70. To accomplish this task, we multiply the numerator and denominator of

Note

If you have had difficulty in the past with addition and subtraction of fractions with different denominators, this is the time to get it straightened out. Go over Example 3 as many times as is necessary for you to understand the process.

each fraction by the factor of the LCD that is not also a factor of its denominator:

Original Fractions		Denominators in Factored Form		Multiply by Factor Needed to Obtain LCD		These Have the Same Value as the Original Fractions
$\dfrac{1}{10}$	$=$	$\dfrac{1}{2 \cdot 5}$	$=$	$\dfrac{1}{2 \cdot 5} \cdot \dfrac{\mathbf{7}}{\mathbf{7}}$	$=$	$\dfrac{7}{70}$
$\dfrac{3}{14}$	$=$	$\dfrac{3}{2 \cdot 7}$	$=$	$\dfrac{3}{2 \cdot 7} \cdot \dfrac{\mathbf{5}}{\mathbf{5}}$	$=$	$\dfrac{15}{70}$

The fraction $\frac{7}{70}$ has the same value as the fraction $\frac{1}{10}$. Likewise, the fractions $\frac{15}{70}$ and $\frac{3}{14}$ are equivalent; they have the same value.

Step 3:   Add numerators and put the result over the LCD:

$$\frac{7}{70} + \frac{15}{70} = \frac{7 + 15}{70} = \frac{22}{70}$$

Step 4:   Reduce to lowest terms:

$$\frac{22}{70} = \frac{11}{35} \qquad \textbf{Divide numerator and denominator by 2}$$

The main idea in adding fractions is to write each fraction again with the LCD for a denominator. Once we have done that, we simply add numerators. The same process can be used to add rational expressions, as the next example illustrates.

 EXAMPLE 4   Subtract $\dfrac{3}{x} - \dfrac{1}{2}$.

SOLUTION

Step 1:   The LCD for x and 2 is $2x$. It is the smallest expression divisible by x and by 2.

Step 2:   To change to equivalent expressions with the denominator $2x$, we multiply the first fraction by $\frac{2}{2}$ and the second by $\frac{x}{x}$:

$$\frac{3}{x} \cdot \frac{\mathbf{2}}{\mathbf{2}} = \frac{6}{2x}$$

$$\frac{1}{2} \cdot \frac{\mathbf{x}}{\mathbf{x}} = \frac{x}{2x}$$

Step 3:   Subtracting numerators of the rational expressions in step 2, we have:

$$\frac{6}{2x} - \frac{x}{2x} = \frac{6 - x}{2x}$$

Step 4:   Since $6 - x$ and $2x$ do not have any factors in common, we cannot reduce any further. Here is the complete solution:

$$\frac{3}{x} - \frac{1}{2} = \frac{3}{x} \cdot \frac{\mathbf{2}}{\mathbf{2}} - \frac{1}{2} \cdot \frac{\mathbf{x}}{\mathbf{x}}$$

$$= \frac{6}{2x} - \frac{x}{2x}$$

$$= \frac{6 - x}{2x}$$

EXAMPLE 5   Add $\dfrac{5}{2x-6} + \dfrac{x}{x-3}$.

SOLUTION   If we factor $2x - 6$, we have $2x - 6 = 2(x - 3)$. We need only multiply the second rational expression in our problem by $\frac{2}{2}$ to have two expressions with the same denominator:

$$\frac{5}{2x-6} + \frac{x}{x-3} = \frac{5}{2(x-3)} + \frac{x}{x-3}$$

$$= \frac{5}{2(x-3)} + \frac{\mathbf{2}}{\mathbf{2}}\left(\frac{x}{x-3}\right)$$

$$= \frac{5}{2(x-3)} + \frac{2x}{2(x-3)}$$

$$= \frac{2x+5}{2(x-3)}$$

EXAMPLE 6   Add $\dfrac{1}{x+4} + \dfrac{8}{x^2-16}$.

SOLUTION   After writing each denominator in factored form, we find that the least common denominator is $(x + 4)(x - 4)$. To change the first rational expression to an equivalent rational expression with the common denominator, we multiply its numerator and denominator by $x - 4$:

$$\frac{1}{x+4} + \frac{8}{x^2-16}$$

$$= \frac{1}{x+4} + \frac{8}{(x+4)(x-4)} \qquad \text{Factor each denominator}$$

$$= \frac{1}{x+4} \cdot \frac{\mathbf{x-4}}{\mathbf{x-4}} + \frac{8}{(x+4)(x-4)} \qquad \begin{array}{l}\text{Change to equivalent} \\ \text{rational expressions}\end{array}$$

$$= \frac{x-4}{(x+4)(x-4)} + \frac{8}{(x+4)(x-4)} \qquad \text{Simplify}$$

$$= \frac{x+4}{(x+4)(x-4)} \qquad \text{Add numerators}$$

$$= \frac{1}{x-4} \qquad \begin{array}{l}\text{Divide out common} \\ \text{factor } x + 4\end{array}$$

Note
In the last step we reduced the rational expression to lowest terms by dividing out the common factor of $x + 4$.

EXAMPLE 7   Add $\dfrac{2}{x^2+5x+6} + \dfrac{x}{x^2-9}$.

SOLUTION

Step 1:   We factor each denominator and build the LCD from the factors:

$$\left.\begin{array}{l} x^2 + 5x + 6 = (x+2)(x+3) \\ x^2 - 9 = (x+3)(x-3) \end{array}\right\} \;\; \mathbf{LCD = (x+2)(x+3)(x-3)}$$

Step 2:   Change to equivalent rational expressions:

$$\frac{2}{x^2+5x+6} = \frac{2}{(x+2)(x+3)} \cdot \frac{\mathbf{(x-3)}}{\mathbf{(x-3)}} = \frac{2x-6}{(x+2)(x+3)(x-3)}$$

$$\frac{x}{x^2-9} = \frac{x}{(x+3)(x-3)} \cdot \frac{\mathbf{(x+2)}}{\mathbf{(x+2)}} = \frac{x^2+2x}{(x+2)(x+3)(x-3)}$$

Step 3: Add numerators of the rational expressions produced in step 2:

$$\frac{2x - 6}{(x + 2)(x + 3)(x - 3)} + \frac{x^2 + 2x}{(x + 2)(x + 3)(x - 3)}$$

$$= \frac{x^2 + 4x - 6}{(x + 2)(x + 3)(x - 3)}$$

The numerator and denominator do not have any factors in common.

 EXAMPLE 8   Subtract $\dfrac{x + 4}{2x + 10} - \dfrac{5}{x^2 - 25}$.

SOLUTION   We begin by factoring each denominator:

$$\frac{x + 4}{2x + 10} - \frac{5}{x^2 - 25} = \frac{x + 4}{2(x + 5)} - \frac{5}{(x + 5)(x - 5)}$$

The LCD is $2(x + 5)(x - 5)$. Completing the solution, we have:

$$= \frac{x + 4}{2(x + 5)} \cdot \frac{(x - 5)}{(x - 5)} + \frac{-5}{(x + 5)(x - 5)} \cdot \frac{2}{2}$$

$$= \frac{x^2 - x - 20}{2(x + 5)(x - 5)} + \frac{-10}{2(x + 5)(x - 5)}$$

$$= \frac{x^2 - x - 30}{2(x + 5)(x - 5)}$$

> **Note**
> In the second step we replaced subtraction by addition of the opposite. There seems to be less chance for error when this is done on longer problems.

To see if this expression will reduce, we factor the numerator into $(x - 6)(x + 5)$:

$$= \frac{(x - 6)\cancel{(x + 5)}}{2\cancel{(x + 5)}(x - 5)}$$

$$= \frac{x - 6}{2(x - 5)}$$

 EXAMPLE 9   Write an expression for the sum of a number and its reciprocal, and then simplify that expression.

SOLUTION   If we let x = the number, then its reciprocal is $\dfrac{1}{x}$. To find the sum of the number and its reciprocal, we add them:

$$x + \frac{1}{x}$$

The first term x can be thought of as having a denominator of 1. Since the denominators are 1 and x, the least common denominator is x.

$$x + \frac{1}{x} = \frac{x}{1} + \frac{1}{x}   \qquad \textbf{Write } x \textbf{ as } \frac{x}{1}$$

$$= \frac{x}{1} \cdot \frac{x}{x} + \frac{1}{x}   \qquad \textbf{The LCD is } x$$

$$= \frac{x^2}{x} + \frac{1}{x}$$

$$= \frac{x^2 + 1}{x}   \qquad \textbf{Add numerators}$$

LINKING OBJECTIVES AND EXAMPLES

Next to each objective we have listed the examples that are best described by that objective.

A	1, 2
B	4–9

GETTING READY FOR CLASS

After reading through the preceding section, respond in your own words and in complete sentences.

1. How do we add two rational expressions that have the same denominator?
2. What is the least common denominator for two fractions?
3. What role does factoring play in finding a least common denominator?
4. Explain how to find a common denominator for two rational expressions.

Problem Set 7.3

Online support materials can be found at www.thomsonedu.com/login

Find the following sums and differences.

1. $\dfrac{3}{x} + \dfrac{4}{x} \quad \dfrac{7}{x}$

2. $\dfrac{5}{x} + \dfrac{3}{x} \quad \dfrac{8}{x}$

3. $\dfrac{9}{a} - \dfrac{5}{a} \quad \dfrac{4}{a}$

4. $\dfrac{8}{a} - \dfrac{7}{a} \quad \dfrac{1}{a}$

5. $\dfrac{1}{x+1} + \dfrac{x}{x+1} \quad 1$

6. $\dfrac{x}{x-3} - \dfrac{3}{x-3} \quad 1$

▶ **7.** $\dfrac{y^2}{y-1} - \dfrac{1}{y-1} \quad y+1$

8. $\dfrac{y^2}{y+3} - \dfrac{9}{y+3} \quad y-3$

9. $\dfrac{x^2}{x+2} + \dfrac{4x+4}{x+2} \quad x+2$

10. $\dfrac{x^2-6x}{x-3} + \dfrac{9}{x-3} \quad x-3$

11. $\dfrac{x^2}{x-2} - \dfrac{4x-4}{x-2} \quad x-2$

12. $\dfrac{x^2}{x-5} - \dfrac{10x-25}{x-5} \quad x-5$

13. $\dfrac{x+2}{x+6} - \dfrac{x-4}{x+6} \quad \dfrac{6}{x+6}$

14. $\dfrac{x+5}{x+2} - \dfrac{x+3}{x+2} \quad \dfrac{2}{x+2}$

15. $\dfrac{y}{2} - \dfrac{2}{y} \quad \dfrac{(y+2)(y-2)}{2y}$

16. $\dfrac{3}{y} + \dfrac{y}{3} \quad \dfrac{9+y^2}{3y}$

17. $\dfrac{1}{2} + \dfrac{a}{3} \quad \dfrac{3+2a}{6}$

18. $\dfrac{2}{3} + \dfrac{2a}{5} \quad \dfrac{10+6a}{15}$

▶ **19.** $\dfrac{x}{x+1} + \dfrac{3}{4} \quad \dfrac{7x+3}{4(x+1)}$

20. $\dfrac{x}{x-3} + \dfrac{1}{3} \quad \dfrac{4x-3}{3(x-3)}$

21. $\dfrac{x+1}{x-2} - \dfrac{4x+7}{5x-10} \quad \dfrac{1}{5}$

22. $\dfrac{3x+1}{2x-6} - \dfrac{x+2}{x-3} \quad \dfrac{1}{2}$

23. $\dfrac{4x-2}{3x+12} - \dfrac{x-2}{x+4} \quad \dfrac{1}{3}$

24. $\dfrac{6x+5}{5x-25} - \dfrac{x+2}{x-5} \quad \dfrac{1}{5}$

25. $\dfrac{6}{x(x-2)} + \dfrac{3}{x} \quad \dfrac{3}{x-2}$

26. $\dfrac{10}{x(x+5)} - \dfrac{2}{x} \quad \dfrac{2}{x+5}$

27. $\dfrac{4}{a} - \dfrac{12}{a^2+3a} \quad \dfrac{4}{a+3}$

28. $\dfrac{5}{a} + \dfrac{20}{a^2-4a} \quad \dfrac{5}{a-4}$

29. $\dfrac{2}{x+5} - \dfrac{10}{x^2-25} \quad \dfrac{2x-20}{(x+5)(x-5)}$

30. $\dfrac{6}{x^2-1} + \dfrac{3}{x+1} \quad \dfrac{3}{x-1}$

31. $\dfrac{x-4}{x-3} + \dfrac{6}{x^2-9} \quad \dfrac{x+2}{x+3}$

32. $\dfrac{x+1}{x-1} - \dfrac{4}{x^2-1} \quad \dfrac{x+3}{x+1}$

33. $\dfrac{a-4}{a-3} + \dfrac{5}{a^2-a-6} \quad \dfrac{a+1}{a+2}$

34. $\dfrac{a+2}{a+1} + \dfrac{7}{a^2-5a-6} \quad \dfrac{a-5}{a-6}$

▶ **35.** $\dfrac{8}{x^2-16} - \dfrac{7}{x^2-x-12} \quad \dfrac{1}{(x+3)(x+4)}$

36. $\dfrac{6}{x^2-9} - \dfrac{5}{x^2-x-6} \quad \dfrac{1}{(x+3)(x+2)}$

37. $\dfrac{4y}{y^2+6y+5} - \dfrac{3y}{y^2+5y+4} \quad \dfrac{y}{(y+5)(y+4)}$

38. $\dfrac{3y}{y^2+7y+10} - \dfrac{2y}{y^2+6y+8} \quad \dfrac{y}{(y+5)(y+4)}$

39. $\dfrac{4x+1}{x^2+5x+4} - \dfrac{x+3}{x^2+4x+3} \quad \dfrac{3(x-1)}{(x+4)(x+1)}$

40. $\dfrac{2x-1}{x^2+x-6} - \dfrac{x+2}{x^2+5x+6} \quad \dfrac{x+1}{(x+3)(x-2)}$

41. $\dfrac{1}{x} + \dfrac{x}{3x+9} - \dfrac{3}{x^2+3x} \quad \dfrac{1}{3}$

42. $\dfrac{1}{x} + \dfrac{x}{2x+4} - \dfrac{2}{x^2+2x} \quad \dfrac{1}{2}$

= Videos available by instructor request

▶ = Online student support materials available at www.thomsonedu.com/login

▶ **43.** Work each problem according to the instructions given.

a. Multiply: $\dfrac{4}{9} \cdot \dfrac{1}{6}$ $\dfrac{2}{27}$

b. Divide: $\dfrac{4}{9} \div \dfrac{1}{6}$ $\dfrac{8}{3}$

c. Add: $\dfrac{4}{9} + \dfrac{1}{6}$ $\dfrac{11}{18}$

d. Multiply: $\dfrac{x+2}{x-2} \cdot \dfrac{3x+10}{x^2-4}$ $\dfrac{3x+10}{(x-2)^2}$

e. Divide: $\dfrac{x+2}{x-2} \div \dfrac{3x+10}{x^2-4}$ $\dfrac{(x+2)^2}{3x+10}$

f. Subtract: $\dfrac{x+2}{x-2} - \dfrac{3x+10}{x^2-4}$ $\dfrac{x+3}{x+2}$

▶ **44.** Work each problem according to the instructions given.

a. Multiply: $\dfrac{9}{25} \cdot \dfrac{1}{15}$ $\dfrac{3}{125}$

b. Divide: $\dfrac{9}{25} \div \dfrac{1}{15}$ $\dfrac{27}{5}$

c. Subtract: $\dfrac{9}{25} - \dfrac{1}{15}$ $\dfrac{22}{75}$

d. Multiply: $\dfrac{3x-2}{3x+2} \cdot \dfrac{15x+6}{9x^2-4}$ $\dfrac{3(5x+2)}{(3x+2)^2}$

e. Divide: $\dfrac{3x-2}{3x+2} \div \dfrac{15x+6}{9x^2-4}$ $\dfrac{(3x-2)^2}{3(5x+2)}$

f. Subtract: $\dfrac{3x+2}{3x-2} - \dfrac{15x+6}{9x^2-4}$ $\dfrac{3x+1}{3x+2}$

Complete the following tables.

45.

Number	Reciprocal	Sum	Sum
x	$\dfrac{1}{x}$	$1+\dfrac{1}{x}$	$\dfrac{x+1}{x}$
1	1	2	2
2	$\frac{1}{2}$	$\frac{3}{2}$	$\frac{3}{2}$
3	$\frac{1}{3}$	$\frac{4}{3}$	$\frac{4}{3}$
4	$\frac{1}{4}$	$\frac{5}{4}$	$\frac{5}{4}$

46.

Number	Reciprocal	Difference	Difference
x	$\dfrac{1}{x}$	$1-\dfrac{1}{x}$	$\dfrac{x-1}{x}$
1	1	0	0
2	$\frac{1}{2}$	$\frac{1}{2}$	$\frac{1}{2}$
3	$\frac{1}{3}$	$\frac{2}{3}$	$\frac{2}{3}$
4	$\frac{1}{4}$	$\frac{3}{4}$	$\frac{3}{4}$

47.

x	$x+\dfrac{4}{x}$	$\dfrac{x^2+4}{x}$	$x+4$
1	5	5	5
2	4	4	6
3	$\frac{13}{3}$	$\frac{13}{3}$	7
4	5	5	8

48.

x	$2x+\dfrac{6}{x}$	$\dfrac{2x^2+6}{x}$	$2x+6$
1	8	8	8
2	7	7	10
3	8	8	12
4	$\frac{19}{2}$	$\frac{19}{2}$	14

Add or subtract as indicated.

49. $1+\dfrac{1}{x+2}$ $\dfrac{x+3}{x+2}$

50. $1-\dfrac{1}{x+2}$ $\dfrac{x+1}{x+2}$

51. $1-\dfrac{1}{x+3}$ $\dfrac{x+2}{x+3}$

52. $1+\dfrac{1}{x+3}$ $\dfrac{x+4}{x+3}$

Maintaining Your Skills

Solve each equation.

53. $2x+3(x-3)=6$ 3

54. $4x-2(x-5)=6$ -2

55. $x-3(x+3)=x-3$ -2

56. $x-4(x+4)=x-4$ -3

57. $7-2(3x+1)=4x+3$ $\frac{1}{5}$

58. $8-5(2x-1)=2x+4$ $\frac{3}{4}$

▶ **Chalkboard Problems**

Problems 43 and 44 can be used to show the connection between arithmetic with frctions and operations with rational expressions. Plus, they review multiplication and division and give us a variety of polynomials to factor.

Solve each quadratic equation.

59. $x^2 + 5x + 6 = 0$ $-2, -3$ **60.** $x^2 - 5x + 6 = 0$ $2, 3$

61. $x^2 - x = 6$ $3, -2$ **62.** $x^2 + x = 6$ $-3, 2$

63. $x^2 - 5x = 0$ $0, 5$ **64.** $x^2 - 6x = 0$ $0, 6$

Getting Ready for the Next Section

Simplify.

65. $6\left(\dfrac{1}{2}\right)$ 3 **66.** $10\left(\dfrac{1}{5}\right)$ 2

67. $\dfrac{0}{5}$ 0 **68.** $\dfrac{0}{2}$ 0

69. $\dfrac{5}{0}$ Undefined   **70.** $\dfrac{2}{0}$ Undefined

71. $1 - \dfrac{5}{2}$ $-\dfrac{3}{2}$ **72.** $1 - \dfrac{5}{3}$ $-\dfrac{2}{3}$

Use the distributive property to simplify.

73. $6\left(\dfrac{x}{3} + \dfrac{5}{2}\right)$ $2x + 15$ **74.** $10\left(\dfrac{x}{2} + \dfrac{3}{5}\right)$ $5x + 6$

75. $x^2\left(1 - \dfrac{5}{x}\right)$ $x^2 - 5x$ **76.** $x^2\left(1 - \dfrac{3}{x}\right)$ $x^2 - 3x$

Solve.

77. $2x + 15 = 3$ -6 **78.** $15 = 3x - 3$ 6

79. $-2x - 9 = x - 3$ -2 **80.** $a^2 - a - 20 = -2a$ $-5, 4$

7.4 Equations Involving Rational Expressions

OBJECTIVES

A Solve equations that contain rational expressions.

The first step in solving an equation that contains one or more rational expressions is to find the LCD for all denominators in the equation. Once the LCD has been found, we multiply both sides of the equation by it. The resulting equation should be equivalent to the original one (unless we inadvertently multiplied by zero) and free from any denominators except the number 1.

EXAMPLE 1   Solve $\dfrac{x}{3} + \dfrac{5}{2} = \dfrac{1}{2}$ for x.

SOLUTION   The LCD for 3 and 2 is 6. If we multiply both sides by 6, we have:

$$6\left(\frac{x}{3} + \frac{5}{2}\right) = 6\left(\frac{1}{2}\right) \qquad \textbf{Multiply both sides by 6}$$

$$6\left(\frac{x}{3}\right) + 6\left(\frac{5}{2}\right) = 6\left(\frac{1}{2}\right) \qquad \textbf{Distributive property}$$

$$2x + 15 = 3$$

$$2x = -12$$

$$x = -6$$

We can check our solution by replacing x with -6 in the original equation:

$$-\frac{6}{3} + \frac{5}{2} \overset{?}{=} \frac{1}{2}$$

$$\frac{1}{2} = \frac{1}{2}$$

Multiplying both sides of an equation containing fractions by the LCD clears the equation of all denominators, because the LCD has the property that all denominators will divide it evenly.

EXAMPLE 2   Solve for x: $\dfrac{3}{x-1} = \dfrac{3}{5}$.

SOLUTION   The LCD for $(x-1)$ and 5 is $5(x-1)$. Multiplying both sides by $5(x-1)$, we have:

$$5(x-1) \cdot \frac{3}{x-1} = 5(x-1) \cdot \frac{3}{5}$$

$$5 \cdot 3 = (x-1) \cdot 3$$

$$15 = 3x - 3$$

$$18 = 3x$$

$$6 = x$$

If we substitute $x = 6$ into the original equation, we have:

$$\frac{3}{6-1} \stackrel{?}{=} \frac{3}{5}$$

$$\frac{3}{5} = \frac{3}{5}$$

The solution set is $\{6\}$.

EXAMPLE 3   Solve $1 - \dfrac{5}{x} = \dfrac{-6}{x^2}$.

SOLUTION   The LCD is x^2. Multiplying both sides by x^2, we have

$$x^2\left(1 - \frac{5}{x}\right) = x^2\left(\frac{-6}{x^2}\right) \qquad \textbf{Multiply both sides by } x^2$$

$$x^2(1) - x^2\left(\frac{5}{x}\right) = x^2\left(\frac{-6}{x^2}\right) \qquad \begin{array}{l}\textbf{Apply distributive property}\\ \textbf{to the left side}\end{array}$$

$$x^2 - 5x = -6 \qquad \textbf{Simplify each side}$$

We have a quadratic equation, which we write in standard form, factor, and solve as we did in Chapter 6.

$$x^2 - 5x + 6 = 0 \qquad \textbf{Standard form}$$

$$(x-2)(x-3) = 0 \qquad \textbf{Factor}$$

$$x - 2 = 0 \quad \text{or} \quad x - 3 = 0 \qquad \textbf{Set factors equal to 0}$$

$$x = 2 \quad \text{or} \quad x = 3$$

The two possible solutions are 2 and 3. Checking each in the original equation, we find they both give true statements. They are both solutions to the original equation:

Check $x = 2$ Check $x = 3$

$$1 - \frac{5}{2} \stackrel{?}{=} \frac{-6}{4} \qquad\qquad 1 - \frac{5}{3} \stackrel{?}{=} \frac{-6}{9}$$

$$\frac{2}{2} - \frac{5}{2} = -\frac{3}{2} \qquad\qquad \frac{3}{3} - \frac{5}{3} = -\frac{2}{3}$$

$$-\frac{3}{2} = -\frac{3}{2} \qquad\qquad -\frac{2}{3} = -\frac{2}{3}$$

EXAMPLE 4 Solve $\dfrac{x}{x^2-9} - \dfrac{3}{x-3} = \dfrac{1}{x+3}$.

SOLUTION The factors of x^2-9 are $(x+3)(x-3)$. The LCD, then, is $(x+3)(x-3)$:

$$(x+3)(x-3) \cdot \dfrac{x}{(x+3)(x-3)} + (x+3)(x-3) \cdot \dfrac{-3}{x-3}$$

$$= (x+3)(x-3) \cdot \dfrac{1}{x+3}$$

$$x + (x+3)(-3) = (x-3)1$$

$$x + (-3x) + (-9) = x - 3$$

$$-2x - 9 = x - 3$$

$$-3x = 6$$

$$x = -2$$

The solution is $x = -2$. It checks when substituted for x in the original equation.

 EXAMPLE 5 Solve $\dfrac{x}{x-3} + \dfrac{3}{2} = \dfrac{3}{x-3}$.

SOLUTION We begin by multiplying each term on both sides of the equation by $2(x-3)$:

$$2(x-3) \cdot \dfrac{x}{x-3} + 2(x-3) \cdot \dfrac{3}{2} = 2(x-3) \cdot \dfrac{3}{x-3}$$

$$2x + (x-3) \cdot 3 = 2 \cdot 3$$

$$2x + 3x - 9 = 6$$

$$5x - 9 = 6$$

$$5x = 15$$

$$x = 3$$

Our only possible solution is $x = 3$. If we substitute $x = 3$ into our original equation, we get:

$$\dfrac{3}{3-3} + \dfrac{3}{2} \stackrel{?}{=} \dfrac{3}{3-3}$$

$$\dfrac{3}{0} + \dfrac{3}{2} = \dfrac{3}{0}$$

Two of the terms are undefined, so the equation is meaningless. What has happened is that we have multiplied both sides of the original equation by zero. The equation produced by doing this is not equivalent to our original equation. We always must check our solution when we multiply both sides of an equation by an expression containing the variable to make sure we have not multiplied both sides by zero.

Our original equation has no solution; that is, there is no real number x such that:

$$\dfrac{x}{x-3} + \dfrac{3}{2} = \dfrac{3}{x-3}$$

The solution set is $\varnothing$.

When the proposed solution to an equation is not actually a solution, it is called an *extraneous solution*. In the previous example, $x = 3$ is an extraneous solution.

EXAMPLE 6   Solve $\dfrac{a + 4}{a^2 + 5a} = \dfrac{-2}{a^2 - 25}$ for a.

SOLUTION   Factoring each denominator, we have:

$$a^2 + 5a = a(a + 5)$$

$$a^2 - 25 = (a + 5)(a - 5)$$

The LCD is $a(a + 5)(a - 5)$. Multiplying both sides of the equation by the LCD gives us:

$$\cancel{a}\cancel{(a + 5)}(a - 5) \cdot \frac{a + 4}{\cancel{a}\cancel{(a + 5)}} = \frac{-2}{\cancel{(a + 5)}\cancel{(a - 5)}} \cdot a\cancel{(a + 5)}\cancel{(a - 5)}$$

$$(a - 5)(a + 4) = -2a$$

$$a^2 - a - 20 = -2a$$

The result is a quadratic equation, which we write in standard form, factor, and solve:

$$a^2 + a - 20 = 0 \qquad \textbf{Add 2}\boldsymbol{a}\textbf{ to both sides}$$

$$(a + 5)(a - 4) = 0 \qquad \textbf{Factor}$$

$$a + 5 = 0 \quad \text{or} \quad a - 4 = 0 \qquad \textbf{Set each factor to 0}$$

$$a = -5 \quad \text{or} \quad a = 4$$

The two possible solutions are -5 and 4. There is no problem with the 4. It checks when substituted for a in the original equation. However, -5 is not a solution. Substituting -5 into the original equation gives:

$$\frac{-5 + 4}{(-5)^2 + 5(-5)} \stackrel{?}{=} \frac{-2}{(-5)^2 - 25}$$

$$\frac{-1}{0} = \frac{-2}{0}$$

This indicates -5 is not a solution. The solution is 4.

LINKING OBJECTIVES AND EXAMPLES

Next to each **objective** we have listed the examples that are best described by that objective.

A   1–6

GETTING READY FOR CLASS

After reading through the preceding section, respond in your own words and in complete sentences.

1. What is the first step in solving an equation that contains rational expressions?
2. Explain how to find the LCD used to clear an equation of fractions.
3. When will a solution to an equation containing rational expressions be extraneous?
4. When do we check for extraneous solutions to an equation containing rational expressions?

Solve the following equations. Be sure to check each answer in the original equation if you multiply both sides by an expression that contains the variable.

1. $\dfrac{x}{3} + \dfrac{1}{2} = -\dfrac{1}{2}$ -3

2. $\dfrac{x}{2} + \dfrac{4}{3} = -\dfrac{2}{3}$ -4

3. $\dfrac{4}{a} = \dfrac{1}{5}$ 20

4. $\dfrac{2}{3} = \dfrac{6}{a}$ 9

5. $\dfrac{3}{x} + 1 = \dfrac{2}{x}$ -1

6. $\dfrac{4}{x} + 3 = \dfrac{1}{x}$ -1

7. $\dfrac{3}{a} - \dfrac{2}{a} = \dfrac{1}{5}$ 5

8. $\dfrac{7}{a} + \dfrac{1}{a} = 2$ 4

9. $\dfrac{3}{x} + 2 = \dfrac{1}{2}$ -2

10. $\dfrac{5}{x} + 3 = \dfrac{4}{3}$ -3

11. $\dfrac{1}{y} - \dfrac{1}{2} = -\dfrac{1}{4}$ 4

12. $\dfrac{3}{y} - \dfrac{4}{5} = -\dfrac{1}{5}$ 5

13. $1 - \dfrac{8}{x} = \dfrac{-15}{x^2}$ $3, 5$

14. $1 - \dfrac{3}{x} = \dfrac{-2}{x^2}$ $1, 2$

15. $\dfrac{x}{2} - \dfrac{4}{x} = -\dfrac{7}{2}$ $-8, 1$

16. $\dfrac{x}{2} - \dfrac{5}{x} = -\dfrac{3}{2}$ $-5, 2$

17. $\dfrac{x-3}{2} + \dfrac{2x}{3} = \dfrac{5}{6}$ 2

18. $\dfrac{x-2}{3} + \dfrac{5x}{2} = 5$ 2

19. $\dfrac{x+1}{3} + \dfrac{x-3}{4} = \dfrac{1}{6}$ 1

20. $\dfrac{x+2}{3} + \dfrac{x-1}{5} = -\dfrac{3}{5}$ -2

21. $\dfrac{6}{x+2} = \dfrac{3}{5}$ 8

22. $\dfrac{4}{x+3} = \dfrac{1}{2}$ 5

23. $\dfrac{3}{y-2} = \dfrac{2}{y-3}$ 5

24. $\dfrac{5}{y+1} = \dfrac{4}{y+2}$ -6

25. $\dfrac{x}{x-2} + \dfrac{2}{3} = \dfrac{2}{x-2}$ $\varnothing$; 2 does not check.

26. $\dfrac{x}{x-5} + \dfrac{1}{5} = \dfrac{5}{x-5}$ $\varnothing$; 5 does not check.

27. $\dfrac{x}{x-2} + \dfrac{3}{2} = \dfrac{9}{2(x-2)}$ 3

28. $\dfrac{x}{x+1} + \dfrac{4}{5} = \dfrac{-14}{5(x+1)}$ -2

29. $\dfrac{5}{x+2} + \dfrac{1}{x+3} = \dfrac{-1}{x^2 + 5x + 6}$ $\varnothing$; -3 does not check.

30. $\dfrac{3}{x-1} + \dfrac{2}{x+3} = \dfrac{-3}{x^2 + 2x - 3}$ -2

31. $\dfrac{8}{x^2 - 4} + \dfrac{3}{x+2} = \dfrac{1}{x-2}$ 0

32. $\dfrac{10}{x^2 - 25} - \dfrac{1}{x-5} = \dfrac{3}{x+5}$ $\varnothing$; 5 does not check.

33. $\dfrac{a}{2} + \dfrac{3}{a-3} = \dfrac{a}{a-3}$ 2; 3 does not check.

34. $\dfrac{a}{2} + \dfrac{4}{a-4} = \dfrac{a}{a-4}$ 2; 4 does not check.

35. $\dfrac{6}{y^2 - 4} = \dfrac{4}{y^2 + 2y}$ -4

36. $\dfrac{2}{y^2 - 9} = \dfrac{5}{y^2 - 3y}$ -5

37. $\dfrac{2}{a^2 - 9} = \dfrac{3}{a^2 + a - 12}$ -1

38. $\dfrac{2}{a^2 - 1} = \dfrac{6}{a^2 - 2a - 3}$ 0

39. $\dfrac{3x}{x-5} - \dfrac{2x}{x+1} = \dfrac{-42}{x^2 - 4x - 5}$ $-6, -7$

40. $\dfrac{4x}{x-4} - \dfrac{3x}{x-2} = \dfrac{-3}{x^2 - 6x + 8}$ $-1, -3$

41. $\dfrac{2x}{x+2} = \dfrac{x}{x+3} - \dfrac{3}{x^2 + 5x + 6}$ -1; -3 does not check.

42. $\dfrac{3x}{x-4} = \dfrac{2x}{x-3} + \dfrac{6}{x^2 - 7x + 12}$ -2; 3 does not check.

43. Solve each equation.

  a. $5x - 1 = 0$ $\dfrac{1}{5}$

  b. $\dfrac{5}{x} - 1 = 0$ 5

  c. $\dfrac{x}{5} - 1 = \dfrac{2}{3}$ $\dfrac{25}{3}$

  d. $\dfrac{5}{x} - 1 = \dfrac{2}{3}$ 3

  e. $\dfrac{5}{x^2} + 5 = \dfrac{26}{x}$ $\dfrac{1}{5}, 5$

44. Solve each equation.

  a. $2x - 3 = 0$ $\dfrac{3}{2}$

  b. $2 - \dfrac{3}{x} = 0$ $\dfrac{3}{2}$

  c. $\dfrac{x}{3} - 2 = \dfrac{1}{2}$ $\dfrac{15}{2}$

  d. $\dfrac{3}{x} - 2 = \dfrac{1}{2}$ $\dfrac{6}{5}$

  e. $\dfrac{1}{x} + \dfrac{3}{x^2} = 2$ $-1, \dfrac{3}{2}$

▶ **Chalkboard Problems**

Problems 43 and 44 have a variety of equations to solve, starting with some equations they have solved in chapter 2. They make an easy transition to equations with rational expressions. With Problems 45 and 46, students start to see the difference between addition and subtraction with rational expressions and solving equations.

= Videos available by instructor request

▶ = Online student support materials available at www.thomsonedu.com/login

▶ **45.** Work each problem according to the instructions given.

    a. Divide: $\dfrac{7}{a^2 - 5a - 6} \div \dfrac{a + 2}{a + 1}$ $\dfrac{7}{(a - 6)(a + 2)}$

    b. Add: $\dfrac{7}{a^2 - 5a - 6} + \dfrac{a + 2}{a + 1}$ $\dfrac{a - 5}{a - 6}$

    c. Solve: $\dfrac{7}{a^2 - 5a - 6} + \dfrac{a + 2}{a + 1} = 2$ Possible 7, −1; only 7

▶ **46.** Work each problem according to the instructions given.

    a. Divide: $\dfrac{6}{x^2 - 9} \div \dfrac{x - 4}{x - 3}$ $\dfrac{6}{(x + 3)(x - 4)}$

    b. Add: $\dfrac{6}{x^2 - 9} + \dfrac{x - 4}{x - 3}$ $\dfrac{x + 2}{x + 3}$

    c. Solve: $\dfrac{6}{x^2 - 9} + \dfrac{x - 4}{x - 3} = \dfrac{3}{4}$ Possible 3, 1; only 1

Maintaining Your Skills

Solve each word problem.

47. Number Problem If twice the difference of a number and 3 were decreased by 5, the result would be 3. Find the number.   7

48. Number Problem If 3 times the sum of a number and 2 were increased by 6, the result would be 27. Find the number.   5

49. Geometry The length of a rectangle is 5 more than twice the width. The perimeter is 34 inches. Find the length and width.   Length 13 inches, width 4 inches

50. Geometry The length of a rectangle is 2 more than 3 times the width. The perimeter is 44 feet. Find the length and width.   Length 17 feet, width 5 feet

Solve each problem. Be sure to show the equation that describes the situation.

51. Number Problem The product of two consecutive even integers is 48. Find the two integers.   6, 8 or −8, −6

52. Number Problem The product of two consecutive odd integers is 35. Find the two integers.   5, 7 or −7, −5

53. Geometry The hypotenuse (the longest side) of a right triangle is 10 inches, and the lengths of the two legs (the other two sides) are given by two consecutive even integers. Find the lengths of the two legs.   6 inches, 8 inches

54. Geometry One leg of a right triangle is 2 more than twice the other. If the hypotenuse is 13 feet, find the lengths of the two legs.   5 feet, 12 feet

Getting Ready for the Next Section

Solve.

55. $\dfrac{1}{x} + \dfrac{1}{2x} = \dfrac{9}{2}$ $\dfrac{1}{3}$

56. $\dfrac{50}{x + 5} = \dfrac{30}{x - 5}$ 20

57. $\dfrac{1}{10} - \dfrac{1}{15} = \dfrac{1}{x}$ 30

58. $\dfrac{15}{x} + \dfrac{15}{x + 20} = 2$ −15, 10

Find the value of $y = -\dfrac{6}{x}$ for the given value of x.

59. $x = -6$ 1   **60.** $x = -3$ 2

61. $x = 2$ −3   **62.** $x = 1$ −6

7.5 Applications

OBJECTIVES

A Solve applications whose solutions depend on solving an equation containing rational expressions.

B Graph an equation involving a rational expression.

In this section we will solve some word problems whose equations involve rational expressions. Like the other word problems we have encountered, the more you work with them, the easier they become.

Also in this section, we will look at the graphs of rational equations. As we did with the graphs of linear equations, we will first build a table by finding values of x and y that satisfy the given equation.

EXAMPLE 1 One number is twice another. The sum of their reciprocals is $\frac{9}{2}$. Find the two numbers.

SOLUTION Let x = the smaller number. The larger then must be $2x$. Their reciprocals are $\frac{1}{x}$ and $\frac{1}{2x}$, respectively. An equation that describes the situation is:

$$\frac{1}{x} + \frac{1}{2x} = \frac{9}{2}$$

We can multiply both sides by the LCD $2x$ and then solve the resulting equation:

$$2x\left(\frac{1}{x}\right) + 2x\left(\frac{1}{2x}\right) = 2x\left(\frac{9}{2}\right)$$

$$2 + 1 = 9x$$

$$3 = 9x$$

$$x = \frac{3}{9} = \frac{1}{3}$$

The smaller number is $\frac{1}{3}$. The other number is twice as large, or $\frac{2}{3}$. If we add their reciprocals, we have:

$$\frac{3}{1} + \frac{3}{2} = \frac{6}{2} + \frac{3}{2} = \frac{9}{2}$$

The solutions check with the original problem.

EXAMPLE 2 A boat travels 30 miles up a river in the same amount of time it takes to travel 50 miles down the same river. If the current is 5 miles per hour, what is the speed of the boat in still water?

SOLUTION The easiest way to work a problem like this is with a table. The top row of the table is labeled with d for distance, r for rate, and t for time. The left column of the table is labeled with the two trips: upstream and downstream. Here is what the table looks like:

	d	r	t
Upstream			
Downstream			

30 miles
50 miles

The next step is to read the problem over again and fill in as much of the table as we can with the information in the problem. The distance the boat travels upstream is 30 miles and the distance downstream is 50 miles. Since we are asked for the speed of the boat in still water, we will let that be x. If the speed of the boat in still water is x, then its speed upstream (against the current) must be $x - 5$, and its speed downstream (with the current) must be $x + 5$. Putting these four quantities into the appropriate positions in the table, we have

	d	r	t
Upstream	30	$x - 5$	
Downstream	50	$x + 5$	

The last positions in the table are filled in by using the equation $t = \dfrac{d}{r}$.

	d	r	t
Upstream	30	$x - 5$	$\dfrac{30}{x - 5}$
Downstream	50	$x + 5$	$\dfrac{50}{x + 5}$

Note

There are two things to note about this problem. The first is that to solve the equation $d = r \cdot t$ for t, we divide each side by r, like this:

$$\frac{d}{r} = \frac{r \cdot t}{r}$$

$$\frac{d}{r} = t$$

The second thing is this: The speed of the boat in still water is the rate at which it would be traveling if there were no current; that is, it is the speed of the boat through the water. Since the water itself is moving at 5 miles per hour, the boat is going 5 miles per hour slower when it travels against the current and 5 miles per hour faster when it travels with the current.

Reading the problem again, we find that the time for the trip upstream is equal to the time for the trip downstream. Setting these two quantities equal to each other, we have our equation:

$$\text{Time (downstream)} = \text{time (upstream)}$$

$$\frac{50}{x + 5} = \frac{30}{x - 5}$$

The LCD is $(x + 5)(x - 5)$. We multiply both sides of the equation by the LCD to clear it of all denominators. Here is the solution:

$$(x + 5)(x - 5) \cdot \frac{50}{x + 5} = (x + 5)(x - 5) \cdot \frac{30}{x - 5}$$

$$50x - 250 = 30x + 150$$

$$20x = 400$$

$$x = 20$$

The speed of the boat in still water is 20 miles per hour.

 EXAMPLE 3 Tina is training for a biathlon. To train for the bicycle portion, she rides her bike 15 miles up a hill and then 15 miles back down the same hill. The complete trip takes her 2 hours. If her downhill speed is 20 miles per hour faster than her uphill speed, how fast does she ride uphill?

SOLUTION Again, we make a table. As in the previous example, we label the top row with distance, rate, and time. We label the left column with the two trips, uphill and downhill.

	d	r	t
Uphill			
Downhill			

Up and back
total time = 2 hours

Next, we fill in the table with as much information as we can from the problem. We know the distance traveled is 15 miles uphill and 15 miles downhill, which allows us to fill in the distance column. To fill in the rate column, we first note that she rides 20 miles per hour faster downhill than uphill. Therefore, if we let x equal her rate uphill, then her rate downhill is $x + 20$. Filling in the table with this information gives us

	d	r	t
Uphill	15	x	
Downhill	15	x + 20	

Since time is distance divided by rate, $t = d/r$, we can fill in the last column in the table.

	d	r	t
Uphill	15	x	$\dfrac{15}{x}$
Downhill	15	x + 20	$\dfrac{15}{x + 20}$

Rereading the problem, we find that the total time (the time riding uphill plus the time riding downhill) is two hours. We write our equation as follows:

$$\text{Time (uphill)} + \text{time (downhill)} = 2$$

$$\frac{15}{x} + \frac{15}{x + 20} = 2$$

We solve this equation for x by first finding the LCD and then multiplying each term in the equation by it to clear the equation of all denominators. Our LCD is $x(x + 20)$. Here is our solution:

$$x(x + 20)\frac{15}{x} + x(x + 20)\frac{15}{x + 20} = 2 \cdot [x(x + 20)]$$

$$15(x + 20) + 15x = 2x(x + 20)$$

$$15x + 300 + 15x = 2x^2 + 40x$$

$$0 = 2x^2 + 10x - 300$$

$$0 = x^2 + 5x - 150 \quad \textbf{Divide both sides by 2}$$

$$0 = (x + 15)(x - 10)$$

$$x = -15 \quad \text{or} \quad x = 10$$

Since we cannot have a negative speed, our only solution is $x = 10$. Tina rides her bike at a rate of 10 miles per hour when going uphill. (Her downhill speed is $x + 20 = 30$ miles per hour.)

Inlet pipe
10 hours
to fill

Outlet pipe

15 hours
to empty

EXAMPLE 4   An inlet pipe can fill a water tank in 10 hours, while an outlet pipe can empty the same tank in 15 hours. By mistake, both pipes are left open. How long will it take to fill the water tank with both pipes open?

SOLUTION   Let $x =$ amount of time to fill the tank with both pipes open.

One method of solving this type of problem is to think in terms of how much of the job is done by a pipe in 1 hour.

1. If the inlet pipe fills the tank in 10 hours, then in 1 hour the inlet pipe fills $\frac{1}{10}$ of the tank.

2. If the outlet pipe empties the tank in 15 hours, then in 1 hour the outlet pipe empties $\frac{1}{15}$ of the tank.

3. If it takes x hours to fill the tank with both pipes open, then in 1 hour the tank is $\frac{1}{x}$ full.

Here is how we set up the equation. *In 1 hour,*

$$\underset{\substack{\text{Amount of water let} \\ \text{in by inlet pipe}}}{\frac{1}{10}} - \underset{\substack{\text{Amount of water let} \\ \text{out by outlet pipe}}}{\frac{1}{15}} = \underset{\substack{\text{Total amount of} \\ \text{water into tank}}}{\frac{1}{x}}$$

The LCD for our equation is $30x$. We multiply both sides by the LCD and solve:

$$30x\left(\frac{1}{10}\right) - 30x\left(\frac{1}{15}\right) = 30x\left(\frac{1}{x}\right)$$

$$3x - 2x = 30$$

$$x = 30$$

It takes 30 hours with both pipes open to fill the tank.

> **Note**
> In solving a problem of this type, we have to assume that the thing doing the work (whether it is a pipe, a person, or a machine) is working at a constant rate; that is, as much work gets done in the first hour as is done in the last hour and any other hour in between.

EXAMPLE 5 Graph the equation $y = \dfrac{1}{x}$.

SOLUTION Since this is the first time we have graphed an equation of this form, we will make a table of values for x and y that satisfy the equation. Before we do, let's make some generalizations about the graph (Figure 1).

First, notice that since y is equal to 1 divided by x, y will be positive when x is positive. (The quotient of two positive numbers is a positive number.) Likewise, when x is negative, y will be negative. In other words, x and y always will have the same sign. Thus, our graph will appear in quadrants I and III only because in those quadrants x and y have the same sign.

Next, notice that the expression $\dfrac{1}{x}$ will be undefined when x is 0, meaning that there is no value of y corresponding to $x = 0$. Because of this, the graph will not cross the y-axis. Further, the graph will not cross the x-axis either. If we try to find the x-intercept by letting $y = 0$, we have

$$0 = \frac{1}{x}$$

x	y
-3	$-\frac{1}{3}$
-2	$-\frac{1}{2}$
-1	-1
$-\frac{1}{2}$	-2
$-\frac{1}{3}$	-3
0	Undefined
$\frac{1}{3}$	3
$\frac{1}{2}$	2
1	1
2	$\frac{1}{2}$
3	$\frac{1}{3}$

But there is no value of x to divide into 1 to obtain 0. Therefore, since there is no solution to this equation, our graph will not cross the x-axis.

To summarize, we can expect to find the graph in quadrants I and III only, and the graph will cross neither axis.

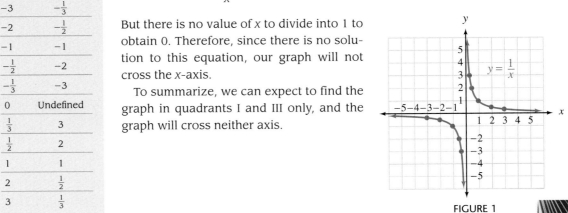

FIGURE 1

EXAMPLE 6   Graph the equation $y = \dfrac{-6}{x}$.

SOLUTION   Since y is -6 divided by x, when x is positive, y will be negative (a negative divided by a positive is negative), and when x is negative, y will be positive (a negative divided by a negative). Thus, the graph (Figure 2) will appear in quadrants II and IV only. As was the case in Example 5, the graph will not cross either axis.

x	y
−6	1
−3	2
−2	3
−1	6
0	Undefined
1	−6
2	−3
3	−2
6	−1

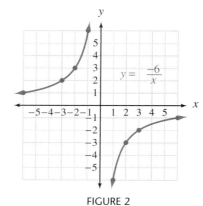

FIGURE 2

GETTING READY FOR CLASS

After reading through the preceding section, respond in your own words and in complete sentences.

1. Write an application problem for which the solution depends on solving the equation $\dfrac{1}{2} + \dfrac{1}{3} = \dfrac{1}{x}$.

2. How does the current of a river affect the speed of a motor boat traveling against the current?

3. How does the current of a river affect the speed of a motor boat traveling in the same direction as the current?

4. What is the relationship between the total number of minutes it takes for a drain to empty a sink and the amount of water that drains out of the sink in 1 minute?

LINKING OBJECTIVES AND EXAMPLES

Next to each **objective** we have listed the examples that are best described by that objective.

A	1–4
B	5, 6

Number Problems

▶ **1.** One number is 3 times as large as another. The sum of their reciprocals is $\frac{16}{3}$. Find the two numbers. $\frac{1}{4}, \frac{3}{4}$

2. If $\frac{3}{5}$ is added to twice the reciprocal of a number, the result is 1. Find the number. 5

3. The sum of a number and its reciprocal is $\frac{13}{6}$. Find the number. $\frac{2}{3}$ and $\frac{3}{2}$

4. The sum of a number and 10 times its reciprocal is 7. Find the number. 2 and 5

5. If a certain number is added to both the numerator and denominator of the fraction $\frac{7}{9}$, the result is $\frac{5}{7}$. Find the number. -2

6. The numerator of a certain fraction is 2 more than the denominator. If $\frac{1}{3}$ is added to the fraction, the result is 2. Find the fraction. $\frac{5}{3}$

7. The sum of the reciprocals of two consecutive even integers is $\frac{5}{12}$. Find the integers. 4, 6

8. The sum of the reciprocals of two consecutive integers is $\frac{7}{12}$. Find the two integers. 3, 4

Motion Problems

9. A boat travels 26 miles up the river in the same amount of time it takes to travel 38 miles down the same river. If the current is 3 miles per hour, what is the speed of the boat in still water? 16 miles per hour

	d	r	t
Upstream			
Downstream			

10. A boat can travel 9 miles up a river in the same amount of time it takes to travel 11 miles down the same river. If the current is 2 miles per hour, what is the speed of the boat in still water? 20 miles per hour

	d	r	t
Upstream			
Downstream			

11. An airplane flying against the wind travels 140 miles in the same amount of time it would take the same plane to travel 160 miles with the wind. If the wind speed is a constant 20 miles per hour, how fast would the plane travel in still air? 300 miles per hour

12. An airplane flying against the wind travels 500 miles in the same amount of time that it would take to travel 600 miles with the wind. If the speed of the wind is 50 miles per hour, what is the speed of the plane in still air? 550 miles per hour

▶ **13.** One plane can travel 20 miles per hour faster than another. One of them goes 285 miles in the same time it takes the other to go 255 miles. What are their speeds? 170 miles per hour; 190 miles per hour

14. One car travels 300 miles in the same amount of time it takes a second car traveling 5 miles per hour slower than the first to go 275 miles. What are the speeds of the cars? 60 miles per hour, 55 miles per hour

15. Tina, whom we mentioned in Example 3 of this section, is training for a biathlon. To train for the running portion of the race, she runs 8 miles each day, over the same course. The first 2 miles of the course is on level ground, while the last 6 miles is downhill. She runs 3 miles per hour slower on level ground than she runs downhill. If the complete course takes 1 hour, how fast does she run on the downhill part of the course? 9 miles per hour

16. Jerri is training for the same biathlon as Tina (Example 3 and Problem 15). To train for the bicycle portion of the race, she rides 24 miles out a straight road, then turns around and rides 24 miles back. The trip out is against the wind, whereas the trip back is with the wind. If she rides 10 miles per hour faster with the wind then she does against the wind, and the complete trip out and back takes 2 hours, how fast does she ride when she rides against the wind? 20 miles per hour

17. To train for the running of a triathlon, Jerri jogs 1 hour each day over the same 9-mile course. Five miles of the course is downhill, whereas the other 4 miles is on level ground. Jerri figures that she runs 2 miles per hour faster downhill than she runs on level ground. Find the rate at which Jerri runs on level ground. 8 miles per hour

18. Travis paddles his kayak in the harbor at Morro Bay, California, where the incoming tide has caused a current in the water. From the point where he enters the water, he paddles 1 mile against the current, then turns around and paddles 1 mile back to where he started. His average speed when paddling with the current is 4 miles per hour faster than his speed against the current. If the complete trip (out and back) takes him 1.2 hours, find his average speed when he paddles against the current. 1 mile per hour

Work Problems

▶ 19. An inlet pipe can fill a pool in 12 hours, while an out-let pipe can empty it in 15 hours. If both pipes are left open, how long will it take to fill the pool. 60 hours

20. A water tank can be filled in 20 hours by an inlet pipe and emptied in 25 hours by an outlet pipe. How long will it take to fill the tank if both pipes are left open? 100 hours

21. A bathtub can be filled by the cold water faucet in 10 minutes and by the hot water faucet in 12 min-utes. How long does it take to fill the tub if both faucets are open? $\frac{60}{11}$ minutes

22. A water faucet can fill a sink in 6 minutes, whereas the drain can empty it in 4 minutes. If the sink is full, how long will it take to empty if both the faucet and the drain are open? 12 minutes

23. A sink can be filled by the cold water faucet in 3 minutes. The drain can empty a full sink in 4 min-utes. If the sink is empty and both the cold water faucet and the drain are open, how long will it take the sink to overflow? 12 minutes

24. A bathtub can be filled by the cold water faucet in 9 minutes and by the hot water faucet in 10 minutes. The drain can empty the tub in 5 minutes. Can the tub be filled if both faucets and the drain are open? Yes

Graph each of the following equations.

25. $y = \dfrac{-4}{x}$

26. $y = \dfrac{4}{x}$

27. $y = \dfrac{8}{x}$

28. $y = \dfrac{-8}{x}$

29. Graph $y = \dfrac{3}{x}$ and $x + y = 4$ on the same coordinate system. At what points do the two graphs intersect? (1, 3) and (3, 1)

30. Graph $y = \dfrac{4}{x}$ and $x - y = 3$ on the same coordinate system. At what points do the two graphs intersect? (4, 1) and (−1, −4)

Maintaining Your Skills

31. Factor out the greatest common factor for $15a^3b^3 - 20a^2b - 35ab^2$. $5ab(3a^2b^2 - 4a - 7b)$

32. Factor by grouping $3ax - 2a + 15x - 10$. $(a + 5)(3x - 2)$

Factor completely.

33. $x^2 - 4x - 12$ $(x - 6)(x + 2)$

34. $4x^2 - 20xy + 25y^2$ $(2x - 5y)^2$

35. $x^4 - 16$ $(x^2 + 4)(x + 2)(x - 2)$

36. $2x^2 + xy - 21y^2$ $(2x + 7y)(x - 3y)$

37. $5x^3 - 25x^2 - 30x$ $5x(x - 6)(x + 1)$

Solve each equation.

38. $x^2 - 9x + 18 = 0$ 3, 6

39. $x^2 - 6x = 0$ 0, 6

40. $8x^2 = -2x + 15$ $\frac{5}{4}, -\frac{3}{2}$

41. $x(x + 2) = 80$ −10, 8

42. **Number Problem** The product of two consecutive even integers is 4 more than twice their sum. Find the two integers. 4, 6 or −2, 0

43. **Geometry** The hypotenuse of a right triangle is 15 inches. One of the legs is 3 inches more than the other. Find the length of the two legs.
9 inches and 12 inches

Getting Ready for the Next Section

Simplify.

44. $\dfrac{1}{2} \div \dfrac{2}{3}$ $\frac{3}{4}$

45. $\dfrac{1}{3} \div \dfrac{3}{4}$ $\frac{4}{9}$

46. $1 + \dfrac{1}{2}$ $\frac{3}{2}$

47. $1 + \dfrac{2}{3}$ $\frac{5}{3}$

48. $y^5 \cdot \dfrac{2x^3}{y^2}$ $2x^3y^3$

49. $y^7 \cdot \dfrac{3x^5}{y^4}$ $3x^5y^3$

50. $\dfrac{2x^3}{y^2} \cdot \dfrac{y^5}{4x}$ $\frac{x^2y^3}{2}$

51. $\dfrac{3x^5}{y^4} \cdot \dfrac{y^7}{6x^2}$ $\frac{x^3y^3}{2}$

Factor.

52. $x^2y + x$ $x(xy + 1)$

53. $xy^2 + y$ $y(xy + 1)$

Reduce.

54. $\dfrac{2x^3y^2}{4x}$ $\frac{x^2y^2}{2}$

55. $\dfrac{3x^5y^3}{6x^2}$ $\frac{x^3y^3}{2}$

56. $\dfrac{x^2 - 4}{x^2 - x - 6}$ $\frac{x - 2}{x - 3}$

57. $\dfrac{x^2 - 9}{x^2 - 5x + 6}$ $\frac{x + 3}{x - 2}$

7.6 Complex Fractions

7.6

OBJECTIVES

A Simplify a complex fraction.

A complex fraction is a fraction or rational expression that contains other fractions in its numerator or denominator. Each of the following is a complex fraction:

$$\dfrac{\frac{1}{2}}{\frac{2}{3}} \qquad \dfrac{x+\frac{1}{y}}{y+\frac{1}{x}} \qquad \dfrac{\frac{a+1}{a^2-9}}{\frac{2}{a+3}}$$

We will begin this section by simplifying the first of these complex fractions. Before we do, though, let's agree on some vocabulary. So that we won't have to use phrases such as "the numerator of the denominator," let's call the numerator of a complex fraction the *top* and the denominator of a complex fraction the *bottom*.

EXAMPLE 1   Simplify $\dfrac{\frac{1}{2}}{\frac{2}{3}}$.

SOLUTION   There are two methods we can use to solve this problem.

Method 1   We can multiply the top and bottom of this complex fraction by the LCD for both fractions. In this case the LCD is 6:

$$\dfrac{\frac{1}{2}}{\frac{2}{3}} = \dfrac{6 \cdot \frac{1}{2}}{6 \cdot \frac{2}{3}} = \dfrac{3}{4}$$

Method 2   We can treat this as a division problem. To divide by $\frac{2}{3}$, we multiply by its reciprocal $\frac{3}{2}$:

$$\dfrac{\frac{1}{2}}{\frac{2}{3}} = \dfrac{1}{2} \cdot \dfrac{3}{2} = \dfrac{3}{4}$$

Using either method, we obtain the same result.

EXAMPLE 2   Simplify:

$$\dfrac{\frac{2x^3}{y^2}}{\frac{4x}{y^5}}$$

SOLUTION

Method 1   The LCD for each rational expression is y^5. Multiplying the top and bottom of the complex fraction by y^5, we have:

$$\dfrac{\frac{2x^3}{y^2}}{\frac{4x}{y^5}} = \dfrac{y^5 \cdot \frac{2x^3}{y^2}}{y^5 \cdot \frac{4x}{y^5}} = \dfrac{2x^3 y^3}{4x} = \dfrac{x^2 y^3}{2}$$

Method 2   To divide by $\dfrac{4x}{y^5}$ we multiply by its reciprocal, $\dfrac{y^5}{4x}$:

$$\frac{\dfrac{2x^3}{y^2}}{\dfrac{4x}{y^5}} = \frac{2x^3}{y^2} \cdot \frac{y^5}{4x} = \frac{x^2 y^3}{2}$$

Again the result is the same, whether we use Method 1 or Method 2.

 EXAMPLE 3   Simplify:

$$\frac{x + \dfrac{1}{y}}{y + \dfrac{1}{x}}$$

SOLUTION   To apply Method 2 as we did in the first two examples, we would have to simplify the top and bottom separately to obtain a single rational expression for both before we could multiply by the reciprocal. It is much easier, in this case, to multiply the top and bottom by the LCD xy:

$$\frac{x + \dfrac{1}{y}}{y + \dfrac{1}{x}} = \frac{xy\left(x + \dfrac{1}{y}\right)}{xy\left(y + \dfrac{1}{x}\right)} \qquad \textbf{Multiply top and bottom by } xy$$

$$= \frac{xy \cdot x + xy \cdot \dfrac{1}{y}}{xy \cdot y + xy \cdot \dfrac{1}{x}} \qquad \textbf{Distributive property}$$

$$= \frac{x^2 y + x}{xy^2 + y} \qquad \textbf{Simplify}$$

We can factor an x from $x^2y + x$ and a y from $xy^2 + y$ and then reduce to lowest terms:

$$= \frac{x\,(\cancel{xy + 1})}{y\,(\cancel{xy + 1})}$$

$$= \frac{x}{y} \qquad\qquad\qquad\qquad\qquad$$

 EXAMPLE 4   Simplify:

$$\frac{1 - \dfrac{4}{x^2}}{1 - \dfrac{1}{x} - \dfrac{6}{x^2}}$$

SOLUTION   The simplest way to simplify this complex fraction is to multiply the top and bottom by the LCD, x^2:

$$\frac{1 - \dfrac{4}{x^2}}{1 - \dfrac{1}{x} - \dfrac{6}{x^2}} = \frac{x^2\left(1 - \dfrac{4}{x^2}\right)}{x^2\left(1 - \dfrac{1}{x} - \dfrac{6}{x^2}\right)} \qquad \textbf{Multiply top and bottom by } x^2$$

$$= \frac{x^2 \cdot 1 - x^2 \cdot \dfrac{4}{x^2}}{x^2 \cdot 1 - x^2 \cdot \dfrac{1}{x} - x^2 \cdot \dfrac{6}{x^2}} \qquad \textbf{Distributive property}$$

$$= \frac{x^2 - 4}{x^2 - x - 6} \qquad \textbf{Simplify}$$

$$= \frac{(x-2)\cancel{(x+2)}}{(x-3)\cancel{(x+2)}} \qquad \textbf{Factor}$$

$$= \frac{x-2}{x-3} \qquad \textbf{Reduce} \qquad$$

In our next example, we find the relationship between a sequence of complex fractions and the numbers in the Fibonacci sequence.

 EXAMPLE 5   Simplify each term in the following sequence, and then explain how this sequence is related to the Fibonacci sequence:

$$1 + \frac{1}{1+1}, \; 1 + \frac{1}{1 + \dfrac{1}{1+1}}, \; 1 + \frac{1}{1 + \dfrac{1}{1 + \dfrac{1}{1+1}}}, \ldots$$

SOLUTION   We can simplify our work somewhat if we notice that the first term $1 + \dfrac{1}{1+1}$ is the larger denominator in the second term and that the second term is the largest denominator in the third term:

First term: $1 + \dfrac{1}{1+1} = 1 + \dfrac{1}{2} = \dfrac{2}{2} + \dfrac{1}{2} = \dfrac{3}{2}$

Second term: $1 + \dfrac{1}{1 + \dfrac{1}{1+1}} = 1 + \dfrac{1}{\dfrac{3}{2}} = 1 + \dfrac{2}{3} = \dfrac{3}{3} + \dfrac{2}{3} = \dfrac{5}{3}$

Third term: $1 + \dfrac{1}{1 + \dfrac{1}{1 + \dfrac{1}{1+1}}} = 1 + \dfrac{1}{\dfrac{5}{3}} = 1 + \dfrac{3}{5} = \dfrac{5}{5} + \dfrac{3}{5} = \dfrac{8}{5}$

Here are the simplified numbers for the first three terms in our sequence:

$$\frac{3}{2}, \frac{5}{3}, \frac{8}{5}, \ldots$$

Recall the Fibonacci sequence:

$$1, 1, 2, 3, 5, 8, 13, 21, \ldots$$

As you can see, each term in the sequence we have simplified is the ratio of two consecutive numbers in the Fibonacci sequence. If the pattern continues in this manner, the next number in our sequence will be $\frac{13}{8}$.

LINKING OBJECTIVES AND EXAMPLES

Next to each **objective** we have listed the examples that are best described by that objective.

A   1–5

GETTING READY FOR CLASS

After reading through the preceding section, respond in your own words and in complete sentences.

1. What is a complex fraction?
2. Explain one method of simplifying complex fractions.
3. How is a least common denominator used to simplify a complex fraction?
4. What types of complex fractions can be rewritten as division problems?

Problem Set 7.6

Online support materials can be found at www.thomsonedu.com/login

Simplify each complex fraction.

1. $\dfrac{\frac{3}{4}}{\frac{1}{8}}$ 6

2. $\dfrac{\frac{1}{3}}{\frac{5}{6}}$ $\frac{2}{5}$

3. $\dfrac{\frac{2}{3}}{\frac{3}{4}}$ $\frac{1}{6}$

4. $\dfrac{\frac{5}{1}}{\frac{1}{2}}$ 10

5. $\dfrac{\frac{x^2}{y}}{\frac{x}{y^3}}$ xy^2

6. $\dfrac{\frac{x^5}{y^3}}{\frac{x^2}{y^8}}$ x^3y^5

7. $\dfrac{\frac{4x^3}{y^6}}{\frac{8x^2}{y^7}}$ $\frac{xy}{2}$

8. $\dfrac{\frac{6x^4}{y}}{\frac{2x}{y^5}}$ $3x^3y^4$

9. $\dfrac{y + \frac{1}{x}}{x + \frac{1}{y}}$ $\frac{y}{x}$

10. $\dfrac{y - \frac{1}{x}}{x - \frac{1}{y}}$ $\frac{y}{x}$

11. $\dfrac{1 + \frac{1}{a}}{1 - \frac{1}{a}}$ $\frac{a+1}{a-1}$

12. $\dfrac{\frac{1}{a} - 1}{\frac{1}{a} + 1}$ $\frac{1-a}{1+a}$

13. $\dfrac{\frac{x+1}{x^2-9}}{\frac{2}{x+3}}$ $\frac{x+1}{2(x-3)}$

14. $\dfrac{\frac{3}{x-5}}{\frac{x+1}{x^2-25}}$ $\frac{3(x+5)}{x+1}$

▶ **15.** $\dfrac{\frac{1}{a+2}}{\frac{1}{a^2-a-6}}$ $a - 3$

16. $\dfrac{\frac{1}{a^2+5a+6}}{\frac{1}{a+3}}$ $\frac{1}{a+2}$

▶ **17.** $\dfrac{1 - \frac{9}{y^2}}{1 - \frac{1}{y} - \frac{6}{y^2}}$ $\frac{y+3}{y+2}$

18. $\dfrac{1 - \frac{4}{y^2}}{1 - \frac{2}{y} - \frac{8}{y^2}}$ $\frac{y-2}{y-4}$

19. $\dfrac{\frac{1}{y} + \frac{1}{x}}{\frac{1}{xy}}$ $x + y$

20. $\dfrac{\frac{1}{xy}}{\frac{1}{y} - \frac{1}{x}}$ $\frac{1}{x-y}$

21. $\dfrac{1 - \frac{1}{a^2}}{1 - \frac{1}{a}}$ $\frac{a+1}{a}$

22. $\dfrac{1 + \frac{1}{a}}{1 - \frac{1}{a^2}}$ $\frac{a}{a-1}$

23. $\dfrac{\frac{1}{10x} - \frac{y}{10x^2}}{\frac{1}{10} - \frac{y}{10x}}$ $\frac{1}{x}$

24. $\dfrac{\frac{1}{2x} + \frac{y}{2x^2}}{\frac{1}{4} + \frac{y}{4x}}$ $\frac{2}{x}$

25. $\dfrac{\frac{1}{a+1} + 2}{\frac{1}{a+1} + 3}$ $\frac{2a+3}{3a+4}$

26. $\dfrac{\frac{2}{a+1} + 3}{\frac{3}{a+1} + 4}$ $\frac{3a+5}{4a+7}$

Although the following problems do not contain complex fractions, they do involve more than one operation. Simplify inside the parentheses first, then multiply.

27. $\left(1 - \dfrac{1}{x}\right)\left(1 - \dfrac{1}{x+1}\right)\left(1 - \dfrac{1}{x+2}\right)$ $\frac{x-1}{x+2}$

28. $\left(1 + \dfrac{1}{x}\right)\left(1 + \dfrac{1}{x + 1}\right)\left(1 + \dfrac{1}{x + 2}\right) \dfrac{x + 3}{x}$

29. $\left(1 + \dfrac{1}{x + 3}\right)\left(1 + \dfrac{1}{x + 2}\right)\left(1 + \dfrac{1}{x + 1}\right) \dfrac{x + 4}{x + 1}$

30. $\left(1 - \dfrac{1}{x + 3}\right)\left(1 - \dfrac{1}{x + 2}\right)\left(1 - \dfrac{1}{x + 1}\right) \dfrac{x}{x + 3}$

31. Simplify each term in the following sequence.

$$2 + \dfrac{1}{2 + 1}, \; 2 + \dfrac{1}{2 + \dfrac{1}{2 + 1}}, \; 2 + \dfrac{1}{2 + \dfrac{1}{2 + \dfrac{1}{2 + 1}}}, \; \ldots$$

$\dfrac{7}{3}, \dfrac{17}{7}, \dfrac{41}{17}$

32. Simplify each term in the following sequence.

$$2 + \dfrac{3}{2 + 3}, \; 2 + \dfrac{3}{2 + \dfrac{3}{2 + 3}}, \; 2 + \dfrac{3}{2 + \dfrac{3}{2 + \dfrac{3}{2 + 3}}}, \; \ldots$$

$\dfrac{13}{5}, \dfrac{41}{13}, \dfrac{121}{41}$

Complete the following tables.

33.

Number x	Reciprocal $\dfrac{1}{x}$	Quotient $\dfrac{x}{\frac{1}{x}}$	Square x^2
1	1	1	1
2	$\dfrac{1}{2}$	4	4
3	$\dfrac{1}{3}$	9	9
4	$\dfrac{1}{4}$	16	16

34.

Number x	Reciprocal $\dfrac{1}{x}$	Quotient $\dfrac{\frac{1}{x}}{x}$	Square x^2
1	1	1	1
2	$\dfrac{1}{2}$	$\dfrac{1}{4}$	4
3	$\dfrac{1}{3}$	$\dfrac{1}{9}$	9
4	$\dfrac{1}{4}$	$\dfrac{1}{16}$	16

35.

Number x	Reciprocal $\dfrac{1}{x}$	Sum $1 + \dfrac{1}{x}$	Quotient $\dfrac{1 + \frac{1}{x}}{\frac{1}{x}}$
1	1	2	2
2	$\dfrac{1}{2}$	$\dfrac{3}{2}$	3
3	$\dfrac{1}{3}$	$\dfrac{4}{3}$	4
4	$\dfrac{1}{4}$	$\dfrac{5}{4}$	5

36.

Number x	Reciprocal $\dfrac{1}{x}$	Difference $1 - \dfrac{1}{x}$	Quotient $\dfrac{1 - \frac{1}{x}}{\frac{1}{x}}$
1	1	0	0
2	$\dfrac{1}{2}$	$\dfrac{1}{2}$	1
3	$\dfrac{1}{3}$	$\dfrac{2}{3}$	2
4	$\dfrac{1}{4}$	$\dfrac{3}{4}$	3

Maintaining Your Skills

Solve each inequality.

37. $2x + 3 < 5$　$x < 1$

38. $3x - 2 > 7$　$x > 3$

39. $-3x \le 21$　$x \ge -7$

40. $-5x \ge -10$　$x \le 2$

41. $-2x + 8 > -4$　$x < 6$

42. $-4x - 1 < 11$　$x > -3$

43. $4 - 2(x + 1) \ge -2$　$x \le 2$

44. $6 - 2(x + 3) \le -8$　$x \ge 4$

Getting Ready for the Next Section

Solve.

45. $21 = 6x$　$\dfrac{7}{2}$

46. $72 = 2x$　36

47. $x^2 + x = 6$　$-3, 2$

48. $x^2 + 2x = 8$　$-4, 2$

OBJECTIVES

A Solve a proportion.

B Solve application problems involving proportions.

A proportion is two equal ratios; that is, if $\dfrac{a}{b}$ and $\dfrac{c}{d}$ are ratios, then:

$$\frac{a}{b} = \frac{c}{d}$$

is a proportion.

Each of the four numbers in a proportion is called a *term* of the proportion. We number the terms as follows:

$$\text{First term} \to \frac{a}{b} = \frac{c}{d} \leftarrow \text{Third term}$$
$$\text{Second term} \to b \quad d \leftarrow \text{Fourth term}$$

The first and fourth terms are called the *extremes,* and the second and third terms are called the *means:*

$$\text{Means} \quad \frac{a}{b} = \frac{c}{d} \quad \text{Extremes}$$

For example, in the proportion:

$$\frac{3}{8} = \frac{12}{32}$$

the extremes are 3 and 32, and the means are 8 and 12.

Means-Extremes Property If a, b, c, and d are real numbers with $b \neq 0$ and $d \neq 0$, then

$$\text{if} \qquad \frac{a}{b} = \frac{c}{d}$$

$$\text{then} \qquad ad = bc$$

In words: In any proportion, the product of the extremes is equal to the product of the means.

This property of proportions comes from the multiplication property of equality. We can use it to solve for a missing term in a proportion.

EXAMPLE 1 Solve the proportion $\dfrac{3}{x} = \dfrac{6}{7}$ for x.

SOLUTION We could solve for x by using the method developed in Section 7.4; that is, multiplying both sides by the LCD $7x$. Instead, let's use our new means-extremes property:

$$\frac{3}{x} = \frac{6}{7} \qquad \textbf{Extremes are 3 and 7; means are } x \textbf{ and 6}$$

$$21 = 6x \qquad \textbf{Product of extremes = product of means}$$

$$\frac{21}{6} = x \qquad \textbf{Divide both sides by 6}$$

$$x = \frac{7}{2} \qquad \textbf{Reduce to lowest terms}$$

EXAMPLE 2   Solve for x: $\dfrac{x+1}{2} = \dfrac{3}{x}$.

SOLUTION   Again, we want to point out that we could solve for x by using the method we used in Section 7.4. Using the means-extremes property is simply an alternative to the method developed in Section 7.4:

$$\frac{x+1}{2} = \frac{3}{x}$$ **Extremes are $x + 1$ and x; means are 2 and 3**

$$x^2 + x = 6$$ **Product of extremes = product of means**

$$x^2 + x - 6 = 0$$ **Standard form for a quadratic equation**

$$(x + 3)(x - 2) = 0$$ **Factor**

$$x + 3 = 0 \quad \text{or} \quad x - 2 = 0$$ **Set factors equal to 0**

$$x = -3 \quad \text{or} \quad x = 2$$

This time we have two solutions: -3 and 2.

EXAMPLE 3   A manufacturer knows that during a production run, 8 out of every 100 parts produced by a certain machine will be defective. If the machine produces 1,450 parts, how many can be expected to be defective?

SOLUTION   The ratio of defective parts to total parts produced is $\frac{8}{100}$. If we let x represent the numbers of defective parts out of the total of 1,450 parts, then we can write this ratio again as $\dfrac{x}{1{,}450}$. This gives us a proportion to solve:

Defective parts in numerator → $\dfrac{x}{1{,}450} = \dfrac{8}{100}$ **Extremes are x and 100; means are 1,450 and 8**

Total parts in denominator $100x = 11{,}600$ **Product of extremes = product of means**

$$x = 116$$

The manufacturer can expect 116 defective parts out of the total of 1,450 parts if the machine usually produces 8 defective parts for every 100 parts it produces.

EXAMPLE 4

The scale on a map indicates that 1 inch on the map corresponds to an actual distance of 85 miles. Two cities are 3.5 inches apart on the map. What is the actual distance between the two cities?

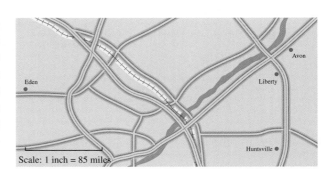

SOLUTION We let x represent the actual distance between the two cities. The proportion is

$$\text{Miles} \longrightarrow \frac{x}{3.5} = \frac{85}{1} \longleftarrow \text{Miles} \atop \longleftarrow \text{Inches}$$

$$x \cdot 1 = 3.5(85)$$

$$x = 297.5 \text{ miles}$$

 EXAMPLE 5 A woman drives her car 270 miles in 6 hours. If she continues at the same rate, how far will she travel in 10 hours?

SOLUTION We let x represent the distance traveled in 10 hours. Using x, we translate the problem into the following proportion:

$$\text{Miles} \longrightarrow \frac{x}{10} = \frac{270}{6} \longleftarrow \text{Miles} \atop \longleftarrow \text{Hours}$$

Notice that the two ratios in the proportion compare the same quantities. That is, both ratios compare miles to hours. In words this proportion says:

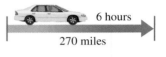

6 hours

270 miles

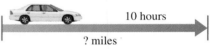

10 hours

? miles

$$x \text{ miles is to } 10 \text{ hours as } 270 \text{ miles is to } 6 \text{ hours}$$

$$\downarrow \qquad\qquad \downarrow \qquad\qquad \downarrow$$

$$\frac{x}{10} \qquad = \qquad \frac{270}{6}$$

Next, we solve the proportion.

$$x \cdot 6 = 10 \cdot 270$$

$$x \cdot 6 = 2{,}700$$

$$\frac{x \cdot \cancel{6}}{\cancel{6}} = \frac{2{,}700}{6}$$

$$x = 450 \text{ miles}$$

If the woman continues at the same rate, she will travel 450 miles in 10 hours.

LINKING OBJECTIVES AND EXAMPLES

Next to each **objective** we have listed the examples that are best described by that objective.

A   1, 2

B   3–5

GETTING READY FOR CLASS

After reading through the preceding section, respond in your own words and in complete sentences.

1. What is a proportion?
2. What are the means and extremes of a proportion?
3. What is the relationship between the means and the extremes in a proportion? (It is called the means-extremes property of proportions.)
4. How are ratios and proportions related?

Solve each of the following proportions.

▶ **1.** $\dfrac{x}{2} = \dfrac{6}{12}$ 1

2. $\dfrac{x}{4} = \dfrac{6}{8}$ 3

3. $\dfrac{2}{5} = \dfrac{4}{x}$ 10

4. $\dfrac{3}{8} = \dfrac{9}{x}$ 24

5. $\dfrac{10}{20} = \dfrac{20}{x}$ 40

6. $\dfrac{15}{60} = \dfrac{60}{x}$ 240

7. $\dfrac{a}{3} = \dfrac{5}{12}$ $\frac{5}{4}$

8. $\dfrac{a}{2} = \dfrac{7}{20}$ $\frac{7}{10}$

9. $\dfrac{2}{x} = \dfrac{6}{7}$ $\frac{7}{3}$

10. $\dfrac{4}{x} = \dfrac{6}{7}$ $\frac{14}{3}$

▶ **11.** $\dfrac{x+1}{3} = \dfrac{4}{x}$ 3, −4

12. $\dfrac{x+1}{6} = \dfrac{7}{x}$ 6, −7

13. $\dfrac{x}{2} = \dfrac{8}{x}$ 4, −4

14. $\dfrac{x}{9} = \dfrac{4}{x}$ 6, −6

15. $\dfrac{4}{a+2} = \dfrac{a}{2}$ 2, −4

16. $\dfrac{3}{a+2} = \dfrac{a}{5}$ 3, −5

17. $\dfrac{1}{x} = \dfrac{x-5}{6}$ 6, −1

18. $\dfrac{1}{x} = \dfrac{x-6}{7}$ 7, −1

Applying the Concepts

19. Baseball A baseball player gets 6 hits in the first 18 games of the season. If he continues hitting at the same rate, how many hits will he get in the first 45 games? 15 hits

20. Basketball A basketball player makes 8 of 12 free throws in the first game of the season. If she shoots with the same accuracy in the second game, how many of the 15 free throws she attempts will she make? 10

21. Mixture Problem A solution contains 12 milliliters of alcohol and 16 milliliters of water. If another solution is to have the same concentration of alcohol in water but is to contain 28 milliliters of water, how much alcohol must it contain? 21 milliliters

22. Mixture Problem A solution contains 15 milliliters of HCl and 42 milliliters of water. If another solution is to have the same concentration of HCl in water but is to contain 140 milliliters of water, how much HCl must it contain? 50 milliliters

23. Nutrition If 100 grams of ice cream contains 13 grams of fat, how much fat is in 350 grams of ice cream? 45.5 grams

24. Nutrition A 6-ounce serving of grapefruit juice contains 159 grams of water. How many grams of water are in 20 ounces of grapefruit juice? 530 grams

25. Map Reading A map is drawn so that every 3.5 inches on the map corresponds to an actual distance of 100 miles. If the actual distance between the two cities is 420 miles, how far apart are they on the map? 14.7 inches

26. Map Reading The scale on a map indicates that 1 inch on the map corresponds to an actual distance of 105 miles. Two cities are 4.5 inches apart on the map. What is the actual distance between the two cities? 472.5 miles

On the map shown here, 0.5 inches on the map is equal to 5 miles. Use the information from the map to work Problems 27 through 30.

27. Map Reading Suppose San Luis Obispo is 1.25 inches from Arroyo Grande on the map. How far apart are the two cities? 12.5 miles

28. Map Reading Suppose San Luis Obispo is 3.4 inches from Paso Robles on the map. How far apart are the two cities? 34 miles

29. Driving Time If Ava drives from Paso Robles to San Luis Obispo in 46 minutes, how long will it take her to drive from San Luis Obispo to Arroyo Grande, if she drives at the same speed? Round to the nearest minute. 17 minutes

30. Driving Time If Brooke drives from Arroyo Grande to San Luis Obispo in 15 minutes, how long will it take her to drive from San Luis Obispo to Paso Robles, if she drives at the same speed? Round to the nearest minute. *41 minutes*

31. Distance A man drives his car 245 miles in 5 hours. At this rate, how far will he travel in 7 hours? *343 miles*

32. Distance An airplane flies 1,380 miles in 3 hours. How far will it fly in 5 hours? *2,300 miles*

33. Improving Your Quantitative Literacy Searching Google News for news articles that involve proportions produces a number of items involving proportions, the top three of which are shown below. Search the Internet for current news articles by searching on the word *proportion.* Then find an article that the material in this section helps you understand.

Callahan Decries Commotion Over Gesture
. . . "I don't use that type of demeanor, and I never have. This is way blown out of **proportion.** I don't know where we get all these . . .
Rapid City Journal, SD, November 2, 2005

Council to Look Again at Inn Plans
. . . meeting months ago. "On paper the size of the inn seemed to be all out of **proportion** with the rest of the park," he said. "It was . . .
Wairarapa Times Age, New Zealand, November 2, 2005

The Dilemma of Halting Clinical Trials Early
. . . What's more, the authors found, a large **pro-portion** of the trials are funded by the drugs and devices industry, and involve drugs with large market potential in . . .
MedPage Today, NJ, November 1, 2005

Trademark of Google, Inc.

Maintaining Your Skills

Reduce to lowest terms.

34. $\dfrac{x^2 - x - 6}{x^2 - 9}$ $\dfrac{x+2}{x+3}$

35. $\dfrac{xy + 5x + 3y + 15}{x^2 + ax + 3x + 3a}$ $\dfrac{y+5}{x+a}$

Multiply or divide, as indicated.

36. $\dfrac{x^2 - 25}{x + 4} \cdot \dfrac{2x + 8}{x^2 - 9x + 20}$ $\dfrac{2(x+5)}{x-4}$

37. $\dfrac{3x + 6}{x^2 + 4x + 3} \div \dfrac{x^2 + x - 2}{x^2 + 2x - 3}$ $\dfrac{3}{x+1}$

Add or subtract, as indicated.

38. $\dfrac{x}{x^2 - 16} + \dfrac{4}{x^2 - 16}$ $\dfrac{1}{x-4}$

39. $\dfrac{2}{x^2 - 1} - \dfrac{5}{x^2 + 3x - 4}$ $\dfrac{-3}{(x+1)(x+4)}$

Getting Ready for the Next Section

Use the formula $y = 5x$ to find y when

40. $x = 4$ 20

41. $x = 3$ 15

Use the formula $y = \dfrac{20}{x}$ to find y when

42. $x = 10$ 2

43. $x = 5$ 4

Use the formula $y = 2x^2$ to find x when

44. $y = 50$ $-5, 5$

45. $y = 72$ $-6, 6$

Use the formula $y = Kx$ to find K when

46. $y = 15$ and $x = 3$ 5

47. $y = 72$ and $x = 4$ 18

Use the formula $y = Kx^2$ to find K when

48. $y = 32$ and $x = 4$ 2

49. $y = 45$ and $x = 3$ 5

OBJECTIVES

A Solve problems involving direct and inverse variation.

Two variables are said to *vary directly* if one is a constant multiple of the other. For instance, y varies directly as x if $y = Kx$, where K is a constant. The constant K is called the constant of variation. The following table gives the relation between direct variation statements and their equivalent algebraic equations.

Statement	Equation (K = constant of variation)
y varies directly as x	$y = Kx$
y varies directly as the square of x	$y = Kx^2$
s varies directly as the square root of t	$s = K\sqrt{t}$
r varies directly as the cube of s	$r = Ks^3$

Any time we run across a statement similar to those in the table, we immediately can write an equivalent expression involving variables and a constant of variation K.

EXAMPLE 1 Suppose y varies directly as x. When y is 15, x is 3. Find y when x is 4.

SOLUTION From the first sentence we can write the relationship between x and y as:

$$y = Kx$$

We now use the second sentence to find the value of K. Since y is 15 when x is 3, we have:

$$15 = K(3) \quad \text{or} \quad K = 5$$

Now we can rewrite the relationship between x and y more specifically as:

$$y = 5x$$

To find the value of y when x is 4 we simply substitute $x = 4$ into our last equation. Substituting:

$$x = 4$$

into:

$$y = 5x$$

we have:

$$y = 5(4)$$
$$y = 20$$

EXAMPLE 2 Suppose y varies directly as the square of x. When x is 4, y is 32. Find x when y is 50.

SOLUTION The first sentence gives us:

$$y = Kx^2$$

Since y is 32 when x is 4, we have:

$$32 = K(4)^2$$
$$32 = 16K$$
$$K = 2$$

The equation now becomes:

$$y = 2x^2$$

When y is 50, we have:

$$50 = 2x^2$$

$$25 = x^2$$

$$x = \pm 5$$

There are two possible solutions, $x = 5$ or $x = -5$.

EXAMPLE 3   The cost of a certain kind of candy varies directly with the weight of the candy. If 12 ounces of the candy cost $1.68, how much will 16 ounces cost?

SOLUTION Let $x =$ the number of ounces of candy and $y =$ the cost of the candy. Then $y = Kx$. Since y is 1.68 when x is 12, we have:

$$1.68 = K \cdot 12$$

$$K = \frac{1.68}{12}$$

$$= 0.14$$

The equation must be:

$$y = 0.14x$$

When x is 16, we have:

$$y = 0.14(16)$$

$$= 2.24$$

The cost of 16 ounces of candy is $2.24.

Inverse Variation

Two variables are said to *vary inversely* if one is a constant multiple of the reciprocal of the other. For example, y varies inversely as x if $y = \dfrac{K}{x}$, where K is a real number constant. Again, K is called the constant of variation. The table that follows gives some examples of inverse variation statements and their associated algebraic equations.

Statement	Equation ($K =$ constant of variation)
y varies inversely as x	$y = \dfrac{K}{x}$
y varies inversely as the square of x	$y = \dfrac{K}{x^2}$
F varies inversely as the square root of t	$F = \dfrac{K}{\sqrt{t}}$
r varies inversely as the cube of s	$r = \dfrac{K}{s^3}$

Every inverse variation statement has an associated inverse variation equation.

EXAMPLE 4   Suppose y varies inversely as x. When y is 4, x is 5. Find y when x is 10.

SOLUTION   The first sentence gives us the relationship between x and y:

$$y = \frac{K}{x}$$

We use the second sentence to find the value of the constant K:

$$4 = \frac{K}{5}$$

or:

$$K = 20$$

We can now write the relationship between x and y more specifically as:

$$y = \frac{20}{x}$$

We use this equation to find the value of y when x is 10. Substituting:

$$x = 10$$

into:

$$y = \frac{20}{x}$$

we have:

$$y = \frac{20}{10}$$

$$y = 2$$

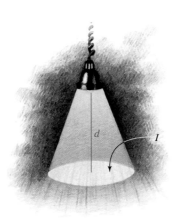

EXAMPLE 5   The intensity (I) of light from a source varies inversely as the square of the distance (d) from the source. Ten feet away from the source the intensity is 200 candlepower. What is the intensity 5 feet from the source?

SOLUTION

$$I = \frac{K}{d^2}$$

Since $I = 200$ when $d = 10$, we have:

$$200 = \frac{K}{10^2}$$

$$200 = \frac{K}{100}$$

$$K = 20,000$$

The equation becomes:

$$I = \frac{20{,}000}{d^2}$$

When $d = 5$, we have

$$I = \frac{20{,}000}{5^2}$$

$$= \frac{20{,}000}{25}$$

$$= 800 \text{ candlepower}$$

LINKING OBJECTIVES AND EXAMPLES

Next to each **objective** we have listed the examples that are best described by that objective.

A   1–5

GETTING READY FOR CLASS

After reading through the preceding section, respond in your own words and in complete sentences.

1. What does it mean when we say "y varies directly with x"?

2. Give an example of a sentence that is a direct variation statement.

3. Translate the equation $y = \dfrac{K}{x}$ into words.

4. Give an example of an everyday situation where one quantity varies inversely with another.

Problem Set 7.8

Online support materials can be found at www.thomsonedu.com/login

For each of the following problems, y varies directly as x.

1. If $y = 10$ when $x = 5$, find y when x is 4.   8

2. If $y = 20$ when $x = 4$, find y when x is 11.   55

▶ **3.** If $y = 39$ when $x = 3$, find y when x is 10.   130

4. If $y = -18$ when $x = 6$, find y when x is 3.   −9

5. If $y = -24$ when $x = 4$, find x when y is −30.   5

6. If $y = 30$ when $x = -15$, find x when y is 8.   −4

7. If $y = -7$ when $x = -1$, find x when y is −21.   −3

8. If $y = 30$ when $x = 4$, find y when x is 7. $\frac{105}{2}$

For each of the following problems, y varies directly as the square of x.

9. If $y = 75$ when $x = 5$, find y when x is 1.   3

10. If $y = -72$ when $x = 6$, find y when x is 3.   −18

11. If $y = 48$ when $x = 4$, find y when x is 9.   243

12. If $y = 27$ when $x = 3$, find x when y is 75.   5, −5

For each of the following problems, y varies inversely with x.

13. If $y = 5$ when $x = 2$, find y when x is 5.   2

14. If $y = 2$ when $x = 10$, find y when x is 4.   5

15. If $y = 2$ when $x = 1$, find y when x is 4. $\frac{1}{2}$

16. If $y = 4$ when $x = 3$, find y when x is 6.   2

17. If $y = 5$ when $x = 3$, find x when y is 15.   1

18. If $y = 12$ when $x = 10$, find x when y is 60.   2

19. If $y = 10$ when $x = 10$, find x when y is 20.   5

20. If $y = 15$ when $x = 2$, find x when y is 6.   5

= Videos available by instructor request

▶ = Online student support materials available at www.thomsonedu.com/login

For each of the following problems, y varies inversely as the square of x.

21. If $y = 4$ when $x = 5$, find y when x is 2.   25

22. If $y = 5$ when $x = 2$, y when x is 6. $\frac{5}{9}$

▶ **23.** If $y = 4$ when $x = 3$, find y when x is 2.   9

24. If $y = 9$ when $x = 4$, find y when x is 3.   16

Applying the Concepts

25. Tension in a Spring The tension t in a spring varies directly with the distance d the spring is stretched. If the tension is 42 pounds when the spring is stretched 2 inches, find the tension when the spring is stretched twice as far.   84 pounds

26. Fill Time The time t it takes to fill a bucket varies directly with the volume g of the bucket. If it takes 1 minute to fill a 4-gallon bucket, how long will it take to fill a 6-gallon bucket? $\frac{3}{2}$ minutes

▶ **27. Electricity** The power P in an electric circuit varies directly with the square of the current I. If $P = 30$ when $I = 2$, find P when $I = 7$. $\frac{735}{2}$ or 367.5

28. Electricity The resistance R in an electric circuit varies directly with the voltage V. If $R = 20$ when $V = 120$, find R when $V = 240$.   40

29. Wages The amount of money M a woman makes per week varies directly with the number of hours h she works per week. If she works 20 hours and earns $157, how much does she make if she works 30 hours? $235.50

30. Volume The volume V of a gas varies directly as the temperature T. If $V = 3$ when $T = 150$, find V when T is 200.   4

31. Weight The weight F of a body varies inversely with the square of the distance d between the body and the center of the Earth. If a man weighs 150 pounds 4,000 miles from the center of the Earth, how much will he weigh at a distance of 5,000 miles from the center of the Earth?   96 pounds

32. Light Intensity The intensity I of a light source varies inversely with the square of the distance d from the source. Four feet from the source, the intensity is 9 footcandles. What is the intensity 3 feet from the source?   16 footcandles

33. Electricity The current I in an electric circuit varies inversely with the resistance R. If a current of 30 amperes is produced by a resistance of 2 ohms, what current will be produced by a resistance of 5 ohms?   12 amperes

34. Pressure The pressure exerted by a gas on the container in which it is held varies inversely with the volume of the container. A pressure of 40 pounds per square inch is exerted on a container of volume 2 cubic feet. What is the pressure on a container whose volume is 8 cubic feet?
10 pounds per square inch

Maintaining Your Skills

Solve each system of equations by the elimination method.

35. $2x + y = 3$
$3x - y = 7$ $(2, -1)$

36. $3x - y = -6$
$4x + y = -8$ $(-2, 0)$

37. $4x - 5y = 1$
$x - 2y = -2$ $(4, 3)$

38. $6x - 4y = 2$
$2x + y = 10$ $(3, 4)$

Solve by the substitution method.

39. $5x + 2y = 7$
$y = 3x - 2$ $(1, 1)$

40. $-7x - 5y = -1$
$y = x + 5$ $(-2, 3)$

41. $2x - 3y = 4$
$x = 2y + 1$ $(5, 2)$

42. $4x - 5y = 2$
$x = 2y - 1$ $(3, 2)$

Chapter 7 SUMMARY

Rational Numbers [7.1]

EXAMPLES

1. We can reduce $\frac{6}{8}$ to lowest terms by dividing the numerator and denominator by their greatest common factor 2:

$$\frac{6}{8} = \frac{2 \cdot 3}{2 \cdot 4} = \frac{3}{4}$$

Any number that can be put in the form $\frac{a}{b}$, where a and b are integers ($b \neq 0$), is called a rational number.

Multiplying or dividing the numerator and denominator of a rational number by the same nonzero number never changes the value of the rational number.

Rational Expressions [7.1]

2. We reduce rational expressions to lowest terms by factoring the numerator and denominator and then dividing out any factors they have in common:

$$\frac{x-3}{x^2-9} = \frac{x-3}{(x-3)(x+3)} = \frac{1}{x+3}$$

Any expression of the form $\frac{P}{Q}$, where P and Q are polynomials ($Q \neq 0$), is a rational expression.

Multiplying or dividing the numerator and denominator of a rational expression by the same nonzero quantity always produces a rational expression equivalent to the original one.

Multiplication [7.2]

3. $\dfrac{x-1}{x^2+2x-3} \cdot \dfrac{x^2-9}{x-2}$

$$= \frac{x-1}{(x+3)(x-1)} \cdot \frac{(x-3)(x+3)}{x-2}$$

$$= \frac{x-3}{x-2}$$

To multiply two rational numbers or two rational expressions, multiply numerators, multiply denominators, and divide out any factors common to the numerator and denominator:

For rational numbers $\dfrac{a}{b}$ and $\dfrac{c}{d}$, $\quad \dfrac{a}{b} \cdot \dfrac{c}{d} = \dfrac{ac}{bd}$

For rational expressions $\dfrac{P}{Q}$ and $\dfrac{R}{S}$, $\quad \dfrac{P}{Q} \cdot \dfrac{R}{S} = \dfrac{PR}{QS}$

Division [7.2]

4. $\dfrac{2x}{x^2-25} \div \dfrac{4}{x-5}$

$$= \frac{2x}{(x-5)(x+5)} \cdot \frac{(x-5)}{4}$$

$$= \frac{x}{2(x+5)}$$

To divide by a rational number or rational expression, simply multiply by its reciprocal:

For rational numbers $\dfrac{a}{b}$ and $\dfrac{c}{d}$, $\quad \dfrac{a}{b} \div \dfrac{c}{d} = \dfrac{a}{b} \cdot \dfrac{d}{c}$

For rational expressions $\dfrac{P}{Q}$ and $\dfrac{R}{S}$, $\quad \dfrac{P}{Q} \div \dfrac{R}{S} = \dfrac{P}{Q} \cdot \dfrac{S}{R}$

Addition [7.3]

5. $\dfrac{3}{x-1} + \dfrac{x}{2}$

$= \dfrac{3}{x-1} \cdot \dfrac{2}{2} + \dfrac{x}{2} \cdot \dfrac{x-1}{x-1}$

$= \dfrac{6}{2(x-1)} + \dfrac{x^2 - x}{2(x-1)}$

$= \dfrac{x^2 - x + 6}{2(x-1)}$

To add two rational numbers or rational expressions, find a common denominator, change each expression to an equivalent expression having the common denominator, then add numerators and reduce if possible:

For rational numbers $\dfrac{a}{c}$ and $\dfrac{b}{c}$, $\dfrac{a}{c} + \dfrac{b}{c} = \dfrac{a+b}{c}$

For rational expressions $\dfrac{P}{S}$ and $\dfrac{Q}{S}$, $\dfrac{P}{S} + \dfrac{Q}{S} = \dfrac{P+Q}{S}$

Subtraction [7.3]

6. $\dfrac{x}{x^2 - 4} - \dfrac{2}{x^2 - 4}$

$= \dfrac{x-2}{x^2 - 4}$

$= \dfrac{\cancel{x-2}}{\cancel{(x-2)}\,(x+2)}$

$= \dfrac{1}{x+2}$

To subtract a rational number or rational expression, simply add its opposite:

For rational numbers $\dfrac{a}{c}$ and $\dfrac{b}{c}$, $\dfrac{a}{c} - \dfrac{b}{c} = \dfrac{a}{c} + \left(\dfrac{-b}{c}\right)$

For rational expressions $\dfrac{P}{S}$ and $\dfrac{Q}{S}$, $\dfrac{P}{S} - \dfrac{Q}{S} = \dfrac{P}{S} + \left(\dfrac{-Q}{S}\right)$

Equations [7.4]

7. Solve $\dfrac{1}{2} + \dfrac{3}{x} = 5$.

$2x\left(\dfrac{1}{2}\right) + 2x\left(\dfrac{3}{x}\right) = 2x(5)$

$x + 6 = 10x$

$6 = 9x$

$x = \dfrac{2}{3}$

To solve equations involving rational expressions, first find the least common denominator (LCD) for all denominators. Then multiply both sides by the LCD and solve as usual. Be sure to check all solutions in the original equation to be sure there are no undefined terms.

Complex Fractions [7.6]

8. $\dfrac{1 - \dfrac{4}{x}}{x - \dfrac{16}{x}} = \dfrac{x\left(1 - \dfrac{4}{x}\right)}{x\left(x - \dfrac{16}{x}\right)}$

$= \dfrac{x - 4}{x^2 - 16}$

$= \dfrac{\cancel{x-4}}{\cancel{(x-4)}\,(x+4)}$

$= \dfrac{1}{x+4}$

A rational expression that contains a fraction in its numerator or denominator is called a complex fraction. The most common method of simplifying a complex fraction is to multiply the top and bottom by the LCD for all denominators.

Ratio and Proportion [7.1, 7.7]

9. Solve for x: $\dfrac{3}{x} = \dfrac{5}{20}$.

$3 \cdot 20 = 5 \cdot x$

$60 = 5x$

$x = 12$

The ratio of a to b is:

$$\dfrac{a}{b}$$

Two equal ratios form a proportion. In the proportion

$$\dfrac{a}{b} = \dfrac{c}{d}$$

a and d are the *extremes*, and b and c are the *means*. In any proportion the product of the extremes is equal to the product of the means.

Direct Variation [7.8]

10. If y varies directly with the square of x, then:
$$y = Kx^2$$

The variable y is said to vary directly with the variable x if $y = Kx$, where K is a real number.

Inverse Variation [7.8]

11. If y varies inversely with the cube of x, then
$$y = \frac{K}{x^3}$$

The variable y is said to vary inversely with the variable x if $y = \dfrac{K}{x}$, where K is a real number.

Chapter 7 Review Test

The problems below form a comprehensive review of the material in this chapter. They can be used to study for exams. If you would like to take a practice test on this chapter, you can use the odd-numbered problems. Give yourself an hour and work as many of the odd-numbered problems as possible. When you are finished, or when an hour has passed, check your answers with the answers in the back of the book. You can use the even-numbered problems for a second practice test.

The numbers in brackets refer to the sections of the text in which similar problems can be found.

Reduce to lowest terms. Also specify any restriction on the variable. [7.1]

1. $\dfrac{7}{14x - 28}$

2. $\dfrac{a + 6}{a^2 - 36}$

3. $\dfrac{8x - 4}{4x + 12}$

4. $\dfrac{x + 4}{x^2 + 8x + 16}$

5. $\dfrac{3x^3 + 16x^2 - 12x}{2x^3 + 9x^2 - 18x}$

6. $\dfrac{x + 2}{x^4 - 16}$

7. $\dfrac{x^2 + 5x - 14}{x + 7}$

8. $\dfrac{a^2 + 16a + 64}{a + 8}$

9. $\dfrac{xy + bx + ay + ab}{xy + 5x + ay + 5a}$

Multiply or divide as indicated. [7.2]

10. $\dfrac{3x + 9}{x^2} \cdot \dfrac{x^3}{6x + 18}$

11. $\dfrac{x^2 + 8x + 16}{x^2 + x - 12} \div \dfrac{x^2 - 16}{x^2 - x - 6}$

12. $(a^2 - 4a - 12)\left(\dfrac{a - 6}{a + 2}\right)$

13. $\dfrac{3x^2 - 2x - 1}{x^2 + 6x + 8} \div \dfrac{3x^2 + 13x + 4}{x^2 + 8x + 16}$

Find the following sums and differences. [7.3]

14. $\dfrac{2x}{2x + 3} + \dfrac{3}{2x + 3}$

15. $\dfrac{x^2}{x - 9} - \dfrac{18x - 81}{x - 9}$

16. $\dfrac{a + 4}{a + 8} - \dfrac{a - 9}{a + 8}$

17. $\dfrac{x}{x + 9} + \dfrac{5}{x}$

18. $\dfrac{5}{4x + 20} + \dfrac{x}{x + 5}$

19. $\dfrac{3}{x^2 - 36} - \dfrac{2}{x^2 - 4x - 12}$

20. $\dfrac{3a}{a^2 + 8a + 15} - \dfrac{2}{a + 5}$

Solve each equation. [7.4]

21. $\dfrac{3}{x} + \dfrac{1}{2} = \dfrac{5}{x}$

22. $\dfrac{a}{a - 3} = \dfrac{3}{2}$

23. $1 - \dfrac{7}{x} = \dfrac{-6}{x^2}$

24. $\dfrac{3}{x + 6} - \dfrac{1}{x - 2} = \dfrac{-8}{x^2 + 4x - 12}$

25. $\dfrac{2}{y^2 - 16} = \dfrac{10}{y^2 + 4y}$

26. Number Problem The sum of a number and 7 times its reciprocal is $\frac{16}{3}$. Find the number. [7.5]

27. Distance, Rate, and Time A boat travels 48 miles up a river in the same amount of time it takes to travel 72 miles down the same river. If the current is 3 miles per hour, what is the speed of the boat in still water? [7.5]

28. Filling a Pool An inlet pipe can fill a pool in 21 hours, whereas an outlet pipe can empty it in 28 hours. If both pipes are left open, how long will it take to fill the pool? [7.5]

Simplify each complex fraction. [7.6]

29. $\dfrac{\dfrac{x + 4}{x^2 - 16}}{\dfrac{2}{x - 4}}$

30. $\dfrac{1 - \dfrac{9}{y^2}}{1 + \dfrac{4}{y} - \dfrac{21}{y^2}}$

31. $\dfrac{\dfrac{1}{a - 2} + 4}{\dfrac{1}{a - 2} + 1}$

32. Write the ratio of 40 to 100 as a fraction in lowest terms. [7.7]

33. If there are 60 seconds in 1 minute, what is the ratio of 40 seconds to 3 minutes? [7.7]

Solve each proportion. [7.7]

34. $\dfrac{x}{9} = \dfrac{4}{3}$

35. $\dfrac{a}{3} = \dfrac{12}{a}$

36. $\dfrac{8}{x - 2} = \dfrac{x}{6}$

Work the following problems involving variation. [7.8]

37. y varies directly as x. If $y = -20$ when $x = 4$, find y when $x = 7$.

38. y varies inversely with x. If $y = 3$ when $x = 2$, find y when $x = 12$.

GROUP PROJECT   Kayak Race

Number of People 2-3

Time Needed 20 minutes

Equipment Paper and pencil

Background In a kayak race, the participants must paddle a kayak 450 meters down a river and then return 450 meters up the river to the starting point (see figure). Susan has deduced correctly that the total time t (in seconds) depends on the speed c (in meters per second) of the water according to the following expression:

$$t = \frac{450}{v + c} + \frac{450}{v - c}$$

where v is the speed of the kayak relative to the water (the speed of the kayak in still water).

Procedure

1. Fill in the following table.

2. If the kayak race were conducted in the still waters of a lake, do you think that the total time of a given par-

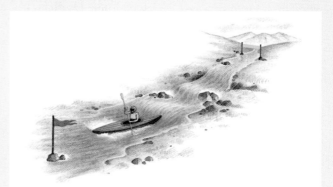

ticipant would be greater than, equal to, or smaller than the time in the river? Justify your answer.

3. Suppose Peter can paddle his kayak at 4.1 meters per second and that the speed of the current is 4.1 meters per second. What will happen when Peter makes the turn and tries to come back up the river? How does this situation show up in the equation for total time?

Time t (seconds)	Speed of Kayak Relative to the Water v (meters/second)	Current of the River c (meters/second)
240		1
300		2
	4	3
	3	1
540	3	
	3	3

Bertrand Russell

Bertrand Russell

Here is a quote taken from the beginning of the first sentence in the book *Principles of Mathematics* by the British philosopher and mathematician, Bertrand Russell.

> Pure Mathematics is the class of all propositions of the form "*p* implies *q*," where *p* and *q* are propositions containing one or more variables . . .

He is using the phrase "*p* implies *q*" in the same way mathematicians use the phrase "If *A*, then *B*." Conditional statements are an introduction to the foundations on which all of mathematics is built.

Write an essay on the life of Bertrand Russell. In the essay, indicate what purpose he had for writing and publishing his book *Principles of Mathematics*. Write in complete sentences and organize your work just as you would if you were writing a paper for an English class.

Roots and Radicals

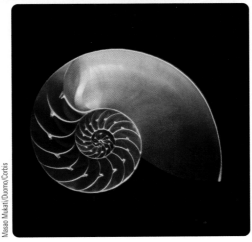

Masao Mukai/Duomo/Corbis

The diagram shown here is called the *spiral of roots*. It is constructed using the Pythagorean theorem, which we introduced in Chapter 6. The spiral of roots gives us a way to visualize positive square roots, one of the topics we will cover in this chapter. If we take each of the diagonals in the spiral of roots and place it above the corresponding whole number on the *x*-axis and then connect the tops of all these segments with a smooth curve, we have the curve shown in Figure 1. Table 1 gives the lengths of the diagonals in the spiral of roots, accurate to the nearest hundredth. That curve in Figure 1 is also the graph of the equation $y = \sqrt{x}$.

TABLE 1	
Approximate Length of Diagonals	
Number	**Positive Square Root**
1	1
2	1.41
3	1.73
4	2
5	2.24
6	2.45
7	2.65
8	2.83
9	3
10	3.16

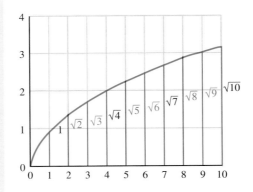

▶ Improve your grade and save time!
Go online to **www.thomsonedu.com/login** where you can
- Watch videos of instructors working through the in-text examples
- Follow step-by-step online tutorials of in-text examples and review questions
- Work practice problems
- Check your readiness for an exam by taking a pre-test and exploring the modules recommended in your Personalized Study plan
- Receive help from a live tutor online through vMentor™

Try it out! Log in with an access code or purchase access at **www.ichapters.com**.

OBJECTIVES

A Find a root of a number.

B Find the root of an expression containing a variable.

C Solve an application problem involving roots.

In Chapter 5 we developed notation (exponents) that would take us from a number to its square. If we wanted the square of 5, we wrote $5^2 = 25$. In this section we will use another type of notation that will take us in the reverse direction—from the square of a number back to the number itself.

In general we are interested in going from a number, say, 49, back to the number we squared to get 49. Since the square of 7 is 49, we say 7 is a square root of 49. The notation we use looks like this:

$$\sqrt{49}$$

Notation In the expression $\sqrt{49}$, 49 is called the *radicand;* $\sqrt{}$ is the *radical sign;* and the complete expression, $\sqrt{49}$, is called the *radical.*

> **DEFINITION** If x represents any positive real number, then the expression $\sqrt{x}$ is the **positive square root** of x. It is the *positive* number we square to get x.
>
> The expression $-\sqrt{x}$ is the **negative square root** of x. It is the negative number we square to get x.

Square Roots of Positive Numbers

Every positive number has two square roots, one positive and the other negative. Some books refer to the positive square root of a number as the principal root.

 EXAMPLE 1   The positive square root of 25 is 5 and can be written $\sqrt{25} = 5$. The negative square root of 25 is -5 and can be written $-\sqrt{25} = -5$.

If we want to consider the negative square root of a number, we must put a negative sign in front of the radical. It is a common mistake to think of $\sqrt{25}$ as meaning either 5 or -5. The expression $\sqrt{25}$ means the *positive* square root of 25, which is 5. If we want the negative square root, we write $-\sqrt{25}$ to begin with.

 EXAMPLES   Find the following roots.

2. $\sqrt{49} = 7$ **7 is the positive number we square to get 49**
3. $-\sqrt{49} = -7$ **−7 is the negative number we square to get 49**
4. $\sqrt{121} = 11$ **11 is the positive number we square to get 121**

 EXAMPLE 5   The positive square root of 17 is written $\sqrt{17}$. The negative square root of 17 is written $-\sqrt{17}$.

We have no other exact representation for the two roots in Example 5. Since 17 itself is not a perfect square (the square of an integer), its two square roots,

$\sqrt{17}$ and $-\sqrt{17}$, are irrational numbers. They have a place on the real number line but cannot be written as the ratio of two integers. The square roots of any number that is not itself a perfect square are irrational numbers.

EXAMPLE 6

Number	Positive Square Root	Negative Square Root	Roots Are
9	3	−3	Rational numbers
36	6	−6	Rational numbers
7	$\sqrt{7}$	$-\sqrt{7}$	Irrational numbers
22	$\sqrt{22}$	$-\sqrt{22}$	Irrational numbers
100	10	−10	Rational numbers

Square Root of Zero

The number 0 is the only real number with one square root. It is also its own square root:

$$\sqrt{0} = 0$$

Square Roots of Negative Numbers

Negative numbers have square roots, but their square roots are not real numbers. They do not have a place on the real number line. We will consider square roots of negative numbers later in the book.

EXAMPLE 7   The expression $\sqrt{-4}$ does not represent a real number since there is no real number we can square and end up with −4. The same is true of square roots of any negative number.

Other Roots

There are many other roots of numbers besides square roots, although square roots seem to be the most commonly used. The cube root of a number is the number we cube (raise to the third power) to get the original number. The cube root of 8 is 2 since $2^3 = 8$. The cube root of 27 is 3 since $3^3 = 27$. The notation for cube roots looks like this:

$$\text{The 3 is called the index} \quad \overset{\rightarrow}{\sqrt[3]{8}} = 2$$

$$\overset{\rightarrow}{\sqrt[3]{27}} = 3$$

We can go as high as we want with roots. The fourth root of 16 is 2 because $2^4 = 16$. We can write this in symbols as $\sqrt[4]{16} = 2$.

Here is a list of the most common roots. They are the roots that will come up most often in the remainder of the book, and they should be memorized.

Square Roots		Cube Roots	Fourth Roots
$\sqrt{1} = 1$	$\sqrt{49} = 7$	$\sqrt[3]{1} = 1$	$\sqrt[4]{1} = 1$
$\sqrt{4} = 2$	$\sqrt{64} = 8$	$\sqrt[3]{8} = 2$	$\sqrt[4]{16} = 2$
$\sqrt{9} = 3$	$\sqrt{81} = 9$	$\sqrt[3]{27} = 3$	$\sqrt[4]{81} = 3$
$\sqrt{16} = 4$	$\sqrt{100} = 10$	$\sqrt[3]{64} = 4$	$\sqrt[4]{256} = 4$
$\sqrt{25} = 5$	$\sqrt{121} = 11$	$\sqrt[3]{125} = 5$	$\sqrt[4]{625} = 5$
$\sqrt{36} = 6$	$\sqrt{144} = 12$		

With even roots—square roots, fourth roots, sixth roots, and so on—we cannot have negative numbers *under* the radical sign. With odd roots, negative numbers under the radical sign do not cause problems.

 EXAMPLES   Find the following roots, if possible.

8. $\sqrt[3]{-8} = -2$ **Because $(-2)^3 = -8$**

9. $\sqrt[3]{-27} = -3$ **Because $(-3)^3 = -27$**

10. $\sqrt{-4}$ **Not a real number since there is no real number whose square is -4**

11. $-\sqrt{4} = -2$ **Because -2 is the negative number we square to get 4**

12. $\sqrt[4]{-16}$ **Not a real number since there is no real number that can be raised to the fourth power to obtain -16**

13. $-\sqrt[4]{16} = -2$ **Because -2 is the negative number we raise to the fourth power to get 16**

Note

At first it may be difficult to see the difference in some of these examples. Generally, we have to be careful with even roots of negative numbers; that is, if the index on the radical is an even number, then we cannot have a negative number under the radical sign. That is why $\sqrt{-4}$ and $\sqrt[4]{-16}$ are not real numbers.

Variables Under the Radical Sign

In this chapter, unless we say otherwise, we will assume that all variables that appear under a radical sign represent positive numbers. That way we can simplify expressions involving radicals that contain variables. Here are some examples.

 EXAMPLE 14   Simplify $\sqrt{49x^2}$.

SOLUTION   We are looking for the expression we square to get $49x^2$. Since the square of 7 is 49 and the square of x is x^2, we can square $7x$ and get $49x^2$:

$$\sqrt{49x^2} = 7x \quad \textbf{Because } (7x)^2 = 49x^2$$

 EXAMPLE 15   Simplify $\sqrt{16a^2b^2}$.

SOLUTION   We want an expression whose square is $16a^2b^2$. That expression is $4ab$:

$$\sqrt{16a^2b^2} = 4ab \quad \textbf{Because } (4ab)^2 = 16a^2b^2$$

 EXAMPLE 16   Simplify $\sqrt[3]{125a^3}$.

SOLUTION   We are looking for the expression we cube to get $125a^3$. That expression is $5a$:

$$\sqrt[3]{125a^3} = 5a \quad \textbf{Because } (5a)^3 = 125a^3$$

EXAMPLE 17 Simplify $\sqrt{x^6}$.

SOLUTION The number we square to obtain x^6 is x^3.

$$\sqrt{x^6} = x^3 \qquad \textbf{Because } (x^3)^2 = x^6$$

 EXAMPLE 18 Simplify $\sqrt[4]{16a^8b^4}$.

SOLUTION The number we raise to the fourth power to obtain $16a^8b^4$ is $2a^2b$.

$$\sqrt[4]{16a^8b^4} = 2a^2b \qquad \textbf{Because } (2a^2b)^4 = 16a^8b^4$$

There are many application problems involving radicals that require decimal approximations. When we need a decimal approximation to a square root, we can use a calculator.

EXAMPLE 19 If you invest P dollars in an account and after 2 years the account has A dollars in it, then the annual rate of return r on the money you originally invested is given by the formula

$$r = \frac{\sqrt{A} - \sqrt{P}}{\sqrt{P}}$$

Suppose you pay $65 for a coin collection and find that the same coins sell for $84 two years later. Find the annual rate of return on your investment.

SOLUTION Substituting $A = 84$ and $P = 65$ in the formula, we have

$$r = \frac{\sqrt{84} - \sqrt{65}}{\sqrt{65}}$$

From either the table of square roots or a calculator, we find that $\sqrt{84} = 9.165$ and $\sqrt{65} = 8.062$. Using these numbers in our formula gives us

$$r = \frac{9.165 - 8.062}{8.062}$$

$$= \frac{1.103}{8.062}$$

$$= 0.137 \text{ or } 13.7\%$$

To earn as much as this in a savings account that compounds interest once a year, you have to find an account that pays 13.7% in annual interest.

The Pythagorean Theorem Again

Now that we have some experience working with square roots, we can rewrite the Pythagorean theorem using a square root. In Figure 1, if triangle ABC is a right triangle with $C = 90°$, then the length of the longest side is the *square root* of the sum of the squares of the other two sides.

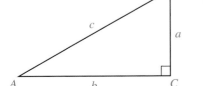

$$c = \sqrt{a^2 + b^2}$$

FIGURE 1

EXAMPLE 20   A tent pole is 8 feet in length and makes an angle of 90° with the ground. One end of a rope is attached to the top of the pole, and the other end of the rope is anchored to the ground 6 feet from the bottom of the pole. Find the length of the rope.

SOLUTION   The diagram in Figure 2 is a visual representation of the situation. To find the length of the pole, we apply the Pythagorean theorem.

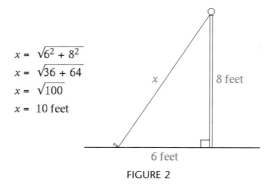

$$x = \sqrt{6^2 + 8^2}$$
$$x = \sqrt{36 + 64}$$
$$x = \sqrt{100}$$
$$x = 10 \text{ feet}$$

FIGURE 2

LINKING OBJECTIVES AND EXAMPLES

Next to each **objective** we have listed the examples that are best described by that objective.

A	1–13
B	14–18
C	19, 20

GETTING READY FOR CLASS

After reading through the preceding section, respond in your own words and in complete sentences.

1. Every real number has two square roots. Explain the notation we use to tell them apart. Use the square roots of 3 for examples.
2. Explain why a square root of −4 is not a real number.
3. We use the notation $\sqrt{17}$ to represent the positive square root of 17. Explain why there isn't a simpler way to express the positive square root of 17.
4. Write a list of the notation and terms associated with radicals.

Problem Set 8.1

Online support materials can be found at www.thomsonedu.com/login

Find the following roots. If the root does not exist as a real number, write "not a real number."

▶ **1.** $\sqrt{9}$ 3

2. $\sqrt{16}$ 4

▶ **3.** $-\sqrt{9}$ −3

4. $-\sqrt{16}$ −4

▶ **5.** $\sqrt{-25}$ Not a real number

6. $\sqrt{-36}$ Not a real number

7. $-\sqrt{144}$ −12

8. $\sqrt{256}$ 16

9. $\sqrt{625}$ 25

10. $-\sqrt{625}$ −25

11. $\sqrt{-49}$ Not a real number

12. $\sqrt{-169}$ Not a real number

▶ **13.** $-\sqrt{64}$ −8

14. $-\sqrt{25}$ −5

15. $-\sqrt{100}$ −10

16. $\sqrt{121}$ 11

17. $\sqrt{1,225}$ 35

18. $-\sqrt{1,681}$ −41

19. $\sqrt[4]{1}$ 1

20. $-\sqrt[4]{81}$ −3

▶ **21.** $\sqrt[3]{-8}$ −2

22. $\sqrt[3]{125}$ 5

23. $-\sqrt[3]{125}$ −5

24. $-\sqrt[3]{-8}$ 2

25. $\sqrt[3]{-1}$ −1

26. $-\sqrt[3]{-1}$ 1

▢ = Videos available by instructor request
▶ = Online student support materials available at www.thomsonedu.com/login

27. $\sqrt[3]{-27}$ -3 **28.** $-\sqrt[3]{27}$ -3

29. $-\sqrt[4]{16}$ -2 **30.** $\sqrt[4]{-16}$ Not a real number

Assume all variables are positive, and find the following roots.

31. $\sqrt{x^2}$ x **32.** $\sqrt{a^2}$ a

▸ **33.** $\sqrt{9x^2}$ $3x$ **34.** $\sqrt{25x^2}$ $5x$

35. $\sqrt{x^2y^2}$ xy **36.** $\sqrt{a^2b^2}$ ab

37. $\sqrt{(a+b)^2}$ $a+b$ **38.** $\sqrt{(x+y)^2}$ $x+y$

39. $\sqrt{49x^2y^2}$ $7xy$ **40.** $\sqrt{81x^2y^2}$ $9xy$

41. $\sqrt[3]{x^3}$ x **42.** $\sqrt[3]{a^3}$ a

43. $\sqrt[3]{8x^3}$ $2x$ **44.** $\sqrt[3]{27x^3}$ $3x$

45. $\sqrt{x^4}$ x^2 **46.** $\sqrt{x^6}$ x^3

▸ **47.** $\sqrt{36a^6}$ $6a^3$ **48.** $\sqrt{64a^4}$ $8a^2$

49. $\sqrt{25a^8b^4}$ $5a^4b^2$ **50.** $\sqrt{16a^4b^8}$ $4a^2b^4$

51. $\sqrt[3]{x^6}$ x^2 **52.** $\sqrt[3]{x^9}$ x^3

53. $\sqrt[3]{27a^{12}}$ $3a^4$ **54.** $\sqrt[3]{8a^6}$ $2a^2$

55. Simplify each expression. Assume all variables are positive.
   a. $\sqrt[4]{16}$ 2 **b.** $\sqrt[4]{x^4}$ x
   c. $\sqrt[4]{x^8}$ x^2 **d.** $\sqrt[4]{16x^8y^{12}}$ $2x^2y^3$

56. Simplify each expression. Assume all variables are positive.
   a. $\sqrt[5]{32}$ 2 **b.** $\sqrt[5]{y^5}$ y
   c. $\sqrt[5]{y^{10}}$ y^2 **d.** $\sqrt[5]{32x^{10}y^{20}}$ $2x^2y^4$

Simplify each expression.

57. a. $\sqrt{9}+\sqrt{16}$ 7 **58. a.** $\sqrt{64}+\sqrt{36}$ 10
   b. $\sqrt{9+16}$ 5 **b.** $\sqrt{64}+\sqrt{36}$ 14
   c. $\sqrt{9\cdot16}$ 12 **c.** $\sqrt{64}\sqrt{36}$ 48
   d. $\sqrt{9}\sqrt{16}$ 12 **d.** $\sqrt{64\cdot36}$ 48

59. a. $\sqrt{25}-\sqrt{16}$ 1 **60. a.** $\sqrt{144+25}$ 13
   b. $\sqrt{25-16}$ 3 **b.** $\sqrt{144}+\sqrt{25}$ 17
   c. $\sqrt{25}\sqrt{16}$ 20 **c.** $\sqrt{144}\sqrt{25}$ 60
   d. $\sqrt{25\cdot16}$ 20 **d.** $\sqrt{144\cdot25}$ 60

61. Use the approximation $\sqrt{5}\approx2.236$ and find approximations for each expression.
   a. $\dfrac{1+\sqrt{5}}{2}$ 1.618
   b. $\dfrac{1-\sqrt{5}}{2}$ -0.618
   c. $\dfrac{1+\sqrt{5}}{2}+\dfrac{1-\sqrt{5}}{2}$ 1

62. Use the approximation $\sqrt{3}\approx1.732$ and find approximations for ech expression.
   a. $\dfrac{1+\sqrt{3}}{2}$ 1.366
   b. $\dfrac{1-\sqrt{3}}{2}$ -0.366
   c. $\dfrac{1+\sqrt{3}}{2}+\dfrac{1-\sqrt{3}}{2}$ 1

63. Evaluate each root.
   a. $\sqrt{9}$ 3
   b. $\sqrt{900}$ 30
   c. $\sqrt{0.09}$ 0.3

64. Evaluate each root.
   a. $\sqrt[3]{27}$ 3
   b. $\sqrt[3]{0.027}$ 0.3
   c. $\sqrt[3]{27,000}$ 30

Simplify each of the following pairs of expressions.

65. $\dfrac{5+\sqrt{49}}{2}$ and $\dfrac{5-\sqrt{49}}{2}$ $6, -1$

66. $\dfrac{3+\sqrt{25}}{4}$ and $\dfrac{3-\sqrt{25}}{4}$ $2, -\dfrac{1}{2}$

67. $\dfrac{2+\sqrt{16}}{2}$ and $\dfrac{2-\sqrt{16}}{2}$ $3, -1$

68. $\dfrac{3+\sqrt{9}}{2}$ and $\dfrac{3-\sqrt{9}}{2}$ $3, 0$

69. We know that the trinomial x^2+6x+9 is the square of the binomial $x+3$; that is,

$$x^2+6x+9=(x+3)^2$$

Use this fact to find $\sqrt{x^2+6x+9}$. $x+3$

70. Use the fact that $x^2+10x+25=(x+5)^2$ to find $\sqrt{x^2+10x+25}$. $x+5$

Applying the Concepts

71. Pendulum Problem The time (in seconds) it takes for the pendulum on a clock to swing through one complete cycle is given by the formula

$$T=\frac{11}{7}\sqrt{\frac{L}{2}}$$

where L is the length (in feet) of the pendulum. Use this formula and a calculator to complete the following table. Round your answers to the nearest hundredth.

Length L (feet)	Time T (seconds)
1	1.11
2	1.57
3	1.92
4	2.22
5	2.48
6	2.72

Height h (feet)	Distance d (miles)
10	4
50	9
90	12
130	14
170	16
190	17

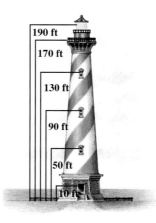

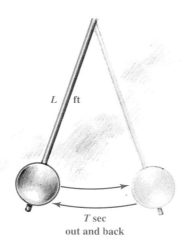

L ft

T sec
out and back

72. Lighthouse Problem The higher you are above the ground, the farther you can see. If your view is unobstructed, then the distance in miles that you can see from h feet above the ground is given by the formula

$$d = \sqrt{\frac{3h}{2}}$$

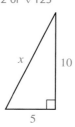

The following figure shows a lighthouse with a door and windows at various heights. The preceding formula can be used to find the distance to the ocean horizon from these heights. Use the formula and a calculator to complete the following table. Round your answers to the nearest whole number.

Find x in each of the following right triangles.

73. 5

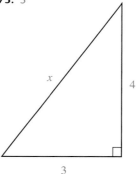

3

4

x

74. 13

5

12

x

75. 11.2 or $\sqrt{125}$

5

10

x

76. 7.1 or $\sqrt{50}$

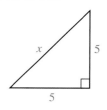

5

5

x

77. Geometry One end of a wire is attached to the top of a 24-foot pole; the other end of the wire is anchored to the ground 18 feet from the bottom of the pole. If the pole makes an angle of 90° with the ground, find the length of the wire. 30 feet

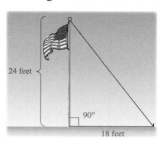

24 feet

90°

18 feet

78. Geometry The screen on a television set is in the shape of a rectangle. If the length is 20 inches and the width is 12 inches, how many inches is it from one corner of the screen to the opposite corner?
23.3 inches

20 inches

12 inches

x

79. Geometry Two children are trying to cross a stream. They want to use a log that goes from one bank to the other. If the left bank is 5 feet higher than the right bank and the stream is 12 feet wide, how long must a log be to just barely reach? 13 feet

Left bank

5 ft

Right bank

12 ft

80. Geometry, Surveying A surveying team wants to calculate how long a straight tunnel through a mountain will be. They find a place where the lines to the two ends of the proposed tunnel form a right angle. The distances are shown. How long will the tunnel be? Give the exact answer, and, if you have a calculator, provide an approximate answer to two decimal places. $\sqrt{5} \approx 2.24$ miles

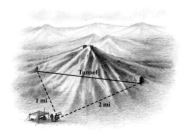

Tunnel

1 mi

2 mi

Maintaining Your Skills

Reduce each rational expression to lowest terms.

81. $\dfrac{x^2 - 16}{x + 4}$ $x - 4$

82. $\dfrac{x - 5}{x^2 - 25}$ $\dfrac{1}{x + 5}$

83. $\dfrac{10a + 20}{5a^2 - 20}$ $\dfrac{2}{a - 2}$

84. $\dfrac{8a - 16}{4a^2 - 16}$ $\dfrac{2}{a + 2}$

85. $\dfrac{2x^2 - 5x - 3}{x^2 - 3x}$ $\dfrac{2x + 1}{x}$

86. $\dfrac{x^2 - 5x}{3x^2 - 13x - 10}$ $\dfrac{x}{3x + 2}$

87. $\dfrac{xy + 3x + 2y + 6}{xy + 3x + ay + 3a}$ $\dfrac{x + 2}{x + a}$

88. $\dfrac{xy + 5x + 4y + 20}{x^2 + bx + 4x + 4b}$ $\dfrac{y + 5}{x + b}$

Getting Ready for the Next Section

Simplify.

89. $3 \cdot \sqrt{16}$ 12

90. $6 \cdot \sqrt{4}$ 12

Factor each of the following numbers into the product of two numbers, one of which is a perfect square. (Remember from Chapter 1, a perfect square is 1, 4, 9, 16, 25, 36, . . ., etc.)

91. 75 $25 \cdot 3$

92. 12 $4 \cdot 3$

93. 50 $25 \cdot 2$

94. 20 $4 \cdot 5$

95. 40 $4 \cdot 10$

96. 18 $9 \cdot 2$

97. Factor x^4 from x^5. $x^4 \cdot x$ **98.** Factor x^2 from x^3. $x^2 \cdot x$

99. Factor $4x^2$ from $12x^2$. $4x^2 \cdot 3$

100. Factor $4x^2$ from $20x^2$. $4x^2 \cdot 5$

101. Factor $25x^2y^2$ from $50x^3y^2$. $25x^2y^2 \cdot 2x$

102. Factor $25x^2y^2$ from $75x^2y^3$. $25x^2y^2 \cdot 3y$

8.2 Properties of Radicals

OBJECTIVES

A Use Property 1 for radicals to simplify a radical expression.

B Use Property 2 for radicals to simplify a radical expression.

C Use a combination of Properties 1 and 2 to simplify a radical expression.

In this section we will consider the first part of what is called *simplified form* for radical expressions. A radical expression is any expression containing a radical, whether it is a square root, a cube root, or a higher root. Simplified form for a radical expression is the form that is easiest to work with. The first step in putting a radical expression in simplified form is to take as much out from under the radical sign as possible. To do this, we first must develop two properties of radicals in general.

Consider the following two problems:

$$\sqrt{9 \cdot 16} = \sqrt{144} = 12$$
$$\sqrt{9} \cdot \sqrt{16} = 3 \cdot 4 = 12$$

Since the answers to both are equal, the original problems also must be equal; that is, $\sqrt{9 \cdot 16} = \sqrt{9} \cdot \sqrt{16}$. We can generalize this property as follows.

> **Property 1 for Radicals** If x and y represent nonnegative real numbers, then it is always true that
> $$\sqrt{xy} = \sqrt{x}\,\sqrt{y}$$
>
> *In words:* The square root of a product is the product of the square roots.

We can use this property to simplify radical expressions.

 EXAMPLE 1   Simplify $\sqrt{20}$.

SOLUTION   To simplify $\sqrt{20}$, we want to take as much out from under the radical sign as possible. We begin by looking for the largest perfect square that is a factor of 20. The largest perfect square that divides 20 is 4, so we write 20 as $4 \cdot 5$:

$$\sqrt{20} = \sqrt{4 \cdot 5}$$

Next, we apply the first property of radicals and write

$$\sqrt{4 \cdot 5} = \sqrt{4}\,\sqrt{5}$$

And since $\sqrt{4} = 2$, we have

$$\sqrt{4}\,\sqrt{5} = 2\sqrt{5}$$

The expression $2\sqrt{5}$ is the simplified form of $\sqrt{20}$ since we have taken as much out from under the radical sign as possible.

Note

Working a problem like the one in Example 1 depends on recognizing the largest perfect square that divides (is a factor of) the number under the radical sign. The set of perfect squares is the set

$$\{1, 4, 9, 16, 25, 36, \dots\}$$

To simplify an expression like $\sqrt{20}$, we first must find the largest number in this set that is a factor of the number under the radical sign.

 EXAMPLE 2   Simplify $\sqrt{75}$.

SOLUTION   Since 25 is the largest perfect square that divides 75, we have

$$\sqrt{75} = \sqrt{25 \cdot 3} \qquad \text{Factor 75 into } 25 \cdot 3$$
$$= \sqrt{25}\,\sqrt{3} \qquad \text{Property 1 for radicals}$$
$$= 5\sqrt{3} \qquad \sqrt{25} = 5$$

The expression $5\sqrt{3}$ is the simplified form for $\sqrt{75}$ since we have taken as much out from under the radical sign as possible.

The next two examples involve square roots of expressions that contain variables. Remember, we are assuming that all variables that appear under a radical sign represent positive numbers.

 EXAMPLE 3   Simplify $\sqrt{25x^3}$.

SOLUTION The largest perfect square that is a factor of $25x^3$ is $25x^2$. We write $25x^3$ as $25x^2 \cdot x$ and apply Property 1:

$$\sqrt{25x^3} = \sqrt{25x^2 \cdot x} \qquad \text{Factor } 25x^3 \text{ into } 25x^2 \cdot x$$
$$= \sqrt{25x^2}\,\sqrt{x} \qquad \text{Property 1 for radicals}$$
$$= 5x\sqrt{x} \qquad \sqrt{25x^2} = 5x$$

 EXAMPLE 4   Simplify $\sqrt{18y^4}$.

SOLUTION The largest perfect square that is a factor of $18y^4$ is $9y^4$. We write $18y^4$ as $9y^4 \cdot 2$ and apply Property 1:

$$\sqrt{18y^4} = \sqrt{9y^4 \cdot 2} \qquad \text{Factor } 18y^4 \text{ into } 9y^4 \cdot 2$$
$$= \sqrt{9y^4}\,\sqrt{2} \qquad \text{Property 1 for radicals}$$
$$= 3y^2\sqrt{2} \qquad \sqrt{9y^4} = 3y^2$$

 EXAMPLE 5   Simplify $3\sqrt{32}$.

SOLUTION We want to get as much out from under $\sqrt{32}$ as possible. Since 16 is the largest perfect square that divides 32, we have:

$$3\sqrt{32} = 3\sqrt{16 \cdot 2} \qquad \text{Factor 32 into } 16 \cdot 2$$
$$= 3\sqrt{16}\,\sqrt{2} \qquad \text{Property 1 for radicals}$$
$$= 3 \cdot 4\sqrt{2} \qquad \sqrt{16} = 4$$
$$= 12\sqrt{2} \qquad 3 \cdot 4 = 12$$

Although we have stated Property 1 for radicals in terms of square roots only, it holds for higher roots as well. If we were to state Property 1 again for cube roots, it would look like this:

$$\sqrt[3]{xy} = \sqrt[3]{x}\,\sqrt[3]{y}$$

 EXAMPLE 6   Simplify $\sqrt[3]{24x^3}$.

SOLUTION Since we are simplifying a cube root, we look for the largest perfect cube that is a factor of $24x^3$. Since 8 is a perfect cube, the largest perfect cube that is a factor of $24x^3$ is $8x^3$.

$$\sqrt[3]{24x^3} = \sqrt[3]{8x^3 \cdot 3} \qquad \text{Factor } 24x^3 \text{ into } 8x^3 \cdot 3$$
$$= \sqrt[3]{8x^3}\,\sqrt[3]{3} \qquad \text{Property 1 for radicals}$$
$$= 2x\sqrt[3]{3} \qquad \sqrt[3]{8x^3} = 2x$$

The second property of radicals has to do with division. The property becomes apparent when we consider the following two problems:

$$\sqrt{\frac{64}{16}} = \sqrt{4} = 2$$

$$\frac{\sqrt{64}}{\sqrt{16}} = \frac{8}{4} = 2$$

Since the answers in each case are equal, the original problems must be also:

$$\sqrt{\frac{64}{16}} = \frac{\sqrt{64}}{\sqrt{16}}$$

Here is the property in general.

Property 2 for Radicals If x and y both represent nonnegative real numbers and $y \neq 0$, then it is always true that

$$\sqrt{\frac{x}{y}} = \frac{\sqrt{x}}{\sqrt{y}}$$

In words: The square root of a quotient is the quotient of the square roots.

Although we have stated Property 2 for square roots only, it holds for higher roots as well.

We can use Property 2 for radicals in much the same way as we used Property 1 to simplify radical expressions.

 EXAMPLE 7 Simplify $\sqrt{\dfrac{49}{81}}$.

SOLUTION We begin by applying Property 2 for radicals to separate the fraction into two separate radicals. Then we simplify each radical separately:

$$\sqrt{\frac{49}{81}} = \frac{\sqrt{49}}{\sqrt{81}} \qquad \textbf{Property 2 for radicals}$$

$$= \frac{7}{9} \qquad \sqrt{49} = 7 \textbf{ and } \sqrt{81} = 9 \qquad$$

 EXAMPLE 8 Simplify $\sqrt[4]{\dfrac{81}{16}}$.

SOLUTION Remember, although Property 2 has been stated in terms of square roots, it holds for higher roots as well. Proceeding as we did in Example 7, we have

$$\sqrt[4]{\frac{81}{16}} = \frac{\sqrt[4]{81}}{\sqrt[4]{16}} \qquad \textbf{Property 2}$$

$$= \frac{3}{2} \qquad \sqrt[4]{81} = 3 \textbf{ and } \sqrt[4]{16} = 2 \qquad$$

EXAMPLE 9   Simplify $\sqrt{\dfrac{50}{49}}$.

SOLUTION   Applying Property 2 for radicals and then simplifying each resulting radical separately, we have

$$\sqrt{\frac{50}{49}} = \frac{\sqrt{50}}{\sqrt{49}} \qquad \text{Property 2 for radicals}$$

$$= \frac{\sqrt{25 \cdot 2}}{7} \qquad \text{Factor } 50 = 25 \cdot 2, \sqrt{49} = 7$$

$$= \frac{\sqrt{25}\,\sqrt{2}}{7} \qquad \text{Property 1 for radicals}$$

$$= \frac{5\sqrt{2}}{7} \qquad \sqrt{25} = 5$$

 EXAMPLE 10   Simplify $\sqrt{\dfrac{12x^2}{25}}$.

SOLUTION   Proceeding as we have in the previous three examples, we use Property 2 for radicals to separate the numerator and denominator into two separate radicals. Then, we simplify each radical separately:

$$\sqrt{\frac{12x^2}{25}} = \frac{\sqrt{12x^2}}{\sqrt{25}} \qquad \text{Property 2 for radicals}$$

$$= \frac{\sqrt{4x^2 \cdot 3}}{5} \qquad \text{Factor } 12x^2 = 4x^2 \cdot 3, \sqrt{25} = 5$$

$$= \frac{\sqrt{4x^2}\,\sqrt{3}}{5} \qquad \text{Property 1 for radicals}$$

$$= \frac{2x\sqrt{3}}{5} \qquad \sqrt{4x^2} = 2x$$

 EXAMPLE 11   Simplify $\sqrt{\dfrac{50x^3y^2}{49}}$.

SOLUTION   We begin by taking the square roots of $50x^3y^2$ and 49 separately and then writing $\sqrt{49}$ as 7:

$$\sqrt{\frac{50x^3y^2}{49}} = \frac{\sqrt{50x^3y^2}}{\sqrt{49}} \qquad \text{Property 2}$$

$$= \frac{\sqrt{50x^3y^2}}{7} \qquad \sqrt{49} = 7$$

To simplify the numerator of this last expression, we determine that the largest perfect square that is a factor of $50x^3y^2$ is $25x^2y^2$. Continuing, we have

$$= \frac{\sqrt{25x^2y^2 \cdot 2x}}{7} \qquad \text{Factor } 50x^3y^2 \text{ into } 25x^2y^2 \cdot 2x$$

$$= \frac{\sqrt{25x^2y^2}\,\sqrt{2x}}{7} \qquad \text{Property 1}$$

$$= \frac{5xy\sqrt{2x}}{7} \qquad \sqrt{25x^2y^2} = 5xy$$

LINKING OBJECTIVES AND EXAMPLES

Next to each **objective** we have listed the examples that are best described by that objective.

A	1–6
B	7, 8
C	9–11

GETTING READY FOR CLASS

After reading through the preceding section, respond in your own words and in complete sentences.

1. Describe Property 1 for radicals.
2. Describe Property 2 for radicals.
3. Explain why this statement is false: "The square root of a sum is the sum of the square roots."
4. Describe how you would apply Property 1 for radicals to $\sqrt{20}$.

Problem Set 8.2

Online support materials can be found at www.thomsonedu.com/login

Use Property 1 for radicals to simplify the following radical expressions as much as possible. Assume all variables represent positive numbers.

1. $\sqrt{8}$ $2\sqrt{2}$
2. $\sqrt{18}$ $3\sqrt{2}$
▶ 3. $\sqrt{12}$ $2\sqrt{3}$
4. $\sqrt{27}$ $3\sqrt{3}$
▶ 5. $\sqrt[3]{24}$ $2\sqrt[3]{3}$
6. $\sqrt[3]{54}$ $3\sqrt[3]{2}$
7. $\sqrt{50x^2}$ $5x\sqrt{2}$
8. $\sqrt{32x^2}$ $4x\sqrt{2}$
9. $\sqrt{45a^2b^2}$ $3ab\sqrt{5}$
10. $\sqrt{128a^2b^2}$ $8ab\sqrt{2}$
11. $\sqrt[3]{54x^3}$ $3x\sqrt[3]{2}$
12. $\sqrt[3]{128x^3}$ $4x\sqrt[3]{2}$
13. $\sqrt{32x^4}$ $4x^2\sqrt{2}$
14. $\sqrt{48x^4}$ $4x^2\sqrt{3}$
▶ 15. $5\sqrt{80}$ $20\sqrt{5}$
16. $3\sqrt{125}$ $15\sqrt{5}$
17. $\frac{1}{2}\sqrt{28x^3}$ $x\sqrt{7x}$
18. $\frac{2}{3}\sqrt{54x^3}$ $2x\sqrt{6x}$
19. $x\sqrt[3]{8x^4}$ $2x^2\sqrt[3]{x}$
20. $x\sqrt[3]{8x^5}$ $2x^2\sqrt[3]{x^2}$
21. $2a\sqrt[3]{27a^5}$ $6a^2\sqrt[3]{a^2}$
22. $3a\sqrt[3]{27a^4}$ $9a^2\sqrt[3]{a}$
23. $\frac{4}{3}\sqrt{45a^3}$ $4a\sqrt{5a}$
24. $\frac{3}{5}\sqrt{300a^3}$ $6a\sqrt{3a}$
▶ 25. $3\sqrt{50xy^2}$ $15y\sqrt{2x}$
26. $4\sqrt{18xy^2}$ $12y\sqrt{2x}$
27. $7\sqrt{12x^2y}$ $14x\sqrt{3y}$
28. $6\sqrt{20x^2y}$ $12x\sqrt{5y}$

Use Property 2 for radicals to simplify each of the following. Assume all variables represent positive numbers.

29. $\sqrt{\dfrac{16}{25}}$ $\frac{4}{5}$
30. $\sqrt{\dfrac{81}{64}}$ $\frac{9}{8}$
31. $\sqrt{\dfrac{4}{9}}$ $\frac{2}{3}$
32. $\sqrt{\dfrac{49}{16}}$ $\frac{7}{4}$
33. $\sqrt[3]{\dfrac{8}{27}}$ $\frac{2}{3}$
34. $\sqrt[3]{\dfrac{64}{27}}$ $\frac{4}{3}$
35. $\sqrt[4]{\dfrac{16}{81}}$ $\frac{2}{3}$
36. $\sqrt[4]{\dfrac{81}{16}}$ $\frac{3}{2}$
37. $\sqrt{\dfrac{100x^2}{25}}$ $2x$
38. $\sqrt{\dfrac{100x^2}{4}}$ $5x$
39. $\sqrt{\dfrac{81a^2b^2}{9}}$ $3ab$
40. $\sqrt{\dfrac{64a^2b^2}{16}}$ $2ab$
41. $\sqrt[3]{\dfrac{27x^3}{8y^3}}$ $\frac{3x}{2y}$
42. $\sqrt[3]{\dfrac{125x^3}{64y^3}}$ $\frac{5x}{4y}$

Use combinations of Properties 1 and 2 for radicals to simplify the following problems as much as possible. Assume all variables represent positive numbers.

43. $\sqrt{\dfrac{50}{9}}$ $\frac{5\sqrt{2}}{3}$
44. $\sqrt{\dfrac{32}{49}}$ $\frac{4\sqrt{2}}{7}$
45. $\sqrt{\dfrac{75}{25}}$ $\sqrt{3}$
46. $\sqrt{\dfrac{300}{4}}$ $5\sqrt{3}$
▶ 47. $\sqrt{\dfrac{128}{49}}$ $\frac{8\sqrt{2}}{7}$
48. $\sqrt{\dfrac{32}{64}}$ $\frac{\sqrt{2}}{2}$
49. $\sqrt{\dfrac{288x}{25}}$ $\frac{12\sqrt{2x}}{5}$
50. $\sqrt{\dfrac{28y}{81}}$ $\frac{2\sqrt{7y}}{9}$
51. $\sqrt{\dfrac{54a^2}{25}}$ $\frac{3a\sqrt{6}}{5}$
52. $\sqrt{\dfrac{243a^2}{49}}$ $\frac{9a\sqrt{3}}{7}$

▨ = Videos available by instructor request
▶ = Online student support materials available at www.thomsonedu.com/login

53. $\dfrac{3\sqrt{50}}{2}$ $\dfrac{15\sqrt{2}}{2}$

54. $\dfrac{5\sqrt{48}}{3}$ $\dfrac{20\sqrt{3}}{3}$

55. $\dfrac{7\sqrt{28y^2}}{3}$ $\dfrac{14y\sqrt{7}}{3}$

56. $\dfrac{9\sqrt{243x^2}}{2}$ $\dfrac{81x\sqrt{3}}{2}$

57. $\dfrac{5\sqrt{72a^2b^2}}{\sqrt{36}}$ $5ab\sqrt{2}$

58. $\dfrac{2\sqrt{27a^2b^2}}{\sqrt{9}}$ $2ab\sqrt{3}$

59. $\dfrac{6\sqrt{8x^2y}}{\sqrt{4}}$ $6x\sqrt{2y}$

60. $\dfrac{5\sqrt{32xy^2}}{\sqrt{25}}$ $4y\sqrt{2x}$

61. Simplify $\sqrt{b^2 - 4ac}$ if
 a. $a = 2, b = 4, c = -3$ $2\sqrt{10}$
 b. $a = 1, b = 1, c = -6$ 5
 c. $a = 1, b = 1, c = -11$ $3\sqrt{5}$
 d. $a = 3, b = 6, c = 2$ $2\sqrt{3}$

62. Simplify $\sqrt{b^2 - 4ac}$ if
 a. $a = -3, b = -4, c = 2$ $2\sqrt{10}$
 b. $a = 1, b = -3, c = 2$ 1
 c. $a = 4, b = 8, c = 1$ $4\sqrt{3}$
 d. $a = 1, b = -4, c = 1$ $2\sqrt{3}$

63. Simplify.
 a. $\sqrt{32x^{10}y^5}$ $4x^5y^2\sqrt{2y}$ **b.** $\sqrt[3]{32x^{10}y^5}$ $2x^3y\sqrt[3]{4xy^2}$
 c. $\sqrt[4]{32x^{10}y^5}$ $2x^2y\sqrt[4]{2x^2y}$ **d.** $\sqrt[5]{32x^{10}y^5}$ $2x^2y$

64. Simplify.
 a. $\sqrt{16x^8y^4}$ $4x^4y^2$ **b.** $\sqrt{16x^4y^8}$ $4x^2y^4$
 c. $\sqrt[3]{16x^8y^4}$ $2x^2y\sqrt[3]{2x^2y}$ **d.** $\sqrt[4]{16x^8y^4}$ $2x^2y$

65. Simplify.
 a. $\sqrt{4}$ 2 **b.** $\sqrt{0.04}$ 0.2
 c. $\sqrt{400}$ 20 **d.** $\sqrt{0.0004}$ 0.02

66. Simplify.
 a. $\sqrt[3]{8}$ 2 **b.** $\sqrt[3]{0.008}$ 0.2
 c. $\sqrt[3]{80}$ $2\sqrt[3]{10}$ **d.** $\sqrt[3]{8,000}$ 20

Use a calculator to help complete the following tables. If an answer needs rounding, round to the nearest thousandth.

67.

x	$\sqrt{x}$	$2\sqrt{x}$	$\sqrt{4x}$
1	1	2	2
2	1.414	2.828	2.828
3	1.732	3.464	3.464
4	2	4	4

68.

x	$\sqrt{x}$	$2\sqrt{x}$	$\sqrt{4x}$
1	1	2	2
4	2	4	4
9	3	6	6
16	4	8	8

69.

x	$\sqrt{x}$	$3\sqrt{x}$	$\sqrt{9x}$
1	1	3	3
2	1.414	4.243	4.243
3	1.732	5.196	5.196
4	2	6	6

70.

x	$\sqrt{x}$	$3\sqrt{x}$	$\sqrt{9x}$
1	1	3	3
4	2	6	6
9	3	9	9
16	4	12	12

Maintaining Your Skills

Multiply or divide as indicated.

71. $\dfrac{8x}{x^2 - 5x} \cdot \dfrac{x^2 - 25}{4x^2 + 4x}$ $\dfrac{2(x + 5)}{x(x + 1)}$

72. $\dfrac{x^2 + 4x}{4x^2} \cdot \dfrac{7x}{x^2 + 5x}$ $\dfrac{7(x + 4)}{4x(x + 5)}$

73. $\dfrac{x^2 + 3x - 4}{3x^2 + 7x - 20} \div \dfrac{x^2 - 2x + 1}{3x^2 - 2x - 5}$ $\dfrac{x + 1}{x - 1}$

74. $\dfrac{x^2 - 16}{2x^2 + 7x + 3} \div \dfrac{2x^2 - 5x - 12}{4x^2 + 8x + 3}$ $\dfrac{x + 4}{x + 3}$

75. $(x^2 - 36)\left(\dfrac{x + 3}{x - 6}\right)$ $(x + 6)(x + 3)$

76. $(x^2 - 49)\left(\dfrac{x + 5}{x + 7}\right)$ $(x - 7)(x + 5)$

Getting Ready for the Next Section

Simplify.

77. $\sqrt{4x^3y^2}$ $2xy\sqrt{x}$

78. $\sqrt{9x^2y^3}$ $3xy\sqrt{y}$

79. $\dfrac{6}{2}\sqrt{16}$ 12

80. $\dfrac{8}{4}\sqrt{9}$ 6

81. $\dfrac{\sqrt{2}}{\sqrt{4}}$ $\dfrac{\sqrt{2}}{2}$

82. $\dfrac{\sqrt{6}}{\sqrt{9}}$ $\dfrac{\sqrt{6}}{3}$

83. $\dfrac{\sqrt[3]{18}}{\sqrt[3]{27}}$ $\dfrac{\sqrt[3]{18}}{3}$

84. $\dfrac{\sqrt[3]{12}}{\sqrt[3]{8}}$ $\dfrac{\sqrt[3]{12}}{2}$

Multiply.

85. $\dfrac{\sqrt{2}}{\sqrt{3}} \cdot \dfrac{\sqrt{3}}{\sqrt{3}}$ $\dfrac{\sqrt{6}}{3}$

86. $\dfrac{\sqrt{y}}{\sqrt{2}} \cdot \dfrac{\sqrt{2}}{\sqrt{2}}$ $\dfrac{\sqrt{2y}}{2}$

87. $\sqrt[3]{3} \cdot \sqrt[3]{9}$ 3

88. $\sqrt[3]{4} \cdot \sqrt[3]{2}$ 2

8.3 Simplified Form for Radicals

OBJECTIVES

A Use both properties of radicals to write a radical expression in simplified form.

B Rationalize the denominator in a radical expression that contains only one term in the denominator.

Radical expressions that are in simplified form are generally easier to work with. A radical expression is in simplified form if it has three special characteristics.

> **DEFINITION** A radical expression is in **simplified form** if
>
> 1. There are no perfect squares that are factors of the quantity under the square root sign, no perfect cubes that are factors of the quantity under the cube root sign, and so on. We want as little as possible under the radical sign.
> 2. There are no fractions under the radical sign.
> 3. There are no radicals in the denominator.

Note

Simplified form for radicals is the form that we work toward when simplifying radicals. The properties of radicals are the tools we use to get us to simplified form.

A radical expression that has these three characteristics is said to be in simplified form. As we will see, simplified form is not always the least complicated expression. In many cases, the simplified expression looks more complicated than the original expression. The important thing about simplified form for radicals is that simplified expressions are easier to work with.

The tools we will use to put radical expressions into simplified form are the properties of radicals. We list the properties again for clarity.

> **Properties of Radicals** If a and b represent any two nonnegative real numbers, then it is always true that
>
> **1.** $\sqrt{a}\,\sqrt{b} = \sqrt{a \cdot b}$
>
> **2.** $\dfrac{\sqrt{a}}{\sqrt{b}} = \sqrt{\dfrac{a}{b}}$ $b \neq 0$
>
> **3.** $\sqrt{a}\,\sqrt{a} = (\sqrt{a})^2 = a$ **This property comes directly from the definition of radicals**

The following examples illustrate how we put a radical expression into simplified form using the three properties of radicals. Although the properties are stated for square roots only, they hold for all roots. [Property 3 written for cube roots would be $\sqrt[3]{a}\,\sqrt[3]{a}\,\sqrt[3]{a} = (\sqrt[3]{a})^3 = a$.]

EXAMPLE 1   Put $\sqrt{\frac{1}{2}}$ into simplified form.

SOLUTION   The expression $\sqrt{\frac{1}{2}}$ is not in simplified form because there is a fraction under the radical sign. We can change this by applying Property 2 for radicals:

$$\sqrt{\frac{1}{2}} = \frac{\sqrt{1}}{\sqrt{2}} \qquad \textbf{Property 2 for radicals}$$

$$= \frac{1}{\sqrt{2}} \qquad \sqrt{\textbf{1}} = \textbf{1}$$

The expression $\frac{1}{\sqrt{2}}$ is not in simplified form because there is a radical sign in the denominator. If we multiply the numerator and denominator of $\frac{1}{\sqrt{2}}$ by $\sqrt{2}$, the denominator becomes $\sqrt{2} \cdot \sqrt{2} = 2$:

$$\frac{1}{\sqrt{2}} = \frac{1}{\sqrt{2}} \cdot \frac{\sqrt{\textbf{2}}}{\sqrt{\textbf{2}}} \qquad \textbf{Multiply numerator and denominator by } \sqrt{2}$$

$$= \frac{\sqrt{2}}{2} \qquad \begin{array}{l} \textbf{1} \cdot \sqrt{\textbf{2}} = \sqrt{\textbf{2}} \\ \sqrt{\textbf{2}} \cdot \sqrt{\textbf{2}} = \sqrt{\textbf{4}} = \textbf{2} \end{array}$$

If we check the expression $\frac{\sqrt{2}}{2}$ against our definition of simplified form for radicals, we find that all three rules hold. There are no perfect squares that are factors of 2. There are no fractions under the radical sign. No radicals appear in the denominator. The expression $\frac{\sqrt{2}}{2}$, therefore, must be in simplified form.

 EXAMPLE 2   Write $\sqrt{\frac{2}{3}}$ in simplified form.

SOLUTION   We proceed as we did in Example 1:

$$\sqrt{\frac{2}{3}} = \frac{\sqrt{2}}{\sqrt{3}} \qquad \textbf{Use Property 2 to separate radicals}$$

$$= \frac{\sqrt{2}}{\sqrt{3}} \cdot \frac{\sqrt{\textbf{3}}}{\sqrt{\textbf{3}}} \qquad \textbf{Multiply by } \frac{\sqrt{\textbf{3}}}{\sqrt{\textbf{3}}} \textbf{ to remove the radical from the denominator}$$

$$= \frac{\sqrt{6}}{3} \qquad \begin{array}{l} \sqrt{\textbf{2}} \cdot \sqrt{\textbf{3}} = \sqrt{\textbf{6}} \\ \sqrt{\textbf{3}} \cdot \sqrt{\textbf{3}} = \sqrt{\textbf{9}} = \textbf{3} \end{array}$$

EXAMPLE 3   Put the expression $\frac{6\sqrt{20}}{2\sqrt{5}}$ into simplified form.

SOLUTION   Although there are many ways to begin this problem, we notice that 20 is divisible by 5. Using Property 2 for radicals as the first step, we can quickly put the expression into simplified form:

$$\frac{6\sqrt{20}}{2\sqrt{5}} = \frac{6}{2}\sqrt{\frac{20}{5}} \qquad \textbf{Property 2 for radicals}$$

$$= 3\sqrt{4} \qquad \frac{\textbf{20}}{\textbf{5}} = \textbf{4}$$

$$= 3 \cdot 2 \qquad \sqrt{\textbf{4}} = \textbf{2}$$

$$= 6$$

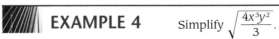

EXAMPLE 4   Simplify $\sqrt{\dfrac{4x^3y^2}{3}}$.

SOLUTION   We begin by separating the numerator and denominator and then taking the perfect squares out of the numerator:

$$\sqrt{\frac{4x^3y^2}{3}} = \frac{\sqrt{4x^3y^2}}{\sqrt{3}} \qquad \textbf{Property 2 for radicals}$$

$$= \frac{\sqrt{4x^2y^2}\,\sqrt{x}}{\sqrt{3}} \qquad \textbf{Property 1 for radicals}$$

$$= \frac{2xy\sqrt{x}}{\sqrt{3}} \qquad \sqrt{4x^2y^2} = 2xy$$

> **Note**
>
> When working with square roots of variable quantities, we will always assume that the variables represent positive numbers. That way we can say that $\sqrt{x^2} = x$.

The only thing keeping our expression from being in simplified form is the $\sqrt{3}$ in the denominator. We can take care of this by multiplying the numerator and denominator by $\sqrt{3}$:

$$\frac{2xy\sqrt{x}}{\sqrt{3}} = \frac{2xy\sqrt{x}}{\sqrt{3}} \cdot \frac{\sqrt{3}}{\sqrt{3}} \qquad \textbf{Multiply numerator and denominator by } \sqrt{3}$$

$$= \frac{2xy\sqrt{3x}}{3} \qquad \sqrt{3} \cdot \sqrt{3} = \sqrt{9} = 3$$

Although the final expression may look more complicated than the original expression, it is in simplified form. The last step is called *rationalizing the denominator*. We have taken the radical out of the denominator and replaced it with a rational number.

EXAMPLE 5   Simplify $\sqrt[3]{\dfrac{2}{3}}$.

SOLUTION   We can apply Property 2 first to separate the cube roots:

$$\sqrt[3]{\frac{2}{3}} = \frac{\sqrt[3]{2}}{\sqrt[3]{3}}$$

To write this expression in simplified form, we must remove the radical from the denominator. Since the radical is a cube root, we will need to multiply it by an expression that will give us a perfect cube under that cube root. We can accomplish this by multiplying the numerator and denominator by $\sqrt[3]{9}$. Here is what it looks like:

$$\frac{\sqrt[3]{2}}{\sqrt[3]{3}} = \frac{\sqrt[3]{2}}{\sqrt[3]{3}} \cdot \frac{\sqrt[3]{9}}{\sqrt[3]{9}}$$

$$= \frac{\sqrt[3]{18}}{\sqrt[3]{27}} \qquad \sqrt[3]{2} \cdot \sqrt[3]{9} = \sqrt[3]{18}$$
$$\qquad\qquad\quad \sqrt[3]{3} \cdot \sqrt[3]{9} = \sqrt[3]{27}$$

$$= \frac{\sqrt[3]{18}}{3} \qquad \sqrt[3]{27} = 3$$

To see why multiplying the numerator and denominator by $\sqrt[3]{9}$ works in this example, you first must convince yourself that multiplying the numerator and denominator by $\sqrt[3]{3}$ would not have worked.

EXAMPLE 6   Simplify $\sqrt[3]{\frac{1}{4}}$.

SOLUTION   We begin by separating the numerator and denominator:

$$\sqrt[3]{\frac{1}{4}} = \frac{\sqrt[3]{1}}{\sqrt[3]{4}} \qquad \textbf{Property 2 for radicals}$$

$$= \frac{1}{\sqrt[3]{4}} \qquad \sqrt[3]{1} = 1$$

To rationalize the denominator, we need to have a perfect cube under the cube root sign. If we multiply the numerator and denominator by $\sqrt[3]{2}$, we will have $\sqrt[3]{4} \cdot \sqrt[3]{2} = \sqrt[3]{8}$ in the denominator:

$$\frac{1}{\sqrt[3]{4}} = \frac{1}{\sqrt[3]{4}} \cdot \frac{\sqrt[3]{2}}{\sqrt[3]{2}} \qquad \textbf{Multiply numerator and denominator by } \sqrt[3]{2}$$

$$= \frac{\sqrt[3]{2}}{\sqrt[3]{8}} \qquad \sqrt[3]{4} \cdot \sqrt[3]{2} = \sqrt[3]{8}$$

$$= \frac{\sqrt[3]{2}}{2} \qquad \sqrt[3]{8} = 2$$

The final expression has no radical sign in the denominator and therefore is in simplified form.

The Spiral of Roots

To visualize the square roots of the positive integers, we can construct the spiral of roots, which we mentioned in the introduction to this chapter. To begin, we draw two line segments, each of length 1, at right angles to each other. Then we use the Pythagorean theorem to find the length of the diagonal. Figure 1 illustrates.

Next, we construct a second triangle by connecting a line segment of length 1 to the end of the first diagonal so that the angle formed is a right angle. We find the length of the second diagonal using the Pythagorean theorem. Figure 2 illustrates this procedure. Continuing to draw new triangles by connecting line segments of length 1 to the end of each new diagonal, so that the angle formed is a right angle, the spiral of roots begins to appear (see Figure 3).

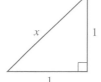

$$x = \sqrt{1^2 + 1^2}$$
$$= \sqrt{2}$$

FIGURE 1

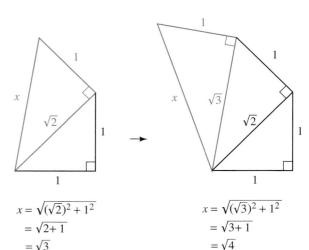

$$x = \sqrt{(\sqrt{2})^2 + 1^2}$$
$$= \sqrt{2+1}$$
$$= \sqrt{3}$$

FIGURE 2

$$x = \sqrt{(\sqrt{3})^2 + 1^2}$$
$$= \sqrt{3+1}$$
$$= \sqrt{4}$$
$$= 2$$

FIGURE 3

LINKING OBJECTIVES AND EXAMPLES

Next to each **objective** we have listed the examples that are best described by that objective.

A   1–6

B   1, 2, 4–6

GETTING READY FOR CLASS

After reading through the preceding section, respond in your own words and in complete sentences.

1. Describe how you would put $\sqrt{\dfrac{1}{2}}$ in simplified form.

2. What is simplified form for an expression that contains a square root?

3. What does it mean to rationalize the denominator in an expression?

4. What is useful about the spiral of roots?

Problem Set 8.3

Online support materials can be found at www.thomsonedu.com/login

Put each of the following radical expressions into simplified form. Assume all variables represent positive numbers.

▶ **1.** $\sqrt{\dfrac{1}{2}}$ $\dfrac{\sqrt{2}}{2}$

2. $\sqrt{\dfrac{1}{5}}$ $\dfrac{\sqrt{5}}{5}$

3. $\sqrt{\dfrac{1}{3}}$ $\dfrac{\sqrt{3}}{3}$

4. $\sqrt{\dfrac{1}{6}}$ $\dfrac{\sqrt{6}}{6}$

5. $\sqrt{\dfrac{2}{5}}$ $\dfrac{\sqrt{10}}{5}$

6. $\sqrt{\dfrac{3}{7}}$ $\dfrac{\sqrt{21}}{7}$

7. $\sqrt{\dfrac{3}{2}}$ $\dfrac{\sqrt{6}}{2}$

8. $\sqrt{\dfrac{5}{3}}$ $\dfrac{\sqrt{15}}{3}$

9. $\sqrt{\dfrac{20}{3}}$ $\dfrac{2\sqrt{15}}{3}$

10. $\sqrt{\dfrac{32}{5}}$ $\dfrac{4\sqrt{10}}{5}$

11. $\sqrt{\dfrac{45}{6}}$ $\dfrac{\sqrt{30}}{2}$

12. $\sqrt{\dfrac{48}{7}}$ $\dfrac{4\sqrt{21}}{7}$

13. $\sqrt{\dfrac{20}{5}}$ 2

14. $\sqrt{\dfrac{12}{3}}$ 2

15. $\dfrac{\sqrt{21}}{\sqrt{3}}$ $\sqrt{7}$

16. $\dfrac{\sqrt{21}}{\sqrt{7}}$ $\sqrt{3}$

17. $\dfrac{\sqrt{35}}{\sqrt{7}}$ $\sqrt{5}$

18. $\dfrac{\sqrt{35}}{\sqrt{5}}$ $\sqrt{7}$

▶ **19.** $\dfrac{10\sqrt{15}}{5\sqrt{3}}$ $2\sqrt{5}$

20. $\dfrac{4\sqrt{12}}{8\sqrt{3}}$ 1

21. $\dfrac{6\sqrt{21}}{3\sqrt{7}}$ $2\sqrt{3}$

22. $\dfrac{8\sqrt{50}}{16\sqrt{2}}$ $\dfrac{5}{2}$

23. $\dfrac{6\sqrt{35}}{12\sqrt{5}}$ $\dfrac{\sqrt{7}}{2}$

24. $\dfrac{8\sqrt{35}}{16\sqrt{7}}$ $\dfrac{\sqrt{5}}{2}$

25. $\sqrt{\dfrac{4x^2y^2}{2}}$ $xy\sqrt{2}$

26. $\sqrt{\dfrac{9x^2y^2}{3}}$ $xy\sqrt{3}$

27. $\sqrt{\dfrac{5x^2y}{3}}$ $\dfrac{x\sqrt{15y}}{3}$

28. $\sqrt{\dfrac{7x^2y}{5}}$ $\dfrac{x\sqrt{35y}}{5}$

▶ **29.** $\sqrt{\dfrac{16a^4}{5}}$ $\dfrac{4a^2\sqrt{5}}{5}$

30. $\sqrt{\dfrac{25a^4}{7}}$ $\dfrac{5a^2\sqrt{7}}{7}$

31. $\sqrt{\dfrac{72a^5}{5}}$ $\dfrac{6a^2\sqrt{10a}}{5}$

32. $\sqrt{\dfrac{12a^5}{5}}$ $\dfrac{2a^2\sqrt{15a}}{5}$

33. $\sqrt{\dfrac{20x^2y^3}{3}}$ $\dfrac{2xy\sqrt{15y}}{3}$

34. $\sqrt{\dfrac{27x^2y^3}{2}}$ $\dfrac{3xy\sqrt{6y}}{2}$

35. $\dfrac{2\sqrt{20x^2y^3}}{3}$ $\dfrac{4xy\sqrt{5y}}{3}$

36. $\dfrac{5\sqrt{27x^3y^2}}{2}$ $\dfrac{15xy\sqrt{3x}}{2}$

37. $\dfrac{6\sqrt{54a^2b^3}}{5}$ $\dfrac{18ab\sqrt{6b}}{5}$

38. $\dfrac{7\sqrt{75a^3b^2}}{6}$ $\dfrac{35ab\sqrt{3a}}{6}$

39. $\dfrac{3\sqrt{72x^4}}{\sqrt{2x}}$ $18x\sqrt{x}$

40. $\dfrac{2\sqrt{45x^4}}{\sqrt{5x}}$ $6x\sqrt{x}$

▶ **41.** $\sqrt[3]{\dfrac{1}{2}}$ $\dfrac{\sqrt[3]{4}}{2}$

42. $\sqrt[3]{\dfrac{1}{4}}$ $\dfrac{\sqrt[3]{2}}{2}$

43. $\sqrt[3]{\dfrac{1}{9}}$ $\dfrac{\sqrt[3]{3}}{3}$

44. $\sqrt[3]{\dfrac{1}{3}}$ $\dfrac{\sqrt[3]{9}}{3}$

45. Rationalize the denominator.

a. $\dfrac{6}{\sqrt{\pi}}$ $\dfrac{6\sqrt{\pi}}{\pi}$

b. $\sqrt{\dfrac{A}{\pi}}$ $\dfrac{\sqrt{A\pi}}{\pi}$

c. $\sqrt[3]{\dfrac{3V}{4\pi}}$ $\dfrac{\sqrt[3]{6V\pi^2}}{2\pi}$

d. $\dfrac{2}{\sqrt[3]{2\pi}}$ $\dfrac{\sqrt[3]{4\pi^2}}{\pi}$

= Videos available by instructor request

▶ = Online student support materials available at www.thomsonedu.com/login

46. Rationalize the denominator.

a. $\dfrac{3}{\sqrt{2}}$ $\dfrac{3\sqrt{2}}{2}$ b. $\sqrt{\dfrac{3}{2}}$ $\dfrac{\sqrt{6}}{2}$

c. $\sqrt[3]{\dfrac{3}{2}}$ $\dfrac{\sqrt[3]{12}}{2}$ d. $\sqrt[3]{\dfrac{3}{4}}$ $\dfrac{\sqrt[3]{6}}{2}$

Use a calculator to help complete the following tables. If an answer needs rounding, round to the nearest thousandth.

47.

x	$\sqrt{x}$	$\dfrac{1}{\sqrt{x}}$	$\dfrac{\sqrt{x}}{x}$
1	1	1	1
2	1.414	.707	.707
3	1.732	.577	.577
4	2	.5	.5
5	2.236	.447	.447
6	2.449	.408	.408

48.

x	$\sqrt{x}$	$\dfrac{1}{\sqrt{x}}$	$\dfrac{\sqrt{x}}{x}$
1	1	1	1
4	2	$\frac{1}{2}$	$\frac{1}{2}$
9	3	$\frac{1}{3}$	$\frac{1}{3}$
16	4	$\frac{1}{4}$	$\frac{1}{4}$
25	5	$\frac{1}{5}$	$\frac{1}{5}$
36	6	$\frac{1}{6}$	$\frac{1}{6}$

49.

x	$\sqrt{x^2}$	$\sqrt{x^3}$	$x\sqrt{x}$
1	1	1	1
2	2	2.828	2.828
3	3	5.196	5.196
4	4	8	8
5	5	11.18	11.18
6	6	14.697	14.697

50.

x	$\sqrt{x^2}$	$\sqrt{x^3}$	$x\sqrt{x}$
1	1	1	1
4	4	8	8
9	9	27	27
16	16	64	64
25	25	125	125
36	36	216	216

51. Spiral of Roots Construct your own spiral of roots by using a ruler. Draw the first triangle by using two 1-inch lines. The first diagonal will have a length of $\sqrt{2}$ inches. Each new triangle will be formed by drawing a 1-inch line segment at the end of the previous diagonal so that the angle formed is 90°.

52. Spiral of Roots Construct a spiral of roots by using the line segments of length 2 inches. The length of the first diagonal will be $2\sqrt{2}$ inches. The length of the second diagonal will be $2\sqrt{3}$ inches.

53. Number Sequence Simplify the terms in the following sequence. The result will be a sequence that gives the lengths of the diagonals in the spiral of roots you constructed in Problem 51.

$$\sqrt{1^2 + 1}, \ \sqrt{(\sqrt{2})^2 + 1}, \ \sqrt{(\sqrt{3})^2 + 1}, \ldots$$

$\sqrt{2}, \ \sqrt{3}, \ \sqrt{4} = 2$

54. Number Sequence Simplify the terms in the following sequence. The result will be a sequence that gives the lengths of the diagonals in the spiral of roots you constructed in Problem 52.

$$\sqrt{2^2 + 4}, \ \sqrt{(2\sqrt{2})^2 + 4}, \ \sqrt{(2\sqrt{3})^2 + 4}, \ldots$$

$2\sqrt{2}, \ 2\sqrt{3}, \ 4$

Maintaining Your Skills

Use the distributive property to combine the following.

55. $3x + 7x$ $10x$

56. $3x - 7x$ $-4x$

57. $15x + 8x$ $23x$

58. $15x - 8x$ $7x$

59. $7a - 3a + 6a$ $10a$

60. $25a + 3a - a$ $27a$

Add or subtract as indicated.

61. $\dfrac{x^2}{x+5} + \dfrac{10x+25}{x+5}$ $x+5$ **62.** $\dfrac{x^2}{x-3} - \dfrac{9}{x-3}$ $x+3$

63. $\dfrac{a}{3} + \dfrac{2}{5}$ $\dfrac{5a+6}{15}$ **64.** $\dfrac{4}{a} + \dfrac{2}{3}$ $\dfrac{12+2a}{3a}$

65. $\dfrac{6}{a^2-9} - \dfrac{5}{a^2-a-6}$ $\dfrac{1}{(a+2)(a+3)}$

66. $\dfrac{4a}{a^2+6a+5} - \dfrac{3a}{a^2+5a+4}$ $\dfrac{a}{(a+4)(a+5)}$

Getting Ready for the Next Section

Combine like terms.

67. $15x + 8x$ $23x$ **68.** $6x + 20x$ $26x$

69. $25y + 3y - y$ $27y$ **70.** $12y + 4y - y$ $15y$

71. $2ab + 5ab$ $7ab$ **72.** $3ab + 7ab$ $10ab$

73. $2xy - 9xy + 50x$ $-7xy + 50x$

74. $2xy - 18xy + 3x$ $-16xy + 3x$

Simplify.

75. $\dfrac{6+2x}{4}$ $\dfrac{3+x}{2}$ **76.** $\dfrac{6-2x}{2}$ $3-x$

8.4 Addition and Subtraction of Radical Expressions

OBJECTIVES

A   Add and subtract similar radical expressions.

To add two or more radical expressions, we apply the distributive property. Adding radical expressions is similar to adding similar terms of polynomials.

 EXAMPLE 1   Combine terms in the expression $3\sqrt{5} - 7\sqrt{5}$.

SOLUTION   The two terms $3\sqrt{5}$ and $7\sqrt{5}$ each have $\sqrt{5}$ in common. Since $3\sqrt{5}$ means 3 times $\sqrt{5}$, or $3 \cdot \sqrt{5}$, we apply the distributive property:

$$3\sqrt{5} - 7\sqrt{5} = (3-7)\sqrt{5} \qquad \textbf{Distributive property}$$
$$= -4\sqrt{5} \qquad \textbf{3 - 7 = -4}$$

Since we use the distributive property to add radical expressions, each expression must contain exactly the same radical.

 EXAMPLE 2   Combine terms in the expression $7\sqrt{2} - 3\sqrt{2} + 6\sqrt{2}$.

SOLUTION $7\sqrt{2} - 3\sqrt{2} + 6\sqrt{2} = (7-3+6)\sqrt{2}$ **Distributive property**

$$= 10\sqrt{2} \qquad \textbf{Addition}$$

In Examples 1 and 2, each term was a radical expression in simplified form. If one or more terms are not in simplified form, we must put them into simplified form and then combine terms, if possible.

> **Rule**   To combine two or more radical expressions, put each expression in simplified form, and then apply the distributive property, if possible.

EXAMPLE 3   Combine terms in the expression $3\sqrt{50} + 2\sqrt{32}$.

SOLUTION   We begin by putting each term into simplified form:

$$3\sqrt{50} + 2\sqrt{32} = 3\sqrt{25}\sqrt{2} + 2\sqrt{16}\sqrt{2} \qquad \text{Property 1 for radicals}$$
$$= 3 \cdot 5\sqrt{2} + 2 \cdot 4\sqrt{2} \qquad \sqrt{25} = 5 \text{ and } \sqrt{16} = 4$$
$$= 15\sqrt{2} + 8\sqrt{2} \qquad \text{Multiplication}$$

Applying the distributive property to the last line, we have

$$15\sqrt{2} + 8\sqrt{2} = (15 + 8)\sqrt{2} \qquad \text{Distributive property}$$
$$= 23\sqrt{2} \qquad 15 + 8 = 23$$

EXAMPLE 4   Combine terms in the expression $5\sqrt{75} + \sqrt{27} - \sqrt{3}$.

SOLUTION

$$5\sqrt{75} + \sqrt{27} - \sqrt{3} = 5\sqrt{25}\sqrt{3} + \sqrt{9}\sqrt{3} - \sqrt{3} \qquad \begin{array}{l}\text{Property 1 for radicals}\\ \sqrt{25} = 5 \text{ and } \sqrt{9} = 3\end{array}$$
$$= 5 \cdot 5\sqrt{3} + 3\sqrt{3} - \sqrt{3}$$
$$= 25\sqrt{3} + 3\sqrt{3} - \sqrt{3} \qquad 5 \cdot 5 = 25$$
$$= (25 + 3 - 1)\sqrt{3} \qquad \text{Distributive property}$$
$$= 27\sqrt{3} \qquad \text{Addition}$$

The most time-consuming part of combining most radical expressions is simplifying each term in the expression. Once this has been done, applying the distributive property is simple and fast.

EXAMPLE 5   Simplify $a\sqrt{12} + 5\sqrt{3a^2}$. (Assume $a > 0$.)

SOLUTION   We simplify each term in the expression by putting it in simplified form for radicals:

$$a\sqrt{12} + 5\sqrt{3a^2} = a\sqrt{4}\sqrt{3} + 5\sqrt{a^2}\sqrt{3} \qquad \text{Property 1 for radicals}$$
$$= a \cdot 2\sqrt{3} + 5 \cdot a\sqrt{3} \qquad \sqrt{4} = 2 \text{ and } \sqrt{a^2} = a$$
$$= 2a\sqrt{3} + 5a\sqrt{3} \qquad \text{Commutative property}$$
$$= (2a + 5a)\sqrt{3} \qquad \text{Distributive property}$$
$$= 7a\sqrt{3} \qquad \text{Addition}$$

EXAMPLE 6   Combine terms in the expression. (Assume $x > 0$.)

$$\sqrt{20x^3} - 3x\sqrt{45x} + 10\sqrt{25x^2}$$

SOLUTION $\sqrt{20x^3} - 3x\sqrt{45x} + 10\sqrt{25x^2} = \sqrt{4x^2}\sqrt{5x} - 3x\sqrt{9}\sqrt{5x} + 10\sqrt{25x^2}$
$$= 2x\sqrt{5x} - 3x \cdot 3\sqrt{5x} + 10 \cdot 5x$$
$$= 2x\sqrt{5x} - 9x\sqrt{5x} + 50x$$

Each term is now in simplified form. The best we can do next is to combine the first two terms. The last term does not have the common radical $\sqrt{5x}$.

$$2x\sqrt{5x} - 9x\sqrt{5x} + 50x = (2x - 9x)\sqrt{5x} + 50x$$
$$= -7x\sqrt{5x} + 50x$$

We have, in any case, succeeded in reducing the number of terms in our original problem.

Our next example involves an expression that is similar to many of the expressions we will find when we solve quadratic equations in Chapter 9.

 EXAMPLE 7   Simplify $\dfrac{6 + \sqrt{12}}{4}$.

SOLUTION   We begin by writing $\sqrt{12}$ as $2\sqrt{3}$:

$$\frac{6 + \sqrt{12}}{4} = \frac{6 + 2\sqrt{3}}{4}$$

Factor 2 from the numerator and denominator and then reduce to lowest terms.

$$\frac{6 + 2\sqrt{3}}{4} = \frac{2(3 + \sqrt{3})}{2 \cdot 2}$$

$$= \frac{3 + \sqrt{3}}{2}$$

Note

Remember,
$\sqrt{12} = \sqrt{4 \cdot 3} = \sqrt{4}\sqrt{3} = 2\sqrt{3}$.

LINKING OBJECTIVES AND EXAMPLES

Next to each **objective** we have listed the examples that are best described by that objective.

A   1–6

GETTING READY FOR CLASS

After reading through the preceding section, respond in your own words and in complete sentences.

1. What are similar radicals?
2. When can we add two radical expressions?
3. What is the first step when adding or subtracting expressions containing radicals?
4. It is not possible to add $\sqrt{2}$ and $\sqrt{3}$. Explain why.

Problem Set 8.4

Online support materials can be found at www.thomsonedu.com/login

In each of the following problems, simplify each term, if necessary, and then use the distributive property to combine terms, if possible.

1. a. $3x + 4x$ $7x$
   b. $3y + 4y$ $7y$
▶ **c.** $3\sqrt{2} + 4\sqrt{2}$ $7\sqrt{2}$
   d. $3\sqrt{5} + 4\sqrt{5}$ $7\sqrt{5}$

2. a. $7x + 2x$ $9x$
   b. $7t + 2t$ $9t$
   c. $7\sqrt{3} + 2\sqrt{3}$ $9\sqrt{3}$
   d. $7\sqrt{x} + 2\sqrt{x}$ $9\sqrt{x}$

3. a. $x + 6x$ $7x$
   b. $t + 6t$ $7t$
   c. $\sqrt{3} + 6\sqrt{3}$ $7\sqrt{3}$
   d. $\sqrt{x} + 6\sqrt{x}$ $7\sqrt{x}$

4. a. $x + 10x$ $11x$
   b. $y + 10y$ $11y$
   c. $\sqrt{2} + 10\sqrt{2}$ $11\sqrt{2}$
   d. $\sqrt{7} + 10\sqrt{7}$ $11\sqrt{7}$

5. $9\sqrt{5} - 7\sqrt{5}$ $2\sqrt{5}$

6. $6\sqrt{7} - 10\sqrt{7}$ $-4\sqrt{7}$

7. $5\sqrt{5} + \sqrt{5}$ $6\sqrt{5}$

8. $\sqrt{6} - 10\sqrt{6}$ $-9\sqrt{6}$

9. $14\sqrt{13} - \sqrt{13}$ $13\sqrt{13}$

10. $-2\sqrt{6} - 9\sqrt{6}$ $-11\sqrt{6}$

11. $-3\sqrt{10} + 9\sqrt{10}$ $6\sqrt{10}$

12. $11\sqrt{11} + \sqrt{11}$ $12\sqrt{11}$

13. $\dfrac{5}{8}\sqrt{5} - \dfrac{3}{7}\sqrt{5}$ $\dfrac{11}{56}\sqrt{5}$

14. $\dfrac{5}{6}\sqrt{11} - \dfrac{7}{9}\sqrt{11}$ $\dfrac{1}{18}\sqrt{11}$

▶ **15.** $\sqrt{8} + 2\sqrt{2}$ $4\sqrt{2}$

16. $\sqrt{20} + 3\sqrt{5}$ $5\sqrt{5}$

17. $3\sqrt{3} - \sqrt{27}$ 0

18. $4\sqrt{5} - \sqrt{80}$ 0

19. $5\sqrt{12} - 10\sqrt{48}$ $-30\sqrt{3}$

20. $3\sqrt{300} - 5\sqrt{27}$ $15\sqrt{3}$

21. $-\sqrt{75} - \sqrt{3}$ $-6\sqrt{3}$

22. $5\sqrt{20} + 8\sqrt{80}$ $42\sqrt{5}$

▶ **23.** $\dfrac{1}{5}\sqrt{75} - \dfrac{1}{2}\sqrt{12}$ 0

24. $\dfrac{1}{2}\sqrt{24} + \dfrac{1}{5}\sqrt{150}$ $2\sqrt{6}$

25. $\dfrac{3}{4}\sqrt{8} + \dfrac{3}{10}\sqrt{75}$ $\dfrac{3}{2}\sqrt{2} + \dfrac{3}{2}\sqrt{3}$

■ = Videos available by instructor request
▶ = Online student support materials available at www.thomsonedu.com/login

26. $\dfrac{5}{6}\sqrt{54} - \dfrac{3}{4}\sqrt{24}$ $\sqrt{6}$

27. $\sqrt{27} - 2\sqrt{12} + \sqrt{3}$ 0

28. $\sqrt{20} + 3\sqrt{45} - \sqrt{5}$ $10\sqrt{5}$

29. $\dfrac{5}{6}\sqrt{72} - \dfrac{3}{8}\sqrt{8} + \dfrac{3}{10}\sqrt{50}$ $\frac{23}{4}\sqrt{2}$

30. $\dfrac{3}{4}\sqrt{24} - \dfrac{5}{6}\sqrt{54} - \dfrac{7}{10}\sqrt{150}$ $-\frac{9}{2}\sqrt{6}$

▶ **31.** $6\sqrt{48} - 2\sqrt{12} + 5\sqrt{27}$ $35\sqrt{3}$

32. $5\sqrt{50} + 8\sqrt{12} - \sqrt{32}$ $21\sqrt{2} + 16\sqrt{3}$

▶ **33.** $9 + 6\sqrt{2} + 2 - 18 - 6\sqrt{2}$ -7

▶ **34.** $4 + 4\sqrt{3} + 3 - 8 - 4\sqrt{3}$ -1

▶ **35.** $4 + 4\sqrt{5} + 5 - 8 - 4\sqrt{5} - 1$ 0

▶ **36.** $1 - 2\sqrt{2} + 2 - 2 + 2\sqrt{2} - 1$ 0

All variables in the following problems represent positive real numbers. Simplify each term, and combine, if possible.

37. $\sqrt{x^3} + x\sqrt{x}$ $2x\sqrt{x}$ **38.** $2\sqrt{x} - 2\sqrt{4x}$ $-2\sqrt{x}$

39. $5\sqrt{3a^2} - a\sqrt{3}$ $4a\sqrt{3}$ **40.** $6a\sqrt{a} + 7\sqrt{a^3}$ $13a\sqrt{a}$

41. $5\sqrt{8x^3} + x\sqrt{50x}$ $15x\sqrt{2x}$ **42.** $2\sqrt{27x^2} - x\sqrt{48}$ $2x\sqrt{3}$

43. $3\sqrt{75x^3y} - 2x\sqrt{3xy}$ $13x\sqrt{3xy}$

44. $9\sqrt{24x^3y^2} - 5x\sqrt{54xy^2}$ $3xy\sqrt{6x}$

45. $\sqrt{20ab^2} - b\sqrt{45a}$ $-b\sqrt{5a}$

46. $4\sqrt{a^3b^2} - 5a\sqrt{ab^2}$ $-ab\sqrt{a}$

47. $9\sqrt{18x^3} - 2x\sqrt{48x}$ $27x\sqrt{2x} - 8x\sqrt{3x}$

48. $8\sqrt{72x^2} - x\sqrt{8}$ $46x\sqrt{2}$

49. $7\sqrt{50x^2y} + 8x\sqrt{8y} - 7\sqrt{32x^2y}$ $23x\sqrt{2y}$

50. $6\sqrt{44x^3y^3} - 8x\sqrt{99xy^3} - 6y\sqrt{176x^3y}$ $-36xy\sqrt{11xy}$

Simplify each expression.

▶ **51.** $\dfrac{6 + 2\sqrt{2}}{2}$ $3 + \sqrt{2}$ ▶ **52.** $\dfrac{6 - 2\sqrt{3}}{2}$ $3 - \sqrt{3}$

▶ **53.** $\dfrac{9 - 3\sqrt{3}}{3}$ $3 - \sqrt{3}$ ▶ **54.** $\dfrac{-8 + 4\sqrt{2}}{2}$ $-4 + 2\sqrt{2}$

55. $\dfrac{8 - \sqrt{24}}{6}$ $\dfrac{4 - \sqrt{6}}{3}$ **56.** $\dfrac{8 + \sqrt{48}}{8}$ $\dfrac{2 + \sqrt{3}}{2}$

57. $\dfrac{6 + \sqrt{8}}{2}$ $3 + \sqrt{2}$ **58.** $\dfrac{4 - \sqrt{12}}{2}$ $2 - \sqrt{3}$

59. $\dfrac{-10 + \sqrt{50}}{10}$ $\dfrac{-2 + \sqrt{2}}{2}$ **60.** $\dfrac{-12 + \sqrt{20}}{6}$ $\dfrac{-6 + \sqrt{5}}{3}$

61. Use the approximation $\sqrt{3} \approx 1.732$ to simplify the following.

   a. $\dfrac{3 + \sqrt{12}}{2}$ 3.232

   b. $\dfrac{3 - \sqrt{12}}{2}$ -0.232

   c. $\dfrac{3 + \sqrt{12}}{2} + \dfrac{3 - \sqrt{12}}{2}$ 3

62. Use the approximation $\sqrt{5} \approx 2.236$ to simplify the following.

   a. $\dfrac{1 + \sqrt{20}}{2}$ 2.736

   b. $\dfrac{1 - \sqrt{20}}{2}$ -1.736

   c. $\dfrac{1 + \sqrt{20}}{2} + \dfrac{1 - \sqrt{20}}{2}$ 1

Use a calculator to help complete the following tables. If an answer needs rounding, round to the nearest thousandth.

63.

x	$\sqrt{x^2 + 9}$	$x + 3$
1	3.162	4
2	3.606	5
3	4.243	6
4	5	7
5	5.831	8
6	6.708	9

64.

x	$\sqrt{x^2 + 16}$	$x + 4$
1	4.123	5
2	4.472	6
3	5	7
4	5.657	8
5	6.403	9
6	7.211	10

65.

x	$\sqrt{x + 3}$	$\sqrt{x} + \sqrt{3}$
1	2	2.732
2	2.236	3.146
3	2.449	3.464
4	2.646	3.732
5	2.828	3.968
6	3	4.182

66.

x	$\sqrt{x + 4}$	$\sqrt{x} + 2$
1	2.236	3
2	2.449	3.414
3	2.646	3.732
4	2.828	4
5	3	4.236
6	3.162	4.449

67. **Comparing Expressions** The following statement is false. Correct the right side to make the statement true.
$$4\sqrt{3} + 5\sqrt{3} = 9\sqrt{6}$$ $9\sqrt{3}$

68. Comparing Expressions The following statement is false. Correct the right side to make the statement true.

$$7\sqrt{5} - 3\sqrt{5} = 4\sqrt{25} \quad 4\sqrt{5}$$

78. $\dfrac{a}{a-3} - \dfrac{a}{2} = \dfrac{3}{a-3}$

Possible solutions 2 and 3; only 2 checks

Maintaining Your Skills

Multiply.

69. $(3x + y)^2 \quad 9x^2 + 6xy + y^2$

70. $(2x - 3y)^2 \quad 4x^2 - 12xy + 9y^2$

71. $(3x - 4y)(3x + 4y) \quad 9x^2 - 16y^2$

72. $(7x + 2y)(7x - 2y) \quad 49x^2 - 4y^2$

Solve each equation.

73. $\dfrac{x}{3} - \dfrac{1}{2} = \dfrac{5}{2} \quad 9$

74. $\dfrac{3}{x} + \dfrac{1}{5} = \dfrac{4}{5} \quad 5$

75. $1 - \dfrac{5}{x} = \dfrac{-6}{x^2} \quad 2, 3$

76. $1 - \dfrac{1}{x} = \dfrac{6}{x^2} \quad 3, -2$

77. $\dfrac{a}{a-4} - \dfrac{a}{2} = \dfrac{4}{a-4}$

Possible solutions 2 and 4; only 2 checks

Getting Ready for the Next Section

Multiply. Assume any variables represent positive numbers.

79. $\sqrt{5}\sqrt{2} \quad \sqrt{10}$

80. $\sqrt{3}\sqrt{2} \quad \sqrt{6}$

81. $\sqrt{5}\sqrt{5} \quad 5$

82. $\sqrt{3}\sqrt{3} \quad 3$

83. $\sqrt{x}\sqrt{x} \quad x$

84. $\sqrt{y}\sqrt{y} \quad y$

85. $\sqrt{5}\sqrt{7} \quad \sqrt{35}$

86. $\sqrt{5}\sqrt{3} \quad \sqrt{15}$

Combine like terms.

87. $5 + 7\sqrt{5} + 2\sqrt{5} + 14 \quad 19 + 9\sqrt{5}$

88. $3 + 5\sqrt{3} + 2\sqrt{3} + 10 \quad 13 + 7\sqrt{3}$

89. $x - 7\sqrt{x} + 3\sqrt{x} - 21 \quad x - 4\sqrt{x} - 21$

90. $x - 6\sqrt{x} + 8\sqrt{x} - 48 \quad x + 2\sqrt{x} - 48$

8.5 Multiplication and Division of Radicals

OBJECTIVES

A Multiply radical expressions.

B Rationalize the denominator in a radical expression that contains two terms in the denominator.

In this section we will look at multiplication and division of expressions that contain radicals. As you will see, multiplication of expressions that contain radicals is very similar to multiplication of polynomials. The division problems in this section are just an extension of the work we did previously when we rationalized denominators.

 EXAMPLE 1 Multiply $(3\sqrt{5})(2\sqrt{7})$.

SOLUTION We can rearrange the order and grouping of the numbers in this product by applying the commutative and associative properties. Following that, we apply Property 1 for radicals and multiply:

$$(3\sqrt{5})(2\sqrt{7}) = (3 \cdot 2)(\sqrt{5}\ \sqrt{7}) \quad \textbf{Commutative and associative properties}$$

$$= (3 \cdot 2)(\sqrt{5 \cdot 7}) \quad \textbf{Property 1 for radicals}$$

$$= 6\sqrt{35} \quad \textbf{Multiplication}$$

In actual practice, it is not necessary to show either of the first two steps, although you may want to show them on the first few problems you work, just to be sure you understand them.

EXAMPLE 2   Multiply $\sqrt{5}(\sqrt{2} + \sqrt{5})$.

SOLUTION

$$\sqrt{5}(\sqrt{2} + \sqrt{5}) = \sqrt{5} \cdot \sqrt{2} + \sqrt{5} \cdot \sqrt{5} \qquad \textbf{Distributive property}$$

$$= \sqrt{10} + 5 \qquad \textbf{Multiplication}$$

 EXAMPLE 3   Multiply $3\sqrt{2}(2\sqrt{5} + 5\sqrt{3})$.

SOLUTION

$$3\sqrt{2}(2\sqrt{5} + 5\sqrt{3}) = 3\sqrt{2} \cdot 2\sqrt{5} + 3\sqrt{2} \cdot 5\sqrt{3} \qquad \textbf{Distributive property}$$

$$= 3 \cdot 2 \cdot \sqrt{2}\,\sqrt{5} + 3 \cdot 5\sqrt{2}\,\sqrt{3} \qquad \textbf{Commutative property}$$

$$= 6\sqrt{10} + 15\sqrt{6}$$

Each item in the last line is in simplified form, so the solution is complete.

 EXAMPLE 4   Multiply $(\sqrt{5} + 2)(\sqrt{5} + 7)$.

SOLUTION   We multiply using the FOIL method that we used to multiply binomials:

$$(\sqrt{5} + 2)(\sqrt{5} + 7) = \underset{\text{F}}{\sqrt{5} \cdot \sqrt{5}} + \underset{\text{O}}{7\sqrt{5}} + \underset{\text{I}}{2\sqrt{5}} + \underset{\text{L}}{14}$$

$$= 5 + 9\sqrt{5} + 14$$

$$= 19 + 9\sqrt{5}$$

We must be careful not to try to simplify further by adding 19 and 9. We can add only radical expressions that have a common radical part; 19 and $9\sqrt{5}$ are not similar.

 EXAMPLE 5   Multiply $(\sqrt{x} + 3)(\sqrt{x} - 7)$.

SOLUTION   Remember, we are assuming that any variables that appear under a radical represent positive numbers.

$$(\sqrt{x} + 3)(\sqrt{x} - 7) = \underset{\text{F}}{\sqrt{x}\,\sqrt{x}} - \underset{\text{O}}{7\sqrt{x}} + \underset{\text{I}}{3\sqrt{x}} - \underset{\text{L}}{21}$$

$$= x - 4\sqrt{x} - 21$$

EXAMPLE 6   Expand and simplify $(\sqrt{3} - 2)^2$.

SOLUTION   Multiplying $\sqrt{3} - 2$ times itself, we have

$$(\sqrt{3} - 2)^2 = (\sqrt{3} - 2)(\sqrt{3} - 2)$$

$$= \sqrt{3}\,\sqrt{3} - 2\sqrt{3} - 2\sqrt{3} + 4$$

$$= 3 - 4\sqrt{3} + 4$$

$$= 7 - 4\sqrt{3}$$

EXAMPLE 7 Multiply $(\sqrt{5} + \sqrt{2})(\sqrt{5} - \sqrt{2})$.

SOLUTION We can apply the formula $(x + y)(x - y) = x^2 - y^2$ to obtain

$$(\sqrt{5} + \sqrt{2})(\sqrt{5} - \sqrt{2}) = (\sqrt{5})^2 - (\sqrt{2})^2$$
$$= 5 - 2$$
$$= 3$$

We also could have multiplied the two expressions using the FOIL method. If we were to do so, the work would look like this:

$$(\sqrt{5} + \sqrt{2})(\sqrt{5} - \sqrt{2}) = \underset{\text{F}}{\sqrt{5}\,\sqrt{5}} - \underset{\text{O}}{\sqrt{2}\,\sqrt{5}} + \underset{\text{I}}{\sqrt{2}\,\sqrt{5}} - \underset{\text{L}}{\sqrt{2}\,\sqrt{2}}$$
$$= 5 - \sqrt{10} + \sqrt{10} - 2$$
$$= 5 - 2$$
$$= 3$$

In either case, the product is 3. Also, the expressions $\sqrt{5} + \sqrt{2}$ and $\sqrt{5} - \sqrt{2}$ are called *conjugates* of each other.

 EXAMPLE 8 Multiply $(\sqrt{a} + \sqrt{b})(\sqrt{a} - \sqrt{b})$.

SOLUTION We can apply the formula $(x + y)(x - y) = x^2 - y^2$ to obtain

$$(\sqrt{a} + \sqrt{b})(\sqrt{a} - \sqrt{b}) = (\sqrt{a})^2 - (\sqrt{b})^2$$
$$= a - b$$

EXAMPLE 9 Rationalize the denominator in the expression

$$\frac{\sqrt{3}}{\sqrt{3} - \sqrt{2}}$$

SOLUTION To remove the two radicals in the denominator, we must multiply both the numerator and denominator by $\sqrt{3} + \sqrt{2}$. That way, when we multiply $\sqrt{3} - \sqrt{2}$ and $\sqrt{3} + \sqrt{2}$, we will obtain the difference of two squares in the denominator:

$$\frac{\sqrt{3}}{\sqrt{3} - \sqrt{2}} = \frac{\sqrt{3}}{(\sqrt{3} - \sqrt{2})} \cdot \frac{(\sqrt{3} + \sqrt{2})}{(\sqrt{3} + \sqrt{2})}$$
$$= \frac{\sqrt{3}\,\sqrt{3} + \sqrt{3}\,\sqrt{2}}{(\sqrt{3})^2 - (\sqrt{2})^2}$$
$$= \frac{3 + \sqrt{6}}{3 - 2}$$
$$= \frac{3 + \sqrt{6}}{1}$$
$$= 3 + \sqrt{6}$$

EXAMPLE 10   Rationalize the denominator in the expression $\dfrac{2}{5 - \sqrt{3}}$.

SOLUTION   We use the same procedure as in Example 9. Multiply the numerator and denominator by the conjugate of the denominator, which is $5 + \sqrt{3}$:

$$\left(\frac{2}{5 - \sqrt{3}}\right)\left(\frac{\mathbf{5 + \sqrt{3}}}{\mathbf{5 + \sqrt{3}}}\right) = \frac{10 + 2\sqrt{3}}{5^2 - (\sqrt{3})^2}$$

$$= \frac{10 + 2\sqrt{3}}{25 - 3}$$

$$= \frac{10 + 2\sqrt{3}}{22}$$

The numerator and denominator of this last expression have a factor of 2 in common. We can reduce to lowest terms by dividing out the common factor 2. Continuing, we have

$$= \frac{\cancel{2}(5 + \sqrt{3})}{\cancel{2} \cdot 11}$$

$$= \frac{5 + \sqrt{3}}{11}$$

The final expression is in simplified form.

 EXAMPLE 11   Rationalize the denominator in the expression
$$\frac{\sqrt{2} + \sqrt{3}}{\sqrt{2} - \sqrt{3}}$$

SOLUTION   We remove the two radicals in the denominator by multiplying both the numerator and denominator by the conjugate of $\sqrt{2} - \sqrt{3}$, which is $\sqrt{2} + \sqrt{3}$:

$$\frac{\sqrt{2} + \sqrt{3}}{\sqrt{2} - \sqrt{3}} = \left(\frac{\sqrt{2} + \sqrt{3}}{\sqrt{2} - \sqrt{3}}\right)\frac{(\mathbf{\sqrt{2} + \sqrt{3}})}{(\mathbf{\sqrt{2} + \sqrt{3}})}$$

$$= \frac{\sqrt{2}\,\sqrt{2} + \sqrt{2}\,\sqrt{3} + \sqrt{3}\,\sqrt{2} + \sqrt{3}\,\sqrt{3}}{(\sqrt{2})^2 - (\sqrt{3})^2}$$

$$= \frac{2 + \sqrt{6} + \sqrt{6} + 3}{2 - 3}$$

$$= \frac{5 + 2\sqrt{6}}{-1}$$

$$= -(5 + 2\sqrt{6}) \text{ or } -5 - 2\sqrt{6}$$

LINKING OBJECTIVES AND EXAMPLES

Next to each **objective** we have listed the examples that are best described by that objective.

A	1–8
B	9–11

GETTING READY FOR CLASS

After reading through the preceding section, respond in your own words and in complete sentences.

1. Describe how you would use the commutative and associative properties to multiply $3\sqrt{5}$ and $2\sqrt{7}$.
2. Explain why $(\sqrt{5} + 7)^2$ is not the same as $5 + 49$.
3. Explain in words how you would rationalize the denominator in the expression $\dfrac{\sqrt{3}}{\sqrt{3} - \sqrt{2}}$.
4. What are conjugates?

Problem Set 8.5

Online support materials can be found at www.thomsonedu.com/login

Perform the following multiplications. All answers should be in simplified form for radical expressions.

1. $\sqrt{3}\,\sqrt{2}$ $\sqrt{6}$
2. $\sqrt{5}\,\sqrt{6}$ $\sqrt{30}$
3. $\sqrt{6}\,\sqrt{2}$ $2\sqrt{3}$
4. $\sqrt{6}\,\sqrt{3}$ $3\sqrt{2}$
▶ 5. $(2\sqrt{3})(5\sqrt{7})$ $10\sqrt{21}$
6. $(3\sqrt{2})(4\sqrt{5})$ $12\sqrt{10}$
7. $(4\sqrt{3})(2\sqrt{6})$ $24\sqrt{2}$
8. $(7\sqrt{6})(3\sqrt{2})$ $42\sqrt{3}$
▶ 9. $(2\sqrt{2})^2$ 8
▶ 10. $(-2\sqrt{3})^2$ 12
▶ 11. $(-2\sqrt{6})^2$ 24
▶ 12. $(5\sqrt{2})^2$ 50
13. $(-1 + 5\sqrt{2} + 1)^2$ 50
14. $(-8 + 2\sqrt{6} + 8)^2$ 24
15. $[2(-3 + \sqrt{2}) + 6]^2$ 8
16. $[2(3 - \sqrt{3}) - 6]^2$ 12
17. $\sqrt{2}(\sqrt{3} - 1)$ $\sqrt{6} - \sqrt{2}$
18. $\sqrt{3}(\sqrt{5} + 2)$ $\sqrt{15} + 2\sqrt{3}$
19. $\sqrt{2}(\sqrt{3} + \sqrt{2})$ $\sqrt{6} + 2$
20. $\sqrt{5}(\sqrt{7} - \sqrt{5})$ $\sqrt{35} - 5$
▶ 21. $\sqrt{3}(2\sqrt{2} + \sqrt{3})$ $2\sqrt{6} + 3$
22. $\sqrt{11}(3\sqrt{2} - \sqrt{11})$ $3\sqrt{22} - 11$
23. $2\sqrt{3}(\sqrt{2} + \sqrt{5})$ $2\sqrt{6} + 2\sqrt{15}$
24. $3\sqrt{2}(\sqrt{3} + \sqrt{2})$ $3\sqrt{6} + 6$
25. $(\sqrt{2} + 1)^2$ $3 + 2\sqrt{2}$
26. $(\sqrt{5} - 4)^2$ $21 - 8\sqrt{5}$
▶ 27. $(\sqrt{x} + 3)^2$ $x + 6\sqrt{x} + 9$
28. $(\sqrt{x} - 4)^2$ $x - 8\sqrt{x} + 16$
29. $\left(\sqrt{a} - \dfrac{1}{2}\right)^2$ $a - \sqrt{a} + \dfrac{1}{4}$
30. $\left(\sqrt{a} + \dfrac{1}{2}\right)^2$ $a + \sqrt{a} + \dfrac{1}{4}$
31. $(\sqrt{5} + 3)(\sqrt{5} + 2)$ $11 + 5\sqrt{5}$
32. $(\sqrt{7} + 4)(\sqrt{7} - 5)$ $-13 - \sqrt{7}$

33. $\left(\sqrt{3} + \dfrac{1}{2}\right)\left(\sqrt{2} + \dfrac{1}{3}\right)$ $\sqrt{6} + \dfrac{1}{3}\sqrt{3} + \dfrac{1}{2}\sqrt{2} + \dfrac{1}{6}$
34. $\left(\sqrt{5} - \dfrac{1}{4}\right)\left(\sqrt{3} + \dfrac{1}{5}\right)$ $\sqrt{15} + \dfrac{1}{5}\sqrt{5} - \dfrac{1}{4}\sqrt{3} - \dfrac{1}{20}$
35. $(\sqrt{x} + 6)(\sqrt{x} - 6)$ $x - 36$
36. $(\sqrt{x} + 7)(\sqrt{x} - 7)$ $x - 49$
37. $\left(\sqrt{a} + \dfrac{1}{3}\right)\left(\sqrt{a} + \dfrac{2}{3}\right)$ $a + \sqrt{a} + \dfrac{2}{9}$
38. $\left(\sqrt{a} + \dfrac{1}{4}\right)\left(\sqrt{a} + \dfrac{3}{4}\right)$ $a + \sqrt{a} + \dfrac{3}{16}$
▶ 39. $(\sqrt{5} - 2)(\sqrt{5} + 2)$ 1
40. $(\sqrt{6} - 3)(\sqrt{6} + 3)$ -3
41. $(2\sqrt{7} + 3)(3\sqrt{7} - 4)$ $30 + \sqrt{7}$
42. $(3\sqrt{5} + 1)(4\sqrt{5} + 3)$ $63 + 13\sqrt{5}$
▶ 43. $(3 + \sqrt{2})^2 - 6(3 + \sqrt{2})$ -7
▶ 44. $(3 - \sqrt{2})^2 - 6(3 - \sqrt{2})$ -7
▶ 45. $(2 + \sqrt{5})^2 - 4(2 + \sqrt{5})$ 1
▶ 46. $(2 + \sqrt{2})^2 - 4(2 + \sqrt{2})$ -2
47. $(7 - \sqrt{5})^2 - 14(7 - \sqrt{5}) + 44$ 0
48. $(5 - \sqrt{7})^2 - 10(5 - \sqrt{7}) - 24$ -42

Rationalize the denominator. All answers should be expressed in simplified form.

▶ 49. $\dfrac{\sqrt{3}}{\sqrt{5} - \sqrt{2}}$ $\dfrac{\sqrt{15} + \sqrt{6}}{3}$
50. $\dfrac{\sqrt{2}}{\sqrt{6} + \sqrt{3}}$ $\dfrac{2\sqrt{3} - \sqrt{6}}{3}$

= Videos available by instructor request

▶ = Online student support materials available at www.thomsonedu.com/login

51. $\dfrac{\sqrt{5}}{\sqrt{5}+\sqrt{2}}$ $\dfrac{5-\sqrt{10}}{3}$ **52.** $\dfrac{\sqrt{7}}{\sqrt{7}-\sqrt{2}}$ $\dfrac{7+\sqrt{14}}{5}$

53. $\dfrac{8}{3-\sqrt{5}}$ $6+2\sqrt{5}$ **54.** $\dfrac{10}{5+\sqrt{5}}$ $\dfrac{5-\sqrt{5}}{2}$

55. $\dfrac{\sqrt{3}+\sqrt{2}}{\sqrt{3}-\sqrt{2}}$ $5+2\sqrt{6}$ **56.** $\dfrac{\sqrt{5}-\sqrt{2}}{\sqrt{5}+\sqrt{2}}$ $\dfrac{7-2\sqrt{10}}{3}$

▶ **57.** Work each problem according to the instructions given.
- **a.** Subtract: $(\sqrt{7}+\sqrt{3})-(\sqrt{7}-\sqrt{3})$ $2\sqrt{3}$
- **b.** Multiply: $(\sqrt{7}+\sqrt{3})(\sqrt{7}-\sqrt{3})$ 4
- **c.** Square: $(\sqrt{7}-\sqrt{3})^2$ $10-2\sqrt{21}$
- **d.** Divide: $\dfrac{\sqrt{7}-\sqrt{3}}{\sqrt{7}+\sqrt{3}}$ $\dfrac{5-\sqrt{21}}{2}$

▶ **58.** Work each problem according to the instructions given.
- **a.** Subtract: $(\sqrt{11}-\sqrt{6})-(\sqrt{11}+\sqrt{6})$ $-2\sqrt{6}$
- **b.** Multiply: $(\sqrt{11}-\sqrt{6})(\sqrt{11}+\sqrt{6})$ 5
- **c.** Square: $(\sqrt{11}-\sqrt{6})^2$ $17-2\sqrt{66}$
- **d.** Divide: $\dfrac{\sqrt{11}-\sqrt{6}}{\sqrt{11}+\sqrt{6}}$ $\dfrac{17-2\sqrt{66}}{5}$

▶ **59.** Work each problem according to the instructions given.
- **a.** Add: $(\sqrt{x}+2)+(\sqrt{x}-2)$ $2\sqrt{x}$
- **b.** Multiply: $(\sqrt{x}+2)(\sqrt{x}-2)$ $x-4$
- **c.** Square: $(\sqrt{x}+2)^2$ $x+4\sqrt{x}+4$
- **d.** Divide: $\dfrac{\sqrt{x}+2}{\sqrt{x}-2}$ $\dfrac{x+4\sqrt{x}+4}{x-4}$

▶ **60.** Work each problem according to the instructions given.
- **a.** Add: $(\sqrt{x}-3)+(\sqrt{x}+3)$ $2\sqrt{x}$
- **b.** Multiply: $(\sqrt{x}-3)(\sqrt{x}+3)$ $x-9$
- **c.** Square: $(\sqrt{x}-3)^2$ $x-6\sqrt{x}+9$
- **d.** Divide: $\dfrac{\sqrt{x}+3}{\sqrt{x}-3}$ $\dfrac{x+6\sqrt{x}+9}{x-9}$

▶ **61.** Work each problem according to the instructions given.
- **a.** Add: $(5+\sqrt{2})+(5-\sqrt{2})$ 10
- **b.** Multiply: $(5+\sqrt{2})(5-\sqrt{2})$ 23
- **c.** Square: $(5+\sqrt{2})^2$ $27+10\sqrt{2}$
- **d.** Divide: $\dfrac{5+\sqrt{2}}{5-\sqrt{2}}$ $\dfrac{27+10\sqrt{2}}{23}$

▶ **62.** Work each problem according to the instructions given.
- **a.** Add: $(2+\sqrt{3})+(2-\sqrt{3})$ 4
- **b.** Multiply: $(2+\sqrt{3})(2-\sqrt{3})$ 1
- **c.** Square: $(2+\sqrt{3})^2$ $7+4\sqrt{3}$

- **d.** Divide: $\dfrac{2+\sqrt{3}}{2-\sqrt{3}}$ $7+4\sqrt{3}$

▶ **63.** Work each problem according to the instructions given.
- **a.** Add: $\sqrt{2}+(\sqrt{6}+\sqrt{2})$ $\sqrt{6}+2\sqrt{2}$
- **b.** Multiply: $\sqrt{2}(\sqrt{6}+\sqrt{2})$ $2+2\sqrt{3}$
- **c.** Divide: $\dfrac{\sqrt{6}+\sqrt{2}}{\sqrt{2}}$ $1+\sqrt{3}$
- **d.** Divide: $\dfrac{\sqrt{2}}{\sqrt{6}+\sqrt{2}}$ $\dfrac{-1+\sqrt{3}}{2}$

▶ **64.** Work each problem according to the instructions given.
- **a.** Add: $\sqrt{5}+(\sqrt{5}+\sqrt{10})$ $2\sqrt{5}+\sqrt{10}$
- **b.** Multiply: $\sqrt{5}(\sqrt{5}+\sqrt{10})$ $5+5\sqrt{2}$
- **c.** Divide: $\dfrac{\sqrt{5}+\sqrt{10}}{\sqrt{5}}$ $1+\sqrt{2}$
- **d.** Divide: $\dfrac{\sqrt{5}}{\sqrt{5}+\sqrt{10}}$ $-1+\sqrt{2}$

65. Work each problem according to the instructions given.
- **a.** Add: $\left(\dfrac{1+\sqrt{5}}{2}\right)+\left(\dfrac{1-\sqrt{5}}{2}\right)$ 1
- **b.** Multiply: $\left(\dfrac{1+\sqrt{5}}{2}\right)\left(\dfrac{1-\sqrt{5}}{2}\right)$ -1

66. Work each problem according to the instructions given.
- **a.** Add: $\left(\dfrac{1+\sqrt{3}}{2}\right)+\left(\dfrac{1-\sqrt{3}}{2}\right)$ 1
- **b.** Multiply: $\left(\dfrac{1+\sqrt{3}}{2}\right)\left(\dfrac{1-\sqrt{3}}{2}\right)$ $-\dfrac{1}{2}$

Applying the Concepts

67. Comparing Expressions The following statement is false. Correct the right side to make the statement true.

$$2(3\sqrt{5}) = 6\sqrt{15} \quad 6\sqrt{5}$$

68. Comparing Expressions The following statement is false. Correct the right side to make the statement true.

$$5(2\sqrt{6}) = 10\sqrt{30} \quad 10\sqrt{6}$$

69. Comparing Expressions The following statement is false. Correct the right side to make the statement true.

$$(\sqrt{3}+7)^2 = 3+49 \quad 52+14\sqrt{3}$$

▶ **Chalkboard Problems**
Problems 57–64 require students to pay attention to instructions and give a nice lead-in to rationalizing the denominator.

70. Comparing Expressions The following statement is false. Correct the right side to make the statement true.

$$(\sqrt{5} + \sqrt{2})^2 = 5 + 2 \quad 7 + 2\sqrt{10}$$

81. $(-9)^2$ 81
82. $(-4)^2$ 16
83. $(\sqrt{x+1})^2$ $x + 1$
84. $(\sqrt{x+2})^2$ $x + 2$
85. $(\sqrt{2x-3})^2$ $2x - 3$
86. $(\sqrt{3x-1})^2$ $3x - 1$
87. $(x+3)^2$ $x^2 + 6x + 9$
88. $(x+2)^2$ $x^2 + 4x + 4$

Maintaining Your Skills

Solve.

Solve each equation.

89. $3a - 2 = 4$ 2
90. $2a - 3 = 25$ 14

71. $x^2 + 5x - 6 = 0$ $-6, 1$
72. $x^2 + 5x + 6 = 0$ $-2, -3$

91. $x + 15 = x^2 + 6x + 9$ $-6, 1$

73. $x^2 - 3x = 0$ $0, 3$
74. $x^2 + 5x = 0$ $0, -5$

92. $x + 8 = x^2 + 4x + 4$ $-4, 1$

Solve each proportion.

Determine whether the given numbers are solutions to the equation.

75. $\dfrac{x}{3} = \dfrac{27}{x}$ $-9, 9$
76. $\dfrac{x}{2} = \dfrac{8}{x}$ $-4, 4$

93. $\sqrt{x+15} = x + 3$
  a. $x = -6$ no
  b. $x = 1$ yes

77. $\dfrac{x}{5} = \dfrac{3}{x+2}$ $3, -5$
78. $\dfrac{x}{2} = \dfrac{4}{x+2}$ $2, -4$

94. $\sqrt{x+8} = x + 2$
  a. $x = -4$ no
  b. $x = 1$ yes

Getting Ready for the Next Section

95. Evaluate $y = 3\sqrt{x}$ for $x = -4, -1, 0, 1, 4, 9, 16$.
  Undefined, Undefined, 0, 3, 6, 9, 12

Simplify.

96. Evaluate $y = \sqrt[3]{x}$ for $x = -27, -8, -1, 0, 1, 8, 27$.
  $-3, -2, -1, 0, 1, 2, 3$

79. 7^2 49
80. 5^2 25

8.6 Equations Involving Radicals

OBJECTIVES

A Solve equations that contain radicals.

B Graph equations that contain radicals.

To solve equations that contain one or more radical expressions, we need an additional property. From our work with exponents we know that if two quantities are equal, then so are the squares of those quantities; that is, for real numbers a and b

$$\text{if} \quad a = b$$
$$\text{then} \quad a^2 = b^2$$

The only problem with squaring both sides of an equation is that occasionally we will change a false statement into a true statement. Let's take the false statement $3 = -3$ as an example.

$$3 = -3 \qquad \textbf{A false statement}$$
$$(3)^2 = (-3)^2 \qquad \textbf{Square both sides}$$
$$9 = 9 \qquad \textbf{A true statement}$$

We can avoid this problem by always checking our solutions if, at any time during the process of solving an equation, we have squared both sides of the equation. Here is how the property is stated.

> **Squaring Property of Equality** We can square both sides of an equation any time it is convenient to do so, as long as we check all solutions in the original equation.

 EXAMPLE 1   Solve for x: $\sqrt{x + 1} = 7$.

SOLUTION   To solve this equation by our usual methods, we must first eliminate the radical sign. We can accomplish this by squaring both sides of the equation:

$$\sqrt{x + 1} = 7$$

$$(\sqrt{x + 1})^2 = 7^2 \qquad \textbf{Square both sides}$$

$$x + 1 = 49$$

$$x = 48$$

To check our solution, we substitute $x = 48$ into the original equation:

$$\sqrt{48 + 1} \stackrel{?}{=} 7$$

$$\sqrt{49} = 7$$

$$7 = 7 \qquad \textbf{A true statement}$$

The solution checks.

 EXAMPLE 2   Solve for x: $\sqrt{2x - 3} = -9$.

SOLUTION   We square both sides and proceed as in Example 1:

$$\sqrt{2x - 3} = -9$$

$$(\sqrt{2x - 3})^2 = (-9)^2 \qquad \textbf{Square both sides}$$

$$2x - 3 = 81$$

$$2x = 84$$

$$x = 42$$

Checking our solution in the original equation, we have

$$\sqrt{2(42) - 3} \stackrel{?}{=} -9$$

$$\sqrt{84 - 3} = -9$$

$$\sqrt{81} = -9$$

$$9 = -9 \qquad \textbf{A false statement}$$

Our solution does not check because we end up with a false statement.

Note

As you can see, when we check $x = 42$ in the original equation, we find that it is not a solution to the equation. Actually, it was apparent from the beginning that the equation had no solution; that is, no matter what x is, the equation

$$\sqrt{2x - 3} = -9$$

never can be true because the left side is a positive number (or zero) for any value of x, and the right side is always negative.

Squaring both sides of the equation has produced what is called an *extraneous solution*. This happens occasionally when we use the squaring property of equality. We can always eliminate extraneous solutions by checking each solution in the original equation.

EXAMPLE 3   Solve for a: $\sqrt{3a - 2} + 3 = 5$.

SOLUTION   Before we can square both sides to eliminate the radical, we must isolate the radical on the left side of the equation. To do so, we add -3 to both sides:

$$\sqrt{3a - 2} + 3 = 5$$

$$\sqrt{3a - 2} = 2 \qquad \textbf{Add } -3 \textbf{ to both sides}$$

$$(\sqrt{3a - 2})^2 = 2^2 \qquad \textbf{Square both sides}$$

$$3a - 2 = 4$$

$$3a = 6$$

$$a = 2$$

Checking $a = 2$ in the original equation, we have

$$\sqrt{3 \cdot 2 - 2} + 3 \overset{?}{=} 5$$

$$\sqrt{4} + 3 = 5$$

$$5 = 5 \qquad \textbf{A true statement}$$

EXAMPLE 4   Solve for x: $\sqrt{x + 15} = x + 3$.

SOLUTION   We begin by squaring both sides:

$$(\sqrt{x + 15})^2 = (x + 3)^2 \qquad \textbf{Square both sides}$$

$$x + 15 = x^2 + 6x + 9$$

We have a quadratic equation. We put it into standard form by adding $-x$ and -15 to both sides. Then we factor and solve as usual:

$$0 = x^2 + 5x - 6 \qquad\qquad \textbf{Standard form}$$

$$0 = (x + 6)(x - 1) \qquad\qquad \textbf{Factor}$$

$$x + 6 = 0 \qquad \text{or} \qquad x - 1 = 0 \qquad \textbf{Set factors equal to 0}$$

$$x = -6 \qquad \text{or} \qquad x = 1$$

We check each solution in the original equation:

Check -6	Check 1
$\sqrt{-6 + 15} \overset{?}{=} -6 + 3$	$\sqrt{1 + 15} \overset{?}{=} 1 + 3$
$\sqrt{9} = -3$	$\sqrt{16} = 4$
$3 = -3$	$4 = 4$
A false statement	A true statement

Since $x = -6$ does not check in the original equation, it cannot be a solution. The only solution is $x = 1$.

EXAMPLE 5 Graph $y = \sqrt{x}$ and $y = \sqrt[3]{x}$.

SOLUTION The graphs are shown in Figures 1 and 2. Notice that the graph of $y = \sqrt{x}$ appears in the first quadrant only because in the equation $y = \sqrt{x}$, x and y cannot be negative.

The graph of $y = \sqrt[3]{x}$ appears in quadrants I and III, since the cube root of a positive number is also a positive number and the cube root of a negative number is a negative number; that is, when x is positive, y will be positive, and when x is negative, y will be negative.

The graphs of both equations will contain the origin since $y = 0$ when $x = 0$ in both equations.

x	y
−4	Undefined
−1	Undefined
0	0
1	1
4	2
9	3
16	4

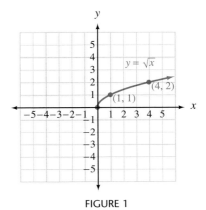

FIGURE 1

x	y
−27	−3
−8	−2
−1	−1
0	0
1	1
8	2
27	3

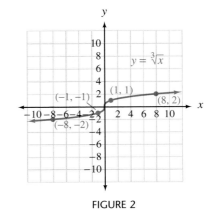

FIGURE 2

LINKING OBJECTIVES AND EXAMPLES

Next to each **objective** we have listed the examples that are best described by that objective.

A   1–4

B   5

GETTING READY FOR CLASS

After reading through the preceding section, respond in your own words and in complete sentences.

1. What is the squaring property of equality?
2. Under what conditions do we obtain extraneous solutions to equations that contain radical expressions?
3. Why do we check our solutions to equations when we have used the squaring property of equality?
4. Why are there no solutions to the equation $\sqrt{2x - 9} = -9$?

Solve each equation by applying the squaring property of equality. Be sure to check all solutions in the original equation.

▶ 1. $\sqrt{x + 1} = 2$ 3

2. $\sqrt{x - 3} = 4$ 19

3. $\sqrt{x + 5} = 7$ 44

4. $\sqrt{x + 8} = 5$ 17

▶ 5. $\sqrt{x - 9} = -6$ ∅

6. $\sqrt{x + 10} = -3$ ∅

7. $\sqrt{x - 5} = -4$ ∅

8. $\sqrt{x + 7} = -5$ ∅

9. $\sqrt{x - 8} = 0$ 8

10. $\sqrt{x - 9} = 0$ 9

11. $\sqrt{2x + 1} = 3$ 4

12. $\sqrt{2x - 5} = 7$ 27

13. $\sqrt{2x - 3} = -5$ ∅

14. $\sqrt{3x - 8} = -4$ ∅

15. $\sqrt{3x + 6} = 2$ $-\frac{2}{3}$

16. $\sqrt{5x - 1} = 5$ $\frac{26}{5}$

17. $2\sqrt{x} = 10$ 25

18. $3\sqrt{x} = 9$ 9

19. $3\sqrt{a} = 6$ 4

20. $2\sqrt{a} = 12$ 36

▶ 21. $\sqrt{3x + 4} - 3 = 2$ 7

22. $\sqrt{2x - 1} + 2 = 5$ 5

23. $\sqrt{5y - 4} - 2 = 4$ 8

24. $\sqrt{3y + 1} + 7 = 2$ ∅

25. $\sqrt{2x + 1} + 5 = 2$ ∅

26. $\sqrt{6x - 8} - 1 = 3$ 4

▶ 27. $\sqrt{x + 3} = x - 3$ Possible 1, 6; only 6

28. $\sqrt{x - 3} = x - 3$ 3, 4

29. $\sqrt{a + 2} = a + 2$ −1, −2

30. $\sqrt{a + 10} = a - 2$ Possible 6, −1; only 6

31. $\sqrt{2x + 9} = x + 5$ −4

32. $\sqrt{x + 6} = x + 4$ Possible −5, −2; only −2

33. Solve each equation.
   a. $\sqrt{y} - 4 = 6$ 100
   b. $\sqrt{y - 4} = 6$ 40
   c. $\sqrt{y} - 4 = -6$ ∅
   d. $\sqrt{y - 4} = y - 6$ Possible 5, 8; only 8

34. Solve each equation.
   a. $\sqrt{2y} + 15 = 7$ ∅
   b. $\sqrt{2y + 15} = 7$ 17
   c. $\sqrt{2y + 15} = y$ Possible −3, 5; only 5
   d. $\sqrt{2y + 15} = y + 6$ Possible −3, −7; only −3

35. Solve each equation.
   a. $x - 3 = 0$ 3
   b. $\sqrt{x} - 3 = 0$ 9
   c. $\sqrt{x - 3} = 0$ 3
   d. $\sqrt{x} + 3 = 0$ ∅
   e. $\sqrt{x} + 3 = 5$ 4
   f. $\sqrt{x} + 3 = -5$ ∅
   g. $x - 3 = \sqrt{5 - x}$ Possible 1, 4; only 4

36. Solve each equation.
   a. $x - 2 = 0$ 2
   b. $\sqrt{x} - 2 = 0$ 4
   c. $\sqrt{x} + 2 = 0$ ∅
   d. $\sqrt{x + 2} = 0$ −2
   e. $\sqrt{x} + 2 = 7$ 25
   f. $x - 2 = \sqrt{2x - 1}$
       Possible 1, 5; only 5

37. **Pendulum Problem** The time (in seconds) it takes for the pendulum on a clock to swing through one complete cycle is given by the formula

$$T = \frac{11}{7}\sqrt{\frac{L}{2}}$$

where L is the length of the pendulum, in feet. The following table was constructed using this formula. Draw a line graph of the information in the table.

Length L (feet)	Time T (seconds)
1	1.11
2	1.57
3	1.92
4	2.22
5	2.48
6	2.72

1 sec

38. **Lighthouse Problem** The higher you are above the ground, the farther you can see. If your view is unobstructed, then the distance in miles that you can see from h feet above the ground is given by the formula

$$d = \sqrt{\frac{3h}{2}}$$

▶ **Chalkboard Problem**
Problem 33 shows almost every situation students will face with the equations in this section.

☐ = Videos available by instructor request
▶ = Online student support materials available at www.thomsonedu.com/login

The following table was constructed using this formula. Draw a line graph of the information in the table.

Height h (feet)	Distance d (miles)
10	3.9
50	8.7
90	11.6
130	14.0
170	16.0
190	16.9

From each equation, complete the given table and sketch each graph.

39. $y = \sqrt{x}$

x	y
0	0
1	1
2	1.4
3	1.7
4	2

40. $y = \sqrt[3]{x}$

x	y
-8	-2
-4	-1.59
-1	-1
0	0
4	1.59
8	2

41. $y = 2\sqrt{x}$

x	y
0	0
1	2
4	4
9	6

42. $y = 3\sqrt[3]{x}$

x	y
-8	-6
-1	-3
0	0
1	3
8	6

43. $y = \sqrt{x} + 2$

x	y
0	2
1	3
2	3.4
4	4
9	5

44. $y = \sqrt[3]{x} + 3$

x	y
-8	1
-1	2
0	3
1	4
8	5

45. Number Problem The sum of a number and 2 is equal to the positive square root of 8 times the number. Find the number. $x + 2 = \sqrt{8x}$; $x = 2$

46. Number Problem The sum of twice a number and 1 is equal to 3 times the positive square root of the number. Find the number. $2x + 1 = 3\sqrt{x}$; $\frac{1}{4}$ and 1

47. Number Problem The difference of a number and 3 is equal to twice the positive square root of the number. Find the number.
$x - 3 = 2\sqrt{x}$; possible solutions 1 and 9; only 9 checks

48. Number Problem The difference of a number and 2 is equal to the positive square root of the number. Find the number. $x - 2 = \sqrt{x}$; Possible solutions 4 and 1; only 4 checks

49. Pendulum Problem The number of seconds T it takes the pendulum of a grandfather clock to swing through one complete cycle is given by the formula

$$T = \frac{11}{7} \sqrt{\frac{L}{2}}$$

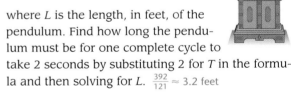

where L is the length, in feet, of the pendulum. Find how long the pendulum must be for one complete cycle to take 2 seconds by substituting 2 for T in the formula and then solving for L. $\frac{392}{121} \approx 3.2$ feet

50. Pendulum Problem How long must the pendulum on a grandfather clock be if one complete cycle is to take 1 second? $\frac{98}{121} \approx 0.8$ foot

Maintaining Your Skills

51. Reduce to lowest terms $\frac{x^2 - x - 6}{x^2 - 9} \cdot \frac{x + 2}{x + 3}$

52. Divide using long division $\frac{x^2 - 2x + 6}{x - 4}$. $x + 2 + \frac{14}{x - 4}$

Perform the indicated operations.

53. $\frac{x^2 - 25}{x + 4} \cdot \frac{2x + 8}{x^2 - 9x + 20}$ $\frac{2(x + 5)}{x - 4}$

54. $\frac{3x + 6}{x^2 + 4x + 3} \div \frac{x^2 + x - 2}{x^2 + 2x - 3}$ $\frac{3}{x + 1}$

55. $\frac{x}{x^2 - 16} + \frac{4}{x^2 - 16}$ $\frac{1}{x - 4}$

56. $\frac{2}{x^2 - 1} - \frac{5}{x^2 + 3x - 4}$ $\frac{-3}{(x + 1)(x + 4)}$

57. $\dfrac{1 - \dfrac{25}{x^2}}{1 - \dfrac{8}{x} + \dfrac{15}{x^2}}$ $\dfrac{x+5}{x-3}$

Solve each equation.

58. $\dfrac{x}{2} - \dfrac{5}{x} = -\dfrac{3}{2}$ $-5, 2$

59. $\dfrac{x}{x^2 - 9} - \dfrac{3}{x - 3} = \dfrac{1}{x + 3}$ -2

60. Speed of a Boat A boat travels 30 miles up a river in the same amount of time it takes to travel 50 miles down the same river. If the current is 5 miles per hour, what is the speed of the boat in still water? 20 miles per hour

61. Filling a Pool A pool can be filled by an inlet pipe in 8 hours. The drain will empty the pool in 12 hours. How long will it take to fill the pool if both the inlet pipe and the drain are open? 24 hours

62. Mixture Problem If 30 liters of a certain solution contains 2 liters of alcohol, how much alcohol is in 45 liters of the same solution? 3 liters

63. y varies directly with x. If $y = 8$ when x is 12, find y when x is 36. 24

Chapter 8 SUMMARY

Roots [8.1]

EXAMPLES

1. The two square roots of 9 are 3 and -3:

$$\sqrt{9} = 3 \quad \text{and} \quad -\sqrt{9} = -3$$

Every positive real number x has two square roots, one positive and one negative. The positive square root is written $\sqrt{x}$. The negative square root of x is written $-\sqrt{x}$. In both cases the square root of x is a number we square to get x. The cube root of x is written $\sqrt[3]{x}$ and is the number we cube to get x.

Notation [8.1]

2. Index   Radical sign

$$\sqrt[3]{24} \longleftarrow \text{Radicand}$$

      Radical

In the expression $\sqrt[3]{8}$, 8 is called the *radicand*, 3 is the *index*, $\sqrt{}$ is called the *radical sign*, and the whole expression $\sqrt[3]{8}$ is called the *radical*.

Properties of Radicals [8.2]

3. a. $\sqrt{3} \cdot \sqrt{2} = \sqrt{3 \cdot 2} = \sqrt{6}$

b. $\dfrac{\sqrt{12}}{\sqrt{3}} = \sqrt{\dfrac{12}{3}} = \sqrt{4} = 2$

c. $\sqrt{5} \cdot \sqrt{5} = (\sqrt{5})^2 = 5$

If a and b represent nonnegative real numbers, then

1. $\sqrt{a}\,\sqrt{b} = \sqrt{ab}$ The product of the square roots is the square root of the product

2. $\dfrac{\sqrt{a}}{\sqrt{b}} = \sqrt{\dfrac{a}{b}} \quad (b \neq 0)$ The quotient of the square roots is the square root of the quotient

3. $\sqrt{a} \cdot \sqrt{a} = (\sqrt{a})^2 = a$ This property shows that squaring and square roots are inverse operations

Simplified Form for Radicals [8.3]

4. Simplify $\sqrt{20}$ and $\sqrt{\dfrac{2}{3}}$.

$$\sqrt{20} = \sqrt{4 \cdot 5} = \sqrt{4}\,\sqrt{5} = 2\sqrt{5}$$

$$\sqrt{\dfrac{2}{3}} = \dfrac{\sqrt{2}}{\sqrt{3}} = \dfrac{\sqrt{2}}{\sqrt{3}} \cdot \dfrac{\sqrt{3}}{\sqrt{3}} = \dfrac{\sqrt{6}}{3}$$

A radical expression is in simplified form if:

1. There are no perfect squares that are factors of the quantity under the square root sign, no perfect cubes that are factors of the quantity under the cube root sign, and so on. We want as little as possible under the radical sign.

2. There are no fractions under the radical sign.

3. There are no radicals in the denominator.

Addition and Subtraction of Radical Expressions [8.4]

5. a. $5\sqrt{7} + 3\sqrt{7} = 8\sqrt{7}$

b. $2\sqrt{18} - 3\sqrt{50}$
$= 2 \cdot 3\sqrt{2} - 3 \cdot 5\sqrt{2}$
$= 6\sqrt{2} - 15\sqrt{2}$
$= -9\sqrt{2}$

We add and subtract radical expressions by using the distributive property to combine terms that have the same radical parts. If the radicals are not in simplified form, we begin by writing them in simplified form and then combining similar terms, if possible.

Multiplication of Radical Expressions [8.5]

6. a. $\sqrt{3}(\sqrt{5} - \sqrt{3}) = \sqrt{15} - 3$
  b. $(\sqrt{7} + 3)(\sqrt{7} - 5)$
  $= 7 - 5\sqrt{7} + 3\sqrt{7} - 15$
  $= -8 - 2\sqrt{7}$

We multiply radical expressions by applying the distributive property or the FOIL method.

Division of Radical Expressions [8.5]

7. $\dfrac{7}{\sqrt{5} - \sqrt{3}}$

$= \dfrac{7}{\sqrt{5} - \sqrt{3}} \cdot \dfrac{\sqrt{5} + \sqrt{3}}{\sqrt{5} + \sqrt{3}}$

$= \dfrac{7\sqrt{5} + 7\sqrt{3}}{2}$

To divide by an expression like $\sqrt{5} - \sqrt{3}$, we multiply the numerator and denominator by its conjugate, $\sqrt{5} + \sqrt{3}$. This process also is called rationalizing the denominator.

Squaring Property of Equality [8.6]

8. Solve $\sqrt{x - 3} = 2$
  $(\sqrt{x - 3})^2 = 2^2$
  $x - 3 = 4$
  $x = 7$
The solution checks in the original equation.

We are free to square both sides of an equation whenever it is convenient, as long as we check all solutions in the original equation. We must check solutions because squaring both sides of an equation occasionally produces extraneous solutions.

! COMMON MISTAKES

1. A very common mistake with radicals is to think of $\sqrt{25}$ as representing both the positive and negative square roots of 25. The notation $\sqrt{25}$ stands for the *positive* square root of 25. If we want the negative square root of 25, we write $-\sqrt{25}$.

2. The most common mistake when working with radicals is to try to apply a property similar to Property 1 for radicals involving addition instead of multiplication. Here is an example:

$$\sqrt{16 + 9} = \sqrt{16} + \sqrt{9} \qquad \textbf{Mistake}$$

Although this example looks like it may be true, it isn't. If we carry it out further, the mistake becomes obvious:

$$\sqrt{16 + 9} \overset{?}{=} \sqrt{16} + \sqrt{9}$$

$$\sqrt{25} = 4 + 3$$

$$5 = 7 \qquad \textbf{False}$$

3. It is a mistake to try to simplify expressions like $2 + 3\sqrt{7}$. The 2 and 3 cannot be combined because the terms they appear in are not similar. Therefore, $2 + 3\sqrt{7} \neq 5\sqrt{7}$. The expression $2 + 3\sqrt{7}$ cannot be simplified further.

The problems below form a comprehensive review of the material in this chapter. They can be used to study for exams. If you would like to take a practice test on this chapter, you can use the odd-numbered problems. Give yourself an hour and work as many of the odd-numbered problems as possible. When you are finished, or when an hour has passed, check your answers with the answers in the back of the book. You can use the even-numbered problems for a second practice test.

Numbers in brackets refer to sections of the text in which similar problems can be found.

Find the following roots. Assume all variables are positive. [8.1]

1. $\sqrt{25}$ **2.** $\sqrt{169}$

3. $\sqrt[3]{-1}$ **4.** $\sqrt[4]{625}$

5. $\sqrt{100x^2y^4}$ **6.** $\sqrt[3]{8a^3}$

Simplify. Assume all variables represent positive numbers. [8.2]

7. $\sqrt{24}$ **8.** $\sqrt{60x^2}$

9. $\sqrt{90x^3y^4}$ **10.** $-\sqrt{32}$

11. $3\sqrt{20x^3y}$

Simplify. Assume all variables represent positive numbers. [8.2]

12. $\sqrt{\dfrac{3}{49}}$ **13.** $\sqrt{\dfrac{8}{81}}$

14. $\sqrt{\dfrac{49}{64}}$ **15.** $\sqrt{\dfrac{49a^2b^2}{16}}$

Simplify. Assume all variables represent positive numbers. [8.2]

16. $\sqrt{\dfrac{80}{49}}$ **17.** $\sqrt{\dfrac{40a^2}{121}}$

18. $\dfrac{5\sqrt{84}}{7}$ **19.** $\dfrac{3\sqrt{120a^2b^2}}{\sqrt{25}}$

20. $\dfrac{-5\sqrt{20x^3y^2}}{\sqrt{144}}$

Write in simplest form. Assume all variables represent positive numbers. [8.3]

21. $\dfrac{2}{\sqrt{7}}$ **22.** $\sqrt{\dfrac{32}{5}}$

23. $\sqrt{\dfrac{5}{48}}$ **24.** $\dfrac{-3\sqrt{60}}{\sqrt{5}}$

25. $\sqrt{\dfrac{32ab^2}{3}}$ **26.** $\sqrt[3]{\dfrac{3}{4}}$

Write in simplest form. Assume all variables represent positive numbers. [8.5]

27. $\dfrac{3}{\sqrt{3}-4}$ **28.** $\dfrac{2}{3+\sqrt{7}}$

29. $\dfrac{3}{\sqrt{5}-\sqrt{2}}$ **30.** $\dfrac{\sqrt{5}}{\sqrt{3}-\sqrt{5}}$

31. $\dfrac{\sqrt{5}-\sqrt{2}}{\sqrt{5}+\sqrt{2}}$ **32.** $\dfrac{\sqrt{x}+3}{\sqrt{x}-3}$

Combine the following expressions. [8.4]

33. $3\sqrt{5}-7\sqrt{5}$ **34.** $3\sqrt{27}-5\sqrt{48}$

35. $-2\sqrt{45}-5\sqrt{80}+2\sqrt{20}$

36. $3\sqrt{50x^2}-x\sqrt{200}$ **37.** $\sqrt{40a^3b^2}-a\sqrt{90ab^2}$

Multiply. Write all answers in simplest form. [8.5]

38. $\sqrt{3}(\sqrt{3}+3)$ **39.** $4\sqrt{2}(\sqrt{3}+\sqrt{5})$

40. $(\sqrt{x}+7)(\sqrt{x}-7)$ **41.** $(2\sqrt{5}-4)(\sqrt{5}+3)$

42. $(\sqrt{x}+5)^2$

Solve each equation. [8.6]

43. $\sqrt{x-3}=3$ **44.** $\sqrt{3x-5}=4$

45. $5\sqrt{a}=20$ **46.** $\sqrt{3x-7}+6=2$

47. $\sqrt{2x+1}+10=8$ **48.** $\sqrt{7x+1}=x+1$

Find x in each of the following right triangles.

49.

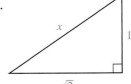

50.

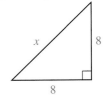

GROUP PROJECT   Unwinding the Spiral of Roots

Number of People 2–3

Time Needed 8–12 minutes

Equipment Pencil, ruler, graph paper, scissors, and tape

Background In this chapter, we used the Spiral of Roots to visualize square roots of positive integers. If we "unwind" the spiral of roots, we can produce the graph of a simple equation on a rectangular coordinate system.

Procedure

1. Carefully cut out each triangle from the Spiral of Roots to the right.

2. Line up the triangles horizontally on the coordinate system shown here so that the side of length 1 is on the x-axis and the hypotenuse is on the left. Note that the first triangle is shown in place, and the outline of the second triangle is next to it. The 1-unit side of each triangle should fit in each of the 1-unit spaces on the x-axis.

3. On the coordinate system, plot a point at the tip of each triangle. Then, connect these points with a smooth line.

4. What is the equation of the line you have just drawn?

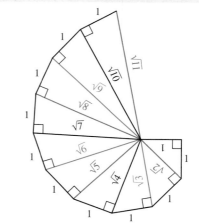

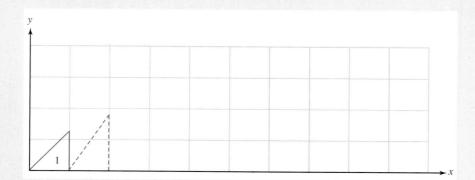

Connections

Although it may not look like it, the three items shown here are related very closely to one another. Your job is to find the connection.

A Continued Fraction

$$1 + \cfrac{1}{1 + \cfrac{1}{1 + \cfrac{1}{1 + \cdots}}}$$

The Fibonacci Sequence

$$1, 1, 2, 3, 5, \ldots$$

The Golden Rectangle

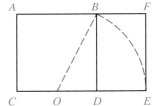

Step 1: The dots in the continued fraction indicate that the pattern shown continues indefinitely. This means that there is no way for us to simplify this expression, as we have simplified the expressions in this chapter. However, we can begin to understand the continued fraction, and what it simplifies to, by working with the following sequence of expressions. Simplify each expression. Write each answer as a fraction, in lowest terms.

$$1 + \cfrac{1}{1 + 1} \qquad 1 + \cfrac{1}{1 + \cfrac{1}{1 + 1}} \qquad 1 + \cfrac{1}{1 + \cfrac{1}{1 + \cfrac{1}{1 + 1}}} \qquad 1 + \cfrac{1}{1 + \cfrac{1}{1 + \cfrac{1}{1 + \cfrac{1}{1 + 1}}}}$$

Step 2: Compare the fractional answers to step 1 with the numbers in the Fibonacci sequence. Based on your observation, give the answer to the following problem, without actually doing any arithmetic.

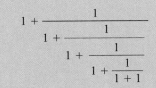

$$1 + \cfrac{1}{1 + \cfrac{1}{1 + \cfrac{1}{1 + \cfrac{1}{1 + 1}}}}$$

Step 3: Continue the sequence of simplified fractions you have written in steps 1 and 2, until you have nine numbers in the sequence. Convert each of these numbers to a decimal, accurate to four places past the decimal point.

Step 4: Find a decimal approximation to the golden ratio $\dfrac{1 + \sqrt{5}}{2}$, accurate to four places past the decimal point.

Step 5: Compare the results in steps 3 and 4, and then make a conjecture about what number the continued fraction would simplify to, if it was actually possible to simplify it.

Quadratic Equations

In the introduction to Chapter 6 we mentioned that hang time for a football depends only on the vertical velocity of the ball. It is also true that the maximum height attained by the ball is dependent only on the initial vertical velocity. The height h of the ball at time t is always given by the equation

$$h = vt - 16t^2$$

where v is the initial velocity of the ball. Table 1 shows the equations that give the height of the ball for different vertical velocities. Figure 1 shows the graph of each of those equations.

TABLE 1

Height of a Football

Initial Vertical Velocity (feet/second)	Height
16	$h = 16t - 16t^2$
32	$h = 32t - 16t^2$
48	$h = 48t - 16t^2$
64	$h = 64t - 16t^2$
80	$h = 80t - 16t^2$

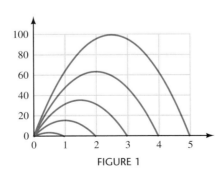

FIGURE 1

> Improve your grade and save time!
> Go online to **www.thomsonedu.com/login** where you can
> - Watch videos of instructors working through the in-text examples
> - Follow step-by-step online tutorials of in-text examples and review questions
> - Work practice problems
> - Check your readiness for an exam by taking a pre-test and exploring the modules recommended in your Personalized Study plan
> - Receive help from a live tutor online through vMentor™
>
> Try it out! Log in with an access code or purchase access at **www.ichapters.com**.

OBJECTIVES

A Solve a quadratic equation by using the square root property.

B Solve an application problem involving a quadratic equation.

Consider the equation $x^2 = 9$. Inspection shows that there are two solutions: $x = 3$ and $x = -3$, the two square roots of 9. Since every positive real number has two square roots, we can write the following property.

> **Square Root Property for Equations** For all positive real numbers b:
> If $a^2 = b$, then $a = \sqrt{b}$ or $a = -\sqrt{b}$

Notation A shorthand notation for

$$a = \sqrt{b} \text{ or } a = -\sqrt{b}$$

is:

$$a = \pm\sqrt{b}$$

which is read "a is plus or minus the square root of b."

We can use the square root property any time we feel it is helpful. We must make sure, however, that we include both the positive and the negative square roots.

EXAMPLE 1   Solve for x: $x^2 = 7$.

Note
This method of solving quadratic equations sometimes is called extraction of roots.

SOLUTION $x^2 = 7$

$x = \pm\sqrt{7}$ **Square root property**

The two solutions are $\sqrt{7}$ and $-\sqrt{7}$.

EXAMPLE 2   Solve for y: $3y^2 = 60$.

SOLUTION   We begin by applying the square root property for equations:

$$3y^2 = 60$$

$y^2 = 20$ **Divide each side by 3**

$y = \pm\sqrt{20}$ **Square root property**

$y = \pm 2\sqrt{5}$ $\sqrt{20} = \sqrt{4 \cdot 5} = \sqrt{4}\,\sqrt{5} = 2\sqrt{5}$

Our two solutions are $2\sqrt{5}$ and $-2\sqrt{5}$. Each of them will yield a true statement when used in place of the variable in the original equation $3y^2 = 60$.

EXAMPLE 3   Solve for a: $(a + 3)^2 = 16$.

SOLUTION   We begin by applying the square root property for equations:

$$(a + 3)^2 = 16$$

$$a + 3 = \pm 4$$

At this point we add -3 to both sides to get:

$$a = -3 \pm 4$$

which we can write as:

$$a = -3 + 4 \quad \text{or} \quad a = -3 - 4$$
$$a = 1 \quad \text{or} \quad a = -7$$

Our solutions are 1 and −7.

 EXAMPLE 4   Solve for x: $(3x - 2)^2 = 25$.

SOLUTION

$$(3x - 2)^2 = 25$$
$$3x - 2 = \pm 5$$

Adding 2 to both sides, we have:

$$3x = 2 \pm 5$$

Dividing both sides by 3 gives us:

$$x = \frac{2 \pm 5}{3}$$

We separate the preceding equation into two separate statements:

$$x = \frac{2 + 5}{3} \quad \text{or} \quad x = \frac{2 - 5}{3}$$

$$x = \frac{7}{3} \quad \text{or} \quad x = \frac{-3}{3} = -1$$

> **Note**
>
> We can solve the equation in Example 4 by factoring (as we did in Section 6.7) if we first expand $(3x - 2)^2$.
>
> $$(3x - 2)^2 = 25$$
> $$9x^2 - 12x + 4 = 25$$
> $$9x^2 - 12x - 21 = 0$$
> $$3(3x^2 - 4x - 7) = 0$$
> $$3(3x - 7)(x + 1) = 0$$
> $$x = \frac{7}{3} \quad \text{or} \quad x = -1$$

 EXAMPLE 5   Solve for y: $(4y - 5)^2 = 6$.

SOLUTION

$$(4y - 5)^2 = 6$$
$$4y - 5 = \pm\sqrt{6}$$
$$4y = 5 \pm \sqrt{6} \qquad \textbf{Add 5 to both sides}$$
$$y = \frac{5 \pm \sqrt{6}}{4} \qquad \textbf{Divide both sides by 4}$$

Since $\sqrt{6}$ is irrational, we cannot simplify the expression further. The solution set is $\left\{ \dfrac{5 + \sqrt{6}}{4}, \dfrac{5 - \sqrt{6}}{4} \right\}$.

 EXAMPLE 6   Solve for x: $(2x + 6)^2 = 8$.

SOLUTION

$$(2x + 6)^2 = 8$$
$$2x + 6 = \pm\sqrt{8}$$
$$2x + 6 = \pm 2\sqrt{2} \qquad \sqrt{8} = \sqrt{4 \cdot 2} = 2\sqrt{2}$$
$$2x = -6 \pm 2\sqrt{2} \qquad \textbf{Add −6 to both sides}$$
$$x = \frac{-6 \pm 2\sqrt{2}}{2} \qquad \textbf{Divide each side by 2}$$

We can reduce the previous expression to lowest terms by factoring a 2 from each term in the numerator and then dividing that 2 by the 2 in the denominator. This is equivalent to dividing each term in the numerator by the 2 in the denominator. Here is what it looks like:

$$x = \frac{\cancel{2}(-3 \pm \sqrt{2})}{\cancel{2}}$$ **Factor a 2 from each term in numerator**

$$x = -3 \pm \sqrt{2}$$ **Divide numerator and denominator by 2**

The two solutions are $-3 + \sqrt{2}$ and $-3 - \sqrt{2}$.

We can check our two solutions in the original equation. Let's check our first solution, $-3 + \sqrt{2}$.

Note
We are showing the check here so you can see that the irrational number $-3 + \sqrt{2}$ is a solution to $(2x + 6)^2 = 8$. Some people don't believe it at first.

When $x = -3 + \sqrt{2}$

the equation $(2x + 6)^2 = 8$

becomes $[2(-3 + \sqrt{2}) + 6]^2 \overset{?}{=} 8$

$(-6 + 2\sqrt{2} + 6)^2 = 8$

$(2\sqrt{2})^2 = 8$

$4 \cdot 2 = 8$

$8 = 8$ **A true statement**

The second solution, $-3 - \sqrt{2}$, checks also.

EXAMPLE 7 If an object is dropped from a height of h feet, the amount of time in seconds it will take for the object to reach the ground (ignoring the resistance of air) is given by the formula

$$h = 16t^2$$

Solve this formula for t.

SOLUTION To solve for t, we apply the square root property:

$h = 16t^2$ **Original formula**

$\pm\sqrt{h} = 4t$ **Square root property**

$\pm\dfrac{\sqrt{h}}{4} = t$ **Divide each side by 4**

Since t represents the time it takes for the object to fall h feet, t will never be negative. Therefore, the formula that gives t in terms of h is:

$$t = \frac{\sqrt{h}}{4}$$

Whenever we are solving an application problem like this one and we obtain a result that includes the $\pm$ sign, we must ask ourselves if the result actually can be negative. If it cannot be, we delete the negative result and use only the positive result.

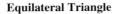

FACTS FROM GEOMETRY

Special Triangles

An *equilateral triangle* (Figure 1) is a triangle with three sides of equal length. If all three sides in a triangle have the same length, then the three interior angles in the triangle must also be equal. Since the sum of the interior angles in a triangle is always 180°, each of the three interior angles in any equilateral triangle must be 60°.

An *isosceles triangle* (Figure 2) is a triangle with two sides of equal length. Angles A and B in the isosceles triangle in Figure 2 are called the *base angles;* they are the angles opposite the two equal sides. In every isosceles triangle, the base angles are equal.

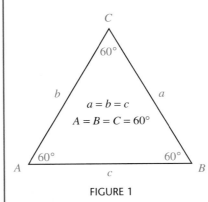

Equilateral Triangle

FIGURE 1

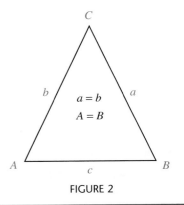

Isosceles Triangle

FIGURE 2

EXAMPLE 8

The lengths of all three sides of an equilateral triangle are 8 centimeters. Find the height of the triangle.

SOLUTION Because the three sides of an equilateral triangle are equal, the height always divides the base into two equal line segments. Figure 3 illustrates this fact. We find the height by applying the Pythagorean theorem.

$$8^2 = x^2 + 4^2$$
$$64 = x^2 + 16$$
$$x^2 = 64 - 16$$
$$x^2 = 48$$
$$x = \sqrt{48}$$
$$x = \sqrt{16 \cdot 3}$$
$$x = 4\sqrt{3} \text{ cm}$$

> **Note**
> A calculator gives a decimal approximation of 6.9 centimeters for the height of the triangle in Example 8.

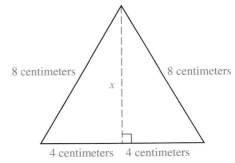

FIGURE 3

EXAMPLE 9 The base of an isosceles triangle is 10 feet, and the length of the two equal sides is 7 feet, as shown in Figure 4. Find the height of the triangle.

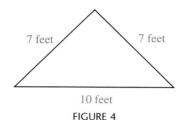

7 feet 7 feet

10 feet

FIGURE 4

SOLUTION Because two sides in every isosceles triangle are equal, the height that is drawn to the base always divides the base into two equal line segments, as shown in Figure 5. This fact allows us to form a right triangle in which two sides are known. We apply the Pythagorean theorem and solve for x.

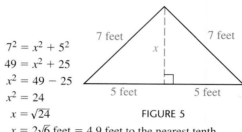

$$7^2 = x^2 + 5^2$$
$$49 = x^2 + 25$$
$$x^2 = 49 - 25$$
$$x^2 = 24$$
$$x = \sqrt{24}$$

FIGURE 5

$x = 2\sqrt{6}$ feet = 4.9 feet to the nearest tenth

LINKING OBJECTIVES AND EXAMPLES

Next to each **objective** we have listed the examples that are best described by that objective.

A 1–6

B 7–9

GETTING READY FOR CLASS

After reading through the preceding section, respond in your own words and in complete sentences.

1. What is the square root property for equations?
2. Describe how you would solve the equation $x^2 = 7$.
3. What does the symbol $\pm$ stand for?
4. What is an isosceles triangle?

Problem Set 9.1

Online support materials can be found at www.thomsonedu.com/login

Answers appear in the Instructor's Edition only.

Solve each of the following equations using the methods learned in this section.

▶ **1.** $x^2 = 9$ ± 3

2. $x^2 = 16$ ± 4

3. $a^2 = 25$ ± 5

4. $a^2 = 36$ ± 6

▶ **5.** $y^2 = 8$ $\pm 2\sqrt{2}$

6. $y^2 = 75$ $\pm 5\sqrt{3}$

7. $2x^2 = 100$ $\pm 5\sqrt{2}$

8. $2x^2 = 54$ $\pm 3\sqrt{3}$

▶ **9.** $3a^2 = 54$ $\pm 3\sqrt{2}$

10. $2a^2 = 64$ $\pm 4\sqrt{2}$

11. $(x + 2)^2 = 4$ $0, -4$

12. $(x - 3)^2 = 16$ $7, -1$

13. $(x + 1)^2 = 25$ $4, -6$

14. $(x + 3)^2 = 64$ $5, -11$

▶ **15.** $(a - 5)^2 = 75$ $5 \pm 5\sqrt{3}$

16. $(a - 4)^2 = 32$ $4 \pm 4\sqrt{2}$

17. $(y + 1)^2 = 50$ $-1 \pm 5\sqrt{2}$

18. $(y - 5)^2 = 27$ $5 \pm 3\sqrt{3}$

19. $(2x + 1)^2 = 25$ $2, -3$

20. $(3x - 2)^2 = 16$ $2, -\frac{2}{3}$

▨ = Videos available by instructor request

21. $(4a - 5)^2 = 36$ $\frac{11}{4}, -\frac{1}{4}$ **22.** $(2a + 6)^2 = 64$ $1, -7$

23. $(3y - 1)^2 = 12$ $\frac{1 \pm 2\sqrt{3}}{3}$ **24.** $(5y - 4)^2 = 12$ $\frac{4 \pm 2\sqrt{3}}{5}$

25. $(6x + 2)^2 = 27$ $\frac{-2 \pm 3\sqrt{3}}{6}$

26. $(8x - 1)^2 = 20$ $\frac{1 \pm 2\sqrt{5}}{8}$

27. $(3x - 9)^2 = 27$ $3 \pm \sqrt{3}$ **28.** $(2x + 8)^2 = 32$ $-4 \pm 2\sqrt{2}$

29. $(3x + 6)^2 = 45$ $-2 \pm \sqrt{5}$ **30.** $(5x - 10)^2 = 75$ $2 \pm \sqrt{3}$

31. $(2y - 4)^2 = 8$ $2 \pm \sqrt{2}$ **32.** $(4y - 6)^2 = 48$ $\frac{3 \pm 2\sqrt{3}}{2}$

33. $\left(x - \frac{2}{3}\right)^2 = \frac{25}{9}$ $\frac{7}{3}, -1$ **34.** $\left(x - \frac{3}{4}\right)^2 = \frac{49}{16}$ $\frac{5}{2}, -1$

35. $\left(x + \frac{1}{2}\right)^2 = \frac{7}{4}$ $\frac{-1 \pm \sqrt{7}}{2}$

36. $\left(x + \frac{1}{3}\right)^2 = \frac{5}{9}$ $\frac{-1 \pm \sqrt{5}}{3}$

37. $\left(a - \frac{4}{5}\right)^2 = \frac{12}{25}$ $\frac{4 \pm 2\sqrt{3}}{5}$

38. $\left(a - \frac{3}{7}\right)^2 = \frac{18}{49}$ $\frac{3 \pm 3\sqrt{2}}{7}$

Since $a^2 + 2ab + b^2$ can be written as $(a + b)^2$, each of the following equations can be solved using our square root method. The first step is to write the trinomial on the left side of the equal sign as the square of a binomial. Solve each equation.

39. $x^2 + 10x + 25 = 7$ $-5 \pm \sqrt{7}$

40. $x^2 + 6x + 9 = 11$ $-3 \pm \sqrt{11}$

41. $x^2 - 2x + 1 = 9$ $4, -2$

42. $x^2 + 8x + 16 = 25$ $1, -9$

43. $x^2 + 12x + 36 = 8$ $-6 \pm 2\sqrt{2}$

44. $x^2 - 4x + 4 = 12$ $2 \pm 2\sqrt{3}$

45. Consider the equation $x^2 = 3$.
 a. Can you solve it by factoring?   Answers will vary
 b. Solve it. $\pm\sqrt{3}$

46. Consider the equation $x^2 - 5 = 0$.
 a. Can you solve it by factoring?   Answers will vary
 b. Solve it. $\pm\sqrt{5}$

47. Consider the equation $(x - 3)^2 = 4$.
 a. Can you solve it by factoring?   Yes
 b. Solve it. $1, 5$

48. Consider the equation $x^2 - 10x + 25 = 1$.
 a. Can you solve it by factoring?   Yes
 b. Solve it. $4, 6$

49. Is $3 + \sqrt{3}$ a solution to $(2x - 6)^2 = 12$?   Yes

50. Is $2 + \sqrt{2}$ a solution to $(3x + 6)^2 = 18$?   No

The next two problems will give you practice with a variety of equations.

51. Solve each equation.
 a. $2x - 1 = 0$ $\frac{1}{2}$
 b. $2x - 1 = 4$ $\frac{5}{2}$
 c. $(2x - 1)^2 = 4$ $-\frac{1}{2}, \frac{3}{2}$
 d. $\sqrt{2x - 1} = 4$ $\frac{25}{2}$
 e. $\frac{1}{2x} - 1 = \frac{1}{4}$ $\frac{2}{5}$

52. Solve each equation.
 a. $x + 5 = 0$ -5
 b. $x + 5 = 8$ 3
 c. $(x + 5)^2 = 8$ $-5 \pm 2\sqrt{2}$
 d. $\sqrt{x} + 5 = 8$ 9
 e. $\frac{1}{x} + 5 = \frac{1}{2}$ $-\frac{2}{9}$

Chalkboard Problems
Problems 51 and 52 give you a variety of problems to solve while introducing the square root method of solving quadratics.

Applying the Concepts

53. Checking Solutions Check the solution
$$x = -1 + 5\sqrt{2}$$
in the equation $(x + 1)^2 = 50$.
$(-1 + 5\sqrt{2} + 1)^2 = (5\sqrt{2})^2 = 50$

54. Checking Solutions Check the solution
$$x = -8 + 2\sqrt{6}$$
in the equation $(x + 8)^2 = 24$.
$(-8 + 2\sqrt{6} + 8)^2 = (2\sqrt{6})^2 = 24$

55. The equation $x^2 - 10x + 22 = 0$ has solutions $5 + \sqrt{3}$ and $5 - \sqrt{3}$. Find:
 a. The sum of the solutions: $(5 + \sqrt{3}) + (5 - \sqrt{3})$ 10
 b. The product of the solutions: $(5 + \sqrt{3})(5 - \sqrt{3})$ 22

56. The equation $x^2 - 14x + 44 = 0$ has solutions $7 + \sqrt{5}$ and $7 - \sqrt{5}$. Find:
 a. The sum of the solutions -14
 b. The product of the solutions 44

57. Geometry The area of a circle with radius r is $A = \pi r^2$. Find the radius if the area is 36π square feet.   6 feet

58. Geometry Solve the formula $A = \pi r^2$ for r. $r = \sqrt{\dfrac{A}{\pi}}$

59. Falling Object A baseball is dropped from the top of a 100-ft building. It will take t seconds to hit the ground, according to the formula $100 = 16t^2$. Find t. $\frac{5}{2}$ sec

60. Ball Toss A ball thrown straight up into the air takes t seconds to reach a height of 25 feet, according to the formula $25 = 16t^2$. Find t. $\frac{5}{4}$ sec

61. Number Problem The square of the sum of a number and 3 is 16. Find the number. (There are two solutions.)   −7, 1

62. Number Problem The square of the sum of twice a number and 3 is 25. Find the number. (There are two solutions.)   1, −4

63. Investing If you invest $100 in an account with interest rate r compounded annually, the amount of money A in the account after 2 years is given by the formula

$$A = 100(1 + r)^2$$

Solve this formula for r. $r = -1 + \dfrac{\sqrt{A}}{10}$

64. Investing If you invest P dollars in an account with interest rate r compounded annually, the amount of money A in the account after 2 years is given by the formula

$$A = P(1 + r)^2$$

Solve this formula for r. $r = -1 + \sqrt{\dfrac{A}{P}}$

65. Geometry The lengths of all three sides of an equilateral triangle are 10 feet. Find the height of the triangle. $5\sqrt{3}$ feet

66. Geometry The lengths of all three sides of an equilateral triangle are 12 meters. Find the height of the triangle. $6\sqrt{3}$ meters

67. Geometry The front of a tent forms an equilateral triangle with sides of 6 feet. Can a person 5 feet 8 inches tall stand up inside the tent?

No; height of tent = $3\sqrt{3} \approx 5.2$ feet
≈ 5 feet 2 inches

68. Geometry The front of a tent forms an equilateral triangle. The tent must be constructed so that a

person 6 feet tall can stand up inside. Find the length of the three sides of the front of the tent.

$4\sqrt{3} \approx 6.9$ feet

69. Geometry The base of an isosceles triangle is 8 feet, and the length of the two equal sides is 5 feet. Find the height of the triangle.   3 feet

70. Geometry An isosceles triangle has a base 8 feet long. If the length of the two equal sides is 6 feet, find the height of the triangle. $2\sqrt{5}$ feet ≈ 4.47 feet

Displacement The displacement, in cubic inches, of a car engine is given by the formula

$$d = \pi \cdot s \cdot c \cdot \left(\frac{1}{2} \cdot b\right)^2$$

where s is the stroke and b is the bore, as shown here, and c is the number of cylinders. Calculate the bore for each of the following cars. Use 3.14 to approximate π.

71. BMW M Coupe Six cylinders, 3.53 inches of stroke, and 192 cubic inches of displacement   3.4 in.

72. Chevrolet Corvette Eight cylinders, 3.62 inches of stroke, and 346 cubic inches of displacement
3.9 in.

Maintaining Your Skills

Multiply.

73. $(x - 5)^2$ $x^2 - 10x + 25$ **74.** $(x + 5)^2$ $x^2 + 10x + 25$

Factor.

75. $x^2 - 12x + 36$ $(x - 6)^2$ **76.** $x^2 + 12x + 36$ $(x + 6)^2$

77. $x^2 + 4x + 4$ $(x + 2)^2$ **78.** $x^2 - 4x + 4$ $(x - 2)^2$

Find the following roots.

79. $\sqrt[3]{8}$ 2   **80.** $\sqrt[3]{27}$ 3

81. $\sqrt[4]{16}$ 2   **82.** $\sqrt[4]{81}$ 3

Getting Ready for the Next Section

Simplify.

83. $\left(\dfrac{1}{2} \cdot 18\right)^2$ 81   **84.** $\left(\dfrac{1}{2} \cdot 8\right)^2$ 16

85. $\left[\dfrac{1}{2}(-2)\right]^2$ 1   **86.** $\left[\dfrac{1}{2}(-10)\right]^2$ 25

87. $\left(\dfrac{1}{2} \cdot 3\right)^2$ $\dfrac{9}{4}$ **88.** $\left(\dfrac{1}{2} \cdot 5\right)^2$ $\dfrac{25}{4}$

89. $\dfrac{2x^2 + 16}{2}$ $x^2 + 8$ **90.** $\dfrac{3x^2 - 3x}{3}$ $x^2 - x$

Factor.

91. $x^2 + 6x + 9$ $(x + 3)^2$ **92.** $x^2 + 12x + 36$ $(x + 6)^2$

93. $y^2 - 3y + \dfrac{9}{4}$ $\left(y - \dfrac{3}{2}\right)^2$

94. $y^2 - 5y + \dfrac{25}{4}$ $\left(y - \dfrac{5}{2}\right)^2$

9.2 Completing the Square

OBJECTIVES

A Solve a quadratic equation by completing the square.

In this section we will develop a method of solving quadratic equations that works whether or not the equation can be factored. Since we will be working with the individual terms of trinomials, we need some new definitions so that we can keep our vocabulary straight.

> **DEFINITION** In the trinomial $2x^2 + 3x + 4$, the first term, $2x^2$, is called the **quadratic term**; the middle term, $3x$, is called the **linear term**; and the last term, 4, is called the **constant term.**

Now consider the following list of perfect square trinomials and their corresponding binomial squares:

$$x^2 + 6x + 9 = (x + 3)^2$$
$$x^2 - 8x + 16 = (x - 4)^2$$
$$x^2 - 10x + 25 = (x - 5)^2$$
$$x^2 + 12x + 36 = (x + 6)^2$$

In each case the coefficient of x^2 is 1. A more important observation, however, stems from the relationship between the linear terms (middle terms) and the constant terms (last terms). Notice that the constant in each case is the square of half the coefficient of x in the linear term; that is,

1. For the first trinomial, $x^2 + 6x + 9$, the last term, 9, is the square of half the coefficient of the middle term: $9 = (\frac{6}{2})^2$.

2. For the second trinomial, $x^2 - 8x + 16$, $16 = [\frac{1}{2}(-8)]^2$.

3. For the trinomial $x^2 - 10x + 25$, it also holds: $25 = [\frac{1}{2}(-10)]^2$.

Check and see that it also works for the final trinomial.

In summary, then, for every perfect square trinomial in which the coefficient of x^2 is 1, the final term is always the square of half the coefficient of the linear term. We can use this fact to build our own perfect square trinomials.

EXAMPLES
Write the correct final term to each of the following expressions so each becomes a perfect square trinomial and then factor.

1. $x^2 - 2x$

SOLUTION The coefficient of the linear term is -2. If we take half of -2, we get -1, the square of which is 1. Adding the 1 as the final term, we have the perfect square trinomial:

$$x^2 - 2x + 1 = (x - 1)^2$$

2. $x^2 + 18x$

SOLUTION Half of 18 is 9, the square of which is 81. If we add 81 at the end, we have:

$$x^2 + 18x + 81 = (x + 9)^2$$

3. $x^2 + 3x$

SOLUTION Half of 3 is $\frac{3}{2}$, the square of which is $\frac{9}{4}$:

$$x^2 + 3x + \frac{9}{4} = \left(x + \frac{3}{2}\right)^2$$

We can use this procedure, along with the method developed in Section 9.1, to solve some quadratic equations.

EXAMPLE 4
Solve $x^2 - 6x + 5 = 0$ by completing the square.

SOLUTION We begin by adding -5 to both sides of the equation. We want just $x^2 - 6x$ on the left side so that we can add on our own final term to get a perfect square trinomial:

$$x^2 - 6x + 5 = 0$$

$$x^2 - 6x \qquad = -5 \qquad \textbf{Add } -5 \textbf{ to both sides}$$

Now we can add 9 to both sides and the left side will be a perfect square:

$$x^2 - 6x + \mathbf{9} = -5 + \mathbf{9}$$

$$(x - 3)^2 = 4$$

The final line is in the form of the equations we solved in Section 9.1:

$$x - 3 = \pm 2$$

$$x = 3 \pm 2 \qquad \textbf{Add 3 to both sides}$$

$$x = 3 + 2 \qquad \text{or} \qquad x = 3 - 2$$

$$x = 5 \qquad \text{or} \qquad x = 1$$

The two solutions are 5 and 1.

The preceding method of solution is called *completing the square*.

Note

The equation in Example 4 can be solved quickly by factoring:

$$x^2 - 6x + 5 = 0$$
$$(x - 5)(x - 1) = 0$$
$$x - 5 = 0 \quad \text{or} \quad x - 1 = 0$$
$$x = 5 \quad \text{or} \quad x = 1$$

The reason we didn't solve it by factoring is we want to practice completing the square on some simple equations.

EXAMPLE 5   Solve by completing the square $2x^2 + 16x - 18 = 0$.

SOLUTION   We begin by moving the constant term to the other side:

$$2x^2 + 16x - 18 = 0$$

$$2x^2 + 16x = 18 \qquad \textbf{Add 18 to both sides}$$

To complete the square, we must be sure the coefficient of x^2 is 1. To accomplish this, we divide both sides by 2:

$$\frac{2x^2}{2} + \frac{16x}{2} = \frac{18}{2}$$

$$x^2 + 8x = 9$$

We now complete the square by adding the square of half the coefficient of the linear term to both sides:

$$x^2 + 8x + \mathbf{16} = 9 + \mathbf{16} \qquad \textbf{Add 16 to both sides}$$

$$(x + 4)^2 = 25$$

$$x + 4 = \pm 5 \qquad \textbf{Square root property}$$

$$x = -4 \pm 5 \qquad \textbf{Add } -4 \textbf{ to both sides}$$

$$x = -4 + 5 \quad \text{or} \quad x = -4 - 5$$

$$x = 1 \quad \text{or} \quad x = -9$$

The solution set arrived at by completing the square is $\{1, -9\}$.

We will now summarize the preceding examples by listing the steps involved in solving quadratic equations by completing the square.

Strategy for Solving a Quadratic Equation by Completing the Square

Step 1: Put the equation in the form $ax^2 + bx = c$. This usually involves moving only the constant term to the opposite side.

Step 2: Make sure the coefficient of the squared term is 1. If it is not 1, simply divide both sides by whatever it is.

Step 3: Add the square of half the coefficient of the linear term to both sides of the equation.

Step 4: Write the left-hand side of the equation as a binomial square, apply the square root property, and solve.

Note

Step 3 is the step at which we actually complete the square.

Here is one final example.

EXAMPLE 6   Solve for y: $3y^2 - 9y + 3 = 0$.

SOLUTION

$$3y^2 - 9y + 3 = 0$$

$$3y^2 - 9y = -3 \qquad \textbf{Add } -3 \textbf{ to both sides}$$

$$y^2 - 3y = -1 \qquad \text{Divide by 3}$$

$$y^2 - 3y + \frac{9}{4} = -1 + \frac{9}{4} \qquad \text{Complete the square}$$

$$\left(y - \frac{3}{2}\right)^2 = \frac{5}{4} \qquad -1 + \frac{9}{4} = -\frac{4}{4} + \frac{9}{4} = \frac{5}{4}$$

$$y - \frac{3}{2} = \pm \frac{\sqrt{5}}{2} \qquad \text{Square root property}$$

$$y = \frac{3}{2} \pm \frac{\sqrt{5}}{2} \qquad \text{Add } \tfrac{3}{2} \text{ to both sides}$$

$$y = \frac{3}{2} + \frac{\sqrt{5}}{2} \qquad \text{or} \qquad y = \frac{3}{2} - \frac{\sqrt{5}}{2}$$

$$y = \frac{3 + \sqrt{5}}{2} \qquad \text{or} \qquad y = \frac{3 - \sqrt{5}}{2}$$

Note

We can use a calculator to get decimal approximations to these solutions. If we use the approximation $\sqrt{5} \approx 2.236$, then

$$\frac{3 + \sqrt{5}}{2} \approx \frac{3 + 2.236}{2}$$

$$= \frac{5.236}{2} = 2.618$$

$$\frac{3 - \sqrt{5}}{2} \approx \frac{3 - 2.236}{2}$$

$$= \frac{0.764}{2} = 0.382$$

The solutions are $\dfrac{3 + \sqrt{5}}{2}$ and $\dfrac{3 - \sqrt{5}}{2}$, which can be written in a shorter form as:

$$\frac{3 \pm \sqrt{5}}{2}$$

GETTING READY FOR CLASS

After reading through the preceding section, respond in your own words and in complete sentences.

1. What kind of equation do we solve using the method of completing the square?
2. Explain in words how you would complete the square on $x^2 - 6x = 5$.
3. What is the first step in completing the square?
4. What is the linear term in a trinomial?

LINKING OBJECTIVES AND EXAMPLES

Next to each **objective** we have listed the examples that are best described by that objective.

A   1–6

Problem Set 9.2

Online support materials can be found at www.thomsonedu.com/login

Give the correct final term for each of the following expressions to ensure that the resulting trinomial is a perfect square trinomial.

1. $x^2 + 6x$ 9

2. $x^2 - 10x$ 25

3. $x^2 + 2x$ 1

4. $x^2 + 14x$ 49

▶ 5. $y^2 - 8y$ 16

6. $y^2 + 12y$ 36

7. $y^2 - 2y$ 1

8. $y^2 - 6y$ 9

▶ 9. $x^2 + 16x$ 64

10. $x^2 - 4x$ 4

11. $a^2 - 3a$ $\frac{9}{4}$

12. $a^2 + 5a$ $\frac{25}{4}$

13. $x^2 - 7x$ $\frac{49}{4}$

14. $x^2 - 9x$ $\frac{81}{4}$

15. $y^2 + y$ $\frac{1}{4}$

16. $y^2 - y$ $\frac{1}{4}$

17. $x^2 - \frac{3}{2}x$ $\frac{9}{16}$

18. $x^2 + \frac{2}{3}x$ $\frac{1}{9}$

☐ = Videos available by instructor request

▶ = Online student support materials available at www.thomsonedu.com/login

Solve each of the following equations by completing the square. Follow the steps given at the end of this section.

19. $x^2 + 4x = 12$ $-6, 2$ **20.** $x^2 - 2x = 8$ $-2, 4$

▶ **21.** $x^2 - 6x = 16$ $-2, 8$ **22.** $x^2 + 12x = -27$ $-9, -3$

23. $a^2 + 2a = 3$ $-3, 1$ **24.** $a^2 - 8a = -7$ $1, 7$

25. $x^2 - 10x = 0$ $0, 10$ **26.** $x^2 + 4x = 0$ $0, -4$

27. $y^2 + 2y - 15 = 0$ $-5, 3$ **28.** $y^2 - 10y - 11 = 0$ $11, -1$

29. $x^2 + 4x - 3 = 0$ $-2 \pm \sqrt{7}$ **30.** $x^2 + 6x + 5 = 0$ $-5, -1$

31. $x^2 - 4x = 4$ $2 \pm 2\sqrt{2}$ **32.** $x^2 + 4x = -1$ $-2 \pm \sqrt{3}$

33. $a^2 = 7a + 8$ $8, -1$ **34.** $a^2 = 3a + 1$ $\dfrac{3 \pm \sqrt{13}}{2}$

35. $4x^2 + 8x - 4 = 0$ $-1 \pm \sqrt{2}$

36. $3x^2 + 12x + 6 = 0$ $-2 \pm \sqrt{2}$

37. $2x^2 + 2x - 4 = 0$ $-2, 1$

38. $4x^2 + 4x - 3 = 0$ $\dfrac{1}{2}, -\dfrac{3}{2}$

39. $4x^2 + 8x + 1 = 0$ $\dfrac{-2 \pm \sqrt{3}}{2}$

40. $3x^2 + 6x + 2 = 0$ $\dfrac{-3 \pm \sqrt{3}}{3}$

41. $2x^2 - 2x = 1$ $\dfrac{1 \pm \sqrt{3}}{2}$

42. $3x^2 - 3x = 1$ $\dfrac{3 \pm \sqrt{21}}{6}$

43. $4a^2 - 4a + 1 = 0$ $\dfrac{1}{2}$

44. $2a^2 + 4a + 1 = 0$ $\dfrac{-2 \pm \sqrt{2}}{2}$

45. $3y^2 - 9y = 2$ $\dfrac{9 \pm \sqrt{105}}{6}$

46. $5y^2 - 10y = 4$ $\dfrac{5 \pm 3\sqrt{5}}{5}$

47. Solve the equation $x^2 - 2x = 0$
 a. By factoring $0, 2$
 b. By completing the square $0, 2$

48. Solve the equation $x^2 + 3x = 40$
 a. By factoring $-8, 5$
 b. By completing the square $-8, 5$

49. Is $x = 3 - \sqrt{2}$ a solution to $x^2 + 6x = -7$?   No.

50. Is $x = 2 + \sqrt{2}$ a solution to $x^2 - 4x = -1$?   No.

51. The equation $x^2 - 6x = 0$ can be solved by completing the square. Is there a faster way to solve it?
 Yes, by factoring.

52. The equation $(11a - 24)^2 = 144$ can be solved by expanding the left side then writing it in standard form and factoring. Is there an easier way?
 Answers will vary.

53. Suppose you solve the equation $x^2 - 6x = 7$ by completing the square and then finding the solutions are -1 and 7. Was there another way to solve the equation?   Yes, by factoring.

54. The solutions to the equation $x^2 - 6x + 7 = 0$ are $3 + \sqrt{2}$ and $3 - \sqrt{2}$. Could you solve the equation by factoring?   No.

There is more than one way to solve each equation below. In each case, give some thought as to which method will be easiest, then solve the equation.

▶ **55.** $(11a - 24)^2 = 144$ $\frac{12}{11}, \frac{36}{11}$ ▶ **56.** $\left(a + \dfrac{3}{4}\right)^2 = \dfrac{9}{16}$ $-\frac{3}{2}, 0$

▶ **57.** $t^2 - 6t = 0$ $0, 6$ ▶ **58.** $2x^2 + 5x = 0$ $-\frac{5}{2}, 0$

▶ **59.** $t^2 - 6t = 7$ $-1, 7$ ▶ **60.** $2x^2 + 5x = 3$ $-3, \frac{1}{2}$

▶ **61.** $t^2 - 6t = -7$ $3 \pm \sqrt{2}$

▶ **62.** $2x^2 + 5x - 1 = 0$ $\dfrac{-5 \pm \sqrt{33}}{4}$

▶ **63.** $(6y - 8)^2 = 20$ $\dfrac{4 \pm \sqrt{5}}{3}$ ▶ **64.** $(9y - 15)^2 = 45$ $\dfrac{5 \pm \sqrt{5}}{3}$

▶ **65.** If one solution to a quadratic equation is $\dfrac{3 - 2\sqrt{5}}{2}$, name another solution. $\dfrac{3 + 2\sqrt{5}}{2}$

▶ **66.** If one solution to a quadratic equation is $\dfrac{7 + 3}{2} = 5$, name another solution. 2

Each diagram below shows a square that is divided into 4 smaller squares and rectangles. Use your knowledge of completing the square to find x in each case.

▶ **67.**

x^2	$4x$
$4x$	16

Total area = 100 cm²
$x = 6$

▶ **68.**

x^2	$2x$
$2x$	4

Total area = 25 cm²
$x = 3$

Applying the Concepts

69. Computer Screen An advertisement for a portable computer indicates it has a 14-inch viewing screen. This means that the diagonal of the screen measures 14 inches. If the ratio of the length to the width of the screen is 3 to 4, we can represent the length with $4x$ and the width with $3x$, and then use the Pythagorean theorem to solve for x. Once we have x, the length will be $4x$ and the width will be $3x$. Find the length and width of this computer screen to the nearest tenth of an inch.

Length = 11.2 in., width = 8.4 in.

70. Television Screen A 25-inch television has a rectangular screen on which the diagonal measures 25 inches. The ratio of the length to the width of the screen is 3 to 4. Let the length equal $4x$ and the width equal $3x$, and then use the Pythagorean theorem to solve for x. Then find the length and width of this television screen.

Length = 20 in., width = 15 in.

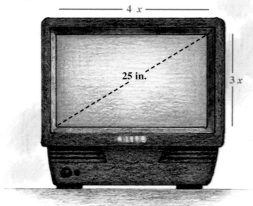

Maintaining Your Skills

Find the value of each expression if $a = 2$, $b = 4$, and $c = -3$.

71. $2a$ 4

72. b^2 16

73. $4ac$ -24

74. $b^2 - 4ac$ 40

75. $\sqrt{b^2 - 4ac}$ $2\sqrt{10}$

76. $-b + \sqrt{b^2 - 4ac}$ $-4 + 2\sqrt{10}$

Put each expression in simplified form for radicals.

77. $\sqrt{12}$ $2\sqrt{3}$

78. $\sqrt{50x^2}$ $5x\sqrt{2}$

79. $\sqrt{20x^2y^3}$ $2xy\sqrt{5y}$

80. $3\sqrt{48x^4}$ $12x^2\sqrt{3}$

81. $\sqrt{\dfrac{81}{25}}$ $\dfrac{9}{5}$

82. $\dfrac{6\sqrt{8x^2y}}{\sqrt{9}}$ $4x\sqrt{2y}$

Getting Ready for the Next Section

Write the quadratic equation in standard form ($ax^2 + bx + c = 0$).

83. $2x^2 = -4x + 3$ $2x^2 + 4x - 3 = 0$

84. $3x^2 = -4x + 2$ $3x^2 + 4x - 2 = 0$

85. $(x - 2)(x + 3) = 5$ $x^2 + x - 11 = 0$

86. $(x - 1)(x + 2) = 4$ $x^2 + x - 6 = 0$

Identify the coefficient of x^2, the coefficient of x, and the constant term.

87. $x^2 - 5x - 6$ $1, -5, -6$

88. $x^2 - 6x + 7$ $1, -6, 7$

89. $2x^2 + 4x - 3$ $2, 4, -3$

90. $3x^2 + 4x - 2$ $3, 4, -2$

Find the value of $b^2 - 4ac$ for the given values of a, b, and c.

91. $a = 1, b = -5, c = -6$ 49

92. $a = 1, b = -6, c = 7$ 8

93. $a = 2, b = 4, c = -3$ 40

94. $a = 3, b = 4, c = -2$ 40

Simplify.

95. $\dfrac{5 + \sqrt{49}}{2}$ 6

96. $\dfrac{5 - \sqrt{49}}{2}$ -1

97. $\dfrac{-4 - \sqrt{40}}{4}$ $\dfrac{-2 - \sqrt{10}}{2}$

98. $\dfrac{-4 + \sqrt{40}}{4}$ $\dfrac{-2 + \sqrt{10}}{2}$

9.3 The Quadratic Formula

OBJECTIVES

A Solve a quadratic equation by using the quadratic formula.

In this section we will derive the quadratic formula. It is one formula that you will use in almost all types of mathematics. We will first state the formula as a theorem and then prove it. The proof is based on the method of completing the square developed in the preceding section.

> **The Quadratic Theorem** For any quadratic equation in the form $ax^2 + bx + c = 0$, where a, b, and c are real numbers and $a \neq 0$, the two solutions are:
>
> $$x = \frac{-b + \sqrt{b^2 - 4ac}}{2a} \quad \text{and} \quad x = \frac{-b - \sqrt{b^2 - 4ac}}{2a}$$

Note

This is one of the few times in the course where we actually get to show a proof. The proof shown here is the reason the quadratic formula looks the way it does. We will use the quadratic formula in every example we do in this section. As you read through the examples, you may find yourself wondering why some parts of the formula are the way they are. If that happens, come back to this proof and see for yourself.

Proof We will prove the theorem by completing the square on:

$$ax^2 + bx + c = 0$$

Adding $-c$ to both sides, we have:

$$ax^2 + bx = -c$$

To make the coefficient of x^2 one, we divide both sides by a:

$$\frac{ax^2}{a} + \frac{bx}{a} = -\frac{c}{a}$$

$$x^2 + \frac{b}{a}x = -\frac{c}{a}$$

Now, to complete the square, we add the square of half of $\frac{b}{a}$ to both sides:

$$x^2 + \frac{b}{a}x + \left(\frac{b}{2a}\right)^2 = -\frac{c}{a} + \left(\frac{b}{2a}\right)^2 \qquad \frac{1}{2} \text{ of } \frac{b}{a} \text{ is } \frac{b}{2a}$$

Let's simplify the right side separately:

$$-\frac{c}{a} + \left(\frac{b}{2a}\right)^2 = -\frac{c}{a} + \frac{b^2}{4a^2}$$

The least common denominator is $4a^2$. We multiply the numerator and denominator of $-\frac{c}{a}$ by $4a$ to give it the common denominator. Then we combine numerators:

$$\frac{4a}{4a}\left(-\frac{c}{a}\right) + \frac{b^2}{4a^2} = -\frac{4ac}{4a^2} + \frac{b^2}{4a^2}$$

$$= \frac{-4ac + b^2}{4a^2}$$

$$= \frac{b^2 - 4ac}{4a^2}$$

Now, back to the equation. We use our simplified expression for the right side:

$$x^2 + \frac{b}{a}x + \left(\frac{b}{2a}\right)^2 = \frac{b^2 - 4ac}{4a^2}$$

$$\left(x + \frac{b}{2a}\right)^2 = \frac{b^2 - 4ac}{4a^2}$$

Applying the square root property, we have:

$$x + \frac{b}{2a} = \pm \frac{\sqrt{b^2 - 4ac}}{2a}$$

$$x = \frac{-b}{2a} \pm \frac{\sqrt{b^2 - 4ac}}{2a} \qquad \text{Add } \frac{-b}{2a} \text{ to both sides}$$

$$x = \frac{-b \pm \sqrt{b^2 - 4ac}}{2a}$$

Note

This formula is called the quadratic formula. You will see it many times if you continue taking math classes. By the time you are finished with this section and the problems in the problem set, you should have it memorized.

Our proof is now complete. What we have is this: If our equation is in the form $ax^2 + bx + c = 0$ (standard form), then the solution can always be found by using the quadratic formula:

$$x = \frac{-b \pm \sqrt{b^2 - 4ac}}{2a}$$

 EXAMPLE 1   Solve $x^2 - 5x - 6 = 0$ by using the quadratic formula.

SOLUTION   To use the quadratic formula, we must make sure the equation is in standard form; identify a, b, and c; substitute them into the formula; and work out the arithmetic.

For the equation $x^2 - 5x - 6 = 0$, $a = 1$, $b = -5$, and $c = -6$:

$$x = \frac{-b \pm \sqrt{b^2 - 4ac}}{2a} = \frac{-(-5) \pm \sqrt{(-5)^2 - 4(1)(-6)}}{2(1)}$$

$$= \frac{5 \pm \sqrt{49}}{2}$$

$$= \frac{5 \pm 7}{2}$$

Note

Whenever the solutions to our quadratic equations turn out to be rational numbers, as in Example 1, it means the original equation could have been solved by factoring. (We didn't solve the equation in Example 1 by factoring because we were trying to get some practice with the quadratic formula.)

$$x = \frac{5 + 7}{2} \qquad \text{or} \qquad x = \frac{5 - 7}{2}$$

$$x = \frac{12}{2} \qquad\qquad\qquad x = -\frac{2}{2}$$

$$x = 6 \qquad\qquad\qquad\quad x = -1$$

The two solutions are 6 and -1.

EXAMPLE 2   Solve for x: $2x^2 = -4x + 3$.

SOLUTION   Before we can identify a, b, and c, we must write the equation in standard form. To do so, we add $4x$ and -3 to each side of the equation:

$$2x^2 = -4x + 3$$

$$2x^2 + 4x - 3 = 0 \qquad \text{**Add $4x$ and -3 to each side**}$$

Now that the equation is in standard form, we see that $a = 2$, $b = 4$, and $c = -3$. Using the quadratic formula we have:

$$x = \frac{-b \pm \sqrt{b^2 - 4ac}}{2a}$$

$$= \frac{-4 \pm \sqrt{4^2 - 4(2)(-3)}}{2(2)}$$

$$= \frac{-4 \pm \sqrt{40}}{4}$$

$$= \frac{-4 \pm 2\sqrt{10}}{4}$$

We can reduce the final expression in the preceding equation to lowest terms by factoring 2 from the numerator and denominator and then dividing it out:

$$x = \frac{\cancel{2}(-2 \pm \sqrt{10})}{\cancel{2} \cdot 2}$$

$$= \frac{-2 \pm \sqrt{10}}{2}$$

Our two solutions are $\dfrac{-2 + \sqrt{10}}{2}$ and $\dfrac{-2 - \sqrt{10}}{2}$.

 EXAMPLE 3   Solve for x: $(x - 2)(x + 3) = 5$.

SOLUTION   We must put the equation into standard form before we can use the formula:

$$(x - 2)(x + 3) = 5$$

$$x^2 + x - 6 = 5 \qquad \textbf{Multiply out the left side}$$

$$x^2 + x - 11 = 0 \qquad \textbf{Add -5 to each side}$$

Now, $a = 1$, $b = 1$, and $c = -11$; therefore:

$$x = \frac{-1 \pm \sqrt{1^2 - 4(1)(-11)}}{2(1)}$$

$$= \frac{-1 \pm \sqrt{45}}{2}$$

$$= \frac{-1 \pm 3\sqrt{5}}{2}$$

The solution set is $\left\{ \dfrac{-1 + 3\sqrt{5}}{2}, \dfrac{-1 - 3\sqrt{5}}{2} \right\}$.

 EXAMPLE 4   Solve $x^2 - 6x = -7$.

SOLUTION   We begin by writing the equation in standard form:

$$x^2 - 6x = -7$$

$$x^2 - 6x + 7 = 0 \qquad \textbf{Add 7 to each side}$$

Using $a = 1$, $b = -6$, and $c = 7$ in the quadratic formula

$$x = \frac{-b \pm \sqrt{b^2 - 4ac}}{2a}$$

we have:

$$x = \frac{-(-6) \pm \sqrt{(-6)^2 - 4(1)(7)}}{2(1)}$$

$$= \frac{6 \pm \sqrt{36 - 28}}{2}$$

$$= \frac{6 \pm \sqrt{8}}{2}$$

$$= \frac{6 \pm 2\sqrt{2}}{2}$$

The two terms in the numerator have a 2 in common. We reduce to lowest terms by factoring the 2 from the numerator and then dividing numerator and denominator by 2:

$$= \frac{\cancel{2}(3 \pm \sqrt{2})}{\cancel{2}}$$

$$= 3 \pm \sqrt{2}$$

The two solutions are $3 + \sqrt{2}$ and $3 - \sqrt{2}$. This time, let's check our solutions in the original equation $x^2 - 6x = -7$.

Checking $x = 3 + \sqrt{2}$, we have:

$$(3 + \sqrt{2})^2 - 6(3 + \sqrt{2}) \stackrel{?}{=} -7$$

$$9 + 6\sqrt{2} + 2 - 18 - 6\sqrt{2} = -7 \qquad \textbf{Multiply}$$

$$11 - 18 + 6\sqrt{2} - 6\sqrt{2} = -7 \qquad \textbf{Add 9 and 2}$$

$$-7 + 0 = -7 \qquad \textbf{Subtraction}$$

$$-7 = -7 \qquad \textbf{A true statement}$$

Checking $x = 3 - \sqrt{2}$, we have:

$$(3 - \sqrt{2})^2 - 6(3 - \sqrt{2}) \stackrel{?}{=} -7$$

$$9 - 6\sqrt{2} + 2 - 18 + 6\sqrt{2} = -7 \qquad \textbf{Multiply}$$

$$11 - 18 - 6\sqrt{2} + 6\sqrt{2} = -7 \qquad \textbf{Add 9 and 2}$$

$$-7 + 0 = -7 \qquad \textbf{Subtraction}$$

$$-7 = -7 \qquad \textbf{A true statement}$$

As you can see, both solutions yield true statements when used in place of the variable in the original equation.

LINKING OBJECTIVES AND EXAMPLES

Next to each **objective** we have listed the examples that are best described by that objective.

A   1–4

GETTING READY FOR CLASS

After reading through the preceding section, respond in your own words and in complete sentences.

1. What is the quadratic formula?
2. Under what circumstances should the quadratic formula be applied?
3. What is standard form for a quadratic equation?
4. What is the first step in solving a quadratic equation using the quadratic formula?

Problem Set 9.3

Online support materials can be found at www.thomsonedu.com/login

Solve the following equations by using the quadratic formula.

1. $x^2 + 3x + 2 = 0$ $-1, -2$ **2.** $x^2 - 5x + 6 = 0$ $2, 3$

▶ **3.** $x^2 + 5x + 6 = 0$ $-2, -3$ **4.** $x^2 - 7x - 8 = 0$ $8, -1$

5. $x^2 + 6x + 9 = 0$ -3 **6.** $x^2 - 10x + 25 = 0$ 5

7. $x^2 + 6x + 7 = 0$ $-3 \pm \sqrt{2}$ **8.** $x^2 - 4x - 1 = 0$ $2 \pm \sqrt{5}$

9. $2x^2 + 5x + 3 = 0$ $-1, -\frac{3}{2}$ **10.** $2x^2 + 3x - 20 = 0$ $\frac{5}{2}, -4$

11. $4x^2 + 8x + 1 = 0$ $\dfrac{-2 \pm \sqrt{3}}{2}$

12. $3x^2 + 6x + 2 = 0$ $\dfrac{-3 \pm \sqrt{3}}{3}$

13. $x^2 - 2x + 1 = 0$ 1 **14.** $x^2 + 2x - 3 = 0$ $-3, 1$

▶ **15.** $x^2 - 5x = 7$ $\dfrac{5 \pm \sqrt{53}}{2}$ **16.** $2x^2 - 6x = 8$ $4, -1$

17. $6x^2 - x - 2 = 0$ $\frac{2}{3}, -\frac{1}{2}$ **18.** $6x^2 + 5x - 4 = 0$ $\frac{1}{2}, -\frac{4}{3}$

19. $(x - 2)(x + 1) = 3$ $\dfrac{1 \pm \sqrt{21}}{2}$

20. $(x - 8)(x + 7) = 5$ $\dfrac{1 \pm 7\sqrt{5}}{2}$

21. $(2x - 3)(x + 2) = 1$ $\dfrac{-1 \pm \sqrt{57}}{4}$

22. $(4x - 5)(x - 3) = 6$ $\dfrac{17 \pm \sqrt{145}}{8}$

23. $2x^2 - 3x = 5$ $\frac{5}{2}, -1$

24. $3x^2 - 4x = 5$ $\dfrac{2 \pm \sqrt{19}}{3}$

25. $2x^2 = -6x + 7$ $\dfrac{-3 \pm \sqrt{23}}{2}$

26. $5x^2 = -6x + 3$ $\dfrac{-3 \pm 2\sqrt{6}}{5}$

▶ **27.** $3x^2 = -4x + 2$ $\dfrac{-2 \pm \sqrt{10}}{3}$

28. $3x^2 = 4x + 2$ $\dfrac{2 \pm \sqrt{10}}{3}$

29. $2x^2 - 5 = 2x$ $\dfrac{1 \pm \sqrt{11}}{2}$

30. $5x^2 + 1 = -10x$ $\dfrac{-5 \pm 2\sqrt{5}}{5}$

31. Solve the equation $2x^3 + 3x^2 - 4x = 0$ by first factoring out the common factor x and then using the quadratic formula. There are three solutions to this equation. $0, \dfrac{-3 \pm \sqrt{41}}{4}$

32. Solve the equation $5y^3 - 10y^2 + 4y = 0$ by first factoring out the common factor y and then using the quadratic formula. $0, \dfrac{5 \pm \sqrt{5}}{5}$

33. To apply the quadratic formula to the equation $3x^2 - 4x = 0$, you have to notice that $c = 0$. Solve the equation using the quadratic formula. $0, \frac{4}{3}$

34. Solve the equation $9x^2 - 16 = 0$ using the quadratic formula. (Notice $b = 0$.) $\frac{4}{3}, -\frac{4}{3}$

35. Solve the following equation by first multiplying both sides by the LCD and then applying the quadratic formula to the result.

$$\frac{1}{2}x^2 - \frac{1}{2}x - \frac{1}{6} = 0 \qquad \dfrac{3 \pm \sqrt{21}}{6}$$

36. Solve the following equation by first multiplying both sides by the LCD and then apply the quadratic formula to the result.

$$\frac{1}{2}y^2 - y - \frac{3}{2} = 0 \qquad 3, -1$$

    = Videos available by instructor request

▶ = Online student support materials available at www.thomsonedu.com/login

37. The equation $(6x - 8)^2 = 32$ can be solved with the quadratic formula. Is there a quicker way?
  Yes, factoring.

38. The equation $2x^2 + 7x = 0$ can be solved with the quadratic formula. Is there a quicker way?
  Yes, factoring.

39. Suppose you solve the equation $x^2 - 10x + 22 = 0$ and find the solutions are $5 \pm \sqrt{3}$. Could you have solved the equation by factoring?   No.

40. Suppose you solve the equation $x^2 - 10x + 16 = 0$ with the quadratic formula and find the solutions are 2 and 8. Does this tell you that factoring would work also?   Yes.

41. For the equation $x^2 - 2x - 1 = 0$
  a. Can it be solved by factoring?   No.
  b. Solve it. $1 \pm \sqrt{2}$

42. For the equation $54x^2 + 111x + 56 = 0$
  a. Can it be solved by factoring?   Answers will vary.
  b. Solve it. $-\frac{8}{9}, -\frac{7}{6}$

43. Is $x = 7 - \sqrt{5}$ a solution to $x^2 - 14x + 44 = 0$?   Yes.

44. Is $x = 5 - \sqrt{7}$ a solution to $x^2 + 10x - 24 = 0$?   No.

▶ **45.** $2t^2 + 7t = 0$ $-\frac{7}{2}, 0$ ▶ **46.** $3a^2 - 8a = 0$ $0, \frac{8}{3}$

▶ **47.** $2t^2 + 7t = -3$ $-3, -\frac{1}{2}$ ▶ **48.** $3a^2 - 8a = 3$ $-\frac{1}{3}, 3$

▶ **49.** $2t^2 + 7t = 3$ $\dfrac{-7 \pm \sqrt{73}}{4}$ ▶ **50.** $3a^2 - 8a = -3$ $\dfrac{4 \pm \sqrt{7}}{3}$

▶ **51.** $(6x - 8)^2 = 32$ $\dfrac{4 \pm 2\sqrt{2}}{3}$

52. a. $(10x - 15)^2 = 50$ $\dfrac{3 \pm \sqrt{2}}{2}$

  b. $2 + \dfrac{7}{t} = -\dfrac{3}{t^2}$ $-\frac{1}{2}, -3$

  c. $3 - \dfrac{8}{a} = \dfrac{3}{a^2}$ $-\frac{1}{3}, 3$

Applying the Concepts

53. Archery Margaret shoots an arrow into the air. The equation for the height (in feet) of the tip of the arrow is $h = 8 + 64t - 16t^2$. To find the time at which the arrow is 56 feet above the ground, we replace h with 56 to obtain

$$56 = 8 + 64t - 16t^2$$

Solve this equation for t to find the times at which the arrow is 56 feet above the ground.
  1 and 3 seconds

54. Coin Toss At the beginning of every football game, the referee flips a coin to see who will kick off. The equation that gives the height (in feet) of the coin tossed in the air is $h = 6 + 32t - 16t^2$. To find the times at which the coin is 18 feet above the ground we substitute 18 for h in the equation, giving us

$$18 = 6 + 32t - 16t^2$$

Solve this equation for t to find the times at which the coin is 18 feet above the ground.
  $\frac{1}{2}$ and 1.5 seconds

Maintaining Your Skills

Add or subtract as indicated.

55. $\dfrac{3}{x-2} + \dfrac{6}{2x-4}$ $\quad \dfrac{6}{x-2}$

56. $\dfrac{6x+5}{5x-25} - \dfrac{x+2}{x-5}$ $\quad \dfrac{1}{5}$

57. $\dfrac{x}{x^2-9} + \dfrac{4}{4x-12}$ $\quad \dfrac{2x+3}{(x+3)(x-3)}$

58. $\dfrac{x+1}{2x-2} - \dfrac{2}{x^2-1}$ $\quad \dfrac{x+3}{2(x+1)}$

59. $\dfrac{2x}{x^2-1} + \dfrac{x}{x^2-3x+2}$ $\quad \dfrac{3x}{(x+1)(x-2)}$

60. $\dfrac{4x}{x^2+6x+5} - \dfrac{3x}{x^2+5x+4}$ $\quad \dfrac{x}{(x+5)(x+4)}$

Getting Ready for the Next Section

Combine.

61. $(3 + 4\sqrt{2}) + (2 - 6\sqrt{2})$ $\quad 5 - 2\sqrt{2}$

62. $(2 + 7\sqrt{2}) + (3 - 5\sqrt{2})$ $\quad 5 + 2\sqrt{2}$

63. $(2 - 5\sqrt{2}) - (3 + 7\sqrt{2}) + (2 - \sqrt{2})$ $\quad 1 - 13\sqrt{2}$

64. $(3 - 8\sqrt{2}) - (2 + 4\sqrt{2}) + (3 - \sqrt{2})$ $\quad 4 - 13\sqrt{2}$

Multiply.

65. $4\sqrt{2}(3 + 5\sqrt{2})$ $\quad 12\sqrt{2} + 40$

66. $5\sqrt{2}(2 + 3\sqrt{2})$ $\quad 10\sqrt{2} + 30$

67. $(3 + 2\sqrt{2})(4 - 3\sqrt{2})$ $\quad -\sqrt{2}$

68. $(2 - 3\sqrt{2})(5 - \sqrt{2})$ $\quad 16 - 17\sqrt{2}$

Rationalize the denominator.

69. $\dfrac{2}{3 + \sqrt{5}}$ $\quad \dfrac{3 - \sqrt{5}}{2}$

70. $\dfrac{2}{5 + \sqrt{3}}$ $\quad \dfrac{5 - \sqrt{3}}{11}$

71. $\dfrac{2 + \sqrt{3}}{2 - \sqrt{3}}$ $\quad 7 + 4\sqrt{3}$

72. $\dfrac{3 + \sqrt{5}}{3 - \sqrt{5}}$ $\quad \dfrac{7 + 3\sqrt{5}}{2}$

9.4 Complex Numbers

OBJECTIVES

A Add and subtract complex numbers.

B Multiply two complex numbers.

C Divide two complex numbers.

To solve quadratic equations such as $x^2 = -4$, we need to introduce a new set of numbers. If we try to solve $x^2 = -4$ using real numbers, we always get no solution. There is no real number whose square is -4.

The new set of numbers is called the *complex numbers* and is based on the following definition.

> **DEFINITION** The **number i** is a number such that $i = \sqrt{-1}$.

The first thing we notice about this definition is that i is not a real number. There are no real numbers that represent the square root of -1. The other observation we make about i is $i^2 = -1$. If $i = \sqrt{-1}$, then, squaring both sides, we must have $i^2 = -1$. The most common power of i is i^2. Whenever we see i^2, we can write it as -1. We are now ready for a definition of complex numbers.

> **DEFINITION** A **complex number** is any number that can be put in the form $a + bi$, where a and b are real numbers and $i = \sqrt{-1}$.

Note

The form $a + bi$ is called *standard form* for complex numbers. The definition in the box indicates that if a number can be written in the form $a + bi$, then it is a complex number.

Example The following are complex numbers:

$$3 + 4i \qquad \frac{1}{2} - 6i \qquad 8 + i\sqrt{2} \qquad \frac{3}{4} - 2i\sqrt{5}$$

Example The number $4i$ is a complex number because $4i = 0 + 4i$.

Example The number 8 is a complex number because $8 = 8 + 0i$.

From the previous example we can see that the real numbers are a subset of the complex numbers because any real number x can be written as $x + 0i$.

Addition and Subtraction of Complex Numbers

We add and subtract complex numbers according to the same procedure we used to add and subtract polynomials: We combine similar terms.

 EXAMPLE 1   Combine $(3 + 4i) + (2 - 6i)$.

SOLUTION

$$(3 + 4i) + (2 - 6i) = (3 + 2) + (4i - 6i) \qquad \textbf{Commutative and}$$
$$\textbf{associative properties}$$
$$= 5 + (-2i) \qquad \textbf{Combine similar terms}$$
$$= 5 - 2i \qquad \text{}$$

 EXAMPLE 2   Combine $(2 - 5i) - (3 + 7i) + (2 - i)$.

SOLUTION

$$(2 - 5i) - (3 + 7i) + (2 - i) = 2 - 5i - 3 - 7i + 2 - i$$
$$= (2 - 3 + 2) + (-5i - 7i - i)$$
$$= 1 - 13i \qquad \text{}$$

Multiplication of Complex Numbers

Multiplication of complex numbers is very similar to multiplication of polynomials. We can simplify many answers by using the fact that $i^2 = -1$.

 EXAMPLE 3   Multiply $4i(3 + 5i)$.

SOLUTION $4i(3 + 5i) = 4i(3) + 4i(5i)$ **Distributive property**
$$= 12i + 20i^2 \qquad \textbf{Multiplication}$$
$$= 12i + 20(-1) \qquad \textbf{\textit{i}}^2 = \textbf{--1}$$
$$= -20 + 12i \qquad \text{}$$

 EXAMPLE 4   Multiply $(3 + 2i)(4 - 3i)$.

SOLUTION

$$(3 + 2i)(4 - 3i) = 3 \cdot 4 + 3(-3i) + 2i(4) + 2i(-3i) \qquad \textbf{FOIL method}$$
$$= 12 - 9i + 8i - 6i^2$$
$$= 12 - 9i + 8i - 6(-1) \qquad \textbf{\textit{i}}^2 = \textbf{--1}$$
$$= (12 + 6) + (-9i + 8i)$$
$$= 18 - i \qquad \text{}$$

Division of Complex Numbers

We divide complex numbers by applying the same process we used to rational-ize denominators.

EXAMPLE 5   Divide $\dfrac{2 + i}{5 + 2i}$.

SOLUTION   The conjugate of the denominator is $5 - 2i$:

$$\left(\frac{2 + i}{5 + 2i}\right)\left(\frac{5 - 2i}{5 - 2i}\right) = \frac{10 - 4i + 5i - 2i^2}{25 - 4i^2}$$

$$= \frac{10 - 4i + 5i - 2(-1)}{25 - 4(-1)} \qquad i^2 = -1$$

$$= \frac{12 + i}{29}$$

When we write our answer in standard form for complex numbers, we get:

$$\frac{12}{29} + \frac{1}{29}i$$

Note

The conjugate of $a + bi$ is $a - bi$. When we multiply complex conjugates the result is always a real number because:

$(a + bi)(a - bi) = a^2 - (bi)^2$
$= a^2 - b^2 i^2$
$= a^2 - b^2(-1)$
$= a^2 + b^2$

which is a real number.

LINKING OBJECTIVES AND EXAMPLES

Next to each objective we have listed the examples that are best described by that objective.

A	1, 2
B	3, 4
C	5

GETTING READY FOR CLASS

After reading through the preceding section, respond in your own words and in complete sentences.

1. What is the number i?

2. What is a complex number?

3. What kind of number will always result when we multiply complex conjugates?

4. Explain how to divide complex numbers.

Problem Set 9.4

Online support materials can be found at www.thomsonedu.com/login

Combine the following complex numbers.

1. $(3 - 2i) + 3i$ $3 + i$

2. $(5 - 4i) - 8i$ $5 - 12i$

▶ **3.** $(6 + 2i) - 10i$ $6 - 8i$

4. $(8 - 10i) + 7i$ $8 - 3i$

5. $(11 + 9i) - 9i$ 11

6. $(12 + 2i) + 6i$ $12 + 8i$

7. $(3 + 2i) + (6 - i)$ $9 + i$

8. $(4 + 8i) - (7 + i)$ $-3 + 7i$

9. $(5 + 7i) - (6 + 8i)$ $-1 - i$

10. $(11 + 6i) - (3 + 6i)$ 8

▶ **11.** $(9 - i) + (2 - i)$ $11 - 2i$

12. $(8 + 3i) - (8 - 3i)$ $6i$

13. $(6 + i) - 4i - (2 - i)$ $4 - 2i$

14. $(3 + 2i) - 5i - (5 + 4i)$ $-2 - 7i$

15. $(6 - 11i) + 3i + (2 + i)$ $8 - 7i$

16. $(3 + 4i) - (5 + 7i) - (6 - i)$ $-8 - 2i$

17. $(2 + 3i) - (6 - 2i) + (3 - i)$ $-1 + 4i$

18. $(8 + 9i) + (5 - 6i) - (4 - 3i)$ $9 + 6i$

Multiply the following complex numbers.

19. $3(2 - i)$ $6 - 3i$

20. $4(5 + 3i)$ $20 + 12i$

▶ **21.** $2i(8 - 7i)$ $14 + 16i$

22. $-3i(2 + 5i)$ $15 - 6i$

23. $(2 + i)(4 - i)$ $9 + 2i$

24. $(6 + 3i)(4 + 3i)$ $15 + 30i$

25. $(2 + i)(3 - 5i)$ $11 - 7i$

26. $(4 - i)(2 - i)$ $7 - 6i$

▶ **27.** $(3 + 5i)(3 - 5i)$ 34

28. $(8 + 6i)(8 - 6i)$ 100

29. $(2 + i)(2 - i)$ 5

30. $(3 + i)(3 - i)$ 10

= Videos available by instructor request

▶ = Online student support materials available at www.thomsonedu.com/login

Divide the following complex numbers.

31. $\dfrac{2}{3-2i}$ $\quad \dfrac{6+4i}{13}$ **32.** $\dfrac{3}{5+6i}$ $\quad \dfrac{15-18i}{61}$

33. $\dfrac{-3i}{2+3i}$ $\quad \dfrac{-9-6i}{13}$ **34.** $\dfrac{4i}{3+i}$ $\quad \dfrac{2+6i}{5}$

35. $\dfrac{2+i}{2-i}$ $\quad \dfrac{3+4i}{5}$ **36.** $\dfrac{3+2i}{3-2i}$ $\quad \dfrac{5+12i}{13}$

37. Work each problem according to the instructions given.
  a. Add: $(2+3i)+(2-3i)$ 4
  b. Multiply: $(2+3i)(2-3i)$ 13
  c. Square: $(2-3i)^2$ $-5-12i$
  d. Divide: $\dfrac{2-3i}{2+3i}$ $\dfrac{-5-12i}{13}$

38. Work each problem according to the instructions given.
  a. Subtract: $(1+i)-(1-i)$ $2i$
  b. Multiply: $(1+i)(1-i)$ 2
  c. Square: $(1+i)^2$ $2i$
  d. Divide: $\dfrac{(1+i)}{(1-i)}$ i

39. Work each problem according to the instructions given.
  a. Add: $6i+(3-i)$ $3+5i$
  b. Multiply: $6i(3-i)$ $6+18i$
  c. Divide: $\dfrac{3-i}{6i}$ $\dfrac{1+3i}{6}$
  d. Divide: $\dfrac{6i}{3-i}$ $\dfrac{-3+9i}{5}$

40. Work each problem according to the instructions given.
  a. Subtract: $7i-(5-4i)$ $-5+11i$
  b. Multiply: $7i(5-4i)$ $28+35i$
  c. Divide: $\dfrac{5-4i}{7i}$ $\dfrac{-4+5i}{7}$
  d. Divide: $\dfrac{7i}{5-4i}$ $\dfrac{-28+35i}{41}$

Simplify.

41. $(3+2i)^2-6(3+2i)$ -13
42. $(5-4i)^2-10(5-4i)$ -41
43. $3(2+i)^2-12(2+i)$ -15
44. $5(1-i)^2-10(1-i)$ -10
45. $(1+i)^2-2(1+i)+2$ 0
46. $(2+2i)^2-4(2+2i)+8$ 0
47. Is $x=3i$ a solution to $x^2+9=0$? Yes.
48. Is $x=-5i$ a solution to $x^2+25=0$? Yes.
49. Solve $x^2=-36$ by applying the square root property for equations. $\pm 6i$

Chalkboard Problems
Problems 37–40 allow you to do all the operations with complex numbers. Each problem sets things up so division follows the other operations very smoothly.

50. Solve $x^2+49=0$ by applying the square root property for equations. $\pm 7i$
51. Is $a=1+i$ a solution to $a^2-2a=-2$? Yes.
52. Is $a=2+2i$ a solution to $a^2-4a=-8$? Yes.
53. Use the FOIL method to multiply $(x+3i)(x-3i)$. x^2+9
54. Use the FOIL method to multiply $(x+5i)(x-5i)$. x^2+25
55. The opposite of i is $-i$. The reciprocal of i is $\dfrac{1}{i}$. Multiply the numerator and denominator of $\dfrac{1}{i}$ by i and simplify the result to see that the opposite of i and the reciprocal of i are the same number.
$\dfrac{1}{i}\cdot\dfrac{i}{i}=\dfrac{i}{i^2}=\dfrac{i}{-1}=-i$
56. If $i^2=-1$, what are i^3 and i^4? (*Hint*: $i^3=i^2\cdot i$.) $-i$ and 1

Maintaining Your Skills

Solve each equation by taking the square root of each side.

57. $(x-3)^2=25$ $8,-2$ **58.** $(x-2)^2=9$ $5,-1$
59. $(2x-6)^2=16$ $5,1$ **60.** $(2x+1)^2=49$ $-4,3$
61. $(x+3)^2=12$ $-3\pm2\sqrt3$ **62.** $(x+3)^2=8$ $-3\pm2\sqrt2$

Put each expression into simplified form for radicals.

63. $\sqrt{\dfrac{1}{2}}$ $\dfrac{\sqrt2}{2}$ **64.** $\sqrt{\dfrac{5}{6}}$ $\dfrac{\sqrt{30}}{6}$
65. $\sqrt{\dfrac{8x^2y^3}{3}}$ $\dfrac{2xy\sqrt{6y}}{3}$ **66.** $\sqrt{\dfrac{45xy^4}{7}}$ $\dfrac{3y^2\sqrt{35x}}{7}$
67. $\sqrt[3]{\dfrac{1}{4}}$ $\dfrac{\sqrt[3]{2}}{2}$ **68.** $\sqrt[3]{\dfrac{2}{3}}$ $\dfrac{\sqrt[3]{18}}{3}$

Getting Ready for the Next Section

Simplify.
69. $\sqrt{36}$ 6 **70.** $\sqrt{49}$ 7
71. $\sqrt{75}$ $5\sqrt3$ **72.** $\sqrt{12}$ $2\sqrt3$

Solve for x.
73. $(x+2)^2=9$ $-5,1$ **74.** $(x-3)^2=9$ $0,6$
75. $\dfrac{1}{10}x^2-\dfrac{1}{5}x=\dfrac{1}{2}$ $1\pm\sqrt6$
76. $\dfrac{1}{18}x^2-\dfrac{2}{9}x=\dfrac{13}{18}$ $2\pm\sqrt{17}$
77. $(2x-3)(2x-1)=4$ $\dfrac{2\pm\sqrt5}{2}$
78. $(x-1)(3x-3)=10$ $\dfrac{3\pm\sqrt{30}}{3}$

9.5 Complex Solutions to Quadratic Equations

OBJECTIVES

A Write square roots of negative numbers as complex numbers.

B Solve quadratic equations whose solutions may be complex numbers.

The quadratic formula tells us that the solutions to equations of the form $ax^2 + bx + c = 0$ are always:

$$x = \frac{-b \pm \sqrt{b^2 - 4ac}}{2a}$$

The part of the quadratic formula under the radical sign is called the *discriminant*:

$$\text{Discriminant} = b^2 - 4ac$$

When the discriminant is negative, we have to deal with the square root of a negative number. We handle square roots of negative numbers by using the definition $i = \sqrt{-1}$. To illustrate, suppose we want to simplify an expression that contains $\sqrt{-9}$, which is not a real number. We begin by writing $\sqrt{-9}$ as $\sqrt{9(-1)}$. Then, we write this expression as the product of two separate radicals: $\sqrt{9}\,\sqrt{-1}$. Applying the definition $i = \sqrt{-1}$ to this last expression, we have

$$\sqrt{9}\,\sqrt{-1} = 3i$$

As you may recall from the previous section, the number $3i$ is called a complex number. Here are some further examples.

 EXAMPLE 1   Write the following radicals as complex numbers:

a. $\sqrt{-4} = \sqrt{4(-1)} = \sqrt{4}\,\sqrt{-1} = 2i$
b. $\sqrt{-36} = \sqrt{36(-1)} = \sqrt{36}\,\sqrt{-1} = 6i$
c. $\sqrt{-7} = \sqrt{7(-1)} = \sqrt{7}\,\sqrt{-1} = i\sqrt{7}$
d. $\sqrt{-75} = \sqrt{75(-1)} = \sqrt{75}\,\sqrt{-1} = 5i\sqrt{3}$

In parts (c) and (d) of Example 1, we wrote i before the radical because it is less confusing that way. If we put i after the radical, it is sometimes mistaken for being under the radical.

Let's see how complex numbers relate to quadratic equations by looking at some examples of quadratic equations whose solutions are complex numbers.

EXAMPLE 2   Solve for x: $(x + 2)^2 = -9$

SOLUTION   We can solve this equation by expanding the left side, putting the results into standard form, and then applying the quadratic formula. It is faster, however, simply to apply the square root property:

$$(x + 2)^2 = -9$$

$$x + 2 = \pm\sqrt{-9} \qquad \textbf{Square root property}$$

$$x + 2 = \pm 3i \qquad \boldsymbol{\sqrt{-9} = \sqrt{9}\,\sqrt{-1} = 3i}$$

$$x = -2 \pm 3i \qquad \textbf{Add } \boldsymbol{-2} \textbf{ to both sides}$$

The solution set contains two complex solutions. Notice that the two solutions are conjugates.

The solution set is $\{-2 + 3i, -2 - 3i\}$.

 EXAMPLE 3   Solve for x: $\frac{1}{10}x^2 - \frac{1}{5}x = -\frac{1}{2}$.

SOLUTION   It will be easier to apply the quadratic formula if we clear the equation of fractions. Multiplying both sides of the equation by the LCD 10 gives us:

$$x^2 - 2x = -5$$

Next, we add 5 to both sides to put the equation into standard form:

$$x^2 - 2x + 5 = 0 \qquad \textbf{Add 5 to both sides}$$

Applying the quadratic formula with $a = 1$, $b = -2$, and $c = 5$, we have:

$$x = \frac{-(-2) \pm \sqrt{(-2)^2 - 4(1)(5)}}{2(1)} = \frac{2 \pm \sqrt{-16}}{2} = \frac{2 \pm 4i}{2}$$

Dividing the numerator and denominator by 2, we have the two solutions:

$$x = 1 \pm 2i$$

The two solutions are $1 + 2i$ and $1 - 2i$.

 EXAMPLE 4   Solve $(2x - 3)(2x - 1) = -4$.

SOLUTION   We multiply the binomials on the left side and then add 4 to each side to write the equation in standard form. From there we identify a, b, and c and apply the quadratic formula:

$$(2x - 3)(2x - 1) = -4$$
$$4x^2 - 8x + 3 = -4 \qquad \textbf{Multiply binomials on left side}$$
$$4x^2 - 8x + 7 = 0 \qquad \textbf{Add 4 to each side}$$

Placing $a = 4$, $b = -8$, and $c = 7$ in the quadratic formula we have:

$$x = \frac{-(-8) \pm \sqrt{(-8)^2 - 4(4)(7)}}{2(4)}$$

$$= \frac{8 \pm \sqrt{64 - 112}}{8}$$

$$= \frac{8 \pm \sqrt{-48}}{8}$$

$$= \frac{8 \pm 4i\sqrt{3}}{8} \qquad \sqrt{-48} = i\sqrt{48} = i\sqrt{16}\,\sqrt{3} = 4i\sqrt{3}$$

Note

It would be a mistake to try to reduce this final expression further. Sometimes first-year algebra students will try to divide the 2 in the denominator into the 2 in the numerator, which is a mistake. Remember, when we reduce to lowest terms, we do so by dividing the numerator and denominator by any factors they have in common. In this case 2 is not a factor of the numerator. This expression is in lowest terms.

To reduce this final expression to lowest terms, we factor a 4 from the numerator and then divide the numerator and denominator by 4:

$$= \frac{\cancel{4}(2 \pm i\sqrt{3})}{\cancel{4} \cdot 2}$$

$$= \frac{2 \pm i\sqrt{3}}{2}$$

LINKING OBJECTIVES AND EXAMPLES

Next to each **objective** we have listed the examples that are best described by that objective.

A 1

B 2–4

GETTING READY FOR CLASS

After reading through the preceding section, respond in your own words and in complete sentences.

1. Describe how you would write $\sqrt{-4}$ in terms of the number i.
2. When would the quadratic formula result in complex solutions?
3. How can we avoid working with fractions when using the quadratic formula?
4. Describe how you would simplify the expression $\dfrac{8 \pm 4i\sqrt{3}}{8}$.

Problem Set 9.5

Online support materials can be found at www.thomsonedu.com/login

Write the following radicals as complex numbers.

1. $\sqrt{-16}$ $4i$
2. $\sqrt{-25}$ $5i$
▶ 3. $\sqrt{-49}$ $7i$
4. $\sqrt{-81}$ $9i$
5. $\sqrt{-6}$ $i\sqrt{6}$
6. $\sqrt{-10}$ $i\sqrt{10}$
7. $\sqrt{-11}$ $i\sqrt{11}$
8. $\sqrt{-19}$ $i\sqrt{19}$
9. $\sqrt{-32}$ $4i\sqrt{2}$
10. $\sqrt{-288}$ $12i\sqrt{2}$
11. $\sqrt{-50}$ $5i\sqrt{2}$
12. $\sqrt{-45}$ $3i\sqrt{5}$
▶ 13. $\sqrt{-8}$ $2i\sqrt{2}$
14. $\sqrt{-24}$ $2i\sqrt{6}$
15. $\sqrt{-48}$ $4i\sqrt{3}$
16. $\sqrt{-27}$ $3i\sqrt{3}$

Solve the following quadratic equations. Use whatever method seems to fit the situation or is convenient for you.

17. $x^2 = 2x - 2$ $1 \pm i$
18. $x^2 = 4x - 5$ $2 \pm i$
19. $x^2 - 4x = -4$ 2
20. $x^2 - 4x = 4$ $2 \pm 2\sqrt{2}$
▶ 21. $2x^2 + 5x = 12$ $\frac{3}{2}, -4$
22. $2x^2 + 30 = 16x$ $3, 5$
23. $(x - 2)^2 = -4$ $2 \pm 2i$
24. $(x - 5)^2 = -25$ $5 \pm 5i$
▶ 25. $\left(x + \dfrac{1}{2}\right)^2 = -\dfrac{9}{4}$ $\dfrac{-1 \pm 3i}{2}$
26. $\left(x - \dfrac{1}{4}\right)^2 = -\dfrac{1}{2}$ $\dfrac{1 \pm 2i\sqrt{2}}{4}$
27. $\left(x - \dfrac{1}{2}\right)^2 = -\dfrac{27}{36}$ $\dfrac{1 \pm i\sqrt{3}}{2}$
28. $\left(x + \dfrac{1}{2}\right)^2 = -\dfrac{32}{64}$ $\dfrac{-1 \pm i\sqrt{2}}{2}$
29. $x^2 + x + 1 = 0$ $\dfrac{-1 \pm i\sqrt{3}}{2}$

30. $x^2 - 3x + 4 = 0$ $\dfrac{3 \pm i\sqrt{7}}{2}$
31. $x^2 - 5x + 6 = 0$ $2, 3$
32. $x^2 + 2x + 2 = 0$ $-1 \pm i$
▶ 33. $\dfrac{1}{2}x^2 + \dfrac{1}{3}x + \dfrac{1}{6} = 0$ $\dfrac{-1 \pm i\sqrt{2}}{3}$
34. $\dfrac{1}{5}x^2 + \dfrac{1}{20}x + \dfrac{1}{4} = 0$ $\dfrac{-1 \pm i\sqrt{79}}{8}$
35. $\dfrac{1}{3}x^2 = -\dfrac{1}{2}x + \dfrac{1}{3}$ $\dfrac{1}{2}, -2$
36. $\dfrac{1}{2}x^2 = -\dfrac{1}{3}x + \dfrac{1}{6}$ $-1, \dfrac{1}{3}$
37. $(x + 2)(x - 3) = 5$ $\dfrac{1 \pm 3\sqrt{5}}{2}$
38. $(x - 1)(x + 1) = 6$ $\pm\sqrt{7}$
39. $(x - 5)(x - 3) = -10$ $4 \pm 3i$
40. $(x - 2)(x - 4) = -5$ $3 \pm 2i$
41. $(2x - 2)(x - 3) = 9$ $\dfrac{4 \pm \sqrt{22}}{2}$
42. $(x - 1)(2x + 6) = 9$ $\dfrac{-2 \pm \sqrt{34}}{2}$

43. What is the fastest method of solving the equation $(9x - 11)^2 = -16$?   Taking the square root of each side.

44. What is the fastest method of solving $2x^2 - 6x = 0$?   Factoring.

45. If the solutions to $x^2 - 4x + 13 = 0$ are $2 \pm 3i$, can you solve the equation by factoring?   No.

46. If the solutions to the equation $2x^2 - 6x = 20$ are -2 and 5, can you solve the equation by factoring?   Yes.

▢ = Videos available by instructor request
▶ = Online student support materials available at www.thomsonedu.com/login

Solve each equation using any method you like. In each case, try to find the most efficient method.

▶ **47.** $(9x - 11)^2 = -16$ $\dfrac{11 \pm 4i}{9}$

▶ **48.** $(13x - 5)^2 = -25$ $\dfrac{5 \pm 5i}{13}$

▶ **49.** $2y^2 - 6y = 0$ $0, 3$ ▶ **50.** $9t^2 - 6t = 0$ $0, \dfrac{2}{3}$

▶ **51.** $2y^2 - 6y = -5$ $\dfrac{3 \pm i}{2}$ ▶ **52.** $9t^2 - 6t = -5$ $\dfrac{1 \pm 2i}{3}$

▶ **53.** $2y^2 - 6y = 20$ $-2, 5$ ▶ **54.** $9t^2 - 6t = 3$ $-\dfrac{1}{3}, 1$

▶ **55.** $(8x - 12)^2 = -36$ $\dfrac{6 \pm 3i}{4}$

▶ **56.** $(6x - 18)^2 = -81$ $\dfrac{6 \pm 3i}{2}$

▶ **57.** $1 - \dfrac{3}{y} = \dfrac{10}{y^2}$ $-2, 5$ ▶ **58.** $3 - \dfrac{2}{t} = \dfrac{1}{t^2}$ $-\dfrac{1}{3}, 1$

59. Intercepts The graph of $y = x^2 - 4x + 5$ is shown in the following figure. As you can see, the graph does not cross the x-axis. If the graph did cross the x-axis, the x-intercepts would be solutions to the equation

$$0 = x^2 - 4x + 5$$

Solve this equation. The solutions will confirm the fact that the graph cannot cross the x-axis. $2 \pm i$

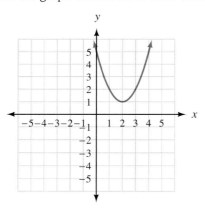

60. Intercepts The graph of $y = x^2 + 2x + 5$ is shown in the following figure. As you can see, the graph does not cross the x-axis. If the graph did cross the x-axis, the x-intercepts would be solutions to the equation

$$0 = x^2 + 2x + 5$$

Solve this equation. $-1 \pm 2i$

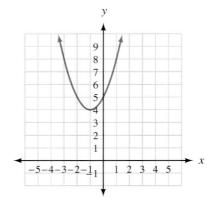

61. Is $x = 2 + 2i$ a solution to the equation $x^2 - 4x + 8 = 0$?   Yes

62. Is $x = 5 + 3i$ a solution to the equation $x^2 - 10x + 34 = 0$?   Yes

63. If one solution to a quadratic equation is $3 + 7i$, what do you think the other solution is? $3 - 7i$

64. If one solution to a quadratic equation is $4 - 2i$, what do you think the other solution is? $4 + 2i$

Maintaining Your Skills

Graph each line.

65. $y = x - 2$ **66.** $y = -x - 2$

67. $2x + 4y = 8$ **68.** $2x - 4y = 8$

Write each term in simplified form for radicals. Then combine similar terms.

69. $3\sqrt{50} + 2\sqrt{32}$ $23\sqrt{2}$

70. $4\sqrt{18} + \sqrt{32} - \sqrt{2}$ $15\sqrt{2}$

71. $\sqrt{24} - \sqrt{54} - \sqrt{150}$ $-6\sqrt{6}$

72. $\sqrt{72} - \sqrt{8} + \sqrt{50}$ $9\sqrt{2}$

73. $2\sqrt{27x^2} - x\sqrt{48}$ $2x\sqrt{3}$

74. $5\sqrt{8x^3} + x\sqrt{50x}$ $15x\sqrt{2x}$

Getting Ready for the Next Section

Add the correct term to each binomial so that it becomes a perfect square trinomial, then factor.

75. $x^2 - 6x$ $x^2 - 6x + 9 = (x - 3)^2$

76. $x^2 - 8x$ $x^2 - 8x + 16, (x - 4)^2$

Use the equation $y = (x + 1)^2 - 3$ to find the value of y when

77. $x = -4$ 6

78. $x = -3$ 1

79. $x = -2$ -2

80. $x = -1$ -3

81. $x = 1$ 1

82. $x = 2$ 6

Graph the ordered pairs.

83. $(1, 2)$

84. $(2, 1)$

85. $(-2, -2)$

86. $(-1, -1)$

87. $(-4, 1)$

88. $(-1, 2)$

89. $(3, -4)$

90. $(1, -2)$

91. $(5, 0)$

92. $(-3, 0)$

9.6 Graphing Parabolas

OBJECTIVES

A Graph a parabola.

In this section we will graph equations of the form $y = ax^2 + bx + c$ and equations that can be put into this form. The graphs of this type of equation all have similar shapes.

We will begin this section by graphing the simplest quadratic equation, $y = x^2$. To get the idea of the shape of this graph, we need to find some ordered pairs that are solutions. We can do this by setting up the following table:

x	$y = x^2$	y

We can choose any convenient numbers for x and then use the equation $y = x^2$ to find the corresponding values for y. Let's use the values $-3, -2, -1, 0, 1, 2,$ and 3 for x and find corresponding values for y. Here is how the table looks when we let x have these values:

x	$y = x^2$	y
-3	$y = (-3)^2 = 9$	9
-2	$y = (-2)^2 = 4$	4
-1	$y = (-1)^2 = 1$	1
0	$y = 0^2 = 0$	0
1	$y = 1^2 = 1$	1
2	$y = 2^2 = 4$	4
3	$y = 3^2 = 9$	9

The table gives us the solutions $(-3, 9)$, $(-2, 4)$, $(-1, 1)$, $(0, 0)$, $(1, 1)$, $(2, 4)$, and $(3, 9)$ for the equation $y = x^2$. We plot each of the points on a rectangular coordinate system and draw a smooth curve between them, as shown in Figure 1.

This graph is called a *parabola*. All equations of the form $y = ax^2 + bx + c$ ($a \neq 0$) produce parabolas when graphed.

Note that the point $(0, 0)$ is called the vertex of the parabola in Figure 1. It is the lowest point on this graph. There are some parabolas that open downward. For those parabolas, the vertex is the highest point on the graph.

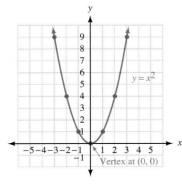

FIGURE 1

EXAMPLE 1 Graph the equation $y = x^2 - 3$.

SOLUTION We begin by making a table using convenient values for x:

x	$y = x^2 - 3$	y
-2	$y = (-2)^2 - 3 = 4 - 3 = 1$	1
-1	$y = (-1)^2 - 3 = 1 - 3 = -2$	-2
0	$y = 0^2 - 3 = -3$	-3
1	$y = 1^2 - 3 = 1 - 3 = -2$	-2
2	$y = 2^2 - 3 = 4 - 3 = 1$	1

The table gives us the ordered pairs $(-2, 1)$, $(-1, -2)$, $(0, -3)$, $(1, -2)$, and $(2, 1)$ as solutions to $y = x^2 - 3$. The graph is shown in Figure 2.

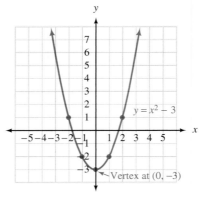

FIGURE 2

EXAMPLE 2 Graph $y = (x - 2)^2$.

SOLUTION

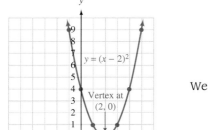

FIGURE 3

x	$y = (x - 2)^2$	y
-1	$y = (-1 - 2)^2 = (-3)^2 = 9$	9
0	$y = (0 - 2)^2 = (-2)^2 = 4$	4
1	$y = (1 - 2)^2 = (-1)^2 = 1$	1
2	$y = (2 - 2)^2 = 0^2 = 0$	0

We can continue the table if we feel more solutions will make the graph clearer.

x	$y = (x-2)^2$	y
3	$y = (3 - 2)^2 = 1^2 = 1$	1
4	$y = (4 - 2)^2 = 2^2 = 4$	4
5	$y = (5 - 2)^2 = 3^2 = 9$	9

Putting the results of the table onto a coordinate system, we have the graph in Figure 3.

When graphing parabolas, it is sometimes easier to obtain the correct graph if you have some idea what the graph will look like before you start making your table.

If you look over the first two examples closely, you will see that the graphs in Figures 2 and 3 have the same shape as the graph of $y = x^2$. The difference is in the position of the vertex. For example, the graph of $y = x^2 - 3$, as shown in Figure 2, looks like the graph of $y = x^2$ with its vertex moved down three units vertically. Similarly, the graph of $y = (x - 2)^2$, shown in Figure 3, looks like the graph of $y = x^2$ with its vertex moved two units to the right.

Without showing the tables necessary to graph them, Figures 4–7 are four more parabolas and their corresponding equations. Can you see the correlation between the numbers in the equations and the position of the graphs on the coordinate systems?

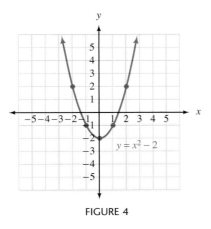

FIGURE 4

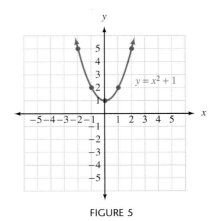

FIGURE 5

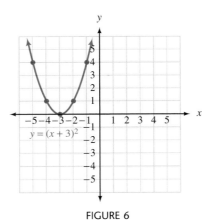

FIGURE 6

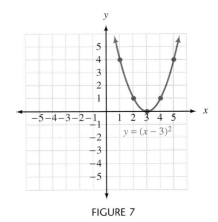

FIGURE 7

For our next example, we will graph a parabola in which the vertex has been moved both vertically and horizontally from the origin.

EXAMPLE 3   Graph $y = (x + 1)^2 - 3$.

SOLUTION

x	$y = (x + 1)^2 - 3$	y
-4	$y = (-4 + 1)^2 - 3 = 9 - 3$	6
-3	$y = (-3 + 1)^2 - 3 = 4 - 3$	1
-1	$y = (-1 + 1)^2 - 3 = 0 - 3$	-3
1	$y = (1 + 1)^2 - 3 = 4 - 3$	1
2	$y = (2 + 1)^2 - 3 = 9 - 3$	6

Graphing the results of the table, we have Figure 8.

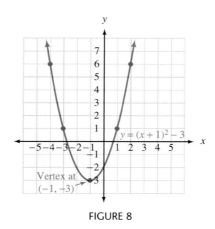

FIGURE 8

We summarize the information shown in the preceding examples and graphs with the following statement.

Graphing Parabolas The graph of each of the following equations is a parabola with the same basic shape as the graph of $y = x^2$.

Equation	Vertex	To obtain the graph, move the graph of $y = x^2$
$y = x^2 + k$	$(0, k)$	k units vertically
$y = (x - h)^2$	$(h, 0)$	h units horizontally
$y = (x - h)^2 + k$	(h, k)	h units horizontally and k units vertically

---

Our last example for this section shows what we do if the equation of the parabola we want to graph is not in the form $y = (x - h)^2 + k$.

EXAMPLE 4   Graph $y = x^2 - 6x + 5$.

SOLUTION   We could graph this equation by making a table of values of x and y as we have previously. However, there is an easier way. If we can write the equation in the form

$$y = (x - h)^2 + k$$

then, to graph it, we simply place the vertex at (h, k) and draw a graph from there that has the same shape as $y = x^2$.

To write our equation in the form given in the preceding equation, we simply complete the square on the first two terms on the right side of the equation, $x^2 - 6x$. We do so by adding 9 to and subtracting 9 from the right side of the equation. This amounts to adding 0 to the equation, so we know we haven't changed its solutions. This is what it looks like:

$$y = (x^2 - 6x\ ) + 5$$

$$y = (x^2 - 6x + \mathbf{9}) + 5 - \mathbf{9}$$

$$y = (x - 3)^2 - 4$$

This final equation has the form we are looking for. Our graph will have the same shape as the graph of $y = x^2$. The vertex of our graph is at $(3, -4)$. The graph is shown in Figure 9.

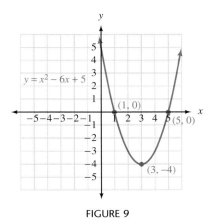

$y = x^2 - 6x + 5$

$(1, 0)$

$(5, 0)$

$(3, -4)$

FIGURE 9

One final note: We can check to see that the x-intercepts in Figure 9 are correct by finding the x-intercepts from our original equation. Remember, the x-intercepts occur when y is 0; that is, to find the x-intercepts, we let $y = 0$ and solve for x:

$$0 = x^2 - 6x + 5$$

$$0 = (x - 5)(x - 1)$$

$$x = 5 \quad \text{or} \quad x = 1$$

LINKING OBJECTIVES AND EXAMPLES

Next to each **objective** we have listed the examples that are best described by that objective.

A   1–4

GETTING READY FOR CLASS

After reading through the preceding section, respond in your own words and in complete sentences.

1. What is a parabola?

2. What is the vertex of a parabola?

3. How do you find the x-intercepts of a parabola?

4. Describe the relationship between the graph of $y = x^2$ and the graph of $y = x^2 - 2$.

Problem Set 9.6

Online support materials can be found at www.thomsonedu.com/login

Graph each of the following equations.

▶ **1.** $y = x^2 - 4$

2. $y = x^2 + 2$

3. $y = x^2 + 5$

4. $y = x^2 - 2$

▶ **5.** $y = (x + 2)^2$

6. $y = (x + 5)^2$

7. $y = (x - 3)^2$

8. $y = (x - 2)^2$

9. $y = (x - 5)^2$

10. $y = (x + 3)^2$

11. $y = (x + 1)^2 - 2$

12. $y = (x - 1)^2 + 2$

13. $y = (x + 2)^2 - 3$

14. $y = (x - 2)^2 + 3$

15. $y = (x - 3)^2 + 2$

16. $y = (x + 4)^2 + 1$

= Videos available by instructor request

▶ = Online student support materials available at www.thomsonedu.com/login

Graph each of the following equations. Begin by completing the square on the first two terms as we did in Example 4 of this section.

17. $y = x^2 + 6x + 5$ **18.** $y = x^2 - 8x + 12$

19. $y = x^2 - 2x - 3$ **20.** $y = x^2 + 2x - 3$

The following equations have graphs that are also parabolas. However, the graphs of these equations will open downward; the vertex of each will be the highest point on the graph. Graph each equation by first making a table of ordered pairs using the given values of x.

21. $y = 4 - x^2$ $x = -3, -2, -1, 0, 1, 2, 3$

22. $y = 3 - x^2$ $x = -3, -2, -1, 0, 1, 2, 3$

23. $y = -1 - x^2$ $x = -2, -1, 0, 1, 2$

24. $y = -2 - x^2$ $x = -2, -1, 0, 1, 2$

25. Graph the line $y = x + 2$ and the parabola $y = x^2$ on the same coordinate system. Name the points where the two graphs intersect.

26. Graph the line $y = x$ and the parabola $y = x^2 - 2$ on the same coordinate system. Name the points where the two graphs intersect.

27. Graph the parabola $y = 2x^2$ and the parabola $y = \frac{1}{2}x^2$ on the same coordinate system.

28. Graph the parabola $y = 3x^2$ and the parabola $y = \frac{1}{3}x^2$ on the same coordinate system.

Applying the Concepts

29. Comet Orbit If a comet approaches the sun with *exactly* the right velocity and direction, its orbit around the sun will be a parabola. (Such a comet only will be seen once and then will leave the solar system forever.) A certain comet's orbit is given by $y = x^2 - 0.25$, where both x and y are measured in millions of miles, and the sun's position is at the origin $(0, 0)$.
   a. Graph the comet's orbit between $x = -2$ and $x = 2$.
   b. How far is the comet from the sun when the comet is at the vertex of its path?
      250,000 miles
   c. How far is the comet from the sun when the comet crosses the x-axis?
      500,000 miles

30. Maximum Profit A company wants to know what price they should charge to make the maximum profit on their inexpensive digital watches. They have discovered that if they sell the watches for $4, they can sell 2 million of them, but if they sell them for $5, they can only sell 1 million.
   a. Let x be the number of *million* watches sold at p dollars per watch. Assume that the relationship between x and p is linear and find an equation that gives that relationship. $x = -p + 6$
   b. Remember that revenue is price times number of items sold: $R = xp$. Solve the equation from part (a) for x and substitute the result into the equation $R = xp$ in place of x.
      $R = p(-p + 6) = -p^2 + 6p$; R is in million of dollars
   c. Graph the equation from part (b) on a coordinate system where the horizontal axis is the p-axis and the vertical axis is the R-axis.
   d. The highest point on the graph will give you the maximum revenue. What is the maximum revenue?   Vertex is $(3, 9)$. Maximum revenue is $9,000,000 at $3 per watch.

Maintaining Your Skills

Find each root.

31. $\sqrt{49}$ 7   **32.** $\sqrt[3]{-8}$ -2

Write in simplified form for radicals.

33. $\sqrt{50}$ $5\sqrt{2}$ **34.** $2\sqrt{18x^2y^3}$ $6xy\sqrt{2y}$

35. $\sqrt{\dfrac{2}{5}}$ $\dfrac{\sqrt{10}}{5}$ **36.** $\sqrt[3]{\dfrac{1}{2}}$ $\dfrac{\sqrt[3]{4}}{2}$

Perform the indicated operations.

37. $3\sqrt{12} + 5\sqrt{27}$ $21\sqrt{3}$ **38.** $\sqrt{3}(\sqrt{3} - 4)$ $3 - 4\sqrt{3}$

39. $(\sqrt{6} + 2)(\sqrt{6} - 5)$ $-4 - 3\sqrt{6}$

40. $(\sqrt{x} + 3)^2$ $x + 6\sqrt{x} + 9$

Rationalize the denominator.

41. $\dfrac{8}{\sqrt{5} - \sqrt{3}}$ $4\sqrt{5} + 4\sqrt{3}$ **42.** $\dfrac{\sqrt{5} - \sqrt{2}}{\sqrt{5} + \sqrt{2}}$ $\dfrac{7 - 2\sqrt{10}}{3}$

Solve for x.

43. $\sqrt{2x - 5} = 3$ 7

44. $\sqrt{x + 15} = x + 3$ Possible solutions -6 and 1; only 1 checks

Solving Quadratic Equations of the Form $(ax + b)^2 = c$ [9.1]

EXAMPLES

1. $(x - 3)^2 = 25$
$x - 3 = \pm 5$
$x = 3 \pm 5$
$x = -2$ or $x = 8$

We can solve equations of the form $(ax + b)^2 = c$ by applying the square root property for equations to write:

$$ax + b = \pm\sqrt{c}$$

Completing the Square [9.2]

2. $x^2 - 6x + 2 = 0$
$x^2 - 6x \quad = -2$
$x^2 - 6x + \mathbf{9} = -2 + \mathbf{9}$
$(x - 3)^2 = 7$
$x - 3 = \pm\sqrt{7}$
$x = 3 \pm \sqrt{7}$

To complete the square on a quadratic equation as a method of solution, we use the following steps:

Step 1: Put the equation in the form $ax^2 + bx = c$. This usually involves moving only the constant term to the opposite side.

Step 2: Make sure the coefficient of the squared term is 1. If it is not 1, simply divide both sides by whatever it is.

Step 3: Add the square of half the coefficient of the linear term to both sides of the equation.

Step 4: Write the left-hand side of the equation as a binomial square, apply the square root property, and solve.

The Quadratic Formula [9.3]

3. If $2x^2 + 3x - 4 = 0$

then $x = \dfrac{-3 \pm \sqrt{9 - 4\,(2)(-4)}}{2(2)}$

$= \dfrac{-3 \pm \sqrt{41}}{4}$

Any equation that is in the form $ax^2 + bx + c = 0$, where $a \neq 0$, has as its solutions:

$$x = \frac{-b \pm \sqrt{b^2 - 4ac}}{2a}$$

The expression under the square root sign, $b^2 - 4ac$, is known as the discriminant. When the discriminant is negative, the solutions are complex numbers.

Complex Numbers [9.4]

4. The numbers $5, 3i, 2 + 4i$, and $7 - i$ are all complex numbers.

Any number that can be put in the form $a + bi$, where $i = \sqrt{-1}$, is called a complex number.

Addition and Subtraction of Complex Numbers [9.4]

5. $(3 + 4i) + (6 - 7i)$
$= (3 + 6) + (4i - 7i)$
$= 9 - 3i$

We add (or subtract) complex numbers by using the same procedure we used to add (or subtract) polynomials: We combine similar terms.

Multiplication of Complex Numbers [9.4]

6. $(2 + 3i)(3 - i)$
$= 6 - 2i + 9i - 3i^2$
$= 6 + 7i + 3$
$= 9 + 7i$

We multiply complex numbers in the same way we multiply binomials. The result, however, can be simplified further by substituting -1 for i^2 whenever it appears.

Division of Complex Numbers [9.4]

7. $\dfrac{3}{2 + 5i} = \dfrac{3}{2 + 5i} \cdot \dfrac{2 - 5i}{2 - 5i}$

$= \dfrac{6 - 15i}{29}$

Division with complex numbers is accomplished with the method for rationalizing the denominator that we developed while working with radical expressions. If the denominator has the form $a + bi$, we multiply both the numerator and the denominator by its conjugate, $a - bi$.

Graphing Parabolas [9.6]

8. Graph $y = x^2 - 2$.

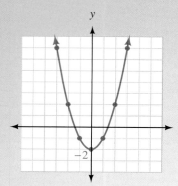

The graph of each of the following equations is a parabola with the same basic shape as the graph of $y = x^2$.

Equation	Vertex	To Obtain the Graph, Move the Graph of $y = x^2$
$y = x^2 + k$	$(0, k)$	k units vertically
$y = (x - h)^2$	$(h, 0)$	h units horizontally
$y = (x - h)^2 + k$	(h, k)	h units horizontally and k units vertically

! COMMON MISTAKES

1. The most common mistake when working with complex numbers is to say $i = -1$. It does not; i is the *square root* of -1, not -1 itself.
2. The most common mistake when working with the quadratic formula is to try to identify the constants a, b, and c before putting the equation into standard form.

The problems below form a comprehensive review of the material in this chapter. They can be used to study for exams. If you would like to take a practice test on this chapter, you can use the odd-numbered problems. Give yourself an hour and work as many of the odd-numbered problems as possible. When you are finished, or when an hour has passed, check your answers with the answers in the back of the book. You can use the even-numbered problems for a second practice test.

The numbers in brackets refer to the sections of the text in which similar problems can be found.

Solve each quadratic equation. [9.1, 9.5]

1. $a^2 = 32$

2. $a^2 = 60$

3. $2x^2 = 32$

4. $(x + 3)^2 = 36$

5. $(x - 2)^2 = 81$

6. $(3x + 2)^2 = 16$

7. $(2x + 5)^2 = 32$

8. $(3x - 4)^2 = 27$

9. $\left(x - \dfrac{2}{3}\right)^2 = -\dfrac{25}{9}$

Solve by completing the square. [9.2]

10. $x^2 + 8x = 4$

11. $x^2 - 8x = 4$

12. $x^2 - 4x - 7 = 0$

13. $x^2 + 4x + 3 = 0$

14. $a^2 = 9a + 3$

15. $a^2 = 5a + 6$

16. $2x^2 + 4x - 6 = 0$

17. $3x^2 - 6x - 2 = 0$

Solve by using the quadratic formula. [9.3, 9.5]

18. $x^2 + 7x + 12 = 0$

19. $x^2 - 8x + 16 = 0$

20. $x^2 + 5x + 7 = 0$

21. $2x^2 = -8x + 5$

22. $3x^2 = -3x - 4$

23. $\dfrac{1}{5}x^2 - \dfrac{1}{2}x = \dfrac{3}{10}$

24. $(2x + 1)(2x - 3) = -6$

Add and subtract the following complex numbers. [9.4]

25. $(4 - 3i) + 5i$

26. $(2 - 5i) - 3i$

27. $(5 + 6i) + (5 - i)$

28. $(2 + 5i) + (3 - 7i)$

29. $(3 - 2i) - (3 - i)$

30. $(5 - 7i) - (6 - 2i)$

31. $(3 + i) - 5i - (4 - i)$

32. $(2 - 3i) - (5 - 2i) + (5 - i)$

Multiply the following complex numbers. [9.4]

33. $2(3 - i)$

34. $-6(4 + 2i)$

35. $4i(6 - 5i)$

36. $(2 - i)(3 + i)$

37. $(3 - 4i)(5 + i)$

38. $(3 - i)^2$

39. $(4 + i)(4 - i)$

40. $(3 + 2i)(3 - 2i)$

Divide the following complex numbers. [9.4]

41. $\dfrac{i}{3 + i}$

42. $\dfrac{i}{2 - i}$

43. $\dfrac{5}{2 + 5i}$

44. $\dfrac{2i}{4 - i}$

45. $\dfrac{-3i}{3 - 2i}$

46. $\dfrac{3 + i}{3 - i}$

47. $\dfrac{4 - 5i}{4 + 5i}$

48. $\dfrac{2 + 3i}{3 - 5i}$

Write the following radicals as complex numbers. [9.5]

49. $\sqrt{-36}$

50. $\sqrt{-144}$

51. $\sqrt{-17}$

52. $\sqrt{-31}$

53. $\sqrt{-40}$

54. $\sqrt{-72}$

55. $\sqrt{-200}$

56. $\sqrt{-242}$

Graph each of the following equations. [9.6]

57. $y = x^2 + 2$

58. $y = (x - 2)^2$

59. $y = (x + 2)^2$

60. $y = (x + 3)^2 - 2$

61. $y = x^2 + 4x + 7$

62. $y = x^2 - 4x + 7$

GROUP PROJECT   Proving the Pythagorean Theorem

Number of People 1–2

Time Needed 5–8 minutes

Equipment pencil, graph paper, and scissors

Background The postage stamp shown here was issued by San Marino, a small state in Europe, in 1982. It depicts the mathematician Pythagoras. As you know, the Pythagorean theorem is used extensively throughout mathematics, and we have used it many times in this book. We can use the diagram shown here to give a visual "proof" of this theorem, similar to the original proof given by the Pythagoreans.

Procedure The shaded triangle shown here is a right triangle.

1. Notice the square representing c^2 has dotted lines making up four additional triangles and a square.

2. Cut out the square labeled c^2, and cut along the dotted lines.

3. Use all the pieces you have cut out to cover up the squares representing a^2 and b^2. Doing so is a visual "proof" that $c^2 = a^2 + b^2$.

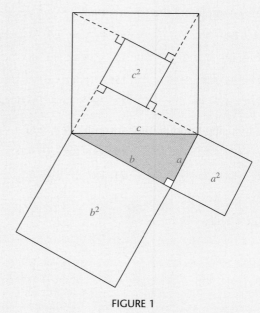

FIGURE 1

A Continued Fraction and the Golden Ratio

You may recall that the continued fraction

$$1 + \cfrac{1}{1 + \cfrac{1}{1 + \cfrac{1}{1 + \cdots}}}$$

is equal to the golden ratio.

The same conclusion can be found with the quadratic formula by using the following conditional statement

$$\text{If } x = 1 + \cfrac{1}{1 + \cfrac{1}{1 + \cfrac{1}{1 + \cdots}}} \text{, then } x = 1 + \frac{1}{x}$$

Work with this conditional statement until you see that it is true. Then solve the equation $x = 1 + \frac{1}{x}$. Write an essay in which you explain in your own words why the conditional statement is true. Then show the details of the solution to the equation $x = 1 + \frac{1}{x}$. The message that you want to get across in your essay is that the continued fraction shown here is actually the same as the golden ratio.

Resources

By gathering resources early in the term, before you need help, the information about these resources will be available to you when they are needed.

INSTRUCTOR

Knowing the contact information for your instructor is very important. You may already have this information from the course syllabus. It is a good idea to write it down again.

Name _____ Office Location _____

Available Hours: M _____ T _____ W _____ TH _____ F _____

Phone Number _____ ext. _____ E-mail Address _____

TUTORING CENTER

Many schools offer tutoring, free of charge to their students. If this is the case at your school, find out when and where tutoring is offered.

Tutoring Location _____ Phone Number _____ ext. _____

Available Hours: M _____ T _____ W _____ TH _____ F _____

COMPUTER LAB

Many schools offer a computer lab where students can use the online resources and software available with their textbook. Other students using the same software and websites as you can be very helpful. Find out where the computer lab at your school is located.

Computer Lab Location _____ Phone Number _____ ext. _____

Available Hours: M _____ T _____ W _____ TH _____ F _____

VIDEO LESSONS

A complete set of video lessons is available to your school. These videos feature the author of your textbook presenting full-length, 15- to 20-minute lessons from every section of your textbook. If you miss class, or find yourself behind, these lessons will prove very useful.

Video Location _____ Phone Number _____ ext. _____

Available Hours: M _____ T _____ W _____ TH _____ F _____

CLASSMATES

Form a study group and meet on a regular basis. When you meet try to speak to each other using proper mathematical language. That is, use the words that you see in the definition and property boxes in your textbook.

Name _____ Phone _____ E-mail _____

Name _____ Phone _____ E-mail _____

Name _____ Phone _____ E-mail _____

Answers to Odd-Numbered Problems and Chapter Reviews

CHAPTER 1

Problem Set 1.1

1. $x + 5 = 14$ **3.** $5y < 30$ **5.** $3y \leq y + 6$ **7.** $\frac{x}{3} = x + 2$ **9.** 9   **11.** 49   **13.** 8   **15.** 64   **17.** 16   **19.** 100   **21.** 121

23. a. 11   **b.** 16   **25. a.** 17   **b.** 42   **27. a.** 30   **b.** 30   **29. a.** 24   **b.** 80   **31.** 27   **33.** 35   **35.** 4   **37.** 37

39. a. 16   **b.** 16   **41. a.** 81   **b.** 41   **43.** 2,345   **45.** 2   **47.** 148   **49.** 36   **51.** 58   **53.** 62   **55.** 100   **57.** 9   **59.** 18

61. 12   **63.** 18   **65.** 42   **67.** 0.17   **69.** 0.22   **71.** 2.3   **73.** 5.55   **75.** 240   **77.** 10   **79.** 3,000   **81.** 2,000

83. 35.33   **85.** 0.75   **87.** 0.33   **89.** 272.73   **91.** 10   **93.** 420   **95. a.** 600 mg   **b.** 231 mg

97.

Activity	Calories Burned in 1 Hour by a 150-Pound Person
Bicycling	374
Bowling	265
Handball	680
Jogging	680
Skiing	544

99. 5   **101.** 10   **103.** 25   **105.** 10

Problem Set 1.2

1–7.

9. $\frac{18}{24}$ **11.** $\frac{12}{24}$ **13.** $\frac{15}{24}$ **15.** $\frac{36}{60}$ **17.** $\frac{22}{60}$ **19.** 2   **21.** 25   **23.** 6   **25.** $-10, \frac{1}{10}, 10$ **27.** $3, -\frac{1}{3}, 3$

29. $-x, \frac{1}{x}, |x|$ **31.** $<$ **33.** $>$ **35.** $>$ **37.** $>$ **39.** $<$ **41.** $<$ **43.** 6   **45.** 22   **47.** 3   **49.** 7

51. 3   **53.** $\frac{8}{15}$ **55.** $\frac{3}{2}$ **57.** 1   **59.** 1   **61. a.** 2   **b.** 4   **c.** 8   **d.** 0.03   **63. a.** 6   **b.** 12   **c.** 24   **d.** 0.09   **65.** $\frac{9}{16}$

67. $\frac{8}{27}$ **69. a.** 4   **b.** 14   **c.** 24   **d.** 34   **71. a.** 12   **b.** 20   **c.** 100   **d.** 1,024   **73. a.** 17   **b.** 25   **c.** 25   **75.** 4 inches; 1 inch²

77. 4.5 inches; 1.125 inches²   **79.** 10.25 centimeters; 5 centimeters²   **81.** $-8, -2$ **83.** $-64°F; -54°F$

85. -100 feet; -105 feet   **87.** Area = 93.5 inches²: perimeter = 39 inches   **89.** 1,387 calories   **91.** 654 more calories

Problem Set 1.3

1. $3 + 5 = 8, 3 + (-5) = -2,$ **3.** $15 + 20 = 35, 15 + (-20) = -5,$
     $-3 + 5 = 2, -3 + (-5) = -8$ $-15 + 20 = 5, -15 + (-20) = -35$

5. 3   **7.** -7 **9.** -14 **11.** -3 **13.** -25 **15.** -12 **17.** -19 **19.** -25 **21.** -8 **23.** -4 **25.** 6   **27.** 6

29. 8   **31.** -4 **33.** -14 **35.** -17 **37.** 4   **39.** 3   **41.** 15   **43.** -8 **45.** 12   **47.** 11   **49.** 10.8   **51.** 980

53. 1.70   **55.** 23, 28   **57.** 30, 35   **59.** 0, -5 **61.** $-12, -18$ **63.** $-4, -8$ **65.** Yes   **67.** $5 + 9 = 14$

69. $[-7 + (-5)] + 4 = -8$ **71.** $[-2 + (-3)] + 10 = 5$ **73.** 3   **75.** -3 **77.** $-12 + 4$ **79.** $10 + (-6) + (-8) = -\$4$

81. $-30 + 40 = 10$

Problem Set 1.4

1. -3 **3.** -6 **5.** 0 **7.** -10 **9.** -16 **11.** -12 **13.** -7 **15.** 35 **17.** 0 **19.** -4 **21.** 4 **23.** -24 **25.** -28
27. 25 **29.** 4 **31.** 7 **33.** 17 **35.** 8 **37.** 4 **39.** 18 **41.** 10 **43.** 17 **45.** 1 **47.** 1 **49.** 27 **51.** -26
53. -2 **55.** 68 **57.** -11.3 **59.** -3.6 **61. a.** -9 **b.** -10 **c.** -3 **63.** $-7 - 4 = -11$ **65.** $12 - (-8) = 20$
67. $-5 - (-7) = 2$ **69.** $[4 + (-5)] - 17 = -18$ **71.** $8 - 5 = 3$ **73.** $-8 - 5 = -13$ **75.** $8 - (-5) = 13$ **77.** 10 **79.** -2
81. $1,500 - 730$ **83.** $-35 + 15 - 20 = -\$40$ **85.** $98 - 65 - 53 = -\$20$ **87.** $\$4,500, \$3,950, \$3,400, \$2,850, \$2,300$; yes
89. 769 feet **91.** 439 feet **93.** 2 seconds **95.** $35°$ **97.** $60°$ **99. a.**
b. 11.5 inches

Day	Plant Height (inches)	Day	Plant Height (inches)
0	0	6	6
1	0.5	7	9
2	1	8	13
3	1.5	9	18
4	3	10	23
5	4		

Problem Set 1.5

1. Commutative **3.** Multiplicative inverse **5.** Commutative **7.** Distributive **9.** Commutative, associative
11. Commutative, associative **13.** Commutative **15.** Commutative, associative **17.** Commutative **19.** Additive inverse
21. $3x + 6$ **23.** $9a + 9b$ **25.** 0 **27.** 0 **29.** 10 **31.** $(4 + 2) + x = 6 + x$ **33.** $x + (2 + 7) = x + 9$ **35.** $(3 \cdot 5)x = 15x$
37. $(9 \cdot 6)y = 54y$ **39.** $\left(\frac{1}{2} \cdot 3\right)a = \frac{3}{2}a$ **41.** $\left(\frac{1}{3} \cdot 3\right)x = x$ **43.** $\left(\frac{1}{2} \cdot 2\right)y = y$ **45.** $\left(\frac{3}{4} \cdot \frac{4}{3}\right)x = x$ **47.** $\left(\frac{6}{5} \cdot \frac{5}{6}\right)a = a$
49. $8x + 16$ **51.** $8x - 16$ **53.** $4y + 4$ **55.** $18x + 15$ **57.** $6a + 14$ **59.** $54y - 72$ **61.** $\frac{3}{2}x - 3$ **63.** $x + 2$
65. $3x + 3y$ **67.** $8a - 8b$ **69.** $12x + 18y$ **71.** $12a - 8b$ **73.** $3x + 2y$ **75.** $4a + 25$ **77.** $6x + 12$ **79.** $14x + 38$
81. $2x + 1$ **83.** $6x - 3$ **85.** $5x + 10$ **87.** $6x + 5$ **89.** $5x + 6$ **91.** $6m - 5$ **93.** $7 + 3x$ **95.** $3x - 2y$ **97.** $0.09x + 180$
99. $0.12x + 60$ **101.** $a + 1$ **103.** $1 - a$ **105.** No **107.** No, not commutative **109.** $8 \div 4 \neq 4 \div 8$
111. $12(2,400 - 480) = \$23,040$; $12(2,400) - 12(480) = \$23,040$ **113.** $3(650 + 225) = \$2,625$; $3(650) + 3(225) = \$2,625$

Problem Set 1.6

1. -42 **3.** -16 **5.** 3 **7.** 121 **9.** 6 **11.** -60 **13.** 24 **15.** 49 **17.** -27 **19.** 6 **21.** 10 **23.** 9 **25.** 45
27. 14 **29.** -2 **31.** 216 **33.** -2 **35.** -18 **37.** 29 **39.** 38 **41.** -5 **43.** 37 **45. a.** 60 **b.** -23 **c.** 17 **d.** -2
47. $-\frac{10}{21}$ **49.** -4 **51.** $\frac{9}{16}$ **53. a.** 27 **b.** 3 **c.** -27 **d.** -27 **55.** 9 **57.** $\frac{25}{4}$ **59.** 4 **61.** $\frac{9}{4}$ **63.** $-8x$ **65.** $42x$
67. x **69.** $-\frac{3}{2}x + 3$ **71.** $-6x + 8$ **73.** $-15x - 30$ **75.** $-12x - 20y$ **77.** $-6x - 10y$ **79.** $-\frac{3}{2}x + 3$ **81.** $-\frac{2}{3}x + 2$
83. $\frac{2}{3}x - 2$ **85.** $-2x + y$ **87. a.** 2 **b.** 1 **c.** 3 **89. a.** 3 **b.** 3 **c.** -4 **91. a.** 12 **b.** 12 **93. a.** 44 **b.** 44 **95.** -25
97. $2(-4x) = -8x$ **99.** -26 **101.** 8 **103.** -80 **105.** $\frac{1}{8}$ **107.** -24 **109.** $\$60$ **111.** $1°F$ **113. a.** $\$544$ **b.** $-\$521$

Problem Set 1.7

1. -2 **3.** -3 **5.** $-\frac{1}{3}$ **7.** 3 **9.** $\frac{1}{7}$ **11.** 0 **13.** 9 **15.** -15 **17.** -36 **19.** $-\frac{1}{4}$ **21.** $\frac{16}{15}$ **23.** $\frac{4}{3}$ **25.** $-\frac{8}{13}$
27. -1 **29.** 1 **31.** $\frac{3}{5}$ **33.** $-\frac{5}{3}$ **35.** -2 **37.** -3 **39.** Undefined **41.** Undefined **43.** 5 **45.** $-\frac{7}{3}$ **47.** -1
49. -7 **51.** $\frac{15}{17}$ **53.** $-\frac{32}{17}$ **55.** $\frac{1}{3}$ **57.** 1 **59.** 1 **61.** -2 **63.** $\frac{9}{7}$ **65.** $\frac{16}{11}$ **67.** -1 **69.** $\frac{7}{20} = 0.35$ **71.** -1 **73. a.** $\frac{3}{2}$ **b.** $\frac{3}{2}$
75. a. $-\frac{5}{7}$ **b.** $-\frac{5}{7}$ **77. a.** 2.618 **b.** 0.382 **c.** 3 **79. a.** 25 **b.** -25 **c.** -25 **d.** -25 **e.** 25 **81. a.** 10 **b.** 0 **c.** -100 **d.** -20
83. $5x + 6$ **85.** $3x + 20$ **87.** $3 + x$ **89.** $3x - 7y$ **91.** 1 **93.** $2 - y$ **95.** 3 **97.** -10 **99.** -3 **101.** -8 **103.** $\$350$
105. Drops $3.5°F$ each hour **107. a.** $\$20,000$ **b.** $\$50,000$ **c.** Yes, the projected revenue for 5,000 email addresses is $\$10,000$, which is $\$5,000$ more than the list costs.

Problem Set 1.8

1. 0, 1 **3.** $-3, -2.5, 0, 1, \frac{3}{2}$ **5.** All **7.** $-10, -8, -2, 9$ **9.** π **11.** T **13.** F **15.** F **17.** T **19.** Composite, $2^4 \cdot 3$
21. Prime **23.** Composite, $3 \cdot 11 \cdot 31$ **25.** $2^4 \cdot 3^2$ **27.** $2 \cdot 19$ **29.** $3 \cdot 5 \cdot 7$ **31.** $2^2 \cdot 3^2 \cdot 5$ **33.** $5 \cdot 7 \cdot 11$ **35.** 11^2
37. $2^2 \cdot 3 \cdot 5 \cdot 7$ **39.** $2^2 \cdot 5 \cdot 31$ **41.** $\frac{7}{11}$ **43.** $\frac{5}{7}$ **45.** $\frac{11}{13}$ **47.** $\frac{14}{15}$ **49.** $\frac{5}{9}$ **51.** $\frac{5}{8}$ **53. a.** -30 **b.** 130 **c.** $-4,000$ **d.** $-\frac{5}{8}$
55. $\frac{2}{9}$ **57.** $\frac{3}{2}$ **59.** 64 **61.** 509 **63. a.** 0.35 **b.** 0.28 **c.** 0.25 **65.** $6^3 = (2 \cdot 3)^3 = 2^3 \cdot 3^3$ **67.** $9^4 \cdot 16^2 = (3^2)^4(2^4)^2 = 2^8 \cdot 3^8$
69. $3 \cdot 8 + 3 \cdot 7 + 3 \cdot 5 = 24 + 21 + 15 = 60 = 2^2 \cdot 3 \cdot 5$ **71.** Irrational numbers **73.** 8, 21, 34

Problem Set 1.9

1. $\frac{2}{3}$ **3.** $-\frac{1}{4}$ **5.** $\frac{1}{2}$ **7.** $\frac{x-1}{3}$ **9.** $\frac{3}{2}$ **11.** $\frac{x+6}{2}$ **13.** $-\frac{3}{5}$ **15.** $\frac{10}{a}$ **17.** $\frac{7}{8}$ **19.** $\frac{1}{10}$ **21.** $\frac{7}{9}$ **23.** $\frac{7}{3}$

25. $\frac{1}{4}$ **27.** $\frac{7}{6}$ **29.** $\frac{19}{24}$ **31.** $\frac{13}{60}$ **33.** $\frac{29}{35}$ **35.** $\frac{949}{1,260}$ **37.** $\frac{13}{420}$ **39.** $\frac{41}{24}$ **41.** $\frac{5}{4}$ **43.** $-\frac{3}{2}$ **45.** $\frac{3}{2}$ **47.** $-\frac{2}{3}$

49. $\frac{7}{3}$ **51.** $\frac{1}{125}$ **53.** 4   **55.** -2 **57.** -18 **59.** -16 **61.** -2 **63.** 2   **65.** -144 **67.** -162 **69.** $\frac{5x+4}{20}$ **71.** $\frac{a+4}{12}$

73. $\frac{9x+4}{12}$ **75.** $\frac{10+3x}{5x}$ **77.** $\frac{3y+28}{7y}$ **79.** $\frac{60+19a}{20a}$ **81.** $\frac{2}{3}x$ **83.** $-\frac{1}{4}x$ **85.** $\frac{14}{15}x$ **87.** $\frac{11}{12}x$ **89.** $\frac{41}{40}x$ **91.** $\frac{x-1}{x}$

93. a. $\frac{1}{4}$ **b.** $\frac{5}{4}$ **c.** $-\frac{3}{8}$ **d.** $-\frac{3}{2}$ **95.** $\frac{1}{3}$ **97.** $\frac{3}{4}$ **99. a.** $\frac{3}{2}$ **b.** $\frac{4}{3}$ **c.** $\frac{5}{4}$ **101. a.** 8 **b.** 7 **c.** 8

CHAPTER 1 REVIEW/TEST

1. $-7+(-10)=-17$ **2.** $(-7+4)+5=2$ **3.** $(-3+12)+5=14$ **4.** $4-9=-5$ **5.** $9-(-3)=12$ **6.** $-7-(-9)=2$
7. $(-3)(-7)-6=15$ **8.** $5(-6)+10=-20$ **9.** $2[(-8)(3x)]=-48x$ **10.** $\frac{-25}{-5}=5$ **11.** 1.8   **12.** -10 **13.** $-6, \frac{1}{6}$
14. $\frac{12}{5}, -\frac{5}{12}$ **15.** -5 **16.** $-\frac{5}{4}$ **17.** 3   **18.** -38 **19.** -25 **20.** 1   **21.** -3 **22.** 22   **23.** -4 **24.** -20 **25.** -30
26. -12 **27.** -24 **28.** 12   **29.** -4 **30.** $-\frac{2}{3}$ **31.** 23   **32.** 47   **33.** -3 **34.** -35 **35.** 32   **36.** 2   **37.** 20
38. -3 **39.** -98 **40.** 70   **41.** 2   **42.** $\frac{17}{2}$ **43.** Undefined   **44.** 3   **45.** Associative   **46.** Multiplicative identity
47. Commutative   **48.** Additive inverse   **49.** Commutative, associative   **50.** Distributive   **51.** $12+x$ **52.** $28a$ **53.** x
54. y **55.** $14x+21$ **56.** $6a-12$ **57.** $\frac{5}{2}x-3$ **58.** $-\frac{3}{2}x+3$ **59.** $-\frac{1}{3}, 0, 5, -4.5, \frac{2}{5}, -3$ **60.** 0, 5   **61.** $\sqrt{7}, \pi$
62. $0, 5, -3$ **63.** $2 \cdot 3^2 \cdot 5$ **64.** $2^3 \cdot 3 \cdot 5 \cdot 7$ **65.** $\frac{173}{210}$ **66.** $\frac{2x+7}{12}$ **67.** -2 **68.** 810   **69.** 8   **70.** 12   **71.** -1
72. $\frac{1}{16}$

CHAPTER 2

Problem Set 2.1

1. $-3x$ **3.** $-a$ **5.** $12x$ **7.** $6a$ **9.** $6x-3$ **11.** $7a+5$ **13.** $5x-5$ **15.** $4a+2$ **17.** $-9x-2$ **19.** $12a+3$
21. $10x-1$ **23.** $21y+6$ **25.** $-6x+8$ **27.** $-2a+3$ **29.** $-4x+26$ **31.** $4y-16$ **33.** $-6x-1$ **35.** $2x-12$
37. $10a+33$ **39.** $4x-9$ **41.** $7y-39$ **43.** $-19x-14$ **45.** 5   **47.** -9 **49.** 4   **51.** 4   **53.** -37 **55.** -41
57. 64   **59.** 64   **61.** 144   **63.** 144   **65.** 3   **67.** 0   **69.** 15   **71.** 6
73. a.

n	1	2	3	4
$3n$	3	6	9	12

b.

n	1	2	3	4
n^3	1	8	27	64

75. 1, 4, 7, 10, . . . an arithmetic sequence   **77.** 0, 1, 4, 9, . . . a sequence of squares   **79.** $-6y+4$ **81.** $0.17x$
83. $2x$ **85.** $5x-4$ **87.** $7x-5$ **89.** $-2x-9$ **91.** $7x+2$ **93.** $-7x+6$ **95.** $7x$ **97.** $-y$ **99.** $10y$ **101.** $0.17x+180$
103. $0.22x+60$ **105.** $a=4; 13x$ **107.** 49   **109.** 40   **111. a.** 42°F **b.** 28°F **c.** $-14°F$ **113. a.** \$37.50 **b.** \$40.00 **c.** \$42.50
115. $0.71G$; \$887.50   **117.** $-\frac{1}{24}$ **119.** $-\frac{7}{9}$ **121.** $-\frac{23}{420}$ **123.** 12   **125.** -3 **127.** -9.7 **129.** $-\frac{5}{4}$ **131.** -3 **133.** $a-12$
135. $-\frac{7}{2}$ **137.** $\frac{51}{40}$ **139.** 7

Problem Set 2.2

1. 11   **3.** 4   **5.** $-\frac{3}{4}$ **7.** -5.8 **9.** -17 **11.** $-\frac{1}{8}$ **13.** -6 **15.** -3.6 **17.** -7 **19.** $-\frac{7}{45}$ **21.** 3   **23.** $\frac{11}{8}$ **25.** 21
27. 7   **29.** 3.5   **31.** 22   **33.** -2 **35.** -16 **37.** -3 **39.** 10   **41.** -12 **43.** 4   **45.** 2   **47.** -5 **49.** -1 **51.** -3
53. 8   **55.** -8 **57.** 2   **59.** 11   **61. a.** $\frac{3}{2}$ **b.** 1 **c.** $-\frac{3}{2}$ **d.** -4 **e.** $\frac{8}{5}$
63. a. 6% **b.** 5% **c.** 2% **d.** 75%   **65.** $x+55+55=180; 70°$ **67.** $18x$ **69.** x **71.** y **73.** x **75.** a **77.** y **79.** x
81. 6   **83.** 6   **85.** -9 **87.** $-\frac{15}{8}$ **89.** -18 **91.** $-\frac{5}{4}$ **93.** $3x$

Problem Set 2.3

1. 2   **3.** 4   **5.** $-\frac{1}{2}$ **7.** -2 **9.** 3   **11.** 4   **13.** 0   **15.** 0   **17.** 6   **19.** -50 **21.** $\frac{3}{2}$ **23.** 12   **25.** -3 **27.** 32
29. -8 **31.** $\frac{1}{2}$ **33.** 4   **35.** 8   **37.** -4 **39.** 4   **41.** -15 **43.** $-\frac{1}{2}$ **45.** 3   **47.** 1   **49.** $\frac{1}{4}$ **51.** -3 **53.** 3
55. 2   **57.** $-\frac{3}{2}$ **59.** $-\frac{3}{2}$ **61.** 1   **63.** 1   **65.** -2 **67.** -2 **69.** $-\frac{4}{5}$ **71.** 1   **73.** $\frac{7}{2}$ **75.** $\frac{5}{4}$ **77.** 40   **79.** $\frac{7}{3}$

81. 200 tickets   **83.** \$1,391 per month   **85.** $6x - 10$ **87.** $\frac{3}{2}x + 3$ **89.** $-x + 2$ **91.** $10x - 43$ **93.** $-3x - 13$
95. $-6y + 4$ **97.** 2   **99.** 6   **101.** 3,000   **103.** $3x - 11$ **105.** $0.09x + 180$ **107.** $-6y + 4$ **109.** $4x - 11$ **111.** $5x$
113. $0.17x$

Problem Set 2.4

1. 3   **3.** -2 **5.** -1 **7.** 2   **9.** -4 **11.** -2 **13.** 0   **15.** 1   **17.** $\frac{1}{2}$ **19.** 7   **21.** 8   **23.** $-\frac{1}{3}$ **25.** $\frac{3}{4}$ **27.** 75
29. 2   **31.** 6   **33.** 8   **35.** 0   **37.** $\frac{3}{7}$ **39.** 1   **41.** 1   **43.** -1 **45.** 6   **47.** $\frac{3}{4}$ **49.** 3   **51.** $\frac{3}{4}$ **53.** 8   **55.** 6
57. -2 **59.** -2 **61.** 2   **63.** -6 **65.** 2   **67.** 20   **69.** 4,000   **71.** 700   **73.** 11   **75.** 7
77. a. $\frac{5}{4} = 1.25$ **b.** $\frac{15}{2} = 7.5$ **c.** $6x + 20$ **d.** 15 **e.** $4x - 20$ **f.** $\frac{45}{2} = 22.5$ **79.** $\frac{3}{2}$ **81.** 4   **83.** 1

85.

x	$3(x + 2)$	$3x + 2$	$3x + 6$
0	6	2	6
1	9	5	9
2	12	8	12
3	15	11	15

87.

a	$(2a + 1)^2$	$4a^2 + 4a + 1$
1	9	9
2	25	25
3	49	49

89. 14   **91.** -3 **93.** $\frac{1}{4}$

95. $\frac{1}{3}$ **97.** $-\frac{3}{2}x + 3$

Problem Set 2.5

1. 100 feet   **3. a.** 2 **b.** $\frac{3}{2}$ **c.** 0   **5. a.** 2 **b.** -3 **c.** 1   **7.** -2 **9.** 1   **11. a.** 2 **b.** 4   **13. a.** 5 **b.** 18
15. a. 20 **b.** 75   **17.** $l = \frac{A}{w}$ **19.** $h = \frac{V}{lw}$ **21.** $a = P - b - c$ **23.** $x = 3y - 1$ **25.** $y = 3x + 6$ **27.** $y = -\frac{2}{3}x + 2$
29. $w = \frac{P - 2l}{2}$ **31.** $v = \frac{h - 16t^2}{t}$ **33.** $h = \frac{A - \pi r^2}{2\pi r}$ **35. a.** $y = -2x - 5$ **b.** $y = 4x - 7$ **37. a.** $y = \frac{3}{4}x + \frac{1}{4}$ **b.** $y = \frac{3}{4}x - 5$
39. a. $y = \frac{3}{5}x + 1$ **b.** $y = \frac{1}{2}x + 2$ **c.** $y = 4x + 3$ **41.** $y = \frac{3}{7}x - 3$ **43.** $y = 2x + 8$
45. a. $\frac{15}{4} = 3.75$ **b.** 17 **c.** $y = -\frac{4}{5}x + 4$ **d.** $x = -\frac{5}{4}y + 5$ **47.** 60°; 150°   **49.** 45°; 135°   **51.** 10   **53.** 240   **55.** 25%   **57.** 35%
59. 64   **61.** 2,000   **63.** 100°C; yes   **65.** 20°C; yes   **67.** $C = \frac{5}{9}(F - 32)$ **69.** 60%   **71.** 26.5%   **73.** 7 meters
75. $\frac{3}{2}$ or 1.5 inches   **77.** 132 feet   **79.** $\frac{2}{9}$ centimeters   **81. a.** 95 **b.** -41 **c.** -95 **d.** 41   **83. a.** 9 **b.** -73 **c.** 9 **d.** -73
85. The sum of 4 and 1 is 5.   **87.** The difference of 6 and 2 is 4.   **89.** The difference of a number and 5 is -12.
91. The sum of a number and 3 is four times the difference of that number and 3.   **93.** $2(6 + 3) = 18$ **95.** $2(5) + 3 = 13$
97. $x + 5 = 13$ **99.** $5(x + 7) = 30$

Problem Set 2.6
Along with the answers to the odd-numbered problems in this problem set and the next, we are including some of the equations used to solve the problems. Be sure that you try the problems on your own before looking to see the correct equations.
1. 8   **3.** $2x + 4 = 14$; 5   **5.** -1 **7.** The two numbers are x and $x + 2$; $x + (x + 2) = 8$; 3 and 5   **9.** 6 and 14
11. Shelly is 39, and Michele is 36   **13.** Evan is 11; Cody is 22   **15.** Barney is 27; Fred is 31   **17.** Lacy is 16; Jack is 32
19. Patrick is 18; Pat is 38   **21.** $s = 9$ inches   **23.** $s = 15$ feet   **25.** 11 feet, 18 feet, 33 feet   **27.** 26 feet, 13 feet, 14 feet
29. Length = 11 inches; width = 6 inches   **31.** Length = 25 inches; width = 9 inches   **33.** Length = 15 feet; width = 3 feet
35. 9 dimes; 14 quarters   **37.** 12 quarters; 27 nickels   **39.** 8 nickels; 17 dimes   **41.** 7 nickels; 10 dimes; 12 quarters
43. 3 nickels; 9 dimes; 6 quarters   **45.** 4 is less than 10.   **47.** 9 is greater than or equal to -5.   **49.** <   **51.** <   **53.** 2
55. 12   **57.** $5x$ **59.** $1.075x$ **61.** $0.09x + 180$ **63.** 6,000   **65.** 30

Problem Set 2.7
1. 5 and 6   **3.** -4 and -5 **5.** 13 and 15   **7.** 52 and 54   **9.** -14 and -16 **11.** 17, 19, and 21   **13.** 42, 44, and 46
15. \$4,000 invested at 8%, \$6,000 invested at 9%   **17.** \$700 invested at 10%, \$1,200 invested at 12%
19. \$500 at 8%, \$1,000 at 9%, \$1,500 at 10%   **21.** 45°, 45°, 90°   **23.** 22.5°, 45°, 112.5°   **25.** 53°, 90°   **27.** 80°, 60°, 40°
29. 16 adult and 22 children's tickets   **31.** 16 minutes   **33.** 39 hours   **35.** They are in offices 7329 and 7331.
37. Kendra is 8 years old and Marissa is 10 years old.   **39.** Jeff   **41.** \$10.38   **43.** Length = 12 meters; width = 10 meters
45. 59°, 60°, 61°   **47.** \$54.00   **49.** Yes   **51.** -3 **53.** -8 **55.** 0   **57.** 8   **59.** -24 **61.** 28   **63.** $-\frac{1}{2}$ **65.** -16
67. -25 **69.** 24   **71. a.** 9 **b.** 3 **c.** -9 **d.** -3 **73. a.** -8 **b.** 8 **c.** 8 **d.** -8 **75.** -2.3125 **77.** $\frac{10}{3}$

Problem Set 2.8

1. $x < 12$

3. $a \le 12$

5. $x > 13$

7. $y \ge 4$

9. $x > 9$

11. $x < 2$

13. $a \le 5$

15. $x > 15$

17. $x < -3$

19. $x \le 6$

21. $x \ge -50$

23. $y < -6$

25. $x < 6$

27. $y \ge -5$

29. $x < 3$

31. $x \le 18$

33. $a < -20$

35. $y < 25$

37. $a \le 3$

39. $x \ge \frac{15}{2}$

41. $x < -1$

43. $y \ge -2$

45. $x < -1$

47. $m \le -6$

49. $x \le -5$

51. $y < -\frac{3}{2}x + 3$ **53.** $y < \frac{2}{5}x - 2$ **55.** $y \le \frac{3}{7}x + 3$

57. $y \le \frac{1}{2}x + 1$ **59. a.** 3 **b.** 2 **c.** No **d.** $x > 2$ **61.** $x < 3$ **63.** $x \ge 3$ **65.** At least 291   **67.** $2x + 6 < 10$; $x < 2$
69. $4x > x - 8$; $x > -\frac{8}{3}$ **71.** $2x + 2(3x) \ge 48$; $x \ge 6$; the width is at least 6 meters.
73. $x + (x + 2) + (x + 4) > 24$; $x > 6$; the shortest side is even and greater than 6 inches.   **75.** $t \ge 100$
77. Lose money if they sell less than 200 tickets; make a profit if they sell more than 200 tickets   **79.** $2x + 1$ **81.** $-2x - 4$
83. $\frac{5}{2}a + \frac{4}{3}$ **85.** $-2a - \frac{5}{2}$ **87.** $\frac{2}{3}x$ **89.** $-\frac{1}{6}x$ **91.** $\frac{11}{12}x$ **93.** $\frac{41}{40}x$ **95.** $x \ge 2$ **97.** $x < 4$ **99.** $-1 \le x$

Problem Set 2.9

1.

3.

5.

7.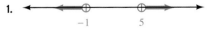

9.

11.

13.

15.

17.

19.

21.

23.

25.

27.

29.

31.

33.

35.

37.

39.

41. $-2 < x < 3$ **43.** $-2 \le x \le 3$ **45. a.** $2x + x > 10$; $x + 10 > 2x$; $2x + 10 > x$ **b.** $\frac{10}{3} < x < 10$

47.   **49.** $4 < x < 5$ **51.** The width is between 3 inches and $\frac{11}{2}$ inches.   **53.** 8   **55.** 24

57. 25%   **59.** 10%   **61.** 80   **63.** 400   **65.** -5 **67.** 5   **69.** 7   **71.** 9   **73.** 6   **75.** $2x - 3$ **77.** $-3, 0, 2$

CHAPTER 2 REVIEW/TEST

1. $-3x$ **2.** $-2x - 3$ **3.** $4a - 7$ **4.** $-2a + 3$ **5.** $-6y$ **6.** $-6x + 1$ **7.** 19   **8.** -11 **9.** -18 **10.** -13 **11.** 8
12. 6   **13.** -8 **14.** $\frac{15}{14}$ **15.** 2   **16.** -5 **17.** -5 **18.** 0   **19.** 12   **20.** -8 **21.** -1 **22.** 1   **23.** 5   **24.** 2   **25.** 0
26. 0   **27.** 10   **28.** 2   **29.** 2   **30.** $-\frac{8}{3}$ **31.** 0   **32.** -5 **33.** 0   **34.** -4 **35.** -8 **36.** 4   **37.** $y = \frac{2}{5}x - 2$

38. $y = \frac{5}{2}x - 5$ **39.** $h = \frac{V}{\pi r^2}$ **40.** $w = \frac{P - 2l}{2}$ **41.** 206.4   **42.** 9%   **43.** 11   **44.** 5 meters; 25 meters

45. \$500 at 9%; \$800 at 10%   **46.** 10 nickels; 5 dimes   **47.** $x > -2$ **48.** $x < 2$ **49.** $a \ge 6$ **50.** $a < -15$

51.

52.

53.

54.

55.

56.

CHAPTER 3

Problem Set 3.1

1–17.

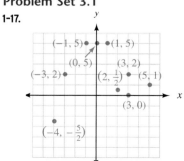

19. $(-4, 4)$ **21.** $(-4, 2)$ **23.** $(-3, 0)$ **25.** $(2, -2)$ **27.** $(-5, -5)$

29. Yes   **31.** No

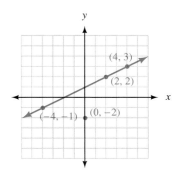

33. Yes   **35.** No

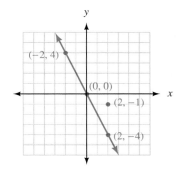

37. Yes   **39.** No

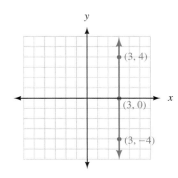

41. No   **43.** No

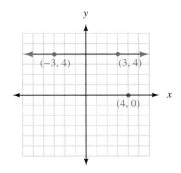

45. a. (5, 40), (10, 80), (20, 160), answers may vary **b.** $320 **c.** 30 hours **d.** No, she should make $280 for working 35 hours.

47.

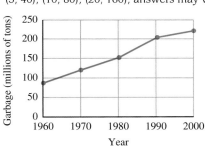

49.

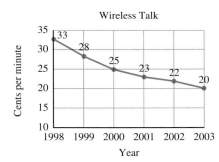

51. (1981, 5), (1985, 20.2), (1990, 34.4), (1995, 44.8), (2000, 65.4), (2004, 99.4)   **53.** $A = (1, 2)$, $B = (6, 7)$

55. $A = (2, 2)$, $B = (2, 5)$, $C = (7, 5)$ **57.** $\dfrac{x+3}{5}$ **59.** $\dfrac{2-a}{7}$ **61.** $\dfrac{1-2y}{14}$ **63.** $\dfrac{x+6}{2x}$ **65.** $\dfrac{x}{6}$ **67.** $\dfrac{8+x}{2x}$

69. a. -3 **b.** 6 **c.** 0 **d.** -4 **71. a.** 4 **b.** 2 **c.** -1 **d.** 9

Problem Set 3.2

1. (0, 6), (3, 0), (6, −6)   **3.** (0, 3), (4, 0), (−4, 6)   **5.** (1, 1), $(\frac{3}{4}, 0)$, (5, 17)   **7.** (2, 13), (1, 6), (0, −1)   **9.** (−5, 4), (−5, −3), (−5, 0)

11.

x	y
1	3
−3	−9
4	12
6	18

13.

x	y
0	0
$-\frac{1}{2}$	−2
−3	−12
3	12

15.

x	y
2	3
3	2
5	0
9	−4

17.

x	y
2	0
3	2
1	−2
−3	−10

19.

x	y
0	−1
−1	−7
−3	−19
$\frac{3}{2}$	8

21. (0, −2) **23.** (1, 5), (0, −2), (−2, −16) **25.** (1, 6), (−2, −12), (0, 0) **27.** (2, −2) **29.** (3, 0), (3, −3)

31. 12 inches
33. a. Yes **b.** No, she should earn $108 for working 9 hours. **c.** No, she should earn $84 for working 7 hours. **d.** Yes
35. a. $375,000 **b.** At the end of six years **c.** No, the crane will be worth $195,000 after nine years. **d.** $600,000
37. 6 **39.** 3 **41.** 6 **43.** 4 **45.** −3 **47.** 2 **49.** 0 **51.** $y = -5x + 4$ **53.** $y = \frac{3}{2}x - 3$

Problem Set 3.3

1.

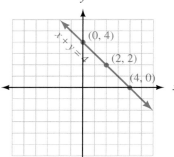

3.

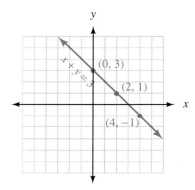

5.

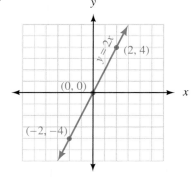

7.

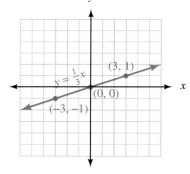

9.

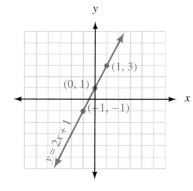

11.

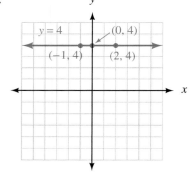

13.

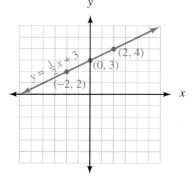

15.

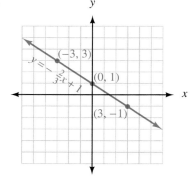

17.

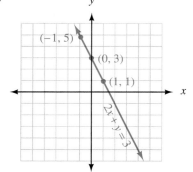

19.

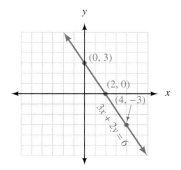

21.

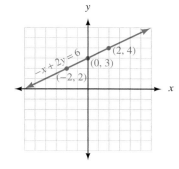

23.

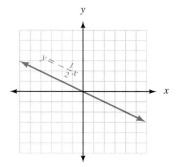

25.

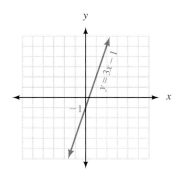

27.

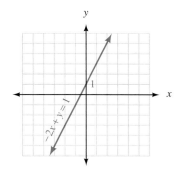

29.

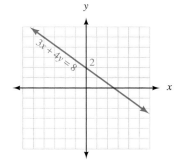

31.

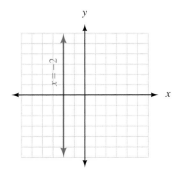

33.

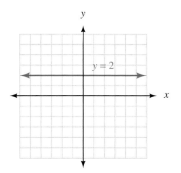

35.

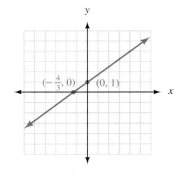

37.

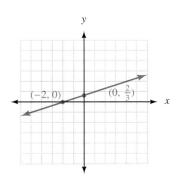

39.

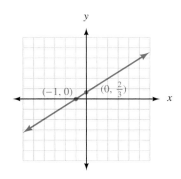

41.

Equation	H, V, and/or O
$x = 3$	V
$y = 3$	H
$y = 3x$	O
$y = 0$	O, H

43.

Equation	H, V, and/or O
$x = -\frac{3}{5}$	V
$y = -\frac{3}{5}$	H
$y = -\frac{3}{5}x$	O
$x = 0$	O, V

45. a. $\frac{5}{2}$ **b.** 5 **c.** 2 **d.**

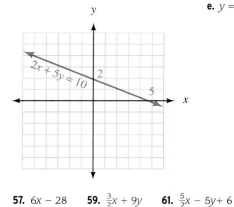

e. $y = -\frac{2}{5}x + 2$

47. $2x + 5$ **49.** $2x - 6$ **51.** $3x + \frac{15}{2}$ **53.** $6x + 9$ **55.** $2x + 50$ **57.** $6x - 28$ **59.** $\frac{3}{2}x + 9y$ **61.** $\frac{5}{2}x - 5y + 6$
63. a. 2 **b.** -3 **65. a.** -4 **b.** 2 **67. a.** 6 **b.** 2

Problem Set 3.4

1.

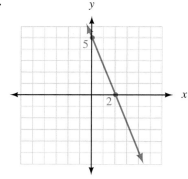

3.

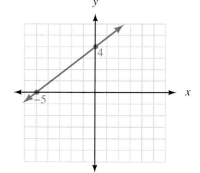

5.

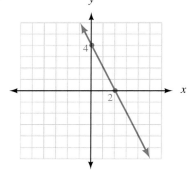

7.

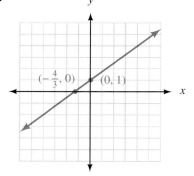

9.

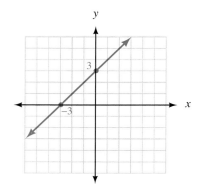

11.

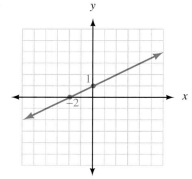

13.

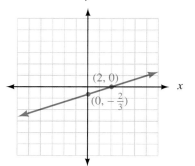

15.

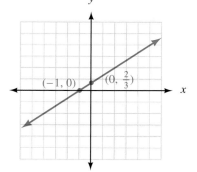

17.

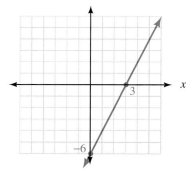

19.

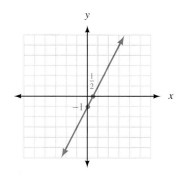

21.

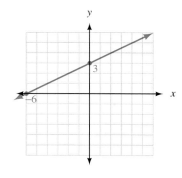

23.

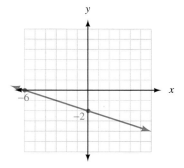

25.

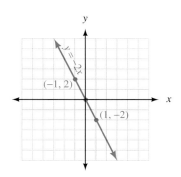

27.

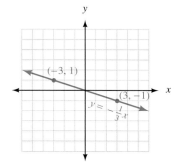

29.

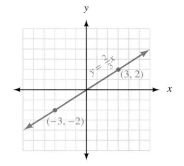

31.

Equation	x-intercept	y-intercept
$3x + 4y = 12$	4	3
$3x + 4y = 4$	$\frac{4}{3}$	1
$3x + 4y = 3$	1	$\frac{3}{4}$
$3x + 4y = 2$	$\frac{2}{3}$	$\frac{1}{2}$

33.

Equation	x-intercept	y-intercept
$x - 3y = 2$	2	$-\frac{2}{3}$
$y = \frac{1}{3}x - \frac{2}{3}$	2	$-\frac{2}{3}$
$x - 3y = 0$	0	0
$y = \frac{1}{3}x$	0	0

35. a. 0 **b.** $-\frac{3}{2}$ **c.** 1 **d.** **e.** $y = \frac{2}{3}x + 1$

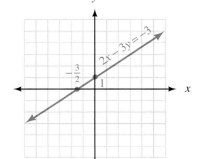

37. x-intercept $= 3$; y-intercept $= 3$

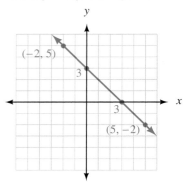

39. x-intercept $= -1$; y-intercept $= -3$

41.

x	y
−2	1
0	−1
−1	0
1	−2

43.

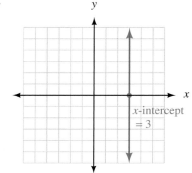

45.

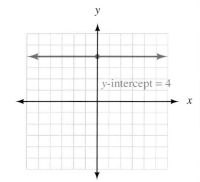

47. a. $10x + 12y = 480$ **b.** x-intercept $= 48$; y-intercept $= 40$ **c.**

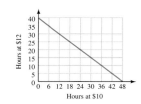

d. 10 hours   **e.** 18 hours

49. 12   **51.** -11 **53.** -6 **55.** 7   **57.** -2 **59.** -1 **61.** 14   **63.** -6 **65. a.** $\frac{3}{2}$ **b.** $\frac{3}{2}$ **67. a.** $\frac{3}{2}$ **b.** $\frac{3}{2}$

Problem Set 3.5

1.

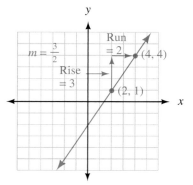

3.

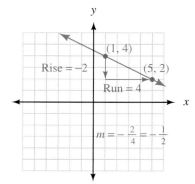

5.

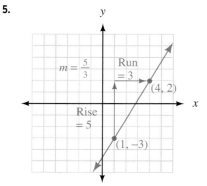

7.

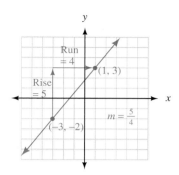

9.

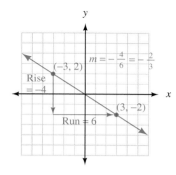

11.

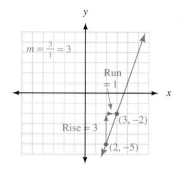

13.

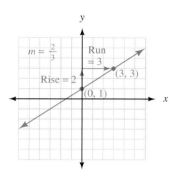

15.

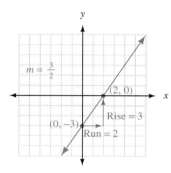

17.

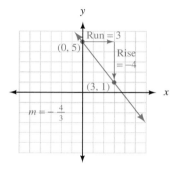

19.

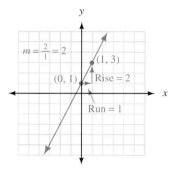

21.

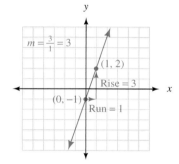

23. Slope = 3, y-intercept = 2

25. Slope = 2, y-intercept = -2

27.

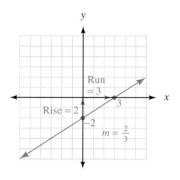

29.

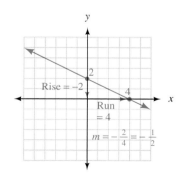

31.

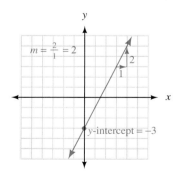

33.

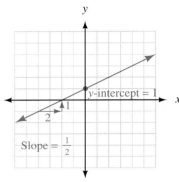

35. 6   **37.**

Equation	Slope
$x = 3$	Undefined
$y = 3$	0
$y = 3x$	3

39.

Equation	Slope
$x = -\frac{2}{3}$	0
$y = -\frac{2}{3}$	Undefined
$y = -\frac{2}{3}x$	$-\frac{2}{3}$

41. Slopes: A, $\frac{1}{2}$; B, 2; C, 5   **43.** 3   **45.** 15   **47.** -6 **49.** 6   **51.** -39 **53.** 4   **55.** 10   **57.** -20 **59.** $y = 2x + 4$
61. $y = -3x + 3$ **63.** $y = \frac{4}{5}x - 4$ **65.** $y = -2x - 5$ **67.** $y = -\frac{2}{3}x + 1$ **69.** $y = \frac{3}{2}x + 1$

Problem Set 3.6

1. $y = \frac{2}{3}x + 1$ **3.** $y = \frac{3}{2}x - 1$ **5.** $y = -\frac{2}{5}x + 3$ **7.** $y = 2x - 4$

9. $m = 2, b = 4$ **11.** $m = -3, b = 3$ **13.** $m = -\frac{3}{2}, b = 3$

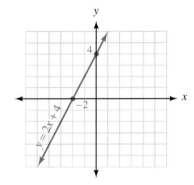   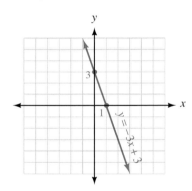   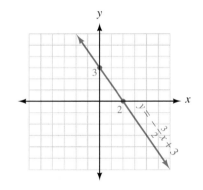

15. $m = \frac{4}{5}, b = -4$ **17.** $m = -\frac{2}{5}, b = -2$

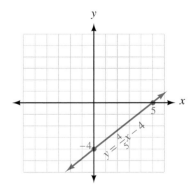   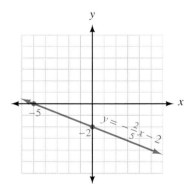

19. $y = 2x - 1$ **21.** $y = -\frac{1}{2}x - 1$ **23.** $y = \frac{3}{2}x - 6$ **25.** $y = -3x + 1$ **27.** $y = x - 2$ **29.** $y = 2x - 3$ **31.** $y = \frac{4}{3}x + 2$
33. $y = -\frac{2}{3}x - 3$ **35.** $m = 3, b = 3, y = 3x + 3$ **37.** $m = \frac{1}{4}, b = -1, y = \frac{1}{4}x - 1$

39. a. $-\frac{5}{2}$ **b.** $y = 2x + 6$ **c.** 6 **d.** 2 **e.**

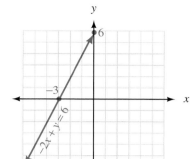

41. $y = -\frac{2}{3}x + 2$ **43.** $y = -\frac{5}{2}x - 5$ **45.** $x = 3$

47. a. \$6,000 **b.** 3 years **c.** Slope $= -3,000$ **d.** \$3,000 **e.** $V = -3,000t + 21,000$

49. a. \$3,000 **b.** 500 shirts **c.** Slope $= 2$ **d.** \$2 **e.** $y = 2x + 1,000$ **51.** 6   **53.** -2 **55.** 4   **57.** -11 **59.** 10   **61.** -38

63. -12 **65.** -18

67.

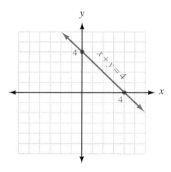

69.

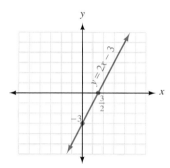

71.

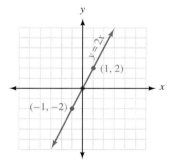

Problem Set 3.7

1.

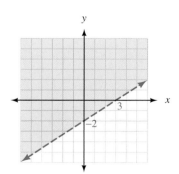

3.

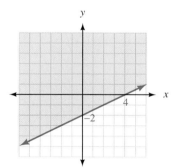

5.

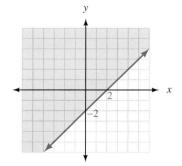

7.

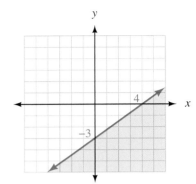

9.

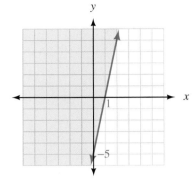

11.

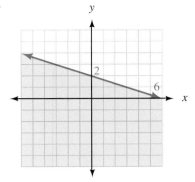

13.

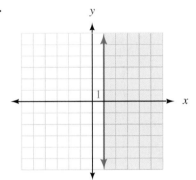

15.

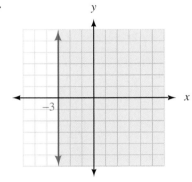

17.

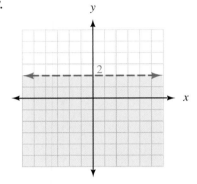

19.

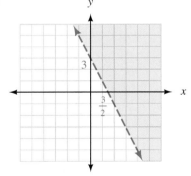

21.

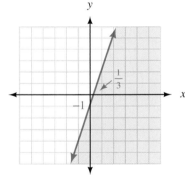

23.

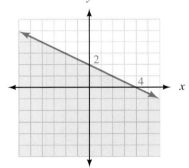

25. a. $y < \frac{8}{3}$ **b.** $y > -\frac{8}{3}$ **c.** $y = -\frac{4}{3}x + 4$ **d.**

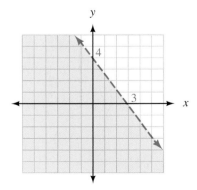

27. a. $y = \frac{2}{5}x + 2$ **b.** $y < \frac{2}{5}x + 2$ **c.** $y > \frac{2}{5}x + 2$ **29.** $-6x + 11$ **31.** -8 **33.** -4 **35.** $w = \dfrac{P - 2l}{2}$

37. 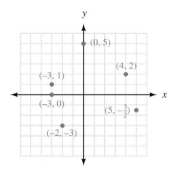   **39.** $y \geq \frac{3}{2}x - 6$ **41.** Width 2 inches, length 11 inches

CHAPTER 3 REVIEW/TEST

1. $(4, -6)$, $(0, 6)$, $(1, 3)$, $(2, 0)$ **2.** $(5, -2)$, $(0, -4)$, $(15, 2)$, $(10, 0)$ **3.** $(4, 2)$, $(2, -2)$, $\left(\frac{9}{2}, 3\right)$ **4.** $(2, 13)$, $\left(-\frac{3}{5}, 0\right)$, $\left(-\frac{6}{5}, -3\right)$

5. $(2, -3)$, $(-1, -3)$, $(-3, -3)$ **6.** $(6, 5)$, $(6, 0)$, $(6, -1)$ **7.** $\left(2, -\frac{3}{2}\right)$ **8.** $\left(-\frac{8}{3}, -1\right)$, $(-3, -2)$

9.-14.   **15.** $(-2, 0)$, $(0, -2)$, $(1, -3)$ **16.** $(-1, -3)$, $(1, 3)$, $(0, 0)$

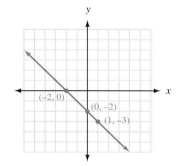

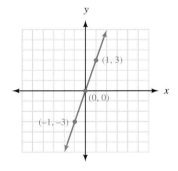

17. $(1, 1)$, $(0, -1)$, $(-1, -3)$ **18.** $(-3, 0)$, $(-3, 5)$, $(-3, -5)$ **19.**

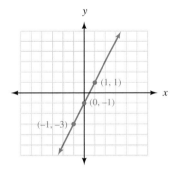

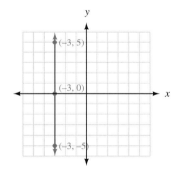

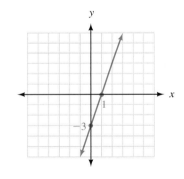

20.   **21.**   **22.**

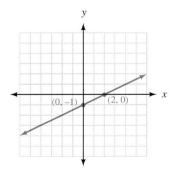

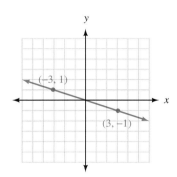

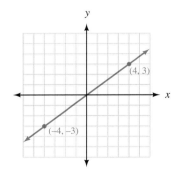

23.

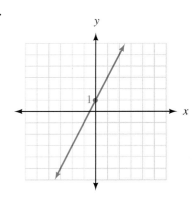

24.

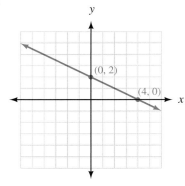

25.

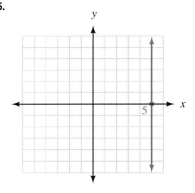

26.

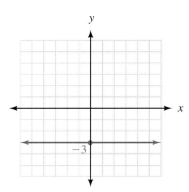

27.

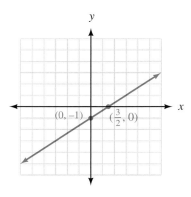

28.

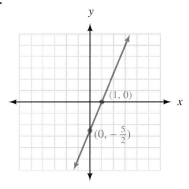

29. x-intercept = 2; y-intercept = -6 **30.** x-intercept = 12; y-intercept = -4 **31.** x-intercept = 3; y-intercept = -3
32. x-intercept = 2; y-intercept = -6 **33.** no x-intercept; y-intercept = -5 **34.** x-intercept = 4; no y-intercept   **35.** $m = 2$
36. $m = -1$ **37.** $m = 2$ **38.** $m = 2$ **39.** $x = 6$ **40.** $x = -25$ **41.** $y = -2x + 2$ **42.** $y = \frac{1}{2}x + 1$ **43.** $y = -\frac{3}{4}x + \frac{1}{4}$

44. $y = 5$ **45.** $y = 3x + 2$ **46.** $y = -x + 6$ **47.** $y = -\frac{1}{3}x + \frac{3}{4}$ **48.** $y = 0$ **49.** $m = 4, b = -1$ **50.** $m = -2, b = -5$

51. $m = -2, b = 3$ **52.** $m = -\frac{5}{2}, b = 4$

53.

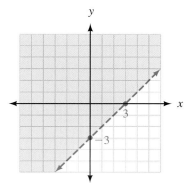

54.

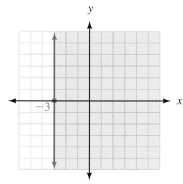

55.

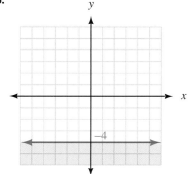

56.

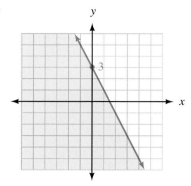

CHAPTER 4

Problem Set 4.1

1.

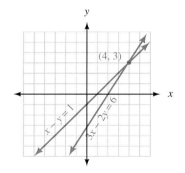

3.

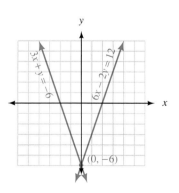

5.

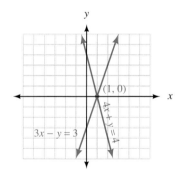

7.

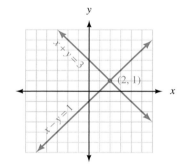

9.

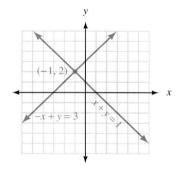

11.

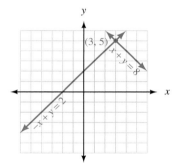

13.

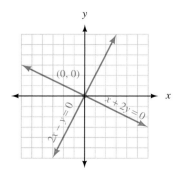

15.

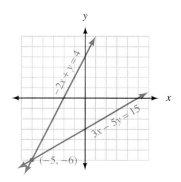

17.

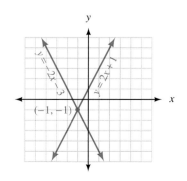

19.

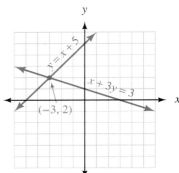

21.

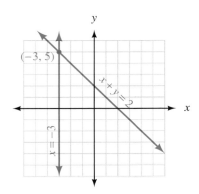

23.

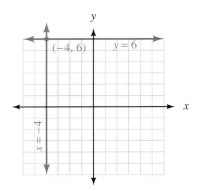

25. ∅

27. Any point on the line

29. a. $4x - 5y$ **b.** 1 **c.** -2
d.   **e.** $(0, -2)$

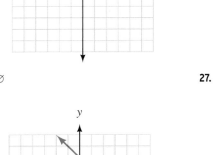

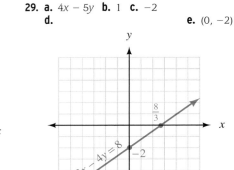

31.

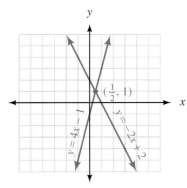

33.

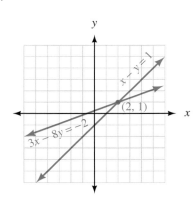

35. a. 25 hours **b.** Gigi's **c.** Marcy's

37. $10x + 75$ **39.** $11x - 35$ **41.** $-x + 58$ **43.** $-0.02x + 6$ **45.** $-0.003x - 28$ **47.** $1.31x - 125$ **49.** $2x$ **51.** $7x$
53. $13x$ **55.** $-12x - 20y$ **57.** $3x + 8y$ **59.** $-4x + 2y$ **61.** 1   **63.** 0   **65.** -5

Problem Set 4.2

1. $(2, 1)$ **3.** $(3, 7)$ **5.** $(2, -5)$ **7.** $(-1, 0)$ **9.** Lines coincide   **11.** $(4, 8)$ **13.** $\left(\frac{1}{5}, 1\right)$ **15.** $(1, 0)$ **17.** $(-1, -2)$
19. $\left(-5, \frac{3}{4}\right)$ **21.** $(-4, 5)$ **23.** $(-3, -10)$ **25.** $(3, 2)$ **27.** $\left(5, \frac{1}{3}\right)$ **29.** $\left(-2, \frac{2}{3}\right)$ **31.** $(2, 2)$ **33.** Lines are parallel; ∅
35. $(1, 1)$ **37.** Lines are parallel; ∅   **39.** $(10, 12)$ **41.** $(6, 8)$ **43.** $(7,000, 8,000)$ **45.** $(3,000, 8,000)$ **47.** $(11, 12)$
49. $(10, 12)$ **51.** slope $= 3$; y-int $= -3$ **53.** slope $= \frac{2}{5}$; y-int $= -5$ **55.** slope $= -1$ **57.** slope $= 2$ **59.** 11   **61.** 9
63. $y = 3x$ **65.** $y = x - 2$ **67.** 1   **69.** 2   **71.** 1   **73.** $x = 3y - 1$ **75.** 1   **77.** 5   **79.** 34.5   **81.** 33.95

Problem Set 4.3
1. $(4, 7)$ **3.** $(3, 17)$ **5.** $\left(\frac{3}{2}, 2\right)$ **7.** $(2, 4)$ **9.** $(0, 4)$ **11.** $(-1, 3)$ **13.** $(1, 1)$ **15.** $(2, -3)$ **17.** $\left(-2, \frac{3}{5}\right)$ **19.** $(-3, 5)$
21. Lines are parallel; $\varnothing$ **23.** $(3, 1)$ **25.** $\left(\frac{1}{2}, \frac{3}{4}\right)$ **27.** $(2, 6)$ **29.** $(4, 4)$ **31.** $(5, -2)$ **33.** $(18, 10)$ **35.** Lines coincide
37. $(8, 4)$ **39.** $(6, 12)$ **41.** $(16, 32)$ **43.** $(10, 12)$ **45. a.** $\frac{25}{4}$ **b.** $y = \frac{4}{5}x - 4$ **c.** $x = y + 5$ **d.** $(5, 0)$
47. a. 1,000 miles **b.** Car **c.** Truck **d.** We are only working with positive numbers.   **49.** 47   **51.** 14   **53.** 70   **55.** 35
57. 5   **59.** 6,540   **61.** 1,760   **63.** 20   **65.** 63   **67.** 53   **69.** 3 and 23   **71.** 15 and 24   **73.** Length = 23 in.; Width = 6 in.
75. 14 nickles; 10 dimes   **77.** $(3, 17)$ **79.** $(7,000, 8,000)$

Problem Set 4.4
As you can see, in addition to the answers to the problems we sometimes have included the system of equations used to solve the problems. Remember, you should attempt the problem on your own before checking your answers or equations.
1. $x + y = 25$ The two numbers   **3.** 3 and 12   **5.** $x - y = 5$ The two numbers   **7.** 6 and 29
$\quad\quad y = x + 5$ are 10 and 15. $\quad\quad\quad\quad\quad\quad\quad\quad x = 2y + 1$ are 9 and 4.
9. Let $x =$ the amount invested at 6% and $y =$ the amount invested at 8%.   **11.** \$2,000 at 6%, \$8,000 at 5%
$\quad\quad\quad x + y = 20,000$ He has \$9,000 at 8%   **13.** 6 nickels, 8 quarters
$\quad 0.06x + 0.08y = 1,380$ and \$11,000 at 6%.
15. Let $x =$ the number of dimes and $y =$ the number of quarters.
$\quad\quad\quad x + y = 21$ He has 12 dimes
$\quad 0.10x + 0.25y = 3.45$ and 9 quarters.
17. Let $x =$ the number of liters of 50% solution and $y =$ the number of liters of 20% solution.
$\quad\quad\quad x + y = 18$ 6 liters of 50% solution
$\quad 0.50x + 0.20y = 0.30(18)$ 12 liters of 20% solution
19. 10 gallons of 10% solution, 20 gallons of 7% solution   **21.** 20 adults, 50 kids   **23.** 16 feet wide, 32 feet long
25. 33 \$5 chips, 12 \$25 chips   **27.** 50 at \$11, 100 at \$20   **29.** $(-2, 2), (0, 3), (2, 4)$ **31.**

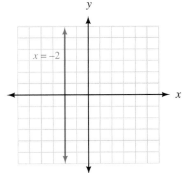

33. 2   **35.** $y = \frac{1}{2}x + 2$ **37.** $y = 2x + 1$

CHAPTER 4 REVIEW/TEST
1. $(4, -2)$ **2.** $(-3, 2)$ **3.** $(3, -2)$ **4.** $(2, 0)$ **5.** $(2, 1)$ **6.** $(-1, -2)$ **7.** $(1, -3)$ **8.** $(-2, 5)$ **9.** Lines coincide
10. $(2, -2)$ **11.** $(1, 1)$ **12.** Lines are parallel; $\varnothing$ **13.** $(-2, -3)$ **14.** $(5, -1)$ **15.** $(-2, 7)$ **16.** $(-4, -2)$ **17.** $(-1, 4)$
18. $(2, -6)$ **19.** Lines are parallel; $\varnothing$ **20.** $(-3, -10)$ **21.** $(1, -2)$ **22.** Lines coincide   **23.** 10, 8   **24.** 24, 8
25. \$4,000 at 4%, \$8,000 at 5%   **26.** \$3,000 at 6%, \$11,000 at 8%   **27.** 10 dimes, 7 nickels   **28.** 9 dimes, 6 quarters
29. 40 liters of 10% solution, 10 liters of 20% solution   **30.** 20 liters of 25% solution, 20 liters of 15% solution

CHAPTER 5

Problem Set 5.1
1. Base 4; exponent 2; 16   **3.** Base 0.3; exponent 2; 0.09   **5.** Base 4; exponent 3; 64   **7.** Base -5; exponent 2; 25
9. Base -2; exponent 3; -8 **11.** Base 3; exponent 4; 81   **13.** Base $\frac{2}{3}$; exponent 2; $\frac{4}{9}$ **15.** Base $\frac{1}{2}$; exponent 4; $\frac{1}{16}$

17. a.

Number x	Square x^2
1	1
2	4
3	9
4	16
5	25
6	36
7	49

b. Either *larger* or *greater* will work.
19. x^9 **21.** y^{30} **23.** 2^{12} **25.** x^{28} **27.** x^{10} **29.** 5^{12} **31.** y^9 **33.** 2^{50} **35.** a^{3x}
37. b^{xy} **39.** $16x^2$ **41.** $32y^5$ **43.** $81x^4$ **45.** $0.25a^2b^2$ **47.** $64x^3y^3z^3$ **49.** $8x^{12}$
51. $16a^6$ **53.** x^{14} **55.** a^{11} **57.** $128x^7$ **59.** $432x^{10}$ **61.** $16x^4y^6$ **63.** $\frac{8}{27}a^{12}b^{15}$
65. x^2 **67.** $4x$ **69.** 2 **71.** $4x$ **73. a.** 32 **b.** 64 **c.** 32 **d.** 64

75.

Number x	Square x^2
−3	9
−2	4
−1	1
0	0
1	1
2	4
3	9

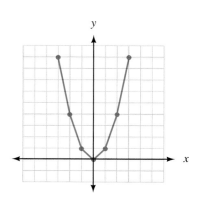

77.

Number x	Square x^2
−2.5	6.25
−1.5	2.25
−0.5	0.25
0	0
0.5	0.25
1.5	2.25
2.5	6.25

79. 4.32×10^4 **81.** 5.7×10^2 **83.** 2.38×10^5 **85.** 2,490 **87.** 352 **89.** 28,000
91. 27 inches³ **93.** 15.6 inches³ **95.** 36 inches³ **97.** Yes, the dimensions could be 2 feet times 3 feet times 7 feet.
99. 6.5×10^8 seconds **101.** \$740,000 **103.** \$180,000 **105.** 219 inches³ **107.** 182 inches³ **109.** 2^7 **111.** $2 \cdot 5^3$
113. $2^4 \cdot 3^2 \cdot 5$ **115.** $2^2 \cdot 5 \cdot 41$ **117.** $2^3 \cdot 3^3$ **119.** $2^3 \cdot 3^3 \cdot 5^3$ **121.** 5^6 **123.** $2^6 \cdot 3^3$ **125.** −3 **127.** 11 **129.** −5 **131.** 5
133. 2 **135.** 6 **137.** 4 **139.** 3

Problem Set 5.2

1. $\frac{1}{9}$ **3.** $\frac{1}{36}$ **5.** $\frac{1}{64}$ **7.** $\frac{1}{125}$ **9.** $\frac{2}{x^3}$ **11.** $\frac{1}{8x^3}$ **13.** $\frac{1}{25y^2}$ **15.** $\frac{1}{100}$

17.

Number x	Square x^2	Power of 2 2^x
−3	9	$\frac{1}{8}$
−2	4	$\frac{1}{4}$
−1	1	$\frac{1}{2}$
0	0	1
1	1	2
2	4	4
3	9	8

19. $\frac{1}{25}$ **21.** x^6 **23.** 64 **25.** $8x^3$ **27.** 6^{10} **29.** $\frac{1}{6^{10}}$ **31.** $\frac{1}{2^8}$ **33.** 2^8

35. $27x^3$ **37.** $81x^4y^4$ **39.** 1 **41.** $2a^2b$ **43.** $\frac{1}{49y^6}$ **45.** $\frac{1}{x^8}$ **47.** $\frac{1}{y^3}$ **49.** x^2 **51.** a^6 **53.** $\frac{1}{y^9}$ **55.** y^{40}

57. $\frac{1}{x}$ **59.** x^9 **61.** a^{16} **63.** $\frac{1}{a^4}$ **65. a.** 32 **b.** 32 **c.** $\frac{1}{32}$ **d.** $\frac{1}{32}$ **67. a.** $\frac{1}{5}$ **b.** $\frac{1}{8}$ **c.** $\frac{1}{x}$ **d.** $\frac{1}{x^2}$

69.

Number x	Power of 2 2^x
-3	$\frac{1}{8}$
-2	$\frac{1}{4}$
-1	$\frac{1}{2}$
0	1
1	2
2	4
3	8

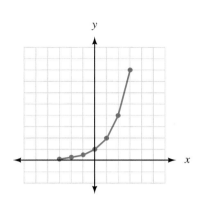

71. 4.8×10^{-3} **73.** 2.5×10^{1} **75.** 9×10^{-6}

77.

Expanded Form	Scientific Notation $n \times 10^r$
0.000357	3.57×10^{-4}
0.00357	3.57×10^{-3}
0.0357	3.57×10^{-2}
0.357	3.57×10^{-1}
3.57	3.57×10^{0}
35.7	3.57×10^{1}
357	3.57×10^{2}
3,570	3.57×10^{3}
35,700	3.57×10^{4}

79. 0.00423   **81.** 0.00008   **83.** 4.2   **85.** 0.002

87. $0.0024 = 2.4 \times 10^{-3}$, $0.0051 = 5.1 \times 10^{-3}$, $0.0096 = 9.6 \times 10^{-3}$, $0.0099 = 9.9 \times 10^{-3}$, $0.0111 = 1.11 \times 10^{-2}$ **89.** 2.5×10^{4}
91. 2.35×10^{5} **93.** 8.2×10^{-4} **95.** 100 inches²; 400 inches²; 4   **97.** x^2; $4x^2$; 4   **99.** 216 inches³; 1,728 inches³; 8
101. x^3; $8x^3$; 8   **103.** $7x$ **105.** $2a$ **107.** $10y$ **109.** 13.5   **111.** 8   **113.** 26.52   **115.** 12   **117.** x^8 **119.** x **121.** $\frac{1}{y^2}$
123. 340

Problem Set 5.3

1. $12x^7$ **3.** $-16y^{11}$ **5.** $32x^2$ **7.** $200a^6$ **9.** $-24a^3b^3$ **11.** $24x^6y^8$ **13.** $3x$ **15.** $\frac{6}{y^3}$ **17.** $\frac{1}{2a}$ **19.** $-\frac{3a}{b^2}$ **21.** $\frac{x^2}{9z^2}$
23.

a	b	ab	$\frac{a}{b}$	$\frac{b}{a}$
10	$5x$	$50x$	$\frac{2}{x}$	$\frac{x}{2}$
$20x^3$	$6x^2$	$120x^5$	$\frac{10x}{3}$	$\frac{3}{10x}$
$25x^5$	$5x^4$	$125x^9$	$5x$	$\frac{1}{5x}$
$3x^{-2}$	$3x^2$	9	$\frac{1}{x^4}$	x^4
$-2y^4$	$8y^7$	$-16y^{11}$	$-\frac{1}{4y^3}$	$-4y^3$

25. 6×10^{8} **27.** 1.75×10^{-1} **29.** 1.21×10^{-6} **31.** 4.2×10^{3}

33. 3×10^{10} **35.** 5×10^{-3} **37.** $8x^2$ **39.** $-11x^5$ **41.** 0 **43.** $4x^3$ **45.** $31ab^2$ **47.**

49. $4x^3$ **51.** $\dfrac{1}{b^2}$ **53.** $\dfrac{6y^{10}}{x^4}$

a	b	ab	a + b
$5x$	$3x$	$15x^2$	$8x$
$4x^2$	$2x^2$	$8x^4$	$6x^2$
$3x^3$	$6x^3$	$18x^6$	$9x^3$
$2x^4$	$-3x^4$	$-6x^8$	$-x^4$
x^5	$7x^5$	$7x^{10}$	$8x^5$

55. 2×10^6 **57.** 1×10^1 **59.** 4.2×10^{-6} **61.** $9x^3$ **63.** $-20a^2$ **65.** $6x^5y^2$ **67.** $x^2y + x$ **69.** $x + y$ **71.** $x^2 - 4$
73. $x^2 - x - 6$ **75.** $x^2 - 5x$ **77.** $x^2 - 8x$ **79. a.** 2 **b.** $-3b$ **c.** $5b^2$ **81. a.** $3xy$ **b.** $2y$ **c.** $-\dfrac{y^2}{2}$ **83.** -8 **85.** 9 **87.** 0
89.   **91.**   **93.** -5 **95.** 6 **97.** 76 **99.** $6x^2$ **101.** $2x$

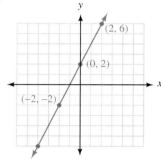
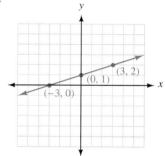

103. $-2x - 9$ **105.** 11

Problem Set 5.4

1. Trinomial, 3   **3.** Trinomial, 3   **5.** Binomial, 1   **7.** Binomial, 2   **9.** Monomial, 2   **11.** Monomial, 0   **13.** $5x^2 + 5x + 9$
15. $5a^2 - 9a + 7$ **17.** $x^2 + 6x + 8$ **19.** $6x^2 - 13x + 5$ **21.** $x^2 - 9$ **23.** $3y^2 - 11y + 10$ **25.** $6x^3 + 5x^2 - 4x + 3$ **27.** $2x^2 - x + 1$
29. $2a^2 - 2a - 2$ **31.** $-\dfrac{1}{9}x^3 - \dfrac{2}{3}x^2 - \dfrac{5}{2}x + \dfrac{7}{4}$ **33.** $-4y^2 + 15y - 22$ **35.** $-2x$ **37.** $4x$ **39.** $x^3 - 27$ **41.** $x^3 + 6x^2 + 12x + 8$
43. $4x - 8$ **45.** $x^2 - 33x + 63$ **47.** $8y^2 + 4y + 26$ **49.** $75x^2 - 150x - 75$ **51.** $12x + 2$ **53.** 4 **55. a.** 0 **b.** 0
57. a. $2{,}200$ **b.** $-1{,}800$ **59.** 18π inches³   **61.** $-15x^2$ **63.** $6x^3$ **65.** $6x^4$ **67.** 5 **69.** -6 **71.** $-20x^2$ **73.** $-21x$
75. $2x$ **77.** $6x - 18$

Problem Set 5.5

1. $6x^2 + 2x$ **3.** $6x^4 - 4x^3 + 2x^2$ **5.** $2a^3b - 2a^2b^2 + 2ab$ **7.** $3y^4 + 9y^3 + 12y^2$ **9.** $8x^5y^2 + 12x^4y^3 + 32x^2y^4$ **11.** $x^2 + 7x + 12$
13. $x^2 + 7x + 6$ **15.** $x^2 + 2x + \dfrac{3}{4}$ **17.** $a^2 + 2a - 15$ **19.** $xy + bx - ay - ab$ **21.** $x^2 - 36$ **23.** $y^2 - \dfrac{25}{36}$ **25.** $2x^2 - 11x + 12$
27. $2a^2 + 3a - 2$ **29.** $6x^2 - 19x + 10$ **31.** $2ax + 8x + 3a + 12$ **33.** $25x^2 - 16$ **35.** $2x^2 + \dfrac{5}{2}x - \dfrac{3}{4}$ **37.** $3 - 10a + 8a^2$
39.

	x	3
x	x^2	$3x$
2	$2x$	6

$(x + 2)(x + 3) = x^2 + 2x + 3x + 6$
$= x^2 + 5x + 6$

41.

	x	x	2
x	x^2	x^2	$2x$
1	x	x	2

$(x + 1)(2x + 2) = 2x^2 + 4x + 2$

43. $a^3 - 6a^2 + 11a - 6$ **45.** $x^3 + 8$ **47.** $2x^3 + 17x^2 + 26x + 9$ **49.** $5x^4 - 13x^3 + 20x^2 + 7x + 5$ **51.** $2x^4 + x^2 - 15$
53. $6a^6 + 15a^4 + 4a^2 + 10$ **55.** $x^3 + 12x^2 + 47x + 60$ **57.** $x^2 - 5x + 8$ **59.** $8x^2 - 6x - 5$ **61.** $x^2 - x - 30$ **63.** $x^2 + 4x - 6$
65. $x^2 + 13$ **67.** $x^2 + 2x - 3$ **69.** $a^2 - 3a + 6$ **71. a.** $x^2 - 1$ **b.** $x^2 + 2x + 1$ **c.** $x^3 + 3x^2 + 3x + 1$ **d.** $x^4 + 4x^3 + 6x^2 + 4x + 1$
73. a. $x^2 - 1$ **b.** $x^2 - x - 2$ **c.** $x^2 - 2x - 3$ **d.** $x^2 - 3x - 4$ **75.** 0 **77.** -5 **79.** $-2, 3$ **81.** Never 0 **83.** a, b
85. $A = x(2x + 5) = 2x^2 + 5x$ **87.** $A = x(x + 1) = x^2 + x$ **89.** $R = (1{,}200 - 100p)p = 1{,}200p - 100p^2$

91.

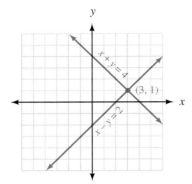

93. $(5, -7)$ **95.** $(3, 17)$ **97.** $800 at 8%, $400 at 10%   **99.** 169 **101.** $-10x$ **103.** 0

105. 0 **107.** $-12x + 16$ **109.** $x^2 + x - 2$ **111.** $x^2 + 6x + 9$

Problem Set 5.6

1. $x^2 - 4x + 4$ **3.** $a^2 + 6a + 9$ **5.** $x^2 - 10x + 25$ **7.** $a^2 - a + \frac{1}{4}$ **9.** $x^2 + 20x + 100$ **11.** $a^2 + 1.6a + 0.64$ **13.** $4x^2 - 4x + 1$
15. $16a^2 + 40a + 25$ **17.** $9x^2 - 12x + 4$ **19.** $9a^2 + 30ab + 25b^2$ **21.** $16x^2 - 40xy + 25y^2$ **23.** $49m^2 + 28mn + 4n^2$
25. $36x^2 - 120xy + 100y^2$ **27.** $x^4 + 10x^2 + 25$ **29.** $a^4 + 2a^2 + 1$ **31.** $y^2 + 3y + \frac{9}{4}$ **33.** $a^2 + a + \frac{1}{4}$ **35.** $x^2 + \frac{3}{2}x + \frac{9}{16}$
37. $t^2 + \frac{2}{5}t + \frac{1}{25}$

39.

x	$(x + 3)^2$	$x^2 + 9$	$x^2 + 6x + 9$
1	16	10	16
2	25	13	25
3	36	18	36
4	49	25	49

41.

a	b	$(a + b)^2$	$a^2 + b^2$	$a^2 + ab + b^2$	$a^2 + 2ab + b^2$
1	1	4	2	3	4
3	5	64	34	49	64
3	4	49	25	37	49
4	5	81	41	61	81

43. $a^2 - 25$ **45.** $y^2 - 1$ **47.** $81 - x^2$ **49.** $4x^2 - 25$ **51.** $16x^2 - \frac{1}{9}$ **53.** $4a^2 - 49$ **55.** $36 - 49x^2$ **57.** $x^4 - 9$
59. $a^4 - 16$ **61.** $25y^8 - 64$ **63.** $2x^2 - 34$ **65.** $-12x^2 + 20x + 8$ **67.** $a^2 + 4a + 6$ **69.** $8x^3 + 36x^2 + 54x + 27$
71. a. 11 **b.** 1 **c.** 11 **73. a.** 1 **b.** 13 **c.** 1 **75.** $x^2 + (x + 1)^2 = 2x^2 + 2x + 1$ **77.** $x^2 + (x + 1)^2 + (x + 2)^2 = 3x^2 + 6x + 5$
79. $a^2 + ab + ba + b^2 = a^2 + 2ab + b^2$ **81.**

	a	b
a	a^2	ab
b	ba	b^2

83.

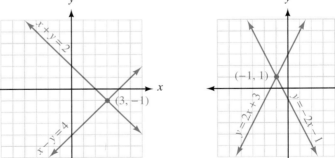

85. $2x^2$ **87.** x^2 **89.** $3x$ **91.** $3xy$ **93.** $5x$ **95.** $\dfrac{a^4}{2b^2}$

Problem Set 5.7

1. $x - 2$ **3.** $3 - 2x^2$ **5.** $5xy - 2y$ **7.** $7x^4 - 6x^3 + 5x^2$ **9.** $10x^4 - 5x^2 + 1$ **11.** $-4a + 2$ **13.** $-8a^4 - 12a^3$ **15.** $-4b - 5a$
17. $-6a^2b + 3ab^2 - 7b^3$ **19.** $-\dfrac{a}{2} - b - \dfrac{b^2}{2a}$ **21.** $3x + 4y$ **23.** $-y + 3$ **25.** $x^2 + 8x - 9$ **27.** $y^2 - 3y + 1$ **29.** $-1 + xy$
31. $-a + 1$ **33.** $x^2 - 3xy + y^2$ **35.** $2 - 3b + 5b^2$ **37.** $-2xy + 1$ **39.** $xy - \dfrac{1}{2}$ **41.** $\dfrac{1}{4x} - \dfrac{1}{2a} + \dfrac{3}{4}$ **43.** $\dfrac{4x^2}{3} + \dfrac{2}{3x} + \dfrac{1}{x^2}$
45. $3a^{3m} - 9a^m$ **47.** $2x^{4m} - 5x^{2m} + 7$ **49.** $3x^2 - x + 6$ **51.** 4 **53.** $x + 5$ **55. a.** 7 **b.** 7 **c.** 23

57. a. 19 **b.** 34 **c.** 19 **59. a.** 2 **b.** 2 **61.** $(7, -1)$ **63.** $(2, 3)$ **65.** $(1, 1)$ **67.** Lines coincide. **69.** $146\frac{20}{27}$ **71.** $2x + 5$
73. $x^2 - 3x$ **75.** $2x^3 - 10x^2$ **77.** $-2x$ **79.** 2

Problem Set 5.8

1. $x - 2$ **3.** $a + 4$ **5.** $x - 3$ **7.** $x + 3$ **9.** $a - 5$ **11.** $x + 2 + \dfrac{2}{x + 3}$ **13.** $a - 2 + \dfrac{12}{a + 5}$ **15.** $x + 4 + \dfrac{9}{x - 2}$

17. $x + 4 + \dfrac{-10}{x + 1}$ **19.** $a + 1 + \dfrac{-1}{a + 2}$ **21.** $x - 3 + \dfrac{17}{2x + 4}$ **23.** $3a - 2 + \dfrac{7}{2a + 3}$ **25.** $2a^2 - a - 3$ **27.** $x^2 - x + 5$

29. $x^2 + x + 1$ **31.** $x^2 + 2x + 4$ **33. a.** 19 **b.** 19 **c.** 5 **35. a.** 7 **b.** $\frac{25}{7}$ **c.** $\frac{25}{7}$ **37.** $430.17 **39.** $290.17 **41.** 5, 20

43. $800 at 8%, $400 at 9% **45.** Eight $5 bills, twelve $10 bills **47.** 10 gallons of 20%, 6 gallons of 60%

CHAPTER 5 REVIEW/TEST

1. -1 **2.** -64 **3.** $\frac{9}{49}$ **4.** y^{12} **5.** x^{30} **6.** x^{35} **7.** 2^{24} **8.** $27y^3$ **9.** $-8x^3y^3z^3$ **10.** $\frac{1}{49}$ **11.** $\frac{4}{x^5}$ **12.** $\frac{1}{27y^3}$

13. a^6 **14.** $\frac{1}{x^4}$ **15.** x^{15} **16.** $\frac{1}{x^5}$ **17.** 1 **18.** -3 **19.** $9x^6y^4$ **20.** $64a^{22}b^{20}$ **21.** $\frac{-1}{27x^3y^6}$ **22.** b **23.** $\frac{1}{x^{35}}$ **24.** $5x^7$

25. $\dfrac{6y^7}{x^5}$ **26.** $4a^6$ **27.** $2x^5y^2$ **28.** 6.4×10^7 **29.** 2.3×10^8 **30.** 8×10^3 **31.** $8a^2 - 12a - 3$ **32.** $-12x^2 + 5x - 14$

33. $-4x^2 - 6x$ **34.** 14 **35.** $12x^2 - 21x$ **36.** $24x^5y^2 - 40x^4y^3 + 32x^3y^4$ **37.** $a^3 + 6a^2 + a - 4$ **38.** $x^3 + 125$

39. $6x^2 - 29x + 35$ **40.** $25y^2 - \frac{1}{25}$ **41.** $a^4 - 9$ **42.** $a^2 - 10a + 25$ **43.** $9x^2 + 24x + 16$ **44.** $y^4 + 6y^2 + 9$ **45.** $-2b - 4a$

46. $-5x^4y^3 + 4x^2y^2 + 2x$ **47.** $x + 9$ **48.** $2x + 5$ **49.** $x^2 - 4x + 16$ **50.** $x - 3 + \dfrac{16}{3x + 2}$ **51.** $x^2 - 4x + 5 + \dfrac{5}{2x + 1}$

CHAPTER 6

Problem Set 6.1

1. $5(3x + 5)$ **3.** $3(2a + 3)$ **5.** $4(x - 2y)$ **7.** $3(x^2 - 2x - 3)$ **9.** $3(a^2 - a - 20)$ **11.** $4(6y^2 - 13y + 6)$ **13.** $x^2(9 - 8x)$
15. $13a^2(1 - 2a)$ **17.** $7xy(3x - 4y)$ **19.** $11ab^2(2a - 1)$ **21.** $7x(x^2 + 3x - 4)$ **23.** $11(11y^4 - x^4)$ **25.** $25x^2(4x^2 - 2x + 1)$
27. $8(a^2 + 2b^2 + 4c^2)$ **29.** $4ab(a - 4b + 8ab)$ **31.** $11a^2b^2(11a - 2b + 3ab)$ **33.** $12x^2y^3(1 - 6x^3 - 3x^2y)$ **35.** $(x + 3)(y + 5)$
37. $(x + 2)(y + 6)$ **39.** $(a - 3)(b + 7)$ **41.** $(a - b)(x + y)$ **43.** $(2x - 5)(a + 3)$ **45.** $(b - 2)(3x - 4)$ **47.** $(x + 2)(x + a)$
49. $(x - b)(x - a)$ **51.** $(x + y)(a + b + c)$ **53.** $(3x + 2)(2x + 3)$ **55.** $(10x - 1)(2x + 5)$ **57.** $(4x + 5)(5x + 1)$ **59.** $(x + 2)(x^2 + 3)$
61. $(3x - 2)(2x^2 + 5)$ **63.** 6 **65.** $3(4x^2 + 2x + 1)$ **67.** $y^2 - 16y + 64$ **69.** $x^2 - 4x - 12$ **71.** $2y^2 - 7y + 12$ **73.** $2x^2 + 7x - 11$
75. $3x + 8$ **77.** $3x^2 + 4x - 5$ **79.** $x^2 - 5x - 14$ **81.** $x^2 - x - 6$ **83.** $x^3 + 27$ **85.** $2x^3 + 9x^2 - 2x - 3$
87. $18x^7 - 12x^6 + 6x^5$ **89.** $x^2 + x + \frac{2}{9}$ **91.** $12x^2 - 10xy - 12y^2$ **93.** $81a^2 - 1$ **95.** $x^2 - 18x + 81$ **97.** $x^3 + 8$

Problem Set 6.2

1. $(x + 3)(x + 4)$ **3.** $(x + 1)(x + 2)$ **5.** $(a + 3)(a + 7)$ **7.** $(x - 2)(x - 5)$ **9.** $(y - 3)(y - 7)$ **11.** $(x - 4)(x + 3)$
13. $(y + 4)(y - 3)$ **15.** $(x + 7)(x - 2)$ **17.** $(r - 9)(r + 1)$ **19.** $(x - 6)(x + 5)$ **21.** $(a + 7)(a + 8)$ **23.** $(y + 6)(y - 7)$
25. $(x + 6)(x + 7)$ **27.** $2(x + 1)(x + 2)$ **29.** $3(a + 4)(a - 5)$ **31.** $100(x - 2)(x - 3)$ **33.** $100(p - 5)(p - 8)$ **35.** $x^2(x + 3)(x - 4)$
37. $2r(r + 5)(r - 3)$ **39.** $2y^2(y + 1)(y - 4)$ **41.** $x^3(x + 2)(x + 2)$ **43.** $3y^2(y + 1)(y - 5)$ **45.** $4x^2(x - 4)(x - 9)$
47. $(x + 2y)(x + 3y)$ **49.** $(x - 4y)(x - 5y)$ **51.** $(a + 4b)(a - 2b)$ **53.** $(a - 5b)(a - 5b)$ **55.** $(a + 5b)(a + 5b)$
57. $(x - 6a)(x + 8a)$ **59.** $(x + 4b)(x - 9b)$ **61.** $(x^2 - 3)(x^2 - 2)$ **63.** $(x - 100)(x + 20)$ **65.** $(x - \frac{1}{2})(x - \frac{1}{2})$ **67.** $(x + 0.2)(x + 0.4)$
69. $x + 16$ **71.** $4x^2 - x - 3$ **73.** $\frac{4}{25}$ **75.** $72a^{12}$ **77.** $\dfrac{1}{16x^2}$ **79.** $\dfrac{a^3}{6b^2}$ **81.** $\dfrac{1}{3x^7y^2}$ **83.** $-21x^4y^5$ **85.** $-20ab^3$
87. $-27a^5b^8$ **89.** $6a^2 + 13a + 2$ **91.** $6a^2 + 7a + 2$ **93.** $6a^2 + 8a + 2$

Problem Set 6.3

1. $(2x + 1)(x + 3)$ **3.** $(2a - 3)(a + 1)$ **5.** $(3x + 5)(x - 1)$ **7.** $(3y + 1)(y - 5)$ **9.** $(2x + 3)(3x + 2)$ **11.** $(2x - 3y)(2x - 3y)$
13. $(4y + 1)(y - 3)$ **15.** $(4x - 5)(5x - 4)$ **17.** $(10a - b)(2a + 5b)$ **19.** $(4x - 5)(5x + 1)$ **21.** $(6m - 1)(2m + 3)$ **23.** $(4x + 5)(5x + 3)$
25. $(3a - 4b)(4a - 3b)$ **27.** $(3x - 7y)(x + 2y)$ **29.** $(2x + 5)(7x - 3)$ **31.** $(3x - 5)(2x - 11)$ **33.** $(5t - 19)(3t - 2)$
35. $2(2x + 3)(x - 1)$ **37.** $2(4a - 3)(3a - 4)$ **39.** $x(5x - 4)(2x - 3)$ **41.** $x^2(3x + 2)(2x - 5)$ **43.** $2a(5a + 2)(a - 1)$
45. $3x(5x + 1)(x - 7)$ **47.** $5y(7y + 2)(y - 2)$ **49.** $a^2(5a + 1)(3a - 1)$ **51.** $3y(4x + 5)(2x - 3)$ **53.** $2y(2x - y)(3x - 7y)$
55. Both equal 25. **57.** $4x^2 - 9$ **59.** $x^4 - 81$ **61.** $12x - 35$ **63.** $9x + 8$ **65.** $16(t - 1)(t - 3)$

67. $h = 2(4 - t)(1 + 8t)$

Time t (seconds)	Height h (feet)
0	8
1	54
2	68
3	50
4	0

69. $6x^3 + 5x^2 - 4x + 3$ **71.** $x^4 + 11x^3 - 6x - 9$ **73.** $2x^5 + 8x^3 - 2x - 7$ **75.** $-8x^5 + 2x^4 + 2x^2 + 12$ **77.** $7x^5y^3$ **79.** $3a^2b^3$
81. $x^2 - 9$ **83.** $x^2 - 25$ **85.** $x^2 - 49$ **87.** $x^2 - 81$ **89.** $4x^2 - 9y^2$ **91.** $x^4 - 16$ **93.** $x^2 + 6x + 9$ **95.** $x^2 + 10x + 25$
97. $x^2 + 14x + 49$ **99.** $x^2 + 18x + 81$ **101.** $4x^2 + 12x + 9$ **103.** $16x^2 - 16xy + 4y^2$

Problem Set 6.4
1. $(x + 3)(x - 3)$ **3.** $(a + 6)(a - 6)$ **5.** $(x + 7)(x - 7)$ **7.** $4(a + 2)(a - 2)$ **9.** Cannot be factored   **11.** $(5x + 13)(5x - 13)$
13. $(3a + 4b)(3a - 4b)$ **15.** $(3 + m)(3 - m)$ **17.** $(5 + 2x)(5 - 2x)$ **19.** $2(x + 3)(x - 3)$ **21.** $32(a + 2)(a - 2)$
23. $2y(2x + 3)(2x - 3)$ **25.** $(a^2 + b^2)(a + b)(a - b)$ **27.** $(4m^2 + 9)(2m + 3)(2m - 3)$ **29.** $3xy(x + 5y)(x - 5y)$ **31.** $(x - 1)^2$
33. $(x + 1)^2$ **35.** $(a - 5)^2$ **37.** $(y + 2)^2$ **39.** $(x - 2)^2$ **41.** $(m - 6)^2$ **43.** $(2a + 3)^2$ **45.** $(7x - 1)^2$ **47.** $(3y - 5)^2$
49. $(x + 5y)^2$ **51.** $(3a + b)^2$ **53.** $\left(y - \frac{3}{2}\right)^2$ **55.** $\left(a + \frac{1}{2}\right)^2$ **57.** $\left(x - \frac{7}{2}\right)^2$ **59.** $\left(x - \frac{3}{8}\right)^2$ **61.** $\left(t - \frac{1}{5}\right)^2$ **63.** $3(a + 3)^2$
65. $2(x + 5y)^2$ **67.** $5x(x + 3y)^2$ **69.** $(x + 3 + y)(x + 3 - y)$ **71.** $(x + y + 3)(x + y - 3)$ **73.** 14   **75.** 25   **77.** $4y^2 - 6y - 3$
79. $12x^5 - 9x^3 + 3$ **81.** $-6x^3 - 4x^2 + 2x$ **83.** $-7x^3 + 4x^2 - 11$ **85.** $x - 2 + \frac{2}{x - 3}$ **87.** $3x - 2 + \frac{9}{2x + 3}$
89. a. 1 **b.** 8 **c.** 27 **d.** 64 **e.** 125   **91. a.** $x^3 - x^2 + x$ **b.** $x^2 - x + 1$ **c.** $x^3 + 1$ **93. a.** $x^3 - 2x^2 + 4x$ **b.** $2x^2 - 4x + 8$ **c.** $x^3 + 8$
95. a. $x^3 - 3x^2 + 9x$ **b.** $3x^2 - 9x + 27$ **c.** $x^3 + 27$ **97. a.** $x^3 - 4x^2 + 16x$ **b.** $4x^2 - 16x + 64$ **c.** $x^3 + 64$
99. a. $x^3 - 5x^2 + 25x$ **b.** $5x^2 - 25x + 125$ **c.** $x^3 + 125$

Problem Set 6.5
1. $(x - y)(x^2 + xy + y^2)$ **3.** $(a + 2)(a^2 - 2a + 4)$ **5.** $(3 + x)(9 - 3x + x^2)$ **7.** $(y - 1)(y^2 + y + 1)$ **9.** $(y - 4)(y^2 + 4y + 16)$
11. $(5h - t)(25h^2 + 5ht + t^2)$ **13.** $(x - 6)(x^2 + 6x + 36)$ **15.** $2(y - 3)(y^2 + 3y + 9)$ **17.** $2(a - 4b)(a^2 + 4ab + 16b^2)$
19. $2(x + 6y)(x^2 - 6xy + 36y^2)$ **21.** $10(a - 4b)(a^2 + 4ab + 16b^2)$ **23.** $10(r - 5)(r^2 + 5r + 25)$ **25.** $(4 + 3a)(16 - 12a + 9a^2)$
27. $(2x - 3y)(4x^2 + 6xy + 9y^2)$ **29.** $\left(t + \frac{1}{3}\right)\left(t^2 - \frac{1}{3}t + \frac{1}{9}\right)$ **31.** $\left(3x - \frac{1}{3}\right)\left(9x^2 + x + \frac{1}{9}\right)$ **33.** $(4a + 5b)(16a^2 - 20ab + 25b^2)$
35. $\left(\frac{1}{2}x - \frac{1}{3}y\right)\left(\frac{1}{4}x^2 + \frac{1}{6}xy + \frac{1}{9}y^2\right)$ **37.** $(a - b)(a^2 + ab + b^2)(a + b)(a^2 - ab + b^2)$ **39.** $(2x - y)(4x^2 + 2xy + y^2)(2x + y)(4x^2 - 2xy + y^2)$
41. $(x - 5y)(x^2 + 5xy + 25y^2)(x + 5y)(x^2 - 5xy + 25y^2)$ **43.** $x = 3y + 4$ **45.** $x = \frac{3}{2}y + 2$ **47.** $y = \frac{1}{2}x + 3$ **49.** $y = \frac{2}{3}x - 4$
51. $2x^5 - 8x^3$ **53.** $3x^4 - 18x^3 + 27x^2$ **55.** $y^3 + 25y$ **57.** $15a^2 - a - 2$ **59.** $4x^4 - 12x^3 - 40x^2$ **61.** $2ab^5 - 8ab^4 + 2ab^3$

Problem Set 6.6
1. $(x + 9)(x - 9)$ **3.** $(x + 5)(x - 3)$ **5.** $(x + 3)^2$ **7.** $(y - 5)^2$ **9.** $2ab(a^2 + 3a + 1)$ **11.** Cannot be factored
13. $3(2a + 5)(2a - 5)$ **15.** $(3x - 2y)^2$ **17.** $4x(x^2 + 4y^2)$ **19.** $2y(y + 5)^2$ **21.** $a^4(a^2 + 4b^2)$ **23.** $(x + 4)(y + 3)$
25. $(x - 3)(x^2 + 3x + 9)$ **27.** $(x + 2)(y - 5)$ **29.** $5(a + b)^2$ **31.** Cannot be factored   **33.** $3(x + 2y)(x + 3y)$ **35.** $(2x + 19)(x - 2)$
37. $100(x - 2)(x - 1)$ **39.** $(x + 8)(x - 8)$ **41.** $(x + a)(x + 3)$ **43.** $a^5(7a + 3)(7a - 3)$ **45.** Cannot be factored
47. $a(5a + 1)(5a + 3)$ **49.** $(x + y)(a - b)$ **51.** $3a^2b(4a + 1)(4a - 1)$ **53.** $5x^2(2x + 3)(2x - 3)$ **55.** $(3x + 41y)(x - 2y)$
57. $2x^3(2x - 3)(4x - 5)$ **59.** $(2x + 3)(x + a)$ **61.** $(y^2 + 1)(y + 1)(y - 1)$ **63.** $3x^2y^2(2x + 3y)^2$ **65.** $16(t - 1)(t - 3)$
67. $(9x + 8)(6x + 7)$ **69.** 1   **71.** 30   **73.** -6 **75.** 3   **77.** 2   **79.** 5   **81.** $-\frac{3}{2}$ **83.** $-\frac{3}{4}$

Problem Set 6.7
1. $-2, 1$ **3.** $4, 5$ **5.** $0, -1, 3$ **7.** $-\frac{2}{3}, -\frac{3}{2}$ **9.** $0, -\frac{4}{3}, \frac{4}{3}$ **11.** $0, -\frac{1}{3}, -\frac{3}{5}$ **13.** $-1, -2$ **15.** $4, 5$ **17.** $6, -4$ **19.** $2, 3$
21. -3 **23.** $4, -4$ **25.** $\frac{3}{2}, -4$ **27.** $-\frac{2}{3}$ **29.** 5 **31.** $4, -\frac{5}{2}$ **33.** $\frac{5}{3}, -4$ **35.** $\frac{7}{2}, -\frac{7}{2}$ **37.** $0, -6$ **39.** $0, 3$ **41.** $0, 4$
43. $0, 5$ **45.** $2, 5$ **47.** $\frac{1}{2}, -\frac{4}{3}$ **49.** $4, -\frac{5}{2}$ **51.** $8, -10$ **53.** $5, 8$ **55.** $6, 8$ **57.** -4 **59.** $5, 8$ **61.** $6, -8$ **63.** $0, -\frac{3}{2}, -4$
65. $0, 3, -\frac{5}{2}$ **67.** $0, \frac{1}{2}, -\frac{5}{2}$ **69.** $0, \frac{3}{5}, -\frac{3}{2}$ **71.** $\frac{1}{2}, \frac{3}{2}$ **73.** $-5, 4$ **75.** $-7, -6$ **77.** $-3, -1$ **79.** $2, 3$ **81.** $-15, 10$
83. $-5, 3$ **85.** $-3, -2, 2$ **87.** $-4, -1, 4$ **89.** $x^2 - 8x + 15 = 0$ **91. a.** $x^2 - 5x + 6 = 0$ **b.** $x^2 - 7x + 6 = 0$ **c.** $x^2 - x - 6 = 0$
93. $\frac{1}{8}$ **95.** x^8 **97.** x^{18} **99.** 5.6×10^{-3} **101.** 5.67×10^9 **103.** $x(x + 1) = 72$ **105.** $x(x + 2) = 99$
107. $x(x + 2) = 5[x + (x + 2)] - 10$ **109.** Bicycle $75, suit $15   **111.** House $2,400, lot $600

Problem Set 6.8

1. Two consecutive even integers are x and $x + 2$; $x(x + 2) = 80$; 8, 10 and -10, -8 **3.** 9, 11 and -11, -9

5. $x(x + 2) = 5(x + x + 2) - 10$; 8, 10 and 0, 2   **7.** 8, 6   **9.** The numbers are x and $5x + 2$; $x(5x + 2) = 24$; 2, 12 and $-\frac{12}{5}$, -10

11. 5, 20 and 0, 0   **13.** Let x = the width; $x(x + 1) = 12$; width 3 inches, length 4 inches

15. Let x = the base; $\frac{1}{2}(x)(2x) = 9$; base 3 inches   **17.** $x^2 + (x + 2)^2 = 10^2$; 6 inches and 8 inches   **19.** 12 meters

21. $1,400 = 400 + 700x - 100x^2$; 200 or 500 items   **23.** 200 videotapes or 800 videotapes   **25.** $R = xp = (1,200 - 100p)p = 3,200$; \$4 or \$8

27. \$7 or \$10   **29. a.** 5 feet **b.** 12 feet   **31. a.** 25 seconds later **b.**

t (sec)	h (feet)
0	100
5	1,680
10	2,460
15	2,440
20	1,620
25	0

33. $200x^{24}$ **35.** x^7 **37.** 8×10^1 **39.** $10ab^2$ **41.** $6x^4 + 6x^3 - 2x^2$ **43.** $9y^2 - 30y + 25$ **45.** $4a^4 - 49$

CHAPTER 6 REVIEW/TEST

1. $10(x - 2)$ **2.** $x^2(4x - 9)$ **3.** $5(x - y)$ **4.** $x(7x^2 + 2)$ **5.** $4(2x + 1)$ **6.** $2(x^2 + 7x + 3)$ **7.** $8(3y^2 - 5y + 6)$
8. $15xy^2(2y - 3x^2)$ **9.** $7(7a^3 - 2b^3)$ **10.** $6ab(b + 3a^2b^2 - 4a)$ **11.** $(x + a)(y + b)$ **12.** $(x - 5)(y + 4)$ **13.** $(2x - 3)(y + 5)$
14. $(5x - 4a)(x - 2b)$ **15.** $(y + 7)(y + 2)$ **16.** $(w + 5)(w + 10)$ **17.** $(a - 6)(a - 8)$ **18.** $(r - 6)(r - 12)$ **19.** $(y + 9)(y + 11)$
20. $(y + 6)(y + 2)$ **21.** $(2x + 3)(x + 5)$ **22.** $(2y - 5)(2y - 1)$ **23.** $(5y + 6)(y + 1)$ **24.** $(5a - 3)(4a - 3)$ **25.** $(2r + 3t)(3r - 2t)$
26. $(2x - 7)(5x + 3)$ **27.** $(n + 9)(n - 9)$ **28.** $(2y + 3)(2y - 3)$ **29.** Cannot be factored.   **30.** $(6y + 11x)(6y - 11x)$
31. $(8a + 11b)(8a - 11b)$ **32.** $(8 + 3m)(8 - 3m)$ **33.** $(y + 10)^2$ **34.** $(m - 8)^2$ **35.** $(8t + 1)^2$ **36.** $(4n - 3)^2$ **37.** $(2r - 3t)^2$
38. $(3m + 5n)^2$ **39.** $2(x + 4)(x + 6)$ **40.** $a(a - 3)(a - 7)$ **41.** $3m(m + 1)(m - 7)$ **42.** $5y^2(y - 2)(y + 4)$ **43.** $2(2x + 1)(2x + 3)$
44. $a(3a + 1)(a - 5)$ **45.** $2m(2m - 3)(5m - 1)$ **46.** $5y(2x - 3y)(3x - y)$ **47.** $4(x + 5)^2$ **48.** $x(2x + 3)^2$ **49.** $5(x + 3)(x - 3)$
50. $3x(2x + 3y)(2x - 3y)$ **51.** $(2a - b)(4a^2 + 2ab + b^2)$ **52.** $(3x + 2y)(9x^2 - 6xy + 4y^2)$ **53.** $(5x - 4y)(25x^2 + 20xy + 16y^2)$
54. $3ab(2a + b)(a + 5b)$ **55.** $x^3(x + 1)(x - 1)$ **56.** $y^4(4y^2 + 9)$ **57.** $4x^3(x + 2y)(3x - y)$ **58.** $5a^2b(3a - b)(2a + 3b)$
59. $3ab^2(2a - b)(3a + 2b)$ **60.** $-2, 5$ **61.** $-\frac{5}{2}, \frac{5}{2}$ **62.** $2, -5$ **63.** $7, -7$ **64.** $0, 9$ **65.** $-\frac{3}{2}, -\frac{2}{3}$ **66.** $0, -\frac{5}{3}, \frac{2}{3}$
67. 10, 12 and -12, -10 **68.** 10, 11 and -11, -10 **69.** 5, 7, and -1, 1   **70.** 5, 15   **71.** 6, 11 and $-\frac{11}{2}$, -12 **72.** 2 inches

CHAPTER 7

Problem Set 7.1

1. a. $\frac{1}{4}$ **b.** $\frac{1}{x - 1}$, $x \neq -1, x \neq 1$ **c.** $\frac{x}{x + 1}$, $x \neq -1, x \neq 1$ **d.** $\frac{x^2 + x + 1}{x + 1}$, $x \neq -1, x \neq 1$ **e.** $\frac{x^2 + x + 1}{x^2}$, $x \neq 0, x \neq 1$

3. $\frac{1}{a + 3}$, $a \neq -3, a \neq 3$ **5.** $\frac{1}{x - 5}$, $x \neq -5, x \neq 5$ **7.** $\frac{(x + 2)(x - 2)}{2}$ **9.** $\frac{2(x - 5)}{3(x - 2)}$, $x \neq 2$ **11.** 2   **13.** $\frac{5(x - 1)}{4}$ **15.** $\frac{1}{x - 3}$

17. $\frac{1}{x + 3}$ **19.** $\frac{1}{a - 5}$ **21.** $\frac{1}{3x + 2}$ **23.** $\frac{x + 5}{x + 2}$ **25.** $\frac{2m(m + 2)}{m - 2}$ **27.** $\frac{x - 1}{x - 4}$ **29.** $\frac{2(2x - 3)}{2x + 3}$ **31.** $\frac{2x - 3}{2x + 3}$

33. $\frac{1}{(x^2 + 9)(x - 3)}$ **35.** $\frac{3x - 5}{(x^2 + 4)(x - 2)}$ **37.** $\frac{2x(7x + 6)}{2x + 1}$ **39.** $\frac{x^2 + xy + y^2}{x + y}$ **41.** $\frac{x^2 - 2x + 4}{x - 2}$ **43.** $\frac{x^2 - 2x + 4}{x - 1}$ **45.** $\frac{x + 2}{x + 5}$

47. $\frac{x + a}{x + b}$ **49. a.** $x^2 - 16$ **b.** $x^2 - 8x + 16$ **c.** $4x^3 - 32x^2 + 64x$ **d.** $\frac{x}{4}$ **51.** $\frac{4}{3}$ **53.** $\frac{4}{5}$ **55.** $\frac{8}{1}$

57.

Checks Written x	Total Cost $2.00 + 0.15x$	Cost per Check $\dfrac{2.00 + 0.15x}{x}$
0	$2.00	Undefined
5	$2.75	$0.55
10	$3.50	$0.35
15	$4.25	$0.28
20	$5.00	$0.25

59. 40.7 miles/hour **61.** 39.25 feet/minute **63.** Johnson: 1.23
Wood: 1.16
Martinez: 1.16
Nomo: 0.98

65. Level ground: 10 minutes/mile or 0.1 mile/minute; downhill: $\frac{20}{3}$ minutes/mile or $\frac{3}{20}$ mile/minute
67. 48 miles/gallon **69.** $5 + 4 = 9$ **71.**

x	$\dfrac{x-3}{3-x}$
-2	-1
-1	-1
0	-1
1	-1
2	-1

The entries are all -1 because the numerator and denominator are opposites.
Or, $\dfrac{x-3}{3-x} = \dfrac{-1(x-3)}{3-x} = -1$.

73.

x	$\dfrac{x-5}{x^2-25}$	$\dfrac{1}{x+5}$
0	$\frac{1}{5}$	$\frac{1}{5}$
2	$\frac{1}{7}$	$\frac{1}{7}$
-2	$\frac{1}{3}$	$\frac{1}{3}$
5	Undefined	$\frac{1}{10}$
-5	Undefined	Undefined

75. Disney, Nike.

Company	P/E
Yahoo	35.33
Google	84.53
Disney	18.63
Nike	17.51
eBay	56.11

77. 0 **79.** $16a^2b^2$ **81.** $19x^4 + 21x^2 - 42$

83. $7a^4b^4 + 9b^3 - 11a^3$ **85.** $\frac{5}{14}$ **87.** $\frac{9}{10}$ **89.** $(x+3)(x-3)$ **91.** $3(x-3)$ **93.** $(x-5)(x+4)$ **95.** $a(a+5)$
97. $(a-5)(a+4)$ **99.** 5.8

Problem Set 7.2

1. 2 **3.** $\frac{x}{2}$ **5.** $\frac{3}{2}$ **7.** $\dfrac{1}{2(x-3)}$ **9.** $\dfrac{4a(a+5)}{7(a+4)}$ **11.** $\dfrac{y-2}{4}$ **13.** $\dfrac{2(x+4)}{x-2}$ **15.** $\dfrac{x+3}{(x-3)(x+1)}$ **17.** 1 **19.** $\dfrac{y-5}{(y+2)(y-2)}$

21. $\dfrac{x^2 + 5x + 25}{x-5}$ **23.** 1 **25.** $\dfrac{a+3}{a+4}$ **27.** $\dfrac{2y-3}{y-6}$ **29.** $\dfrac{x-1}{(x-2)(x^2-x+1)}$ **31.** $\frac{3}{2}$

33. a. $\frac{4}{13}$ **b.** $\dfrac{x+1}{x^2+x+1}$ **c.** $\dfrac{x-1}{x^2+x+1}$ **d.** $\dfrac{x+1}{x-1}$ **35.** $2(x-3)$ **37.** $(x+2)(x+1)$ **39.** $-2x(x-5)$ **41.** $\dfrac{2(x+5)}{x(y+5)}$

43. $\dfrac{2x}{x-y}$ **45.** $\dfrac{1}{x-2}$ **47.** $\frac{1}{5}$ **49.** $\frac{1}{100}$ **51.** 2.7 miles **53.** 742 miles/hour **55.** 0.45 mile/hour **57.** 8.8 miles/hour

59. a. 21.3 kilometers/hour **b.** 13.2 miles/hour **61.** 3 **63.** $\frac{11}{4}$ **65.** $9x^2$ **67.** $6ab^2$ **69.** $\frac{4}{5}$ **71.** $\frac{11}{35}$ **73.** $-\frac{4}{35}$

75. $2x - 6$ **77.** $x^2 - x - 20$ **79.** $\dfrac{1}{x-3}$ **81.** $\dfrac{x-6}{2(x-5)}$ **83.** $x^2 - x - 30$

Problem Set 7.3

1. $\frac{7}{x}$ **3.** $\frac{4}{a}$ **5.** 1 **7.** $y+1$ **9.** $x+2$ **11.** $x-2$ **13.** $\dfrac{6}{x+6}$ **15.** $\dfrac{(y+2)(y-2)}{2y}$ **17.** $\dfrac{3+2a}{6}$ **19.** $\dfrac{7x+3}{4(x+1)}$

21. $\frac{1}{5}$ **23.** $\frac{1}{3}$ **25.** $\dfrac{3}{x-2}$ **27.** $\dfrac{4}{a+3}$ **29.** $\dfrac{2x-20}{(x-5)(x+5)}$ **31.** $\dfrac{x+2}{x+3}$ **33.** $\dfrac{a+1}{a+2}$ **35.** $\dfrac{1}{(x+3)(x+4)}$

37. $\dfrac{y}{(y+5)(y+4)}$ **39.** $\dfrac{3(x-1)}{(x+4)(x+1)}$ **41.** $\dfrac{1}{3}$ **43. a.** $\dfrac{2}{27}$ **b.** $\dfrac{8}{3}$ **c.** $\dfrac{11}{18}$ **d.** $\dfrac{3x+10}{(x-2)^2}$ **e.** $\dfrac{(x+2)^2}{3x+10}$ **f.** $\dfrac{x+3}{x+2}$

45.

Number x	Reciprocal $\dfrac{1}{x}$	Sum $1+\dfrac{1}{x}$	Sum $\dfrac{x+1}{x}$
1	1	2	2
2	$\dfrac{1}{2}$	$\dfrac{3}{2}$	$\dfrac{3}{2}$
3	$\dfrac{1}{3}$	$\dfrac{4}{3}$	$\dfrac{4}{3}$
4	$\dfrac{1}{4}$	$\dfrac{5}{4}$	$\dfrac{5}{4}$

47.

x	$x+\dfrac{4}{x}$	$\dfrac{x^2+4}{x}$	$x+4$
1	5	5	5
2	4	4	6
3	$\dfrac{13}{3}$	$\dfrac{13}{3}$	7
4	5	5	8

49. $\dfrac{x+3}{x+2}$ **51.** $\dfrac{x+2}{x+3}$ **53.** 3 **55.** -2 **57.** $\dfrac{1}{5}$ **59.** $-2, -3$ **61.** $3, -2$ **63.** $0, 5$ **65.** 3 **67.** 0 **69.** Undefined

71. $-\dfrac{3}{2}$ **73.** $2x+15$ **75.** x^2-5x **77.** -6 **79.** -2

Problem Set 7.4

1. -3 **3.** 20 **5.** -1 **7.** 5 **9.** -2 **11.** 4 **13.** 3, 5 **15.** $-8, 1$ **17.** 2 **19.** 1 **21.** 8 **23.** 5
25. Possible solution 2, which does not check; $\varnothing$ **27.** 3 **29.** Possible solution -3, which does not check; $\varnothing$ **31.** 0
33. Possible solutions 2 and 3, but only 2 checks; 2 **35.** -4 **37.** -1 **39.** $-6, -7$
41. Possible solutions -3 and -1, but only -1 checks; -1 **43. a.** $\dfrac{1}{5}$ **b.** 5 **c.** $\dfrac{25}{3}$ **d.** 3 **e.** $\dfrac{1}{5}$, 5

45. a. $\dfrac{7}{(a-6)(a+2)}$ **b.** $\dfrac{a-5}{a-6}$ **c.** Possible answers 7, -1; only 7 checks **47.** 7 **49.** Length 13 inches, width 4 inches

51. 6, 8, or $-8, -6$ **53.** 6 inches, 8 inches **55.** $\dfrac{1}{3}$ **57.** 30 **59.** 1 **61.** -3

Problem Set 7.5

1. $\dfrac{1}{4}$, $\dfrac{3}{4}$ **3.** $x+\dfrac{1}{x}=\dfrac{13}{6}$; $\dfrac{2}{3}$ and $\dfrac{3}{2}$ **5.** -2 **7.** $\dfrac{1}{x}+\dfrac{1}{x+2}=\dfrac{5}{12}$; 4 and 6

9. Let x = the speed of the boat in still water.

	d	r	t
Upstream	26	$x-3$	$\dfrac{26}{x-3}$
Downstream	38	$x+3$	$\dfrac{38}{x+3}$

The equation is $\dfrac{26}{x-3}=\dfrac{38}{x+3}$; $x=16$ miles per hour.

11. 300 miles/hour **13.** 170 miles/hour, 190 miles/hour **15.** 9 miles/hour **17.** 8 miles/hour

19. $\dfrac{1}{12}-\dfrac{1}{15}=\dfrac{1}{x}$; 60 hours **21.** $\dfrac{60}{11}$ minutes **23.** 12 minutes

25.

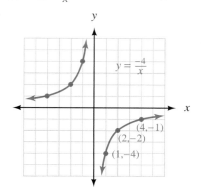

27.

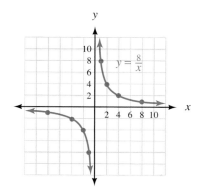

29.

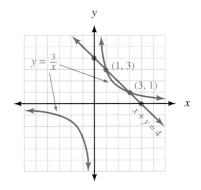

31. $5ab(3a^2b^2 - 4a - 7b)$ **33.** $(x - 6)(x + 2)$ **35.** $(x^2 + 4)(x + 2)(x - 2)$ **37.** $5x(x - 6)(x + 1)$ **39.** $0, 6$ **41.** $-10, 8$

43. 9 inches and 12 inches   **45.** $\frac{4}{9}$ **47.** $\frac{5}{3}$ **49.** $3x^5y^3$ **51.** $\frac{x^3y^3}{2}$ **53.** $y(xy + 1)$ **55.** $\frac{x^3y^3}{2}$ **57.** $\frac{x + 3}{x - 2}$

Problem Set 7.6

1. 6 **3.** $\frac{1}{6}$ **5.** xy^2 **7.** $\frac{xy}{2}$ **9.** $\frac{y}{x}$ **11.** $\frac{a + 1}{a - 1}$ **13.** $\frac{x + 1}{2(x - 3)}$ **15.** $a - 3$ **17.** $\frac{y + 3}{y + 2}$ **19.** $x + y$ **21.** $\frac{a + 1}{a}$ **23.** $\frac{1}{x}$

25. $\frac{2a + 3}{3a + 4}$ **27.** $\frac{x - 1}{x + 2}$ **29.** $\frac{x + 4}{x + 1}$ **31.** $\frac{7}{3}, \frac{17}{7}, \frac{41}{17}$

33.

Number	Reciprocal	Quotient	Square
x	$\dfrac{1}{x}$	$\dfrac{x}{\frac{1}{x}}$	x^2
1	1	1	1
2	$\frac{1}{2}$	4	4
3	$\frac{1}{3}$	9	9
4	$\frac{1}{4}$	16	16

35.

Number	Reciprocal	Sum	Quotient
x	$\dfrac{1}{x}$	$1 + \dfrac{1}{x}$	$\dfrac{1 + \frac{1}{x}}{\frac{1}{x}}$
1	1	2	2
2	$\frac{1}{2}$	$\frac{3}{2}$	3
3	$\frac{1}{3}$	$\frac{4}{3}$	4
4	$\frac{1}{4}$	$\frac{5}{4}$	5

37. $x < 1$ **39.** $x \geq -7$ **41.** $x < 6$ **43.** $x \leq 2$ **45.** $\frac{7}{2}$ **47.** $-3, 2$

Problem Set 7.7

1. 1 **3.** 10 **5.** 40 **7.** $\frac{5}{4}$ **9.** $\frac{7}{3}$ **11.** $3, -4$ **13.** $4, -4$ **15.** $2, -4$ **17.** $6, -1$ **19.** 15 hits   **21.** 21 milliliters

23. 45.5 grams   **25.** 14.7 inches   **27.** 12.5 miles   **29.** 17 minutes   **31.** 343 miles   **35.** $\frac{y + 5}{x + a}$ **37.** $\frac{3}{x + 1}$

39. $\frac{-3}{(x + 1)(x + 4)}$ **41.** 15 **43.** 4 **45.** $-6, 6$ **47.** 18 **49.** 5

Problem Set 7.8

1. 8 **3.** 130 **5.** 5 **7.** -3 **9.** 3 **11.** 243 **13.** 2 **15.** $\frac{1}{2}$ **17.** 1 **19.** 5 **21.** 25 **23.** 9 **25.** 84 pounds
27. $\frac{735}{2}, 367.5$ **29.** $\$235.50$ **31.** 96 pounds   **33.** 12 amperes   **35.** $(2, -1)$ **37.** $(4, 3)$ **39.** $(1, 1)$ **41.** $(5, 2)$

CHAPTER 7 REVIEW/TEST

1. $\frac{1}{2(x - 2)}, x \neq 2$ **2.** $\frac{1}{a - 6}, a \neq -6, 6$ **3.** $\frac{2x - 1}{x + 3}, x \neq -3$ **4.** $\frac{1}{x + 4}, x \neq -4$ **5.** $\frac{3x - 2}{2x - 3}, x \neq \frac{3}{2}, -6, 0$ **6.** $\frac{1}{(x - 2)(x^2 + 4)}, x \neq \pm 2$

7. $x - 2, x \neq -7$ **8.** $a + 8, a \neq -8$ **9.** $\frac{y + b}{y + 5}, x \neq -a, y \neq -5$ **10.** $\frac{x}{2}$ **11.** $\frac{x + 2}{x - 4}$ **12.** $(a - 6)^2$ **13.** $\frac{x - 1}{x + 2}$ **14.** 1 **15.** $x - 9$

16. $\frac{13}{a + 8}$ **17.** $\frac{x^2 + 5x + 45}{x(x + 9)}$ **18.** $\frac{4x + 5}{4(x + 5)}$ **19.** $\frac{1}{(x + 6)(x + 2)}$ **20.** $\frac{a - 6}{(a + 5)(a + 3)}$ **21.** 4 **22.** 9 **23.** $1, 6$

24. Possible solution 2, which does not check; $\varnothing$ **25.** 5; -4 does not check   **26.** 3 and $\frac{7}{3}$ **27.** 15 miles/hour   **28.** 84 hours
29. $\frac{1}{2}$ **30.** $\dfrac{y+3}{y+7}$ **31.** $\dfrac{4a-7}{a-1}$ **32.** $\frac{2}{5}$ **33.** $\frac{2}{9}$ **34.** 12   **35.** $-6, 6$ **36.** $-6, 8$ **37.** -35 **38.** $\frac{1}{2}$

CHAPTER 8

Problem Set 8.1
1. 3   **3.** -3 **5.** Not a real number   **7.** -12 **9.** 25   **11.** Not a real number   **13.** -8 **15.** -10 **17.** 35   **19.** 1
21. -2 **23.** -5 **25.** -1 **27.** -3 **29.** -2 **31.** x **33.** $3x$ **35.** xy **37.** $a+b$ **39.** $7xy$ **41.** x **43.** $2x$ **45.** x^2
47. $6a^3$ **49.** $5a^4b^2$ **51.** x^2 **53.** $3a^4$ **55. a.** 2   **b.** x **c.** x^2 **d.** $2x^2y^3$ **57. a.** 7   **b.** 5   **c.** 12   **d.** 12
59. a. 1   **b.** 3   **c.** 20   **d.** 20   **61. a.** 1.618   **b.** -0.618 **c.** 1   **63. a.** 3   **b.** 30   **c.** 0.3   **65.** 6, -1 **67.** 3, -1 **69.** $x+3$
71.

Length l (feet)	Time t (seconds)
1	1.11
2	1.57
3	1.92
4	2.22
5	2.48
6	2.72

73. 5   **75.** $\sqrt{125} \approx 11.2$ **77.** 30 feet   **79.** 13 feet

81. $x-4$ **83.** $\dfrac{2}{a-2}$ **85.** $\dfrac{2x+1}{x}$ **87.** $\dfrac{x+2}{x+a}$ **89.** 12   **91.** $25 \cdot 3$ **93.** $25 \cdot 2$ **95.** $4 \cdot 10$ **97.** $x^4 \cdot x$ **99.** $4x^2 \cdot 3$
101. $25x^2y^2 \cdot 2x$

Problem Set 8.2
1. $2\sqrt{2}$ **3.** $2\sqrt{3}$ **5.** $2\sqrt[3]{3}$ **7.** $5x\sqrt{2}$ **9.** $3ab\sqrt{5}$ **11.** $3x\sqrt[3]{2}$ **13.** $4x^2\sqrt{2}$ **15.** $20\sqrt{5}$ **17.** $x\sqrt{7x}$ **19.** $2x^2\sqrt[3]{x}$
21. $6a^2\sqrt[3]{a^2}$ **23.** $4a\sqrt{5a}$ **25.** $15y\sqrt{2x}$ **27.** $14x\sqrt{3y}$ **29.** $\frac{4}{5}$ **31.** $\frac{2}{3}$ **33.** $\frac{2}{3}$ **35.** $\frac{2}{3}$ **37.** $2x$ **39.** $3ab$ **41.** $\dfrac{3x}{2y}$
43. $\dfrac{5\sqrt{2}}{3}$ **45.** $\sqrt{3}$ **47.** $\dfrac{8\sqrt{2}}{7}$ **49.** $\dfrac{12\sqrt{2x}}{5}$ **51.** $\dfrac{3a\sqrt{6}}{5}$ **53.** $\dfrac{15\sqrt{2}}{2}$ **55.** $\dfrac{14y\sqrt{7}}{3}$ **57.** $5ab\sqrt{2}$ **59.** $6x\sqrt{2y}$
61. a. $2\sqrt{10}$ **b.** 5   **c.** $3\sqrt{5}$ **d.** $2\sqrt{3}$ **63. a.** $4x^5y^2\sqrt{2y}$ **b.** $2x^3y\sqrt[3]{4xy^2}$ **c.** $2x^2y\sqrt[4]{2x^2y}$ **d.** $2x^2y$ **65. a.** 2   **b.** 0.2   **c.** 20   **d.** 0.02
67.

x	$\sqrt{x}$	$2\sqrt{x}$	$\sqrt{4x}$
1	1	2	2
2	1.414	2.828	2.828
3	1.732	3.464	3.464
4	2	4	4

69.

x	$\sqrt{x}$	$3\sqrt{x}$	$\sqrt{9x}$
1	1	3	3
2	1.414	4.243	4.243
3	1.732	5.196	5.196
4	2	6	6

71. $\dfrac{2(x+5)}{x(x+1)}$ **73.** $\dfrac{x+1}{x-1}$ **75.** $(x+6)(x+3)$ **77.** $2xy\sqrt{x}$ **79.** 12   **81.** $\dfrac{\sqrt{2}}{2}$ **83.** $\dfrac{\sqrt[3]{18}}{3}$ **85.** $\dfrac{\sqrt{6}}{3}$ **87.** 3

Problem Set 8.3
1. $\dfrac{\sqrt{2}}{2}$ **3.** $\dfrac{\sqrt{3}}{3}$ **5.** $\dfrac{\sqrt{10}}{5}$ **7.** $\dfrac{\sqrt{6}}{2}$ **9.** $\dfrac{2\sqrt{15}}{3}$ **11.** $\dfrac{\sqrt{30}}{2}$ **13.** 2   **15.** $\sqrt{7}$ **17.** $\sqrt{5}$ **19.** $2\sqrt{5}$ **21.** $2\sqrt{3}$ **23.** $\dfrac{\sqrt{7}}{2}$
25. $xy\sqrt{2}$ **27.** $\dfrac{x\sqrt{15y}}{3}$ **29.** $\dfrac{4a^2\sqrt{5}}{5}$ **31.** $\dfrac{6a^2\sqrt{10a}}{5}$ **33.** $\dfrac{2xy\sqrt{15y}}{3}$ **35.** $\dfrac{4xy\sqrt{5y}}{5}$ **37.** $\dfrac{18ab\sqrt{6b}}{5}$ **39.** $18x\sqrt{x}$ **41.** $\dfrac{\sqrt[3]{4}}{2}$
43. $\dfrac{\sqrt[3]{3}}{3}$ **45. a.** $\dfrac{6\sqrt{\pi}}{\pi}$ **b.** $\dfrac{\sqrt{A\pi}}{\pi}$ **c.** $\dfrac{\sqrt[3]{6V\pi^2}}{2\pi}$ **d.** $\dfrac{\sqrt[3]{4\pi^2}}{\pi}$

47.

x	$\sqrt{x}$	$\dfrac{1}{\sqrt{x}}$	$\dfrac{\sqrt{x}}{x}$
1	1	1	1
2	1.414	.707	.707
3	1.732	.577	.577
4	2	.5	.5
5	2.236	.447	.447
6	2.449	.408	.408

49.

x	$\sqrt{x^2}$	$\sqrt{x^3}$	$x\sqrt{x}$
1	1	1	1
2	2	2.828	2.828
3	3	5.196	5.196
4	4	8	8
5	5	11.18	11.18
6	6	14.697	14.697

51.

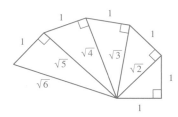

53. $\sqrt{2}, \sqrt{3}, \sqrt{4} = 2$ **55.** $10x$ **57.** $23x$ **59.** $10a$ **61.** $x + 5$ **63.** $\dfrac{5a + 6}{15}$ **65.** $\dfrac{1}{(a + 2)(a + 3)}$ **67.** $23x$ **69.** $27y$
71. $7ab$ **73.** $-7xy + 50x$ **75.** $\dfrac{3 + x}{2}$

Problem Set 8.4
1. a. $7x$ **b.** $7y$ **c.** $7\sqrt{2}$ **d.** $7\sqrt{5}$ **3. a.** $7x$ **b.** $7t$ **c.** $7\sqrt{3}$ **d.** $7\sqrt{x}$ **5.** $2\sqrt{5}$ **7.** $6\sqrt{5}$ **9.** $13\sqrt{13}$ **11.** $6\sqrt{10}$ **13.** $\frac{11}{56}\sqrt{5}$
15. $4\sqrt{2}$ **17.** 0 **19.** $-30\sqrt{3}$ **21.** $-6\sqrt{3}$ **23.** 0 **25.** $\frac{3}{2}\sqrt{2} + \frac{3}{2}\sqrt{3}$ **27.** 0 **29.** $\frac{23}{4}\sqrt{2}$ **31.** $35\sqrt{3}$ **33.** -7 **35.** 0
37. $2x\sqrt{x}$ **39.** $4a\sqrt{3}$ **41.** $15x\sqrt{2x}$ **43.** $13x\sqrt{3xy}$ **45.** $-b\sqrt{5a}$ **47.** $27x\sqrt{2x} - 8x\sqrt{3x}$ **49.** $23x\sqrt{2y}$ **51.** $3 + \sqrt{2}$
53. $3 - \sqrt{3}$ **55.** $\dfrac{4 - \sqrt{6}}{3}$ **57.** $3 + \sqrt{2}$ **59.** $\dfrac{-2 + \sqrt{2}}{2}$ **61. a.** 3.232 **b.** -0.232 **c.** 3

63.

x	$\sqrt{x^2 + 9}$	$x + 3$
1	3.162	4
2	3.606	5
3	4.243	6
4	5	7
5	5.831	8
6	6.708	9

65.

x	$\sqrt{x + 3}$	$\sqrt{x} + \sqrt{3}$
1	2	2.732
2	2.236	3.146
3	2.449	3.464
4	2.646	3.732
5	2.828	3.968
6	3	4.182

67. $9\sqrt{3}$ **69.** $9x^2 + 6xy + y^2$ **71.** $9x^2 - 16y^2$ **73.** 9

75. $2, 3$ **77.** Possible solutions are 2 and 4; only 2 checks. **79.** $\sqrt{10}$ **81.** 5 **83.** x **85.** $\sqrt{35}$ **87.** $19 + 9\sqrt{5}$
89. $x - 4\sqrt{x} - 21$

Problem Set 8.5
1. $\sqrt{6}$ **3.** $2\sqrt{3}$ **5.** $10\sqrt{21}$ **7.** $24\sqrt{2}$ **9.** 8 **11.** 24 **13.** 50 **15.** 8 **17.** $\sqrt{6} - \sqrt{2}$ **19.** $\sqrt{6} + 2$ **21.** $2\sqrt{6} + 3$
23. $2\sqrt{6} + 2\sqrt{15}$ **25.** $3 + 2\sqrt{2}$ **27.** $x + 6\sqrt{x} + 9$ **29.** $a - \sqrt{a} + \frac{1}{4}$ **31.** $11 + 5\sqrt{5}$ **33.** $\sqrt{6} + \frac{1}{3}\sqrt{3} + \frac{1}{2}\sqrt{2} + \frac{1}{6}$
35. $x - 36$ **37.** $a + \sqrt{a} + \frac{2}{9}$ **39.** 1 **41.** $30 + \sqrt{7}$ **43.** -7 **45.** 1 **47.** 0 **49.** $\dfrac{\sqrt{15} + \sqrt{6}}{3}$ **51.** $\dfrac{5 - \sqrt{10}}{3}$
53. $6 + 2\sqrt{5}$ **55.** $5 + 2\sqrt{6}$ **57. a.** $2\sqrt{3}$ **b.** 4 **c.** $10 - 2\sqrt{21}$ **d.** $\dfrac{5 - \sqrt{21}}{2}$

59. a. $2\sqrt{x}$ **b.** $x - 4$ **c.** $x + 4\sqrt{x} + 4$ **d.** $\dfrac{x + 4\sqrt{x} + 4}{x - 4}$ **61. a.** 10 **b.** 23 **c.** $27 + 10\sqrt{2}$ **d.** $\dfrac{27 + 10\sqrt{2}}{23}$

63. a. $\sqrt{6} + 2\sqrt{2}$ **b.** $2 + 2\sqrt{3}$ **c.** $1 + \sqrt{3}$ **d.** $\dfrac{-1 + \sqrt{3}}{2}$ **65. a.** 1 **b.** -1 **67.** $6\sqrt{5}$ **69.** $52 + 14\sqrt{3}$ **71.** $-6, 1$ **73.** $0, 3$
75. $-9, 9$ **77.** $3, -5$ **79.** 49 **81.** 81 **83.** $x + 1$ **85.** $2x - 3$ **87.** $x^2 + 6x + 9$ **89.** 2 **91.** $-6, 1$ **93. a.** no **b.** yes
95. Undefined, Undefined, 0, 3, 6, 9, 12

Problem Set 8.6

1. 3 **3.** 44 **5.** ∅ **7.** ∅ **9.** 8 **11.** 4 **13.** ∅ **15.** $-\frac{2}{3}$ **17.** 25 **19.** 4 **21.** 7 **23.** 8 **25.** ∅

27. Possible solutions are 1 and 6; only 6 checks. **29.** $-1, -2$ **31.** -4 **33. a.** 100 **b.** 40 **c.** ∅ **d.** Possible 5, 8; only 8 checks

35. a. 3 **b.** 9 **c.** 3 **d.** ∅ **e.** 4 **f.** ∅ **d.** Possible 1, 4; only 4 checks

37.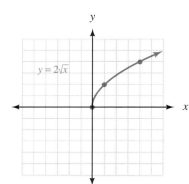

39.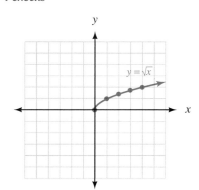

41.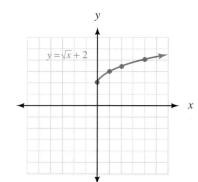

43.

45. $x + 2 = \sqrt{8x}$; $x = 2$ **47.** $x - 3 = 2\sqrt{x}$; possible solutions are 1 and 9; only 9 checks. **49.** $\frac{392}{121} \approx 3.2$ feet **51.** $\frac{x + 2}{x + 3}$

53. $\frac{2(x + 5)}{x - 4}$ **55.** $\frac{1}{x - 4}$ **57.** $\frac{x + 5}{x - 3}$ **59.** -2 **61.** 24 hours **63.** 24

CHAPTER 8 REVIEW/TEST

1. 5 **2.** 13 **3.** -1 **4.** 5 **5.** $10xy^2$ **6.** $2a$ **7.** $2\sqrt{6}$ **8.** $2x\sqrt{15}$ **9.** $3xy^2\sqrt{10x}$ **10.** $-4\sqrt{2}$ **11.** $6x\sqrt{5xy}$

12. $\frac{\sqrt{3}}{7}$ **13.** $\frac{2\sqrt{2}}{9}$ **14.** $\frac{7}{8}$ **15.** $\frac{7ab}{4}$ **16.** $\frac{4\sqrt{5}}{7}$ **17.** $\frac{2a\sqrt{10}}{11}$ **18.** $\frac{10\sqrt{21}}{7}$ **19.** $\frac{6ab\sqrt{30}}{5}$ **20.** $\frac{-5xy\sqrt{5x}}{6}$ **21.** $\frac{2\sqrt{7}}{7}$

22. $\frac{4\sqrt{10}}{5}$ **23.** $\frac{\sqrt{15}}{12}$ **24.** $-6\sqrt{3}$ **25.** $\frac{4b\sqrt{6a}}{3}$ **26.** $\frac{\sqrt[3]{6}}{2}$ **27.** $\frac{-3\sqrt{3} - 12}{13}$ **28.** $3 - \sqrt{7}$ **29.** $\sqrt{5} + \sqrt{2}$ **30.** $\frac{-\sqrt{5} - 5}{2}$

31. $\frac{7 - 2\sqrt{10}}{3}$ **32.** $\frac{x + 6\sqrt{x} + 9}{x - 9}$ **33.** $-4\sqrt{5}$ **34.** $-11\sqrt{3}$ **35.** $-22\sqrt{5}$ **36.** $5x\sqrt{2}$ **37.** $-ab\sqrt{10a}$ **38.** $3 + 3\sqrt{3}$

39. $4\sqrt{6} + 4\sqrt{10}$ **40.** $x - 49$ **41.** $2\sqrt{5} - 2$ **42.** $x + 10\sqrt{x} + 25$ **43.** 12 **44.** 7 **45.** 16 **46.** No solution

47. No solution **48.** 0, 5 **49.** $\sqrt{3}$ **50.** $8\sqrt{2}$

CHAPTER 9

Problem Set 9.1

1. ± 3 **3.** ± 5 **5.** $\pm 2\sqrt{2}$ **7.** $\pm 5\sqrt{2}$ **9.** $\pm 3\sqrt{2}$ **11.** $0, -4$ **13.** $4, -6$ **15.** $5 \pm 5\sqrt{3}$ **17.** $-1 \pm 5\sqrt{2}$ **19.** $2, -3$

21. $\frac{11}{4}, -\frac{1}{4}$ **23.** $\frac{1 \pm 2\sqrt{3}}{3}$ **25.** $\frac{-2 \pm 3\sqrt{3}}{6}$ **27.** $3 \pm \sqrt{3}$ **29.** $-2 \pm \sqrt{5}$ **31.** $2 \pm \sqrt{2}$ **33.** $\frac{7}{3}, -1$ **35.** $\frac{-1 \pm \sqrt{7}}{2}$

37. $\dfrac{4 \pm 2\sqrt{3}}{5}$ **39.** $-5 \pm \sqrt{7}$ **41.** $4, -2$ **43.** $-6 \pm 2\sqrt{2}$ **45. a.** Answers will vary **b.** $\pm\sqrt{3}$ **47. a.** Yes **b.** $1, 5$

49. Yes   **51. a.** $\frac{1}{2}$ **b.** $\frac{5}{2}$ **c.** $-\frac{1}{2}, \frac{3}{2}$ **d.** $\frac{25}{2}$ **e.** $\frac{2}{5}$ **53.** $(-1 + 5\sqrt{2} + 1)^2 = (5\sqrt{2})^2 = 50$ **55. a.** 10 **b.** 22 **57.** 6 ft   **59.** $\frac{5}{2}$ sec

61. $-7, 1$ **63.** $r = -1 + \dfrac{\sqrt{A}}{10}$ **65.** $5\sqrt{3}$ feet

67. No, tent height is $3\sqrt{3}$ feet ≈ 5.20 feet, which is less than 5 feet 8 inches ($5\frac{8}{12}$ feet ≈ 5.67 feet)   **69.** 3 feet   **71.** 3.4 inches

73. $x^2 - 10x + 25$ **75.** $(x - 6)^2$ **77.** $(x + 2)^2$ **79.** 2   **81.** 2   **83.** 81   **85.** 1   **87.** $\frac{9}{4}$ **89.** $x^2 + 8$ **91.** $(x + 3)^2$
93. $\left(y - \frac{3}{2}\right)^2$

Problem Set 9.2
1. 9   **3.** 1   **5.** 16   **7.** 1   **9.** 64   **11.** $\frac{9}{4}$ **13.** $\frac{49}{4}$ **15.** $\frac{1}{4}$ **17.** $\frac{9}{16}$ **19.** $-6, 2$ **21.** $-2, 8$ **23.** $-3, 1$ **25.** $0, 10$

27. $-5, 3$ **29.** $-2 \pm \sqrt{7}$ **31.** $2 \pm 2\sqrt{2}$ **33.** $8, -1$ **35.** $-1 \pm \sqrt{2}$ **37.** $-2, 1$ **39.** $\dfrac{-2 \pm \sqrt{3}}{2}$ **41.** $\dfrac{1 \pm \sqrt{3}}{2}$ **43.** $\frac{1}{2}$

45. $\dfrac{9 \pm \sqrt{105}}{6}$ **47. a.** $0, 2$ **b.** $0, 2$ **49.** No   **51.** Yes, by factoring   **53.** Yes, by factoring   **55.** $\frac{12}{11}, \frac{36}{11}$ **57.** $0, 6$

59. $-1, 7$ **61.** $3 \pm \sqrt{2}$ **63.** $\dfrac{4 \pm \sqrt{5}}{3}$ **65.** $\dfrac{3 + 2\sqrt{5}}{2}$ **67.** $x = 6$ **69.** Length $= 11.2$ inches, width $= 8.4$ inches

71. 4   **73.** -24 **75.** $2\sqrt{10}$ **77.** $2\sqrt{3}$ **79.** $2xy\sqrt{5y}$ **81.** $\frac{9}{5}$ **83.** $2x^2 + 4x - 3 = 0$ **85.** $x^2 + x - 11 = 0$ **87.** $1, -5, -6$

89. $2, 4, -3$ **91.** 49   **93.** 40   **95.** 6   **97.** $\dfrac{-2 - \sqrt{10}}{2}$

Problem Set 9.3
1. $-1, -2$ **3.** $-2, -3$ **5.** -3 **7.** $-3 \pm \sqrt{2}$ **9.** $-1, -\frac{3}{2}$ **11.** $\dfrac{-2 \pm \sqrt{3}}{2}$ **13.** 1   **15.** $\dfrac{5 \pm \sqrt{53}}{2}$ **17.** $\frac{2}{3}, -\frac{1}{2}$

19. $\dfrac{1 \pm \sqrt{21}}{2}$ **21.** $\dfrac{-1 \pm \sqrt{57}}{4}$ **23.** $\frac{5}{2}, -1$ **25.** $\dfrac{-3 \pm \sqrt{23}}{2}$ **27.** $\dfrac{-2 \pm \sqrt{10}}{3}$ **29.** $\dfrac{1 \pm \sqrt{11}}{2}$ **31.** $0, \dfrac{-3 \pm \sqrt{41}}{4}$ **33.** $0, \frac{4}{3}$

35. $\dfrac{3 \pm \sqrt{21}}{6}$ **37.** Yes, by factoring   **39.** No   **41. a.** No **b.** $1 \pm \sqrt{2}$ **43.** Yes   **45.** $-\frac{7}{2}, 0$ **47.** $-3, -\frac{1}{2}$ **49.** $\dfrac{-7 \pm \sqrt{73}}{4}$

51. $\dfrac{4 \pm 2\sqrt{2}}{3}$ **53.** 1 second and 3 seconds   **55.** $\dfrac{6}{x - 2}$ **57.** $\dfrac{2x + 3}{(x + 3)(x - 3)}$ **59.** $\dfrac{3x}{(x + 1)(x - 2)}$ **61.** $5 - 2\sqrt{2}$

63. $1 - 13\sqrt{2}$ **65.** $12\sqrt{2} + 40$ **67.** $-\sqrt{2}$ **69.** $\dfrac{3 - \sqrt{5}}{2}$ **71.** $7 + 4\sqrt{3}$

Problem Set 9.4
1. $3 + i$ **3.** $6 - 8i$ **5.** 11   **7.** $9 + i$ **9.** $-1 - i$ **11.** $11 - 2i$ **13.** $4 - 2i$ **15.** $8 - 7i$ **17.** $-1 + 4i$ **19.** $6 - 3i$

21. $14 + 16i$ **23.** $9 + 2i$ **25.** $11 - 7i$ **27.** 34   **29.** 5   **31.** $\dfrac{6 + 4i}{13}$ **33.** $\dfrac{-9 - 6i}{13}$ **35.** $\dfrac{3 + 4i}{5}$

37. a. 4 **b.** 13 **c.** $-5 - 12i$ **d.** $\dfrac{-5 - 12i}{13}$ **39. a.** $3 + 5i$ **b.** $6 + 18i$ **c.** $-\dfrac{1 + 3i}{6}$ **d.** $\dfrac{-3 + 9i}{5}$ **41.** -13 **43.** -15 **45.** 0

47. Yes   **49.** $\pm 6i$ **51.** Yes   **53.** $x^2 + 9$ **55.** $\dfrac{1}{i} \cdot \dfrac{i}{i} = \dfrac{i}{i^2} = \dfrac{i}{-1} = -i$ **57.** $8, -2$ **59.** $5, 1$ **61.** $-3 \pm 2\sqrt{3}$ **63.** $\dfrac{\sqrt{2}}{2}$

65. $\dfrac{2xy\sqrt{6y}}{3}$ **67.** $\dfrac{\sqrt[3]{2}}{2}$ **69.** 6   **71.** $5\sqrt{3}$ **73.** $-5, 1$ **75.** $1 \pm \sqrt{6}$ **77.** $\dfrac{2 \pm \sqrt{5}}{2}$

Problem Set 9.5
1. $4i$ **3.** $7i$ **5.** $i\sqrt{6}$ **7.** $i\sqrt{11}$ **9.** $4i\sqrt{2}$ **11.** $5i\sqrt{2}$ **13.** $2i\sqrt{2}$ **15.** $4i\sqrt{3}$ **17.** $1 \pm i$ **19.** 2   **21.** $\frac{3}{2}, -4$ **23.** $2 \pm 2i$

25. $\dfrac{-1 \pm 3i}{2}$ **27.** $\dfrac{1 \pm i\sqrt{3}}{2}$ **29.** $\dfrac{-1 \pm i\sqrt{3}}{2}$ **31.** $2, 3$ **33.** $\dfrac{-1 \pm i\sqrt{2}}{3}$ **35.** $\frac{1}{2}, -2$ **37.** $\dfrac{1 \pm 3\sqrt{5}}{2}$ **39.** $4 \pm 3i$

41. $\dfrac{4 \pm \sqrt{22}}{2}$ **43.** Taking the square root of each side   **45.** No   **47.** $\dfrac{11 \pm 4i}{9}$ **49.** $0, 3$ **51.** $\dfrac{3 \pm i}{2}$ **53.** $-2, 5$

55. $\dfrac{6 \pm 3i}{4}$ **57.** $-2, 5$ **59.** $2 + i$ and $2 - i$ **61.** yes   **63.** $3 - 7i$

65.

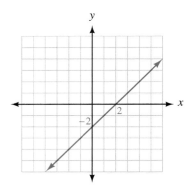

67.

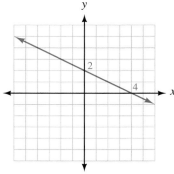

69. $23\sqrt{2}$ **71.** $-6\sqrt{6}$ **73.** $2x\sqrt{3}$

75. $x^2 - 6x + 9 = (x - 3)^2$ **77.** 6 **79.** -2 **81.** 1

83.-91.

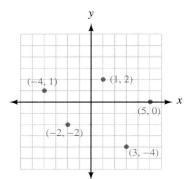

Problem Set 9.6

1.

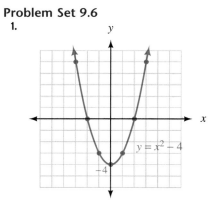

3.

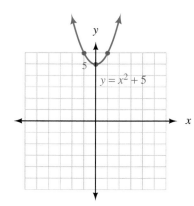

5.

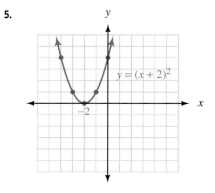

7.

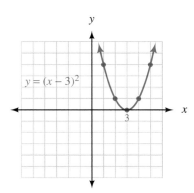

$y = (x - 3)^2$

3

9.

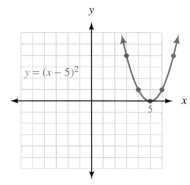

$y = (x - 5)^2$

5

11.

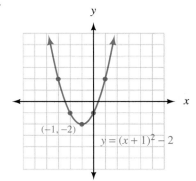

$(-1, -2)$

$y = (x + 1)^2 - 2$

13.

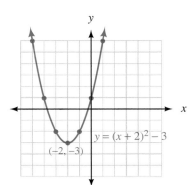

$y = (x + 2)^2 - 3$

$(-2, -3)$

15.

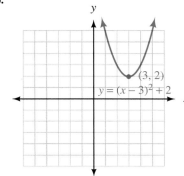

$(3, 2)$

$y = (x - 3)^2 + 2$

17.

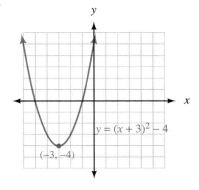

$y = (x + 3)^2 - 4$

$(-3, -4)$

19.

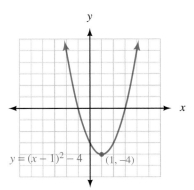

$y = (x - 1)^2 - 4$ $(1, -4)$

21.

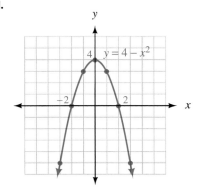

4   $y = 4 - x^2$

-2 2

23.

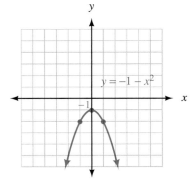

$y = -1 - x^2$

-1

25.

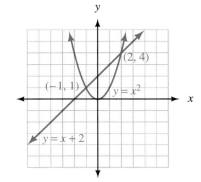

$(2, 4)$

$(-1, 1)$ $y = x^2$

$y = x + 2$

27.

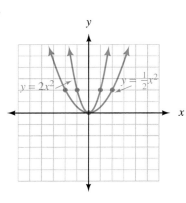

29. a.

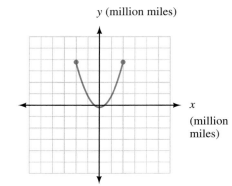

b. 250,000 miles **c.** 500,000 miles

31. 7 **33.** $5\sqrt{2}$ **35.** $\dfrac{\sqrt{10}}{5}$ **37.** $21\sqrt{3}$ **39.** $-4 - 3\sqrt{6}$ **41.** $4\sqrt{5} + 4\sqrt{3}$ **43.** 7

CHAPTER 9 REVIEW/TEST

1. $\pm 4\sqrt{2}$ **2.** $\pm 2\sqrt{15}$ **3.** ± 4 **4.** $-9, 3$ **5.** $-7, 11$ **6.** $-2, \frac{2}{3}$ **7.** $\dfrac{-5 \pm 4\sqrt{2}}{2}$ **8.** $\dfrac{4 \pm 3\sqrt{3}}{3}$ **9.** $\frac{2}{3} \pm \frac{5}{3}i$

10. $-4 \pm 2\sqrt{5}$ **11.** $4 \pm 2\sqrt{5}$ **12.** $2 \pm \sqrt{11}$ **13.** $-3, -1$ **14.** $\dfrac{9 \pm \sqrt{93}}{2}$ **15.** $-1, 6$ **16.** $-3, 1$ **17.** $\dfrac{3 \pm \sqrt{15}}{3}$ **18.** $-4, -3$

19. 4 **20.** $\dfrac{-5 \pm i\sqrt{3}}{2}$ **21.** $\dfrac{-4 \pm \sqrt{26}}{2}$ **22.** $\dfrac{-3 \pm i\sqrt{39}}{6}$ **23.** $-\frac{1}{2}, 3$ **24.** $\dfrac{1 \pm i\sqrt{2}}{2}$ **25.** $4 + 2i$ **26.** $2 - 8i$ **27.** $10 + 5i$

28. $5 - 2i$ **29.** $-i$ **30.** $-1 - 5i$ **31.** $-1 - 3i$ **32.** $2 - 2i$ **33.** $6 - 2i$ **34.** $-24 - 12i$ **35.** $20 + 24i$ **36.** $7 - i$

37. $19 - 17i$ **38.** $8 - 6i$ **39.** 17 **40.** 13 **41.** $\dfrac{1 + 3i}{10}$ **42.** $\dfrac{-1 + 2i}{5}$ **43.** $\dfrac{10 - 25i}{29}$ **44.** $\dfrac{-2 + 8i}{17}$ **45.** $\dfrac{6 - 9i}{13}$

46. $\dfrac{4 + 3i}{5}$ **47.** $\dfrac{-9 - 40i}{41}$ **48.** $\dfrac{-9 + 19i}{34}$ **49.** $6i$ **50.** $12i$ **51.** $i\sqrt{17}$ **52.** $i\sqrt{31}$ **53.** $2i\sqrt{10}$ **54.** $6i\sqrt{2}$ **55.** $10i\sqrt{2}$

56. $11i\sqrt{2}$

57.

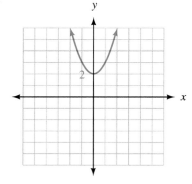

58.

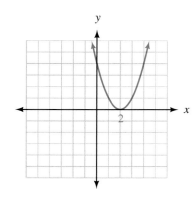

59.

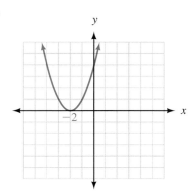

60.

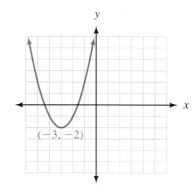

61.

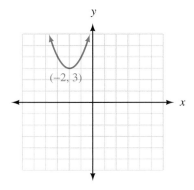

62.

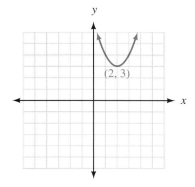

Index

Answers to Selected Even-Numbered Problems

CHAPTER 2

Problem Set 2.8

2.
     -8 0

4. 0 7

6. 0 13

8. 0 9

10. -5 0

12. 0 7

14. 0 4

16. 0 7

18. -3 0

20. 0 3

22. 0 9

24. -8 0

26. 0 7

28. -7 0

30. -3 0

32. 0 12

34. -15 0

36. 0 18

38. 0 3

40. 0 2

42. 0 2

44. 0 $\frac{15}{4}$

46. -3 0

48. 0 5

50. 0 7

Problem Set 2.9

2. ![number line with closed dots at −2 and −1, segment between, arrow left]
    −2 −1

4. ![number line with open dots at 1 and 5, segment between]
    1   5

6. ![number line with open dot at 0 and closed dot at 7]
    0   7

8. ![number line with open dots at 2 and 4, arrows both ways]
    2   4

10. ![number line with closed dots at 2 and 4]
    2   4

12. ![number line with open dots at −1 and 5, arrows both ways]
    −1 5

14. ![number line closed dots at −1 and 3]
    −1 3

16. ![number line closed dots at −5 and 0]
    −5 0

18. ![number line open dots at −4 and 8]
    −4 8

20. ![number line closed dots at −1 and 3]
    −1 3

22. ![number line open dot −5 closed dot 6]
    −5 6

24. ![number line open dots at −3 and 6, arrows both ways]
    −3 6

26. ![number line open dots at −1 and 1, arrows both ways]
    −1 1

28. ![number line open dots at −2/3 and 9, arrow left]
    $-\frac{2}{3}$ 9

30. ![number line closed dots at −2 and 1]
    −2 1

32. ![number line open dots at −1 and 2]
    −1 2

34. ![number line closed dots at −2 and 5]
    −2 5

36. ![number line closed dots at 1 and 3]
    1   3

38. ![number line open dots at −1/2 and 5/3]
    $-\frac{1}{2}$ $\frac{5}{3}$

40. ![number line closed dots at −1 and 2]
    −1 2

48. ![number line closed dots at 10 and 130]
    10   130

CHAPTER 3

Problem Set 3.1

Even-numbered problems 2-18.

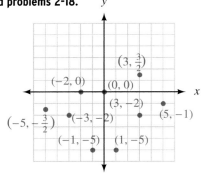

48.

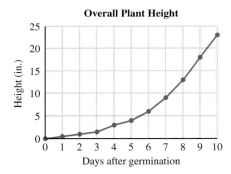

50.

Problem Set 3.3

2.
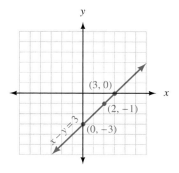

4.
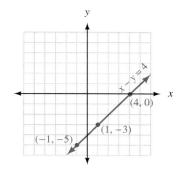

6.
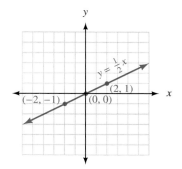

8.
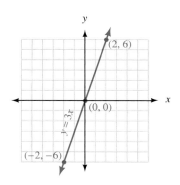

10.
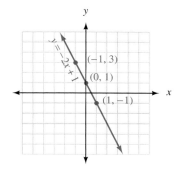

12.
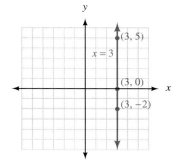

14.
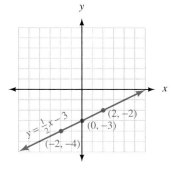

16.
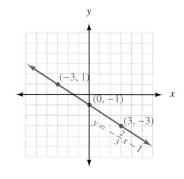

18.
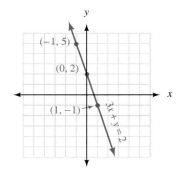

20.
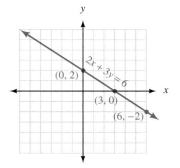

22.
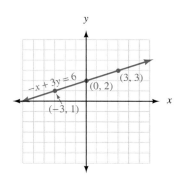

24.
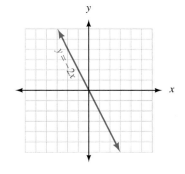

26.

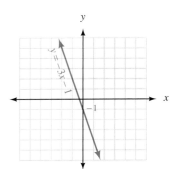

28.

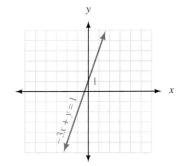

30.

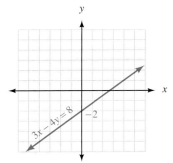

32.

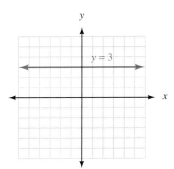

34.

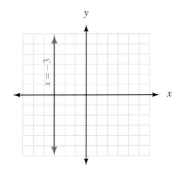

36.

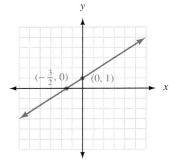

38.

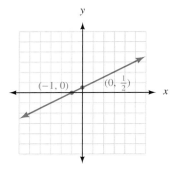

40.

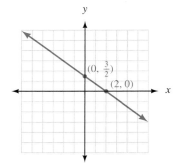

Problem Set 3.4

2.

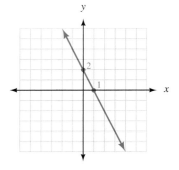

4.

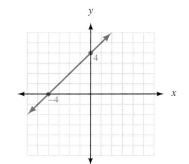

6.

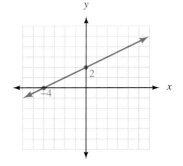

8.

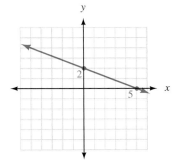

10.

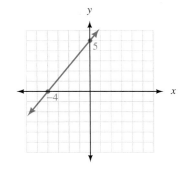

12.

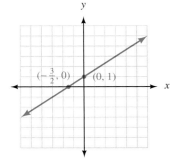

14.

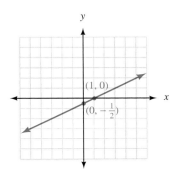

16.

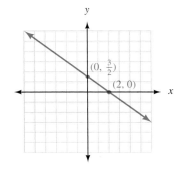

18.

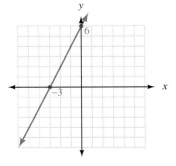

20.

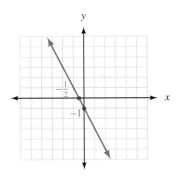

22.

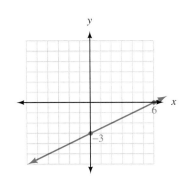

24.

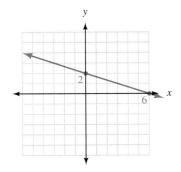

26.

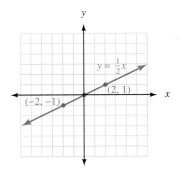

28.

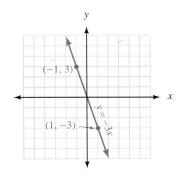

30.

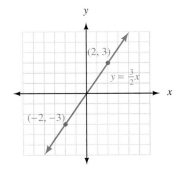

44.

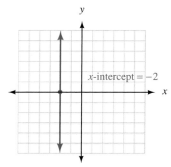

x-intercept = −2

46.

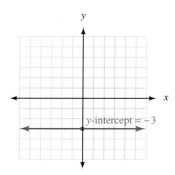

y-intercept = −3

Problem Set 3.5

2.

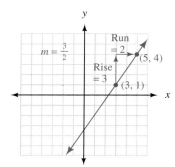

$m = \frac{3}{2}$ Run = 2 (5, 4) Rise = 3 (3, 1)

4.

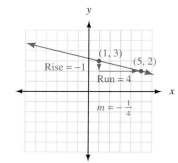

(1, 3) (5, 2) Rise = −1 Run = 4 $m = -\frac{1}{4}$

6.

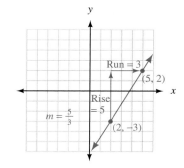

Run = 3 (5, 2) Rise = 5 $m = \frac{5}{3}$ (2, −3)

8.

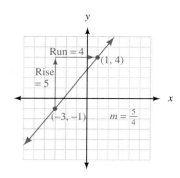

Run = 4 (1, 4) Rise = 5 (−3, −1) $m = \frac{5}{4}$

10.

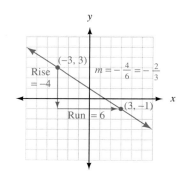

(−3, 3) $m = -\frac{4}{6} = -\frac{2}{3}$ Rise = −4 (3, −1) Run = 6

12.

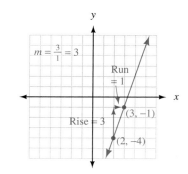

$m = \frac{3}{1} = 3$ Run = 1 Rise = 3 (3, −1) (2, −4)

14.

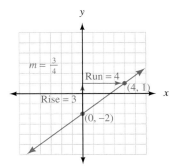

16.

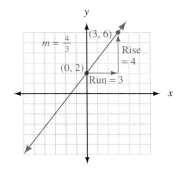

18.

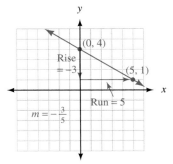

20.

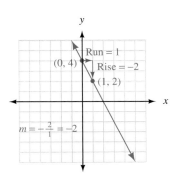

22.

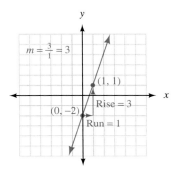

28.

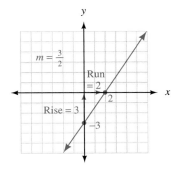

30.

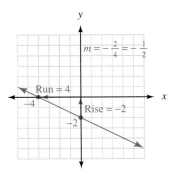

32.

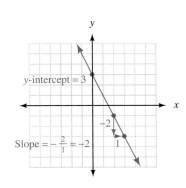

34.

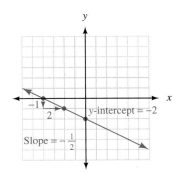

Problem Set 3.6

10.

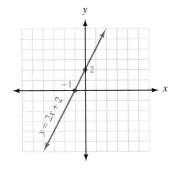

12.

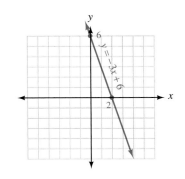

14.

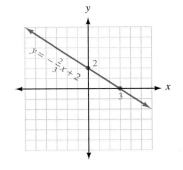

16.

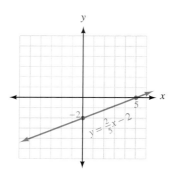

18.

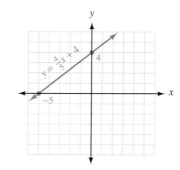

40.

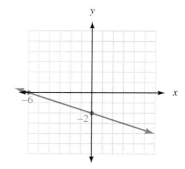

68.

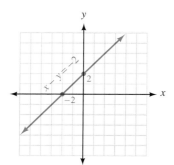

70.

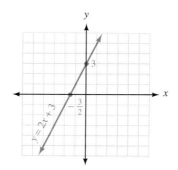

72.

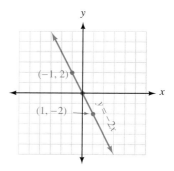

Problem Set 3.7

2.

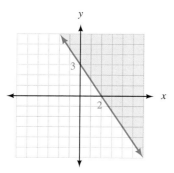

4.

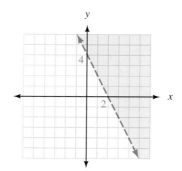

6.

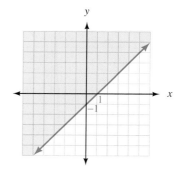

8.

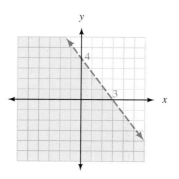

10.

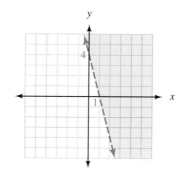

12.

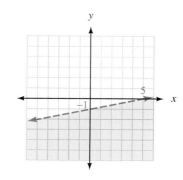

14.

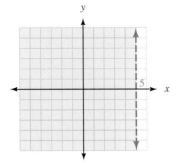

16.

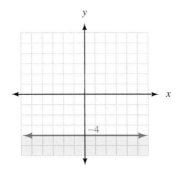

18.

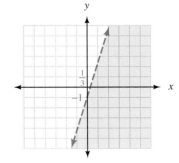

20.

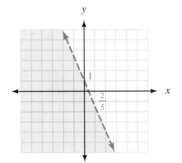

22.

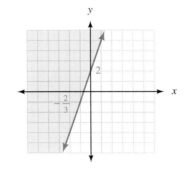

24.

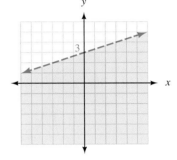

26.

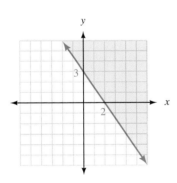

36.

38.

CHAPTER 4

Problem Set 4.1

2.

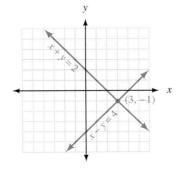

4.

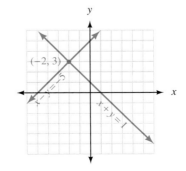

6.

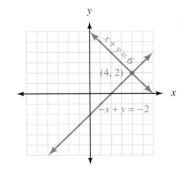

8.

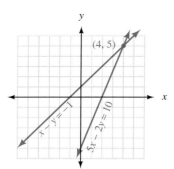

10.

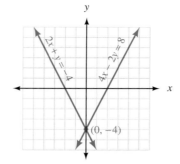

12.

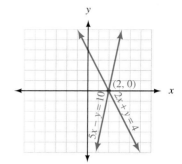

14.

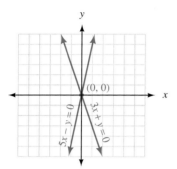

16.

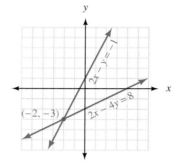

18.

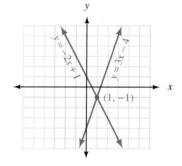

20.

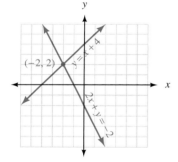

22.

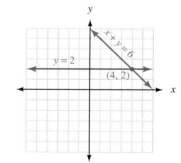

24.

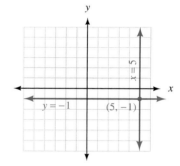

26.

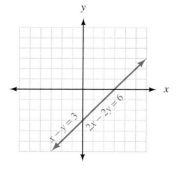

28.

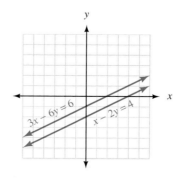

30.

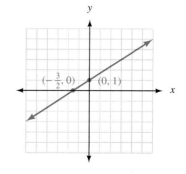

32.

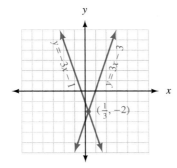

42.

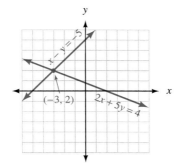

Problem Set 4.4

30.

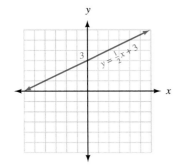

32.

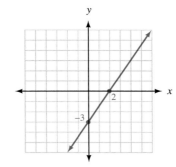

38.

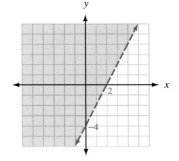

CHAPTER 5

Problem Set 5.3

90.

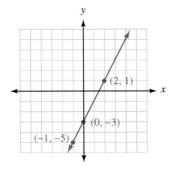

92.

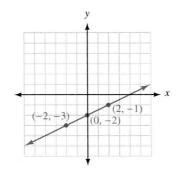

Problem Set 5.6

82.

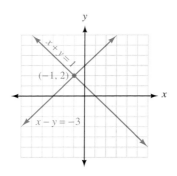

84.

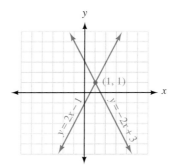

CHAPTER 7

Problem Set 7.5

26.

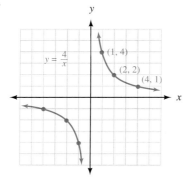

28.

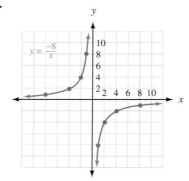

30.

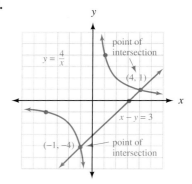

CHAPTER 8

Problem Set 8.3

52.

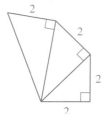

CHAPTER 9

Problem Set 9.5

66.

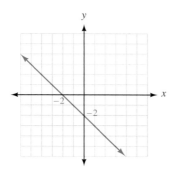

68.

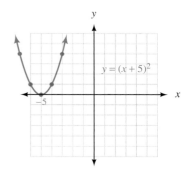

Problem Set 9.6

2.

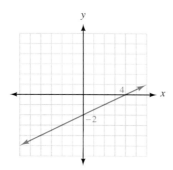

4.

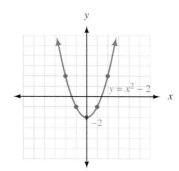

6.

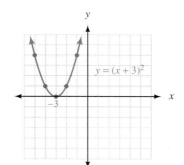

8.

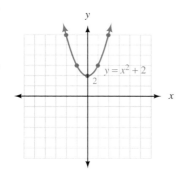

10.

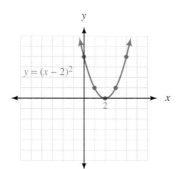

12.

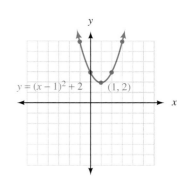

14.

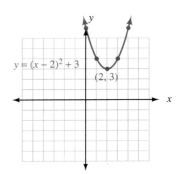

$y = (x - 2)^2 + 3$

$(2, 3)$

16.

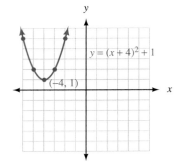

$y = (x + 4)^2 + 1$

$(-4, 1)$

18.

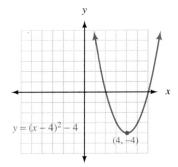

$y = (x - 4)^2 - 4$

$(4, -4)$

20.

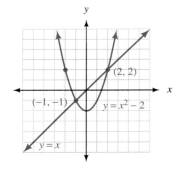

$y = (x + 1)^2 - 4$

$(-1, -4)$

22.

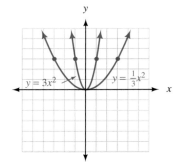

$y = 3 - x^2$

24.

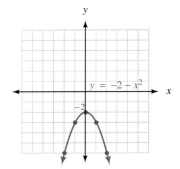

$y = -2 - x^2$

26.

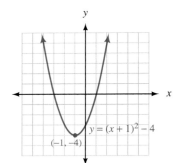

$(2, 2)$

$(-1, -1)$

$y = x^2 - 2$

$y = x$

28.

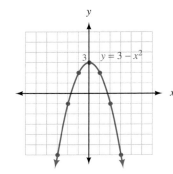

$y = 3x^2$

$y = \frac{1}{3}x^2$

30.

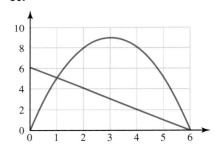